Lecture Notes in Computer Science 16372

Founding Editors

Gerhard Goos
Juris Hartmanis

The series Lecture Notes in Computer Science (LNCS), including its subseries Lecture Notes in Artificial Intelligence (LNAI) and Lecture Notes in Bioinformatics (LNBI), has established itself as a medium for the publication of new developments in computer science and information technology research, teaching, and education.

LNCS enjoys close cooperation with the computer science R & D community, the series counts many renowned academics among its volume editors and paper authors, and collaborates with prestigious societies. Its mission is to serve this international community by providing an invaluable service, mainly focused on the publication of conference and workshop proceedings and postproceedings. LNCS commenced publication in 1973.

Ratna Dutta · Luca De Feo ·
Sugata Gangopadhyay

Editors

Progress in Cryptology –
INDOCRYPT 2025

26th International Conference on Cryptology in India
Bhubaneshwar, India, December 14–17, 2025
Proceedings

 Springer

Editors
Ratna Dutta
Indian Institute of Technology
Kharagpur, West Bengal, India

Luca De Feo
IBM Research Europe
Rüschlikon, Switzerland

Sugata Gangopadhyay
Indian Institute of Technology
Roorkee, Uttarakhand, India

ISSN 0302-9743   ISSN 1611-3349 (electronic)
Lecture Notes in Computer Science
ISBN 978-3-032-13300-7   ISBN 978-3-032-13301-4 (eBook)
https://doi.org/10.1007/978-3-032-13301-4

This Springer imprint is published by the registered company Springer Nature Switzerland AG
The registered company address is: Gewerbestrasse 11, 6330 Cham, Switzerland

If disposing of this product, please recycle the paper.

Foreword

We are delighted to welcome you to Indocrypt 2025, the 26th International Conference on Cryptology in India. This year, the conference was jointly organized by the International Institute of Information Technology, Bhubaneswar (IIIT Bhubaneswar) and the Cryptology Research Society of India (CRSI). The event took place in the vibrant city of Bhubaneswar, Odisha, from December 14–17, 2025.

Following the successful Silver Jubilee edition of Indocrypt held in Chennai in 2024, we were honored to carry forward this esteemed legacy and host the 2025 edition in Odisha for the very first time. As General Chairs, we take immense pride in bringing together the global cryptology community to this region rich in cultural heritage and academic promise.

Since its inception in 2000 under the leadership of Bimal Roy, Indocrypt has grown to become one of the most prominent international venues for cryptology research. Over the years, the CRSI has played a pivotal role in advancing cryptographic research and nurturing academic and industry collaborations across the country. We gratefully acknowledge the enduring contributions and guidance of Bimal Roy and R. Balasubramanian, whose vision continues to shape the Indocrypt series.

The journey of Indocrypt from 2000 to 2024 has mirrored the dramatic evolution of global information security itself. The Foundational Era (2000–2005) focused primarily on establishing rigorous academic standards and mastering the cryptanalysis of symmetric primitives like Block Ciphers and Hash Functions, ensuring India's work stood alongside the global debate following the AES selection. The mid-decade brought a strategic pivot. Around 2006, the conference seriously began to confront the Quantum and Implementation Challenge, providing a key platform for early research into Post-Quantum Cryptography (PQC)—long before the NIST standardization process formalized the field. Papers on lattice-based schemes and code-based cryptography, foundational to our work today, were nurtured here.

This forward-looking perspective ensured that when the Applied and Strategic Cryptology Pivot (2015–2020) hit with the rise of FinTech and distributed ledger technologies, Indocrypt was ready. The focus shifted dramatically to Privacy-Enhancing Technologies (PETs), particularly the practical implementation challenges and security analysis of Homomorphic Encryption (HE) and Zero-Knowledge Proofs (ZKPs). The research presented in this period was directly relevant to India's burgeoning digital identity and cloud service sectors, driving the adoption of robust, privacy-preserving techniques.

The work presented at Indocrypt is rarely purely theoretical; it has consistently addressed serious analysis areas that directly impact the security domain. From early Side-Channel Attack (SCA) and Fault Injection analyses that hardened our national smart card and secure hardware standards, to the current focus on formal security models and AI-driven cryptanalysis, the conference proceedings serve as a crucial document for both academic innovation and practical defense strategy.

The current landscape, heavily influenced by India's National Quantum Mission (NQM), makes the Indocrypt platform more vital than ever. The research accepted here in 2025 will guide the essential national strategy of Cryptographic Migration, focusing on benchmarking, hybrid deployment, and formal verification of PQC candidates like Kyber and Dilithium. Furthermore, we are eager to showcase the newest frontier: the intersection of Artificial Intelligence and Cryptology, covering the use of ML to other strengthen or break cryptographic implementations.

The Programme Committee Chairs for Indocrypt 2025 — Luca De Feo, Ratna Dutta, and Sugata Gangopadhyay curated a strong, diverse, and intellectually rich academic programme that reflected this strategic depth. Attendees enjoyed the distinguished invited talks, engaging tutorials, and a refereed selection of cutting-edge research papers across a wide range of topics in cryptology, including theoretical foundations, new cryptographic primitives, cryptanalysis, cryptographic protocols, post-quantum cryptography, and more. We believe this programme offered deep insights into both foundational and emerging areas, benefiting seasoned researchers and early-career scholars alike.

Our host institute, IIIT Bhubaneswar, is a premier center for research and education in information technology and allied disciplines. Supported by the Government of Odisha, the institute is known for its world-class infrastructure, vibrant academic ecosystem, and a strong emphasis on innovation and research excellence.

We express our heartfelt gratitude to all colleagues, collaborators, and institutional partners who worked tirelessly to make this event a success. We were pleased to welcome all delegates and participants and hope you enjoyed stimulating academic discussions, meaningful networking, and enriching experiences during your time in Bhubaneswar.

We are confident that Indocrypt 2025 built upon its 25-year history to contribute significantly to the advancement of cryptologic research and foster new collaborations that benefit both the academic and professional communities and society at large.

December 2025   Indivar Gupta
                                                                       Debasish Jena

Preface

It is our pleasure and honour to invite you to peruse the proceedings of the International Conference on Cryptology in India, Indocrypt 2025. The conference was held from 14–17 December 2025 in Bhubaneswar, India. It was held under the aegis of the Cryptology Research Society of India (CRSI), and organized by the International Institute of Information Technology, Bhubaneswar (IIIT Bhubaneswar), generously supported by the ISEA-III project under MeitY, Govt. of India.

We were fortunate to receive the help of 73 brilliant researchers, who graciously agreed to be part of the Technical Program Committee (TPC). Our call for papers generated 83 full-paper submissions. Based upon a rigorous process and discussions among the TPC members, we selected 19 papers, which were finally accepted based on double-blind (at least 3) reviews and extensive discussions among the TPC members. The acceptance rate was 22.89%. The accepted papers were organized under the following themes: Foundations, Post-Quantum Cryptography, Quantum Cryptography, Symmetric-Key Cryptography, Secret Sharing and Cloud Security.

As part of the technical program, we also solicited inspiring keynote speeches from Shivam Bhasin (Nanyang Technological University, Singapore), Debdeep Mukhopadhyay (Indian Institute of Technology Kharagpur, India), and Shuichi Katsumata (PQShield Ltd & AIST, Japan). We are very thankful to several people and organizations who played a huge supporting role behind the successful organization of this conference. The following list is our humble attempt to acknowledge their service and support. First and foremost, we would like to thank our sponsors, IIIT Bhubaneswar. We are very thankful for the service of the entire organization committee for their hard work through the last few months of event organization. Last but not least, we would like to thank our General Chairs, Debasish Jena (IIIT Bhubaneswar, India), Indivar Gupta (Defence Research and Development Organisation, India), the Advisory Committee, Ashish Ghosh (Director, IIIT Bhubaneswar, India), Bimal Roy (President, CRSI & Former Director, Indian Statistical Institute, Kolkata), Pradeep Kumar Raut (CEO, OCAC, India), Vishal Kumar Dev (Indian Administrative Service) (Principal Secretary, E&IT Department, Govt. of Odisha, India), for guidance and motivation, as well as our students and scholars including Rakesh Kumar (ISI Kolkata, India), Pratima Jana (IIT Kharagpur, India), and Manas Jana (IIT Kharagpur, India) who helped in many ways to make the conference a success.

We lastly thank the authors who submitted and presented their research at the conference and the scholars who attended these talks in person or remotely.

December 2025

Ratna Dutta
Luca De Feo
Sugata Gangopadhyay

Organization

General Chairs

Debasish Jena	IIIT Bhubaneswar, India
Indivar Gupta	DRDO, India

Program Committee Chairs

Ratna Dutta	IIT Kharagpur, India
Luca De Feo	IBM Research Europe, Switzerland
Sugata Gangopadhyay	IIT Roorkee, India

Advisory Committee

Ashish Ghosh	IIIT Bhubaneswar, India
Bimal Kumar Roy	ISI Kolkata, India
Pradeep Kumar Raut	OCAC, India
Vishal Kumar Dev	Govt. of Odisha, India

Organizing Chairs

Srichandan Sobhanayak	IIIT Bhubaneswar, India
Deepak Kumar Rout	IIIT Bhubaneswar, India
Bharati Mishra	IIIT Bhubaneswar, India
Shanta Kumari Sunanda	IIIT Bhubaneswar, India
Umamani Subudhi	IIIT Bhubaneswar, India
Rakesh Kumar	ISI Kolkata, India
Saroj Kumar Panigrahy	VIT - AP University, India
Sambit Mohapatra	IIIT Bhubaneswar, India
Arpit Sourav Mohapatra	IIIT Bhubaneswar, India
Pradeep Kumar Raut	Govt. of Odisha, India

Finance and Advisory Committee

Sabyasachi Dash   IIIT Bhubaneswar, India
Arun Kumar Sahoo   IIIT Bhubaneswar, India
Rakesh Kumar   ISI Kolkata, India

Publicity Chairs

Pradyut Kumar Biswal   IIIT Bhubaneswar, India
Ajaya Kumar Dash   IIIT Bhubaneswar, India
Lipika Das   IIIT Bhubaneswar, India

Publication Co-chairs

Pratima Jana   IIT Kharagpur, India
Manas Jana   IIT Kharagpur, India

Industry Chair

Kaliprasd Vittal   Informatica, India

Program Committee

Aditi Gangopadhyay   IIT Roorkee, India
Ajith Suresh   Technology Innovation Institute, UAE
Ana Salagean   Loughborough University, UK
Andre Esser   Technology Innovation Institute, UAE
Anubhab Baksi   Lund University, Sweden
Anupam Chattopadhyay   Nanyang Technological University, Singapore
Arka Rai Choudhuri   NTT Research, USA
Arpita Maitra   IAI, TCG CREST, Kolkata, India
Avishek Adhikari   Presidency University, India
Ayantika Chatterjee   IIT Kharagpur, India
Bhupendra Singh   CAIR, DRDO, India
Chaoyun Li   University of Surrey, UK
Chinmoy Biswas   University of Calgary, Canada
Christina Boura   IRIF, Université Paris Cité, France

Debapriya Basu Roy	IIT Kanpur, India
Deepak Kumar Dalai	National Institute of Science Education and Research, Bhubaneswar, India
Dipanwita Roy Chowdhury	IIT Kharagpur, India
Dirmanto Jap	Nanyang Technological University, Singapore
Divya Ravi	University of Amsterdam, Netherlands
Douglas Stebila	University of Waterloo, Canada
Francisco Rodríguez-Henríquez	Technology Innovation Institute, UAE
Haoyang Wang	Shanghai Jiao Tong University, China
Indivar Gupta	SAG DRDO, India
Jason LeGrow	Virginia Polytechnic Institute and State University, USA
Jayaprakash Kar	LNM Institute of Information Technology, India
Kazuhiko Minematsu	NEC, Japan
Lilya Budaghyan	University of Bergen, Norway
Mahabir Prasad Jhanwar	Ashoka University, India
Mahsweta Sarkar	San Diego State University, USA
Mainack Mondal	IIT Kharagpur, India
Marine Minier	LORIA, France
Meltem Turan	National Institute of Standards and Technology, USA
Mridul Nandi	Indian Statistical Institute, Kolkata, India
Mustafa Khairallah	Nanyang Technological University, Singapore
Nicolas Sendrier	Inria, France
Nilanjan Datta	IAI, TCG CREST, Kolkata, India
Nishanth Chandran	Microsoft Research, India
Pantelimon Stanica	Naval Postgraduate School, USA
Prasanna Ravi	Nanyang Technological University, Singapore
Pratish Datta	NTT Research, USA
Raghvendra Singh Rohit	IIT Roorkee, India
Rajat Sadhukhan	IIT Roorkee, India
Rei Ueno	Kyoto University, Japan
Saibal Pal	SAG, DRDO, India
Sambuddho	IIIT Delhi, India
Sanjit Chatterjee	IISc Bangalore, India
Santanu Sarkar	IIT Madras, India
Sarani Bhattacharya	IIT Kharagpur, India
Sartaj ul Hassan	IIT Jammu, India
Satrajit Ghosh	IIT Kharagpur, India
Sherman Chow	Chinese University of Hong Kong, China
Sihem Mesnager	Universities of Paris VIII and Sorbonne Paris Nord, France

Sofia Celi	Brave, Portugal
Somitra Sanadhya	IIT Jodhpur, India
Souradyuti Paul	IIT Bhilai, India
Sourav Mukhopadhyay	IIT Kharagpur, India
Srikanta Patnaik	IIMT Bhubaneswar, India
Sruthi Sekar	IIT Bombay, India
Stjepan Picek	Radboud University, Netherlands
Subhamoy Maitra	Indian Statistical Institute Kolkata, India
Subhranil Dutta	University of St Gallen, Switzerland
Subidh Ali	IIT Bhilai, India
Sumanta Sarkar	University of Warwick, UK
Tanmoy Kanti Das	National Institute of Technology, Raipur, India
Tapas Pal	Karlsruhe Institute of Technology, Germany
Vireshwar Kumar	IIT Delhi, India
Yixin Shen	Inria, University of Rennes, IRISA, France
Takashima Katsuyuki	Waseda University, Japan
Steven Duong	University of Wollongong, Australia

Additional Reviewers

Amaury Pouly	Hyunji Kim	Shashank Singh
Amit Jana	Igor Semaev	Sikhar Patranabis
Amit K. Awasthi	Indranil Sengupta	Steven Galbraith
Andes Y. L. Kei	Jack P. K. Ma	Suchetana Goswami
Angshuman Karmakar	Javier Verbel	Suman Dutta
Animesh Singh	José LuisImaña	Sumanta Chakraborty
Anindya Ganguly	Marco Calderini	Sumit Kumar Pandey
Arindam Mukherjee	Massimo Ostuzzi	Suprava Roy
Avijit Dutta	Nabanita Chakraborty	Surbhi Shaw
Bimal Mandal	Prasanna R. Mishra	Swagata Sasmal
Debadrita Talapatra	Prem Laxman Das	Tapas Pandit
Debranjan Pal	Atul Chaturvedi	Travis Morrison
Elena Kirshanova	Sanajit Patra	Trevor Yap
Florias Papadopoulos	Satyam Kumar	Ying-Yu Pan
Haradhan Ghosh	Sayan Das	Yogesh Kumar

Invited Talks

Cracking Secrets Beyond the Dataset: Revisiting Deep Learning in Side-Channel Analysis

Shivam Bhasin

Temasek Lab and National Integrated Center for Evaluation, Nanyang Technological University, Singapore

Abstract. Side-channel analysis (SCA) continues to challenge the security foundations of modern computing systems by exploiting subtle leakages from physical phenomena such as timing, power consumption, and electromagnetic emanations. Over the past decade, the intersection of SCA and deep learning has opened an exciting new chapter in the field. Deep Learning–based Side-Channel Analysis (DLSCA) has shown impressive capabilities—often succeeding where classical methods struggle, and performing well even against countermeasures like hiding and masking. However, despite these advances, most evaluations still rely on simplified datasets collected under ideal conditions. This raises the question: how well do deep models perform when faced with realistic, noisy targets that challenge these assumptions?

This talk revisits DLSCA from a critical and exploratory perspective. The talk begins with a brief overview of the field, its progress, and key challenges. We then move beyond conventional benchmarks to explore the challenges of applying DLSCA to high-performance platforms such as the Raspberry Pi 4B, powered by a multi-core out-of-order ARM Cortex-A72 processor. Finally, we examine *blind side-channel analysis*, in which attackers operate without access to plaintexts or ciphertexts—a fundamental assumption in conventional SCA.

Through these explorations, the talk encourages reflection on the role of deep learning in side-channel analysis, highlighting both its potential and its limitations as the field moves toward more realistic attack scenarios.

"Every Contact Leaves a Trace": Microarchitecture Leakages in Modern Computing Systems

Debdeep Mukhopadhyay

Indian Institute of Technology Kharagpur, India

Abstract. Modern-day computing systems have advanced microarchitectural elements, which are often invisible to the programmer implementing cryptographic algorithms. While these components aim at enabling high-performance of the computing devices, they often capture traces of secret data which cryptography tries to hide. At Secured Embedded Architecture Laboratory (SEAL), Indian Institute of Technology Kharagpur, India, we have been engaging in active research on such micro-architectural attacks for over a decade. This talk offers a gist of our experience with developing side-channels when process isolation boundaries (including state-of-the-art trusted architectures) are violated at the microarchitectural level. We present our research experience across several microarchitectural aspects (like branch prediction, speculative vulnerabilities, Model Specific Registers, Out-of-Band System Management (relevant for Trusted Architectures like Intel TDX) as well as across various threat models. We also extend our talk to discuss our findings in the context of now-more-hot GPU architectures, commercialized by companies like Nvidia, and present our findings on the leakages exhibited by these platforms. Interestingly, we show that these platforms on which our coveted machine learning codes run also leaves significant traces which can lead to piracy of "our precious" machine learning models.

The talk also emphasizes the importance of close collaborations between academic research groups and industry. Such collaborations not only help better align expectations wrt. impactful research outputs on both sides, but also help drive multifaceted improvements to security in general. Conversations on both sides have often led to a better understanding of how the security landscape evolves in the presence of side-channels. Our experience can hopefully offer a foundation to not only encourage more such academia/industry collaborations, but also to spread widespread awareness of the potency of micro-architectural side-channels (even on Trusted Architectures). This will enable us to bolster the fact that security can only be achieved by design, and not by obscurity!

Taking Post-Quantum Cryptography from Theory to Practice: A Case Study with Signal

Shuichi Katsumata

PQShield and AIST, Japan

Abstract. Not long ago, post-quantum cryptography (PQC) was a topic mainly discussed by academic researchers. However, as governmental bodies and industry began to realize the massive potential threat posed by quantum computers, the narrative shifted. With NIST publishing five FIPS PQC standards for KEMs and digital signatures mid-2024, the transition to PQC has become highly active. While most classical cryptography can *theoretically* be replaced by PQC, this does not mean our job as researchers is complete. An illustrative example is the TLS protocol: making it fully post-quantum has proven to be a nightmare for practitioners due to the massive overhead incurred by the new post-quantum KEMs and digital signatures.

In this keynote, we use Signal as a motivating example where theory alone was not enough. We will walk through our recently deployed *Triple Ratchet* protocol and provide a preview of the *RingXKEM* protocol, which, when combined, yields a fully post-quantum Signal protocol. We explain why known theoretical results were insufficient and detail the pragmatic choices, accounting for real-world bandwidth and memory restrictions, that were needed to make the protocol practical.

Contents

Classical Algorithms

On the Classical Hardness of the Semidirect Discrete Logarithm Problem
in Finite Groups 3
    Mohammad Ferry Husnil Arif and Muhammad Imran

High-Performance FPGA Implementation of a Recursive Modular
Karatsuba Multiplier over $GF(2^m)$ 21
    Ruby Kumari, Sumeet Saurav, and Abhijit Karmakar

Symmetric-Key Cryptography

Improved Modeling for Substitution Boxes with Negative Samples
and Beyond 45
    Debranjan Pal, Anubhab Baksi, Surajit Mandal, and Santanu Sarkar

Refined Linear Approximations for ARX Ciphers and Their Application
to ChaCha 70
    Yurie Okada, Atsuki Nagai, and Atsuko Miyaji

Zero-Knowledge and Interactive Proofs

BOIL: Proof-Carrying Data from Accumulation of Correlated Holographic
IOPs 95
    Tohru Kohrita, Maksim Nikolaev, and Javier Silva

Rejection-Free Framework of Zero-Knowledge Proof Based
on Hint-MLWE 119
    Antoine Douteau and Adeline Roux-Langlois

COMPASS: A Compact PASS-Lineage Accumulator with Succinct Proofs 145
    Tao-Hsiang Chang, Jen-Chieh Hsu, Hao-Yi Hsu, Raylin Tso,
    and Masahiro Mambo

Isogeny-based Cryptography

Smooth Twins for Cryptographic Applications from Pell Equations 173
    Daniel Berger

Hardened CTIDH: Dummy-Free and Deterministic CTIDH 194
  Gustavo Banegas, Andreas Hellenbrand, and Matheus Saldanha

Key-Updatable Identity-Based Signature Schemes 216
  Tobias Guggemos and Farzin Renan

Beyond Sequential Walks: Parallelizing the GA-Dlog Problem 239
  Sudeshna Karmakar, Abul Kalam, and Santanu Sarkar

Multivariate and Lattice-based Cryptography

Multivariate Encryptions with LL' Perturbations: - Is it Possible to Repair
HFE in Encryption? When 0 Makes a Difference- 263
  Jacques Patarin and Pierre Varjabedian

Efficient Identity-Based Inner Product Functional Encryption from RLWE 285
  Anushree Belel and Junji Shikata

Module Lattice Based Constant-Size Group Signature with Verifier Local
Revocation and Backward Unlinkability 312
  Komal Pursharthi and Dheerendra Mishra

Quantum Cryptography

New Results in Quantum Analysis of LED: Featuring One and Two Oracle
Attacks 339
  *Siyi Wang, Kyungbae Jang, Anubhab Baksi, Sumanta Chakraborty,
  Bryan Lee, Anupam Chattopadhyay, and Hwajeong Seo*

One-Time Memories Secure Against Depth-Bounded Quantum Circuits 365
  Kyosuke Sekii and Takashi Nishide

Practically Implementable Minimal Universal Gate Sets for Multi-qudit
Systems with Cryptographic Validation 390
  *Anisha Dutta, Sayantan Chakraborty, Chandan Goswami,
  and Avishek Adhikari*

Secret Sharing and Cloud Security

Traceable Bottom-Up Secret Sharing and Law and Order on Community
Social Key Recovery 417
  Rittwik Hajra, Subha Kar, Pratyay Mukherjee, and Soumit Pal

Beyond Confidentiality: Framing-Resistant Secure Vault Schemes 442
  Meghna Sengupta

Author Index 461

Classical Algorithms

On the Classical Hardness
of the Semidirect Discrete Logarithm
Problem in Finite Groups

Mohammad Ferry Husnil Arif[1] and Muhammad Imran[1,2]

[1] Universitas Indonesia, Depok City, Indonesia
muhammad.imran03@ui.ac.id
[2] University of Birmingham, Birmingham, UK

Abstract. The semidirect discrete logarithm problem (SDLP) in finite groups was proposed as a foundation for post-quantum cryptographic protocols, based on the belief that its non-abelian structure would resist quantum attacks. However, recent results have shown that SDLP in finite groups admits efficient quantum algorithms, undermining its quantum resistance. This raises a fundamental question: does the SDLP offer any computational advantages over the standard discrete logarithm problem (DLP) against classical adversaries? In this work, we investigate the classical hardness of SDLP across different finite group platforms. We establish that the group-case SDLP can be reformulated as a generalized discrete logarithm problem, enabling adaptation of classical algorithms to study its complexity. We present a concrete adaptation of the Baby-Step Giant-Step algorithm for SDLP, achieving time and space complexity $O(\sqrt{r})$ where r is the period of the underlying cycle structure. Through theoretical analysis and experimental validation in SageMath, we demonstrate that the classical hardness of SDLP is highly platform-dependent and does not uniformly exceed that of standard DLP. In finite fields $\mathbb{F}_p^*$, both problems exhibit comparable complexity. Surprisingly, in elliptic curves $E(\mathbb{F}_p)$, the SDLP becomes trivial due to the bounded automorphism group, while in elementary abelian groups $\mathbb{F}_p^n$, the SDLP can be harder than DLP, with complexity varying based on the eigenvalue structure of the automorphism. Our findings reveal that the non-abelian structure of semidirect products does not inherently guarantee increased classical hardness, suggesting that the search for classically hard problems for cryptographic applications requires more careful consideration of the underlying algebraic structures.

Keywords: Semidirect discrete logarithm problem · Post-quantum cryptography · Classical algorithms · Group-based cryptography

1   Introduction

The presumed intractability of the discrete logarithm problem (DLP) has long been a cornerstone of modern cryptographic security, particularly through its application in the Diffie-Hellman key exchange protocol. Numerous cryptographic schemes, including key exchange, digital signatures, and public-key encryption, rely on the hardness of the DLP. However, the advent of quantum computing, particularly Shor's algorithm [15], has fundamentally altered the security landscape. Shor's algorithm demonstrates that the DLP can be solved efficiently in polynomial time when implemented on a sufficiently large quantum computer, rendering classical cryptographic schemes based on the DLP vulnerable to quantum attacks.

In response to this existential threat, researchers have sought alternative approaches to construct cryptographic protocols resistant to quantum adversaries. One prominent direction is the development of new DLP variants designed to circumvent Shor's algorithm. Since the applicability of Shor's algorithm relies crucially on the algebraic structures of the underlying group, two general strategies have emerged.

The first strategy considers the DLP analogue in algebraic objects with less structure. One approach considers the DLP in finite commutative semigroups, where the absence of full group properties was initially thought to obstruct Shor's algorithm. This is based on the observation that Shor's algorithm to compute a from the elements $g, h = g^a \in G$, for a finite commutative group G, requires the evaluation of the oracle $f(x, y) = g^x h^{-y}$. Therefore, in the case of semigroup, the oracle is not well-defined. However, Childs and Ivanyos [5] demonstrated that a modified version of Shor's algorithm can efficiently solve the DLP in finite commutative semigroups, undermining its viability as a quantum-resistant alternative.

Moreover, another approach considers fully unstructured sets, leading to the proposal of the DLP analogue in commutative group actions, which appears to be the most promising approach for constructing quantum-resistant cryptographic protocols. This problem, known as the *vectorization problem*, is an instance of constructive membership testing in the orbits of commutative permutation groups on large finite sets. The framework was originally introduced by Couveignes [6] and has since become a central problem in isogeny-based cryptography, as exemplified by CSIDH [4].

The second strategy relies on increasing the complexity of the algebraic structures underlying the DLP. First, we have the DLP analogue in finite semidirect product groups which first appears in [10]. The problem is also known as the semidirect discrete logarithm problem (SDLP). The SDLP in the semidirect product groups $G \rtimes \mathrm{End}(G)$ for a finite group G can be defined as follows.

Problem 1. Given $g \in G, \sigma \in \mathrm{End}(G)$, and $h = \prod_{i=0}^{t-1} \sigma^i(g)$ for some integer t, determine t.

The SDLP can be defined at different levels of generality. In its most general form, the problem considers a finite semigroup G with $\sigma \in \mathrm{End}(G)$. When G is restricted to be a finite group, we obtain the *group-base case*, and when we further require σ to be an automorphism, we have the *group case* (or *full group case*).

The SDLP was believed to resist against Shor's algorithm and thus several post-quantum cryptographic protocols have been proposed based on the formulation of Diffie-Hellman key exchange follows from the SDLP in finite groups. Suppose two parties, Alice and Bob, agree on a public group G, an element $g \in G$, and an endomorphism $\sigma \in \mathrm{End}(G)$. Then they can arrive at the same $G-$element as follows.

1. Alice picks a random positive integer x and computes $(g, \sigma)^x = (A, \sigma^x)$. Then, Alice sends $A = \prod_{i=0}^{x-1} \sigma^i(g)$ to Bob.
2. Bob also picks a random positive integer y, computes $(g, \sigma)^y = (B, \sigma^y)$ and sends $B = \prod_{i=0}^{y-1} \sigma^i(g)$ to Alice.
3. Alice computes its shared key $K_A = A\sigma^x(B)$.
4. Bob computes its shared key $K_B = B\sigma^y(A)$.

Note that $K_A = K_B$, as the following calculation shows.

$$
\begin{aligned}
A\sigma^x(B) &= \prod_{i=0}^{x-1} \sigma^i(g) \prod_{i=0}^{y-1} \sigma^{x+i}(g) = \prod_{i=0}^{x+y-1} \sigma^i(g) \\
&= \prod_{i=0}^{y-1} \sigma^i(g) \prod_{i=0}^{x-1} \sigma^{y+i}(g) \\
&= B\sigma^y(A).
\end{aligned}
$$

However, it was recently shown in [1,11] that the SDLP in finite groups can be reduced to some instances that can be solved efficiently by Shor's algorithm and well-known generalizations.

More general variant of the DLP is proposed within the framework of general finite non-abelian groups. This was first introduced by Stickel in [17]. The DLP analogue in general finite non-abelian groups G is defined as follows.

Problem 2. Given $u, v, w, x \in G$ such that $w = u^a x v^b$ for some integer a and b, determine a and b.

The problem serves as the basis for Stickel's Diffie-Hellman-like key exchange protocol and has also been adapted in tropical cryptography, notably first proposed by Grigoriev and Shpilrain in [9].

Since the SDLP in finite groups is efficiently solvable by quantum algorithms, this raises a question whether this problem is even considerably harder than the standard DLP or asymptotically living in the same classical complexity classes. The reduction from group-base to group case is classical and efficient, suggesting that the additional structure of the semidirect product might not provide significant classical hardness benefits. However, unlike the DLP analogue in general non-abelian groups, the SDLP appears quite different from the standard DLP. Therefore, it was not clear how to adopt the standard classical algorithms for DLP to solve the SDLP.

The first result of this work establishes that the SDLP in finite groups can be equivalently expressed in the form of the DLP analogue in general non-abelian groups in Problem 2, which is a form that more naturally generalizes the standard DLP. This reformulation allows us to adapt the *Baby-step Giant-step* algorithm to the SDLP in finite groups, which we implement in SageMath. Finally, we provide a detailed complexity comparison of these algorithms for the standard DLP and the SDLP across several classes of finite groups, including the multiplicative group of finite fields $\mathbb{F}_p^*$, the group of rational points of elliptic curves $E(\mathbb{F}_p)$ over finite fields $\mathbb{F}_p$, and elementary abelian groups $\mathbb{F}_p^n$.

In this work, we focus specifically on the group setting for several reasons. First, classical algorithms like Baby-Step Giant-Step are most naturally formulated for groups where inverses exist. Second, and most importantly, [11] established an efficient classical reduction from the group-base case to the group case, allowing us to restrict our attention to the case where σ is an automorphism without loss of generality for group-based instances.

2   Classical Algorithms for the SDLP

In this section, we systematically develop classical algorithmic approaches for solving the semidirect discrete logarithm problem (SDLP). Our exposition follows a deliberate progression through several key components. First, we establish a reduction from the general group-base case SDLP to the group case, demonstrating that we can focus our algorithmic efforts on the latter without loss of generality. Next, we analyze the structural properties of the group case SDLP, particularly examining the cycle structure and period bounds that constrain potential solutions. We then demonstrate a critical reduction from the SDLP to a generalized discrete logarithm problem (GDLP) formulation, which enables us to adapt classical DLP algorithms to the SDLP context. Building on this foundation, we present a modified Baby-step Giant-step algorithm specifically adapted for the SDLP and analyze its complexity in terms of the period of the underlying cycle structure. This comprehensive treatment establishes a framework for comparing the computational difficulty of the SDLP relative to the standard DLP across various group structures, which we explore in subsequent sections.

2.1   Reduction from Group-Base Case to Group Case

The reduction follows directly from [11]. Let G be a finite group with σ an endomorphism of G. To reduce the problem to the automorphism case, consider $K = \cap_{i=0}^{\infty} \sigma^i(G)$ and let k_0 be the smallest non-negative integer such that $K = \sigma^{k_0}(G)$. It is obvious that $k_0 \leq \lceil \log |G| \rceil$. So we can just choose k as the upper bound of $\lceil \log |G| \rceil$ and get $K = \sigma^k(G)$. Then the restriction of σ to K is an automorphism.

Let t be solution for SDLP(G, σ) given by g and h, so we get $h = \prod_{i=0}^{t-1} \sigma^i(g)$. Moreover

$$\sigma^k(h) = \sigma^k \left(\prod_{i=0}^{t-1} \sigma^i(g) \right) = \prod_{i=0}^{t-1} \sigma^i \left(\sigma^k(g) \right).$$

It follows that t also is the solution for $\mathrm{SDLP}(K, \sigma|_K)$ given by $\sigma^k(g)$ and $\sigma^k(h)$. This gives a classical polynomial-time reduction from the group-base case SDLP to the group case SDLP. Throughout the rest of this work we will focus only on the group case SDLP.

2.2   Group Case SDLP

The structure of the semidirect product group provides important insights into the properties of the SDLP. Following the approach in [2], we first introduce some key concepts to understand the cycle structure.

Definition 1. *Let G be a finite group. Fix some $(g, \sigma) \in G \rtimes \mathrm{Aut}(G)$. For any $x \in \mathbb{Z}$, the function $s_{g,\sigma} : \mathbb{Z} \to G$ is defined as the group element such that*

$$(g, \sigma)^x = (s_{g,\sigma}(x), \sigma^x).$$

Recall the binary operation in the semidirect product $G \rtimes \mathrm{Aut}(G)$ is defined as $(g_1, \phi_1)(g_2, \phi_2) = (g_1 \phi_1(g_2), \phi_1 \phi_2)$ for any $(g_1, \phi_1), (g_2, \phi_2) \in G \rtimes \mathrm{Aut}(G)$.

Definition 2. *Let G be a finite group and $(g, \sigma) \in G \rtimes \mathrm{Aut}(G)$. The set*

$$\mathcal{X}_{g,\sigma} := \{s_{g,\sigma}(i) : i \in \mathbb{Z}\}.$$

is called the cycle of (g, σ), and its size is called the period of (g, σ).

The work of [2] gives some important result. Let r be the period of $\mathcal{X}_{g,\sigma}$. Then we have $\mathcal{X}_{g,\sigma} = \{1, g, \cdots, s_{g,\sigma}(r-1)\}$. Consequently, the solution of $\mathrm{SDLP}(G, \sigma)$ will be unique up to modulo r. Also, every k such that $s_{g,\sigma}(k) = 1$ we will have $r \mid k$. Let n be of the order σ. Because $(g, \sigma) \in G \rtimes \langle \sigma \rangle$, we have $s_{g,\sigma}(n|G|) = 1$. This means r is one of the factors of $n|G|$. But we can actually find much lower bound than this value for most case platform groups that we will study.

Define fixed-point subgroup $G^\sigma := \{g \in G \mid \sigma(g) = g\}$. We show that $s_{g,\sigma}(n) \in G^\sigma$.

Theorem 1. *Let G be a finite group and σ is automorphism with order n. Then $s_{g,\sigma}(n) \in G^\sigma$.*

Proof. Suppose $g' := s_{g,\sigma}(n)$. We have

$$\sigma(g') = \sigma(s_{g,\sigma}(n)) = \prod_{i=1}^{n} \sigma^i(g) = \prod_{i=0}^{n-1} \sigma^i(g) = s_{g,\sigma}(n) = g'.$$

Now we are ready to show the main result.

Theorem 2. *Let G be a finite group and σ is automorphism with order n. Then $s_{g,\sigma}(\mathrm{ord}(s_{g,\sigma}(n)) \cdot n) = 1$.*

Proof. Since σ has order n, we have $\sigma^{in} = \text{id}$ for all $i \in \mathbb{Z}$. By [2, Theorem 1], we have

$$s_{g,\sigma}(kn) = \prod_{i=0}^{k-1} \sigma^{in}(s_{g,\sigma}(n)) = \prod_{i=0}^{k-1} s_{g,\sigma}(n) = s_{g,\sigma}(n)^k.$$

Setting $k = \text{ord}(s_{g,\sigma}(n))$ gives $s_{g,\sigma}(\text{ord}(s_{g,\sigma}(n)) \cdot n) = s_{g,\sigma}(n)^{\text{ord}(s_{g,\sigma}(n))} = 1$.

In many of the group cases considered, $\text{ord}(s_{g,\sigma}(n))$ is substantially smaller than $|G|$, resulting in a considerably tighter bound on r.

2.3   Reduction from SDLP to GDLP

To efficiently solve the SDLP using classical methods, it is helpful to reformulate the problem in terms of a more general group-theoretic framework. In this subsection, we show that every instance of the group-case SDLP can be equivalently expressed as a GDLP in the semidirect product group $G \rtimes \text{Aut}(G)$, where the group operation incorporates the action of automorphisms.

Theorem 3. *Let G be a finite group and $\sigma \in \text{Aut}(G)$. Given $g, h \in G$ such that*

$$h = \prod_{i=0}^{t-1} \sigma^i(g) = s_{g,\sigma}(t),$$

then the SDLP instance defined by (G, σ, g, h) is equivalent to solving the following GDLP: find $t \in \mathbb{Z}$ such that $w = u^t v^t$ for

$$u = (g, \sigma), \quad v = (1, \sigma^{-1}), \quad w = (h, id) \in G \rtimes Aut(G).$$

Proof. Recall the group operation in the semidirect product $G \rtimes \text{Aut}(G)$ is defined by:

$$(g_1, \phi_1) \cdot (g_2, \phi_2) = (g_1 \cdot \phi_1(g_2), \phi_1 \circ \phi_2).$$

Therefore, we have $u^t = (g, \sigma)^t = (s_{g,\sigma}(t), \sigma^t)$ and $v^t = (1, \sigma^{-1})^t = (1, \sigma^{-t})$. Thus,

$$u^t v^t = (s_{g,\sigma}(t), \sigma^t) \cdot (1, \sigma^{-t}) = (s_{g,\sigma}(t), \text{id}) = (h, \text{id}) = w.$$

Therefore, solving SDLP is equivalent to finding t such that $u^t v^t = w$ in the group $G \rtimes \text{Aut}(G)$.

This reduction is significant for two main reasons:

1. Algorithmic leverage: It allows us to view the SDLP through the lens of group exponentiation in a non-abelian setting, opening the door to adapting known discrete logarithm algorithms (e.g., Baby-Step Giant-Step, Pollard's rho) to the semidirect product framework.
2. Structural clarity: The decomposition of SDLP into GDLP isolates the non-linear part of the problem (the recursive application of automorphisms) into a group-theoretic form that is more amenable to collision-based methods.

Moreover, this reformulation highlights the similarity between SDLP and the Stickel-type discrete log problems in non-abelian groups, as introduced in [17]. In both cases, the logarithmic exponent is recovered not through simple group exponentiation, but through the interaction of two operations—here captured by u^t and v^t respectively—that encode algebraic structure beyond commutativity.

This equivalence sets the stage for Sect. 2.4, where we present a concrete adaptation of the Baby-Step Giant-Step algorithm to solve this GDLP efficiently. The insight from Theorem 3 is key to this adaptation, as it ensures that every SDLP instance can be encoded as a GDLP in a unified and structurally sound manner.

2.4 Baby-Step Giant-Step Adaptation

The Baby-step Giant-step (BSGS) algorithm, originally introduced by Shanks [14], is a classical collision-based method for solving the standard DLP in cyclic groups. In this section, we present a generalized adaptation of the BSGS algorithm designed to solve the SDLP, using the reduction to the GDLP established in Theorem 3.

Let $u = (g, \sigma)$, $v = (1, \sigma^{-1})$, and $w = (h, \mathrm{id}) \in G \rtimes \mathrm{Aut}(G)$, where $g, h \in G$, and $\sigma \in \mathrm{Aut}(G)$. The SDLP then reduces to finding $t \in \mathbb{Z}$ such that $w = u^t v^t$.

Let r be the period of the cycle $X_{g,\sigma}$, which bounds the search space for t. Set $m = \lceil \sqrt{r} \rceil$. Then, any $t < r$ can be uniquely expressed in the form $t = im + j$ for integers $0 \leq i, j < m$.

We now derive the core identity exploited by the adapted BSGS algorithm. From the group structure of $G \rtimes \mathrm{Aut}(G)$, we have:

$$w = u^t v^t = u^{im+j} v^{im+j} = u^{im} u^j v^j v^{im},$$

which can be rearranged as:

$$u^j v^j = u^{-im} w v^{-im}.$$

This identity implies that a collision between the two sides over their respective ranges yields a solution for t. Thus, we construct two sets

1. Baby steps: $\{u^j v^j : 0 \leq j < m\}$.
2. Giant steps: $\{u^{-im} w v^{-im} : 0 \leq i < m\}$

A match between an element from each set yields

$$u^j v^j = u^{-im} w v^{-im} \implies t = im + j.$$

Algorithm 1. BSGS for GDLP

Require: $u, v, w \in G \rtimes \mathrm{Aut}(G)$ and r is period of $\mathcal{X}_{g,\sigma}$
Ensure: t such that $u^t v^t = w$
 1: Initialize empty hash table $\mathcal{T}$
 2: $m \leftarrow \lceil \sqrt{r} \rceil$
 3: $temp \leftarrow (1, \mathrm{id})$
 4: **for** $j = 0$ **to** $m - 1$ **do**
 5:     $\mathcal{T}[temp] \leftarrow j$ $\triangleright$ Baby Step
 6:     $temp \leftarrow u \cdot temp \cdot v$
 7: **end for**
 8: Precompute u^{-m}, v^{-m}
 9: $temp \leftarrow w$
10: **for** $i = 0$ **to** $m - 1$ **do**
11:     **if** $temp \in \mathcal{T}$ **then** $\triangleright$ Giant Step
12:         $j \leftarrow \mathcal{T}[temp]$
13:         **return** $i \cdot m + j$
14:     **else**
15:         $temp \leftarrow u^{-m} \cdot temp \cdot v^{-m}$
16:     **end if**
17: **end for**

Theorem 4. *Let $u, v, w \in G \rtimes \mathrm{Aut}(G)$ be such that $u^t v^t = w$ and r is period of $\mathcal{X}_{g,\sigma}$. Assuming u^{-1} and v^{-1} are efficiently computable, we can find t with runtime $O(\sqrt{r})$ and space $O(\sqrt{r})$ using Algorithm 1.*

Proof. The algorithm correctness follows from the equality $u^j v^j = u^{-im} w v^{-im}$ when $t = im + j$. Since we know that $0 \leq t < r$, and we have partitioned this range with $0 \leq j < m$ and $0 \leq i < m$ where $m = \lceil \sqrt{r} \rceil$, we will find a solution if one exists. The algorithm achieves $O(\sqrt{r})$ time complexity and $O(\sqrt{r})$ space complexity because it performs exactly m group operations in the baby step phase while storing these values in a hash table, followed by at most m group operations in the giant step phase with constant-time hash table lookups.

When applying this algorithm to SDLP instances via the reduction in Theorem 3, we have $u = (g, \sigma)$, $v = (1, \sigma^{-1})$, and $w = (h, \mathrm{id})$. In this case, computing $u^{-1} = (\sigma^{-1}(g^{-1}), \sigma^{-1})$ and $v^{-1} = (1, \sigma)$ requires efficient computation of σ^{-1}. For the platform groups studied in Sect. 3.1, 3.2, and 3.3, this requirement is satisfied as we discuss in the respective subsections.

This algorithm forms the computational foundation for our experimental analysis in later sections. Its performance, as we will see, depends heavily on the structure of the underlying platform group and the properties of the automorphism σ. The period r plays a crucial role in determining the efficiency of solving SDLP via this method.

3   Specific Platform Groups

In this section, we analyze the computational complexity of both the discrete logarithm problem (DLP) and the SDLP across several fundamental platform groups that are commonly employed in cryptographic applications. Our investigation focuses on three platform groups: the multiplicative group of finite fields $\mathbb{F}_p^*$, elliptic curves over finite fields $E(\mathbb{F}_p)$, and elementary abelian groups $\mathbb{F}_p^n$.

As established in Sect. 2, we focus specifically on the group case of the SDLP where σ is an automorphism, leveraging the polynomial-time reduction from the group-base case. For each platform, we derive complexity bounds for both problems, examining how the inherent algebraic structure influences the computational difficulty. Our analysis reveals that the relative hardness between DLP and SDLP varies dramatically depending on the underlying group structure, with some platforms favoring one problem over the other.

The results of our analysis are summarized in Table 1, which provides a comprehensive comparison of asymptotic complexities for the group case SDLP.

Table 1. Complexity comparison of DLP and SDLP across different platform groups

Platform Group	DLP	SDLP (group case)
$\mathbb{F}_p^*$	$e^{((64/9)^{\frac{1}{3}}+o(1))(\ln p)^{\frac{1}{3}}(\ln\ln p)^{\frac{2}{3}}}$	$\mathcal{O}(p)^3$
$E(\mathbb{F}_p)$	$\mathcal{O}(\sqrt{p})$	$\mathcal{O}(1)$
$\mathbb{F}_p^n$	$\mathcal{O}(1)$	$\mathcal{O}(p^{n/2})$ (no eigenvalue 1)
		$\mathcal{O}(p^{\frac{n}{2}+1})$ (with eigenvalue 1)

[3]Precise complexity: $\mathcal{O}(\sqrt{(p-1)\cdot\gcd(k-1,p-1)})$ for automorphism $x \mapsto x^k$. For random $k \in \mathbb{Z}_{p-1}^*$, $\gcd(k-1,p-1)$ is typically small on average

These findings reveal several interesting patterns. The relationship between DLP and SDLP complexity is highly platform-dependent: while both problems exhibit comparable difficulty in finite fields $\mathbb{F}_p^*$, the SDLP becomes trivial in elliptic curves but can be harder in elementary abelian groups.

3.1   Multiplicative Group of Finite Field

Let $\mathbb{F}_p$ be a finite field of prime order p. The multiplicative group $\mathbb{F}_p^*$ is cyclic of order $p-1$, which completely determines its automorphism structure. Since any automorphism of a cyclic group is uniquely determined by where it sends a generator, and must map the generator to another generator to maintain bijectivity, we have $\mathrm{Aut}(\mathbb{F}_p^*) \cong \mathbb{Z}_{p-1}^*$.

Specifically, each automorphism of $\mathbb{F}_p^*$ has the form $\phi_k : x \mapsto x^k$ where $k \in \mathbb{Z}_{p-1}^*$ (i.e., $\gcd(k,p-1) = 1$). The requirement that $\gcd(k,p-1) = 1$ ensures that ϕ_k is bijective: if g is a generator of $\mathbb{F}_p^*$, then g^k is also a generator if and only if k is coprime to $p-1$. This characterization of automorphisms will be crucial for analyzing the SDLP complexity in this group.

For computational purposes, we note that computing σ^{-1} is efficient in this setting. Given an automorphism $\phi_k : x \mapsto x^k$ where $k \in \mathbb{Z}_{p-1}^*$, the inverse automorphism is $\phi_{k^{-1}} : x \mapsto x^{k^{-1} \bmod (p-1)}$. The modular inverse $k^{-1} \bmod (p-1)$ can be computed in $O(\log p)$ time using the Extended Euclidean Algorithm, making σ^{-1} efficiently computable as required by Algorithm 1.

Now we are ready to compare the standard DLP and the SDLP for the multiplicative group of finite fields $\mathbb{F}_p^*$.

DLP. For the multiplicative group of finite fields $\mathbb{F}_p^*$, there are several classical algorithms available for solving the DLP. The generic approaches include BSGS [14] and Pollard's rho algorithm [13], both with time complexity $\mathcal{O}(\sqrt{p-1}) = \mathcal{O}(\sqrt{p})$.

However, the best classical algorithm for solving the DLP is the Number Field Sieve (NFS), which was originally developed for integer factorization [3,12] and later adapted to the discrete logarithm problem [8].

For a prime p, the NFS algorithm achieves a subexponential time complexity of $e^{((64/9)^{\frac{1}{3}} + o(1))(\ln p)^{\frac{1}{3}} (\ln \ln p)^{\frac{2}{3}}}$. The algorithm works by constructing number fields and using relations between elements to build a system of linear equations in a large but sparse matrix. The solution to this system gives the discrete logarithm.

SDLP. The semidirect product $\mathbb{F}_p^* \rtimes \mathrm{Aut}(\mathbb{F}_p^*)$ has elements of the form (a, ϕ_k) where $a \in \mathbb{F}_p^*$ and $\phi_k \in \mathrm{Aut}(\mathbb{F}_p^*)$. The group operation is defined as:

$$(a, \phi_k) \cdot (b, \phi_l) = (a \cdot \phi_k(b), \phi_k \circ \phi_l) = (a \cdot b^k, \phi_{kl \bmod p-1}).$$

The identity element is $(1, \phi_1)$, and the inverse of (a, ϕ_k) is $(\phi_{k^{-1}}(a^{-1}), \phi_{k^{-1}})$ where k^{-1} is computed modulo $p - 1$.

To analyze the SDLP complexity, we apply Theorem 1 and Theorem 2. Let $n = \mathrm{ord}(k)$ as an element in $\mathbb{Z}_{p-1}^*$. By Theorem 1, we have $a' := s_{a,\phi_k}(n) \in (\mathbb{F}_p^*)^{\phi_k}$, the fixed-point subgroup under ϕ_k. We need to compute the size of this subgroup:

$$(\mathbb{F}_p^*)^{\phi_k} = \{x \in \mathbb{F}_p^* \mid \phi_k(x) = x\} = \{x \in \mathbb{F}_p^* \mid x^k = x\} = \{x \in \mathbb{F}_p^* \mid x^{k-1} = 1\}.$$

The size of this subgroup is precisely $|(\mathbb{F}_p^*)^{\phi_k}| = \gcd(k-1, p-1)$, since these are exactly the elements whose order divides $k - 1$ in the cyclic group $\mathbb{F}_p^*$. Therefore, $\mathrm{ord}(a') \leq \gcd(k-1, p-1)$. By Theorem 2, the period r of the cycle $\mathcal{X}_{a,\phi_k}$ divides $\mathrm{ord}(a') \cdot n$, giving us the bound:

$$r \leq \mathrm{ord}(a') \cdot n \leq \gcd(k-1, p-1) \cdot \phi(p-1).$$

In the worst case, when $\gcd(k-1, p-1)$ is maximized, we have $r = \mathcal{O}(p \cdot \phi(p-1)) = \mathcal{O}(p^2)$. Applying Algorithm 1, the time complexity of SDLP becomes $\mathcal{O}(\sqrt{p^2}) = \mathcal{O}(p)$.

Remark 1. It is worth noting that SDLP in $\mathbb{F}_p^*$ generally admits more efficient algorithms using Algorithm 1 if k chosen randomly because value of $\gcd(k -$

$1, p-1)$ will be small on average the value of common divisor of random number $k - 1$ and $p - 1$. In practice as later will shown on Sect. 4, we will have time complexity $\mathcal{O}\left(\sqrt{\phi(p-1)}\right) = \mathcal{O}(\sqrt{p})$ as compared to $\mathcal{O}(p)$.

3.2   Elliptic Curve

Let E be an elliptic curve over the finite field $\mathbb{F}_p$ where $p > 3$. The curve can be given by the Weierstrass equation $y^2 = x^3 + ax + b$ with $a, b \in \mathbb{F}_p$ and $4a^3 + 27b^2 \neq 0$. The set $E(\mathbb{F}_p)$ of $\mathbb{F}_p$-rational points forms an abelian group under the chord-and-tangent law.

The endomorphism ring $\text{End}(E)$ consists of all group homomorphisms from E to itself that are defined over $\overline{\mathbb{F}}_p$. The automorphism group $\text{Aut}(E)$ comprises all invertible elements in $\text{End}(E)$. A fundamental result states that $|\text{Aut}(E)|$ divides 24, with the exact order determined by the j-invariant of E (see [16]):

$|\text{Aut}(E)|$	$j(E)$	$\text{char}(\mathbb{F}_p)$
2	$j(E) \neq 0, 1728$	any
4	$j(E) = 1728$	$p \neq 2, 3$
6	$j(E) = 0$	$p \neq 2, 3$
12	$j(E) = 0$ or 1728	$p = 3$
24	$j(E) = 0$ or 1728	$p = 2$

This bounded size of $\text{Aut}(E)$ is crucial for our analysis of the SDLP complexity on elliptic curves. Additionally, the automorphisms of elliptic curves have explicit descriptions and bounded order (at most 24), making their inverses readily computable.

DLP. Unlike the DLP in multiplicative groups of finite fields, the best known classical algorithms for solving the (Elliptic Curve Discrete Log Problem) ECDLP are generic algorithms such as Baby-Step Giant-Step [14] and Pollard's rho algorithms [13], both with time complexity $\mathcal{O}(\sqrt{\#E(\mathbb{F}_p)}) = \mathcal{O}(\sqrt{p})$ (By Hasse Bound). This suggests that the ECDLP is genuinely harder than the DLP in finite fields of comparable size, which is a key reason for the popularity of elliptic curve cryptography.

SDLP. The semidirect product $E(\mathbb{F}_p) \rtimes \text{Aut}(E(\mathbb{F}_p))$ has elements of the form (P, α) where $P \in E(\mathbb{F}_p)$ and $\alpha \in \text{Aut}(E)$. The group operation in this semidirect product is defined as:

$$(P, \alpha) \cdot (Q, \beta) = (P + \alpha(Q), \alpha \circ \beta).$$

The identity element is $(\mathcal{O}, [1])$. The inverse of (P, α) is $(-\alpha^{-1}(P), \alpha^{-1})$.

Instead using Theorem 2, we can show that the bound r is constant using the fact that α is element of endomorphism ring $\text{End}(E)$.

Theorem 5. *Let $P \in E(\mathbb{F}_p)$ and $\alpha \in \mathrm{Aut}(E)$ is non-trivial automorphism with order n, then $s_{P,\alpha}(n) = \mathcal{O}$*

Proof. We have that

$$s_{P,\alpha}(n) = P + \alpha(P) + \cdots + \alpha^{n-1}(P) = ([1] + \alpha + \cdots + \alpha^{n-1})(P).$$

Also notice that

$$\alpha^n = [1]$$
$$\alpha^n - [1] = [0]$$
$$(\alpha - [1])(\alpha^{n-1} + \cdots + \alpha + [1]) = [0]$$
$$\implies \alpha^{n-1} + \cdots + \alpha + [1] = [0]$$

By assumption α is non-trivial automorphism. Finally, we have $([1] + \alpha + \cdots + \alpha^{n-1})(P) = [0]P = \mathcal{O}$.

As a result, by Theorem 5 we have the period r of the cycle $\mathcal{X}_{P,\alpha}$ divide n which is at most 24. So the complexity for solving SDLP is $\mathcal{O}(1)$.

3.3   Elementary Abelian Group

An elementary abelian p-group of rank n is a group isomorphic to $\mathbb{F}_p^n$, where every non-identity element has order p. As $\mathbb{F}_p^n$ is an n-dimensional vector space over $\mathbb{F}_p$, its automorphism group is isomorphic to $\mathrm{GL}_n(\mathbb{F}_p)$, the general linear group of $n \times n$ invertible matrices over $\mathbb{F}_p$. Under this isomorphism, each automorphism $\sigma \in \mathrm{Aut}(\mathbb{F}_p^n)$ corresponds to a matrix $A_\sigma \in \mathrm{GL}_n(\mathbb{F}_p)$ such that $\sigma(v) = A_\sigma v$ for all $v \in \mathbb{F}_p^n$. A well-known result states that the maximum order of an element in $\mathrm{GL}_n(\mathbb{F}_p)$ is $p^n - 1$ [7], achieved by companion matrices of primitive polynomials of degree n over $\mathbb{F}_p$.

Under this isomorphism, computing σ^{-1} for an automorphism σ corresponds to computing the inverse of the matrix $A_\sigma \in \mathrm{GL}_n(\mathbb{F}_p)$. Matrix inversion can be performed in $O(n^3)$ operations over $\mathbb{F}_p$ using Gaussian elimination, which is polynomial time. Therefore, σ^{-1} is efficiently computable as required by Algorithm 1.

DLP. For elementary abelian groups, the discrete logarithm problem takes a particularly simple form. Given $v \in \mathbb{F}_p^n$ and $tv \in \mathbb{F}_p^n$ for some scalar t, we need to determine t. Unlike the situation in general groups, where no direct formula exists, in a vector space structure like $\mathbb{F}_p^n$, we can solve this problem directly.

Specifically, since v and tv are both vectors, we can select any non-zero component of v, say $v_i \neq 0$, and compute $t = (tv)_i \cdot (v_i)^{-1} \mod p$. This direct computation can be performed in constant time, making the complexity $\mathcal{O}(1)$. This is substantially more efficient than the DLP in other group structures such as $\mathbb{F}_p^*$ or $E(\mathbb{F}_p)$.

SDLP. The semidirect product $\mathbb{F}_p^n \rtimes \mathrm{Aut}(\mathbb{F}_p^n)$ has elements of the form (v, A) where $v \in \mathbb{F}_p^n$ and $A \in \mathrm{GL}_n(\mathbb{F}_p)$, with the group operation:

$$(v_1, A_1) \cdot (v_2, A_2) = (v_1 + A_1 v_2, A_1 A_2).$$

The identity element is $(0, I)$ where I is the identity matrix in $\mathrm{GL}_n(\mathbb{F}_p)$, and the inverse of (v, A) is $(-A^{-1}v, A^{-1})$.

For the semidirect discrete logarithm problem in elementary abelian groups, we need to analyze how the complexity depends on the properties of the automorphism matrix. As established earlier, for a matrix $A \in \mathrm{GL}_n(\mathbb{F}_p)$ the maximum order is $p^n - 1$.

The complexity of SDLP varies based on whether the automorphism matrix A has an eigenvalue of 1. When A has no eigenvalue equal to 1, the fixed-point subgroup $(\mathbb{F}_p^n)^A = \{v \in \mathbb{F}_p^n \mid Av = v\}$ contains only the zero vector, since $Av = v$ if and only if $v = 0$. Let $\ell = \mathrm{ord}(A)$. By Theorem 1, we have $s_{v,A}(\ell) \in (\mathbb{F}_p^n)^A$. Since the only element in this fixed-point subgroup is the zero vector, we conclude that $s_{v,A}(\ell) = 0$. By Theorem 2, this implies that the period r of the cycle $\mathcal{X}_{v,A}$ divides $\mathrm{ord}(s_{v,A}(\ell)) \cdot \ell = \mathrm{ord}(0) \cdot \ell = \ell$, which is at most $p^n - 1$. Using Algorithm 1, we achieve a complexity of $\mathcal{O}(\sqrt{r}) = \mathcal{O}(\sqrt{p^n - 1}) = \mathcal{O}(p^{n/2})$.

Otherwise, when A has an eigenvalue of 1, the fixed-point subgroup $(\mathbb{F}_p^n)^A$ is non-trivial. With high probability $s_{v,A}(\ell) \neq 0$. But since every non-zero element in $\mathbb{F}_p^n$ has order p, by Theorem 2 we have $s_{v,A}(p\ell) = 0$. The worst-case bound for the period r becomes $p \cdot \mathrm{ord}(A) \leq p \cdot (p^n - 1)$. Using Algorithm 1, the complexity increases to $\mathcal{O}(\sqrt{p \cdot (p^n - 1)}) = \mathcal{O}(p^{\frac{n}{2}+1})$.

4   Simulation Results

To validate our theoretical analysis and provide practical insights into the relative hardness of the DLP and SDLP across different platform groups, we implemented the algorithms in SageMath and conducted systematic experiments. Our implementation is publicly available at https://github.com/swusjask/classical-sdlp. We used the Baby-Step Giant-Step algorithm for all test cases to ensure a fair comparison of the inherent structural properties of each group, even though more efficient algorithms exist for specific platforms (such as the Number Field Sieve for finite fields).

Our experiments measured both the theoretical bounds (group order for DLP, cycle period for SDLP) and the actual computation time required by the BSGS algorithm. For finite fields and elliptic curves, we tested primes ranging from 2^{30} to 2^{37} to cover cryptographically relevant sizes. For elementary abelian groups, we fixed the dimension $n = 3$ and varied the prime from 2^6 to 2^{12}, specifically examining the impact of eigenvalue structure on SDLP complexity.

4.1   Finite Fields and Elliptic Curves

We also compared the DLP and SDLP across both finite fields $\mathbb{F}_p^*$ and elliptic curves $E(\mathbb{F}_p)$ for primes p ranging from 2^{30} to 2^{37}, focusing on cryptographically relevant sizes. Figure 1 shows both the comparison of group orders/cycle periods and their corresponding computation times.

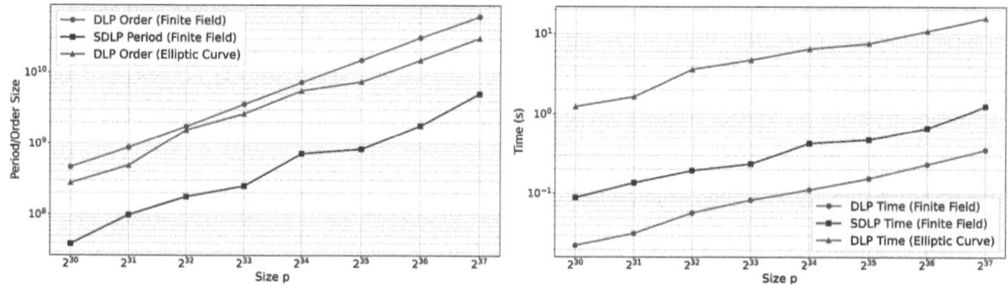

Fig. 1. Comparison of order/period sizes and computation times for DLP and SDLP in finite fields and elliptic curves.

Several interesting observations emerge from this experiment. The orders for both DLP in finite fields $(p-1)$ and DLP in elliptic curves (approximately p by Hasse's Theorem) grow similarly, as expected. However, the period for SDLP in finite fields is consistently smaller than the group order, often by a significant factor. This is because the period is bounded by $\phi(p-1)$ multiplied by some constant.

Despite this smaller period size, the SDLP in finite fields consistently requires more computation time than the standard DLP, demonstrating that the additional complexity of group operations in the semidirect product outweighs the advantage of a smaller search space. The DLP in elliptic curves shows the highest computation time despite having a similar order to the finite field case, reflecting the more complex nature of elliptic curve group operations.

For elliptic curves, we have excluded the SDLP from these charts because, as demonstrated in Sect. 3.2, the cycle period is bounded by the order of the automorphism, which is at most 24, making the SDLP trivially solvable in constant time.

4.2   Elementary Abelian Groups

For elementary abelian groups, we fixed the dimension $n = 3$ and varied the prime p from 2^6 to 2^{12}. We chose to fix $n = 3$ because the problem size grows exponentially with increasing dimension, making experiments with larger dimensions computationally expensive. We specifically examined the impact of the eigenvalue structure of the automorphism matrix on the SDLP complexity. Figure 2 shows both the cycle period lengths and computation times, comparing matrices with and without eigenvalue 1.

The experimental results strongly support our theoretical analysis from Sect. 3.3. Our theory predicts that matrices with eigenvalue 1 have larger periods than those without: specifically, the period can reach $p \cdot \operatorname{ord}(A)$ when eigenvalue

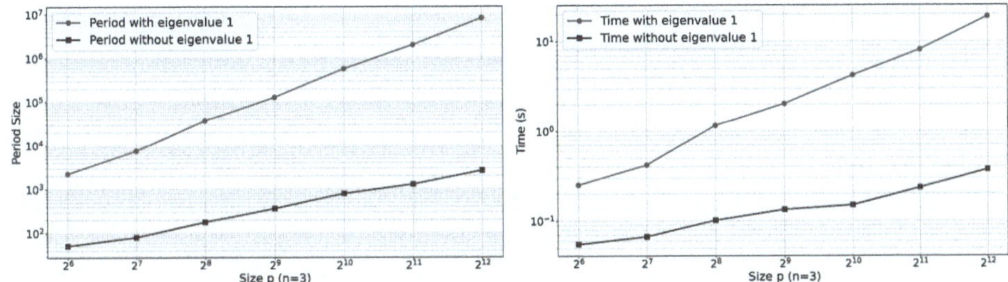

Fig. 2. Period sizes and computation times for BSGS algorithm on elementary abelian groups ($n = 3$) with varying p, comparing matrices with and without eigenvalue 1.

1 is present, compared to at most ord(A) when it is absent. The experiments confirm this distinction - for matrices without eigenvalue 1, the period remains bounded by the order of the matrix, which is at most $p^n - 1$. In contrast, matrices with eigenvalue 1 exhibit periods that can be larger by a factor of p, reaching close to $p \cdot (p^n - 1)$ in many cases. This validates our theoretical complexity bounds of $\mathcal{O}(p^{\frac{n}{2}+1})$ for matrices with eigenvalue 1 versus $\mathcal{O}(p^{\frac{n}{2}})$ for those without. This difference in period length directly translates to computational time, as shown in the right graph of Fig. 2. The gap between the two cases widens as p increases, with matrices having eigenvalue 1 requiring significantly more computation time.

5   Conclusion

In this work, we investigated the classical hardness of the semidirect discrete logarithm problem in finite groups, a problem proposed as a foundation for post-quantum cryptographic protocols. Our investigation reveals that the SDLP does not offer uniform classical hardness advantages over the standard DLP, with the relative difficulty being highly dependent on the underlying platform group.

Our main findings are:

1. **Platform-Dependent Hardness:** The classical complexity of SDLP varies dramatically across different group structures:
    - In finite fields $\mathbb{F}_p^*$, SDLP has worse worst-case complexity than DLP ($O(p)$ vs $O(\sqrt{p})$). However, in practice with random automorphisms, SDLP instances often have significantly smaller periods due to small values of $\gcd(k-1, p-1)$, making them easier to solve despite the worse theoretical bound.
    - In elliptic curves $E(\mathbb{F}_p)$, the SDLP degenerates to a trivial problem solvable in constant time, making it strictly easier than the corresponding DLP.
    - In elementary abelian groups $\mathbb{F}_p^n$, the situation reverses: DLP is trivial while SDLP can require up to $O(p^{n/2+1})$ operations when the automorphism matrix has eigenvalue 1, or $O(p^{n/2})$ when it does not.

2. **Structural Insights:** Our analysis revealed that the non-abelian nature of semidirect products introduces fundamental constraints. We showed that the SDLP can be reformulated as a generalized discrete logarithm problem in the form $u^t v^t = w$, which enabled systematic analysis using adapted classical algorithms.

3. **Implications for Cryptographic Design:** The assumption that more complex algebraic structures automatically yield harder computational problems is not supported by our findings. The bounded automorphism groups in elliptic curves and the direct solvability in vector spaces demonstrate that additional structure can sometimes facilitate rather than hinder attacks.

Our experimental validation confirms these theoretical insights, with implementations demonstrating the predicted complexity behaviors across all tested platforms. Notably, the presence of eigenvalue 1 in automorphism matrices of elementary abelian groups increases the problem difficulty by exactly the factor predicted by our analysis. Interestingly, while SDLP sometimes has larger asymptotic complexity than DLP, our experiments reveal that the actual cycle periods in SDLP are often significantly smaller than the corresponding group orders in DLP. For instance, in finite fields, even though both problems have comparable worst-case complexity, the SDLP period is typically much smaller than $p-1$ due to small values of $\gcd(k-1, p-1)$ in practice, which can make SDLP instances somewhat easier on average despite similar theoretical bounds.

An important observation about the non-abelian structure is that it creates fundamental obstacles for adapting other classical DLP algorithms. For instance, Pollard's rho algorithm [13], which achieves constant space complexity for standard DLP, cannot be directly adapted to SDLP. In the standard setting, Pollard's rho generates a pseudorandom sequence of group elements of the form $g^{\alpha_i} h^{\beta_i}$ until a collision is detected: $g^{\alpha_i} h^{\beta_i} = g^{\alpha_j} h^{\beta_j}$. Since $h = g^x$ for the unknown discrete logarithm x, this collision yields $g^{\alpha_i + x\beta_i} = g^{\alpha_j + x\beta_j}$, which in a cyclic group of order n gives us $\alpha_i + x\beta_i \equiv \alpha_j + x\beta_j \pmod{n}$. Rearranging, we obtain $x(\beta_i - \beta_j) \equiv \alpha_j - \alpha_i \pmod{n}$, allowing us to solve for x when $\gcd(\beta_i - \beta_j, n) = 1$.

However, in the semidirect product setting with our GDLP formulation $u^t v^t = w$, the non-commutativity prevents this approach. A natural attempt would be to generate sequences of the form $u^a v^a w^b$, but upon finding a collision, we cannot rearrange terms to isolate the unknown t. The abelian property $g^\alpha h^\beta = h^\beta g^\alpha$ that enables the algebraic manipulation in standard Pollard's rho simply does not hold in semidirect products. Moreover, if the semidirect product were abelian—which would enable such rearrangements—the condition $(g_1, \phi_1)(g_2, \phi_2) = (g_2, \phi_2)(g_1, \phi_1)$ would force all automorphisms to act trivially. To see this, consider any $g \in G$ and $\phi \in \mathrm{Aut}(G)$, and let (g, id) commute with (e, ϕ). This gives us $(g, \mathrm{id})(e, \phi) = (e, \phi)(g, \mathrm{id})$, which expands to $(g, \phi) = (\phi(g), \phi)$. Therefore $g = \phi(g)$ for all $g \in G$, meaning ϕ is the identity automorphism. With only trivial automorphisms, the SDLP reduces to standard DLP since $h = \prod_{i=0}^{t-1} \sigma^i(g) = \prod_{i=0}^{t-1} g = g^t$. This structural constraint suggests that memory-efficient algorithms for SDLP may require fundamentally new approaches beyond adapting existing DLP methods.

These results contribute to the broader understanding of group-based cryptographic hardness assumptions in the post-quantum era. While SDLP may resist quantum attacks in certain settings, our work shows that it does not provide reliable classical hardness advantages across all platforms. This underscores the importance of careful platform selection and thorough classical cryptanalysis when designing cryptographic schemes based on novel group-theoretic problems.

Future research directions include:

- Developing novel constant-space algorithms for SDLP that respect the non-abelian structure of semidirect products, potentially achieving better space-time tradeoffs than our BSGS adaptation.
- Investigating whether other algebraic structures beyond semidirect products might provide more uniform classical hardness guarantees across different platforms.
- Exploring the theoretical limits of classical hardness in non-abelian group settings, particularly understanding when additional structure helps versus hinders cryptanalysis.

The balance between structure and hardness revealed in this work highlights the importance of thorough cryptanalysis when proposing new post-quantum primitives based on group-theoretic problems.

Acknowledgments. Muhammad Imran is supported by EPSRC through grant number EP/V011324/1.

References

1. Battarbee, C., et al.: On the semidirect discrete logarithm problem in finite groups. In: Chung, K.M., Sasaki, Y. (eds.) ASIACRYPT 2024. LNCS, vol. 15491, pp. 330–357. Springer, Singapore (2024). https://doi.org/10.1007/978-981-96-0944-4_11
2. Battarbee, C., Kahrobaei, D., Perret, L., Shahandashti, S.F.: SPDH-sign: towards efficient, post-quantum group-based signatures. Cryptology ePrint Archive (2023)
3. Buhler, J.P., Lenstra, H.W., Pomerance, C.: Factoring integers with the number field sieve. In: Lenstra, A.K., Lenstra, H.W. (eds.) The development of the number field sieve. LNM, vol. 1554, pp. 50–94. Springer, Heidelberg (1993). https://doi.org/10.1007/BFb0091539
4. Castryck, W., Lange, T., Martindale, C., Panny, L., Renes, J.: CSIDH: an efficient post-quantum commutative group action. In: Peyrin, T., Galbraith, S. (eds.) ASIACRYPT 2018. LNCS, vol. 11274, pp. 395–427. Springer, Cham (2018). https://doi.org/10.1007/978-3-030-03332-3_15
5. Childs, A.M., Ivanyos, G.: Quantum computation of discrete logarithms in semigroups. J. Math. Cryptol. **8**(4), 405–416 (2014)
6. Couveignes, J.M.: Hard homogeneous spaces. Cryptology ePrint Archive (2006)
7. Darafsheh, M.: Order of elements in the groups related to the general linear group. Finite Fields Appl. **11**(4), 738–747 (2005). https://doi.org/10.1016/j.ffa.2004.12.003, https://www.sciencedirect.com/science/article/pii/S1071579704000760

8. Gordon, D.M.: Discrete logarithms in $gf (p)$ using the number field sieve. SIAM J. Discret. Math. **6**(1), 124–138 (1993). https://doi.org/10.1137/0406010
9. Grigoriev, D., Shpilrain, V.: Tropical cryptography. Comm. Algebra **42**(6), 2624–2632 (2014)
10. Habeeb, M., Kahrobaei, D., Koupparis, C., Shpilrain, V.: Public key exchange using semidirect product of (semi)groups. In: Jacobson, M., Locasto, M., Mohassel, P., Safavi-Naini, R. (eds.) ACNS 2013. LNCS, vol. 7954, pp. 475–486. Springer, Heidelberg (2013). https://doi.org/10.1007/978-3-642-38980-1_30
11. Imran, M., Ivanyos, G.: Efficient quantum algorithms for some instances of the semidirect discrete logarithm problem. Designs Codes Cryptogr. 1–19 (2024)
12. Lenstra, A.K., Lenstra, H.W., Manasse, M.S., Pollard, J.M.: The number field sieve. In: Proceedings of the Twenty-Second Annual ACM Symposium on Theory of Computing, STOC 1990, pp. 564–572. Association for Computing Machinery, New York (1990). https://doi.org/10.1145/100216.100295
13. Pollard, J.M.: Monte Carlo methods for index computation (mod$p$). Math. Comput. **32**(143), 918–924 (1978). http://www.jstor.org/stable/2006496
14. Shanks, D.: Class number, a theory of factorization, and genera. In: Proceedings of Symposia in Pure Mathematics, pp. 415–440 (1971). https://doi.org/10.1090/pspum/020/0316385
15. Shor, P.W.: Algorithms for quantum computation: discrete logarithms and factoring. In: Proceedings 35th Annual Symposium on Foundations of Computer Science, pp. 124–134. IEEE (1994)
16. Silverman, J.H.: The Arithmetic of Elliptic Curves, vol. 106. Springer, New York (2009). https://doi.org/10.1007/978-0-387-09494-6
17. Stickel, E.: A new method for exchanging secret keys. In: Third International Conference on Information Technology and Applications (ICITA 2005), vol. 2, pp. 426–430. IEEE (2005)

High-Performance FPGA Implementation of a Recursive Modular Karatsuba Multiplier over $GF(2^m)$

Ruby Kumari[1,2]($\boxtimes$) , Sumeet Saurav[1,2] , and Abhijit Karmakar[1]

[1] CSIR - Central Electronics Engineering Research Institute (CSIR-CEERI), Pilani 333031, Rajastha, India
`sumeet.ceeri@csir.res.in`
[2] Academy of Scientific and Innovative Research (AcSIR), Ghaziabad 201002, India
`ruby.ceeri20a@acsir.res.in`

Abstract. Efficient finite field multiplication over $GF(2^m)$ is essential for high-performance elliptic curve cryptography (ECC) on resource-constrained hardware. This paper presents a recursive modular Karatsuba multiplier for $GF(2^m)$ that integrates modular reduction within a 5-stage stagewise recursive pipeline, achieving one result per clock cycle. The proposed architecture reduces logic depth, balances pipeline stages, and minimizes resource usage (875 LUTs and 1,371 registers on a Xilinx Artix-7 FPGA), while maintaining a low critical path delay (4.2 ns). Experimental results demonstrate significant improvements in area–delay product (ADP) compared to state-of-the-art multipliers. The design is scalable to larger NIST binary fields (e.g., B-233, B-283, B-409, B-571) and supports energy-efficient operation, making it suitable for high-speed, low-power ECC accelerators in embedded and IoT systems.

Keywords: Finite field arithmetic · Karatsuba algorithm · Polynomial multiplier · NIST irreducible polynomial

1 Introduction

Finite field arithmetic [1] is fundamental to cryptography, error-control coding [2], lightweight ciphers for encryption [3] and VLSI testing [4]. In $GF(2^m)$, addition is efficiently implemented using XOR gates, whereas multiplication remains computationally intensive, significantly impacting area, power, and performance in hardware implementations. Polynomial multiplication underpins ECC [5], coding theory, and homomorphic authentication [6], with larger operand sizes needed for modern security levels [7].

Conventional Schoolbook Multiplication (SBM) exhibits quadratic complexity $O(n^2)$, while the Karatsuba method (KM) reduces it to $O(n^{\log_2 3}) \approx O(n^{1.585})$ by recursively partitioning operands [8], replacing one multiplication with multiple additions [9]. In ECC, $GF(2^m)$ is attractive for its balance of performance and

R. Dutta et al. (Eds.): INDOCRYPT 2025, LNCS 16372, pp. 21–41, 2026.
https://doi.org/10.1007/978-3-032-13301-4_2

security, but multiplication requires reduction modulo an irreducible polynomial. Prior FPGA implementations [10,11] and optimizations such as digit-serial [12], parallel feedback [13], and hybrid Karatsuba approaches [14,15] highlight trade-offs among area, delay, and complexity. Despite extensive FPGA-based Karatsuba multiplier architectures [9,13,16–18], achieving low area, high throughput, and low power simultaneously for large-field ECC remains challenging.

To address these challenges, we propose a Pipeline-Based Recursive Modular Karatsuba Multiplier for $GF(2^{163})$. The design integrates modular reduction directly into the multiplier and employs a 5-stage stagewise recursive mapping, achieving low hardware overhead, reduced critical path delay, and fully stagewise recursive mapping throughput of one result per cycle. It is implemented on a Xilinx Artix-7 FPGA and evaluated against state-of-the-art designs in terms of LUT/register utilization, timing, and area-delay product (ADP).

The key contributions of this work are:

- An **integrated recursive modular Karatsuba multiplier** that combines multiplication and modular reduction at the architectural level, sustaining one-product-per-cycle throughput. This integration eliminates redundant modular stages, reduces logic depth, and lowers the area-delay product compared to prior non-integrated designs.
- A **pipeline-aware recursive mapping** with five balanced stages, achieving constant throughput with bounded latency and minimal register overhead. Balanced stagewise recursion ensures uniform combinational delay, improving hardware utilization and enabling high-frequency operation.
- A **hardware-efficient realization** on a Xilinx Artix-7 FPGA, showing improved area–delay product (ADP) and power efficiency compared to earlier non-integrated Karatsuba–Montgomery multipliers. FPGA results confirm lower LUT use and shorter critical paths due to integrated modular reduction.
- **Scalability and broader applicability**, including support for larger NIST binary fields (B-233, B-283, B-409, B-571). The stagewise recursive pipeline allows replication and adjustment of lower-level submodules while maintaining one-result-per-cycle throughput. The architecture is algorithm-agnostic and can also support other multiplication schemes (Classical, Toom–Cook, Montgomery), making it suitable for lightweight ECC, post-quantum accelerators, and RISC-V embedded cryptographic extensions.

Collectively, these contributions make the proposed architecture a hardware-efficient foundation for high-performance cryptographic primitives in embedded and IoT systems, showing clear gains in resource use, throughput, and flexibility.

The remainder of this paper is organized as follows. Section 2 reviews the theoretical foundations of Karatsuba multiplication. Section 3 presents the proposed architecture design and discusses performance and device utilization. Section 4 reports the results, analysis, and comparative evaluation with existing methods. Finally, Sect. 5 concludes the paper.

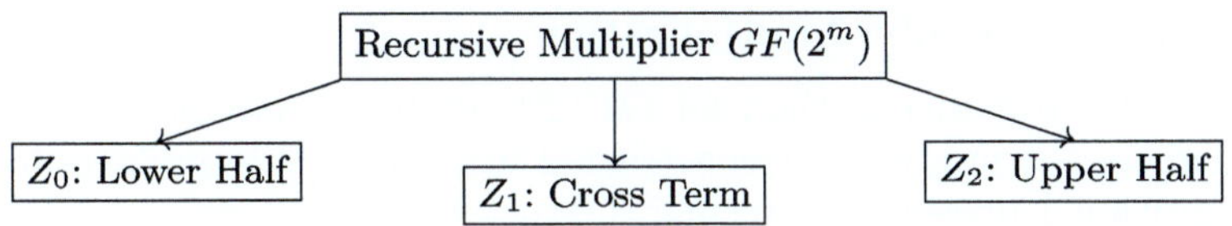

Fig. 1. Recursive decomposition of the Karatsuba multiplier over $GF(2^m)$.

2 Background

Efficient arithmetic in binary extension fields $\mathrm{GF}(2^m)$ is a cornerstone of modern cryptography, like elliptic curve cryptography (ECC) and related finite-field applications. Multiplication over $\mathrm{GF}(2^m)$ dominates the computational cost in many cryptographic protocols, making both area and speed critical considerations in hardware implementations. Various algorithms exist for performing finite-field multiplication, with the Karatsuba algorithm standing out due to its sub-quadratic complexity.

In this work, we focus on FPGA-based implementation of the Karatsuba algorithm combined with modular reduction. Before describing our proposed architecture, we briefly review the theoretical foundation of Karatsuba multiplication, the basics of $\mathrm{GF}(2^m)$ arithmetic, and prior hardware implementations. These discussions establish the context and highlight the limitations of existing approaches, motivating our recursive modular integration.

2.1 Theoretical Foundation of the Karatsuba Algorithm

The Karatsuba algorithm reduces the complexity of n-bit multiplication from $O(n^2)$ to $O(n^{\log_2 3}) \approx O(n^{1.585})$. For two n-bit operands [19,20]

$$X = X_1 \cdot 2^m + X_0, \quad Y = Y_1 \cdot 2^m + Y_0, \quad m = \left\lfloor \tfrac{n}{2} \right\rfloor,$$

the product expands as

$$\begin{aligned} Z = XY &= (X_1 2^m + X_0)(Y_1 2^m + Y_0) \\ &= Z_2 \cdot 2^{2m} + Z_1 \cdot 2^m + Z_0, \end{aligned} \tag{1}$$

with

$$\begin{aligned} Z_2 &= X_1 Y_1, \\ Z_0 &= X_0 Y_0, \\ Z_1 &= (X_1 + X_0)(Y_1 + Y_0) - Z_2 - Z_0. \end{aligned}$$

Thus, four partial multiplications are reduced to three, enabling recursive decomposition and sub-quadratic growth,

As shown in Fig. 1, the multiplier decomposes into three sub-blocks: Z_0 (lower half), Z_1 (cross term), and Z_2 (upper half). Each block may be recursively instantiated until the base word size is reached.

2.2 Binary Karatsuba Multiplication

The modular Karatsuba method for $GF(2^m)$ directly extends the principle of Eq. (1), where the product is reconstructed from three sub-products instead of four. Algorithm 1 presents the recursive structure.

Algorithm Parameters. The function `KaratsubaMod` takes the following parameters:

- A, B: m-bit operands to be multiplied.
- $P(x)$: irreducible reduction polynomial of degree m, used to ensure the result remains in $GF(2^m)$.
- w: base multiplier width, defining the recursion threshold. If $m \leq w$, multiplication is performed by a hardware base multiplier.

At each recursion level, operands are split into halves, recursive products Z_0, Z_1, Z_2 are computed, and the result is reconstructed as in Eq. (1), followed by modular reduction. This ensures that the number of recursive calls grows sub-quadratically while maintaining correctness within the field. The recursion depth is $\lceil \log_2(m/w) \rceil$, and the total cost scales as $O(m^{1.585})$, which is significantly more efficient than the classical $O(m^2)$ method for large operand sizes.

Algorithm 1 outlines the recursive modular Karatsuba multiplier for $GF(2^m)$:

- **Base Case:** If $m \leq w$, compute $C = A \cdot B \bmod P(x)$ using a hardware base multiplier.
- **Operand Splitting:** Split operands into halves:

$$A \to (A_H, A_L), \quad B \to (B_H, B_L).$$

- **Recursive Products:** Compute

$$Z_0 = A_L \cdot B_L \bmod P(x), \quad Z_2 = A_H \cdot B_H \bmod P(x),$$

$$Z_1 = (A_H \oplus A_L)(B_H \oplus B_L) \oplus Z_0 \oplus Z_2.$$

- **Reconstruction:** Form the result:

$$C = (Z_2 \ll m) \oplus (Z_1 \ll m/2) \oplus Z_0 \bmod P(x).$$

Algorithm 1 Recursive Modular Karatsuba Multiplier

Require: A, B: m-bit operands; $P(x)$: irreducible polynomial; w: base width
Ensure: $C = A \cdot B \bmod P(x)$
    if $m \leq w$ **then**
        $C \leftarrow \text{BaseMultiplier}(A, B, P)$
    else
        Split $A \to A_H, A_L, B \to B_H, B_L$
        $Z_0 \leftarrow \text{KaratsubaMod}(A_L, B_L, P, w)$
        $Z_2 \leftarrow \text{KaratsubaMod}(A_H, B_H, P, w)$
        $Z_1 \leftarrow \text{KaratsubaMod}(A_H \oplus A_L, B_H \oplus B_L, P, w) \oplus Z_0 \oplus Z_2$
        $C \leftarrow (Z_2 \ll m) \oplus (Z_1 \ll m/2) \oplus Z_0 \bmod P$
    end if

This recursive structure reduces the number of sub-multiplications from four to three at each level, integrating modular reduction directly within the recursion. As a result, the algorithm achieves $O(m^{1.585})$ complexity, making it highly suitable for finite-field operations in elliptic curve cryptography (ECC) and other large-field cryptographic applications.

2.3   Advantages of Modular Karatsuba Multiplier

The modular Karatsuba multiplier offers several benefits over conventional finite field multipliers, especially for cryptographic applications:

- **Reduced Computational Complexity:** Lowers asymptotic complexity from $O(n^2)$ to $O(n^{\log_2 3}) \approx O(n^{1.585})$, enabling efficient operations on large operands [21].
- **Hardware Efficiency:** Recursive structure supports pipelining and optimal sizing of base multipliers, balancing area and performance [22,23].
- **Scalability and Flexibility:** Recursion depth and base multiplier width can be adjusted for various field sizes, supporting multiple elliptic curve standards [9].
- **Optimized Memory Access:** Recursive partitioning reduces intermediate storage and memory bandwidth, improving local memory utilization [23].
- **Power Efficiency:** Fewer operations and stagewise recursive mappingd data flow reduce switching activity and dynamic power, suitable for embedded/IoT systems [24].
- **Enhanced Cryptographic Performance:** Faster multiplication accelerates ECC operations such as point multiplication, ECDSA, and ECDH [25].
- **Integrated Modular Reduction:** Direct inclusion of the irreducible polynomial keeps results within field bounds, eliminating separate reduction steps [26].

3   Proposed Method for Architecture Design

The proposed Karatsuba multiplication circuit employ a recursive divide-and-conquer strategy. Figure 2 illustrates the 2-bit multiplier, which computes $a_1a_0 \times b_1b_0$ using three AND gates and four XOR gates to generate the 3-bit output $c_2c_1c_0$. Figure 3 extends this approach with three 2-bit Karatsuba blocks ("2×2 KA") and six XOR gates to process 4-bit operands, yielding the result $c_6c_5c_4c_3c_2c_1c_0$. The portion of the right side circuit indicate Overlap logic that combines partial products, while delay elements ensure synchronization.

Both designs demonstrate Karatsuba's key advantage reducing multiplications while retaining necessary additions in the recursive structure.

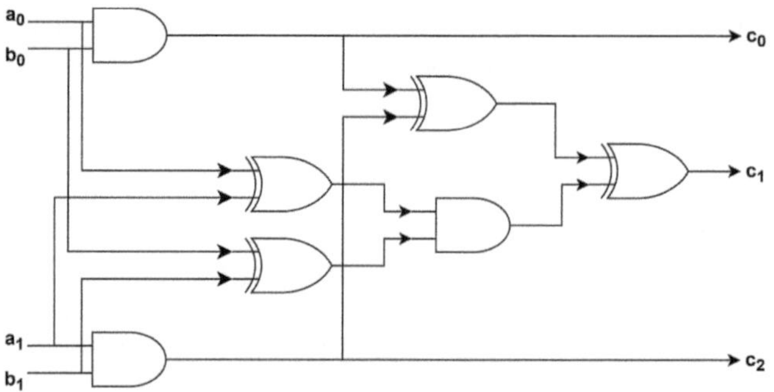

Fig. 2. Logic diagram of a 2-bit modular Karatsuba multiplier.

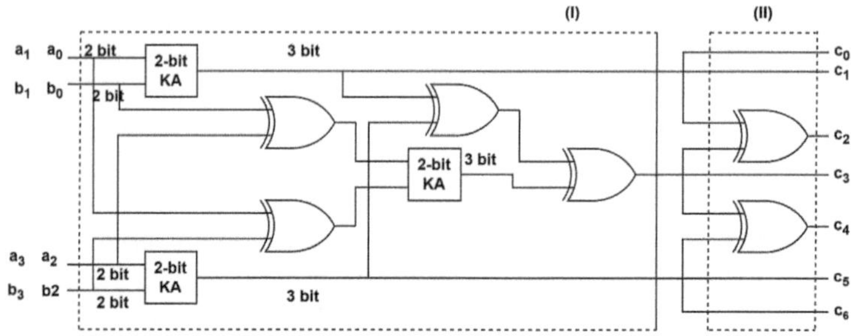

Fig. 3. Logic diagram of a 4-bit modular Karatsuba multiplier.

3.1 Sub-Quadratic Complexity

Let $T(m)$ denote the AND count for m-bit operands:

$$T(m) = 3\,T(m/2) + c\,m,$$

where $c\,m$ accounts for XOR operations in the cross-term. By the Master Theorem ($a = 3$, $b = 2$, $f(m) = O(m)$):

$$T(m) = \Theta(m^{\log_2 3}) \approx \Theta(m^{1.585}).$$

Similarly, XOR count $X(m) = \Theta(m^{\alpha})$, with $\alpha \approx 1.63$, due to additional recombination operations.

Figures 4 and 5 depict the process of calculating the sub-quadratic complexity of the recursive Karatsuba multiplier for $\mathrm{GF}(2^{163})$, and the stagewise recursive mapping partitions operands into progressively smaller subblocks, with each stage performing XOR/AND operations for partial products and registers ensuring continuous data flow. The recursive decomposition principle of the Karatsuba multiplier, shown in Fig. 6, can be efficiently pipelined to enhance throughput while preserving the sub-quadratic computational complexity. Each recursion level performs three concurrent sub-multiplications over half-sized operands,

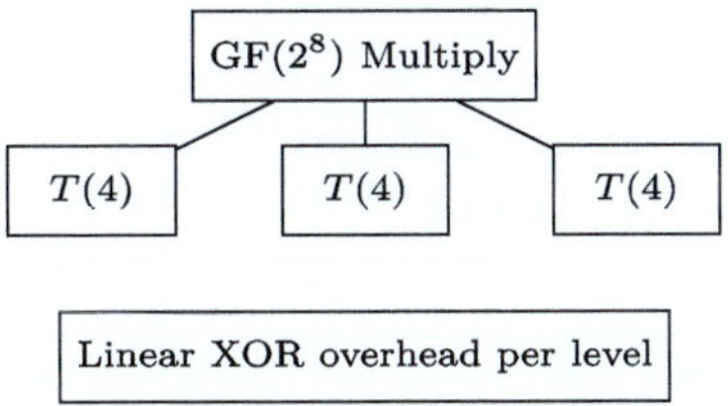

Fig. 4. Recursion tree: each node makes 3 calls per level with linear XOR overhead, yielding sub-quadratic growth.

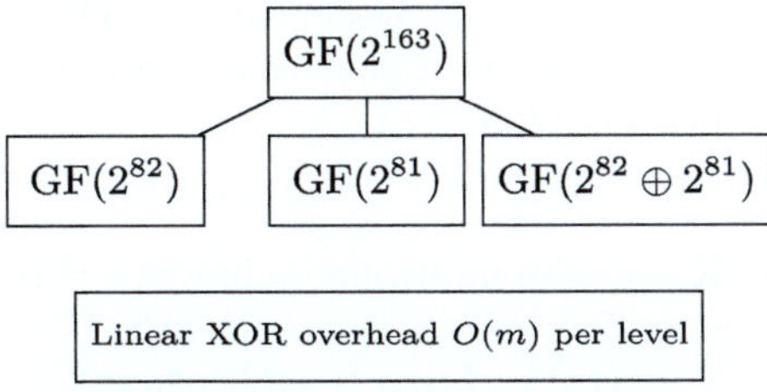

Fig. 5. Recursion tree for $GF(2^{163})$ Karatsuba multiplication. Each node represents a sub-multiplication, showing three recursive calls per level. Linear XOR overhead per level adds $O(m)$, resulting in sub-quadratic growth: $T(m) = 3T(m/2) + O(m)$.

which are structurally independent and hence suitable for parallel execution within a pipeline stage.

Let the operand length be m and the corresponding delay for one multiplication stage be D_m. The pipelined implementation partitions the computation into $L = \log_2 m$ recursive stages, where each stage i performs three parallel multiplications of size $\frac{m}{2^i}$, followed by XOR-based recombination with $O(m/2^i)$ complexity. The total latency T_L and throughput Θ can be expressed as:

$$T_L = \sum_{i=0}^{L-1} D_{m/2^i} + O(m), \quad \Theta = \frac{1}{D_{\text{stage}}}.$$

Since the depth of the pipeline grows logarithmically with operand size, the asymptotic delay reduces to:

$$T_L = O(\log_2 m), \quad \text{and} \quad \Theta = O(1),$$

assuming each stage is balanced in terms of combinational delay. The recurrence relation for the pipelined version remains identical to the classical form:

$$T(m) = 3T(m/2) + O(m),$$

but with concurrent execution of the three sub-operations, the effective latency per recursive level reduces from sequential $3T(m/2)$ to parallel $T(m/2)$. Thus,

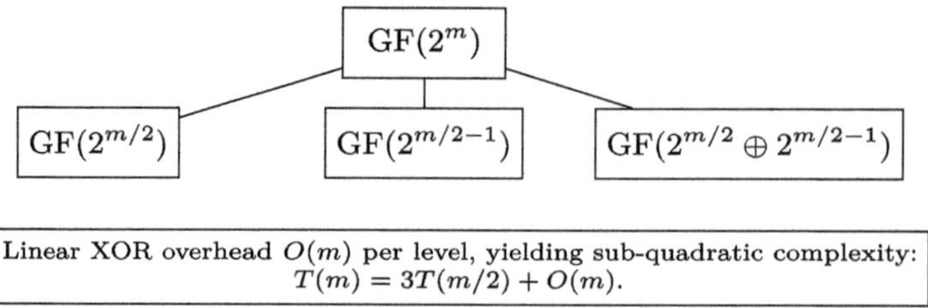

Fig. 6. Generalized recursive structure of the Karatsuba multiplier over GF(2^m). Each node represents a sub-multiplication resulting from recursive decomposition into three halves. The recursion continues until the base field (e.g., GF(2^8)) is reached. At each level, XOR and linear combination operations contribute $O(m)$ overhead, leading to overall sub-quadratic complexity $O(m^{\log_2 3})$.

the pipeline-optimized Karatsuba multiplier achieves a theoretical time complexity of:

$$T_{\mathrm{pipe}}(m) = T(m/2) + O(\log_2 m),$$

while maintaining the same hardware complexity order $O(m^{\log_2 3})$. In hardware realization, each node in Fig. 6 can be mapped to an independent pipeline stage, enabling continuous data flow and one multiplication result per clock cycle after the pipeline is filled.

3.2 Recursive Modular Karatsuba Multiplier: Complexity, Pipeline, and Security

The proposed Recursive Modular Karatsuba Multiplier (RMKM), illustrated in Fig. 6, employs recursive decomposition of m-bit operands into three sub-multiplications: GF($2^{m/2}$), GF($2^{m/2-1}$), and GF($2^{m/2} \oplus 2^{m/2-1}$). This follows the Karatsuba identity:

$$A(x) \cdot B(x) = P_2 \cdot x^m + (P_1 + P_2 + P_0) \cdot x^{m/2} + P_0,$$

where

$$P_0 = A_0 \cdot B_0, \quad P_2 = A_1 \cdot B_1, \quad P_1 = (A_0 + A_1)(B_0 + B_1),$$

and linear XOR-based combinations contribute $O(m)$ per recursion level. The recurrence relation is therefore $T(m) = 3T(m/2) + O(m)$, leading to sub-quadratic complexity $T(m) = O(m^{\log_2 3}) \approx O(m^{1.585})$. The recursion continues until a base field (e.g., GF(2^8)) is reached, where conventional multiplication is used.

Pipelining: The RMKM architecture maps each recursion level to a dedicated pipeline stage, allowing three sub-multiplications to execute concurrently. Let

$D_{m/2^i}$ be the delay of stage i. With $L = \log_2 m$ recursive stages, the total pipeline latency is:

$$T_L = \sum_{i=0}^{L-1} D_{m/2^i} + O(m),$$

while the throughput after filling the pipeline is one multiplication per cycle:

$$\Theta = \frac{1}{D_{\text{stage}}}.$$

Parallel execution of sub-multiplications reduces effective latency per recursive level from $3T(m/2)$ (sequential) to $T(m/2)$ (parallel), achieving an effective pipelined delay of $T_{\text{pipe}}(m) = T(m/2) + O(\log_2 m)$ without increasing hardware complexity.

Side-Channel and Security Considerations: The uniform and regular computation in each pipeline stage minimizes timing variations, while integrated modular reduction prevents leakage of intermediate results. Balanced logic paths and independent sub-multipliers further reduce susceptibility to SPA/DPA attacks. Additionally, masking or blinding can be applied to sub-multipliers to enhance resistance to higher-order attacks.

Although optimized for $GF(2^{163})$ in NIST ECC, the RMKM design is scalable to other binary fields (e.g., $GF(2^{233})$, $GF(2^{283})$) and can serve as a core arithmetic engine for a variety of cryptographic primitives, including elliptic curve integrated encryption scheme (ECIES), pairing-based cryptography, and post-quantum schemes. Its pipelined recursive structure enables high throughput with low area and energy overhead on both FPGAs and ASICs, making it well-suited for embedded and IoT security applications.

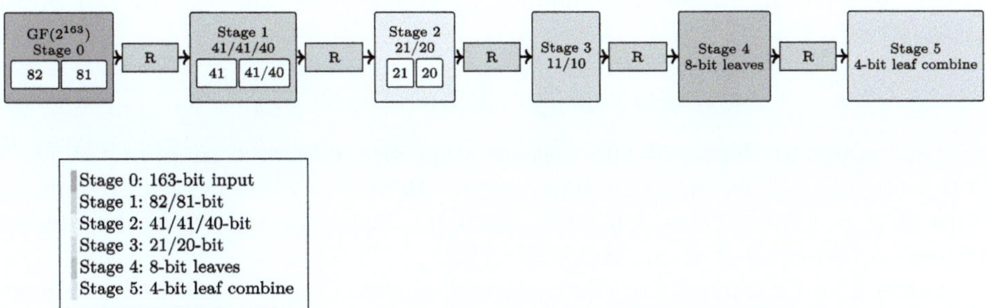

Fig. 7. Proposed 6-stage Recursive Modular Karatsuba multiplier for $GF(2^{163})$ down to 4-bit leaf multiplier, showing staged bit-width partitions, and registers.

3.3   Scalability and Depth of Recursion

Figure 7 illustrates the recursive Karatsuba multiplier for $GF(2^{163})$, partitioned down to 4-bit leaf multipliers with intermediate stagewise recursive mapping registers for high throughput.

Stagewise Partition Mapping: Figure 7 illustrates the 6-stage stagewise recursive mappingd architecture. The stages process operands hierarchically:

- **Stage 0:** Split 163-bit input into 82- and 81-bit subblocks; perform top-level XORs.
- **Stage 1:** Process 82/81-bit blocks, halving into 41- and 41/40-bit subblocks.
- **Stage 2:** Operate on 21- and 20-bit subblocks.
- **Stage 3:** Divide 21- and 20-bit blocks into 11- and 10-bit subblocks.
- **Stage 4:** Base 8-bit leaf multipliers compute fundamental products.
- **Stage 5:** 4-bit leaf combination stage finalizes the multiplication.

Registers between stages maintain one result per cycle throughput. The critical path propagates through all six stages, giving a total latency of six cycles. Color-coded stages indicate pipeline levels: blue for input, green to red for intermediate decomposition, cyan for leaf multipliers. This structure balances hardware utilization, clock frequency, and throughput for high-performance cryptographic multiplication.

3.4   Stagewise Space Complexity of $GF(2^{163})$ Karatsuba Multiplier

The recursive Karatsuba algorithm decomposes an m-bit multiplication into three sub-multiplications per level. Let the operand width at recursion level k be $n_k = \lceil 163/2^{k+1} \rceil$, with 3^k sub-multiplications at level k. Using a schoolbook model, each sub-multiplication requires n_k^2 AND gates and roughly $3n_k$ XOR gates, giving

$$\text{ANDs}(k) = 3^k n_k^2, \quad \text{XORs}(k) \approx 3^{k+1} n_k.$$

The recursion depth of the 6-stage stagewise recursive mapping is $D = \lceil \log_2(163/w) \rceil = 6$ levels, assuming a base width of $w = 4$ bits per leaf. Recursion stops at $k = 5$ (base case: 3–6 bits). With full stagewise partitions, the design achieves a throughput of one result per cycle.

Tables 1 and 2 summarize the total logical gates and instantiated hardware resources, respectively, for the 6-stage stagewise recursive mapping achieving one result per cycle throughput.

Design Trade-offs: Finer-grain staging improves timing but increases area and register usage, whereas merging stages reduces registers at the expense of longer combinational paths. The proposed 6-stage stagewise recursive mapping with 4-bit leaf combination strikes a balance between critical path, area, and

Table 1. Aggregated AND/XOR counts per stagewise recursive stage (all sub-multiplications included).

Stage k	Bit-width(s)	AND	XOR
0	82/81	6724	163
1	41/41/40	5043	82/81
2	21/20	3969	41/40
3	11/10	3267	21/20
4	6-bit leaves	2916	11/(8×3)
5	3-bit leaf	2187	8
Total		24106	411

Table 2. Instantiated AND/XOR counts per stagewise recursive stage (single multiplier per stage).

Stage k	Bit-width(s)	AND	XOR
0	82/81	6724	246
1	41/41/40	1681	123
2	21/20	441	63
3	11/10	121	33
4	6-bit leaves	36	18
5	3-bit leaf	9	9
Total		9012	492

throughput, making it best suited for ECC over $GF(2^{163})$ on larger or different operand sizes.

The multiplier achieves a latency of 6 cycles, a throughput of one result per cycle, and sub-quadratic growth in hardware resources. This balanced stage allocation enables high operating frequency, demonstrating the design's efficiency in the next result Sect. 4.

This recursion directly maps to the six-stage mapping shown in Fig. 8: Stage S0 performs the initial 163-bit split into 82/81-bit blocks; Stage S1 splits 82/81-bit operands into 41/41/40-bit subblocks; Stage S2 processes 21/20-bit partitions; Stage S3 handles 11/10-bit subblocks; Stage S4 reduces to 8-bit leaves and combines partial products; and Stage S5 computes the 4-bit base multiplications at the leaf nodes. Registers between stages ensure a one-result-per-cycle throughput, while each stage bounds the combinational delay, enabling high-frequency operation.

Figures 8 and 9 depict the proposed stagewise recursive mapping of the Hybrid Karatsuba Multiplier. In contrast to conventional Karatsuba architectures that perform multiplication and modular reduction as two distinct processes, the proposed framework integrates modular reduction within each recursive stage (S0–S5). This integration significantly reduces data transfer

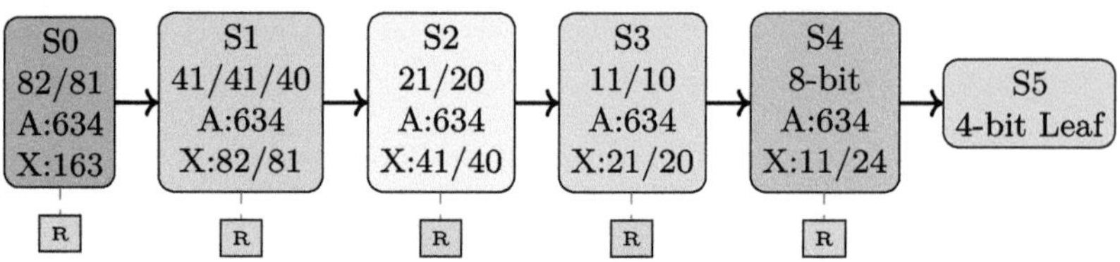

Fig. 8. Compact six-stage Karatsuba recursive mapping for $GF(2^{163})$ with per-stage XOR/AND counts. Gray registers ensure a one-result-per-cycle throughput with a 6-cycle latency. (Color figure online)

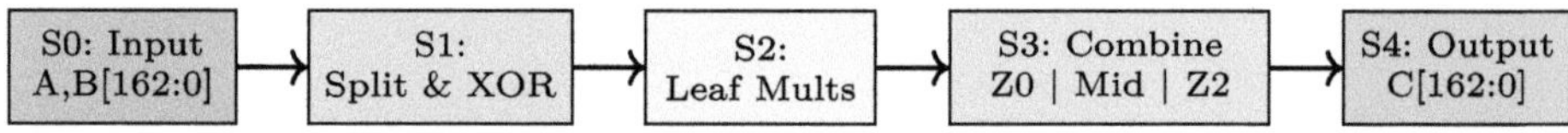

Fig. 9. 5-stage stagewise pipeline mapping $GF(2^{163})$ Karatsuba multiplier. Stage 3 shows intermediate results (Z0, Mid, Z2). Latency = 5 clocks; throughput = 1 product/clock.

overhead and improves the utilization of partial products. Each stage is balanced through pipeline registers (shown in gray), allowing a uniform propagation delay across the datapath and enabling a one-output-per-cycle throughput after initial latency. The recursive inter-stage structure ensures parallelism between sub-block multiplications and modular reduction, which represents a major architectural innovation compared to prior non-integrated Karatsuba or classical multipliers.

Figure 9 illustrates the proposed 5-stage stagewise recursive mapping, where input operands are registered, split/XORed, processed through leaf multiplications, and combined via Karatsuba before $GF(2^{163})$ reduction. The architecture sustains one result per cycle after a 5-cycle latency.

The recursive modular Karatsuba multiplier (Algorithm 1) is implemented using a 5-stage stagewise recursive mapping to maximize throughput on FPGA. Operands are split hierarchically, and each stage of the stagewise recursive mapping performs a portion of the recursive multiplication: Stage 0 splits inputs and computes top-level XORs; Stages 1–3 perform recursive Karatsuba multiplications of subblocks; Stage 4 executes the final combination and modular reduction. This mapping allows one multiplication result per clock cycle once the stagewise recursive mapping is filled, while maintaining the reduced computational complexity of $O(n^{\log_2 3})$ and integrating modular reduction within the recursion. Consequently, the architecture achieves high throughput, low latency, and efficient resource utilization suitable for $GF(2^{163})$ arithmetic in cryptographic applications.

Figure 10 illustrates the cycle-by-cycle behavior of the 5-stage pipelined Karatsuba multiplier. Each new pair of 163-bit operands enters Stage 0 in every clock cycle. After passing through the pipeline stages (Split, High-part multiplication, Low-part multiplication, Cross-term multiplication, and Recombination),

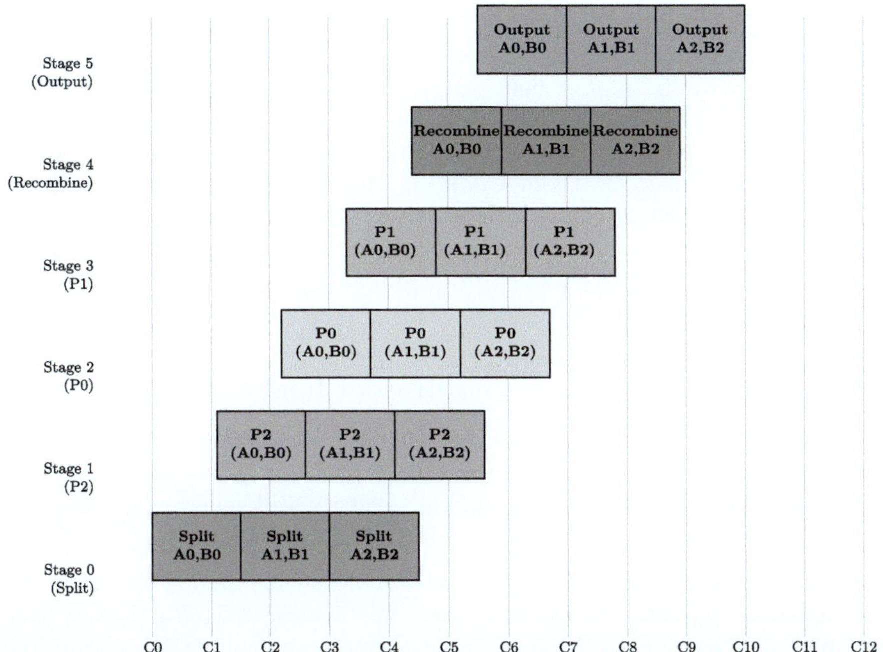

Fig. 10. Timing diagram for 163-bit Karatsuba multiplier. Each new input enters Stage 0 every cycle; after a latency of 5 cycles, one product exits per cycle.

the result emerges at the output after a latency of five cycles. Once the pipeline is filled, the architecture achieves a throughput of one multiplication result per cycle. This demonstrates the advantage of pipelining, where multiple operand pairs are processed concurrently in different stages of the multiplier, significantly improving performance over a non-pipelined design.

Discussion. Figure 10 illustrates the pipeline timing sequence of the proposed architecture. Each recursive stage operates in a synchronized fashion with balanced pipeline registers to maintain uniform stage delays. The diagram highlights the overlapping execution of partial multiplication and modular reduction phases, demonstrating how recursion-level pipelining improves both frequency and throughput. Once the pipeline is filled, a new modular product is obtained every clock cycle, achieving full hardware utilization and throughput efficiency compared to traditional non-pipelined designs. This timing behavior validates the effectiveness of the proposed integrated mapping strategy and its impact on sustained high-speed operation.

Scalability and Extension. Although the proposed pipelined recursive architecture is demonstrated for the NIST B-163 field, the same stagewise mapping and pipeline strategy can be efficiently extended to larger binary fields such as B-233, B-283, B-409, and B-571 by appropriately modifying the operand segmentation and recursion depth. Since the Karatsuba structure recursively decomposes operands, higher field sizes only require proportional replication of the lower-level submodules and pipeline registers. Moreover, the pipelined framework is algorithm-independent and can be adapted to other multiplication schemes such as Classical, Toom–Cook, or Montgomery multipliers with minimal changes in control and modular reduction logic. This scalability confirms the general-purpose and reconfigurable nature of the proposed architecture, making it suitable for a wide spectrum of elliptic curve and post-quantum cryptographic applications over $GF(2^m)$.

3.5   Security Considerations and Broader Implications

While the primary focus of this work is achieving a hardware-efficient and high-throughput Karatsuba multiplier over $GF(2^m)$, the proposed architecture also provides significant advantages in terms of security and generalization.

Side-Channel Resistance. The recursive modular Karatsuba structure inherently balances pipeline stages and data operations, reducing instantaneous switching variations and mitigating potential power and timing side-channel leakage. Integration of modular reduction within each recursion level eliminates separate data-dependent reduction steps, further enhancing uniformity in hardware activity. Although explicit countermeasures are not implemented in this work, the design provides a foundation for future protection using techniques such as operand masking, dual-rail precharge logic, or randomized pipeline scheduling.

Scalability to Other Fields. Although evaluated for the NIST B-163 curve, the stagewise pipeline and recursive mapping are fully scalable to larger binary fields such as B-233, B-283, B-409, and B-571. Scaling requires only adjustment of recursion depth and replication of lower-level submodules, while maintaining balanced stage delays and one-result-per-cycle throughput.

Broader Applicability. Beyond elliptic curve cryptography, the proposed pipeline framework is algorithm-agnostic and can be adapted to other multiplication schemes, including classical, Toom–Cook, or Montgomery multipliers. Its integrated modular reduction at each recursion level makes the architecture a versatile building block for efficient arithmetic accelerators in embedded cryptographic systems, IoT security cores, and lightweight post-quantum cryptography.

4   Results and Analysis

The proposed modular Karatsuba multipliers were evaluated on Xilinx FPGAs, comparing resource utilization, timing, and area-delay efficiency across multiple operand sizes.

4.1   Resource Utilization and Timing Performance

Table 3 presents the resource usage and timing of the modular Karatsuba multiplier for operand sizes from 4 to 128 bits. LUT and FF usage increase gradually with operand size. For instance, 64-bit operands require 2,413 LUTs and 580 FFs. The critical path delay grows from 3.22 ns (4-bit) to 5.25 ns (128-bit) due to integrated modular reduction, while eliminating separate post-processing stages keeps overall latency low.

Table 3. Resource Utilization of non-pipeline Karatsuba Multiplier

Operand	LUT	FF	Delay (ns)	ADP
4	63	37	3.22	202.86
8	145	73	3.22	466.9
16	361	145	3.25	1173.25
32	909	289	4.03	3663.27
64	2413	580	4.45	10737.85
128	6641	1159	5.25	34865.25

The **Area-Delay Product (ADP)** shows that while small operands have higher ADP per bit, larger operands scale efficiently, reflecting good scalability of the modular Karatsuba approach.

4.2   Proposed 5-Stage Recursive Modular Karatsuba Multiplier

The **Recursive-Based Modular Karatsuba Multiplier** for $GF(2^{163})$ was implemented on a Xilinx Artix-7 FPGA (xc7a200tff1156-1). Table 4 summarizes resource usage, power, and timing performance.

The results indicate high hardware efficiency, with less than 1% of FPGA resources consumed. The critical path delay of 4.2 ns supports a maximum operating frequency above 200 MHz, and full throughput of one result per cycle. Over 80% of dynamic power is consumed by I/O switching; in practical deployments with BRAM or AXI/UART interfacing, I/O power would be significantly reduced.

The recursive modular Karatsuba multiplier (Algorithm 1) is implemented using a 5-stage stagewise recursive mapping to maximize throughput on FPGA. Each stage maps to a portion of the recursion: Stage 0 splits operands and computes top-level XORs, Stages 1–3 perform recursive Karatsuba multiplications on subblocks, and Stage 4 executes the final combination and modular reduction. Table 4 shows that this mapping achieves high hardware efficiency, utilizing only 875 LUTs (0.65%) and 1,371 registers (0.51%), while 491 IOBs (98%) handle 163-bit operands. Dynamic and static power are 0.185 W and 0.131 W, respectively, for a total of 0.316 W. The critical path delay of 4.2 ns enables $f_{\max} > 200$ MHz,

Table 4. FPGA Implementation Results of 5-Stage pipeline $GF(2^{163})$ Karatsuba Multiplier

Metric	Value	Observation
Slice LUTs	875/134,600 (0.65%)	Low logic overhead
Slice Registers	1371/269,200 (0.51%)	Lightweight
IOBs	491/500 (98%)	Due to 163-bit operands
BUFGCTRL	1/32 (3.1%)	Single global clock
Dynamic Power	0.185 W	I/O dominates (0.154 W)
Static Power	0.131 W	Baseline device leakage
Total Power	0.316 W	Efficient overall
Critical Delay	4.2 ns	$f_{\max} > 200$ MHz
Latency	5 cycles	One per stage
Throughput	1 result/cycle	Stagewise Pipeline

and the fully recursive modular Karatsuba multiplier design provides a latency of 5 cycles with one result per cycle. These results confirm that integrating the recursive Karatsuba computation into a 5-stage design yields a low-area, high-throughput, and energy-efficient multiplier suitable for FPGA-based cryptographic applications.

Figure 11 illustrates the comparative FPGA implementation results of various $GF(2^m)$ multipliers, highlighting LUTs, Slice utilization, and critical path delay. The proposed 5-stage Recursive-Based modular Karatsuba multiplier (PROP) demonstrates significantly lower LUT and Slice usage compared to prior designs while achieving the shortest critical path of 4.2 ns. Classical (CLS) and Yang [29] designs consume considerably more LUTs and exhibit higher delays, whereas digit-serial or modular-reduction integrated designs (CVS, IMN) show intermediate performance. The recursive modular Karatsuba multiplie architecture of the proposed design enables one result per clock cycle, maintaining high throughput alongside low area and delay. Overall, the proposed multiplier achieves superior area–delay–throughput efficiency, confirming its suitability for high-speed, resource-constrained cryptographic applications.

4.3   Comparison with State-of-the-Art Designs

Table 5 compares the proposed design with classical and other state-of-the-art $GF(2^m)$ multipliers. The 5-stage recursive architecture reduces LUT and slice utilization, improves critical path delay, and achieves single-cycle throughput, outperforming both different Operand size and previous Karatsuba-based designs.

Overall, integrating stagewise recursive mapping stages in the modular Karatsuba multiplier enables **significant speedup and area efficiency**, making it suitable for high-performance cryptographic hardware.

Classical [5] on Virtex-7, Mastrovito [27], and Montgomery multipliers [28] on Virtex-5 FPGAs consume significantly more resources and exhibit higher ADP

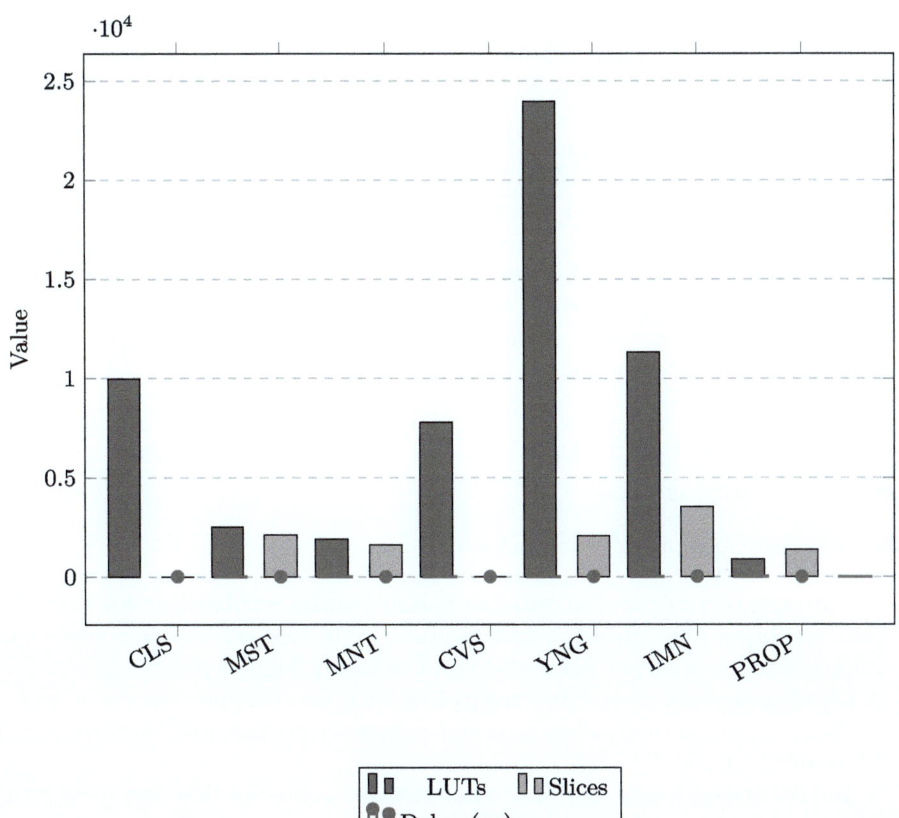

Fig. 11. Comparative FPGA implementation results of $GF(2^m)$ multipliers. Short labels correspond to references: CLS [5], MST [27], MNT [28], CVS [13], YNG [29], IMN [30], PROP: Proposed design.

Table 5. Comparison of the Proposed Recursive-Based Modular Karatsuba Multiplier with State-of-the-Art $GF(2^m)$ Multipliers ($m \leq 163$).

Design/Ref	Device	m	LUTs	Slices	Delay (ns)	ADP	Reduction
Classical [5]	Virtex-7	163	9,982	–	18.129	180,963.68	Yes
Mastrovito [27]	Virtex-5	163	2,500	2,100	8.33	20,825	No
Montgomery [28]	Virtex-5	163	1,900	1,600	7.14	13,566	No
Samanta [31]	Spartan-3E	8	62	36	13.95	865	No
Cuevas [13]	Virtex-5	163	7,786	–	5.468	–	Yes
Yang [29]	Virtex-6	163	23,977	2,061	11.695	–	Yes
Imana [30]	Artix-7	163	11,320	3,532	21.33	75,337	Yes
Proposed Design	Artix-7	163	**875**	**1,371**	**4.2**	**3,675**	Yes

values, providing only half-cycle throughput. Other Karatsuba-based designs, such as Cuevas [13] (Virtex-5) and Yang [29](Virtex-6), achieve comparable throughput but at the cost of higher logic utilization. The Imana [30] design on Artix-7, although high-speed, consumes over 11,320 LUTs and 3,500 slices. for the Prior small-field designs [31] use minimal logic but lack integrated modular reduction and scalability.

In contrast, the proposed Recursive-Based multiplier consumes only 875 LUTs and 1,371 slices, achieving a critical path delay of 4.2 ns and **fully pipeline stagewise throughput of one result per cycle**. This demonstrates that integrating recursive stages within the modular Karatsuba architecture provides the best trade-off between speed, area, and throughput, making it highly suitable for hardware-efficient cryptographic applications.

4.4   Contributions, Future Work

Key Contributions

The main contributions of this work are:

- An **integrated recursive modular Karatsuba multiplier** for $GF(2^{163})$ that combines multiplication and modular reduction within each stage, sustaining one product per clock cycle and reducing logic depth.
- A **pipeline-aware recursive mapping** with five balanced stages, ensuring uniform combinational delay, minimal register overhead, and improved area–delay product (ADP).
- A **hardware-efficient FPGA realization**, achieving low resource usage (875 LUTs, 1,371 registers) and a critical path delay of 4.2 ns, demonstrating superior ADP and power efficiency compared to prior non-integrated Karatsuba and Montgomery multipliers.
- **Scalability and broader applicability**, [15] supporting larger NIST binary fields (B-233, B-283, B-409, B-571) and adaptable to other multiplication schemes (Toom–Cook, Montgomery),
  making it suitable for lightweight ECC, post-quantum accelerators, and RISC-V embedded cryptographic cores.

Future Work

Potential directions for extending this work include:

- Extension to larger NIST fields (B-233, B-283, B-409, B-571) and variable operand sizes, while maintaining pipeline balance and throughput.
- Incorporation of side-channel countermeasures, such as masking or dual-rail logic, to enhance security against power and timing attacks.
- ASIC implementation for further optimization of area, power, and operating frequency.

- Adaptation of the stagewise pipeline framework to other multiplication algorithms (Toom–Cook, Montgomery) and post-quantum cryptographic primitives.
- Exploration of dynamic pipeline scheduling and operand-level parallelism to further increase throughput in high-performance embedded cryptographic systems.

Overall, this work presents a scalable, high-performance, and energy-efficient hardware framework for finite-field multiplication, forming the basis for secure and flexible cryptographic accelerators.

5   Conclusion

This paper presented a recursive modular Karatsuba multiplier with stage-wise pipelining, achieving high throughput, balanced stage delays, and energy-efficient operation on FPGA. By integrating modular reduction within each recursive stage, the design reduces combinational depth and resource usage, outperforming prior non-integrated architectures. The architecture is scalable to larger fields, supports flexible operand sizes, and provides a versatile building block for high-speed cryptographic applications.

Specifically, we proposed a **5-stage pipeline-based modular Karatsuba multiplier** for $GF(2^{163})$, optimized for FPGA-based cryptography. The integrated modular reduction and fully pipelined mapping enable low hardware overhead, reduced critical path delay, and a result per clock cycle. Implementation on a Xilinx Artix-7 FPGA demonstrates that the multiplier consumes only 875 LUTs and 1,371 registers, with a critical path of 4.2 ns, achieving superior area–delay efficiency. Comparative analysis with classical and state-of-the-art $GF(2^m)$ multipliers confirms that the recursive approach offers an optimal trade-off among speed, area, and throughput.

Overall, the design is highly suitable for resource-constrained, high-speed cryptographic systems and can be extended to larger fields or integrated with ECC cores and hardware cryptographic systems to enable secure communications.

Acknowledgment. This research was supported through CSIR-HRDG, with support and resources provided by CSIR-CEERI and the AcSIR academic program.

References

1. Deschamps, J.-P.: Hardware Implementation of Finite-Field Arithmetic. McGraw-Hill, Inc., Boston (2009)
2. Willam, S.: Cryptography and Network Security: Principles and Practice (Global Edition-). Pearson Education, Boston (2022)
3. Kumari, R., Pandey, J.G., Karmakar, A.: An rtl implementation of the data encryption standard (des). arXiv preprint arXiv:2301.05530 (2023)

4. Inoguchi, Y.: Outline of the ultra fine grained parallel processing by fpga. In: Proceedings. Seventh International Conference on High Performance Computing and Grid in Asia Pacific Region, 2004, pp. 434–441. IEEE (2004)
5. Kumari, R., Rout, T., Saini, B., Pandey, J.G., Karmakar, A.: An efficient hardware implementation of elliptic curve point multiplication over gf (2 m) on fpga. In: International Symposium on VLSI Design and Test, pp. 257–271. Springer, Heidelberg (2023)
6. Chen, H., Laine, K., Player, R., Xia, Y.: High-precision arithmetic in homomorphic encryption. In: Smart, N.P. (ed.) CT-RSA 2018. LNCS, vol. 10808, pp. 116–136. Springer, Cham (2018). https://doi.org/10.1007/978-3-319-76953-0_7
7. Zheng, Z., Liu, F., Tian, K.: An unbounded fully homomorphic encryption scheme based on ideal lattices and Chinese remainder theorem. arXiv preprint arXiv:2301.12060 (2023)
8. Hankerson, D., Menezes, A.J., Vanstone, S.: Guide to Elliptic Curve Cryptography. Springer, Heidelberg (2004)
9. Huang, J ., Lee, J., Li, H.: A fast fpga implementation of tate pairing in cryptography over binary field. In: Security and Management, pp. 3–9 (2008)
10. Arish, S., Sharma, R.K: An efficient floating point multiplier design for high speed applications using karatsuba algorithm and urdhva-tiryagbhyam algorithm. In: 2015 International Conference on Signal Processing and Communication (ICSC), pp. 303–308. IEEE (2015)
11. Prema, C., Mainkanda Babu, C.S.: Enhanced high speed modular multiplier using karatsuba algorithm. In: 2013 International Conference on Computer Communication and Informatics, pp. 1–5. IEEE (2013)
12. Lee, T.Y., Liu, M.J., Fan, C.C., Tsai, C.C., Wu, H.: Low complexity digit-serial multiplier over gf (2ˆ m) using karatsuba technology. In: 2013 Seventh International Conference on Complex, Intelligent, and Software Intensive Systems, pp. 461–466. IEEE (2013)
13. Cuevas-Farfan, E., et al.: Karatsuba-ofman multiplier with integrated modular reduction for (2) m gf. Adv. Electr. Comput. Eng. **13**(2), 1–5 (2013)
14. Renita, J., Asokan, N., et al.: Implementation and performance analysis of elliptic curve cryptography using an efficient multiplier. J. Semiconductor Technol. Sci. **22**(2), 53–60 (2022)
15. Kumari, R., Purohit, G., Karmakar, A.: Efficient hardware implementation of modular multiplier over gf (2ˆ m) on fpga. arXiv preprint arXiv:2506.09464 (2025)
16. Chen, H., Jiang, Y., Jin, B.: Scalable karatsuba multiplier over finite field gf (2 m). In: Informatics and Management Science I, pp. 79–84. Springer, Heidelberg (2012). https://doi.org/10.1007/978-1-4471-4802-9_11
17. Lee, T.-Y., Liu, M.-J., Huang, C.-H., Fan, C.-C., Tsai, C.-C., Haixia, W.: Design of a digit-serial multiplier over gf (2 m) using a karatsuba algorithm. J. Chin. Inst. Eng. **42**(7), 602–612 (2019)
18. Cheng, Y.: Space-efficient karatsuba multiplication for multi-precision integers. arXiv preprint arXiv:1605.06760 (2016)
19. Wikipedia contributors. Karatsuba algorithm—wikipedia (2025). https://en.wikipedia.org/wiki/Karatsuba_algorithm. Accessed 10 Sept 2025
20. MIT OpenCourseWare. Lecture 11: Integer arithmetic, karatsuba multiplication (2011). URL https://ocw.mit.edu/courses/6-006-introduction-to-algorithms-fall-2011/MIT6_006F11_lec11.pdf. Accessed 10 Sept 2025
21. Zhang, J.: Efficiency of large integer multiplication algorithms: a comparative study of traditional methods and karatsuba's algorithm. Appl. Comput. Eng. **69**, 30–36 (2024)

22. Chow, G.C.T., Eguro, K., Luk, W., Leong, P.: A karatsuba-based montgomery multiplier. In: Proceedings of the 2010 International Conference on Field Programmable Logic and Applications, pp. 1–6 (2010)
23. Jain, R., Pahwa, K., Pandey, N.: Booth-encoded karatsuba: a novel hardware-efficient multiplier. In: Proceedings of the 2021 International Conference on VLSI Design, pp. 1–6 (2021)
24. Kabin, I., Dyka, Z., Kreiser, D., Langendoerfer, P.: Unified field multiplier for ecc: inherent resistance against horizontal sca attacks. arXiv preprint arXiv:2201.01147 (2022)
25. Pogue, T.E., Nicolici, N.: Karatsuba matrix multiplication and its efficient custom hardware implementations. arXiv preprint arXiv:2501.08889 (2025)
26. Liu, R., Li, S.: A design and implementation of montgomery modular multiplier. In: 2019 IEEE International Symposium on Circuits and Systems (ISCAS), pp. 1–4. IEEE (2019)
27. Petra, N., De Caro, D., Strollo, A.G.M.: A novel architecture for galois fields gf (2^ m) multipliers based on mastrovito scheme. IEEE Trans. Comput. **56**(11), 1470–1483 (2007)
28. Mentens, N., Ors, S.B., Preneel, B., Vandewalle, J.: An fpga implementation of a montgomery multiplier over gf (2^ m). Comput. Inf. **23**(5–6), 487–499 (2004)
29. Yang, Y., Chao, W., Li, Z., Yang, J.: Efficient fpga implementation of modular multiplication based on montgomery algorithm. Microprocess. Microsyst. **47**, 209–215 (2016)
30. Imaña, J.L.: High-speed polynomial basis multipliers over $gf(2^m)$ for special pentanomials. IEEE Trans. Circ. Syst. I: Reg. Pap. **63**(1), 58–69 (2015)
31. Samanta, J., Sultana, R., Bhaumik, J.: Fpga based modified karatsuba multiplier. In: Proceedings of International Conference VLSI Signal Processing (ICVSP), vol. 10, p. 12 (2014)

Symmetric-Key Cryptography

Improved Modeling for Substitution Boxes with Negative Samples and Beyond

Debranjan Pal[1], Anubhab Baksi[2], Surajit Mandal[3(✉)], and Santanu Sarkar[3]

[1] Indian Institute of Technology Kanpur, Kanpur, India
debranjan.crl@gmail.com
[2] Lund University, Lund, Sweden
anubhab.baksi@eit.lth.se
[3] Indian Institute of Technology Madras, Chennai, India
surajit1810@gmail.com, sarkar.santanu.bir1@gmail.com

Abstract. It is a common practice for symmetric-key ciphers is to encode a cryptanalysis problem as an instance of the Mixed Integer Linear Programming (MILP) and then run the instance with an efficient solver. For this purpose, it is essential to model the components in a way that is compatible with the MILP formulation while preserving the characteristics of the cipher. In this work, we look at the problem of efficiently encoding a substitution box (SBox for short). More specifically, we take the evaluation tables, namely, the Difference Distribution Table (DDT), the Linear Approximation Table (LAT), the Division Property Table (DPT) and the Boomerang Connectivity Table (BCT).

In the current stage, the only model proposed/used in the literature is to look for the positive samples (i.e., non-zero) entries in the evaluation table and encode for that (this ultimately leads to the minimum number of active SBoxes for the target cipher). In our first contribution, we experiment with the complementary concept of using the negative samples (i.e., treating the zero entries as non-zero entries and vice versa), finally a proper recovery procedure is performed to get back to the original problem.

In our second contribution, we observe that the top row and column of each table (which are always taken into consideration by the past researchers) are actually redundant, as those are inherently taken care of through additional constraints at another level. We thus propose a simplified model that removes those row and column.

We show the efficacy of our proposals by furnishing results on the evaluation tables on multiple SBoxes. Our proposals often reduce the number of constraints for certain SBox evaluation tables (depending on the properties of the table) from the state-of-the-art results.

Keywords: Block Cipher · Substitution Box · Substitution Box Evaluation Table · Mixed Integer Linear Programming

© The Author(s), under exclusive license to Springer Nature Switzerland AG 2026
R. Dutta et al. (Eds.): INDOCRYPT 2025, LNCS 16372, pp. 45–69, 2026.
https://doi.org/10.1007/978-3-032-13301-4_3

1   Introduction

Over the past few years, multiple tools to find the security bounds for ciphers have emerged. These tools take the problem description of the classical attacks, such as the differential or linear attacks, as a constraint satisfaction or objective optimization problem. Not only does it take away the tedious manual effort, but it is also capable of fine-tuning the bounds (thereby finding a more optimal bound). Naturally, there has been a great deal of interest in this research area, as evidenced by the recent publications. Generally, two types of solvers are used: Satisfiability (SAT) solvers like Cryptominisat and Mixed Integer Linear Programming (MILP) solvers like Gurobi or CPLEX are used, though occasionally we have seen Constraint Programming (CP) or Quadratically Constrained Linear Programming or Simple Theorem Prover (STP) [25] as the tool of choice.

In fact, it is highly important to evaluate the security of classical attacks. This is typically done by the cipher designers (in the paper proposing a new cipher) and independent researchers interested in that cipher.

While automated solvers undeniably reduce the workload for the cipher designers and analyzers alike, there is the problem of formatting the problem (related to the bounds of the cipher) to a form compatible with the solver. When designing new ciphers, it is standard practice to prioritize resistance against classical cryptanalysis. One crucial aspect of assessing a cipher's security against classical attacks is determining the minimum number of active SBoxes (Sect. 2.1) within the cipher[1]. Consequently, calculating the smallest number of active SBoxes for block ciphers [27,37] has garnered significant attention in cryptographic research.

1.1   Overview of Prior Works

In this part, we briefly describe the major works in this field. The basic problem is to take an SBox evaluation table (*Difference Distribution Table* (DDT), *Linear Approximation Table* (LAT), *Division Property Table* (DPT), and *Boomerang Connectivity Table* (BCT)) and convert the information in the said table to a system of linear constraints (so that the constraints can be used in an MILP problem formulation). On top of that, a heuristic idea is to minimize the number of constraints [1,2,17,24,29,34] (it is believed that the lower number of constraints will lead to faster runtime of the overall MILP instance), and this poses an extra challenge. Therefore, the fundamental challenges in this context are:

1. Construct an appropriate set of linear constraints that describes the given SBox evaluation table.
2. Identify a minimal subset of the said constraints that still preserves the SBox evaluation table.

---

[1] This method does not work on ciphers with linear structures, such as DEFAULT [5, Chapter 8] or BAKSHEESH [6].

To address these challenges, researchers have proposed various techniques. Sun et al. [34,35] propose a method to convert from the SBox evaluation tables to linear constraints. Sun et al. [35] introduce a greedy method to eliminate redundant inequalities. In 2017, Sasaki and Todo [31] introduce an innovative algorithm designed to automatically reduce the size of descriptive models. In 2020, Boura and Coggia [18] develop a technique for generating new inequalities based on existing ones. Their experimental results showcase the ability to reduce inequalities compared to previous approaches. In 2017, Abdelkhalek et al. [1] present the first MILP model tailored for 8-bit SBoxes. They focus on the product-of-sum representation and employ the Espresso algorithm to generate the associated inequalities.

1.2   Contribution

Despite recent progress, we observe that it is further possible to optimize the results from the previous works. In this paper, we present two novel approaches to further optimize the state-of-the-art methods. In summary, we propose improved models for SBoxes that are suitable for MILP-aided cryptanalysis to find the minimum number of active SBoxes. Our approaches are general and can be applied to any SBox evaluation table like DDT, LAT, DPT and BCT. We only restrict ourselves to bijective SBoxes in this paper for simplicity, though further generalization for non-bijective SBoxes is possible. For additional details, one may refer to the extended version available in [28].

Our key contributions can be summarized as:

- **Switching to negative samples.** We propose reformulating the SBox constraint system to model the absence of specific transitions or correlations (i.e., the negative samples in the SBox evaluation tables). In this way, we find the complement model (i.e., the maximum number of inactive SBoxes), from which we convert to the actual problem (i.e., the minimum number of active SBoxes). If the solution from the MILP solver, say I, represents the maximum number of SBoxes that can be inactive, then to determine the minimal number of active SBoxes, this value I is subtracted from the total number of SBoxes in the cipher.
- **Exclusion of 0th row/column.** In the survey of literature, we observe that all the papers take the redundant top row and column (that correspond to the zero-to-zero transitions) in the SBox evaluation tables. However, this can be modeled through other means (the constraints which encode whether or not the SBox is active) automatically detect such conditions.

By applying the above two approaches, we get the combined best result in terms of minimum number of linear constraints for the following SBoxes: GIFT, PRESENT, PRINCE, RECTANGLE, LILLIPUT, MIBS, MIDORI S0, MINALPHER, TWINE, ASCON, FIDES-5, KECCAK, SC2000-5, FIDES-6, SC2000-6.

1.3   Organization

- In Sect. 2 we give the necessary prerequisites, including definitions of the SBox and the cryptographic property tables used in the analysis, such as the *Difference Distribution Table* (DDT), *Linear Approximation Table* (LAT), *Division Property Table* (DPT), and *Boomerang Connectivity Table* (BCT).
- Section 3 reviews related work in the field, discussing previous methods for generating and reducing inequalities for SBox modeling.
- In Sect. 4 we introduce our first proposed idea, the concept of complementary variable encoding by focusing on modeling negative transitions (impossible propagation) to simplify the MILP model.
- We present our second idea, which involves excluding redundant zero-to-zero transitions (the first row and column) in the SBox evaluation tables to optimize the modeling process further in Sect. 5.
- Section 6 presents our experimental results, where we compare our proposed methods to traditional approaches by looking at the number of inequalities they produce across different SBoxes and property tables.
- Section 7 concludes the paper by summarizing our findings and discussing their implications for the automated security evaluation of ciphers.
- Additionally, Appendix A provides some relevant examples. These include description of our proposed methods applied to specific SBoxes and property tables. Indices to the SBoxes used/analyzed in our experiments are provided in Appendix B.

2   Prerequisites

2.1   Substitution Box (SBox)

A substitution box, (SBox for short), denoted S, is a mapping represented by $S : \mathbb{F}_2^n \to \mathbb{F}_2^n$. Any operation on an SBox can be depicted as a transformation from an input x to an output y, where both x and y belong to the finite field $\mathbb{F}_2^n$.

Consider the $2n$-bit vector $(x_0, x_1, \ldots, x_{n-1}, y_0, y_1, \ldots, y_{n-1})$ over $\mathbb{F}_2^{2n}$ where we indicate through x the input characteristic to the SBox and through y the output characteristic from the SBox. Depending on the property of interest, the input-output characteristics can be differential propagation (in the context of differential attack) or linear masking (in the context of linear attack), and so on.

2.2   Difference Distribution Table (DDT)

The differential behavior of an SBox is usually characterized by its *Difference Distribution Table* (DDT). For an n-bit input and an m-bit output SBox $S : \mathbb{F}_2^n \to \mathbb{F}_2^m$, the DDT is a $2^n \times 2^m$ matrix whose entry at position $(\Delta x, \Delta y)$ is defined as $\mathrm{DDT}[\Delta x, \Delta y] = \#\{x \in \mathbb{F}_2^n \mid S(x) \oplus S(x \oplus \Delta x) = \Delta y\}$. Here, $\Delta x \in \mathbb{F}_2^n$ denotes an input difference and $\Delta y \in \mathbb{F}_2^m$ an output difference, while

$\oplus$ represents bitwise XOR. Each row of the DDT corresponds to a fixed input difference and sums to 2^n, reflecting that every input x contributes exactly one output difference. The first row (for $\Delta x = 0$) is trivial, with $\text{DDT}[0,0] = 2^n$ and all other entries equal to zero. The maximum value among the nontrivial entries, called the *differential uniformity* of the SBox, determines its resistance against differential cryptanalysis [12,14,23].

2.3 Linear Approximation Table (LAT)

A *Linear Approximation Table* (LAT) is constructed to measure the biases of linear approximations between the input and output bits of a cryptographic function, such as an SBox. For an n-input, m-output function $F : \{0,1\}^n \to \{0,1\}^m$, we are interested in linear combinations of its inputs and outputs. Let $a \in \{0,1\}^n$ be a binary vector representing a linear combination of input bits, and $b \in \{0,1\}^m$ bes also a binary vector representing a linear combination of output bits. The linear approximation is expressed as: $a \cdot x \oplus b \cdot F(x) = 0$, where: $a \cdot x = \bigoplus_{i=1}^{n} a_i x_i$ (XOR of the selected input bits), $b \cdot F(x) = \bigoplus_{j=1}^{m} b_j F(x_j)$ (XOR of the selected output bits). Now, the bias measures how often this approximation holds true compared to random guessing. We define: $P(a,b) = \Pr(a \cdot x \oplus b \cdot F(x) = 0)$, where x is chosen uniformly at random from $\{0,1\}^n$. The bias $\epsilon(a,b)$ is, $\epsilon(a,b) = P(a,b) - \frac{1}{2}$.

The entry $\text{LAT}[a,b]$ in the table is computed as, $\text{LAT}[a,b] = N \cdot \epsilon(a,b)$, where $N = 2^n$ is the total number of possible inputs. Alternatively, $\text{LAT}[a,b] = \#\{x \in \{0,1\}^n : a \cdot x \oplus b \cdot F(x) = 0\} - \frac{N}{2}$.

2.4 Division Property Table (DPT)

Let d represent the algebraic degree of the SBox. It is possible to select 2^{d+1} inputs to the SBox such that the XOR sum of the corresponding SBox outputs equals 0.

Let $\pi_u : \mathbb{F}_2^n \to \mathbb{F}_2$ be a bit-product function defined for any $u \in \mathbb{F}_2^n$ and $x \in \mathbb{F}_2^n$, where the i-th bits of u and x are denoted as u_i and x_i, respectively. The function $\pi_u(x)$ is defined as: $\pi_u(x) := \prod_{i=1}^{n} x_i^{u_i}$.

Let X be a multiset whose elements are values in $\mathbb{F}_2^n$, and k be an integer between 0 and n. If the multiset X satisfies the division property D_n^k, the following two conditions hold: the parity of $\pi_u(x)$ is always even for all $x \in X$ and for any u with a Hamming weight less than k. Also, the parity of $\pi_u(x)$ becomes indeterminate for any u with a Hamming weight greater than or equal to k. One approach is to examine the presence of monomials in the algebraic normal forms (ANFs) of $x \mapsto \pi_v(S(x))$ for $v \in \mathbb{F}_2^n$. This is represented by the set $P_S(u)$, defined as $\bigcup_{w \in \text{Succ}(u)} Q_S(w)$, where $Q_S(w) = \{v \in \mathbb{F}_2^n : \pi_v(S(x)) \text{ contains } \pi_w(x)\}$, and $\text{Succ}(u) = \{x \in \mathbb{F}_2^n : u \preceq x\}$, which forms an affine subspace of dimension $n - \text{wt}(u)$, with $\text{wt}(u)$ being the Hamming weight of u. The table representation of $P_S(u)$ for all u provides valuable insight into the resistance of the function against division-property-based attacks.

2.5   Boomerang Connectivity Table (BCT)

The *Boomerang Connectivity Table* (BCT) is a cryptanalysis tool used to evaluate the probability of successfully connecting two differential characteristics in a boomerang-style attack, particularly when the middle part of the cipher consists of a single SBox layer. In a boomerang attack, the cipher is often considered a composition of sub-ciphers. The attack's success depends on how the differential characteristics of these parts interact at their boundary. The BCT provides a systematic way to measure this interaction.

It analyzes the formation of a boomerang quartet, a set of four values (x_1, x_2, x_3, x_4) at the input of the SBox layer. These values must satisfy two conditions simultaneously:

1. The input difference from the first part of the cipher is Δ_i, such that $x_1 \oplus x_2 = \Delta_i$, and we check if $x_3 \oplus x_4 = \Delta_i$.
2. The output difference required by the second part of the cipher is ∇_o, such that $S(x_1) \oplus S(x_3) = \nabla_o$ and $S(x_2) \oplus S(x_4) = \nabla_o$.

The BCT calculates the probability of these conditions holding true for all possible pairs of (Δ_i, ∇_o).

For an n-bit invertible SBox, the BCT is a $2^n \times 2^n$ table. Each entry in the table, for a given input difference Δ_i and output difference ∇_o, stores a count. It represents the number of initial values x that will result in a valid boomerang quartet.

Each entry of $\mathcal{T}_{BCT}(\Delta_i, \nabla_o)$ is calculated by the following formula, which counts the number of successful inputs: $\mathcal{T}_{BCT}(\Delta_i, \nabla_o) = \#\{x \in \{0,1\}^n | S^{-1}(S(x) \oplus \nabla_o) \oplus S^{-1}(S(x \oplus \Delta_i) \oplus \nabla_o) = \Delta_i\}$. To find an entry, one must iterate through every possible input x (from 0 to $2^n - 1$) and check if the condition in the formula is met. The final value is the total count. The probability of forming a right quartet is this count divided by 2^n.

2.6   Classification of Samples

Initially, to construct a model for an SBox, we adopt a binary classification perspective of its evaluation tables. In this viewpoint, each entry in the SBox evaluation table is treated as a data point that is to be assigned to one of two classes (though, one may note that the DPT is already a binary table). Entries corresponding to positive values are categorized into positive classes, indicating feasible transitions contributing to potential differential or linear characteristics. Conversely, entries that take the value zero are placed into the negative class, since they represent impossible transitions that must be excluded from any valid characteristic. This classification provides a clean separation between feasible and infeasible behaviors of the SBox, enabling us to systematically design linear constraints that capture its propagation properties within a MILP framework.

Example 1. We illustrate the two encodings on the DDT of the chosen 3-bit SBox S, 05276134. We first convert the DDT (which is given in Table 1a) into a binary table as given in Table 1b. Here, we use '+' to denote positive samples and '−' for negative samples.   □

Table 1. ABC SBox (conversion from DDT to binary DDT)

(a) DDT

Δx \ Δy	0	1	2	3	4	5	6	7
0	8	0	0	0	0	0	0	0
1	0	0	0	0	0	4	0	4
2	0	0	4	0	0	4	0	0
3	0	0	4	0	0	0	0	4
4	0	2	0	2	2	0	2	0
5	0	2	0	2	2	0	2	0
6	0	2	0	2	2	0	2	0
7	0	2	0	2	2	0	2	0

(b) Binary DDT

Δx \ Δy	0	1	2	3	4	5	6	7
0	+	−	−	−	−	−	−	−
1	−	−	−	−	+	−	+	−
2	−	−	+	−	−	+	−	−
3	−	−	+	−	−	−	−	+
4	−	+	−	+	+	−	+	−
5	−	+	−	+	+	−	+	−
6	−	+	−	+	+	−	+	−
7	−	+	−	+	+	−	+	−

2.7   Hierarchy of Structures

We classify the modeling of an SBox into a hierarchy of structures based on the depth of integration of its evaluation tables. This allows us to systematically build MILP models that capture the cryptographic properties of the SBox at varying levels of detail and complexity. Within the context, only the *primary* and *secondary* structures are discussed. However, further hierarchical levels are needed to formulate for the entire MILP model that encodes for multiple rounds of a cipher where each round consists of multiple SBoxes in parallel (*tertiary*) and makes use of MILP solver-friendly optimization in the final hierarchy (*quaternary*), both of which are kept out of scope. Refer to Appendix A for examples of primary and secondary structures.

Primary Structure. In the primary structure, the focus is on directly modeling the propagation properties of an SBox using its classical cryptographic evaluation table, such as the DDT, (absolute) LAT, DPT, and BCT (and possibly more). These tables capture the cryptographic behavior of the SBox under various cryptanalytic methods (differential, linear, algebraic, and boomerang analyses, respectively).

Here, every entry of the DDT, LAT, DPT, or BCT is classified into one of two categories, namely positive (i.e., non-zero transition) class and negative (i.e., zero transition) class. Traditionally, the feasible transitions (positive samples) are solely used in the primary structure to construct the corresponding MILP model. By treating the modeling task as a binary classification problem, constraints are generated to separately capture the conditions for positive transitions and negative transitions. The primary structure, therefore, enriches the MILP framework with discriminative power, as it systematically distinguishes between valid and impossible propagations (i.e., negative samples).

Secondary Structure. Moving to the secondary structure, the objective shifts from the binary classification viewpoint to integration for an SBox. Auxiliary variables are introduced to represent the state of each SBox, indicating whether it is active (participating in propagation) or inactive (not participating). Thus, the secondary structure encapsulates the primary structure's classification into a slightly more generalized framework that tracks the activity status of each SBox.

3   Related Works

In this part, we assume that the cryptographic table from the SBox is provided (the table elucidates a binary classification problem). Now we want to formulate that table into a collection of linear constraints so that it can be used in an MILP formulation.

3.1   Representation of an SBox Using Convex Hull (CH)

The *Convex Hull* (CH) of a set P of points is the smallest convex set that contains all the points in P. The process involves calculating the H-representation, which is achieved by determining the cutting planes that collectively form the convex hull. Specifically, we apply the implementation from Sage[2] to extract the H-representation. Consequently, the H-representation yields a system of w linear constraints, taking the form, $A \cdot [x_0, \ldots, x_{n-1}, y_0, \ldots, y_{n-1}]^\top \leq b^\top$. Here, A is the $w \times 2n$ matrix and $b = (b_0, b_1, b_2, \ldots, b_{2n-1})$. Due to the innate implementation in Sage, both matrix A and vector b consist solely of integers. Each linear constraint serves to invalidate certain points, associated with unattainable differential propagation.

3.2   Modeling Using Conditional Differential Behavior

The logical relationship $(x_0, \ldots, x_{m-1}) = (\delta_0, \ldots, \delta_{m-1}) \in \{0,1\}^m \subseteq \mathbb{Z}^m$, which implies $y = \delta \in \{0,1\} \subseteq \mathbb{Z}$, can be effectively captured through the following linear constraint: $\sum_{i=0}^{m-1} (-1)^{\delta_i} x_i + (-1)^{\delta+1} y - \delta + \sum_{i=0}^{m-1} \delta_i \geq 0$
Here, let us consider (x_0, x_1, x_2, x_3) and $(y_0, y1, y_2, y_3)$ as the variables for input and output differences of a 4-bit SBox. Suppose there's an impossible propagation in the DDT of the SBox: $(1010) \rightarrow (0111)$, indicating that the input difference (1010) does not propagate to (0111). Utilizing the relationship noted earlier, the linear constraint that asserts the impossible point (i.e., negative sample) $(1001, 0111)$ is expressed as: $-x_0 + x_1 - x_2 + x_3 + y_0 - y_1 - y_2 - y_3 + 4 \geq 0$.

For an SBox, if there are n impossible paths evident in a DDT, at most n linear constraints are necessary to accurately model the DDT. However, it is possible to reduce the value of n by merging existing constraints, thereby generating new ones.

---

[2] https://www.sagemath.org/.

As an illustration, consider two impossible propagations (i.e., negative samples): $(1010) \rightarrow (0111)$ and $(1010) \rightarrow (0110)$. A combined linear constraint: $-x_0 + x_1 - x_2 + x_3 + y_0 - y_1 - y_2 \geq -3$.
effectively eliminates both of these impossible points simultaneously. This consolidation of constraints helps streamline the representation of impossible propagations and enhances the efficiency of the system.

3.3   Selecting Best Constraints from Convex Hull

When applying an MILP model, obtaining a feasible solution does not guarantee a valid differential path. Our objective is to reduce the number of active SBoxes over a broader range, with the optimal value being less than or equal to the actual minimum count of active SBoxes. To achieve this, we seek linear constraints that can effectively prune parts of the MILP model while preserving the valid differential characteristics within the region. Various algorithms are proposed for reducing the number of constraints in an SBox representation.

Greedy Algorithm Based Approach. The discrete points within the H-Representation (convex hull) often generate a substantial number of constraints. An effective strategy involves identifying the best valid constraints that maximize the elimination of impossible differential patterns within the feasible region of the convex hull. Sun et al. [35] present this greedy approach.

Modeling via Selection of Random Subset of Constraints. The generation of an extensive set of new constraints [17] by randomly adding k constraints often results in redundancy, as they are likely to cover the entire space $\{0,1\}^m$. However, when k hyperplanes from the H-representation share a vertex within the $\{0,1\}^m$ cube, combining the corresponding constraints can potentially yield a new constraint $I_{Q_{new}}$. Importantly, this new hyperplane represented by $I_{Q_{new}}$ should intersect with the cube at least once. In such cases, $I_{Q_{new}}$ has the potential to invalidate a distinct set of impossible points within the H-representation compared to the older constraints.

MILP-Based Reduction Algorithm. Sasaki and Todo [31] introduce a method that employs a standard MILP solver to derive a minimized set of constraints. Initially, they identify all the impossible differential points within the DDT of an SBox. These points serve as the foundation for generating impossible patterns. For each impossible pattern generated, the algorithm assesses which subset of constraints can effectively render the pattern invalid. This step pinpoints the crucial constraints required for pattern invalidation. Subsequently, they formulate a constraint specifying that each impossible pattern must be made infeasible within the valid region by at least one constraint. An MILP problem is then constructed to minimize the overall set of constraints while adhering to the constraints established in the prior steps. The formulated MILP problem is solved, yielding a minimized set of constraints that effectively reduces the number of impossible patterns.

4 Our First Idea: Switching Positive and Negative Samples

In this part, we describe our first idea, where we swap the roles of the positive and negative samples from the given SBox evaluation table. The basic concept is that we now treat the positive samples as negative and vice versa. Thus, we actually model the complementary problem, then construct the overarching MILP problem, and then, from its solution, we revert back to the original problem (by taking the complement again). We show that, for certain SBox evaluation tables, this method is in fact beneficial in reducing the number of linear constraints.

To demonstrate the basic concept of this, we present a suitable example. Let $f_1(X_3, X_2, X_1, X_0)$ be a set of 4 elements in $\mathbb{F}_2^4$, defined as: $f_1 = \{0000, 0011, 1100, 1111\}$ that represents the non-zero elements.

The goal is to find a minimal representation of the set f_1. Figure 1a shows the set f_1 represented in a Karnaugh map. The minimized Boolean expression after applying the Karnaugh map method (over 1's), $f_1^1(X_3, X_2, X_1, X_0) = X_3'X_2'X_1'X_0' + X_3'X_2'X_1X_0 + X_3X_2X_1'X_0' + X_3X_2X_1X_0$ (where, each X_i' denotes the complement of binary value X_i). We can also apply 0's to find the minimized expression in another way (see Fig. 1b). So, the minimized Boolean expression applying the Karnaugh map method (over 0's), $f_1^0(X_3, X_2, X_1, X_0) = (X_3 + X_2')(X_3' + X_2)(X_1 + X_0')(X_1' + X_0)$. We can notice that when using the 0's (i.e., negative samples) map, the simple product of sum terms, which also contains fewer literals than the 1's (i.e., positive samples) map of product of sum.

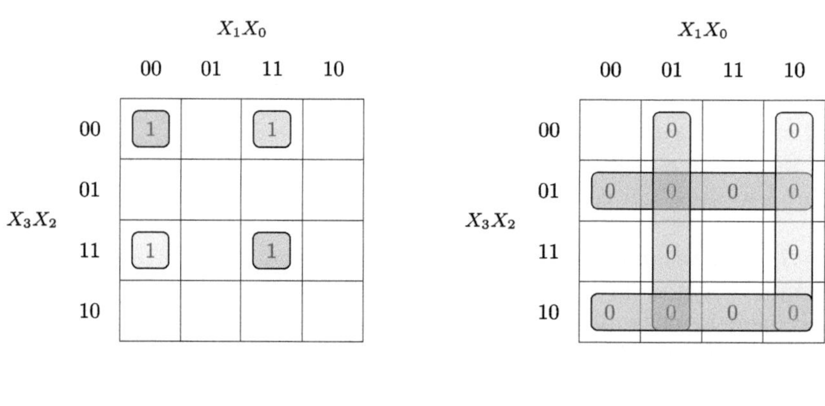

(a) Non-zero elements   (b) Zero elements

Fig. 1. K-Map for f_1

Suppose $f_2(X_3, X_2, X_1, X_0)$ be a set of 14 elements in $\mathbb{F}_2^4$ in another example, where $f_2 = \{0000, 0001, \ 0010, 0011, 0100, 0101, 0110,$

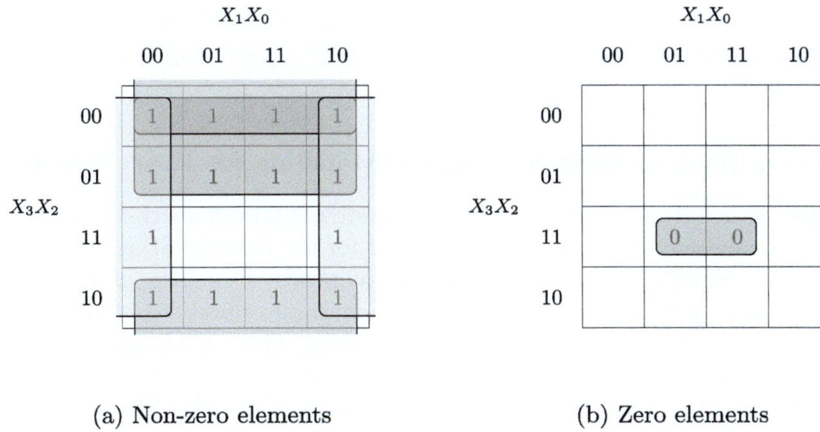

(a) Non-zero elements (b) Zero elements

Fig. 2. K-Map for f_2

0111, 1000, 1001, 1010, 1011, 1100, 1110}. Figure 2a presents the representation of the set f_2 using a Karnaugh map. By applying the Karnaugh map minimization procedure over the 1's, the simplified Boolean expression is, $f_2^1(X_3, X_2, X_1, X_0) = X_0' + X_2' + X_3'$. Alternatively, minimization can also be performed using the 0's (see Fig. 2b), yielding, $f_2^0(X_3, X_2, X_1, X_0) = X_0' + X_2' + X_3'$. It is worth noting that minimization with respect to the 0's leads directly to a product-of-sums form, which contains the same number of literals as the product-of-sums expression derived from the 1's map. However, obtaining the minimal expression is generally more straightforward when employing the 0's map.

We can extend this approach to the modeling of SBoxes in cryptanalysis-based MILP formulations by altering the constraint-generation process derived from the LAT, DDT, DPT, and BCT. Typically, in classical SBox modeling, the binary variables representing input-output patterns are assigned a value of 1 to indicate the presence of an active bit or a valid approximation/differential, and the resulting constraints are constructed to enforce the propagation constraints. However, in certain scenarios, such as when searching for impossible differentials, zero-correlation linear approximations, or other structural properties, it is beneficial to invert this encoding, assign 0 to indicate an active or valid position, and 1 to indicate inactivity or invalid propagation. By doing so, we essentially reformulate the constraint system so that the constraints capture the absence of specific transitions or correlations rather than their presence. This "complementary variable encoding" can simplify the MILP model when the cryptanalytic objective is to identify patterns of non-propagation or to detect impossible transitions.

This inversion of bit semantics allows the evaluation table to derive constraints to be directly reused by simply flipping the binary mapping while preserving their structural relationship to the underlying combinatorial properties of the SBox. This approach is particularly advantageous when integrating SBox constraints into multi-round block cipher models, where representing zero-propagation conditions efficiently can reduce solver complexity and improve runtime performance.

Switching Back to Positive Samples: Modeling for Exact Problem

One may notice that the model we discuss here actually finds the formulation of the complementary problem, that is, it looks for the inactive SBoxes (as it treats the negative samples of an evaluation table as positive samples, and vice-versa). However, since our objective is to find the minimum number of active SBoxes, we need to adjust the MILP problem formulation. One idea is as follows.

After we model the negative samples as the positive samples (and vice-versa), we make the MILP problem as "maximize number of inactive SBoxes". Then we get the maximum possible number of SBoxes that can be inactive (say, the result from the MILP solver is I). Then, we subtract that from the total number of SBoxes. Say, for a cipher with 32 SBoxes at one round and tested for 10 rounds, we have in total $32 \times 10 = 320$ SBoxes. Therefore, the minimal number of active SBoxes for our case will be $320 - I$.

Aside from that, another idea is to change the indicator variable that captures if the SBox is active or not, to reflect the complementary nature. For example, if we have an indicator variable S that is 1 when the SBox is active and 0 otherwise, we can replace it with $1 - S$. The objective function will remain the same, i.e., "minimize number of active SBoxes". However, since we have changed the indicator variable, the MILP solver will now find the minimum number of active SBoxes directly.

5   Our Second Idea: Exclusion of Trivial Transitions in SBox Modeling

In the construction of DDT, LAT, DPT, and BCT, the trivial zero transitions are included. These entries correspond to cases where at least one of the input and output characteristics is zero. Since the goal of SBox modeling is to capture meaningful propagation patterns that reflect the cryptographic strength or weakness of the design, such trivial transitions can be considered redundant.

By deliberately excluding these trivial transitions, the modeling process becomes more efficient. Specifically, the MILP representation, which relies on generating sets of constraints for feasible and infeasible transitions, can be significantly simplified. The exclusion of trivial cases results in a direct reduction of the number of constraints required, since no modeling effort is spent on transitions that do not affect the differential, linear, division, or boomerang propagation properties of the cipher. This reduction not only decreases computational overhead but also improves solver performance when searching for optimal or impossible trails.

Moreover, this selective elimination aligns with the broader principle of focusing on cryptographically significant events in the analysis. By ignoring

Table 2. Minimum number of linear constraints for DDT

(a) Including 0th row and 0th column

SBox	Size	DDT (Zero, Non-zero) [4]	Full Convex Hull (Sage) [35]	Espresso	Sun et al. (constraints) in Asiacrypt'14 [35]	Randomized [4,29]	Sasaki and Todo [31] (2017)	Boura and Coggia* [18]	Subset Addition [30] $k=2$	$k=3$
GIFT		157, 99	73, 237	31, 33	34, 24	31, 22	30, 21	21, 17	21, 17	21, 17
KLEIN		150, 106	158, 311	44, 43	33, 22	30, 22	28, 21	23, 17	23, 17	23, 17
LILLIPUT		150, 106	183, 324	47, 45	31, 27	30, 26	29, 23	26, 19	26, 20	25, 19
MIBS		150, 106	173, 378	46, 47	32, 27	30, 24	29, 23	26, 20	26, 20	26, 20
MIDORI S0		159, 97	142, 239	40, 47	32, 23	30, 25	27, 21	26, 16	26, 17	26, 16
MIDORI S1		150, 106	166, 367	51, 56	29, 26	28, 24	28, 22	26, 20	26, 20	26, 20
MINALPHER		150, 106	140, 338	46, 50	26, 24	26, 24	26, 22	21, 19	21, 19	21, 18
PICCOLO	4	159, 97	72, 202	26, 31	27, 23	26, 24	21, 21	21, 16	21, 16	21, 16
PRESENT		159, 97	115, 237	38, 36	35, 22	32, 22	30, 21	21, 17	21, 17	20, 17
PRIDE		159, 97	74, 194	26, 31	30, 22	28, 22	26, 21	21, 16	21, 17	21, 17
PRINCE		150, 106	151, 300	47, 51	31, 26	31, 26	29, 22	25, 19	25, 19	25, 18
RECTANGLE		159, 97	70, 267	28, 30	25, 25	24, 23	24, 21	19, 17	20, 17	20, 16
SKINNY		159, 97	72, 202	26, 31	28, 24	27, 24	26, 21	21, 16	21, 16	21, 16
TWINE		150, 106	183, 324	47, 45	33, 23	30, 25	29, 23	26, 19	26, 20	25, 19
ASCON		707, 317	216, 2415	77, 59	66, 51	64, 48	63, 40	48, 32	48, 31	47, 31
FIDES-5		527, 497	768, 910	122, 124	101, 96	96, 90	84, 79	64, 61	64, 64	63, 62
KECCAK	5	707, 314	182, 316	72, 46	56, 50	52, 47	52, 46	37, 34	37, 36	32, 36
ICEPOLE		687, -	347, -	117, -	80, -	77, -	72, -	57, -	59, -	57, -
SC2000-5		527, 497	800, 908	120, 123	98, 99	95, 92	84, 82	66, 64	66, 65	65, 63
APN		2079, 2017	5003, 195	-, 298	233, -	229, -	-, -	-, 167	-, 163	-, -
FIDES-6	6	-, 2017	-, 223	-, 489	-, -	-, -	-, -	-, 180	-, 184	-, -
SC2000-6		-, 1954	-, 241	-, 612	-, -	-, -	-, -	-, 214	-, 89	-, -

*: Our implementation of Boura and Coggia's method [18]

Each cell indicates (negative samples, positive samples)

(b) Excluding 0th row and 0th column

SBox	Size	DDT (Zero, non-zero) [4]	Full Convex Hull (Sage) [35]	Sun et al. (constraints) in Asiacrypt'14 [35]	Randomized [4,29]	Sasaki and Todo [31] (2017)	Boura and Coggia* [18]	Subset Addition [30] $k=2$	$k=3$
GIFT		98, 127	119, 141	27, 26	27, 24	25, 22	18, 17	18, 17	18, 17
KLEIN		120, 105	208, 281	29, 23	27, 22	25, 20	21, 18	21, 18	21, 18
LILLIPUT		120, 105	225, 297	28, 23	27, 22	24, 21	20, 19	20, 19	20, 19
MIBS		120, 105	213, 320	28, 26	27, 25	25, 23	22, 20	22, 20	21, 20
MIDORI S0		129, 96	164, 217	33, 23	30, 22	27, 20	24 ,17	24, 17	24, 17
MIDORI S1		120, 105	209, 334	26, 25	25, 25	23, 22	21, 20	21, 20	21, 20
MINALPHER		120, 105	214, 320	27, 24	26, 24	25, 22	19, 19	19, 18	19, 18
PICCOLO	4	129, 96	106, 186	25, 23	25, 22	24, 20	19, 15	19, 16	19, 15
PRESENT		129, 96	170, 233	32, 21	30, 20	26, 18	18, 17	18, 15	18, 15
PRIDE		129, 96	107, 170	28, 21	27, 21	25, 21	20, 17	20, 19	20, 19
PRINCE		120, 105	202, 270	27, 26	27, 25	25, 22	22, 18	22, 19	22, 18
RECTANGLE		129, 96	104, 158	24, 24	22, 24	20, 22	16, 16	16, 17	16, 17
SKINNY		129, 96	106, 186	28, 24	27, 23	24, 20	19, 15	19, 16	19, 15
TWINE		120, 105	225, 297	28, 25	27, 24	24, 21	20, 19	20, 19	20, 19
ASCON		645, 316	273, 1739	65, 48	63, 45	59, 40	46, -	46, 31	45, -
FIDES-5	5	465, 496	1034, 729	93, 95	84, 87	78, 80	61, -	61, -	59, -
KECCAK		645, 316	229, 308	52, 53	50, 49	47, 45	33, -	32, -	31, -
SC2000-5		465, 496	1091, 747	92, 95	87, 90	76, 82	61, -	61, -	59, -
APN		1953, 2016	6112, 5006	230, 237	225, 230	-, -	-, -	-, -	-, -
FIDES-6	6	1953, 2016	8200, 7012	257, 275	248, 263	-, -	-, -	-, -	-, -
SC2000-6		2016, 1953	9816, 11238	292, 294	286, 284	-, -	-, -	-, -	-, -

*: Our implementation of Boura and Coggia's method [18]

Each cell indicates (negative samples, positive samples)

Table 3. Minimum number of linear constraints for LAT

(a) Including 0th row and 0th column

SBox	Size	LAT (Zero, Non-zero) [4]	Full Convex Hull (Sage) [35]	Espresso	Chosen Convex Hull (Deterministic) [35]	Chosen [4,29]	Sasaki,[31]	Boura, Coggia* [18]	Subset Addition [30]	
									$k=2$	$k=3$
GIFT		123, 133	50, 211	19, 27	19, 18	19, 18	19, 16	15, 13	15, 14	15, 13
KLEIN		105, 151	285, 133	45,45	27, 31	27, 29	24, 27	18, 23	18, 23	18, 23
LILLIPUT		105, 151	239, 126	42, 41	26, 32	26, 32	24, 32	19, 22	19, 22	18, 22
MIBS		105, 151	347, 124	44, 43	28, 36	26, 35	23, 35	19, 26	19, 26	19, 26
MIDORI S0		123, 133	195, 92	39, 38	31, 28	29, 28	25, 25	19, 16	19, 18	19, 17
MIDORI S1		105, 151	338, 104	43, 40	25, 32	25, 31	23, 29	18, 22	18, 22	18, 22
MINALPHER		105, 151	304, 143	42, 44	25, 32	25, 31	21, 28	18, 23	18, 23	18, 23
PICCOLO	4	123, 133	64, 76	21, 19	22, 17	21, 16	20, 16	14, 11	14, 11	14, 11
PRESENT		123, 133	169, 137	35, 36	27, 29	25, 26	23, 24	17, 13	17, 17	17, 17
PRIDE		123, 133	67, 72	21, 19	21, 16	21, 16	21, 15	15, 11	15, 12	15, 11
PRINCE		105, 151	298, 114	42, 39	25, 32	25, 31	23, 29	18, 20	18, 21	18, 20
RECTANGLE		123, 133	55, 186	21, 25	22, 17	22, 16	21, 15	18, 12	19, 12	18, 12
SKINNY		123, 133	64, 76	21, 19	21, 16	21, 16	20, 16	14, 11	14, 11	14, 11
TWINE		105, 151	239, 126	42, 41	27, 32	26, 32	24, 32	19, 22	19, 22	18, 22
ASCON		647, 377	304, 1613	86, 86	83, 70	79, 69	75, 58	-, 44	52, 44	50, -
FIDES-5		527, 497	765, 889	122, 123	94, 96	94, 90	83, 77	-, 62	66, 63	64, 62
ICEPOLE	5	497, -	857, -	192, -	102, -	102, -	90, -	80, 64	84, -	81, -
KECCAK		647, 377	137, 236	67, 66	69, 67	67, 66	67, 65	37, 38	42, 40	37, 39
SC2000-5		527, 497	801, 917	122, 124	100, 93	99, 88	84, 78	68, 63	68, 64	67, 63
APN		-, 3025	-, 3869	172, 179	-, 144	-, 134	-, 114	-, 88	-, 89	-, -
FIDES-6	6	-, 3025	-, 2699	203, 253	-, 185	-, 177	-, 159	-, -	-, 119	-, -
SC2000-6		-, 3214	-, 1053	-, 449	-, 320	-, 315	-, 307	-, -	-, 274	-, -

★: Our implementation of Boura and Coggia's method [18]

Each cell indicates (negative samples, positive samples)

(b) Excluding 0th row and 0th column

SBox	Size	LAT (Zero, Non-zero) [4]	Full Convex Hull (Sage) [35]	Sun et al. (constraints) in Asiacrypt'14 [35]	Randomized [4,29]	Sasaki and Todo [31] (2017)	Boura and Coggia* [18]	Subset Addition [30]	
								$k=2$	$k=3$
GIFT		93, 132	74, 116	18, 19	18, 18	17, 15	13, 12	13, 13	13, 12
KLEIN		75, 150	413, 92	23, 28	22, 27	19, 27	15, 23	-, 23	-, 23
LILLIPUT		75, 150	376, 87	19, 31	19, 31	18, 31	15, 22	15, 21	15, 21
MIBS		75, 150	497, 86	23, 35	22, 35	19, 35	16, 26	16, 25	16, 25
MIDORI S0		93, 132	248, 82	25, 25	23, 25	20, 25	16, 16	17, 16	16, 17
MIDORI S1		75, 150	449, 76	24, 30	22, 30	19, 29	15, 22	15, 22	15, 22
MINALPHER		75, 150	438, 100	23, 29	21, 28	19, 27	16, 23	16, 22	16, 22
PICCOLO	4	93, 132	103, 62	19, 15	18, 15	17, 15	12, 11	12, 11	11, 10
PRESENT		93,132	276, 93	20, 28	19, 24	18, 16	15, 16	15, 17	15, 17
PRIDE		93, 132	110, 59	19, 13	19, 13	18, 13	12, 10	12, 9	11, 9
PRINCE		75, 150	379, 84	22, 30	21, 30	18, 29	15, 19	15, 19	15, 10
RECTANGLE		93, 132	84, 90	21, 19	21, 18	20, 16	13, 12	14, 12	13, 11
SKINNY		93, 132	103, 62	19, 15	18, 15	17, 15	12, 11	12, 11	11, 10
TWINE		75, 150	376, 86	20, 32	19, 32	18, 31	15, 22	15, 21	15, 21
ASCON		585, 376	390, 1107	76, 68	73, 65	69, 60	-, -	50, 45	47, -
FIDES-5		465, 495	1034, 716	90, 91	87, 84	76, 77	-, -	62, -	61, -
KECCAK	5	585, 376	164, 228	66, 66	64, 65	62, 64	-, -	41, -	36, -
SC2000-5		465, 496	1085, 760	92, 89	87, 87	77, 79	-, -	63, -	62, -
APN		945, 3024	4734, 2951	105, 141	99, 134	88, 114	-, -	-, -	-, -
FIDES-6	6	945, 3024	17147, 2392	115, 182	110, 172	-, 159	-, -	-, -	-, -
SC2000-6		-, -	-, 942	-, 307	-, -	-, -	-, -	-, -	-, -

★: Our implementation of Boura and Coggia's method [18]

Each cell indicates (negative samples, positive samples)

non-contributing transitions, the SBox model preserves all essential information needed for security evaluation while discarding unnecessary complexity. The gain here is twofold: a reduction in the mathematical burden of constraint generation and an improvement in the clarity of the cryptographic structure being analyzed. Thus, ignoring trivial transitions is not merely a simplification, but a targeted optimization that enhances both the efficiency and precision of SBox modeling.

6   Results

We have the approach to all the earlier approaches for SBox modeling. To make the explanation clear, we consider both types of constraints generated from zero and non-zero transitions in LAT/DDT/DPT/BCT. Our results section has mainly two parts, each having four tables. The first part (Tables 2a, 3a, 4a, 5a) provides the number of constraints for different SBox evaluation tables. The second part (Tables 2b, 3b, 4b, 5b) explains the number of constraints for different SBox evaluation tables excluding the first row and first column, that is, ignoring the trivial transitions. Each resultant table describes the number of constraints required for modeling different 4, 5, and 6-bit SBoxes applying different SBox modeling approaches. Each cell of any table has one pair of values, namely the number of constraints required for applying negative sampling and positive sampling (in this order). We have marked the red color for improved results in each algorithm. Blue color for the best result, including all algorithms, provided that we get improved results for negative sampling compared to the positive sampling approach. For those we do not get a result, we mark with the '-' symbol.

Our analysis yields superior results for a wide range of ciphers by utilizing the zero transition property detailed in the DDT Table 2a. Specifically, applying this to the Sage approach led to improvements for almost all SBoxes tested. Additionally, we recorded advancements for KECCAK using the subset addition method [30] and for SC2000-5 in relation to Sun et al.'s approach [35].

DDT Excluding Trivial Transitions. Table 2b compares the constraints generated from zero and non-zero transitions (i.e., positive and negative samples) in DDT for different modeling, ignoring the trivial transitions. This approach leads to significant improvements for almost every existing approach for a wide range of ciphers, including RECTANGLE, FIDES-5, KECCAK, SC2000-5, FIDES-6, and SC2000-6 when applied to the DDT as given in Table 2b.

LAT. Table 3a presents a comparison of different SBox modeling algorithms for LAT under both positive and negative sampling. In this case, we observe improved results for SBoxes used in KLEIN, LILLIPUT, MIBS, MIDORI S1, MINALPHER, PRINCE, and TWINE across most of the modeling approaches.

LAT Excluding Trivial Transitions. Table 3b compares the constraints generated from zero and non-zero transitions in LAT for different modeling, ignoring the trivial transitions. Here we obtain improved results for SBoxes like

Table 4. Minimum number of linear constraints for DPT

(a) Including 0th row and 0th column

SBox	Size	DPT (Zero, Non-zero) [4]	Full Convex Hull (Sage) [35]	Sun et al. (constraints) in Asiacrypt'14 [35]	Randomized [4, 29]	Sasaki and Todo [31] (2017)	Boura and Coggia* [18]	Subset Addition [30]	
								$k = 2$	$k = 3$
GIFT		208, 48	50, 213	31, 14	31, 14	31, 13	23, 9	23, 9	23, 9
KLEIN		206, 50	50, 117	21, 11	20, 10	20, 9	15, 7	15, 7	13, 6
LILLIPUT		210, 46	49, 122	20, 11	20, 11	20, 9	14, 7	14, 7	12, 7
MIBS		204, 52	51, 84	20, 9	19, 9	19, 8	14, 7	14, 7	12, 7
MIDORI S0		209, 47	50, 45	19, 6	19, 6	19, 6	14, 5	14, 5	12, 5
MIDORI S1		208, 48	47, 75	20, 10	19, 9	19, 9	13, 7	13, 7	12, 7
MINALPHER		206, 50	50, 134	21, 11	20, 10	20, 9	15, 7	15, 7	14, 7
PICCOLO	4	211, 45	47, 98	27, 13	26, 13	26, 12	20, 8	20, 8	20, 7
PRESENT		210, 46	48, 103	22, 10	20, 10	20, 9	15, 7	15, 7	13, 7
PRIDE		213, 43	47, 108	25, 11	24, 11	24, 10	18, 7	18, 7	17, 6
PRINCE		207, 49	48, 86	20, 12	20, 11	20, 10	14, 8	14, 9	12, 8
RECTANGLE		208, 48	52, 197	31, 15	31, 14	31, 13	24, 9	24, 9	23, 9
SKINNY		211, 45	47, 98	27, 13	26, 13	26, 12	20, 8	20, 8	20, 7
TWINE		210, 46	49, 122	21, 11	20, 10	20, 9	14, 7	14, 7	12, 7
ASCON		835, 189	158, 1393	99, 22	99, 22	99, 20	72, 13	72, 15	67, 15
FIDES-5		825, 199	173, 2012	87, 23	83, 23	81, 20	56, -	56, -	54, -
KECCAK	5	888, 136	111, 412	91, 21	91, 19	91, 17	61, -	61, 15	61, 15
ICEPOLE		873, 151	121, 18	31, 8	31, 8	26, 8	16, -	16, 7	16, 6
SC2000-5		883, 141	174, 1719	68, 24	67, 24	65, 19	60, -	60, -	55, -
APN		3652, 444	439, 66461	211, 46	207, 44	202, -	-, -	174, -	171, -
FIDES-6	6	3678,-	403, -	178, -	176, -	173, -	-, -	152, -	149, -
SC2000-6		3734, -	204, -	51, -	49, -	45, -	-, -	33, -	28, -

*: Our implementation of Boura and Coggia's method [18]

Each cell indicates (negative samples, positive samples)

(b) Excluding 0th row and 0th column

SBox	Size	DPT (Zero, Non-zero) [4]	Full Convex Hull (Sage) [35]	Sun et al. (constraints) in Asiacrypt'14 [35]	Randomized [4, 29]	Sasaki and Todo [31] (2017)	Boura and Coggia* [18]	Subset Addition [30]	
								$k = 2$	$k = 3$
GIFT		177, 48	49, 213	21, 14	21, 14	21, 12	16, 8	16, 8	17, 8
KLEIN		175, 50	68, 117	8, 9	8, 8	7, 7	6, 6	6, 6	6, 5
LILLIPUT		179, 46	54, 122	11, 10	10, 9	9, 8	7, 6	7, 6	7, 6
MIBS		173, 52	62, 84	8, 7	8, 7	7, 7	6, 6	6, 6	5, 6
MIDORI S0		178, 47	49, 45	9, 5	9, 5	8, 5	7, 4	7, 4	7, 4
MIDORI S1		177, 48	51, 75	8, 9	8, -	7, 8	6, 6	6, 6	6, 6
MINALPHER		175, 50	56, 134	10, 9	10, 9	9, 8	7, 6	7, 6	6, 6
PICCOLO	4	180, 45	69, 98	17, 12	17, 12	16, 11	14, 7	14, 7	13, 7
PRESENT		179, 46	48, 103	9, 9	9, 9	9, 8	7, 6	7, 6	6, 6
PRIDE		182, 43	77, 108	13, 10	13, 10	13, 9	10, 7	10, 7	10, 6
PRINCE		176, 49	50, 86	9, 10	9, 10	8, 9	7, 7	7, 7	7, 7
RECTANGLE		177, 48	52, 197	21, 14	21, 13	21, 11	16, 8	16, 9	16, 8
SKINNY		180, 45	69, 98	17, 13	17, 12	16, 11	14, 7	14, 7	13, 7
TWINE		179, 46	54, 122	11, 10	11, 9	9, 8	7, 6	7, 6	7, 6
ASCON		772, 189	178, 1393	86, 22	82, 20	81, 18	-, -	58, 15	53, -
FIDES-5		762, 199	171, 2012	73, 21	69, 22	65, 18	-, -	47, -	45, -
KECCAK	5	825, 136	128, 412	81, 19	81, 18	81, 15	-, -	56, -	56, -
SC2000-5		820, 141	217, 1719	47, 23	47, 23	46, 18	-, -	45, -	45, -
APN		3525, 444	592, 66461	179, -	178, -	176, -	-, -	153, -	152, -
FIDES-6	6	3551, 418	763, -	148, -	144, -	144, -	-, -	132, -	132, -
SC2000-6		3607, 362	174, -	17, -	17, -	16, -	-, -	-, -	-, -

*: Our implementation of Boura and Coggia's method [18]

Each cell indicates (negative samples, positive samples)

Table 5. Minimum number of linear constraints for BCT

(a) Including 0th row and 0th column

SBox	Size	BCT (Zero, Non-Zero) [4]	Full Convex Hull (Sage) [35]	Espresso	Sun et al. (constraints) in Asiacrypt'14 [35]	Randomized [4, 29]	Sasaki and Todo [31] (2017)	Boura and Coggia* [18]	Subset Addition [30] $k = 2$	$k = 3$
GIFT		102, 154	173, 104	32, 35	25, 26	24, 26	23, 25	17, 19	17, 17	16, 17
KLEIN		122, 144	246, 164	43, 45	30, 31	28, 27	24, 25	20, 23	20, 19	20, 19
LILLIPUT		120, 136	225, 210	46, 45	29, 29	28, 28	26, 26	21, 22	21, 19	21, 19
MIBS		120, 136	213, 220	46, 47	29, 30	29, 28	26, 26	22, 23	22, 20	21, 20
MIDORI S0		107, 149	127, 146	37, 41	26, 25	24, 24	23, 22	18, 18	18, 16	17, 15
MIDORI S1		105, 151	242, 159	45, 51	25, 33	24, 32	22, 20	18, 28	20, 24	18, 24
MINALPHER		122, 144	215, 182	50, 49	30, 30	29, 30	27, 20	21, 26	21, 22	21, 22
PICCOLO	4	103, 153	139, 83	27, 25	24, 23	23, 22	21, 22	16, 16	16, 14	15, 14
PRESENT		107, 149	239, 147	39, 40	26, 30	26, 28	23, 28	18, 24	18, 21	17, 20
PRIDE		103, 153	115, 70	25, 24	25, 26	23, 25	22, 24	16, 16	16, 16	16, 16
PRINCE		105, 151	269, 159	45, 51	27, 31	25, 31	23, 29	19, 23	19, 21	19, 20
RECTANGLE		107, 149	162, 103	31, 29	26, 25	25, 23	21, 22	15, 18	15, 16	15, 16
SKINNY		103, 153	139, 83	27, 25	23, 22	23, 22	21, 22	16, 16	16, 14	-, 14
TWINE		120, 136	225, 210	46, 45	30, 29	28, 27	26, 26	21, 23	21, 19	21, 19
ASCON		445, 579	1154, 666	89, 101	57, 64	54, 60	47, 54	-, 42	38, 40	-, 38
FIDES-5	5	465, 559	1034, 815	123, 137	94, 97	93, 93	78, 83	61, 66	-, 63	-, 62
KECCAK		445, 579	1230, 435	67, 65	40, 58	41, 56	36, 55	-, 41	-, 45	-, 43
SC2000-5		465, 559	1091, 831	121, 137	94, 100	92, 99	77, 84	-, 70	-, 65	-, 64
APN		1953, 2143	6112, 4716	286, 301	236, 252	230, 243	-, -	-, -	-, -	-, -
FIDES-6	6	1953, 2143	8200, 6290	481, 487	260, 286	-, 278	-, -	-, -	-, -	-, -
SC2000-6		1980, 2206	12384, 7158	609, 609	278, 314	-, 305	-, -	-, -	-, -	-, -

*: Our implementation of Boura and Coggia's method [18]

Each cell indicates (negative samples, positive samples)

(b) Excluding 0th row and 0th column

SBox	Size	BCT (zero, non-zero) [4]	Full Convex Hull (Sage) [35]	Sun et al. (constraints) in Asiacrypt'14 [35]	Randomized [4, 29]	Sasaki and Todo [31] (2017)	Boura and Coggia* [18]	Subset Addition [30] $k = 2$	$k = 3$
GIFT		102, 123	173, 143	24, 21	21, 21	21, 21	17, 17	17, 17	16, 17
KLEIN		112, 113	246, 226	28, 25	25, 24	23, 23	20, 19	20, 19	20, 19
LILLIPUT		120, 105	225, 297	26, 25	23, 23	21, 21	20, 19	20, 19	20, 19
MIBS		120, 105	213, 320	29, 26	26, 25	23, 23	22, 20	22, 20	21, 20
MIDORI S0		107, 118	127, 162	25, 23	20, 20	20, 20	17, 16	17, 16	17, 15
MIDORI S1		105, 120	242, 220	22, 29	29, 28	27, 27	17, 24	18, 24	17, 24
MINALPHER		112, 113	215, 249	29, 26	27, 26	25, 25	20, 22	20, 22	20, 22
PICCOLO	4	103, 122	139, 86	22, 24	23, 24	21, 21	15, 14	15, 14	15, 14
PRESENT		107, 118	239, 180	25, 27	28, 27	24, 24	18, 21	18, 21	17, 20
PRIDE		103, 122	115, 88	22, 25	22, 22	22, 22	16, 14	16, 16	15, 16
PRINCE		105, 120	269, 221	25, 29	28, 27	25, 25	19, 21	19, 21	19, 20
RECTANGLE		107, 118	162, 126	24, 22	21, 21	19, 19	15, 16	15, 16	15, 16
SKINNY		103, 122	139, 86	22, 24	24, 24	21, 21	15, 14	15, 14	15, 14
TWINE		120, 105	225, 297	28, 25	24, 23	21, 21	20, 19	20, 19	20, 19
ASCON		445, 516	1154, 712	54, 60	57, 56	50, 50	-, -	37, 40	35, 38
FIDES-5	5	465, 496	1034, 729	94, 94	87, 89	80, 80	-, -	61, 63	59, 62
KECCAK		445, 516	1230, 552	40, 64	65, 64	55, 55	-, -	32, 45	-, 43
SC2000-5		465, 496	1091, 747	93, 96	89, 91	82, 82	-, -	-, 65	-, 64
APN		1953, 2016	6112, 5006	225, 242	239, 239	-, -	-, -	-, -	-, -
FIDES-6	6	1953, 2016	8200, 7012	256, 269	269, -	-, -	-, -	-, -	-, -
SC2000-6		1890, 2079	12384, 8568	275, -	296, -	-, -	-, -	-, -	-, -

*: Our implementation of Boura and Coggia's method [18]

Each cell indicates (negative samples, positive samples)

`KLEIN, LILLIPUT, MIBS, MIDORI S1, MINALPHER, PRESENT, PRINCE, TWINE` and `FIDES-6` for most of the modeling algorithms.

DPT. Table 4a presents a comparative analysis of different S-box modeling techniques for DPT under both positive and negative sampling strategies. The results indicate that improvements are observed primarily in the Sage-generated outcomes, while across all evaluated ciphers, the negative sampling approach consistently demonstrates superior performance.

DPT Excluding Trivial Transitions. Table 4b compares the constraints derived from zero and non-zero transitions in the DPT across different modeling approaches, excluding the trivial transitions. In this setting, improvements occur only in the Sage-generated results, while for most ciphers the negative sampling strategy yields better performance. In particular, `PRINCE` shows strong results when evaluated with the greedy [35], randomized greedy [30] and MILP-aided [31] approaches.

BCT. Table 5a describes the comparisons of different SBox modeling algorithms for BCT against positive and negative sampling. Here we obtain improved results for SBoxes like `GIFT, KLEIN, MIBS, MIDORI S0, MIDORI S1, MINALPHER, PRESENT, PRINCE, PRIDE, ASCON, FIDES-5,` and `KECCAK` for some of the modeling algorithms. Here, all the algorithms provide better results with respect to negative sampling.

BCT Excluding Trivial Transitions. Table 5b compares the constraints generated from zero and non-zero transitions in BCT for different modeling, ignoring the trivial transitions. Here we obtain improved results for SBoxes like `GIFT, KLEIN, MIBS, MIDORI S0, MIDORI S1, MINALPHER, PRESENT, PRINCE, PRIDE, ASCON, FIDES-5,` and `KECCAK` for some of the modeling algorithms. Here, greedy [35], Boura and Coggia's method [18], subset addition [30] algorithms have better performance with respect to negative sampling.

Finally, applying our two ideas, we get the best results in terms of the minimum number of linear constraints for the following SBoxes, categorized by SBox evaluation tables.

- **DDT:** `FIDES-5, KECCAK, SC2000-5.`
- **LAT:** `KLEIN, LILLIPUT, MIBS, MIDORI S0, MINALPHER, TWINE, KECCAK, SC2000-5, FIDES-5.`
- **DPT:** `MIBS.`
- **BCT:** `GIFT, PRESENT, PRINCE, RECTANGLE, MIDORI S0, ASCON, FIDES-5, KECCAK, FIDES-6, SC2000-6.`

7   Conclusion

In this paper, we introduce two efficient strategies for modeling SBoxes in MILP-based cryptanalysis. Instead of modeling possible transitions, we create a model based on the impossible transitions (i.e., negative samples). We further stream-line by excluding the redundant top row and left column (that correspond to trivial zero-to-zero transition).

Our experimental results, presented across a wide range of 4, 5, and 6-bit SBoxes; and various property tables like the DDT, LAT, DPT and BCT. We confirm that these methods successfully produce a smaller set of linear constraints than traditional modeling. By reducing the number of constraints, these models are expected to improve the efficiency of automated cryptanalysis, ultimately providing a more powerful tool for the security evaluation of modern ciphers.

A   Exemplary Constraints for Evaluation Tables

For better clarity, some examples of the generated linear constraints are presented here. All variables are binary, and x's (respectively, y's) denote the input (respectively, output) characteristics.

Notice that Examples 2, 3, 4, 5 and 6 correspond to the primary structure; and Examples 7 and 8 correspond to the secondary structure (see Sect. 2.7 for the relevant details). The switching of positive and negative samples is handled in the secondary structure setting (it cannot be captured in the primary structure), and similarly the exclusion of zero-to-zero transitions are automatically managed by the auxiliary constraints used in the secondary structure.

Example 2 (Sun et al.'s [35] ABC SBox (DDT) with positive samples using convex hull (Sage)).

```
 1 +1x2 +0x1 -2x0 +1y2 +2y1 +1y0 >= 0      11 +1x2 +2x1 +2x0 -1y2 +0y1 -1y0 >= 0
 2 +1x2 +2x1 -2x0 +1y2 +0y1 +1y0 >= 0      12 +0x2 +0x1 -1x0 +0y2 +0y1 +0y0 >= 1
 3 +1x2 +0x1 +2x0 -1y2 -2y1 -1y0 >= 2      13 +3x2 -2x1 -2x0 +1y2 +4y1 +1y0 >= 0
 4 +3x2 +2x1 +4x0 -1y2 -2y1 -1y0 >= 0      14 -1x2 +0x1 +0x0 +1y2 +0y1 +1y0 >= 0
 5 +0x2 +0x1 +0x0 +0y2 +1y1 +0y0 >= 0      15 +0x2 +1x1 +0x0 +0y2 +0y1 +0y0 >= 0
 6 +0x2 -1x1 +0x0 +0y2 +0y1 +0y0 >= 1      16 +1x2 +2x1 +0x0 +1y2 -2y1 +1y0 >= 0
 7 +1x2 +0x1 +0x0 -1y2 +0y1 +1y0 >= 0      17 +1x2 +0x1 +0x0 +1y2 +0y1 -1y0 >= 0
 8 -1x2 +0x1 +0x0 -1y2 +0y1 -1y0 >= 2      18 +0x2 +0x1 +0x0 +0y2 -1y1 +0y0 >= 1
 9 +0x2 +0x1 +1x0 +0y2 +0y1 +0y0 >= 0      19 +1x2 -1x1 -1x0 +0y2 +1y1 +0y0 >= 1
10 +1x2 -2x1 +0x0 +1y2 +2y1 +1y0 >= 0
```

□

Example 3. (Greedy Approach [35] on **LILLIPUT** *(LAT without 0th row and 0th column): Negative samples).*

```
 1 +2x3 +1x2 +7x1 -5x0 +3y3 +4y2 -1y1 -7y0 >= 6       8 -4x3 -1x2 -5x1 -2x0 +4y3 -2y2 +3y1 -1y0 >= 10
 2 +2x3 +4x2 +1x1 +4x0 -2y3 -1y2 -3y1 +1y0 >= 1       9 +4x3 +1x2 -2x1 +2x0 +3y3 -1y2 -3y1 +1y0 >= 2
 3 -3x3 -1x2 -1x1 -1x0 +3y3 +1y2 +2y1 +2y0 >= 2      10 +2x3 -1x2 -2x1 -1x0 -2y3 +0y2 +1y1 +1y0 >= 4
 4 +1x3 -2x2 +1x1 +2x0 -3y3 -1y2 +1y1 -1y0 >= 4      11 -1x3 +2x2 +1x1 -2x0 -1y3 -2y2 +1y1 +1y0 >= 4
 5 -1x3 +2x2 -5x1 +1x0 +2y3 +5y2 -2y1 +3y0 >= 3      12 +2x3 +1x2 +4x1 +1x0 -3y3 +3y2 +1y1 -1y0 >= 0
 6 +4x3 -1x2 +1x1 -3x0 +1y3 +5y2 +6y1 +3y0 >= -2     13 -2x3 -1x2 -1x1 +1x0 -2y3 -1y2 -1y1 +2y0 >= 6
 7 -2x3 +1x2 +4x1 +2x0 +3y3 -4y2 -1y1 +1y0 >= 3      14 -2x3 -1x2 -2x1 +2x0 +1y3 +3y2 -2y1 +1y0 >= 4
```

```
15 +0x3 -1x2 +1x1 +0x0 -1y3 -1y2 +0y1 -1y0 >= 3    18 +1x3 -1x2 -1x1 -2x0 +2y3 +1y2 +3y1 -2y0 >= 3
16 -2x3 +2x2 -1x1 +1x0 -2y3 +1y2 +2y1 -1y0 >= 4    19 -1x3 -1x2 +1x1 -2x0 -1y3 +1y2 -2y1 -1y0 >= 6
17 +1x3 +2x2 -2x1 +0x0 -1y3 -1y2 -2y1 -2y0 >= 6
```

$\square$

Example 4. (Randomized greedy approach [4, 30] on **PRINCE***(DPT without 0th row and 0th column): Negative samples).*

```
1 +1x3 +1x2 +1x1 +2x0 +2y3 +2y2 +2y1 +3y0 >= -5    6 -1x3 -1x2 -1x1 +1x0 -1y3 +1y2 +1y1 -1y0 >= 4
2 +1x3 +0x2 +0x1 +0x0 +1y3 +2y2 +1y1 +1y0 >= -2    7 -1x3 -1x2 +1x1 -1x0 +1y3 +1y2 -1y1 -1y0 >= 4
3 +0x3 +1x2 +0x1 +0x0 +1y3 +1y2 +2y1 +2y0 >= -2    8 -1x3 -1x2 +1x1 -1x0 +1y3 -1y2 +1y1 -1y0 >= 4
4 +1x3 +1x2 +2x1 +1x0 +2y3 +4y2 +4y1 +3y0 >= -6    9 -1x3 -1x2 -1x1 -1x0 -1y3 -1y2 -1y1 -1y0 >= 7
5 +0x3 +0x2 +0x1 +1x0 +2y3 +1y2 +1y1 +2y0 >= -2
```

$\square$

Example 5. (MILP [31] on **RECTANGLE***(DDT without 0th row and 0th column): Negative samples).*

```
 1 +1x3 +1x2 +0x1 +2x0 -1y3 -2y2 -1y1 -2y0 >= 4    12 +2x3 -1x2 -1x1 +1x0 -2y3 -1y2 -1y1 -2y0 >= 6
 2 +0x3 +1x2 +1x1 -1x0 +0y3 -1y2 -1y1 +1y0 >= 2    13 +0x3 -1x2 -1x1 +0x0 +2y3 +1y2 +1y1 +2y0 >= 0
 3 +1x3 -1x2 +0x1 -1x0 +1y3 -1y2 +1y1 +0y0 >= 2    14 -1x3 -1x2 +1x1 +1x0 +1y3 +1y2 +2y1 +2y0 >= 0
 4 +1x3 +1x2 +2x1 +1x0 +0y3 +0y2 -1y1 -1y0 >= 0    15 -1x3 -1x2 +1x1 -2x0 -1y3 -1y2 +2y1 -2y0 >= 6
 5 -2x3 -1x2 -2x1 -1x0 +2y3 +1y2 -1y1 -1y0 >= 6    16 +0x3 +2x2 -2x1 +0x0 -1y3 +1y2 +1y1 +2y0 >= 1
 6 +1x3 -1x2 -1x1 +1x0 +2y3 +1y2 +1y1 +1y0 >= 0    17 +1x3 -2x2 +1x1 -2x0 -1y3 +1y2 +2y1 -1y0 >= 4
 7 -1x3 +0x2 +0x1 -1x0 +1y3 +1y2 +1y1 +0y0 >= 1    18 +0x3 +0x2 -1x1 +0x0 -1y3 -1y2 +1y1 +1y0 >= 2
 8 -1x3 -1x2 +1x1 +0x0 -1y3 +0y2 -1y1 +1y0 >= 3    19 -1x3 -1x2 -1x1 +1x0 -1y3 +1y2 +1y1 -1y0 >= 4
 9 -1x3 +1x2 +1x1 +0x0 +0y3 +0y2 +1y1 -1y0 >= 1    20 +1x3 +1x2 -1x1 -2x0 -1y3 +2y2 -1y1 -1y0 >= 4
10 -1x3 +0x2 -1x1 +1x0 +1y3 -1y2 +0y1 -1y0 >= 3
11 +0x3 +1x2 -1x1 +0x0 +1y3 -1y2 -1y1 +1y0 >= 2
```

$\square$

Example 6. (Subset Addition ($k = 2$) [30] on **MIDORI S1** *(BCT without 0th row and 0th column): Negative samples).*

```
 1 -5x3 +3x2 -2x1 -1x0 -5y3 +3y2 -2y1 -1y0 >= 11    11 +5x3 -2x2 -4x1 +4x0 +3y3 +2y2 +1y1 -4y0 >= 5
 2 -2x3 -1x2 +3x1 -2x0 +3y3 +1y2 -2y1 -3y0 >= 7     12 -4x3 +2x2 -3x1 +1x0 -3y3 +3y2 -1y1 -2y0 >= 9
 3 +3x3 +3x2 -1x1 -3x0 +2y3 +1y2 -2y1 +2y0 >= 3     13 -1x3 -2x2 -2x1 -1x0 +2y3 -3y2 -2y1 -3y0 >= 11
 4 +3x3 +3x2 +3x1 +2x0 +2y3 +2y2 +4y1 +3y0 >= -7    14 -1x3 -4x2 -2x1 -5x0 -2y3 -3y2 -3y1 -5y0 >= 20
 5 -3x3 +2x2 +1x1 -3x0 -3y3 +2y2 +1y1 -3y0 >= 9     15 -1x3 -1x2 +2x1 +3x0 +1y3 +2y2 -1y1 +2y0 >= 0
 6 +3x3 +1x2 -2x1 -3x0 -2y3 -1y2 +3y1 -2y0 >= 7     16 -2x3 -3x2 +1x1 +3x0 -1y3 -3y2 +3y1 -2y0 >= 8
 7 -1x3 -2x2 +3x1 -3x0 -3y3 -3y2 -1y1 +2y0 >= 10    17 -3x3 -4x2 +1x1 -2x0 +5y3 -1y2 +2y1 -2y0 >= 8
 8 +2x3 -3x2 +1x1 -3x0 -4y3 -5y2 -2y1 -1y0 >= 14    18 -1x3 -2x2 -2x1 +2x0 -1y3 -2y2 -2y1 +2y0 >= 8
 9 +3x3 +2x2 +1x1 -4x0 +5y3 -2y2 -4y1 +4y0 >= 5
10 +1x3 +2x2 -1x1 +2x0 -1y3 -1y2 +2y1 +3y0 >= 0
```

$\square$

Example 7 (Positive sample handling in secondary structure). In this case, suppose S is a binary variable which indicates if the SBox is active (with respect to positive samples). We present two methods to handle positive samples in secondary structure, which are based on the models from [35] and [4], respectively.

The objective function is to maximize S. For constructing the secondary structure, we start from the primary structure constraints, and then modify them depending on the chosen method. We start with the primary structure constraints, then modify accordingly to convert to the secondary structure.

Primary structure:

```
 1 -2x3 +1x2 +0x1 -2x0 +2y3 +2y2 +1y1 -1y0 >= 3    11 +2x3 +2x2 -2x1 +3x0 -3y3 -1y2 -4y1 +4y0 >= 6
 2 +2x3 +0x2 +1x1 -1x0 +2y3 -2y2 +2y1 -2y0 >= 3    12 +1x3 -1x2 +0x1 +1x0 +1y3 +0y2 -2y1 -1y0 >= 3
 3 +1x3 +3x2 +2x1 -3x0 -2y3 +2y2 +2y1 +3y0 >= 1    13 +1x3 -3x2 -1x1 -2x0 +2y3 +3y2 +0y1 +3y0 >= 3
 4 +1x3 +3x2 +3x1 +2x0 -1y3 +0y2 -2y1 -2y0 >= 2    14 -2x3 -2x2 +0x1 +1x0 +1y3 +0y2 +2y1 +2y0 >= 2
 5 -1x3 +2x2 +0x1 -2x0 +2y3 -2y2 -2y1 +1y0 >= 5    15 -2x3 -2x2 +2x1 +1x0 -1y3 +1y2 -1y1 +2y0 >= 4
 6 -1x3 +2x2 +1x1 -2x0 -2y3 -1y2 +2y1 +1y0 >= 4    16 -1x3 -2x2 -1x1 -2x0 -2y3 +1y2 -1y1 -2y0 >= 9
 7 +1x3 -2x2 +3x1 -3x0 -3y3 -3y2 -1y1 -2y0 >= 11   17 -2x3 +0x2 -2x1 +1x0 -1y3 +1y2 +2y1 -2y0 >= 5
 8 +3x3 -1x2 -1x1 -3x0 -2y3 +1y2 +2y1 +2y0 >= 4    18 -1x3 +2x2 -3x1 -1x0 -1y3 -3y2 -2y1 +3y0 >= 8
 9 +2x3 +2x2 -1x1 -2x0 -2y3 -1y2 -1y1 -1y0 >= 6
10 +4x3 +2x2 +2x1 +4x0 -1y3 -1y2 +2y1 -2y0 >= 0
```

Secondary structure following Sun et al.'s [35] method (5 auxiliary constraints):

```
 1 x0 + x1 + x2 + x3 - S >= 0                       14 +1x3 -2x2 +3x1 -3x0 -3y3 -3y2 -1y1 -2y0 >= 11
 2                                                  15 +3x3 -1x2 -1x1 -3x0 -2y3 +1y2 +2y1 +2y0 >= 4
 3 x0 - S <= 0                                      16 +2x3 +2x2 -1x1 -2x0 -2y3 -1y2 -1y1 -1y0 >= 6
 4 x1 - S <= 0                                      17 +4x3 +2x2 +2x1 +4x0 -1y3 -1y2 +2y1 -2y0 >= 0
 5 x2 - S <= 0                                      18 +2x3 +2x2 -2x1 +3x0 -3y3 -1y2 -4y1 +4y0 >= 6
 6 x3 - S <= 0                                      19 +1x3 -1x2 +0x1 +1x0 +1y3 +0y2 -2y1 -1y0 >= 3
 7                                                  20 +1x3 -3x2 -1x1 -2x0 +2y3 +3y2 +0y1 +3y0 >= 3
 8 -2x3 +1x2 +0x1 -2x0 +2y3 +2y2 +1y1 -1y0 >= 3     21 -2x3 -2x2 +0x1 +1x0 +1y3 +0y2 +2y1 +2y0 >= 2
 9 +2x3 +0x2 +1x1 -1x0 +2y3 -2y2 +2y1 -2y0 >= 3     22 -2x3 -2x2 +2x1 +1x0 -1y3 +1y2 -1y1 +2y0 >= 4
10 +1x3 +3x2 +2x1 -3x0 -2y3 +2y2 +2y1 +3y0 >= 1     23 -1x3 -2x2 -1x1 -2x0 -2y3 +1y2 -1y1 -2y0 >= 9
11 +1x3 +3x2 +3x1 +2x0 -1y3 +0y2 -2y1 -2y0 >= 2     24 -2x3 +0x2 -2x1 +1x0 -1y3 +1y2 +2y1 -2y0 >= 5
12 -1x3 +2x2 +0x1 -2x0 +2y3 -2y2 -2y1 +1y0 >= 5     25 -1x3 +2x2 -3x1 -1x0 -1y3 -3y2 -2y1 +3y0 >= 8
13 -1x3 +2x2 +1x1 -2x0 -2y3 -1y2 +2y1 +1y0 >= 4
```

Secondary structure following Baksi's [4] method (1 auxiliary constraint):

```
 1 x0 + x1 + x2 + x3 + y0 + y1 + y2 + y3 - 8S <=    12 +4x3 +2x2 +2x1 +4x0 -1y3 -1y2 +2y1 -2y0 - 8S
     0                                                   >= 0
 2                                                  13 +2x3 +2x2 -2x1 +3x0 -3y3 -1y2 -4y1 +4y0 - 8S
 3 -2x3 +1x2 +0x1 -2x0 +2y3 +2y2 +1y1 -1y0 - 8S          >= 6
     >= 3                                           14 +1x3 -1x2 +0x1 +1x0 +1y3 +0y2 -2y1 -1y0 - 8S
 4 +2x3 +0x2 +1x1 -1x0 +2y3 -2y2 +2y1 -2y0 - 8S          >= 3
     >= 3                                           15 +1x3 -3x2 -1x1 -2x0 +2y3 +3y2 +0y1 +3y0 - 8S
 5 +1x3 +3x2 +2x1 -3x0 -2y3 +2y2 +2y1 +3y0 - 8S          >= 3
     >= 1                                           16 -2x3 -2x2 +0x1 +1x0 +1y3 +0y2 +2y1 +2y0 - 8S
 6 +1x3 +3x2 +3x1 +2x0 -1y3 +0y2 -2y1 -2y0 - 8S          >= 2
     >= 2                                           17 -2x3 -2x2 +2x1 +1x0 -1y3 +1y2 -1y1 +2y0 - 8S
 7 -1x3 +2x2 +0x1 -2x0 +2y3 -2y2 -2y1 +1y0 - 8S          >= 4
     >= 5                                           18 -1x3 -2x2 -1x1 -2x0 -2y3 +1y2 -1y1 -2y0 - 8S
 8 -1x3 +2x2 +1x1 -2x0 -2y3 -1y2 +2y1 +1y0 - 8S          >= 9
     >= 4                                           19 -2x3 +0x2 -2x1 +1x0 -1y3 +1y2 +2y1 -2y0 - 8S
 9 +1x3 -2x2 +3x1 -3x0 -3y3 -3y2 -1y1 -2y0 - 8S          >= 5
     >= 11                                          20 -1x3 +2x2 -3x1 -1x0 -1y3 -3y2 -2y1 +3y0 - 8S
10 +3x3 -1x2 -1x1 -3x0 -2y3 +1y2 +2y1 +2y0 - 8S          >= 8
     >= 4
11 +2x3 +2x2 -1x1 -2x0 -2y3 -1y2 -1y1 -1y0 - 8S
     >= 6
```

The following observations can be made regarding the two methods. Baksi's method [4] takes less number of auxiliary constraints. It requires only 1 auxiliary constraint, even though it modifies all the original constraints by inserting an additional term involving S. Sun et al.'s method [35], in contrast, requires more auxiliary constraints ($n + 1$ for an n-bit SBox).

Further, both methods [4,35] inherently deal with 0th row and 0th column issues. For instance, in Sun et al.'s method, the first auxiliary constraint ensures that when all x_i are zero, S must be set to 0 (inactive). The next n auxiliary constraints ensure that if any x_i is non-zero, S can be set to 1 (active); and for bijective SBoxes the same constraints also ensure that at least one of the y_i is non-zero. $\square$

Example 8 (Negative sample handling in secondary structure). S is a binary variable which indicates if the SBox is active (with respect to negative samples), following the secondary structure modeling from [35] (the alternate method from [4] is omitted here for brevity). Similar to Example 7, the objective function remains the same, to maximize S.

```
 1 x0 + x1 + x2 + x3 - (1 - S) >= 0           14 +1x3 -2x2 +3x1 -3x0 -3y3 -3y2 -1y1 -2y0 >= 11
 2                                            15 +3x3 -1x2 -1x1 -3x0 -2y3 +1y2 +2y1 +2y0 >= 4
 3 x0 - (1 - S) <= 0                          16 +2x3 +2x2 -1x1 -2x0 -2y3 -1y2 -1y1 -1y0 >= 6
 4 x1 - (1 - S) <= 0                          17 +4x3 +2x2 +2x1 +4x0 -1y3 -1y2 +2y1 -2y0 >= 0
 5 x2 - (1 - S) <= 0                          18 +2x3 +2x2 -2x1 +3x0 -3y3 -1y2 -4y1 +4y0 >= 6
 6 x3 - (1 - S) <= 0                          19 +1x3 -1x2 +0x1 +1x0 +1y3 +0y2 -2y1 -1y0 >= 3
 7                                            20 +1x3 -3x2 -1x1 -2x0 +2y3 +3y2 +0y1 +3y0 >= 3
 8 -2x3 +1x2 +0x1 -2x0 +2y3 +2y2 +1y1 -1y0 >= 3  21 -2x3 -2x2 +0x1 +1x0 +1y3 +0y2 +2y1 +2y0 >= 2
 9 +2x3 +0x2 +1x1 -1x0 +2y3 -2y2 +2y1 -2y0 >= 3  22 -2x3 -2x2 +2x1 +1x0 -1y3 +1y2 -1y1 +2y0 >= 4
10 +1x3 +3x2 +2x1 -3x0 -2y3 +2y2 +2y1 +3y0 >= 1  23 -1x3 -2x2 -1x1 -2x0 -2y3 +1y2 -1y1 -2y0 >= 9
11 +1x3 +3x2 +3x1 +2x0 -1y3 +0y2 -2y1 -2y0 >= 2  24 -2x3 +0x2 -2x1 +1x0 -1y3 +1y2 +2y1 -2y0 >= 5
12 -1x3 +2x2 +0x1 -2x0 +2y3 -2y2 -2y1 +1y0 >= 5  25 -1x3 +2x2 -3x1 -1x0 -1y3 -3y2 -2y1 +3y0 >= 8
13 -1x3 +2x2 +1x1 -2x0 -2y3 -1y2 +2y1 +1y0 >= 4
```

However, an alternate version of this model can be constructed by simply running for the minimum number of active negative samples instead of maximum number of active positive samples. In that case, no auxiliary constraints are needed, and the objective function in that case would be to minimize S. Finally, to revert back to the original maximization problem, one has to subtract that from the total number of SBoxes. $\square$

B   Reference to SBoxes

The following SBoxes are used in our experiments:

- ASCON [20]
- FIDES-5 [13]
- GIFT [8]
- ICEPOLE [26]
- KECCAK [11]
- KLEIN [21]
- LILLIPUT [10]
- MIBS [22]
- MIDORI [7]
- MINALPHER [19]
- PICCOLO [32]
- PRESENT [15]
- PRIDE [3]
- PRINCE [16]
- RECTANGLE [38]
- SKINNY [9]
- TWINE [36]
- SC2000 [33]

References

1. Abdelkhalek, A., Sasaki, Y., Todo, Y., Tolba, M., Youssef, A.M.: MILP modeling for (large) s-boxes to optimize probability of differential characteristics. IACR Trans. Symmetric Cryptol. **2017**(4), 99–129 (2017). https://doi.org/10.13154/tosc.v2017.i4.99-129

2. Aitipamula, A., Pal, D., Chowdhury, D.R.: Efficient and optimized modeling of s-boxes. In: Progress in Cryptology - AFRICACRYPT 2025 - 16th International Conference on Cryptology in Africa, Rabat, Morocco, July 21–23, 2025, Proceedings. Lecture Notes in Computer Science, vol. 15651, pp. 441–469. Springer (2025). https://doi.org/10.1007/978-3-031-97260-7_20

3. Albrecht, M.R., Driessen, B., Kavun, E.B., Leander, G., Paar, C., Yalçin, T.: Block ciphers - focus on the linear layer (feat. PRIDE). In: Garay, J.A., Gennaro, R. (eds.) Advances in Cryptology - CRYPTO 2014 - 34th Annual Cryptology Conference, Santa Barbara, CA, USA, August 17–21, 2014, Proceedings, Part I. Lecture Notes in Computer Science, vol. 8616, pp. 57–76. Springer (2014). https://doi.org/10.1007/978-3-662-44371-2_4

4. Baksi, A.: New insights on differential and linear bounds using mixed integer linear programming (full version). Cryptology ePrint Archive, Report 2020/1414 (2020). https://eprint.iacr.org/2020/1414

5. Baksi, A.: DEFAULT: Cipher-Level Resistance Against Differential Fault Attack. Springer, Singapore (2022). https://doi.org/10.1007/978-981-16-6522-6_8

6. Baksi, A., et al.: Baksheesh: similar yet different from gift. Cryptology ePrint Archive, Paper 2023/750 (2023). https://eprint.iacr.org/2023/750

7. Banik, S., Bogdanov, A., Isobe, T., Shibutani, K., Hiwatari, H., Akishita, T., Regazzoni, F.: Midori: a block cipher for low energy. In: Iwata, T., Cheon, J.H. (eds.) Advances in Cryptology - ASIACRYPT 2015 - 21st International Conference on the Theory and Application of Cryptology and Information Security, Auckland, New Zealand, November 29 – December 3, 2015, Proceedings, Part II. Lecture Notes in Computer Science, vol. 9453, pp. 411–436. Springer (2015). https://doi.org/10.1007/978-3-662-48800-3_17

8. Banik, S., Pandey, S.K., Peyrin, T., Sasaki, Y., Sim, S.M., Todo, Y.: Gift: a small present. Cryptology ePrint Archive, Report 2017/622 (2017). https://eprint.iacr.org/2017/622

9. Beierle, C., et al.: The SKINNY family of block ciphers and its low-latency variant MANTIS. IACR Cryptology ePrint Archive **2016**, 660 (2016). http://eprint.iacr.org/2016/660

10. Berger, T.P., Francq, J., Minier, M., Thomas, G.: Extended generalized feistel networks using matrix representation to propose a new lightweight block cipher: lilliput. IEEE Trans. Comput. **65**(7), 2074–2089 (2016). https://doi.org/10.1109/TC.2015.2468218

11. Bertoni, G., Daemen, J., Peeters, M., Assche, G.V.: Sponge functions. In: Ecrypt Hash Workshop (May 2007) (2007). https://keccak.team/files/CSF-0.1.pdf

12. Biham, E., Shamir, A.: Differential Cryptanalysis of the Data Encryption Standard. Springer-Verlag, Berlin, Heidelberg (1993). https://doi.org/10.1007/978-1-4613-9314-6

13. Bilgin, B., Bogdanov, A., Knežević, M., Mendel, F., Wang, Q.: FIDES: lightweight authenticated cipher with side-channel resistance for constrained hardware. In: Bertoni, G., Coron, J.-S. (eds.) CHES 2013. LNCS, vol. 8086, pp. 142–158. Springer, Heidelberg (2013). https://doi.org/10.1007/978-3-642-40349-1_9

14. Biryukov, A., Kushilevitz, E.: From differential cryptanalysis to ciphertext-only attacks. In: Krawczyk, H. (ed.) CRYPTO 1998. LNCS, vol. 1462, pp. 72–88. Springer, Heidelberg (1998). https://doi.org/10.1007/BFb0055721
15. Bogdanov, A., et al.: PRESENT: an ultra-lightweight block cipher. In: Paillier, P., Verbauwhede, I. (eds.) CHES 2007. LNCS, vol. 4727, pp. 450–466. Springer, Heidelberg (2007). https://doi.org/10.1007/978-3-540-74735-2_31
16. Borghoff, J., et al.: Prince - a low-latency block cipher for pervasive computing applications (full version). Cryptology ePrint Archive, Report 2012/529 (2012). https://ia.cr/2012/529
17. Boura, C., Coggia, D.: Efficient milp modelings for sboxes and linear layers of SPN ciphers. IACR Trans. Symmetric Cryptol. **2020**(3), 327–361 (2020). https://tosc.iacr.org/index.php/ToSC/article/view/8705
18. Boura, C., Coggia, D.: Efficient MILP modelings for sboxes and linear layers of SPN ciphers. IACR Trans. Symmetric Cryptol. **2020**(3), 327–361 (2020). https://doi.org/10.13154/tosc.v2020.i3.327-361
19. Chakraborti, A., Datta, N., Nandi, M.: Practical fault attacks on minalpher: how to recover key with minimum faults? In: Security, Privacy, and Applied Cryptography Engineering - 7th International Conference, SPACE 2017, Goa, India, December 13–17, 2017, Proceedings. Lecture Notes in Computer Science, vol. 10662, pp. 111–132. Springer (2017). https://doi.org/10.1007/978-3-319-71501-8_7
20. Dobraunig, C., Eichlseder, M., Mendel, F., Schläffer, M.: ASCON v1.2: lightweight authenticated encryption and hashing. J. Cryptol. **34**(3), 1–42 (2021). https://doi.org/10.1007/s00145-021-09398-9
21. Gong, Z., Nikova, S., Law, Y.W.: KLEIN: a new family of lightweight block ciphers. In: Juels, A., Paar, C. (eds.) RFID. Security and Privacy, pp. 1–18. Springer Berlin Heidelberg, Berlin, Heidelberg (2012). https://doi.org/10.1007/978-3-642-25286-0_1
22. Izadi, M., Sadeghiyan, B., Sadeghian, S.S., Khanooki, H.A.: MIBS: a new lightweight block cipher. In: Garay, J.A., Miyaji, A., Otsuka, A. (eds.) CANS 2009. LNCS, vol. 5888, pp. 334–348. Springer, Heidelberg (2009). https://doi.org/10.1007/978-3-642-10433-6_22
23. Lai, X., Massey, J.L., Murphy, S.: Markov ciphers and differential cryptanalysis. In: Advances in Cryptology - EUROCRYPT 1991, Workshop on the Theory and Application of Cryptographic Techniques, Brighton, UK, April 8–11, 1991, Proceedings, pp. 17–38 (1991). https://doi.org/10.1007/3-540-46416-6_2
24. Li, T., Sun, Y.: Superball: a new approach for MILP modelings of boolean functions. IACR Trans. Symmetric Cryptol. **2022**(3), 341–367 (2022)
25. Liu, Yu., et al.: STP models of optimal differential and linear trail for S-box based ciphers. Sci. China Inf. Sci. **64**(5), 1–3 (2021). https://doi.org/10.1007/s11432-018-9772-0
26. Morawiecki, P., et al.: ICEPOLE: high-speed, hardware-oriented authenticated encryption. Cryptology ePrint Archive, Paper 2014/266 (2014). https://eprint.iacr.org/2014/266
27. Mouha, N., Wang, Q., Gu, D., Preneel, B.: Differential and linear cryptanalysis using mixed-integer linear programming. In: Wu, C., Yung, M., Lin, D. (eds.) Information Security and Cryptology - 7th International Conference, Inscrypt 2011, Beijing, China, November 30 – December 3, 2011. Revised Selected Papers. Lecture Notes in Computer Science, vol. 7537, pp. 57–76. Springer (2011). https://doi.org/10.1007/978-3-642-34704-7_5

28. Pal, D., Baksi, A., Mandal, S., Sarkar, S.: Improved modeling for substitution boxes with negative samples and beyond (extended version). Cryptology ePrint Archive, Paper 2025/1925 (2025). https://eprint.iacr.org/2025/1925
29. Pal, D., Chandratreya, V.P., Chowdhury, D.R.: Efficient algorithms for modeling sboxes using MILP. CoRR arXiv:2306.02642 (2023)
30. Pal, D., Chandratreya, V.P., Chowdhury, D.R.: New techniques for modeling sboxes: an MILP approach. In: Cryptology and Network Security - 22nd International Conference, CANS 2023, Augusta, GA, USA, October 31 – November 2, 2023, Proceedings. Lecture Notes in Computer Science, vol. 14342, pp. 318–340. Springer (2023). https://doi.org/10.1007/978-981-99-7563-1_15
31. Sasaki, Y., Todo, Y.: New algorithm for modeling s-box in MILP based differential and division trail search. In: Innovative Security Solutions for Information Technology and Communications - 10th International Conference, SecITC 2017, Bucharest, Romania, June 8–9, 2017, Revised Selected Papers. Lecture Notes in Computer Science, vol. 10543, pp. 150–165. Springer (2017). https://doi.org/10.1007/978-3-319-69284-5_11
32. Shibutani, K., Isobe, T., Hiwatari, H., Mitsuda, A., Akishita, T., Shirai, T.: Piccolo: an ultra-lightweight blockcipher. In: Cryptographic Hardware and Embedded Systems - CHES 2011 - 13th International Workshop, Nara, Japan, September 28 – October 1, 2011. Proceedings, pp. 342–357 (2011). https://doi.org/10.1007/978-3-642-23951-9_23
33. Shimoyama, T., et al.: The block cipher SC2000. In: Matsui, M. (ed.) Fast Software Encryption, 8th International Workshop, FSE 2001 Yokohama, Japan, April 2–4, 2001, Revised Papers. Lecture Notes in Computer Science, vol. 2355, pp. 312–327. Springer (2001). https://doi.org/10.1007/3-540-45473-X_26
34. Sun, S., et al.: Towards finding the best characteristics of some bit-oriented block ciphers and automatic enumeration of (related-key) differential and linear characteristics with predefined properties. Cryptology ePrint Archive, Paper 2014/747 (2014). https://eprint.iacr.org/2014/747
35. Sun, S., Hu, L., Wang, P., Qiao, K., Ma, X., Song, L.: Automatic security evaluation and (related-key) differential characteristic search: application to SIMON, PRESENT, LBlock, DES(L) and other bit-oriented block ciphers. In: Sarkar, P., Iwata, T. (eds.) ASIACRYPT 2014. LNCS, vol. 8873, pp. 158–178. Springer, Heidelberg (2014). https://doi.org/10.1007/978-3-662-45611-8_9
36. Suzaki, T., Minematsu, K., Morioka, S., Kobayashi, E.: TWINE: a lightweight block cipher for multiple platforms. In: Knudsen, L.R., Wu, H. (eds.) Selected Areas in Cryptography, pp. 339–354. Springer, Berlin Heidelberg, Berlin, Heidelberg (2013). https://doi.org/10.1007/978-3-642-35999-6_22
37. Wu, S., Wang, M.: Security evaluation against differential cryptanalysis for block cipher structures. Cryptology ePrint Archive, Paper 2011/551 (2011). https://eprint.iacr.org/2011/551
38. Zhang, W., Bao, Z., Lin, D., Rijmen, V., Yang, B., Verbauwhede, I.: RECTANGLE: a bit-slice ultra-lightweight block cipher suitable for multiple platforms. IACR Cryptol. ePrint Arch 84 (2014)

Refined Linear Approximations for ARX Ciphers and Their Application to ChaCha

Yurie Okada[1]([✉])[iD], Atsuki Nagai[2], and Atsuko Miyaji[1][iD]

[1] Graduate School of Engineering, The University of Osaka, Suita, Japan
`yurie.okada@cy2sec.comm.eng.osaka-u.ac.jp`, `miyaji@comm.eng.osaka-u.ac.jp`
[2] KDDI Corporation, Chiyoda City, Japan
`at-nagai@kddi.com`

Abstract. ARX-based ciphers such as Salsa20 and ChaCha achieve high performance using only modular addition, rotation, and XOR. While ARX constructions are widely deployed in practice, linear and differential-linear cryptanalysis often reveal non-negligible biases in their reduced-round variants. Previous work has shown that a 7-round distinguisher on ChaCha is feasible, requiring about 2^{214} operations and relying on a linear approximation with a theoretical bias of 2^{-53}. However, such theoretical approximations significantly deviate from experimental observations. In this work, we resolve these discrepancies by introducing new fundamental linear approximations for two consecutive additions over three independent variables. We rigorously derive the exact probabilities of these approximations, demonstrating that the conventional independence assumption leads to systematic errors in bias estimation. Applying our theorem to ChaCha, we refine the probabilities of key approximations used in previous attacks. Our refined estimates closely match experimentally observed biases, reducing the gap between theory and practice. These results provide a more accurate foundation for future differential-linear cryptanalysis of ChaCha and other ARX-based designs.

Keywords: ARX · Piling-up Lemma · ChaCha · linear approximation

1 Introduction

ARX Ciphers. In today's advanced information society, symmetric-key cryptography plays a vital role in securing digital communications, transactions, and data exchanges, thereby protecting both stored and transmitted data. Among symmetric primitives, stream ciphers such as Salsa20 [3] and ChaCha [10] have been widely deployed in hardware and software products, making them especially important in practice.

Two common design paradigms for stream and block ciphers are Substitution-Permutation Networks (SPNs) and ARX (Addition, Rotation,

© The Author(s), under exclusive license to Springer Nature Switzerland AG 2026
R. Dutta et al. (Eds.): INDOCRYPT 2025, LNCS 16372, pp. 70–91, 2026.
https://doi.org/10.1007/978-3-032-13301-4_4

XOR) structures. ARX ciphers rely solely on modular addition, bitwise rotation, and exclusive OR, achieving both simplicity and efficiency. Modular addition introduces nonlinearity, while rotation enhances diffusion. ARX constructions are appealing for their high performance and resistance to many standard cryptanalytic techniques; however, they may exhibit noticeable biases in differential and linear approximations, particularly in reduced-round instances.

In the cryptanalysis of ARX designs, linear approximations of additions are fundamental. For estimating linear biases, Wallén's method [15] for analyzing carry propagation in modular addition is essential. Let $\Theta_i(x, y)$ denote the ith carry bit of the modular addition $x + y$. In particular, the following linear approximations [6,15] are frequently used:

$$\Pr[\Theta_i(x, y) = y_{i-1}] = \tfrac{1}{2}\left(1 + \tfrac{1}{2}\right), \quad i > 0, \tag{1}$$

$$\Pr[\Theta_i(x, y) \oplus \Theta_{i-1}(x, y) = 0] = \tfrac{1}{2}\left(1 + \tfrac{1}{2}\right), \quad i > 0. \tag{2}$$

For more advanced analysis, partial linear approximations in lower rounds are combined into longer approximations across several rounds, where Matsui's Piling-up Lemma [12] is often applied. This approach constructs theoretical estimates of linear biases by combining local approximations and then validates them against experimental measurements. It is crucial that theoretical bias estimates closely reflect empirical values. For example, Nyberg et al. [13] refined the bias estimation of linear approximations in the ARX cipher SNOW 2.0, and more generally, similar refinements have been investigated for block ciphers [14].

Related Works on ChaCha. ChaCha [10], proposed by Bernstein in 2008 as a redesign of Salsa20, was later adopted by the Internet Engineering Task Force (IETF) for TLS 1.3, establishing itself as a cornerstone of secure communications. Compared to its predecessor, ChaCha provides better diffusion and stronger resistance to cryptanalysis. Numerous security analyses have since been carried out. Table 1 summarizes the number of rounds, time complexity, and data complexity of existing distinguishers and key-recovery attacks.

Theoretical bias estimates are used to derive complexities in [4,5], whereas experimental bias measurements are employed in [2,6,7,17]. In parallel, key-recovery attacks based on Probabilistic Neutral Bits (PNBs) [1] have been explored. These have been enhanced through techniques such as syncopation, optimized right-pair search strategies [2,4,5,8,16], and more recently bit puncturing [9], which reduces complexity by selectively omitting state bits.

The first application of differential-linear cryptanalysis to ChaCha was introduced by Maitra et al. [4], who analyzed its internal structure and proposed a distinguisher by transforming one-bit output differences into multi-bit XOR expressions. This line of research was extended by Coutinho et al. [5], who incorporated two-bit output differences to construct more effective linear approximations and demonstrated an improved 6-round attack using theoretical bias estimation. Later works [2,6,7,17] extended the number of covered rounds by leveraging experimentally observed biases in partial approximations. Notably, [7] presented a 7-round distinguisher with complexity 2^{214}, which remains the best

Table 1. Time and data complexity of attacks against reduced-round ChaCha

Rounds	Time	Data	Type	Linear Round	References
4	2^6	2^6	Distinguisher	–	[4]
4.5	2^{12}	2^{12}	Distinguisher	–	[4]
5	2^{16}	2^{16}	Distinguisher	(3, 5)	[4]
6	2^{116}	2^{116}	Distinguisher	(3, 6)	[4]
	2^{75}	2^{75}	Distinguisher	(3, 6)	[5]
	2^{51}	2^{51}	Distinguisher	(3, 6)	[6]
	$2^{127.5}$	$2^{37.5}$	Key Recovery	–	[4]
	$2^{102.2}$	2^{56}	Key Recovery	–	[5]
	$2^{61.4}$	2^{51}	Key Recovery	–	[9]
	$2^{57.4}$	$2^{55.7}$	Key Recovery	–	[9]
7	2^{224}	2^{224}	Distinguisher	(3, 7)	[6]
	2^{214}	2^{214}	Distinguisher	(3, 7)	[7]
	$2^{120.9}$	$2^{120.9}$	Distinguisher	(4, 7)	[17]
	$2^{237.7}$	2^{96}	Key Recovery	–	[4]
	$2^{231.9}$	2^{50}	Key Recovery	–	[5]
	$2^{218.92}$	$2^{87.18}$	Key Recovery	–	[8]
	$2^{216.9}$	$2^{68.9}$	Key Recovery	–	[16]
	$2^{210.3}$	$2^{103.3}$	Key Recovery	–	[16]
	$2^{210.68}$	$2^{110.81}$	Key Recovery	–	[2]
	$2^{148.2}$	$2^{127.7}$	Key Recovery	–	[9]
	$2^{154.2}$	$2^{102.9}$	Key Recovery	–	[9]
7.25	$2^{244.85}$	$2^{93.24}$	Key Recovery	–	[8]
7.5	$2^{251.54}$	$2^{251.54}$	Distinguisher	(3.5, 7.5)	[2]

known against ChaCha. However, although discrepancies between theoretical
and experimental bias values have been pointed out, no prior work has attempted
to refine the theoretical estimates themselves. To illustrate these discrepancies
in more detail, consider the structure of the distinguisher attack against ChaCha
in [7]. It combines differential cryptanalysis over rounds 1–3 with linear cryptanalysis over rounds 3–7. To describe the resulting approximations, we use the
following notation: the jth bit of the ith word in the mth round is denoted by
$x_{i,j}^{(m)}$, the jth bit of the ith word in the $(\frac{s}{2})^{\text{th}}$ round by $x_{i,j}^{[s]}$, and the XOR of
bits $j_1, j_2, \cdots, j_n$ of word i by $x_i[j_1, j_2, \cdots, j_n]$. The linear approximation for
rounds 3–7 (Eq. (3)) consists of the approximation for rounds 3–6 (Eq. (4)) and
that for rounds 6–7.

$$x_{3,0}^{[6]} \oplus x_{4,0}^{[6]} = x_0^{[14]}[0, 3, 4, 7, 8, 11, 12, 14, 15, 18, 20, 27, 28]$$

$$\oplus\ x_1^{[14]}[0, 5, 7, 8, 10, 14, 15, 16, 22, 23, 24, 25, 27, 30, 31] \oplus x_2^{[14]}[7, 9, 10, 16, 19, 25, 26]$$

$$\oplus\ x_3^{[14]}[6, 7, 8, 24] \oplus x_4^{[14]}[0, 2, 3, 5, 18, 22, 23, 27] \oplus x_5^{[14]}[1, 2, 9, 10, 13, 14, 18, 21, 22, 25, 29]$$

$$\oplus\ x_6^{[14]}[0, 2, 3, 7, 10, 11, 13, 14, 19, 22, 23, 25, 27, 31] \oplus x_7^{[14]}[1, 2, 13, 25, 26, 30, 31]$$

$$\oplus\ x_8^{[14]}[8, 11, 13, 20, 25, 27, 28, 30, 31] \oplus x_9^{[14]}[2, 3, 6, 7, 11, 14, 15, 18, 23, 27]$$

$$\oplus\ x_{10}^{[14]}[0, 3, 4, 6, 8, 12, 13, 14, 18, 20, 23, 25, 27, 28]$$

$$\oplus\ x_{11}^{[14]}[6, 14, 15, 18, 19, 23, 24, 27] \oplus x_{12}^{[14]}[3, 4, 6, 11, 13, 22, 23, 24, 26, 27, 30, 31]$$

$$\oplus\ x_{13}^{[14]}[1, 2, 6, 7, 8, 13, 14, 16, 18, 20, 22, 23, 24, 25, 26]$$

$$\oplus\ x_{14}^{[14]}[0, 7, 13, 14, 15, 16, 17, 18, 23, 24] \oplus x_{15}^{[14]}[16, 25, 26]. \tag{3}$$

$$x_{3,0}^{[6]} \oplus x_{4,0}^{[6]} = x_0^{[12]}[0, 16] \oplus x_1^{[12]}[0, 6, 7, 11, 12, 22, 23] \oplus x_2^{[12]}[0, 6, 7, 8, 16, 18, 19, 24]$$

$$\oplus\ x_4^{[12]}[7, 13, 19] \oplus x_{5,7}^{[12]} \oplus x_6^{[12]}[7, 13, 14, 19] \oplus x_7^{[12]}[6, 7, 14, 15, 26] \oplus x_8^{[12]}[0, 7, 8, 19, 31]$$

$$\oplus\ x_9^{[12]}[0, 6, 12, 26] \oplus x_{10,0}^{[12]} \oplus x_{11}^{[12]}[6, 7] \oplus x_{12}^{[12]}[0, 11, 12, 19, 20, 30, 31]$$

$$\oplus\ x_{13}^{[12]}[0, 14, 15, 24, 26, 27] \oplus x_{14}^{[12]}[8, 25, 26] \oplus x_{15,24}^{[12]}, \tag{4}$$

We expand the right-hand side of Eq. (4) from round 6 to round 7. In this process, we divide the right-hand side according to the inputs of the round function, and present the expanded equations for each as Groups I–IV in [7].

Group I

$$x_0^{[12]}[0, 16] \oplus x_4^{[12]}[7, 13, 19] \oplus x_8^{[12]}[0, 7, 8, 19, 31] \oplus x_{12}^{[12]}[0, 11, 12, 19, 20, 30, 31]$$

$$= x_0^{[14]}[0, 3, 4, 7, 8, 11, 12, 14, 15, 18, 20, 27, 28] \oplus x_4^{[14]}[0, 2, 3, 5, 18, 22, 23, 27]$$

$$\oplus\ x_8^{[14]}[8, 11, 13, 20, 25, 27, 28, 30, 31] \oplus x_{12}^{[14]}[3, 4, 6, 11, 13, 22, 23, 24, 26, 27, 30, 31]$$

Group II

$$x_1^{[12]}[0, 6, 7, 11, 12, 22, 23] \oplus x_{5,7}^{[12]} \oplus x_9^{[12]}[0, 6, 12, 26] \oplus x_{13}^{[12]}[0, 14, 15, 24, 26, 27]$$

$$= x_1^{[14]}[0, 5, 7, 8, 10, 14, 15, 16, 22, 23, 24, 25, 27, 30, 31]$$

$$\oplus\ x_5^{[14]}[1, 2, 9, 10, 13, 14, 18, 21, 22, 25, 29] \oplus x_9^{[14]}[2, 3, 6, 7, 11, 14, 15, 18, 23, 27]$$

$$\oplus\ x_{13}^{[14]}[1, 2, 6, 7, 8, 13, 14, 16, 18, 20, 22, 23, 24, 25, 26]$$

Group III

$$x_2^{[12]}[0, 6, 7, 8, 16, 18, 19, 24] \oplus x_6^{[12]}[7, 13, 14, 19] \oplus x_{10,0}^{[12]} \oplus x_{14}^{[12]}[8, 25, 26]$$

$$= x_2^{[14]}[7, 9, 10, 16, 19, 25, 26] \oplus x_6^{[14]}[0, 2, 3, 7, 10, 11, 13, 14, 19, 22, 23, 25, 27, 31]$$

$$\oplus\ x_{10}^{[14]}[0, 3, 4, 6, 8, 12, 13, 14, 18, 20, 23, 25, 27, 28]$$

$$\oplus\ x_{14}^{[14]}[0, 7, 13, 14, 15, 16, 17, 18, 23, 24]$$

Group IV

$$x_7^{[12]}[6,7,14,15,26] \oplus x_{11}^{[12]}[6,7] \oplus x_{15,24}^{[12]}$$
$$= x_3^{[14]}[6,7,8,24] \oplus x_7^{[14]}[1,2,13,25,26,30,31] \oplus x_{11}^{[14]}[6,14,15,18,19,23,24,27]$$
$$\oplus x_{15}^{[14]}[16,25,26]$$

Significant discrepancies between theoretical and experimental biases have been observed for Eq. (4) and Groups I–III [6,7]. Theoretical biases for Eq. (4) and Groups I, II, and III are estimated at 2^{-8}, 2^{-12}, 2^{-14}, and 2^{-15}, respectively. In contrast, the corresponding experimental values are $2^{-7.17}$, $2^{-11.70}$, $2^{-12.31}$, and $2^{-12.813}$ [6,7]. These partial approximations were first introduced in [6], with Groups II and III later refined in [7]. Nevertheless, even after these refinements, a substantial gap between theoretical and experimental biases remains.

Our First Contribution: Fundamental Linear Approximation. To resolve the discrepancies discussed above, we introduce new fundamental linear approximations beyond Eqs. (1) and (2). These approximations are not limited to ChaCha but are broadly applicable to ARX-based ciphers that include the same additions in their structure.

In previous linear cryptanalysis, linear bias was estimated under the assumption that two additions are independent, by combining Eqs. (1) and (2) with Matsui's piling-up lemma. However, since the two consecutive additions are not independent, significant discrepancies arise between these theoretical probabilities and the experimental results. We consider two consecutive additions of three variables x, y, z, namely $x + y$ and $(x + y) + z$, as illustrated in Fig. 1. From this setting we derive three new linear relations:

$$\Theta_i(x,y) \oplus \Theta_i(x+y,z) = x_{i-1} \oplus z_{i-1}, \tag{5}$$
$$\Theta_{i-1}(x,y) \oplus \Theta_i(x,y) \oplus \Theta_{i-1}(x+y,z) \oplus \Theta_i(x+y,z) = 0, \tag{6}$$
$$\Theta_i(x,y) \oplus \Theta_{i-1}(x,y) \oplus \Theta_i(x+y,z) = z_{i-1}. \tag{7}$$

The properties of these relations are summarized as follows:

1. Equation (5): The XOR of the i-th carries of the two consecutive additions equals $x_{i-1} \oplus z_{i-1}$.
2. Equation (6): The XOR of the four carries at positions $i - 1$ and i in both additions equals zero.
3. Equation (7): The XOR of the two carries at positions $i - 1$ and i in the first addition and the i-th carry in the second addition equals z_{i-1}.

In existing approaches, the probabilities that Eqs. (5)–(7) hold were computed as follows:

$$\Pr[\text{Eq. (5)}] = \Pr[\text{Eq. (6)}] = \Pr[\text{Eq. (7)}] = \tfrac{1}{2}\left(1 + \tfrac{1}{4}\right). \tag{8}$$

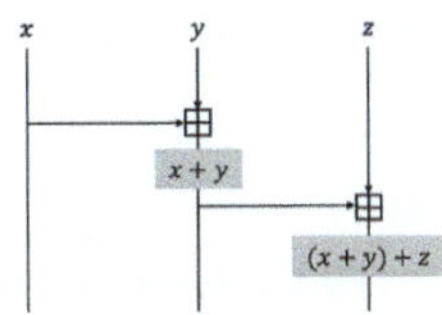

Fig. 1. Addition of three independent elements

However, since these two consecutive additions are not independent, the use of the piling-up lemma causes discrepancies between the experimental and theoretical biases. This work rigorously derives the *exact* theoretical probabilities of Eqs. (5)–(7), summarized in the following theorem.

Theorem 1. *Let x, y, and z be independent variables. For the i-th carry in the binary additions $x + y$ and $(x + y) + z$, the following holds for all $i \geq 1$:*

$$\Pr[\text{Eq. (5)}] = \tfrac{2}{3} - \tfrac{1}{6} \cdot \left(\tfrac{1}{4}\right)^{i-1} \approx \tfrac{2}{3} = \tfrac{1}{2}\left(1 + \tfrac{1}{3}\right), \quad i \gg 1, \tag{9}$$

$$\Pr[\text{Eq. (6)}] = \tfrac{2}{3} - \tfrac{1}{6} \cdot \left(\tfrac{1}{4}\right)^{i-1} \approx \tfrac{2}{3} = \tfrac{1}{2}\left(1 + \tfrac{1}{3}\right), \quad i \gg 1, \tag{10}$$

$$\Pr[\text{Eq. (7)}] = \tfrac{7}{12} + \tfrac{1}{6} \cdot \left(\tfrac{1}{4}\right)^{i-1} \approx \tfrac{7}{12} = \tfrac{1}{2}\left(1 + \tfrac{1}{6}\right), \quad i \gg 1. \tag{11}$$

Our Second Contribution: Application to ChaCha. We now return to the discrepancies observed in the distinguisher attack against ChaCha in [7]. By applying Theorem 1 to Eq. (4) and Groups I–IV, we obtain refined probabilities for the corresponding linear approximations, as given in the following lemmas.

Lemma 1. *The probability that Eq. (4) holds for rounds 3–6 is*

$$\Pr[\text{Eq. (4)}] = \tfrac{1}{2}\left(1 + \tfrac{1}{2^{6} \cdot 3}\right).$$

Lemma 2. *The probabilities that* Group I $-$ III *hold are*

$$\Pr[\text{Group I}] = \tfrac{1}{2}\left(1 + \tfrac{1}{2^{7} \cdot 3^{3}}\right), \quad \Pr[\text{Group II}] = \tfrac{1}{2}\left(1 + \tfrac{1}{2^{6} \cdot 3^{4}}\right), \quad \Pr[\text{Group III}] = \tfrac{1}{2}\left(1 + \tfrac{1}{2^{8} \cdot 3^{3}}\right).$$

Combining Lemmas 1 and 2 with Group IV from [6,7], we refine the bias of Eq. (3), as presented in Theorem 2. Since no discrepancy was observed between the experimental and theoretical values for Group IV, no further refinement was necessary.

Theorem 2. *The probability that Eq. (3) holds for rounds 3–7 is*

$$\Pr[\text{Eq. (3)}] = \tfrac{1}{2}\left(1 + \tfrac{1}{2^{31} \cdot 3^{11}}\right) \approx \tfrac{1}{2}\left(1 + \tfrac{1}{2^{48.43}}\right).$$

Table 2 compares our refined theoretical biases with experimental results and with the existing estimates in [7]. The experiments were conducted by ourselves, and the error is calculated as (experimental value $-$ theoretical value)/theoretical value. The results show that our refined probabilities match

Table 2. Comparison of theoretical and experimental linear biases and errors between [7] and this work

Round		Experimental	[7]		This work	
			Theoretical	Error	Theoretical	Error
3–6	Eq. (4)	$2^{-7.17}$	2^{-8}	0.781	$2^{-6} \cdot 3^{-1} \approx 2^{-7.58}$	0.336
6–7	Group 1	$2^{-11.70}$	2^{-12}	0.232	$2^{-7} \cdot 3^{-3} \approx 2^{-11.75}$	0.039
	Group 2	$2^{-12.28}$	2^{-14}	2.304	$2^{-6} \cdot 3^{-4} \approx 2^{-12.34}$	0.045
	Group 3	$2^{-12.84}$	2^{-15}	3.473	$2^{-8} \cdot 3^{-3} \approx 2^{-12.75}$	−0.057

the experimental biases significantly more closely than the earlier theoretical estimates.

The structure of this paper is as follows. Section 2 introduces linear approximations for ARX-based ciphers, the differential-linear cryptanalysis that forms the basis of this study, and the structure of ChaCha. Section 3 reviews existing linear approximations of ChaCha. Section 4 presents and proves a new theorem on carries in non-independent additions of three variables. Section 5 applies this theorem to the linear approximations of ChaCha derived in [7]. Finally, Sect. 6 concludes the paper.

2   Preliminary

This section provides background for this work. We first review the fundamental linear approximations and their integration in ARX ciphers, then discuss differential-linear cryptanalysis, and finally describe the structure of ChaCha. Table 3 summarizes the notations used in this paper.

Table 3. Notation

Notation	Description
$+$	32-bit modular addition
$\oplus$	bitwise XOR
$\lll i$	rotation by i bits to the left
$X^{(r)}$	state matrix after application of r round functions
$X^{[s]}$	state matrix after application of $\frac{s}{2}$ round functions
$x_{i,j}^{(r)}$	j^{th} bit of ith word in the rth round
$x_i^{(r)}[j_1, j_2, \cdots, j_n]$	bitwise XOR of $x_{i,j_1}^{(r)}$, $x_{i,j_1}^{(r)}$, $\cdots$ and $x_{i,j_n}^{(r)}$
$x_{i,j}^{[s]}$	j^{th} bit of ith word in the $\frac{s}{2}^{th}$ round
b w.p. a	b with probability a

2.1   A Review of Differential-Linear Cryptanalysis

Differential-linear cryptanalysis builds on linear approximations by describing an addition $x + y$ of two numbers as a combination of XOR and carry, $\Theta(x, y)$:

$$x + y = x \oplus y \oplus \Theta(x, y). \tag{12}$$

The i-th bit of $\Theta(x, y)$ is denoted by $\Theta_i(x, y)$, with $\Theta_0(x, y) = 0$. Fundamental linear approximations such as Eqs. (1) and (2) are frequently used in constructing linear relations. To extend these into multi-round linear approximations, the following lemma is commonly applied to combine several approximations.

Lemma 3 (Piling-up Lemma (Lemma 3 of [12])). *Let $X_i(1 \leq i \leq n)$ be independent random variables taking value 0 with probability p_i and 1 with probability $1 - p_i$. Then $\Pr[X_1 \oplus X_2 \oplus \cdots \oplus X_n = 0]$ is*

$$\tfrac{1}{2} + 2^{n-1} \prod_{i=1}^{n} \left(p_i - \tfrac{1}{2}\right).$$

Differential-linear cryptanalysis, proposed by S. Langford and M.E. Hellman in 1994 [11], is a two-step technique that combines differential analysis with linear approximations:

STEP 1: Perform a differential analysis between input and output differences.

STEP 2: Identify linear approximations holding on the output difference from STEP 1 and apply linear analysis.

The differential-linear bias is defined as follows. Let σ and σ' denote linear combinations of two states $X(0)$ and $X'(0)$ after r rounds of transformation. Similarly, let ρ and ρ' denote linear combinations of two states after R rounds of transformation, where $r < R$:

$$\Delta\sigma = \sigma \oplus \sigma' = \bigoplus_u \Delta x^{(r)}_{p_u, q_u}, \quad \rho = \bigoplus_v x^{(R)}_{p_v, q_v}, \quad \rho' = \bigoplus_v x'^{(R)}_{p_v, q_v}.$$

Here, p_u and q_u indicate the word and bit positions of differing bits in σ and σ', while p_v and q_v indicate positions of terms involved in the linear approximation.

The differential bias ϵ_d and the linear bias ϵ_L are defined as

$$\Pr[\Delta\sigma = 0] = \tfrac{1}{2}(1 + \epsilon_d), \qquad \Pr[\sigma = \rho] = \tfrac{1}{2}(1 + \epsilon_L).$$

The differential-linear bias after R rounds, γ, can then be computed as

$$\begin{aligned}
\Pr[\Delta\rho = 0] &= \Pr[\Delta\sigma = 0] \cdot \Pr[\Delta\sigma = \Delta\rho] + \Pr[\Delta\sigma = 1] \cdot \Pr[\Delta\sigma = \bar{\Delta}\rho] \\
&= \tfrac{1}{2}(1 + \epsilon_d) \cdot \tfrac{1}{2}(1 + \epsilon_L^2) + \tfrac{1}{2}(1 - \epsilon_d) \cdot \tfrac{1}{2}(1 - \epsilon_L^2) \\
&= \tfrac{1}{2}(1 + \epsilon_d \cdot \epsilon_L^2),
\end{aligned}$$

using the relation

$$\Pr[\Delta\sigma = \Delta\rho] = \Pr[\sigma = \rho] \cdot \Pr[\sigma' = \rho'] + \Pr[\sigma = \bar{\rho}] \cdot \Pr[\sigma' = \bar{\rho}'] = \tfrac{1}{2}(1 + \epsilon_L^2).$$

Thus, the resulting differential-linear bias is given by $\gamma = \epsilon_d \cdot \epsilon_L^2$.

2.2 Structure of ChaCha

ChaCha utilizes three operations: 32-bit modular addition, bitwise exclusive OR (XOR), and shift rotation. A single word consists of 4 bytes, and an initial state matrix is constructed from 16 words derived from the secret key, block counter, nonce, and constants. Each round consists of four QuarterRound functions, which process four words at a time. The words in the 4×4 matrix are processed either column-wise or diagonally, with the pattern repeating every two rounds. The initial state matrix of ChaCha is give as follows:

$$
X^{(0)} = \begin{pmatrix} x_0^{(0)} & x_1^{(0)} & x_2^{(0)} & x_3^{(0)} \\ x_4^{(0)} & x_5^{(0)} & x_6^{(0)} & x_7^{(0)} \\ x_8^{(0)} & x_9^{(0)} & x_{10}^{(0)} & x_{11}^{(0)} \\ x_{12}^{(0)} & x_{13}^{(0)} & x_{14}^{(0)} & x_{15}^{(0)} \end{pmatrix} = \begin{pmatrix} c_0 & c_1 & c_2 & c_3 \\ k_0 & k_1 & k_2 & k_3 \\ k_4 & k_5 & k_6 & k_7 \\ t_0 & v_0 & v_1 & v_2 \end{pmatrix},
$$

which takes four predefined constants $c_0 = 0x61707865, c_1 = 0x3320646e, c_2 = 0x79622d32, c_3 = 0x6b206574$, 256-bit key $k_0 \sim k_7$, 96-bit nonce $v_0 \sim v_2$ and 32-bit counter t_0. The QuaterRound Function (QRF) has the following operations denoted by $QR_{ChaCha}(x_a^{(m)}, x_b^{(m)}, x_c^{(m)}, x_d^{(m)})$ (See Fig. 2.):

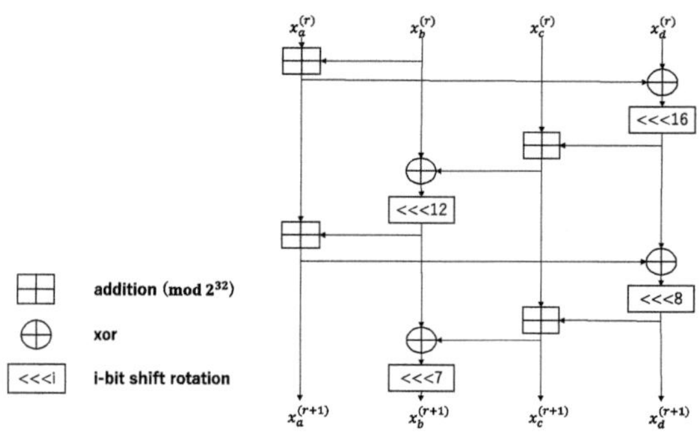

Fig. 2. One QR_{ChaCha}

$$
\begin{cases}
x_{a'}^{(m)} = x_a^{(m)} + x_b^{(m)}; \; x_{d'}^{(m)} = x_d^{(m)} \oplus x_{a'}^{(m)}; \; x_{d''}^{(m)} = x_{d'}^{(m)} \lll 16; \\
x_{c'}^{(m)} = x_c^{(m)} + x_{d''}^{(m)}; \; x_{b'}^{(m)} = x_b^{(m)} \oplus x_{c'}^{(m)}; \; x_{b''}^{(m)} = x_{b'}^{(m)} \lll 12; \\
x_a^{(m+1)} = x_{a'}^{(m)} + x_{b''}^{(m)}; \; x_{d'''}^{(m)} = x_{d''}^{(m)} \oplus x_a^{(m+1)}; \; x_d^{(m+1)} = x_{d'''}^{(m)} \lll 8; \\
x_c^{(m+1)} = x_{c'}^{(m)} + x_d^{(m+1)}; \; x_{b'''}^{(m)} = x_{b''}^{(m)} \oplus x_c^{(m+1)}; \; x_b^{(m+1)} = x_{b'''}^{(m)} \lll 7;
\end{cases}
\tag{13}
$$

Using Eq. (12), the QRF Eq. (13) can be rewritten as follows:

$$x_{b,i}^{(m-1)} = x_{b,i+19}^{(m)} \oplus x_c^{(m)}[i, i+12] \oplus x_{d,i}^{(m)} \oplus \Theta_i(x_{c'}^{(m-1)}, x_d^{(m)}),$$

$$x_{a,i}^{(m-1)} = x_{a,i}^{(m)} \oplus x_b^{(m)}[i+7, i+19] \oplus x_{c,i+12}^{(m)} \oplus x_{d,i}^{(m)}$$

$$\oplus \Theta_i(x_{c'}^{(m-1)}, x_d^{(m)}) \oplus \Theta_i(x_a^{(m-1)}, x_b^{(m-1)}) \oplus \Theta_i(x_{a'}^{(m-1)}, x_{b''}^{(m-1)}), \qquad (14)$$

$$x_{c,i}^{(m-1)} = x_{a,i}^{(m)} \oplus x_{c,i}^{(m)} \oplus x_d^{(m)}[i, i+8] \oplus \Theta_i(x_{c'}^{(m-1)}, x_d^{(m)}) \oplus \Theta_i(x_c^{(m-1)}, x_{d''}^{(m-1)}),$$

$$x_{d,i}^{(m-1)} = x_a^{(m)}[i, i+16] \oplus x_{b,i+7}^{(m)} \oplus x_{c,i}^{(m)} \oplus x_{d,i+24}^{(m)} \oplus \Theta_i(x_{a'}^{(m-1)}, x_{b''}^{(m-1)}).$$

There are two types of rounds, column-round and diagonal-round correspond to the odd and even round, respectively. In the column-round, the following four QRF is applied: $QR_{ChaCha}(x_0^{(m)}, x_4^{(m)}, x_8^{(m)}, x_{12}^{(m)})$, $QR_{ChaCha}(x_1^{(m)}, x_5^{(m)}, x_9^{(m)}, x_{13}^{(m)})$, $QR_{ChaCha}(x_2^{(m)}, x_6^{(m)}, x_{10}^{(m)}, x_{14}^{(m)})$ and $QR_{ChaCha}(x_3^{(m)}, x_7^{(m)}, x_{11}^{(m)}, x_{15}^{(m)})$. In the diagonal-round, the following four QRF is applied: $QR_{ChaCha}(x_0^{(m)}, x_5^{(m)}, x_{10}^{(m)}, x_{15}^{(m)})$, $QR_{ChaCha}(x_1^{(m)}, x_6^{(m)}, x_{11}^{(m)}, x_{12}^{(m)})$, $QR_{ChaCha}(x_2^{(m)}, x_7^{(m)}, x_8^{(m)}, x_{13}^{(m)})$, and $QR_{ChaCha}(x_3^{(m)}, x_4^{(m)}, x_9^{(m)}, x_{14}^{(m)})$. The keystream block Z is generated by 32-bit addition of $X^{(0)}$ and $X^{(20)}$. $X^{(20)}$ is obtained by alternately repeating column-round and diagonal-round. Then, the ciphertext c is obtained by XOR operation of the keystream Z and the plaintext m: $Z = X^{(0)}+X^{(20)}$, $c = Z \oplus m$.

We also use the SubRound Function (SRF) [7], defined as follows:

$$(x_a^{[s]}, x_b^{[s]}, x_c^{[s]}, x_d^{[s]}) = SR(x_a^{[s-1]}, x_b^{[s-1]}, x_c^{[s-1]}, x_d^{[s-1]}, r_1, r_2),$$

$$x_a^{[s]} = x_a^{[s-1]} + x_b^{[s-1]}, x_d^{[s]} = (x_d^{[s-1]} \oplus x_a^{[s]}) \lll r_1, \qquad (15)$$

$$x_c^{[s]} = x_c^{[s-1]} + x_d^{[s]}, x_b^{[s]} = (x_b^{[s-1]} \oplus x_c^{[s]}) \lll r_2.$$

The SRF Eq. (15) is transformed as follows:

$$x_{b,i}^{[s-1]} = x_{b,i+r_2}^{[s]} \oplus x_{c,i}^{[s]}, \; x_{c,i}^{[s-1]} = x_{c,i}^{[s]} \oplus x_{d,i}^{[s]} \oplus \Theta_i(x_c^{[s-1]}, x_d^{[s]}),$$

$$x_{d,i}^{[s-1]} = x_{a,i}^{[s]} \oplus x_{d,i+r_1}^{[s]}, \; x_{a,i}^{[s-1]} = x_{a,i}^{[s]} \oplus x_{b,i+r_2}^{[s]} \oplus \Theta_i(x_a^{[s-1]}, x_b^{[s-1]}). \qquad (16)$$

Two SRFs correspond to One QRF, $X(s) = X[2s]$.

3 Related Works on Linear Approximation for ChaCha

This section reviews the linear approximations related to the proof of the 37 round linear approximation improved in Sect. 5, as well as linear approximations derived from the structure of SRF and the $m-1$-round single-bit and m-round multi-bit linear approximations in [4,7], and [6]. Using Eqs. (1), (2), and (14),the following Lemma 4 is derived.

Lemma 4. *Lemma 9 of [4] combined with Lemma 6 of [6]? For one active input bit in round $m-1$ and multiple active output bits in round $m \geq 1$, the following holds for $i > 0$.*

$$x_{b,i}^{(m-1)} = x_{b,i+19}^{(m)} \oplus x_{c,i}^{(m)} \oplus x_{c,i+12}^{(m)} \oplus x_{d,i}^{(m)} \oplus x_{d,i-1}^{(m)}, \ w.p. \tfrac{1}{2}\left(1+\tfrac{1}{2}\right), \quad (17)$$

$$\begin{aligned}
x_{a,i}^{(m-1)} = {} & x_{a,i}^{(m)} \oplus x_{b,i+7}^{(m)} \oplus x_{b,i+19}^{(m)} \oplus x_{c,i+12}^{(m)} \oplus x_{d,i}^{(m)} \\
& \oplus x_{b,i+6}^{(m)} \oplus x_{b,i+18}^{(m)} \oplus x_{c,i+11}^{(m)} \oplus x_{d,i-1}^{(m)}, \ w.p. \tfrac{1}{2}\left(1+\tfrac{1}{2^3}\right),
\end{aligned} \quad (18)$$

$$\begin{aligned}
x_{c,i}^{(m-1)} = {} & x_{a,i}^{(m)} \oplus x_{c,i}^{(m)} \oplus x_{d,i}^{(m)} \oplus x_{d,i+8}^{(m)} \oplus x_{a,i-1}^{(m)} \\
& \oplus x_{d,i+7}^{(m)} \oplus x_{d,i-1}^{(m)}, \ w.p. \tfrac{1}{2}\left(1+\tfrac{1}{2^2}\right),
\end{aligned} \quad (19)$$

$$\begin{aligned}
x_{d,i}^{(m-1)} = {} & x_{a,i}^{(m)} \oplus x_{a,i+16}^{(m)} \oplus x_{b,i+7}^{(m)} \oplus x_{c,i}^{(m)} \\
& \oplus x_{d,i+24}^{(m)} \oplus x_{c,i-1}^{(m)} \oplus x_{b,i+6}^{(m)}, \ w.p. \tfrac{1}{2}\left(1+\tfrac{1}{2}\right).
\end{aligned} \quad (20)$$

The following Lemma 5 is a reformulation of Lemmas 5–7 from [7], based on Eqs. (1) and (2).

Lemma 5. *Lemma 5–7 of [7] For one or two active input bits in subround $s-1$ and multiple output bits in subround s, the following linear approximations hold with probability $\frac{1}{2}(1+\frac{1}{2})$ with rotation distances r_1 and r_2, when $i > 0$*

$$\begin{aligned}
x_{c,i}^{[s-1]} &= x_{c,i}^{[s]} \oplus x_{d,i}^{[s]} \oplus x_{d,i-1}^{[s]}, \\
x_{a,i}^{[s-1]} &= x_{a,i}^{[s]} \oplus x_{b,i+r_2}^{[s]} \oplus x_{c,i}^{[s]} \oplus x_{b,i+r_2-1}^{[s]} \oplus x_{c,i-1}^{[s]}, \\
x_{c,i}^{[s-1]} \oplus x_{c,i-1}^{[s-1]} &= x_{c,i}^{[s]} \oplus x_{d,i}^{[s]} \oplus x_{c,i-1}^{[s]} \oplus x_{d,i-1}^{[s]}, \\
x_{a,i}^{[s-1]} \oplus x_{a,i-1}^{[s-1]} &= x_{a,i}^{[s]} \oplus x_{b,i+r_2}^{[s]} \oplus x_{c,i}^{[s]} \oplus x_{a,i-1}^{[s]} \oplus x_{b,i+r_2-1}^{[s]} \oplus x_{c,i-1}^{[s]}, \\
x_{c,i}^{[s-1]} \oplus x_{c,i-1}^{[s-1]} &= x_{c,i}^{[s]} \oplus x_{d,i}^{[s]}, \\
x_{a,i}^{[s-1]} \oplus x_{a,i-1}^{[s-1]} &= x_{a,i}^{[s]} \oplus x_{b,i+r_2}^{[s]} \oplus x_{c,i}^{[s]}.
\end{aligned} \quad (21)$$

Some linear approximations in Lemma 9 of [6] are represented in the following Lemma 6, which combines each two or more $(m-1)$-round to m-round linear approximations into each one linear equation.

Lemma 6. *Lemma 9 of [6] For multiple active input bits in round $m-1$ and multiple output bits in round m, the following linear approximations hold with probability $\frac{1}{2}(1+\frac{1}{2^k})$.*

$$\begin{aligned}
& x_a^{(m-1)}[i-1,i] \oplus x_{b,i}^{(m-1)} \\
& = x_a^{(m)}[i-1,i] \oplus x_b^{(m)}[i+6,i+7,i+18] \oplus x_c^{(m)}[i,i+11] \oplus x_d^{(m)}[i-2,i-1] \ (k=3), \\
& x_a^{(m-1)}[i-1,i] \oplus x_{c,i}^{(m-1)} \\
& = x_b^{(m)}[i+6,i+7,i+18,i+19] \oplus x_c^{(m)}[i,i+11,i+12] \oplus x_d^{(m)}[i-2,i-1,i+7,i+8] \ (k=4), \\
& x_a^{(m-1)}[i-1,i] \oplus x_{b,i}^{(m-1)} \oplus x_{c,i}^{(m-1)} \\
& = x_a^{(m)}[i-2,i] \oplus x_b^{(m)}[i+6,i+7,i+18] \oplus x_c^{(m)}[i-1,i,i+11] \oplus x_d^{(m)}[i+6,i+7] \ (k=3), \\
& x_a^{(m-1)}[i-1,i] \\
& = x_a^{(m)}[i-1,i] \oplus x_b^{(m)}[i+6,i+7,i+18,i+19] \oplus x_c^{(m)}[i+11,i+12] \oplus x_d^{(m)}[i-1,i] \ (k=3).
\end{aligned}$$

The computational complexity of 2^{214} in [7] is derived from the bias of the linear approximation for rounds 3–6 and four partial linear approximations for rounds 6–7, categorized into Groups 14. The linear approximation for rounds 3–6 is described in Lemma 7 (originally Lemma 10 in [6]). The four partial linear approximations corresponding to Groups 1–4 are given in Lemmas 8. The combined linear approximation for rounds 3–7 is then described in Lemma 9 (originally Lemma 8 in [7]).

Lemma 7. *Lemma 10 of [6]*
The probabilities for the linear approximation for 3–6 rounds is as follows:

$$\Pr[\text{Eq. (4)}] = \tfrac{1}{2}\left(1 + \tfrac{1}{2^8}\right). \tag{22}$$

Lemma 8. *(Lemma 8 of [7] and Lemma 11 of [6])*
The probabilities for the linear approximations in Groups $\mathrm{I - IV}$ *are as follows:*

$$\Pr[\text{Group I}] = \tfrac{1}{2}\left(1 + \tfrac{1}{2^{12}}\right), \quad \Pr[\text{Group II}] = \tfrac{1}{2}\left(1 + \tfrac{1}{2^{14}}\right),$$
$$\Pr[\text{Group III}] = \tfrac{1}{2}\left(1 + \tfrac{1}{2^{15}}\right), \quad \Pr[\text{Group IV}] = \tfrac{1}{2}\left(1 + \tfrac{1}{2^4}\right).$$

Lemma 9. *Lemma 8 of [7]*
The probability for the linear approximations for 3–7 rounds is as follows:

$$\Pr[\text{Eq. (3)}] = \tfrac{1}{2}\left(1 + \tfrac{1}{2^{53}}\right) \tag{23}$$

4 Newly Derived Fundamental Linear Approximations for ARX Ciphers

Analysis of stream ciphers often begins with approximations of a single addition of two elements such as Eqs (1) and (2) in Sect. 1. Not only approximations of a single addition but also those of two consecutive additions $x + y$ and $(x + y) + z$ of three elements x, y, and z, (as shown in Fig. 1), such as Eqs. (5)–(7), which hold for all $i \geq 1$, are frequently used in stream ciphers? For example, in [6,7], both Eq. (4) and Groups I-III rely on Eqs. (5)–(7). Equation (4) and Group III use Eq. (5); Group I uses Eqs.(5) and (7); and Group II uses Eqs.(5) and (6). Interestingly, the probabilities of these three equations are equal to Eq. (8) when they are evaluated by applying Lemma 3 under the assumption that the two additions are independent. In this section, we refine the linear approximations in Theorem 1 without relying on Lemma 3.

4.1 Proof of Theorem 1

This subsection provides a proof of Theorem 1. We now rigorously derive the exact theoretical probabilities of Eqs. (5)–(7) in Theorem 1.

Equations (9), (10), and (11) all rely on $\Pr[\Theta_i(x,y) \oplus \Theta_i(x+y,z) = 0]$. Define $F_i(x,y,z) := \Theta_i(x,y) \oplus \Theta_i(x+y,z)$; we prove by induction that

$$q_i = \Pr[F_i(x,y,z) = 0] = \tfrac{2}{3} \cdot \left(\tfrac{1}{4}\right)^i + \tfrac{1}{3} \quad (i \geq 0). \tag{24}$$

Note that $1 - q_i = \Pr[F_i(x,y,z) = 1]$. Table 4 shows the combinations of $(\Theta_i(x,y), \Theta_i(x+y,z))$ and (x_i, y_i, z_i) for which $\Theta_{i+1}(x,y) \oplus \Theta_{i+1}(x+y,z) = 0$ holds. For example, when $(\Theta_i(x,y), \Theta_i(x+y,z)) = (0,1)$, the cases $(x_i, y_i, z_i) = (0,0,0)$ and $(1,1,1)$ satisfy $\Theta_{i+1}(x,y) \oplus \Theta_{i+1}(x+y,z) = 0$.

Table 4. Combinations of $(\Theta_i(x,y), \Theta_i(x+y,z))$ and (x_i, y_i, z_i) that satisfy $\Theta_{i+1}(x,y) \oplus \Theta_{i+1}(x+y,z) = 0$

$(\Theta_i(x,y), \Theta_i(x+y,z))$	(x_i, y_i, z_i)
(0, 0)	(0, 0, 0), (0, 0, 1), (0, 1, 0), (1, 0, 0)
(0, 1)	(0, 0, 0), (1, 1, 1)
(1, 0)	(0, 0, 0), (1, 1, 1)
(1, 1)	(0, 1, 1), (1, 0, 1), (1, 1, 0), (1, 1, 1)

We first prove the case $i = 0$. Since no carry occurs from $i = 0$, we always have $(\Theta_0(x,y), \Theta_0(x+y,z)) = (0,0)$. Therefore, we obtain $q_0 = 1 = \tfrac{2}{3} + \tfrac{1}{3}$; thus Eq. (24) holds for $i = 0$.

Next, assume Eq. (24) holds for some $i \geq 0$. We prove the case $i+1$ by considering separately $F_i(x,y,z) = 0$ and $F_i(x,y,z) = 1$. When $F_i(x,y,z) = 0$, i.e. $(\Theta_i(x,y), \Theta_i(x+y,z)) \in \{(0,0),(1,1)\}$, Table 4 shows there are 8 cases with $F_{i+1}(x,y,z) = 0$. When $F_i(x,y,z) = 1$, i.e. $(\Theta_i(x,y), \Theta_i(x+y,z)) \in \{(0,1),(1,0)\}$, Table 4 shows there are 4 cases with $F_{i+1}(x,y,z) = 0$. Hence,

$$\begin{aligned} q_{i+1} &= q_i \cdot \Pr[F_{i+1}(x,y,z) = 0 \mid F_i(x,y,z) = 0] \\ &\quad + (1 - q_i) \cdot \Pr[F_{i+1}(x,y,z) = 0 \mid F_i(x,y,z) = 1] \\ &= q_i \cdot \tfrac{8}{16} + (1 - q_i) \cdot \tfrac{4}{16} = \tfrac{1}{4} + \tfrac{1}{4}q_i = \tfrac{2}{3} \cdot \left(\tfrac{1}{4}\right)^{i+1} + \tfrac{1}{3}. \end{aligned}$$

Thus, Eq. (24) also holds for $i+1$, and by induction, it holds for all $i \geq 0$.

Using q_i, we prove Eqs. (9), (10), and (11). Table 5 shows the combinations of $(\Theta_{i-1}(x,y), \Theta_{i-1}(x+y,z), x_{i-1}, y_{i-1}, z_{i-1})$ that satisfy Eqs. (5), (6), and (7).

Table 5. Combinations of $(\Theta_{i-1}(x,y), \Theta_{i-1}(x+y,z))$ and $(x_{i-1}, y_{i-1}, z_{i-1})$ with Eqs. (5), (6), and (7)

	$(\Theta_{i-1}(x,y), \Theta_{i-1}(x+y,z))$	$(x_{i-1}, y_{i-1}, z_{i-1})$
Equation (5)	(0, 0)	(0, 0, 0), (0, 1, 0), (0, 1, 1), (1, 1, 0)
	(0, 1)	(0, 0, 0), (0, 0, 1), (0, 1, 1), (1, 0, 0), (1, 1, 0), (1, 1, 1)
	(1, 0)	(0, 0, 0), (0, 0, 1), (0, 1, 1), (1, 0, 0), (1, 1, 0), (1, 1, 1)
	(1, 1)	(0, 0, 1), (1, 0, 0), (1, 0, 1), (1, 1, 1)
Equation (6)	(0, 0)	(0, 0, 0), (0, 0, 1), (0, 1, 0), (1, 0, 0)
	(0, 1)	(0, 0, 1), (0, 1, 0), (0, 1, 1), (1, 0, 0), (1, 0, 1), (1, 1, 0)
	(1, 0)	(0, 0, 1), (0, 1, 0), (0, 1, 1), (1, 0, 0), (1, 0, 1), (1, 1, 0)
	(1, 1)	(0, 1, 1), (1, 0, 1), (1, 1, 0), (1, 1, 1)
Equation (7)	(0, 0)	(0, 0, 0), (0, 1, 0), (0, 1, 1), (1, 0, 0), (1, 0, 1), (1, 1, 1)
	(0, 1)	(0, 0, 0), (0, 0, 1), (1, 0, 1), (0, 1, 1)
	(1, 0)	(0, 0, 0), (0, 0, 1), (0, 1, 1), (1, 0, 1)
	(1, 1)	(0, 0, 0), (0, 1, 0), (0, 1, 1), (1, 0, 0), (1, 0, 1), (1, 1, 1)

For the i-th bit, we derive the probability that each equation holds by considering separately the cases $F_{i-1}(x, y, z) = 0$ and $F_{i-1}(x, y, z) = 1$.

Equation (9). We determine the probability that Eq. (5) holds for $i \geq 1$. When $F_{i-1}(x, y, z) = 0$, i.e. $(\Theta_{i-1}(x, y), \Theta_{i-1}(x+y, z)) \in \{(0,0), (1,1)\}$, Table 5 shows that there are 8 cases that satisfy Eq. (5). When $F_{i-1}(x, y, z) = 1$, i.e. $(\Theta_{i-1}(x, y), \Theta_{i-1}(x+y, z)) \in \{(0,1), (1,0)\}$, Table 5 shows that there are 12 cases that satisfy Eq. (5). From the above results and the previously derived q_i, the probability that Eq. (5) holds can be obtained as follows.

$$\begin{aligned}
\Pr[\text{Eq. (5)}] &= q_{i-1} \cdot \Pr[\text{Eq. (5)} \mid F_{i-1}(x, y, z) = 0] \\
&\quad + (1 - q_{i-1}) \cdot \Pr[\text{Eq. (5)} \mid F_{i-1}(x, y, z) = 1] \\
&= q_{i-1} \cdot \tfrac{8}{16} + (1 - q_{i-1}) \cdot \tfrac{12}{16} = \tfrac{3}{4} - \tfrac{1}{4} q_{i-1} = \tfrac{2}{3} - \tfrac{1}{6} \cdot \left(\tfrac{1}{4}\right)^{i-1}
\end{aligned}$$

Equation (10). We determine the probability that Eq. (6) holds for $i \geq 1$. When $F_{i-1}(x, y, z) = 0$, Table 5 shows that there are 8 cases that satisfy Eq. (6). When $F_{i-1}(x, y, z) = 1$, Table 5 shows that there are 12 cases that satisfy Eq. (6). From the above results and q_i, the probability that Eq. (6) holds can be obtained as follows.

$$\begin{aligned}
\Pr[\text{Eq. (6)}] &= q_{i-1} \cdot \Pr[\text{Eq. (6)} \mid F_{i-1}(x, y, z) = 0] \\
&\quad + (1 - q_{i-1}) \cdot \Pr[\text{Eq. (6)} \mid F_{i-1}(x, y, z) = 1] \\
&= q_{i-1} \cdot \tfrac{8}{16} + (1 - q_{i-1}) \cdot \tfrac{12}{16} = \tfrac{3}{4} - \tfrac{1}{4} q_{i-1} = \tfrac{2}{3} - \tfrac{1}{6} \cdot \left(\tfrac{1}{4}\right)^{i-1}
\end{aligned}$$

Equation (11). We determine the probability that Eq. (7) holds for $i \geq 1$. When $F_{i-1}(x, y, z) = 0$, Table 5 shows that there are 12 cases that satisfy Eq. (7). When $F_{i-1}(x, y, z) = 1$, Table 5 shows that there are 8 cases that satisfy Eq. (7). From the above results and q_i, the probability that Eq. (7) holds can be obtained as follows.

$$\begin{aligned}
\Pr[\text{Eq. (7)}] &= q_{i-1} \cdot \Pr[\text{Eq. (7)} \mid F_{i-1}(x, y, z) = 0] \\
&\quad + (1 - q_{i-1}) \cdot \Pr[\text{Eq. (7)} \mid F_{i-1}(x, y, z) = 1] \\
&= q_{i-1} \cdot \tfrac{12}{16} + (1 - q_{i-1}) \cdot \tfrac{8}{16} = \tfrac{1}{2} + \tfrac{1}{4} q_{i-1} = \tfrac{7}{12} + \tfrac{1}{6} \cdot \left(\tfrac{1}{4}\right)^{i-1}
\end{aligned}$$

$\square$

The following corollary is obtained in the same manner as Theorem 1, by generalizing the right-hand sides of Eqs. (5) and (7).

Corollary 1. *Let x, y, and z be independent variables. For the i-th carry in the binary additions $x + y$ and $(x + y) + z$, the following relations hold for all $i \geq 1$:*

$$\Pr[\Theta_i(x, y) \oplus \Theta_i(x + y, z) = \alpha \oplus \beta] = \tfrac{2}{3} - \tfrac{1}{6} \cdot \left(\tfrac{1}{4}\right)^{i-1}, \tag{25}$$

$$\Pr[\Theta_i(x, y) \oplus \Theta_{i-1}(x, y) \oplus \Theta_i(x + y, z) = \gamma] = \tfrac{7}{12} + \tfrac{1}{6} \cdot \left(\tfrac{1}{4}\right)^{i-1}. \tag{26}$$

where $(\alpha, \beta) \in \{(x_{i-1}, y_{i-1}), (x_{i-1}, z_{i-1}), (y_{i-1}, z_{i-1})\}$ and $\gamma \in \{x_{i-1}, y_{i-1}, z_{i-1}\}$.

4.2   Experiment

To experimentally verify Theorem 1, we evaluated the probabilities that Eqs. (5)–(7) hold by performing 10,000 trials of additions of 10–bit values x, y, z. Table 6 shows the theoretical probabilities, the experimental probabilities, and the experimental errors, all rounded to four decimal places.

Similarly, Corollary 1 was verified experimentally in the same manner. In this case, Eqs. (25) and (26) were tested separately, each corresponding to the right-hand side expressions (one of which coincides with Theorem 1). Table 7 compares the theoretical and experimental probabilities of Corollary 1, again showing values rounded to four decimal places.

In all cases, the experimental errors were negligible, confirming the correctness of both Theorem 1 and Corollary 1.

Table 6. Probabilities of Eqs. (5)(7) in Theorem 1: theory, experiment, and error

Eq.		i								
		1	2	3	4	5	6	7	8	9
(5)	Theoretical	0.5	0.625	0.6563	0.6641	0.6660	0.6665	0.6666	0.6667	0.6667
	Experimental	0.4997	0.6196	0.6583	0.668	0.663	0.6629	0.6728	0.6697	0.6754
	Error	−0.0006	−0.0086	0.0031	0.0059	−0.0045	−0.0054	0.0093	0.0046	0.0131
(6)	Theoretical	0.5	0.625	0.6563	0.6641	0.6660	0.6665	0.6666	0.6667	0.6667
	Experimental	0.4926	0.6164	0.6599	0.6594	0.6712	0.6651	0.6623	0.6597	0.6577
	Error	−0.0148	−0.0138	0.0056	−0.0070	0.0078	−0.0021	−0.0065	−0.0104	−0.0134
(7)	Theoretical	0.75	0.625	0.5938	0.5859	0.5840	0.5835	0.5834	0.5833	0.5833
	Experimental	0.7478	0.6403	0.5915	0.5823	0.5911	0.5828	0.5777	0.5923	0.5820
	Error	−0.0029	0.0245	−0.0038	−0.0062	0.0122	−0.0012	−0.0097	0.0154	−0.0023

Table 7. Probabilities of Eqs. (25) and (26) in Corollary 1: theory, experiment, and error

Eq.	RHS		i								
			1	2	3	4	5	6	7	8	9
(25)	$(\alpha \oplus \beta)$	Theoretical	0.5	0.625	0.6563	0.6641	0.6660	0.6665	0.6666	0.6667	0.6667
	$x_{i-1} \oplus y_{i-1}$	Experimental	0.5081	0.6132	0.6572	0.6671	0.6677	0.6652	0.6721	0.6673	0.6709
		Error	0.0162	−0.0189	0.0014	0.0046	0.0025	−0.0020	0.0082	0.0010	0.0064
	$y_{i-1} \oplus z_{i-1}$	Experimental	0.5014	0.6270	0.6564	0.6665	0.6743	0.6669	0.6594	0.6685	0.6630
		Error	0.0028	0.0032	0.0002	0.0037	0.0124	0.0006	−0.0108	0.0028	−0.0055
(26)	γ	Theoretical	0.75	0.625	0.5938	0.5859	0.5840	0.5835	0.5834	0.5833	0.5833
	x_{i-1}	Experimental	0.7485	0.6326	0.5940	0.5878	0.5797	0.5840	0.5766	0.5739	0.5944
		Error	−0.0020	0.0122	0.0004	0.0032	−0.0073	0.0009	−0.0116	−0.0162	0.0190
	y_{i-1}	Experimental	0.7536	0.6238	0.5818	0.5918	0.5894	0.5764	0.5762	0.5868	0.5840
		Error	0.0048	−0.0019	−0.0201	0.0100	0.0093	−0.0122	−0.0123	0.0059	0.0011

5   Application of Theorem 1 to the ChaCha Permutation

In this section, we refine the theoretical biases of the 3–7 round linear approximations used in [7] by applying Theorem 1, with the goal of reducing the discrepancies between theoretical and experimental values shown in Table 2.

First, the 3–7 round linear approximation is divided into the 3–6 round linear approximation Eq. (4) and the 6–7 round linear approximations Groups I-IV. It is known that Group IV does not exhibit any discrepancy. Next, Eq. (4) is further separated into the 3–5 round and 5–6 round approximations. By comparing their theoretical and experimental values, we found that Eq. (27), corresponding to the 5–6 round approximation, causes discrepancies. Similarly, by decomposing Groups I–III into finer sub-approximations, we identified the following equations as sources of discrepancies: Eqs. (28)–(30) in Group I, Eqs. (31)–(34) in Group II, and Eqs. (35)-(37) in Group III. For these equations, Theorem 1 is applied to derive more precise theoretical biases.

$$x_{11,12}^{[10]} = x_1^{[12]}[11,12] \oplus x_{11,12}^{[12]} \oplus x_{12}^{[12]}[11,12,19,20], \tag{27}$$

$$x_{0,16}^{[12]} = x_{0,16}^{[14]} \oplus x_4^{[14]}[2,3,22,23] \oplus x_8^{[14]}[27,28] \oplus x_{12}^{[14]}[15,16], \tag{28}$$

$$x_{4,7}^{[12]} \oplus x_8^{[12]}[7,8] = x_0^{[14]}[7,8] \oplus x_{4,26}^{[14]} \oplus x_8^{[14]}[8,19] \oplus x_{12}^{[14]}[7,8,15,16], \tag{29}$$

$$x_{8,31}^{[12]} = x_0^{[14]}[30,31] \oplus x_{8,31}^{[14]} \oplus x_{12}^{[14]}[6,7,30,31], \tag{30}$$

$$x_1^{[12]}[11,12] \oplus x_{9,12}^{[12]} = x_{1,11}^{[14]} \oplus x_5^{[14]}[18,19,31] \oplus x_9^{[14]}[11,12,24] \oplus x_{13}^{[14]}[19,20], \tag{31}$$

$$x_1^{[12]}[22,23] = x_1^{[14]}[22,23] \oplus x_5^{[14]}[9,10,29,30] \oplus x_9^{[14]}[2,3] \oplus x_{13}^{[14]}[22,23], \tag{32}$$

$$x_1^{[12]}[6,7] \oplus x_{5,7}^{[12]} \oplus x_{9,6}^{[12]} = x_1^{[14]}[5,7] \oplus x_5^{[14]}[13,14,25] \oplus x_9^{[14]}[6,7,18] \oplus x_{13}^{[14]}[13,14], \tag{33}$$

$$x_{9,26}^{[12]} = x_1^{[14]}[25,26] \oplus x_{9,26}^{[14]} \oplus x_{13}^{[14]}[1,2,25,26], \tag{34}$$

$$x_{2,16}^{[12]} = x_{2,16}^{[14]} \oplus x_6^{[14]}[2,3,22,23] \oplus x_{10}^{[14]}[27,28] \oplus x_{14}^{[14]}[15,16], \tag{35}$$

$$x_2^{[12]}[6,7] \oplus x_{6,7}^{[12]} = x_{2,7}^{[14]} \oplus x_6^{[14]}[13,14] \oplus x_{10}^{[14]}[6,7], \tag{36}$$

$$x_2^{[12]}[18,19] \oplus x_{6,19}^{[12]} = x_{2,19}^{[14]} \oplus x_6^{[14]}[25,26] \oplus x_{10}^{[14]}[18,19]. \tag{37}$$

Equations (27)–(37) are obtained as compositions of linear approximations for two consecutive additions, as illustrated in Fig. 1. Table 8 shows, for each approximation, the two underlying linear approximations of additions, together with the corresponding fundamental linear approximations Eqs. (9), (10), and (11) in Theorem 1.

Accordingly, for the compositions of these dependent linear approximations, we apply Theorem 1, whereas for the compositions of independent approximations, we use Lemma 3. The resulting probabilities that Eqs. (27)–(37) hold are presented in Lemma 10.

Table 8. Two Additions where Theorem 1 Applied

Eq	Two Additions	Applied Equation in Theorem 1
(27)	$x_{11,12}^{[11]} = x_{11,12}^{[10]} + x_{12,12}^{[11]}, x_{11,12}^{[12]} = x_{11,12}^{[11]} + x_{12,12}^{[12]}$	(9)
(28)	$x_{0,16}^{[13]} = x_{0,16}^{[12]} + x_{4,16}^{[12]}, x_{0,16}^{[14]} = x_{0,16}^{[13]} + x_{4,16}^{[13]}$	(9)
(29)	$x_{8,8}^{[13]} = x_{8,8}^{[12]} + x_{12,8}^{[13]}, x_{8,8}^{[14]} = x_{8,8}^{[13]} + x_{12,8}^{[14]}$	(11)
(30)	$x_{8,31}^{[13]} = x_{8,31}^{[12]} + x_{12,31}^{[13]}, x_{8,31}^{[14]} = x_{8,31}^{[13]} + x_{12,31}^{[14]}$	(9)
(31)	$x_{1,12}^{[13]} = x_{1,12}^{[12]} + x_{5,12}^{[12]}, x_{1,12}^{[14]} = x_{1,12}^{[13]} + x_{5,12}^{[13]}$	(9)
(32)	$x_{1,23}^{[13]} = x_{1,23}^{[12]} + x_{5,23}^{[12]}, x_{1,23}^{[14]} = x_{1,23}^{[13]} + x_{5,23}^{[13]}$	(10)
(33)	$x_{1,7}^{[13]} = x_{1,7}^{[12]} + x_{5,7}^{[12]}, x_{1,7}^{[14]} = x_{1,7}^{[13]} + x_{5,7}^{[13]}$	(10)
(34)	$x_{9,26}^{[13]} = x_{9,26}^{[12]} + x_{13,26}^{[13]}, x_{9,26}^{[14]} = x_{9,26}^{[13]} + x_{13,26}^{[14]}$	(9)
(35)	$x_{2,16}^{[13]} = x_{2,16}^{[12]} + x_{6,16}^{[12]}, x_{2,16}^{[14]} = x_{2,16}^{[13]} + x_{6,16}^{[13]}$	(9)
(36)	$x_{2,7}^{[13]} = x_{2,7}^{[12]} + x_{6,7}^{[12]}, x_{2,7}^{[14]} = x_{2,7}^{[13]} + x_{6,7}^{[13]}$	(9)
(37)	$x_{2,19}^{[13]} = x_{2,19}^{[12]} + x_{6,19}^{[12]}, x_{2,19}^{[14]} = x_{2,19}^{[13]} + x_{6,19}^{[13]}$	(9)

Lemma 10. *Equations (27)–(37) hold with the following probabilities.*

$$\Pr[\text{Eq. (27)}] = \tfrac{1}{2}\left(1 + \tfrac{1}{3}\right), \quad \Pr[\text{Eq. (28)}] = \tfrac{1}{2}\left(1 + \tfrac{1}{6}\right), \quad \Pr[\text{Eq. (29)}] = \tfrac{1}{2}\left(1 + \tfrac{1}{6}\right),$$
$$\Pr[\text{Eq. (30)}] = \tfrac{1}{2}\left(1 + \tfrac{1}{3}\right), \quad \Pr[\text{Eq. (31)}] = \tfrac{1}{2}\left(1 + \tfrac{1}{6}\right), \quad \Pr[\text{Eq. (32)}] = \tfrac{1}{2}\left(1 + \tfrac{1}{6}\right),$$
$$\Pr[\text{Eq. (33)}] = \tfrac{1}{2}\left(1 + \tfrac{1}{6}\right), \quad \Pr[\text{Eq. (34)}] = \tfrac{1}{2}\left(1 + \tfrac{1}{3}\right), \quad \Pr[\text{Eq. (35)}] = \tfrac{1}{2}\left(1 + \tfrac{1}{6}\right),$$
$$\Pr[\text{Eq. (36)}] = \tfrac{1}{2}\left(1 + \tfrac{1}{3}\right), \quad \Pr[\text{Eq. (37)}] = \tfrac{1}{2}\left(1 + \tfrac{1}{3}\right).$$

Proof. For simplicity, we prove only the probability that Eq. (27) holds. The other equations can be proved in the same manner. Consider two linear approximations $x^{[11]}_{11,12} = x^{[10]}_{11,12} + x^{[11]}_{12,12}$ and $x^{[12]}_{11,12} = x^{[11]}_{11,12} + x^{[12]}_{12,12}$, which can be explicitly expressed using carries as follows:

$$x^{[10]}_{11,12} = x^{[11]}_{11,12} \oplus x^{[11]}_{12,12} \oplus \Theta_{12}(x^{[10]}_{11}, x^{[11]}_{12}),$$
$$x^{[11]}_{11,12} = x^{[12]}_{11,12} \oplus x^{[12]}_{12,12} \oplus \Theta_{12}(x^{[11]}_{11}, x^{[12]}_{12}).$$

Combining these two equations, we obtain

$$x^{[10]}_{11,12} = x^{[11]}_{12,12} \oplus x^{[12]}_{11,12} \oplus x^{[12]}_{12,12} \oplus \Theta_{12}(x^{[10]}_{11}, x^{[11]}_{12}) \oplus \Theta_{12}(x^{[11]}_{11}, x^{[12]}_{12}). \tag{38}$$

Since $x^{[11]}_{11} = x^{[10]}_{11} + x^{[11]}_{12}$, we can apply Eq. (9) of Theorem 1 to the carries and obtain

$$\Pr[\Theta_{12}(x^{[10]}_{11}, x^{[11]}_{12}) \oplus \Theta_{12}(x^{[11]}_{11}, x^{[12]}_{12}) = x^{[11]}_{12,11} \oplus x^{[12]}_{12,11}] \approx \tfrac{1}{2}\left(1 + \tfrac{1}{3}\right). \tag{39}$$

The remaining terms $x^{[11]}_{12,12}$ and $x^{[12]}_{12,12}$ can be further expanded with probability 1 as follows:

$$x^{[11]}_{12,11} = x^{[12]}_{1,11} \oplus x^{[12]}_{12,19}, \quad x^{[11]}_{12,12} = x^{[12]}_{1,12} \oplus x^{[12]}_{12,20}. \tag{40}$$

Finally, by applying Lemma 3 to combine Eqs. (38), (39), and (40), we obtain $\Pr[\text{Eq. (27)}] = \tfrac{1}{2}\left(1 + \tfrac{1}{3}\right)$. $\qquad\square$

Table 9. Biases and Errors before/after refinement of Eqs. (27)(37)

Round	Group	Eq.	Experimental	[6,7]		This work	
				Theoretical	Error	Theoretical	Error
5–6	Eq. (4)	(27)	$0.333358 \approx 2^{-1.585}$	2^{-2}	0.333432	$\frac{1}{3}$	0.000074
6-7	Group I	(28)	$0.166536 \approx 2^{-2.586}$	2^{-3}	0.332288	$\frac{1}{6}$	−0.000784
		(29)	$0.166524 \approx 2^{-2.586}$	2^{-2}	−0.333904	$\frac{1}{6}$	−0.000856
		(30)	$0.333272 \approx 2^{-1.585}$	2^{-2}	0.333088	$\frac{1}{3}$	−0.000184
	Group II	(31)	$0.166748 \approx 2^{-2.584}$	2^{-3}	0.333984	$\frac{1}{6}$	0.000488
		(32)	$0.166443 \approx 2^{-2.586}$	2^{-3}	0.331544	$\frac{1}{6}$	−0.001342
		(33)	$0.167328 \approx 2^{-2.579}$	2^{-3}	0.338624	$\frac{1}{6}$	0.003968
		(34)	$0.333313 \approx 2^{-1.585}$	2^{-2}	0.333252	$\frac{1}{3}$	−0.000061
	Group III	(35)	$0.166443 \approx 2^{-2.587}$	2^{-3}	0.331544	$\frac{1}{6}$	−0.001342
		(36)	$0.333374 \approx 2^{-1.585}$	2^{-2}	0.333496	$\frac{1}{3}$	0.000122
		(37)	$0.333618 \approx 2^{-1.584}$	2^{-2}	0.334472	$\frac{1}{3}$	0.000854

Table 9 presents the theoretical values, experimental results, and experimental errors of Eqs. (27)–(37) shown in Lemma 10, together with the results of [6,7]. From the table, it can be seen that the new Lemma 10 reproduces the experimental results more precisely.

Proof of Lemma 1

Equation (4) is constructed by combining Eq. (27) with the following two linear approximations:

$$x_{3,0}^{[6]} \oplus x_{4,0}^{[6]}$$
$$= x_{1,0}^{[10]} \oplus x_{3,0}^{[10]} \oplus x_{4,26}^{[10]} \oplus x_{7}^{[10]}[7,19] \oplus x_{8}^{[10]}[7,19] \oplus x_{9,0}^{[10]} \oplus x_{11,12}^{[10]}$$
$$\oplus x_{12}^{[10]}[6,7] \oplus x_{13}^{[10]}[0,8] \oplus x_{15,0}^{[10]}, \ w.p. \ \tfrac{1}{2}\left(1+\tfrac{1}{2}\right), \tag{41}$$
$$x_{1,0}^{[10]} \oplus x_{3,0}^{[10]} \oplus x_{4,26}^{[10]} \oplus x_{7}^{[10]}[7,19] \oplus x_{8}^{[10]}[7,19] \oplus x_{9,0}^{[10]}$$
$$\oplus x_{11,12}^{[10]} \oplus x_{12}^{[10]}[6,7] \oplus x_{13}^{[10]}[0,8] \oplus x_{15,0}^{[10]}$$
$$= x_{0}^{[12]}[0,16] \oplus x_{1}^{[12]}[0,6,7,22,23] \oplus x_{2}^{[12]}[0,6,7,8,16,18,19,24]$$
$$\oplus x_{4}^{[12]}[7,13,19] \oplus x_{5,7}^{[12]} \oplus x_{6}^{[12]}[7,13,14,19] \oplus x_{7}^{[12]}[6,7,14,15,26]$$
$$\oplus x_{8}^{[12]}[0,7,8,19,31] \oplus x_{9}^{[12]}[0,6,12,26] \oplus x_{10,0}^{[12]} \oplus x_{11}^{[12]}[6,7,12]$$
$$\oplus x_{12}^{[12]}[0,30,31] \oplus x_{13}^{[12]}[0,14,15,24,26,27]$$
$$\oplus x_{14}^{[12]}[8,25,26] \oplus x_{15,24}^{[12]}, \ w.p. \ \tfrac{1}{2}\left(1+\tfrac{1}{2^5}\right). \tag{42}$$

By experiments, we confirmed that Eq. (41) and Eq. (42) incur no experimental error (i.e., their empirical correlations match the stated values), whereas the only source of experimental deviation in the construction is Eq. (27). By

Lemma 10, the correlation contributed by Eq. (27) is $\frac{1}{3}$. Under the (standard) independence assumptions required by Lemma 3, the overall correlation of $2^{-1} \cdot 2^{-5} \cdot \frac{1}{3} = \frac{1}{2^6 \cdot 3}$. Therefore, $\Pr[\text{Eq. (4)}] = \frac{1}{2}\left(1 + \frac{1}{2^6 \cdot 3}\right)$, which proves the claim. $\qquad\square$

Proof of Lemma 2

For simplicity, we prove only the probability that Group I holds. Groups I is constructed by combining Eqs. (28)–(30) with the following linear approximation:

$$x_{0,0}^{[12]} \oplus x_4^{[12]}[7, 13, 19] \oplus x_8^{[12]}[0, 19] \oplus x_{12}^{[12]}[0, 11, 12, 19, 20, 30, 31]$$

$$= x_0^{[14]}[0, 3, 4, 11, 12, 14, 15, 16, 18, 20, 27, 28, 30, 31]$$

$$\oplus\, x_4^{[14]}[0, 5, 18, 26, 27] \oplus x_8^{[14]}[11, 13, 19, 20, 25, 30]$$

$$\oplus\, x_{12}^{[14]}[3, 4, 8, 11, 13, 22, 23, 24, 26, 27] \ w.p. \ \tfrac{1}{2}\left(1 + \tfrac{1}{2^5}\right), \tag{43}$$

By experiments, we confirmed that Eq. (43) incur no experimental error, whereas the onlysource of experimental deviation in the construction is Eqs. (28)–(30). By Lemma 10, Eqs. (28), (29), and (30) contribute correlations of $\frac{1}{6}$, $\frac{1}{6}$, and $\frac{1}{3}$, respectively. Under the independence assumptions required by Lemma 3, the overall correlation of $2^{-5} \cdot \frac{1}{6} \cdot \frac{1}{6} \cdot \frac{1}{3} = \frac{1}{2^7 \cdot 3^3}$. Therefore, $\Pr[\text{Group I}] = \frac{1}{2}\left(1 + \frac{1}{2^7 \cdot 3^3}\right)$, which proves the claim. $\qquad\square$

Proof of Theorem 2

The linear approximation in Theorem 2 consists of the partial linear approximations Eq. (4) and Groups I–IV. By applying Lemma 3 to combine Eq. (4) and Groups I–III refined in Lemmas 1 and 2 with the remaining approximations including Group IV in Lemma 8, we derive Theorem 2.

6   Conclusion

In this paper, we addressed the discrepancies between theoretical and experimental bias estimates in the linear and differential-linear cryptanalysis of ARX-based ciphers, with a particular focus on ChaCha. We introduced new fundamental linear approximations for two consecutive additions over three independent variables and rigorously derived their exact probabilities. Our analysis demonstrated that the conventional assumption of independence between consecutive additions introduces systematic errors, leading to theoretical values that diverge significantly from experimental measurements.

By applying our theorem to ChaCha, we refined the probabilities of key linear approximations used in existing distinguishers. In particular, we improved the theoretical estimates of Eq. (4) and Groups I–III, which had shown large deviations in previous work. Our refined approximations yield theoretical biases that are much closer to the experimental results, thereby reducing the gap between theory and practice. This not only strengthens the validity of existing attacks but also provides a more reliable framework for evaluating the security of ChaCha against differential-linear cryptanalysis.

The results presented here demonstrate the importance of carefully analyzing dependencies in ARX-based designs and highlight that seemingly small refinements in bias estimation can lead to a significant improvement in the accuracy of cryptanalytic evaluations. Future work includes applying the proposed framework to other ARX ciphers, such as Salsa20 and variants of ChaCha, as well as exploring its integration into key-recovery scenarios. We believe that the methodology developed in this paper will serve as a solid foundation for further advances in the cryptanalysis of ARX-based primitives.

Acknowledgements. This work is partially supported by JSPS KAKENHI Grant Number JP21H03443 and SECOM Science and Technology Foundation.

References

1. Aumasson, J.-P., Fischer, S., Khazaei, S., Meier, W., Rechberger, C.: New features of latin dances: analysis of salsa, ChaCha, and rumba. In: Nyberg, K. (ed.) FSE 2008. LNCS, vol. 5086, pp. 470–488. Springer, Heidelberg (2008). https://doi.org/10.1007/978-3-540-71039-4_30
2. Bellini, E., Gerault, D., Grados, J., Makarim, R.H., Peyrin, T.: Boosting differential-linear cryptanalysis of ChaCha7 with MILP. IACR Trans. Symmetric Cryptol. (2023)
3. Bernstein, D.J.: Salsa20 specification. eSTREAM Project algorithm description (2005). http://www.ecrypt.eu.org/stream/salsa20pf.html
4. Choudhuri, A.R., Maitra, S.: Significantly improved multi-bit differentials for reduced round salsa and ChaCha. IACR Trans. Symmetric Cryptol. **2016**(2), 261–287 (2016)
5. Coutinho, M., Souza Neto, T.C.: New multi-bit differentials to improve attacks against ChaCha. Cryptology ePrint Archive (2020)
6. Coutinho, M., Souza Neto, T.C.: Improved linear approximations to ARX ciphers and attacks against ChaCha. In: Canteaut, A., Standaert, F.-X. (eds.) EUROCRYPT 2021. LNCS, vol. 12696, pp. 711–740. Springer, Cham (2021). https://doi.org/10.1007/978-3-030-77870-5_25
7. Coutinho, M., et al.: Latin dances reloaded: improved cryptanalysis against salsa and ChaCha, and the proposal of Forró. J. Cryptol. **36**(3), 18 (2023)
8. Dey, S., Garai, H.K., Sarkar, S., Sharma, N.K.: Enhanced differential-linear attacks on reduced round ChaCha. IEEE Trans. Inf. Theory **69**(8), 5318–5336 (2023)
9. Flórez-Gutiérrez, A., Todo, Y.: Improved cryptanalysis of ChaCha: beating PNBs with Bit puncturing. In: Fehr, S., Fouque, P.A. (eds.) EUROCRYPT 2025. LNCS, vol. 15061, pp. 427–457. Springer, Cham (2025). https://doi.org/10.1007/978-3-031-91107-1_15
10. Bernstein, D.J.: ChaCha, a variant of Salsa20. In: Workshop Record of SASC, vol. 8, pp. 3–5 (2008)
11. Langford, S.K., Hellman, M.E.: Differential-linear cryptanalysis. In: Desmedt, Y.G. (ed.) CRYPTO 1994. LNCS, vol. 839, pp. 17–25. Springer, Heidelberg (1994). https://doi.org/10.1007/3-540-48658-5_3
12. Matsui, M.: Linear cryptanalysis method for DES cipher. In: Helleseth, T. (ed.) EUROCRYPT 1993. LNCS, vol. 765, pp. 386–397. Springer, Heidelberg (1994). https://doi.org/10.1007/3-540-48285-7_33

13. Nyberg, K., Wallén, J.: Improved linear distinguishers for SNOW 2.0. In: Robshaw, M. (ed.) FSE 2006. LNCS, vol. 4047, pp. 144–162. Springer, Heidelberg (2006). https://doi.org/10.1007/11799313_10
14. Röck, A., Nyberg, K.: Exploiting linear hull in Matsui's algorithm 1 (extended version). Cryptology ePrint Archive (2011)
15. Wallén, J.: Linear approximations of addition modulo 2^n. In: Johansson, T. (ed.) FSE 2003. LNCS, vol. 2887, pp. 261–273. Springer, Heidelberg (2003). https://doi.org/10.1007/978-3-540-39887-5_20
16. Wang, S., Liu, M., Hou, S., Lin, D.: Moving a step of ChaCha in syncopated rhythm. In: Handschuh, H., Lysyanskaya, A. (eds.) CRYPTO 2023. LNCS vol. 14083, pp. 273–304. Springer, Cham (2023). https://doi.org/10.1007/978-3-031-38548-3_10
17. Watanabe, R., Ghafoori, N., Miyaji, A.: Improved differential-linear cryptanalysis of reduced rounds of ChaCha. In: Kim, H., Youn, J. (eds.) WISA 2023. LNCS, vol. 14402, pp. 269–281. Springer, Singapore (2024). https://doi.org/10.1007/978-981-99-8024-6_21

Zero-Knowledge and Interactive Proofs

BOIL: Proof-Carrying Data from Accumulation of Correlated Holographic IOPs

Tohru Kohrita, Maksim Nikolaev$^{(\boxtimes)}$, and Javier Silva

=nil; Foundation, Limassol, Cyprus
`maksim.n@mailbox.org`, `javier.silva@nil.foundation`

Abstract. In this paper, we present a batching technique for oracles corresponding to codewords of a Reed–Solomon code. This protocol is inspired by the round function of the STIR protocol (CRYPTO 2024). Using this oracle batching protocol, we propose a construction of a practically efficient accumulation scheme, which we call BOIL. Our accumulation scheme can be initiated with an arbitrary correlated holographic IOP, leading to a new class of PCD constructions.

Keywords: Split Accumulation · IVC · PCD · IOPP · Proof Systems

1 Introduction

Proof-carrying data (PCD) [17] is a cryptographic primitive that enables the dynamic compilation of a distributed computation, where each message is augmented by a short proof verifying that some local condition is met. An ideal PCD construction achieves this goal with minimal additional computation or communication overhead. A specific instance of PCD is incrementally verifiable computation (IVC) [46], a cryptographic primitive that allows one to produce a proof of the correctness of an iterative computation in an incremental fashion.

PCD via Recursive Composition. Early constructions of IVC and PCD relied on recursive composition, where each prover attaches a proof to each outgoing message, attesting that all previous local conditions have been met [6,7,18]. To achieve this, the constructions included a circuit representation of the verifier algorithm within a recursive statement. As a result, PCD was primarily limited to SNARKs – proof systems where the verifier's description is asymptotically smaller than the overall size of the statement being verified.

PCD via Accumulation. Later works [9,10,12,13] showed that it is often possible to avoid the complete recursive proof verification at each iteration. Instead, the most expensive part of the verification can be accumulated outside the recursive statement, and checked only once at the very end of the distributed computation. Verifying the proof of the correctness of the accumulation is usually much simpler than a full proof verification. The folding approach can be considered

R. Dutta et al. (Eds.): INDOCRYPT 2025, LNCS 16372, pp. 95–118, 2026.
https://doi.org/10.1007/978-3-032-13301-4_5

an extreme version of accumulation schemes for constructing IVC [11,21,34]. While a very small recursive overhead distinguishes constructions using folding, the accumulation approach allows using a broader class of NARKs as a basic block of the PCD scheme.

However, accumulation and folding schemes usually require the use of additively homomorphic commitment schemes. Therefore, this entire series of results is not compatible with (S)NARKs relying on code-based polynomial commitments, which are not homomorphic. However, even when using a full in-circuit verifier for recursion [18], such SNARKs perform excellently and are widely used in practice [30,42,44,45]. The work of [14] was the first attempt in trying to close this gap, showing how to build an accumulation scheme by postponing the code proximity test of such non-homomorphic constructions to the end. However, their work has significant shortcomings, discussed below in more detail, which our new construction addresses.

1.1   Our Contributions

Oracle Batching Protocol. In this paper, we consider SNARKs that are built using a compilation of a Polynomial-IOP and a proximity proof for a linear code. This approach is widely adopted in the industry. Among the reasons, we can highlight the following.

- No need for a trusted ceremony and trapdoors to generate parameters.
- Such systems do not require public key assumptions.
- The system parameters can be adjusted to alter the efficiency trade-off between the prover and the verifier.
- Protocols for code proximity tests, for example, FRI, keep all arithmetic in the same field for prover and verifier, so no field switching is necessary.

However, the polynomial commitment schemes derived from proximity proofs are not an additive like KZG [29] or Pedersen [39], which prevents a direct application of standard accumulation results.

The primary goal of FRI [3] (or any other proximity test) is to distinguish, by querying a function $f : \mathrm{D} \to \mathbb{F}$ at a few locations, whether f coincides with the evaluation of some polynomial of degree less than $d < |D|$ on the domain D, or whether it is far in relative Hamming distance from the evaluation of any low-degree polynomial. The batched version of the protocol [4,28] allows one to perform a proximity test for several functions $f_1, \ldots, f_n$ at once. However, to build more efficient IVC/PCD schemes, we need a reduction for multiple functions that does not require a full-fledged proximity test. Informally, this allows the incremental aggregation of new functions in batches, during the long-running computation. The problem of finding such a reduction was partially solved in [14], but with a limitation on the number of recursive steps allowed. The question of finding a more general solution remained open.

Our first contribution is a batching oracle protocol heavily inspired by the recent STIR protocol [1]. Suppose that we have n functions $f_1, \ldots, f_n : D \to \mathbb{F}$, for some $D \subset \mathbb{F}$. The verifier has oracle access to them, and the prover claims that f_i is δ-close to a low-degree polynomial, for all $i = 1, \ldots, n$. The protocol will produce a function $f_{\mathsf{new}} : D' \to \mathbb{F}$, for some other $D' \subset \mathbb{F}$, and reduce the n claims above to the single claim that f_{new} is δ-close to a low-degree polynomial.

This protocol does not require a proximity test, so it is much more efficient than batched FRI. Thus, it can be a base block for various designs such as a split accumulator or a linear combination scheme for PCS [9]. We consider the first construction in detail. The second follows implicitly, but the formalization is out of the scope of this paper.

Split Accumulation for IOPP. The new oracle batching protocol opens the door to more efficient aggregation and recursive composition of proofs. Informally speaking, instead of performing a low-degree proximity test on each iteration of recursion, we can use the oracle batching protocol, thereby deferring this expensive test to the very end of the computation.

In the context of IVC/PCD constructions, an important performance metric is the recursive overhead. In the case of SNARKs based on the low-degree proximity test for Reed–Solomon codes, the recursive statement usually includes many hash invocations. The paper [18] formally shows how such SNARKs can be used to construct IVC/PCD. The described approach requires representing the verifier as a circuit, in which the lion's share is occupied by hash operations, even when using SNARK-friendly hash functions such as Poseidon [27]. The proximity test makes the largest contribution to this size: $O(\lambda \cdot c_\delta \log^2 d)$ hash invocations, where λ is a security parameter, d is the corresponding polynomial degree and c_δ depends on the proximity parameter δ.

The paper [14] proposes an alternative construction of the IVC/PCD based on a linearity test and exclusively symmetric assumptions. Their linearity test has a better asymptotic estimate of $O(\lambda \cdot c_\delta \log d)$ hash invocations. However, its use entails the problem of distance decay: with each iteration, the provable distance increases, so a much smaller proximity parameter δ/T must be used, where T is the number of iterations. For practically significant parameters, this negates the advantage since $O(\lambda \cdot c_{\delta/T} \log d)$ can be comparable to or even larger than $O(\lambda \cdot c_\delta \log^2 d)$. Our technique, which we call BOIL (**B**atching **O**racles for **I**OPP from **L**inearity), allows us to get the best of both worlds – logarithmic complexity and the absence of the distance decay problem. It is a theoretical and practical improvement over previous results.

We aim to show that BOIL can be used with various proof systems. Therefore, for formalization, we use the abstraction of correlated Holographic IOPs [16]. Roughly speaking, a δ-correlated Holographic IOP (HIOP) [8] is a holographic proof system in which the verifier has access to an oracle that checks whether the requested polynomial is close to the Reed–Solomon code. This model captures many protocols (PLONK [25], Plonky2 [44], RISC Zero [45], Redshift [30]) and gives us the flexibility we need. We show that we can build a split accumulation scheme, given a round-by-round (RBR) knowledge sound δ-correlated HIOP, the Oracle batching protocol, and a RBR sound proximity test. The result of [12] directly yields the IVC/PCD construction.

1.2 Our Techniques

As noted above, running a proximity test for the Reed–Solomon code $\mathsf{RS}[\mathbb{F}, D, d]$ within a circuit is expensive, due to the large amount of hashes. Thus, we want to defer as much to the end of the recursive computation as we can, to minimize the size of the in-circuit verifier in the final IVC/PCD construction.

To achieve this, we want to recursively aggregate all these proximity checks for Reed–Solomon into a single one, and perform a proximity test only once at the end; hence we are interested in an oracle batching technique. This was achieved in [14], using a simpler oracle batching technique. However, it imposed some limitations on their parameter choices, which hindered the practicality of the scheme.

Oracle Batching. The construction in [14] achieves the goal above using a technique that can be thought of as a single round of the FRI proximity test [3]: given functions $f_1, \ldots, f_n : \ D \to \mathbb{F}$, these are combined into a single function $f_{\mathsf{new}} : D \to \mathbb{F}$ by means of a random linear combination, with randomness supplied by the verifier. Informally, if an honestly computed f_{new} is δ-close to $\mathsf{RS}[\mathbb{F}, D, d]$, then with high probability so are $f_1, \ldots, f_n$. The prover sends an oracle for the purported f_{new}. The verifier ensures consistency between $f_1, \ldots, f_n$ and f_{new} by means of spot checks at random points in D. This process can be iterated upon, and at the end the verifier simply checks proximity of the final aggregated function to $\mathsf{RS}[\mathbb{F}, D, d]$. This can be performed directly or by means of a fully fledged proximity test, like FRI or STIR.

This construction has two significant limitations. One limitation is that it requires operating within the unique decoding radius $(1 - \rho)/2$ of the Reed–Solomon code, where ρ is the code rate. Indeed, as shown in [14], the use of a larger decoding radius leads to a possible attack where a malicious prover starts from a codeword that should be rejected and, after several accumulation steps, transitions to an accepting codeword. This leads to parameters that are inefficient in practice: the smaller the decoding radius is, the more queries are necessary to guarantee the same level of soundness in the spot checks phase. The other issue is that the distance guarantee degrades with subsequent iterations of this procedure. In more detail, the protocol only checks that

- f_{new} is δ-close to $\mathsf{RS}[\mathbb{F}, D, d]$ by the final proximity test,
- the honest folding of $f_1, \ldots, f_n$ is δ-close to f_{new} by the spot checks.

Hence, overall we can only guarantee that $f_1, \ldots, f_n$ are 2δ-close to the code. More generally, over k recursion steps, we can only guarantee $k\delta$-closeness to the code. From a theoretical point of view, this means that we can only build bounded-depth accumulation from this technique (although [14] shows that this is enough to build PCD). From a practical point of view, this makes us into even more expensive choices of the code's proximity parameter, namely $(1 - \delta)/2k$.

These two problems are solved if we try to use a round from STIR [1] instead of a round from FRI as our batching technique. It starts exactly as the previous approach, but includes additional steps at the end. Namely, it adds an out-of-domain check and quotienting to disambiguate the decoding of f_{new}.

This is very reminiscent of the technique often used to turn FRI into a polynomial commitment scheme [30]. Informally, this ensures that, with high probability, there is only one valid choice of codeword, even if we are in the list decoding regime. Thus, the out-of-domain sample binds the malicious prover to a single codeword, eliminating the possibility of the attack mentioned above. Moreover, it also gets rid of the distance decay issue: we can directly prove that if a (possibly dishonest) f_{new} is δ-close to the code, then $f_1, \ldots, f_n$ are δ-close to the code. This allows us to work with the much better proximity parameter $1 - \sqrt{\rho}$ (or $1 - \rho$, under the so-called Reed–Solomon decoding conjectures [4, Conjecture 8.4]).

Split Accumulation. In order to obtain a PCD scheme, we follow the overall strategy of [14]. In particular, it suffices to show that, for a given NARK, its verifier relation admits a split accumulation scheme.

In our setting, the SNARK verification procedure is naturally divided into two parts:

- the Polynomial-IOP–related checks, and
- the verification of a proximity proof.

Additionally, for the latter, we now have an oracle batching protocol. Hence, a natural idea is to split the verifier relation accordingly.

Although the idea is conceptually straightforward, its formalization is technically involved. The main difficulty arises because the relation for the proximity check is defined over oracles, so we cannot simply decouple it. We must first obtain its plain (non-oracle) version. Substituting oracles with vector commitments introduces another subtlety: the underlying vectors may be partially undefined, since proximity testing doesn't guarantee full codeword membership.

Our starting point for constructing a NARK is the framework of δ-correlated IOPs introduced in [8]. Such IOPs are equipped with an oracle that checks for δ-correlated agreement. Functions $f_1, \ldots, f_n : D \to \mathbb{F}$ are said to be in δ-correlated agreement if each agrees with some codeword in $\mathrm{RS}[\mathbb{F}, D, d]$ on at least a $(1 - \delta)$ fraction of points in D, and the agreement subset of D is the same for all f_i. This notion captures an IOP that combines a polynomial check with a proximity (or correlated-agreement) check as part of its final verification.

In Sect. 4, we introduce a series of technical transformations, which will later be used to define the split accumulation scheme in Sect. 5. Below we provide an informal overview:

- Transformation B takes as input a δ-correlated IOP that checks agreement for n functions and converts it into a δ-correlated IOP that checks proximity to an RS code for a single function. Such a transformation can be directly derived from the oracle batching protocol.
- The paper [8] shows that a δ-correlated IOP for a relation $\mathcal{R}$ can be combined with a standard IOP for correlated agreement to produce a (regular) IOP for

$\mathcal{R}$, while preserving soundness. However, we cannot reuse this transformation directly, since we require two separate verifiers: V_1 for the Polynomial-IOP checks, and V_2 for the proximity-proof verification. To achieve this split verifier, we introduce a modified transformation T.

- Once we obtain a regular IOP, we can apply the BCS transformation [5] to derive a NARK in the random oracle model. Since our verifier is split, we require a modified version of the BCS transformation. Moreover, the verifier V_2 must account for potentially undefined entries in its input vector. To handle these cases, we define the transformations BCST and $\text{BCST}^{\perp}$, respectively.

We show that, under reasonable conditions, these transformations also preserve knowledge soundness (Lemmas 6 and 5). Hence, we can define

$$\text{NARK} = \text{BCST}(\text{T}(\text{B}(\Pi))),$$

and its relaxed version

$$\text{NARK}^{\perp} = \text{BCST}^{\perp}(\text{T}(\text{B}(\Pi))),$$

which serve as the basis for constructing the split accumulation scheme, formally described in Sect. 5.

This results in a construction with an accumulator verifier that does not need to run a full proximity test on each accumulation step. Thus, we avoid the issue of running a full proximity test inside of the final PCD's prover circuit. Moreover, because of the STIR-based oracle batching technique, our construction does not inherit the parameter limitations as in [14].

1.3   Related Works

Comparison with Folding-Based Constructions. Nova-like folding schemes [32,34,35] allow efficient IVC constructions but are based on the use of R1CS or CCS arithmetization. Our construction allows the use of the Plonkish arithmetization [23,24], which is very expressive and widely used in practice.

Another important difference is that the EC-based IVC/PCD designs use cycles of curves [11,21,34]. This makes formalization somewhat more complicated [33,37]. Moreover it requires the use of non-native arithmetic, which significantly increases the recursive overhead. By avoiding cycles of curves, we get a conceptually more straightforward construction.

Finally, IVC constructions are usually neither succinct nor zero knowledge: the size of the state that the prover must maintain is proportional to the size of the circuit. This significantly complicates the parallelization of proof generation for a distributed computation. In our design, this state size can be significantly reduced using the Oracle batching protocol. In particular, for a circuit represented by a matrix of $N \times M$ elements, the state size will be proportional to one column, i.e. N elements. In this aspect, we obtain some properties of end-to-end IVC schemes [41].

PCD from Solely Symmetric-Key Assumptions. The authors of [14] built a bounded depth accumulator without using a proximity test as in [18] and without public key assumptions. Our work improves this result and shows that building a full-fledged split accumulator in this setting is possible.

A Note on Concurrent Work. After the work-in-progress version of our results was presented publicly, two independent preprints, [15] and [43], were published. Both concurrent works use the idea of a STIR-based oracle batching protocol but for slightly different purposes.

The work of [43] develops the DEEP Commitment notion, allowing for aggregating STARK proofs. The author considers the use case when knowledge extraction is not required.

The results of [15] are more closely related to ours. Specifically, the authors introduce a many-to-one reduction for Reed–Solomon proximity claims, which is reminiscent of our oracle batching protocol. Building on this reduction, they present two accumulation schemes, one for Polynomial IOPs and one for R1CS.

In contrast, our work targets the construction of an accumulation scheme for correlated Holographic IOPs. As established in [8], this model captures a wide range of practical and performant proof systems based on univariate polynomials and proximity testing, including PLONK, Risc0, and Plonky2. Consequently, our results enable security level estimations for accumulation schemes instantiated over such systems with essentially no additional analysis.

Furthermore, we present concrete optimizations designed to minimize prover time and reduce the size of the accumulator. In particular, we demonstrate that for practical parameters of Holographic IOPs, the accumulator size can be orders of magnitude smaller than the witness size.

Finally, another contribution that distinguishes our work is the provision of a proof-of-concept implementation together with experimental evaluations. This demonstrates not only the theoretical but also the practical importance of our constructions.

2   Preliminaries

2.1   Reed–Solomon Codes

Let $\mathbb{F}$ be a finite field, and $D \subset \mathbb{F}$. Given a function $f : D \to \mathbb{F}$, we denote by $\hat{f}$ the lowest-degree polynomial in $\mathbb{F}[X]$ that extends f.

Let $\mathsf{RS}[D, d]$ be the *Reed–Solomon code* over $D \subset \mathbb{F}$ with degree bound $d \mid \#D$, that is, $\mathsf{RS}[D, d] = \{f : D \to \mathbb{F} \mid \deg \hat{f} < d\}$. We might write $\mathsf{RS}[\mathbb{F}, D, d]$ if we want to make the field explicit, but most of the time we ignore it for simplicity. The *code rate* is defined as $\rho = d/\#D$.

Given $f, g : D \to \mathbb{F}$, we denote by $\Delta(f, g)$ the *relative Hamming distance* between them Similarly, for a set $S \subset \mathbb{F}^D$, we denote $\Delta(f, S) = \min_{g \in S}\{\Delta(f, g)\}$. We define the *list decoding* of a function $f : D \to \mathbb{F}$ as $\mathsf{List}(f, d, \delta) = \{g \in \mathsf{RS}[D, d] \mid \Delta(f, g) \leq \delta\}$.

We say that $\mathsf{RS}[D, d]$ is (δ, ℓ)-*list decodable* if $\mathsf{List}(f, d, \delta) \leq \ell$, $\forall f : D \to \mathbb{F}$.

Folding Preserves Correlated Agreement For $\delta \geq 0$, we say that $f_1, \ldots, f_n : D \to \mathbb{F}$ have δ-*corellated agreement in* $\mathsf{RS}[D, d]$ if there exist

- a set $S \subset D$, with size $\#S/\#D \geq (1 - \delta)$, and
- codewords $g_1, \ldots, g_n \in \mathsf{RS}[D, d]$,

such that $f_{i|S} = g_{i|S}$, $\forall i = 1, \ldots, n$. In particular, we have $\Delta(f_i, \mathsf{RS}[D, d]) < \delta$, $\forall i = 1, \ldots, n$.

Let $f_1, \ldots, f_n : D \to \mathbb{F}$ and $\alpha \in \mathbb{F}$. We define:

$$\mathsf{Fold}_\alpha(f_1, \ldots, f_n) = \sum_{i=1}^{n} f_i \alpha^i.$$

Note that it is defined over the same domain as each individual function.

Lemma 1 ([4], Theorems 4.1 and 5.1 and [1], Theorem 4.1). *Let* $d \in \mathbb{N}, f_1, \ldots, f_n : D \to \mathbb{F}, \rho = d/\#D, \delta \in (0, 1 - \sqrt{\rho})$. *Define the error term*

$$\varepsilon_{\mathsf{fold}} = \varepsilon_{\mathsf{fold}}(d, \rho, \delta, n) = \begin{cases} \frac{(n-1)\cdot d}{\rho \cdot \#\mathbb{F}} & \text{if } 0 < \delta \leq \frac{1-\rho}{2}. \\ \frac{(n-1)\cdot d^2}{\#\mathbb{F}\cdot\left(2\cdot\min\{1-\sqrt{\rho}-\delta, \frac{\rho}{20}\}\right)^7} & \text{if } \frac{1-\rho}{2} < \delta < 1 - \rho. \end{cases}$$

Suppose that $f_1, \ldots, f_n$ *do not have* δ-*corellated agreement in* $RS[D, d]$. *Then*

$$\Pr\left[\alpha \leftarrow \mathbb{F} : \Delta(\mathsf{Fold}_\alpha(f_1, \ldots, f_n), RS[D, d]) \leq \delta\right] \leq \varepsilon_{\mathsf{fold}},$$

Out-of-Domain Sampling.

Lemma 2 ([1], Lemma 4.5). *Let* $f : D \to \mathbb{F}, d, s \in \mathbb{N}, \delta \in [0, 1]$. *Define the error term*

$$\varepsilon_{\mathsf{out}} = \binom{\ell}{2} \cdot \left(\frac{d - 1}{\#\mathbb{F} - \#D}\right)^s \leq \frac{d^s \cdot \ell^2}{2 \cdot (\#\mathbb{F} - \#D)^s}.$$

If $RS[D, d]$ *is* (δ, ℓ)-*list decodable, then*

$$\Pr\left[x_1, \ldots, x_s \leftarrow \mathbb{F} \setminus D : \begin{array}{c} \exists u, u' \in \mathsf{List}(f, d, \delta) \ s. \ t. \\ u \neq u' \wedge u(x_i) = u'(x_i) \\ \forall i = 1, \ldots, s \end{array}\right] \leq \varepsilon_{\mathsf{out}}.$$

Quotienting. Let $f : D \to \mathbb{F}$ and $\boldsymbol{x}, \boldsymbol{y} \in \mathbb{F}^q$, with $D \cap \boldsymbol{x} = \emptyset$. We define $\mathsf{Quotient}(f, \boldsymbol{x}, \boldsymbol{y}) : D \to \mathbb{F}$ as follows. Let $\hat{p} \in \mathbb{F}[X]$ such that $\deg p < q$ and $p(x_j) = y_j$ for $j = 1, \ldots, q$. Then

$$\mathsf{Quotient}(f, \boldsymbol{x}, \boldsymbol{y})(x) = \frac{f(x) - \hat{p}(x)}{\prod_{j=1}^{q}(x - x_j)}.$$

Lemma 3 ([1], Lemma 4.4). *Let* $f : D \to \mathbb{F}, d \in \mathbb{N}, \delta \in (0, 1), \boldsymbol{x}, \boldsymbol{y} \in \mathbb{F}^q$ *with* $D \cap \boldsymbol{x} = \emptyset$ *and* $q < d$. *Suppose that for every* $u \in \mathsf{List}(f, d, \delta)$, *there exists* $j \in \{1, \ldots, q\}$ *such that* $\hat{u}(x_j) \neq y_j$. *Then* $\Delta(\mathsf{Quotient}(f, \boldsymbol{x}, \boldsymbol{y}), RS[D, d - q]) > \delta$.

Degree Correction. Let $f : D \to \mathbb{F}$ and $d, d^* \in \mathbb{N}, r \in \mathbb{F}$ with $0 \leq d \leq d^*$. We define $\mathsf{DegCor} : D \to \mathbb{F}$ as follows:

$$\mathsf{DegCor}(d^*, r, f, d)(x) = f(x) \cdot \left(\sum_{\ell=0}^{d^*-d} (rx)^\ell \right).$$

Lemma 4 ([1], Lemma 4.13). *Let* $f : D \to \mathbb{F}, d, d^* \in \mathbb{N}, r \in \mathbb{F}$ *with* $0 \leq d \leq d^*$. *Let* $\rho = d^*/\#D, \delta \in (0, \min\{1 - \sqrt{\rho}, 1 - \rho - 1/\#D\})$. *Suppose that* $\Delta(f, \mathsf{RS}[D, d]) > \delta$. *Then*

$$\Pr\left[r \leftarrow \mathbb{F} : \Delta(\mathsf{DegCor}(d^*, r, f, d), \mathsf{RS}[D, d^*]) \leq \delta\right] \leq \varepsilon_{\mathsf{corr}},$$

where $\varepsilon_{\mathsf{corr}} = \varepsilon_{\mathsf{fold}}(d^*, \rho, \delta, d^* + 1 - d)$ *is the error term defined in Lemma 1.*

2.2 Interactive Oracle Proofs

We denote oracle access to a function f by $\boxed{f}$. The result of a query at a point x is denoted by $\boxed{f(x)}$.

We follow the definitions of Holographic IOPs from [18]. We consider *indexed* relations $\mathcal{R} = \{(\mathfrak{i}, \mathbb{x}, \mathbb{w})\}$. In the context of verifiable computation, $\mathfrak{i}$ is an index that defines the computation, $\mathbb{x}$ is the statement that contains the public inputs, and $\mathbb{w}$ is the witness that contains the secret inputs. In many modern general-purpose proof systems, these take polynomial form. We define the *indexed language* $\mathcal{L}_{\mathfrak{i}} = \{\mathbb{x} \mid \exists \mathbb{w} \text{ s. t. } (\mathfrak{i}, \mathbb{x}, \mathbb{w}) \in \mathcal{R}\}$.

The work of [8] considers *oracle relations*, that is, relations in which $\mathbb{w}$ may contain some functions, and $\mathbb{x}$ contains oracles to those functions. This is convenient to frame proximity proofs as IOPs, so we follow this approach. Thus, all definitions below apply to both regular relations and oracle relations, unless it is specified otherwise.

A *Holographic Interactive Oracle Proof (HIOP)* for an indexed relation $\mathcal{R}$ is a tuple $\Pi = (\mathcal{I}, \mathcal{P}, \mathcal{V})$, where $\mathcal{I}$ is a PT algorithm, and $\mathcal{P}, \mathcal{V}$ are stateful PPT algorithms. The prover and verifier engage interactively. The record of communications between them is called a *Transcript*, and we denote it by $\pi \leftarrow \langle \mathcal{P}(\mathfrak{i}, \mathbb{x}, \mathbb{w}), \mathcal{V}^{\mathcal{I}(\mathfrak{i})}(\mathbb{x}) \rangle$. At the end, the verifier examines the transcript and outputs a bit. We denote this by $0/1 \leftarrow \mathcal{V}^{\mathcal{I}(\mathfrak{i})}(\mathbb{x}, \pi)$.

We use the notions of round-by-round soundness and round-by-round knowledge soundness as formulated in [8, Definitions 3.12 and 3.13].

From IOPs to Non-interactive Arguments in the ROM. An IOP can be seen as a generalization of both IPs and PCPs, both of which can be transformed into non-interactive arguments, via the Fiat–Shamir transformation [22,40] and the CS proofs construction [36,46], respectively. Thus, it is natural that a combination of these techniques yields a generic transformation from IOPs to non-interactive arguments. This was formalized as the BCS transformation [5].

2.3 δ-Correlated IOPs

A common strategy to build SNARKs is to combine a polynomial IOP with a proximity test for a Reed–Solomon code, and the BCS transformation. We recall the notion of δ-correlated HIOPs, introduced in [8], which provides a framework for some such polynomial IOPs, e.g. Plonk [26]. We start by considering a certain type of oracle relations relative to a fixed Reed–Solomon code.

Definition 1. *An oracle relation $\mathcal{R}$ is a $(\mathbb{F}, D, d)$-polynomial oracle relation if the oracles in statements in $\mathcal{R}$ correspond to codewords in $RS[\mathbb{F}, D, d]$.*

In particular, we consider the following strict $(\mathbb{F}, D, d)$-polynomial oracle relation.

$$
\mathsf{CoAgg}(\delta) = \left\{ \begin{pmatrix} \mathbb{i} \\ \mathbb{x} \\ \mathbb{w} \end{pmatrix} = \begin{pmatrix} \mathbb{F}, D, d, \delta, n \\ \boxed{f_1, \ldots, f_n} \\ f_1, \ldots, f_n \end{pmatrix} \ \middle| \ \begin{array}{c} \delta, n > 0, f_i : D \to \mathbb{F}, \\ \Delta(f_i, RS[\mathbb{F}, D, d]) \le \delta \quad \forall i \in [n] \\ \text{(with } \delta\text{-correlated agreement)} \end{array} \right\}.
$$

Additionally, let $\mathsf{OCoAgg}(\delta)$ be a function that works as follows. It receives as input $(\mathbb{i}, \mathbb{x})$, with δ being the proximity parameter in $\mathbb{i}$. Then, it outputs 1 if and only if $(\mathbb{i}, \mathbb{x}, \mathbb{w}) \in \mathsf{CoAgg}(\delta)$.

Given a statement $\mathbb{x}$ and a (potentially partial) transcript τ, let $\boxed{\mathsf{F}(\mathbb{x}, \tau)}$ denote the set of oracles that have appeared so far in either of them. We denote the set of functions behind these oracles by $\mathsf{F}(\mathbb{x}, \tau)$.

Definition 2. *Let $\delta \geq 0$. A HIOP $\Pi = (\mathcal{I}, \mathcal{P}, \mathcal{V})$ for a $(\mathbb{F}, D, d)$-polynomial oracle relation $\mathcal{R}$ is δ-correlated if the following hold:*

- *$\mathcal{V}$ has oracle access to $\mathsf{OCoAgg}(\delta)$.*
- *Let τ denote the transcript up to the last round of interaction. For the last round:*
  - *$\mathcal{V}$ sends $\zeta \leftarrow S \subset \mathbb{F}$ (or an extension of $\mathbb{F}$).*
  - *$\mathcal{P}$ sends evaluations of functions in $\mathsf{F}(\mathbb{x}, \tau)$.*
- *$\mathcal{V}$'s final check consists of the following:*
  - *Assert whether the evaluations sent by the prover in the final round are roots of a certain multivariate polynomial, determined from $\mathbb{i}, \mathbb{x}, \tau$.*
  - *Check that a set of maps $\{\mathsf{Quotient}(f_i, x_{i,\zeta}, f_i(x_{i,\zeta})) \mid f_i \in \mathsf{F}(\mathbb{x}, \tau)\}_{i=1}^{r}$ has δ-correlated agreement in $RS[D, d-1]$, using OCoAgg on their oracles.*

In our construction, we will need to deal with slightly more general protocols. The case with $d' = d - 1$ below encompasses δ-correlated IOPs. We will later encounter protocols with $d' = d$.

Definition 3. *Let $\delta \geq 0$. A HIOP $\Pi = (\mathcal{I}, \mathcal{P}, \mathcal{V})$ for a $(\mathbb{F}, D, d, d')$-polynomial oracle relation $\mathcal{R}$ is called a Semi-δ-correlated HIOP if the following hold:*

- *$\mathcal{V}$ has oracle access to $\mathsf{OCoAgg}(\delta)$.*
- *$\mathcal{V}$'s final check consists of the following:*
  - *Assert whether the evaluations sent by the prover in the final round are roots of a certain multivariate polynomial, determined from $\mathbb{i}, \mathbb{x}, \tau$.*
  - *Using OCoAgg, check δ-correlated agreement in $RS[D, d']$ of a single set of oracles.*

From δ-correrlated IOPs to regular IOPs. One of the main results from [8] is that one can turn a 0-correlated HIOP Π for $\mathcal{R}$ into a δ-correlated IOP for $\mathcal{R}$, and then combine it with a HIOP for the $\mathsf{CoAgg}(\delta)$ relation to produce a standard HIOP for $\mathcal{R}$.

Given a 0-correlated HIOP Π, the transformation can be instantiated by using as Π_{CA} any correlated agreement protocol, like the batch variants of FRI [3], STIR [1] or WHIR [2].[1] Moreover, one can just use the trivial check in which the verifier reads the whole function by querying the oracles at every position. Afterwards, the BCS transformation can be applied, yielding a non-interactive argument in the ROM.

2.4 Split Accumulation in the ROM

We follow the split accumulation definitions from [12], adapted to our setting. Let $\mathcal{R} = \{(\mathsf{qi}, \mathsf{qx}, \mathsf{qw})\}$ be a relation. A *split accumulation scheme* for $\mathcal{R}$ is a tuple of algorithms $\mathsf{SA} = (\mathsf{I}, \mathsf{P}, \mathsf{V}, \mathsf{D})$ with the following syntax:

- $\mathsf{I}(\mathsf{qi})$ outputs the index-specific prover key pk, verifier key vk, and decider key dk.
- $\mathsf{P}\left(\mathsf{pk}, (\mathsf{qx}_i, \mathsf{qw}_i)_{i=1}^n, (\mathsf{acc}_j)_{j=1}^m\right)$ outputs an accumulator $\mathsf{acc} = (\mathsf{acc.x}, \mathsf{acc.w})$, and a proof π_{acc} of correct accumulation. We consider $\mathsf{acc.x}$ and $\mathsf{acc.w}$ the short part and long part of the accumulator, respectively.
- $\mathsf{V}\left(\mathsf{vk}, (\mathsf{qx})_{i=1}^n, (\mathsf{acc}_j.\mathsf{x})_{j=1}^m, \mathsf{acc.x}, \pi_{\mathsf{acc}}\right)$ outputs $0/1$. Note that it only accesses the short part of accumulators.
- $\mathsf{D}(\mathsf{dk}, \mathsf{acc})$ outputs $0/1$. Unlike V, the decider has access to a full accumulator.

A *split accumulation scheme in the ROM* is a split accumulation scheme in which P, V have access to the same random oracle $\mathcal{H}$.

Split accumulators are particularly useful if we can build them for the verifier relation of a NARK for circuit satisfiability. More precisely, let $\mathcal{R} = \{(\mathbb{i}, \mathbb{x}, \mathbb{w})\}$ be the relation for circuit satisfability, and let $\mathsf{ARG} = (\mathcal{I}, \mathcal{P}, \mathcal{V})$ be a NARK for $\mathcal{R}$, and let us write $\mathsf{qi} = \mathbb{i}$, $\mathsf{qx} = (\mathbb{x}, \pi.\mathbb{x})$, $\mathsf{qw} = \pi.\mathbb{w}$. Then, we define the relation $\mathcal{R}_\mathcal{V}$ such that $(\mathsf{qi}, \mathsf{qx}, \mathsf{qw}) \in \mathcal{R}_\mathcal{V} \iff 1 \leftarrow \mathcal{V}(\mathsf{vk}_{\mathsf{NARK}}, \mathbb{x}, (\pi.\mathbb{x}, \pi.\mathbb{w}))$ where $\mathsf{vk}_{\mathsf{NARK}}$ is the NARK verifier key obtained from $\mathcal{I}(\mathbb{i})$.

Suppose that there is a split accumulation scheme SA for $\mathcal{R}_\mathcal{V}$. Then, we can use ARG and SA to build PCD schemes [12, Theorem 5.3].

3 Oracle Batching

3.1 The Core Interactive Protocol

Suppose that we have n functions $f_1, \ldots, f_n : D \to \mathbb{F}$. The verifier has oracle access to them, and the prover claims that f_i is δ-close to a low-degree polynomial, for all $i = 1, \ldots, n$. The following protocol will produce a function f_{new}, and

---

[1] In fact, one can see these protocols as HIOPs for CoAgg, with $n = 1$ in the non-batch case.

reduce the n claims above to the single claim that f_{new} is δ-close to a low-degree polynomial. The function f_{new} is defined over some other domain D' such that $D \cap D' = \emptyset$. The domains D, D' can be chosen as the two cosets of degree 2^s of the group of roots of unity of degree 2^{s+1}.

In our description below, we start from $f_i \in \mathsf{RS}[D, d]$ (or close) for $i = 1, \ldots, n$, and end up with $f_{\text{new}} \in \mathsf{RS}[D']$ (or close), where t is a parameter that determines the number of verifier queries. Because we want to iteratively apply this protocol to aggregate incoming sets of functions of degree d, we will apply some degree correction to f_{new}, to bring it back to degree d.

We describe the interactive protocol in Algorithm 1, and denote by $f_{\text{new}} \leftarrow \mathsf{OB}(f_1, \ldots, f_n)$ its execution with input oracles to $f_1, \ldots, f_n$ and output an oracle to f_{new}.[2]

Algorithm 1. The oracle batching (OB) interactive protocol

1: The Verifier samples $\alpha \leftarrow \mathbb{F}$

2: The Prover computes $f_{\text{fold}} = \widehat{\mathsf{Fold}}_\alpha(f_1, \ldots, f_n)_{|D'}$ and sends oracle $\boxed{f_{\text{fold}}}$

3: The Verifier samples $x_1^{\text{out}}, \ldots, x_s^{\text{out}} \leftarrow \mathbb{F} \setminus D'$

4: The Prover sends $y_j^{\text{out}} = \hat{f}_{\text{fold}}(x_j^{\text{out}}) \quad \forall j = 1, \ldots, s$

5: The Verifier samples $x_1, \ldots, x_t \leftarrow D, r \leftarrow \mathbb{F}$ and defines

$$\boxed{f_{\text{quot}}} = \mathsf{Quotient}(\boxed{f_{\text{fold}}}, \boldsymbol{x}, \boldsymbol{y}), \text{ where}$$

$$\boldsymbol{x} = (x_1^{\text{out}}, \ldots, x_s^{\text{out}}, x_1 \ldots, x_t) \text{ and } \boldsymbol{y} = (y_1^{\text{out}}, \ldots, y_s^{\text{out}}, y_1 \ldots, y_t)$$

$$y_j = \sum_{i=1}^{n} \boxed{f_i(x_j)} \cdot \alpha^i \quad \forall j = 1, \ldots, t$$

$$\boxed{f_{\text{new}}} = \mathsf{DegCor}(d, r, \boxed{f_{\text{quot}}}, d - (s + t))$$

Proposition 1. *Let $f_1, \ldots, f_n : D \to \mathbb{F}$, and let $d \in \mathbb{N}$. Suppose that, for all $i = 1, \ldots, n$, we have that $f_i \in RS[D, d]$. Then*

$$\Pr\left[f_{\text{new}} \leftarrow \mathsf{OB}(f_1, \ldots, f_n) : f_{\text{new}} \in RS[D', d]\right] = 1.$$

The proposition is conceptually straightforward; however, a formal proof is included in our extended preprint [31].

3.2   A Proximity Proof from Oracle Batching

Given the oracle batching protocol above, $\mathsf{OB} = (\mathcal{P}_{\mathsf{OB}}, \mathcal{V}_{\mathsf{OB}})$, it is trivial to turn it into an IOP $\Pi_{\mathsf{OB}} = (\mathcal{I}, \mathcal{P}, \mathcal{V})$ for the $\mathsf{CoAgg}(\delta)$ relation, i.e. a batched proximity check:

---

[2] Note that $\mathsf{Quotient}(f_{\text{fold}}, \boldsymbol{x}, \boldsymbol{y})$ is well defined, since f_{fold} is defined over D', whereas $x_1^{\text{out}}, \ldots, x_s^{\text{out}} \in \mathbb{F} \setminus D', x_1, \ldots, x_t \in D$, and $D \cap D' = \emptyset$.

- $\mathcal{I}$ chooses the parameters of a Reed–Solomon code $\mathsf{RS}[\mathbb{F}, D, d]$, the proximity parameter δ, and the number n of functions in correlated agreement.
- $\mathcal{P}$ is the same as prover $\mathcal{P}_{\mathsf{OB}}$.
- $\mathcal{V}$ runs $\mathcal{V}_{\mathsf{OB}}$, and then checks directly whether $\Delta(f_{\mathsf{new}}, \mathsf{RS}[D', d]) \leq \delta$, accepting if and only if this check passes.

By itself, this protocol is not very useful because of the expensive verifier, but can fit nicely into the accumulator construction, where the expensive part of checking f_{new} is relegated to the decider.

Proposition 2. *Let $\delta \in (0, \min\{1-\sqrt{\rho}, 1-\rho-1/\#D\})$. The IOP Π_{OB} described above has round-by-round soundness error $\varepsilon_{\mathsf{rbr}} = \max\{\varepsilon_{\mathsf{fold}}, \varepsilon_{\mathsf{out}}, \varepsilon_{\mathsf{corr}}, (1-\delta)^t\}$, where $\varepsilon_{\mathsf{fold}}, \varepsilon_{\mathsf{out}}, \varepsilon_{\mathsf{corr}}$ are as defined in Lemmas 1, 2 and 4, respectively, t is the query parameter in the protocol, and δ is specified on $\mathfrak{i}$.*

We present a formal proof in the extended version of this paper [31].

4 NARKs from Correlated HIOP

In this section, we often deal with non-interactive versions of protocols obtained using the BCS transform or its modifications. In this case, the prover computes the Merkle-tree root for every oracle message and uses it as a short commitment (following the syntax of [19], the Merkle commitment scheme is defined by three algorithms $\mathsf{MT.Commit}, \mathsf{MT.Open}, \mathsf{MT.Check}$). For the sake of clarity, we will use $\lceil f \rceil$ as shorthand for the part of the output of the non-interactive prover corresponding to a (possibly virtual) oracle $\boxed{f}$. In particular, $\lceil f \rceil$ contains

- The Merkle-tree root $\mathsf{rt} = \mathsf{MT.Commit}(\mathsf{g}|_{\mathsf{D}})$ for some function $g : D \to \mathbb{F}$. If $\boxed{f}$ is not virtual then $g = f$,
- (d^*, r, d) if $\boxed{f} = \mathsf{DegCor}(d^*, r, \boxed{g}, d)$,
- $(\boldsymbol{x}, \boldsymbol{y})$ if $\boxed{f} = \mathsf{Quotient}(\boxed{g}, \boldsymbol{x}, \boldsymbol{y})$.

We also denote by $\mathcal{C}heck$ the verification that all **non-empty** value-position pairs of $(v_i, \mathsf{pos}_i)_{i=1}^t$ with the corresponding Merkle-tree authentication paths $(\mathsf{ap}_i)_{i=1}^t$ are consistent with the description of the oracle.

- $\mathcal{C}heck((v_i, \mathsf{pos}_i)_{i=1}^t, (\mathsf{ap}_i)_{i=1}^t, \lceil f \rceil) \to 0/1$:
   1. for $i = 1, \ldots, t$:
   2.   Parse value $g_i = g(\mathsf{pos}_i)$ from ap_i.
   3.   Given g_i and $\lceil f \rceil$, calculate $f(\mathsf{pos}_i)$.
   4. $b_i = \Big((f(\mathsf{pos}_i) = v_i) \wedge (\mathsf{MT.Check}(\lceil f \rceil.\mathsf{rt}, \mathsf{pos}_i, g_i, \mathsf{ap}_i) = 1) \Big) \vee$
          $\vee ((v_i = \perp) \wedge (\mathsf{ap}_i = \perp))$.
   5. **return** $\wedge_{i=1}^t b_i$.

To begin, we present a slight modification of transformation that allows us to compile a δ-correlated HIOP and IOPP into a classical HIOP ([8, Theorem 4.6]). The main difference is that we allow as an input a semi-δ-correlated HIOP (Definition 3) and split the Verifier into two parts. The first one makes his final decision before the interactive part of the second one starts. This way, we can separate the part of the proof (in the non-interactive version) for independent verification. This transformation simplifies our construction of the split accumulator. Let $\delta \geq \delta_0$.

Definition 4. *The transformation* $\mathsf{T}[\delta, \delta_0]$ *takes as input a public coin generalized δ-correlated HIOP* $\Pi = (\mathsf{I}, \mathsf{P}, \mathsf{V})$ *for indexed polynomial oracle relation* $\mathcal{R} = (\mathtt{i}, \mathtt{x}, \mathtt{w})$, *HIOP* $\Pi_{\mathsf{CA}} = (\mathsf{I}_{\mathsf{CA}}, \mathsf{P}_{\mathsf{CA}}, \mathsf{V}_{\mathsf{CA}})$ *for polynomial oracle relation* $\mathsf{CoAgg}(\delta_0)$ *and outputs a plain HIOP* $(\mathcal{I}, \mathcal{P}, \mathcal{V})$ *defined below.*

- *$\mathcal{I}$ outputs encodings $\Pi.\mathsf{I}(\mathtt{i})$ and $\mathtt{i}_{\mathsf{CA}}(\mathtt{i})$, where the latter contains the parameters of a Reed–Solomon code $\mathsf{RS}[F, D, d]$, the proximity parameter δ_0, and the number n of functions in correlated agreement.*
- *$\mathcal{P}(\mathtt{i}, \mathtt{x}, \mathtt{w})$ is a pair of interactive algorithms $\mathcal{P}_1, \mathcal{P}_2$, where*
  - *$\mathcal{P}_1$ is the same as the prover $\Pi.\mathsf{P}$,*
  - *$\mathcal{P}_2$ is the same as the prover $\Pi_{\mathsf{CA}}.\mathsf{P}_{\mathsf{CA}}$.*
  First, the algorithm $\mathcal{P}_1(\mathtt{i}, \mathtt{x}, \mathtt{w})$ is executed. Let $\boldsymbol{f} = (f_1, \ldots, f_n)$ be the words on which the verifier would call the oracle $\mathsf{OCoAgg}(\delta)$ during its final decision process. Then, the algorithm $\mathcal{P}_2(\mathtt{i}_{\mathsf{CA}}, \boxed{\boldsymbol{f}}, \boldsymbol{f})$ is executed.
- *$\mathcal{V}^{\mathcal{I}(\mathtt{i})}(\mathtt{x})$ is a pair of interactive algorithms $\mathcal{V}_1, \mathcal{V}_2$, where*
  - *$\mathcal{V}_1$ is the same as the verifier $\Pi.\mathsf{V}$ except that it does not call oracle OCoAgg. Instead, upon completion, it sends an additional "dummy" message.*
  - *$\mathcal{V}_2$ is the same as the verifier $\Pi_{\mathsf{CA}}.\mathsf{V}_{\mathsf{CA}}$.*
  Verifier launches $\mathcal{V}_1^{\mathcal{I}(\mathtt{i})}(\mathtt{x})$, then $\mathcal{V}_2^{\mathtt{i}_{\mathsf{CA}}}(\boxed{\boldsymbol{f}})$. $\mathcal{V}$ accepts if and only if $\mathcal{V}_1 \wedge \mathcal{V}_2$.

To obtain the corresponding non-interactive algorithm, we introduce a modification of the BCS transformation that reflects the two-component format of the verifier obtained as a result of the T transformation.

Definition 5. *The transformation* BCST *takes as input a public coin HIOP* $\Pi' = \mathsf{T}[\delta, \delta_0](\Pi, \Pi_{\mathsf{CA}})$ *and outputs the preprocessing non-interactive argument in the ROM* $(\mathbf{I}, \mathbf{P}, \mathbf{V} = (\mathbf{V_1}, \mathbf{V_2}))$, *defined below.*

- *$\mathbf{I}^{\mathcal{H}}(\mathtt{i}) \rightarrow (\mathsf{pk}, \mathsf{vk})$. The algorithm is the same as what we would get by applying the BCS transformation. For simplicity, we write the $\mathtt{i}_{\mathsf{CA}}$ parameters directly into the verification vk and proving pk keys. We denote corresponding deterministic extraction algorithm as $\mathsf{Parse}(\mathsf{vk}/\mathsf{pk})) \rightarrow \mathtt{i}_{\mathsf{CA}}$.*
- *$\mathbf{P}^{\mathcal{H}}(\mathsf{pk}, \mathtt{x}, \mathtt{w}) \rightarrow \pi = (\pi.\mathtt{x}, \pi.\mathtt{w})$. We make the following changes compared to the prover obtained using the BCS transform.*
  - *If the prover sends several oracles in one round, a separate commitment is calculated for each in the non-interactive version.*

- *Non-oracle values are written directly to the proof. These include polynomial evaluations provided by the prover, parameters for calculating virtual oracle values, etc.*
- *The output is divided into two parts, corresponding to the execution of $\Pi'.\mathcal{P}_1$ and $\Pi'.\mathcal{P}_2$ respectively. The intermediate root hash σ_{k_1} is written to the first part of the proof so that $\Pi'.\mathcal{V}_1$ can perform the check independently of the second part of the proof.*

Then the proof π has the form

$$\pi.\mathrm{x} = \left((\mathsf{m}_1, \ldots, \mathsf{m}_{k_1}), (\mathsf{ap}_1, \ldots, \mathsf{ap}_{q_1}), \sigma_{k_1}\right),$$
$$\pi.\mathrm{w} = \left((\mathsf{m}_1^{\mathsf{CA}}, \ldots, \mathsf{m}_{k_2}^{\mathsf{CA}}), (\mathsf{ap}_1^{\mathsf{CA}}, \ldots, \mathsf{ap}_{q_2}^{\mathsf{CA}}), \sigma_{k_2}^{\mathsf{CA}}\right),$$

where k_1 and k_2 are the number of oracle messages sent by $\Pi'.\mathcal{P}_1$ and $\Pi'.\mathcal{P}_2$, respectively, q_1 and q_2 are the number of requests to these oracles and the index made by $\Pi'.\mathcal{V}_1$ and $\Pi'.\mathcal{V}_2$, respectively. Messages $\mathsf{m}_i, \mathsf{m}_j^{\mathsf{CA}}$ contain the roots of Merkle-trees and non-oracle values.

- $\mathbf{V}^{\mathcal{H}}(\mathsf{vk}, \mathrm{x}, \pi) = (\mathbf{V}_1^{\mathcal{H}}, \mathbf{V}_2^{\mathcal{H}})(\mathsf{vk}, \mathrm{x}, \pi) \to \{0, 1\}$. *We split the verifier that we obtained using the BCS transformation into two parts.*
  - $\mathbf{V}_1^{\mathcal{H}}(\mathsf{vk}, \mathrm{x}, \pi.\mathrm{x})$ *verifies the first part of the proof, makes a decision, and saves the state* $\mathsf{aux} = (\boxed{f_1, \ldots, f_n}, \sigma_{k_1})$. *We denote the deterministic oracle extraction algorithm as* $\mathsf{ParseDesc}(\mathsf{vk}, \pi.\mathrm{x})$.
  - $\mathbf{V}_2^{\mathcal{H}}(\mathsf{aux}, \mathsf{vk}, \mathrm{x}, \pi.\mathrm{w})$ *uses the state to continue verifying the rest of the proof. In particular, it checks the consistency of the Merkle-tree paths provided in the proof with the corresponding* $\boxed{f_i}$.

$\mathbf{V}$ *accepts if and only if*
1. $\mathbf{V}_1^{\mathcal{H}}(\mathsf{vk}, \mathrm{x}, \pi.\mathrm{x}) = 1$
2. $\mathbf{V}_2^{\mathcal{H}}(\mathsf{aux}, \mathsf{vk}, \mathrm{x}, \pi.\mathrm{w}) = 1$, *where* $\mathsf{aux} = (\mathsf{ParseDesc}(\mathsf{vk}, \pi.\mathrm{x}), \sigma_{k_1})$

Now, let us consider a trivial version of the proximity test. Although such a test is of no practical interest, it helps us conceptualize the split accumulation scheme. Let $\Pi_{\mathsf{TRV}} = (\mathsf{I}, \mathsf{P}, \mathsf{V})$ is a HIOP for the polynomial oracle relation $\mathsf{CoAgg}(\delta)$ (Sect. 2.3).

- I outputs $\mathbb{i}_{\mathsf{CA}}$, which contains the parameters of a Reed–Solomon code $\mathsf{RS}[F, D, d]$, the proximity parameter δ_0, and the number n of functions in correlated agreement.
- V checks directly (by reading the oracle $\boxed{f}$ entirely) whether $(\mathbb{i}_{\mathsf{CA}}, \boxed{f}, f) \in \mathsf{CoAgg}(\delta_0)$, accepting if this check passes.

Suppose that we use the transformation T together with the HIOP Π_{TRV} for some δ_0-correlated HIOP Π. Then, in the non-interactive version of the result $\mathsf{BCST}(\mathsf{T}[\delta, \delta_0](\Pi, \Pi_{\mathsf{TRV}}))$, the second part of the proof $\pi.\mathrm{w}$ will contain only value-authentication path pairs $(v_i, \mathsf{ap}_i^{\mathsf{CA}})_{i=1}^{n \cdot |D|}$. This is a consequence of the fact that the prover $\Pi_{\mathsf{TRV}}.\mathsf{P}$ does not send additional oracle messages and the verifier $\Pi_{\mathsf{TRV}}.\mathsf{V}$ does not need additional randomness. The job of the verifier $\mathbf{V}_2$ is

- to check the authentication paths ap_i with respect to the $\boxed{f_1, \ldots, f_n}$ contained in aux, and the oracle $\mathcal{H}$. We denote the deterministic values and authentication paths extraction algorithms as $\mathsf{ParseEval}(\mathsf{vk}, \pi.\mathsf{w}) \to \mathsf{v_i}$ and $\mathsf{ParseAP}(\mathsf{vk}, \pi.\mathsf{w}) \to \mathsf{ap}_i$, respectively. Then:
  - $\mathsf{CheckAP}(\boxed{f}, \mathsf{vk}, \pi.\mathsf{w}) \to 0/1$:
    1. $v_i \leftarrow \mathsf{ParseEval}(\mathsf{vk}, \pi.\mathsf{w})$
    2. $\mathsf{ap}_i \leftarrow \mathsf{ParseAP}(\mathsf{vk}, \pi.\mathsf{w})$
    3. **return** $\mathit{Check}((v_i, i), (\mathsf{ap}_i), \boxed{f})$
- to check that the resulting vectors of values is δ_0-close to the $\mathsf{RS}[\mathbb{F}, D, d]$, and make the final decision.

Therefore, $\mathbf{V_2}$ only needs a simplified interface (without x and σ_{k_1}):

$$\mathbf{V_2}_{\mathcal{H}}(\mathsf{aux}', \mathsf{vk}, \pi.\mathsf{w}), \quad \text{where } \mathsf{aux}' = \mathsf{ParseDesc}(\mathsf{vk}, \pi.\mathsf{x}).$$

Since we only check for proximity to the code, some values may differ from the expected ones. Therefore, we will also allow some authentication paths to be missing. To reflect this, we will introduce another modification of the BCS transformation (Definition 6) for the special case of Π_{TRV}.

Definition 6. *The transformation* $\mathsf{BCST}^\perp$ *takes as input a public coin HIOP* $\Pi' = \mathsf{T}[\delta, \delta_0](\Pi, \Pi_{\mathsf{TRV}})$ *and outputs the preprocessing non-interactive argument in the ROM* $(\mathbf{I}, \mathbf{P}, \mathbf{V} = (\mathbf{V_1}, \mathbf{V_2}))$. *The transformation* $\mathsf{BCST}^\perp$ *is identical to* BCST *except for the algorithm* $\mathbf{V_2}$, *which can now accept the symbol* $\perp$ *instead of the authentication path* ap_i. *In this case, when checking the proximity to the* $\mathsf{RS}[\mathbb{F}, D, d]$, *it considers the corresponding value to be* $\perp$.

Lemma 5. *Let* $\mathcal{H} : \{0,1\}^* \to \{0,1\}^k$ *be a random oracle, and let* $Q \in \mathbb{N}$ *be a bound on the number of queries to* $\mathcal{H}$. *Let* Π_0 *be a semi-δ-correlated HIOP for an indexed regular (non-oracle) relation, and* $\Pi' = \mathsf{T}[\delta, \delta_0](\Pi_0, \Pi_{\mathsf{TRV}})$ *has round-by-round knowledge error* κ. *Then* $\Pi = \mathsf{BCST}(\Pi')$ *has knowledge error against* Q-*query adversaries*

$$Q\kappa + 3(Q^2 + 1)/2^k, \tag{1}$$

and $\Pi^\perp = \mathsf{BCST}^\perp(\Pi')$ *has knowledge error against* Q-*query adversaries*

$$Q\kappa + 3(Q^2 + 1)/2^k. \tag{2}$$

We present a formal proof in the extended version of this paper [31].

Finally, let us consider the last ingredient. We denote by $\mathsf{OCoAgg}^{(k)}$ the oracle for the verification algorithm of the correlation agreement for a set of k functions. Then, we present a transformation from a δ-correlated HIOP, during which the verifier calls $\mathsf{OCoAgg}^{(n)}$, to a δ-correlated HIOPk during which the verifier calls $\mathsf{OCoAgg}^{(1)}$.

Definition 7. *The transformation* B *takes as input a public coin δ-correlated HIOP* $\Pi = (\mathsf{I}, \mathsf{P}, \mathsf{V}^{\mathsf{OCoAgg}^{(n)}})$ *for indexed polynomial oracle relation* $\mathcal{R} = (\mathtt{i}, \mathtt{x}, \mathtt{w})$ *and outputs a semi-δ-correlated HIOP* $\mathsf{B}(\Pi) = (\mathsf{I_B}, \mathsf{P_B}, \mathsf{V_B}^{\mathsf{OCoAgg}^{(1)}})$ *for the same relation* $\mathcal{R}$. *Let* $\mathsf{OB} = (\mathcal{P}_{\mathsf{OB}}, \mathcal{V}_{\mathsf{OB}})$ *be as defined in Sect. 3.*

- $\mathsf{I_B}$ *is same as the indexer* I,
- $\mathsf{P_B}(\mathtt{i}, \mathtt{x}, \mathtt{w})$ *is a pair of interactive algorithms* $\mathcal{P}_1, \mathcal{P}_2$, *where*
  - $\mathcal{P}_1$ *is the same as the prover* $\Pi.\mathsf{P}$,
  - $\mathcal{P}_2$ *is the same as the prover* $\mathsf{OB}.\mathcal{P}_{\mathsf{OB}}$.

  First, the algorithm $\mathcal{P}_1(\mathtt{i}, \mathtt{x}, \mathtt{w})$ *is executed. Let* $\boldsymbol{f} = (f_1, \dots, f_n)$ *be the words on which the verifier would call the oracle* $\mathsf{OCoAgg}^{(n)}(\delta)$ *during its final decision process. Then, the algorithm* $\mathcal{P}_2(\mathtt{i}_{\mathsf{CA}}, \boxed{\boldsymbol{f}}, \boldsymbol{f})$ *is executed.*

- $\mathsf{V_B}^{\mathcal{I}(\mathtt{i}), \mathsf{OCoAgg}^{(1)}}(\mathtt{x})$ *is a pair of interactive algorithms* $\mathcal{V}_1, \mathcal{V}_2$, *where*
  - $\mathcal{V}_1$ *is the same as verifier* $\Pi.\mathsf{V}$ *except that it does not call oracle* OCoAgg. *Instead, upon completion, it sends an additional "dummy" message.*
  - $\mathcal{V}_2$ *is the same as verifier* $\mathsf{OB}.\mathcal{V}_{\mathsf{OB}}$, *except that, upon completion,* $\mathcal{V}_2$ *queries* $\mathsf{OCoAgg}^{(1)}(\boxed{f_{\mathsf{new}}})$.

  The verifier launches $\mathcal{V}_1^{\mathcal{I}(\mathtt{i})}(\mathtt{x})$, *then* $\mathcal{V}_2^{\mathtt{i}_{\mathsf{CA}}}(\boxed{\boldsymbol{f}})$. $\mathcal{V}$ *accepts if and only if* $\mathcal{V}_1 \wedge \mathcal{V}_2$.

The next lemma is a version of [8, Lemma 4.8] and its proof is verbatim. For completeness, the proof is spelled out in the extended version of this paper [31].

Lemma 6. *Let* Π *be a RBR knowledge sound δ-correlated HIOP with knowledge error* κ_Π. *Then,* $\mathsf{HIOP} := \mathsf{T}[\delta, \delta_0](\mathsf{B}(\Pi), \Pi_{\mathsf{TRV}})$ *is a RBR knowledge sound HIOP with knowledge error* $\kappa = \max\{\kappa_\Pi, \epsilon_{\mathsf{OB}}\}$ *for all* $\delta_0 \leq \delta$. *Here,* ϵ_{OB} *is the RBR soundness error of* Π_{OB}.

5   Split Accumulation and PCD

Let us recall our notations.

- $\mathsf{ParseDesc}(\mathsf{vk}/\mathsf{pk}, \pi.\mathtt{x})$ parses the proof and outputs a description $\boxed{f}$ for which the proximity test is performed.
- $\mathsf{ParseEval}(\mathsf{pk}, \pi.\mathtt{w})$ and $\mathsf{ParseAP}(\mathsf{pk}, \pi.\mathtt{w})$ parse the proof and output a vector of evaluations v_i and corresponding auth paths ap_i, respectively.
- $\mathsf{CheckAP}(\boxed{f}, \mathsf{pk}, \pi.\mathtt{w})$ parses the proof and verifies authentication paths with respect to $\boxed{f}$.

Let Π be a RBR knowledge sound δ-correlated HIOP for an indexed regular (non-oracle) relation $\mathcal{R} = (\mathtt{i}, \mathtt{x}, \mathtt{w})$ and

$$\mathsf{NARK} = \mathsf{BCST}^{\perp}(\mathsf{T}[\delta, 0](\mathsf{B}(\Pi), \Pi_{\mathsf{TRV}})) = (\mathsf{I}, \mathsf{P}, \mathsf{V} = (\mathsf{V}_1, \mathsf{V}_2)).$$

This is a NARK by Lemmas 5 and 6.

Input: $\mathsf{pk}_{\mathsf{SA}}$, $(\mathsf{qx}_i, \mathsf{qw}_i)_{i=1}^n$, $(\mathsf{acc}_j)_{j=1}^m$
Output: $\mathsf{acc} = (\mathsf{acc.x}, \mathsf{acc.w})$, π_{acc}

$(\mathsf{pk}_{\mathsf{NARK}}, \mathsf{pk}_{\mathsf{NOB}}) \leftarrow \mathsf{Parse}(\mathsf{pk}_{\mathsf{SA}})$
for $i \in [n]$ **do**
   $(\mathbb{x}_i, \pi_i.\mathbb{x}, \pi_i.\mathbb{w}) \leftarrow \mathsf{Parse}(\mathsf{qx}_i, \mathsf{qw}_i)$
   $\boxed{f_i} \leftarrow \mathsf{ParseDesc}(\mathsf{pk}_{\mathsf{NARK}}, \pi_i.\mathbb{x})$
   $f_i \leftarrow \mathsf{ParseEval}(\mathsf{pk}_{\mathsf{NARK}}, \pi_i.\mathbb{w})$
end for
for $j \in [n+1, n+m]$ **do**
   $\boxed{f_j} \leftarrow \mathsf{ParseDesc}(\mathsf{pk}_{\mathsf{SA}}, \mathsf{acc}_j.\mathbb{x})$
   $f_j \leftarrow \mathsf{ParseEval}(\mathsf{pk}_{\mathsf{SA}}, \mathsf{acc}_j.\mathbb{w})$
end for
$\mathbb{x}' \leftarrow \boxed{f_1, \ldots, f_{n+m}}$
$\mathbb{w}' \leftarrow (f_1, \ldots, f_{n+m})$
$\mathsf{pf}_{\mathsf{NOB}} \leftarrow \mathsf{NOB.P}^{\mathcal{H}}(\mathsf{pk}_{\mathsf{NOB}}, \mathbb{x}', \mathbb{w}')$
$(\pi_{\mathsf{NOB}}.\mathbb{x}, \pi_{\mathsf{NOB}}.\mathbb{w}) \leftarrow \mathsf{Parse}(\mathsf{pf}_{\mathsf{NOB}})$
$\mathsf{acc.x} \leftarrow \mathsf{ParseDesc}(\mathsf{pk}_{\mathsf{NOB}}, \pi_{\mathsf{NOB}}.\mathbb{x})$
$\mathsf{acc.w} \leftarrow \pi_{\mathsf{NOB}}.\mathbb{w}$
$\pi_{\mathsf{acc}} \leftarrow \pi_{\mathsf{NOB}}.\mathbb{x}$
return $(\mathsf{acc}, \pi_{\mathsf{acc}})$

Input: $\mathsf{vk}_{\mathsf{SA}}$, $(\mathsf{qx}_i)_{i=1}^n$, $(\mathsf{acc}_j.\mathbb{x})_{j=1}^m$, π_{acc}, $\mathsf{acc.x}$
Output: $0/1$

$(\mathsf{vk}_{\mathsf{NARK}}, \mathsf{vk}_{\mathsf{NOB}}) \leftarrow \mathsf{Parse}(\mathsf{vk}_{\mathsf{SA}})$
for $i \in [n]$ **do**
   $(\mathbb{x}_i, \pi_i.\mathbb{x}) \leftarrow \mathsf{Parse}(\mathsf{qx}_i)$
   $v_i \leftarrow \mathsf{NARK.V}_1^{\mathcal{H}}(\mathsf{vk}_{\mathsf{NARK}}, \mathbb{x}_i, \pi_i.\mathbb{x})$
   $\boxed{f_i} \leftarrow \mathsf{ParseDesc}(\mathsf{vk}_{\mathsf{NARK}}, \pi_i.\mathbb{x})$
end for
for $j \in [n+1, n+m]$ **do**
   $\boxed{f_j} \leftarrow \mathsf{ParseDesc}(\mathsf{vk}_{\mathsf{SA}}, \mathsf{acc}_j.\mathbb{x})$
end for
$\mathbb{x}' \leftarrow \boxed{f_1, \ldots, f_{n+m}}$
$v_{n+1} \leftarrow \mathsf{NOB.V}_1^{\mathcal{H}}(\mathsf{vk}_{\mathsf{NOB}}, \mathbb{x}', \pi_{\mathsf{acc}})$
$v_{n+2} \leftarrow (\mathsf{ParseDesc}(\mathsf{vk}_{\mathsf{SA}}, \mathsf{acc.x}) =$
$= \mathsf{ParseDesc}(\mathsf{vk}_{\mathsf{NOB}}, \pi_{\mathsf{acc}}))$
return $\bigwedge_{i=1}^{n+2} v_i$

Fig. 1. Algorithms $\mathsf{SA.P}^{\mathcal{H}}$ (on the left) and $\mathsf{SA.V}^{\mathcal{H}}$ (on the right).

In this section, we build a split accumulation scheme SA for the non-interactive argument system NARK. We will use the relation $\mathcal{R}_{\mathcal{V}}$ defined below:

$$\mathcal{R}_{\mathcal{V}} = \left\{ \begin{pmatrix} \mathsf{qi} \\ \mathsf{qx} \\ \mathsf{qw} \end{pmatrix} = \begin{pmatrix} \mathtt{i} \\ (\mathbb{x}, \pi.\mathbb{x}) \\ \pi.\mathbb{w} \end{pmatrix} \;\middle|\; \begin{array}{l} (\mathsf{pk}_{\mathsf{NARK}}, \mathsf{vk}_{\mathsf{NARK}}) \leftarrow \mathsf{NARK.I}^{\mathcal{H}}(\mathtt{i}), \\ \mathsf{NARK.V}^{\mathcal{H}}(\mathsf{vk}_{\mathsf{NARK}}, \mathbb{x}, (\pi.\mathbb{x}, \pi.\mathbb{w})) = 1 \end{array} \right\}.$$

We will also define a non-interactive version of the oracle batching protocol, a key block in our design.

$$\mathsf{NOB} = \mathsf{BCST}^{\perp}(\mathsf{T}[\delta, 0](\mathsf{OB}, \Pi_{\mathsf{TRV}})).$$

The accumulator is represented by a short part $\mathsf{acc.x}$, containing the description of the oracle $\boxed{f}$, and a witness part $\mathsf{acc.w}$, including authentication paths to the vector of the evaluations of function f. In practice, all the authentication paths in witness part can be represented by a single hash code corresponding to the root of the Merkle-tree. The introduced notations make calls $\mathsf{ParseDesc}(\mathsf{acc.x})$ and $\mathsf{ParseEval}(\mathsf{acc.w})$ legitimate.

Formally, the split accumulation scheme SA is represented by a tuple of algorithms $(\mathsf{SA.P}^{\mathcal{H}}, \mathsf{SA.V}^{\mathcal{H}}, \mathsf{SA.I}^{\mathcal{H}}, \mathsf{SA.D}^{\mathcal{H}})$, presented in Figs. 2 and 1.

<table>
<tr><td>

Input: qi
Output: $\mathsf{pk_{SA}, vk_{SA}, dk_{SA}}$

$(\mathsf{pk_{NARK}, vk_{NARK}}) \leftarrow \mathsf{NARK.I}^{\mathcal{H}}(\mathsf{qi})$
$\mathsf{i_{NOB}} \leftarrow \mathtt{Parse}(\mathsf{vk_{NARK}})$
$(\mathsf{pk_{NOB}, vk_{NOB}}) \leftarrow \mathsf{NOB.I}^{\mathcal{H}}(\mathsf{i_{NOB}})$
$\mathsf{pk_{SA}} \leftarrow \mathsf{pk_{NARK}} \,\|\, \mathsf{pk_{NOB}}$
$\mathsf{vk_{SA}} \leftarrow \mathsf{vk_{NARK}} \,\|\, \mathsf{vk_{NOB}}$
$\mathsf{dk_{SA}} \leftarrow \mathsf{vk_{NOB}}$

</td><td>

Input: $\mathsf{dk_{SA}, acc}$
Output: $0/1$

$\mathsf{vk_{NOB}} \leftarrow \mathtt{Parse}(\mathsf{dk_{SA}})$
$\boxed{f} \leftarrow \mathtt{ParseDesc}(\mathsf{acc.x})$
$v \leftarrow \mathsf{NOB.V}_2^{\mathcal{H}}(\boxed{f}, \mathsf{vk_{NOB}}, \mathsf{acc.w})$
return v

</td></tr>
</table>

Fig. 2. Algorithms $\mathsf{SA.I}^{\mathcal{H}}$ (on the left) and $\mathsf{SA.D}^{\mathcal{H}}$ (on the right).

Theorem 1. SA *is complete.*

Following [14], it is enough to show the knowledge soundness relative to the relaxed verifier and decider. In our terms, this means considering specified property relative to the following systems $\mathsf{NARK}^{\perp} = \mathsf{BCST}^{\perp}(\mathsf{T}[\delta, \delta](\mathsf{B}(\Pi), \Pi_{\mathsf{TRV}}))$ and $\mathsf{NOB}^{\perp} = \mathsf{BCST}^{\perp}(\mathsf{T}[\delta, \delta](\mathsf{OB}, \Pi_{\mathsf{TRV}}))$.

Theorem 2. SA *is knowledge sound with respect to* $\mathsf{NARK}^{\perp}$ *and* $\mathsf{NOB}^{\perp}$.

We present formal proofs of Theorems 1 and 2 in the extended version of this paper [31].

Given NARK and the split accumulation scheme SA, we can use [12, Theorem 5.3] to build a PCD scheme. Recall that PCD allows us to show the correctness of a distributed computation. More precisely, given a compliance predicate ϕ, PCD enables untrusted provers to demonstrate that if a computational node uses local input data z_{loc}, received messages $z_1, \ldots, z_t$, and outputs z_{out}, then $\phi(z_{\mathsf{out}}, z_{\mathsf{loc}}, z_1, \ldots, z_t) = 1$.

Informally, the PCD prover takes as input $z_{\mathsf{loc}}, z_{\mathsf{out}}$ and messages $(z_i)_{i=1}^t$ augmented with corresponding PCD proofs $(\pi_i^{\mathsf{PCD}})_{i=1}^t$, where $\pi_i^{\mathsf{PCD}} = ((\pi_i.\mathbb{x}, \pi_i.\mathbb{w}),$ $(\mathsf{acc}_i.\mathbb{x}, \mathsf{acc}_i.\mathbb{w}))$. It accumulates $((z_i, \pi_i.\mathbb{x}), \pi_i.\mathbb{w}, (\mathsf{acc}_i.\mathbb{x}, \mathsf{acc}_i.\mathbb{w}))_{i=1}^t$ to obtain a new accumulator acc and accumulation proof π_{acc}. Finally, the prover uses NARK to generate a proof π_{NARK} for the following statement (presented as a circuit):

1. $\phi(z_{\mathsf{out}}, z_{\mathsf{loc}}, (z_i)_{i=1}^t) = 1$,
2. $\mathsf{SA.V}((z_i, \pi_i.\mathbb{x})_{i=1}^t, (\mathsf{acc}_i.\mathbb{x})_{i=1}^t, \mathsf{acc.x}, \pi_{\mathsf{acc}}) = 1$.

The PCD prover outputs the new proof $\pi^{\mathsf{PCD}} = (\pi_{\mathsf{NARK}}, \mathsf{acc})$.

6 Experimental Evaluation and Discussion

This section discusses possible optimizations of the BOIL protocol and presents an experimental evaluation.

Metrics and Methodology. We implemented[3] the components of the BOIL protocol using the Plonky2 proof system. This choice was motivated by several factors. Plonky2 is highly optimized for recursive proof composition, and despite recent progress in folding schemes such as Nova [34], it remains a practical preference in many applications [20,38]. Importantly, Plonky2 uses a FRI-based commitment scheme and provides a ready-made verifier circuit, unlike Plonky3, which simplifies integration with our protocol.

We conducted experiments using our implementation on a consumer-grade laptop (Apple M2 MacBook Pro). For the baseline, we used Plonky2 with the following parameters: Reed–Solomon codes with a rate of $1/8$, the number of proof-of-work bits set to 0, and a FRI reduction strategy of $(4, 5)$. This reduction strategy means that at the i-th iteration of the FRI commit phase, the domain size is reduced by a factor of 2^4. Once the polynomial's degree falls to 2^5 or less, it is sent directly to the verifier.

Next, we replaced the FRI with the OB protocol and measured the relevant performance metrics for various values of the polynomial degree d.

Selection of Security Parameters. The results and theoretical estimates presented in Sect. 5 provide an explicit method for selecting parameters of OB protocol that ensure a proven level of completeness and soundness for the resulting PCD scheme. Following the approach used in Plonky2 and other practical implementations, we selected parameters under the assumption of the Reed–Solomon decoding conjecture [4, Conjecture 8.4]. To target 128-bit security, we choose the number of sampled points t such that $t \cdot \log(1/\rho) \geq 128$. For our configuration, we set $s = 2$ and $t = 43$.

Degree Homogenization. In practice, it is useful to think of the accumulator as a vector of $s + t$ polynomials of degree $d - 1$: $f_{\mathsf{new}}^{(i)} = \mathsf{Quotient}(f_{\mathsf{fold}}, \boldsymbol{x}_i, \boldsymbol{y}_i)$, where $\boldsymbol{x} = (x_1^{\mathsf{out}}, \ldots, x_s^{\mathsf{out}}, x_1, \ldots, x_t)$ and $\boldsymbol{y} = (y_1^{\mathsf{out}}, \ldots, y_s^{\mathsf{out}}, y_1, \ldots, y_t)$. Note that we can compute all $s + t$ evaluations of $\boxed{f_{\mathsf{new}}^{(i)}}$ at a given point g using just a single query to $\boxed{f_{\mathsf{fold}}}$. As a result, we can still efficiently verify that the linear combination $\sum_{i=1}^{s+t} \beta^i \cdot f_{\mathsf{new}}^{(i)}$ is close to a Reed–Solomon code for a randomly chosen β, while avoiding the need to interpolate the values in $\boldsymbol{y}$. This reduces both the verification complexity and the size of the corresponding circuit.

Furthermore, recall that the final step of HIOP involves checking whether the set of maps $\mathsf{Quotient}(g_i, \zeta_i, g_i(\zeta_i))$ has a δ-correlated agreement. Since all input and output polynomials of OB protocol have degree $d - 1$, the use of DegCor is unnecessary in this setting.

Proving Time and State Size. The execution times of the OB protocol and FRI are comparable (see Table 1), with the difference being negligible due to their minor contribution to the overall proving time. The vast majority of the remaining time is spent committing to all witness elements.

---

[3] https://github.com/smilesmirk/Boil.

Table 1. Prover's performance and proof sizes

	Proving Time, sec			Size, MB	
	FRI	OB	Total	Witness	Accumulator
$\log(d) = 18$	0.49	0.68	≈ 16	270	4
$\log(d) = 19$	1.38	1.51	≈ 150	540	8
$\log(d) = 20$	3.87	3.83	≈ 2100	1080	16

By using the OB protocol both in the accumulator scheme and in the NARK, we obtain a significantly smaller PCD proof size. Specifically, for a Plonkish circuit represented by a $d \times M$ matrix (with $M = 135$ in Plonky2), the proof size becomes proportional to the size of a single column – i.e., d elements. In practice, this leads to a reduction in proof size by several orders of magnitude. This compression enables more efficient parallelization of computation. In contrast, folding-based schemes generally require storing the entire witness as part of the state.

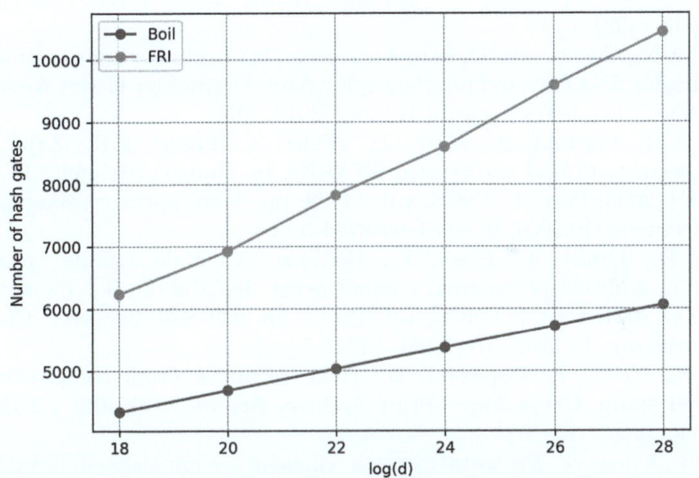

Fig. 3. Size of the verification circuit.

Recursive Circuit Size. Finally, we estimated the size of the verification circuit in terms of the number of hash gates. As shown in Fig. 3, the use of the BOIL protocol yields a noticeable reduction in circuit size for practically relevant parameter settings.

References

1. Arnon, G., Chiesa, A., Fenzi, G., Yogev, E.: STIR: Reed-solomon proximity testing with fewer queries. In: Reyzin, L., Stebila, D. (eds.) CRYPTO 2024, Part X. LNCS, vol. 14929, pp. 380–413. Springer, Cham (2024). https://doi.org/10.1007/978-3-031-68403-6_12

2. Arnon, G., Chiesa, A., Fenzi, G., Yogev, E.: Whir: Reed–solomon proximity testing with super-fast verification. Cryptology ePrint Archive (2024)

3. Ben-Sasson, E., Bentov, I., Horesh, Y., Riabzev, M.: Fast reed-solomon interactive oracle proofs of proximity. In: Chatzigiannakis, I., Kaklamanis, C., Marx, D., Sannella, D. (eds.) ICALP 2018, LIPIcs, vol. 107, pp. 1–17. Schloss Dagstuhl (2018). https://doi.org/10.4230/LIPIcs.ICALP.2018.14

4. Ben-Sasson, E., Carmon, D., Ishai, Y., Kopparty, S., Saraf, S.: Proximity gaps for reed-solomon codes. In: 61st FOCS, pp. 900–909. IEEE Computer Society Press (2020). https://doi.org/10.1109/FOCS46700.2020.00088

5. Ben-Sasson, E., Chiesa, A., Spooner, N.: Interactive oracle proofs. In: Hirt, M., Smith, A.D. (eds.) TCC 2016-B, Part II, LNCS, vol. 9986, pp. 31–60. Springer, Berlin (2016). https://doi.org/10.1007/978-3-662-53644-5_2

6. Ben-Sasson, E., Chiesa, A., Tromer, E., Virza, M.: Scalable zero knowledge via cycles of elliptic curves. In: Garay, J.A., Gennaro, R. (eds.) CRYPTO 2014, Part II, LNCS, vol. 8617, pp. 276–294. Springer, Berlin (2014). https://doi.org/10.1007/978-3-662-44381-1_16

7. Bitansky, N., Canetti, R., Chiesa, A., Tromer, E.: Recursive composition and bootstrapping for SNARKs and proof-carrying data. Cryptology ePrint Archive, Report 2012/095 (2012). https://eprint.iacr.org/2012/095

8. Block, A.R., Garreta, A., Katz, J., Thaler, J., Tiwari, P.R., Zajac, M.: Fiat-shamir security of FRI and related SNARKs. In: Guo, J., Steinfeld, R. (eds.) ASIACRYPT 2023, Part II, LNCS, vol. 14439, pp. 3–40. Springer, Singapore (2023). https://doi.org/10.1007/978-981-99-8724-5_1

9. Boneh, D., Drake, J., Fisch, B., Gabizon, A.: Halo infinite: proof-carrying data from additive polynomial commitments. In: Malkin, T., Peikert, C. (eds.) CRYPTO 2021, Part I, LNCS, vol. 12825, pp. 649–680. Springer, Cham (2021). https://doi.org/10.1007/978-3-030-84242-0_23

10. Bowe, S., Grigg, J., Hopwood, D.: Halo: recursive proof composition without a trusted setup. Cryptology ePrint Archive, Report 2019/1021 (2019). https://eprint.iacr.org/2019/1021

11. Bünz, B., Chen, B.: Protostar: generic efficient accumulation/folding for special-sound protocols. In: Guo, J., Steinfeld, R. (eds.) ASIACRYPT 2023, Part II, LNCS, vol. 14439, pp. 77–110. Springer, Singapore (2023). https://doi.org/10.1007/978-981-99-8724-5_3

12. Bünz, B., Chiesa, A., Lin, W., Mishra, P., Spooner, N.: Proof-carrying data without succinct arguments. In: Malkin, T., Peikert, C. (eds.) CRYPTO 2021, Part I, LNCS, vol. 12825, pp. 681–710. Springer, Cham (2021). https://doi.org/10.1007/978-3-030-84242-0_24

13. Bünz, B., Chiesa, A., Mishra, P., Spooner, N.: Proof-carrying data from accumulation schemes. Cryptology ePrint Archive, Report 2020/499 (2020). https://eprint.iacr.org/2020/499

14. Bünz, B., Mishra, P., Nguyen, W., Wang, W.: Accumulation without homomorphism. Cryptology ePrint Archive, Report 2024/474 (2024). https://eprint.iacr.org/2024/474

15. Bünz, B., Mishra, P., Nguyen, W., Wang, W.: Arc: accumulation for reed–solomon codes (2024). https://eprint.iacr.org/2024/1731, publication info: Preprint
16. Chen, B., Bünz, B., Boneh, D., Zhang, Z.: HyperPlonk: plonk with linear-time prover and high-degree custom gates. In: Hazay, C., Stam, M. (eds.) EURO-CRYPT 2023, Part II, LNCS, vol. 14005, pp. 499–530. Springer, Cham (2023). https://doi.org/10.1007/978-3-031-30617-4_17
17. Chiesa, A.: Proof-carrying data (2010). https://dspace.mit.edu/handle/1721.1/61151. accepted: 2011-02-23T14:20:59Z Journal Abbreviation: PCD
18. Chiesa, A., Ojha, D., Spooner, N.: Fractal: post-quantum and transparent recursive proofs from holography. In: Canteaut, A., Ishai, Y. (eds.) EUROCRYPT 2020, Part I, LNCS, vol. 12105, pp. 769–793. Springer, Cham (2020). https://doi.org/10.1007/978-3-030-45721-1_27
19. Chiesa, A., Yogev, E.: Building cryptographic proofs from hash functions (2024). https://github.com/hash-based-snargs-book
20. Deng, S., Du, B.: zkTree: a zk recursion tree with ZKP membership proofs. Cryptology ePrint Archive, Report 2023/208 (2023). https://eprint.iacr.org/2023/208
21. Eagen, L., Gabizon, A.: ProtoGalaxy: efficient ProtoStar-style folding of multiple instances. Cryptology ePrint Archive, Report 2023/1106 (2023). https://eprint.iacr.org/2023/1106
22. Fiat, A., Shamir, A.: How to prove yourself: practical solutions to identification and signature problems. In: Odlyzko, A.M. (ed.) CRYPTO'86, LNCS, vol. 263, pp. 186–194. Springer, Berlin (1987). https://doi.org/10.1007/3-540-47721-7_12
23. Gabizon, A., Williamson, Z.J.: plookup: a simplified polynomial protocol for lookup tables. Cryptology ePrint Archive, Report 2020/315 (2020). https://eprint.iacr.org/2020/315
24. Gabizon, A., Williamson, Z.J.: Proposal: the turbo-plonk program syntax for specifying snark programs (2020). https://docs.zkproof.org/pages/standards/accepted-workshop3/proposal-turbo_plonk.pdf
25. Gabizon, A., Williamson, Z.J., Ciobotaru, O.: PLONK: permutations over Lagrange-bases for oecumenical noninteractive arguments of knowledge. Cryptology ePrint Archive, Report 2019/953 (2019). https://eprint.iacr.org/2019/953
26. Gabizon, A., Williamson, Z.J., Ciobotaru, O.: Plonk: permutations over lagrange-bases for oecumenical noninteractive arguments of knowledge. Cryptology ePrint Archive (2019)
27. Grassi, L., Khovratovich, D., Rechberger, C., Roy, A., Schofnegger, M.: Poseidon: a new hash function for zero-knowledge proof systems. In: Bailey, M., Greenstadt, R. (eds.) USENIX Security 2021, pp. 519–535. USENIX Association (2021)
28. Haböck, U.: A summary on the FRI low degree test. Cryptology ePrint Archive, Report 2022/1216 (2022). https://eprint.iacr.org/2022/1216
29. Kate, A., Zaverucha, G.M., Goldberg, I.: Constant-size commitments to polynomials and their applications. In: Abe, M. (ed.) ASIACRYPT 2010, LNCS, vol. 6477, pp. 177–194. Springer, Berlin (2010). https://doi.org/10.1007/978-3-642-17373-8_11
30. Kattis, A.A., Panarin, K., Vlasov, A.: RedShift: transparent SNARKs from list polynomial commitments. In: Yin, H., Stavrou, A., Cremers, C., Shi, E. (eds.) ACM CCS 2022, pp. 1725–1737. ACM Press (2022). https://doi.org/10.1145/3548606.3560657
31. Kohrita, T., Nikolaev, M., Silva, J.: BOIL: proof-carrying data from accumulation of correlated holographic IOPs. https://eprint.iacr.org/2024/1993

32. Kothapalli, A., Setty, S.: SuperNova: proving universal machine executions without universal circuits. Cryptology ePrint Archive, Report 2022/1758 (2022). https://eprint.iacr.org/2022/1758
33. Kothapalli, A., Setty, S.: CycleFold: folding-scheme-based recursive arguments over a cycle of elliptic curves. Cryptology ePrint Archive, Report 2023/1192 (2023). https://eprint.iacr.org/2023/1192
34. Kothapalli, A., Setty, S., Tzialla, I.: Nova: recursive zero-knowledge arguments from folding schemes. In: Dodis, Y., Shrimpton, T. (eds.) CRYPTO 2022, Part IV, LNCS, vol. 13510, pp. 359–388. Springer, Cham (2022). https://doi.org/10.1007/978-3-031-15985-5_13
35. Kothapalli, A., Setty, S.T.V.: HyperNova: recursive arguments for customizable constraint systems. In: Reyzin, L., Stebila, D. (eds.) CRYPTO 2024, Part X, LNCS, vol. 14929, pp. 345–379. Springer, Cham (2024). https://doi.org/10.1007/978-3-031-68403-6_11
36. Micali, S.: Computationally sound proofs. SIAM J. Comput. **30**(4), 1253–1298 (2000)
37. Nguyen, W., Boneh, D., Setty, S.: Revisiting the nova proof system on a cycle of curves. Cryptology ePrint Archive, Report 2023/969 (2023). https://eprint.iacr.org/2023/969
38. Papamanthou, C., Srinivasan, S., Gailly, N., Hishon-Rezaizadeh, I., Salumets, A., Golemac, S.: Reckle trees: updatable merkle batch proofs with applications. Cryptology ePrint Archive, Report 2024/493 (2024). https://eprint.iacr.org/2024/493
39. Pedersen, T.P.: Non-interactive and information-theoretic secure verifiable secret sharing. In: Feigenbaum, J. (ed.) CRYPTO'91, LNCS, vol. 576, pp. 129–140. Springer, Berlin, Heidelberg (1992). https://doi.org/10.1007/3-540-46766-1_9
40. Pointcheval, D., Stern, J.: Security proofs for signature schemes. In: Maurer, U.M. (ed.) EUROCRYPT'96, LNCS, vol. 1070, pp. 387–398. Springer, Berlin, Heidelberg (1996). https://doi.org/10.1007/3-540-68339-9_33
41. Soukhanov, L.: Reverie: an end-to-end accumulation scheme from cyclefold. Cryptology ePrint Archive, Report 2023/1888 (2023). https://eprint.iacr.org/2023/1888
42. StarkWare: ethSTARK documentation. Cryptology ePrint Archive, Report 2021/582 (2021). https://eprint.iacr.org/2021/582
43. Szepieniec, A.: DEEP commitments and their applications (2024). https://eprint.iacr.org/2024/1752. publication info: Preprint
44. Team, P.Z.: Plonky2: fast recursive arguments with PLONK and FRI. https://github.com/mir-protocol/plonky2/tree/main/plonky2
45. Team, R.Z.: Zero-knowledge verifiable general computing platform based on zk-STARKs and the RISC-V microarchitecture. https://github.com/risc0
46. Valiant, P.: Incrementally verifiable computation or proofs of knowledge imply time/space efficiency. In: Canetti, R. (ed.) TCC 2008, LNCS, vol. 4948, pp. 1–18. Springer, Berlin (2008). https://doi.org/10.1007/978-3-540-78524-8_1

Rejection-Free Framework
of Zero-Knowledge Proof Based
on Hint-MLWE

Antoine Douteau[✉] and Adeline Roux-Langlois

Université Caen Normandie, ENSICAEN, CNRS, Normandie Univ,
GREYC UMR6072, 14000 Caen, France
`antoine.douteau@unicaen.fr`, `adeline.roux-langlois@cnrs.fr`

Abstract. Commit-and-prove zero-knowledge proofs are a generalized version of zero-knowledge protocols that permit proving relations over the committed elements in addition testifying to its knowledge of the initial message. For example, the existing framework (LNP, Crypto22) allow a user to prove that the secret element committed satisfies quadratic relations with bounded norm (ℓ_2 or ℓ_∞). Security of these frameworks, regarding the zero knowledge property, is mainly assumed by the use of rejection sampling introduced by Lyubashevsky (Asiacrypt09). The main problems with rejection sampling are non-constant time execution and the cost of protecting this step from side-channel attacks.

Our contribution is a new framework of proof for zero-knowledge property that proves knowledge and quadratic relations over lattices without basing the security over rejection sampling. The security of our framework is based on the recent Hint-MLWE (KLSS, Crypto23) assumption. This variant of MLWE gives additional hints about the secret in addition to the original input, and is shown to be as hard as its associated MLWE instance when secrets follow discrete Gaussian distributions.

Keywords: Lattices · Zero-Knowledge Protocol · Commit-and-prove · Hint-MLWE · Rejection sampling

1 Introduction

Lattices have been a well-studied and cryptanalysed mathematical tools that have made it possible to build several optimised, efficient and resistant cryptographic constructions for current standards, but also for future advanced primitives. Originally, fundamental lattice assumptions such as the Closest Vector Problem or the Shortest Vector Problem are NP-hard [4, 27] but the mathematical aspect of lattices inspired relaxed assumptions more cryptographic-oriented, known as the Learning with Errors (LWE) problem [45] and its dual Short Integer Solution (SIS) from [3]. They reduce to these problems but also benefit to be "average-case" unlike the fundamental assumptions. In a global way in cryptography, lack of structure and redundancy can have a negative impact on efficiency and on having reasonable schemes in both size and running time. This is why

© The Author(s), under exclusive license to Springer Nature Switzerland AG 2026
R. Dutta et al. (Eds.): INDOCRYPT 2025, LNCS 16372, pp. 119–144, 2026.
https://doi.org/10.1007/978-3-032-13301-4_6

the constructions often exploit Module Lattices originally defined in [32] that work over polynomials. This special structure is used on ML-DSA and ML-KEM [14,22] as ML stands for Module Lattices, over their assumptions Module-LWE and Module-SIS, variants of LWE and SIS.

Signatures over Lattices. The first framework of signature scheme over lattices [26] exploits a private trapdoor, allowing the signer to invert a one-way function, usually called Hash-and-Sign. The second main framework relies on the Fiat-Shamir paradigm [25] which performs a transformation of a zero-knowledge protocol to a non-interactive instance, secured on the Random Oracle Model (ROM). The idea is to prove knowledge of a specific element, and being zero-knowledge means that the execution of the protocol does not give any additional information on the secret element. The FSp transposes this proof in a single execution without interactivity between a prover and a verifier. The transcript generated defines the signature; the final verification steps are identical.

Rejection Sampling. One of the initial ideas over lattices was considered by Lyubashevsky in [33] in the FS with aborts idea: it allows the prover to send various elements strongly related to the secret committed into the transcript. To ensure the zero-knowledge property, those elements need to be carefully chosen after passing a rejection step successfully. The method is called rejection sampling as those temporary elements need to be resampled in case of failure. The main advantage is that it allows the prover to control the distribution of the leaks. Since the proving phase aborts with a non-negligible probability, it does not run in constant time. This method is used in several constructions over lattices, isogenies and error-correcting codes [11,12,17,22,36].

Commit-and-Prove System of Proofs. In order to be useful in practice and permit more than just proving the knowledge of a secret element, the notion of zero-knowledge commit-and-prove proofs [15] extends the definition of zero-knowledge proofs and allows proving additional information and relations about a committed element. This system of proofs can be modified regarding the Fiat-Shamir (with aborts) paradigm in order to construct efficient advanced signature schemes. The recent framework [36] generalizes a method of construction of this protocol in order to additionnaly prove polynomial relations over the secret and norm bounds of those structure. This allows for example to prove knowledge of a MSIS element or a MLWE secret. The framework also comes with different applications (verifiable encryption, group signature) and is used (a posteriori) in various constructions [1,13,35,42] to take advantage of its versatility. The overall security of the commitment scheme relies on module assumptions, and the zero-knowledge property is established using different optimized rejection sampling mechanisms [21,33,37].

Constraints of Rejection Sampling. In a general manner, the rejection sampling method is theoretically proven to be secure, but several flaws have been discovered in [7,18,23], affecting most of the actual rejection-based constructions over both lattices and isogenies constructions. A major disadvantage is that the proving phase does not run in constant time, making implementation and security difficult, as in [28], which tackles earlier proposals for error-correcting code systems HQC and BIKE [2,6]. Those attacks are called *side-channel attacks* and enable an attacker the ability to extract information by analyzing the physical device running the algorithm.

One of the main methods for protecting against side-channel attacks is to use secret-sharing also called *masking* [30,44]. This countermeasure implies a decomposition of an element in different shares. The masking countermeasure needs to be used over all the functions and variables that involve secret elements, including the rejection sampling. The masking of rejection sampling has been studied in depth in [8,16,23,41], as the leakage of execution information from the proving phase would conceptually contradict the goal of rejection sampling. Despite it has been optimized, masking is particularly expensive: in the recent paper [16] that analyze the second level of security of ML-DSA, the action of masking an execution multiplies the running time by at least 33 (for the first level of security). In a further comparison, the masked rejection sampling step alone takes $3\times$ longer than the total duration of a single (rejection-free) unmasked instance of ML-DSA.

Rejection sampling can also be an issue regarding some constructions with multiple executions of a protocol, such as for the Polynomial Commitment from [29] where its rejection rate would increase exponentially. Then, another perspective would be to completely remove the use of rejection sampling without significantly affecting the overall size of the construction in which it is used.

1.1   Contribution

Our contribution is to build a new commit-and-prove framework of proof for zero-knowledge property on lattices, without performing rejection sampling, based on the new construction from [29]. The security of our framework is based on the Hint-MLWE assumption, introduced in [31]. This variant of MLWE gives additional hints about the secret, and is shown to be as hard as its associated MLWE instance when secret elements follow discrete Gaussian distributions.

Our key idea is to exploit this new assumption in the [36] framework over the new Modified-Ajtai commitment scheme from [29], as it only focused on constructing a polynomial commitment using the Hint-MLWE assumption, we generalize their idea in order to construct and implement a concrete commit-and-prove framework of proof. In a first step, we modify the framework of [36] with the special Gaussianisation process from [29], allowing the proof of knowledge of a secret vector of polynomials s without any restriction on their bounds, unlike the original [36] framework. In a second step, our framework gives to the prover the ability to prove multiple quadratic and linear relations over both $\mathcal{R}_p$ and $\mathbb{Z}_p$ exploiting the automorphism σ_{-1} without additionally committing to it. We also provide the implementation of our modified framework based on the recent LaZer Library [38]. As the structure is relatively close to the original paper [36], the code structure complies with LaZer standards and has been implemented in such a way that it can be used in addition to the original library. As the main contribution lies in the elimination of rejection sampling, which implies a modification of the execution times of the different functions, we add a concrete comparison between instances of [36] using [38] and our implementation on similar parameter sets in Table 1. We provide a static analysis of the complexity, which shows that an instance of the proving phase is at most 4 times longer than a single [36], resulting in faster executions, as the rejection rate increases, but slower (less than 4 times) for other phases. It also affects the size of the transcripts, which are ≈ 3 times larger. Our construction provides an alternative of [36] that runs in constant time. The code is available in the attached file rejection-free-framework-under-Hint-MLWE.zip and in this repository.

Table 1. The first part of the Table explains the size of the transcripts in kilobytes. In the second part, the results show our benchmarks for the mean running time of each phase. The 5 different sets of parameters are detailed in Table 2. In the comparison, the running time of the Prove part of [36] using the [38] implementation takes into account several executions due to rejection sampling.

Size of the transcripts (in kilobytes KB)						
Protocol	Protocol/Phase	Param. 1	Param. 2	Param. 3	Param. 4	Param. 5
[36]	**Opening proof**	13.22	20.77	30.53	×	41.05
	+ 3 Quad.	13.93	21.86	31.31	×	41.68
	+ 3 Eval.	15.16	22.96	32.09	×	42.93
Ours	**Opening proof**	24.23	45.69	103.19	37.33	96.52
	+ 3 Quad.	24.83	46.76	103.95	38.08	97.12
	+ 3 Eval.	25.42	47.82	104.70	38.83	97.72
Mean times (in million cycles)						
[36]	**KeyGen**	7.63	10.32	19.32	×	39.35
	Commit	7.29	20.89	31.33	×	39.42
	Prove	324.22	352.22	4289.68	×	4462.29
	Verify	18.54	39.17	1744.10	×	1629.12
Ours	**KeyGen**	19.01	47.04	66.30	29.55	62.01
	Commit	19.86	86.54	102.47	45.73	67.15
	Prove	59.45	134.26	4248.67	220.98	3961.24
	Verify	41.74	96.79	3897.43	180.31	3624.34

1.2 Technical Overview

The initial idea of our construction is to provide a rejection-free framework of zero-knowledge proofs based on lattice assumptions. We start from [36] based on the Ajtai commitment scheme on a module lattice $\mathcal{R}_p = \mathbb{Z}_p[X]/\langle\phi_d\rangle$ with its modulo p and a power of 2 degree: d. The original construction proves knowledge of the message $\mathbf{s}_1 \in \mathcal{R}_p^{m_1}$, along with a randomness $\mathbf{s}_2 \in \mathcal{R}_p^{m_2}$ by computing the commitment $\mathbf{t_A} = \mathbf{A}_1\mathbf{s}_1 + \mathbf{A}_2\mathbf{s}_2$ for uniformly sampled matrices $\mathbf{A}_1 \in \mathcal{R}_p^{n\times m_1}$ and $\mathbf{A}_2 \in \mathcal{R}_p^{n\times m_2}$. Considering an Hermite Normal Form of the matrix $\mathbf{A}_2$ i.e. with an identity structure such that $\mathbf{A}_2 = \begin{bmatrix} \mathbf{I}_n \mid \mathbf{A}_2' \end{bmatrix}$ generates the same lattice with a faster sampling, inducing a MLWE structure. This commitment scheme is relevant only if the norm of $\mathbf{s}_i$ is bounded, as its binding security is based on the MSIS assumption. The hiding security is provided assuming a hard instance of MLWE with the second component involving the secret randomness $\mathbf{s}_2$. The opening of the Ajtai commitment, coming from [34], needs to provide to the verifier the responses of the form $c\mathbf{s}_i + \mathbf{y}_i$ for a challenge c. Those elements can leak information on $\mathbf{s}_i$ which is avoided using rejection sampling. We want to modify the commitment scheme in order to prevent the use of rejection sampling by having non-leaking responses. The recent assumption Hint-MLWE [31] is aimed at resolving this problem, saying that the responses (no matter their distribution) do not leak any information about $\mathbf{s}_i$, in the specific case where $\mathbf{s}_i$ follows a Gaussian distribution. This was used in [29] where the authors construct the Modified-Ajtai commitment scheme with an opening satisfying the zero-knowledge property over Hint-MLWE. In order to prove knowledge of an element $\mathbf{s}_1 \in \mathcal{R}_p^{m_1}$, they simply redefined the Ajtai commitment for a Gaussian encoded element $\tilde{\mathbf{s}}_1 \in \mathcal{R}_p^{km_1}$ (linked to $\mathbf{s}_1$) by transmitting $\mathbf{t_A} = \tilde{\mathbf{A}}_1\tilde{\mathbf{s}}_1 + \mathbf{A}_2\mathbf{s}_2$. The overall security is also based on MSIS and MLWE, only the zero-knowledge property of the opening is modified from the original version by being shown over the hardness of Hint-MLWE instead of different instances of rejection sampling. This is done by the "Gaussianisation" process that encodes $\mathbf{s}_1$ in base b (i.e. $\mathbf{s}_1 = \sum_{i\in[k-1]}\hat{\mathbf{s}}_{1,i}b^i$). This process in done coordonate-wise (over the vector, and over the coefficients of the polynomials). Proceeding over the modulo p, the process can be made probabilistic as in [39, Theorem 4.1] and can be seen as $\hat{\mathbf{s}} = \mathbf{G}_{b,\sigma}^{-1}\mathbf{s} \bmod p$ for a positive σ such that $\mathbf{G}_b\hat{\mathbf{s}} = \mathbf{s} \bmod p$ (deterministic) by denoting the gadget vector $\mathbf{G}_b := \mathbf{I}_{m_1} \otimes \mathbf{g}_{b,k}^\top := \mathbf{I}_{m_1} \otimes \begin{bmatrix} 1 \ b \cdots b^{k-1} \end{bmatrix}$.

In order to provide a general framework, the authors of [36] showed that the Ajtai commitment can be launched with additional BDLOP commitments from [9] of the form $\mathbf{t_B} = \mathbf{B}\mathbf{s}_2 + \mathbf{m}$ for unbounded messages $\mathbf{m}$ with the same randomness $\mathbf{s}_2$. This commitment is called ABDLOP. As the commitment size is linear in the size of the message, the BDLOP part was intended to be used carefully. Another important point of the Modified-Ajtai commitment scheme is that the committed encodings are not bounded. In our case, we adapt this construction and join to it the same BDLOP construction forming the Rejection-Free ABDLOP commitment scheme. With the Gaussianisation, we commit to elements only by the Ajtai part; the (expensive) BDLOP part will only be used dynamically (used to commit elements during the execution), as both norm bounded and unbounded elements can be initially committed using the (rejection-free) Ajtai part of the commitment scheme. As in the original ABDLOP, the overall opening process is restricted to the one the Ajtai part by computing $\mathbf{m} = \mathbf{t_B} - \mathbf{Bs}_2$. We also adapt the original opening process for a different challenge space $\mathcal{C}$. This involves enlarging some parameters and slightly modifying the security proofs.

Proving only the knowledge of an element $\mathbf{s}$ is not optimal. Commit-and-prove protocol such as the framework [36] additionally allows proving that a secret satisfies a bunch of relations. The prover can ensure quadratic relations over $\mathbf{s}$ but also over an automorphism $\sigma(\mathbf{s})$ (without additionally committing to it). In a second step, they also provide the ability to the prover to prove norm bounds of these relations. Unlike [36], we do not commit the initial message but a randomized encoded version of it, therefore we must modify the framework in consequence and show new adapted proofs of completeness. In [36], they used the fact that a quadratic relation regarding a message $\mathbf{s}$ can be viewed as the result of the function $f(\mathbf{s}) = \mathbf{s}^\top \mathbf{R}_2 \mathbf{s} + \mathbf{r}_1^\top \mathbf{s} + r_0$. We adapt it to a new relation $\tilde{f}$ over the randomized encoding $\tilde{\mathbf{s}}$ such that $\tilde{f}(\tilde{\mathbf{s}}) = 0$ implies that $f(\mathbf{s}) = 0$. As we have the relation $\mathbf{s} = \left[\mathbf{I}_n \otimes \mathbf{g}_{b,k}^\top \right] \cdot \tilde{\mathbf{s}}$ over $\mathcal{R}_p$, we obtain the equality $\tilde{f}(\tilde{\mathbf{s}}) = \tilde{\mathbf{s}}^\top \cdot \left(\mathbf{R}_2 \otimes \left[\mathbf{g}_{b,k}^\top \otimes \mathbf{g}_{b,k} \right] \right) \cdot \tilde{\mathbf{s}} + \left[\mathbf{r}_1 \otimes \mathbf{g}_{b,k}^\top \right] \cdot \tilde{\mathbf{s}} + r_0 = f(\mathbf{s}) \mod p$.

In [36], their protocol allowed relations over $\mathbf{s}$ and any $\mathcal{R}$-automorphism of it with the only the commitment of $\mathbf{s}$. In our case we only consider the conjugate automorphism $\sigma_{-1} : X \longmapsto X^{-1}$, which has an important impact. The relevance is simple and comes from an easy tool from [36, Lemma 2.4]; the constant coefficient of the trace application $\mathrm{tr} : \sum_{i \in [n]} \sigma_{-1}(\mathbf{a}_i) \cdot \mathbf{b}_i$ is equal to $\langle \mathbf{a}, \mathbf{b} \rangle \in \mathbb{Z}_p$ for $\mathbf{a}, \mathbf{b} \in \mathcal{R}^n$. Then, it permits to prove relations over $\mathbb{Z}_p$ and not just over $\mathcal{R}_p$ by using the same mechanism as in [36] that allows to prove that a relation has a zero constant coefficient. In our protocol, as for the quadratic relations, this is implied from the encoding protocol as we have $\sigma_{-1}(\tilde{\mathbf{s}}) = \widetilde{\sigma_{-1}(\mathbf{s})}$. Putting everything together, we can launch a protocol of proof of knowledge with quadratic relations over $\mathcal{R}_p$ and/or $\mathbb{Z}_p$, exploiting [36, Section 4].

1.3   Limitations

The original version of the protocol permits to prove relations over the secret and one of its automorphisms. In our case, we only specify the case of σ_{-1} because we exploit an inherent structure of this automorphism. A generalization regardless of the automorphism may be useful but will involve modifying the Gaussianisation method and inevitably our entire construction. Along with the first limitation, intrinsically dependent on the Gaussianisation method, the protocol imposes a proof environment $\mathcal{R}_p$ where $p = b^k + 1$ a prime modulus. Together with our protocol, our challenge space restricts the choice of k to the specific case of $k = 2$. Finding a new challenge space that does not restrict ourselves to this choice of k together with the automorphism invariant condition $c = \sigma_{-1}(c)$ is an interesting open question. As the bounds of the security assumptions necessary depend on the choices of b (and k), it restricts the choice of p to be a prime higher than $\approx 2^{50}$. The last limitations is that we are not able to prove approximate norm bounds of relations over the secrets as it was build in the [36] framework. At least, it is still possible to apply the [36, Section 5] strategy by recomputing the secret from the encoding and perform rejection sampling. It would be interesting to prove those relations using the pros of our protocol in order to obtain a fully rejection-free proof.

1.4   Related Works

As already mentioned, [36] constructs a generalized framework for proving knowledge of an element $\mathbf{s}$ such that $f(\mathbf{s}) = 0$ for every $f \in \mathcal{F}$. They prove the zero-knowledge property assuming various instances of rejection sampling, without considering potential side-channels attacks neither the cost of protecting it with masking. In our generated framework, we provide a rejection-free alternative of [36] inspired by techniques from [29]. In this paper, the authors construct a polynomial commitment allowing a prover to commit to a polynomial f and convincing a prover that $y = f(x)$ for public y and x without revealing f. They based their construction on a homomorphic decoding protocol. In our work, we exploit their techniques, but in order to commit to an element x and convince a prover that $f(x) = 0$ for a public relation f without revealing x.

We identify in the literature different alternatives for removing the rejection sampling [10,19,20]. In the first article, the authors propose an alternative to the Lyubashevsky Signature Scheme [33] by restricting the construction to deal with both secret keys and masks being distributed uniformly over a hypercube. They also conjecture that all the rejection sampling steps do not have the same impact and that it may be difficult to remove abortion steps while maintaining the correctness of the protocol. In the other papers, the authors exploits Gaussian convolution distributions: by choosing compensating covariance matrices, they manage to construct hints that do not leak information even without rejection sampling. These papers only focused on constructing a rejection-free signature scheme and not a commit-and-prove framework.

2   Preliminaries

2.1   Notations and Definitions

Notations. For $n \in \mathbb{N}$, we define $[n] := \{0, \ldots, n-1\}$. Lower-case v letters mean elements of $\mathcal{R}$, $\vec{v}$ are the same elements viewed as vectors (of size d) of coefficients. Bold lower-case $\mathbf{v}$ letters denote column-vectors with coefficients being elements of $\mathcal{R}$, and bold upper-case $\mathbf{A}$ letters for matrices with coefficients in $\mathcal{R}$. We write $x \xleftarrow{\$} S$ when x is uniformly sampled from a finite set S.

Number Theory. We denote $\mathbb{Z}_q := \mathbb{Z}/q\mathbb{Z}$, $\mathcal{K} := \mathbb{Q}[X]/\langle \phi_d(X) \rangle$ where $\phi_d(X)$ denotes the d-cyclotomic polynomial (we restrict the choice of d being a power of 2), and $\mathcal{R}$ its ring of integers $\mathcal{R} := \mathbb{Z}[X]/\langle \phi_d(X) \rangle$. Also we denote $\mathcal{R}_q := \mathcal{R}/q\mathcal{R}$. We specify the notation of the automorphism $\sigma := \sigma_{-1} \in \mathsf{Aut}(\mathcal{R}) : X \mapsto X^{-1}$.

Norms. For a vector of polynomials $\mathbf{f} = (f_0, \ldots, f_{n-1}) \in \mathcal{R}^n$, the norm is the norm of each coefficient: $\|\mathbf{f}\|_q := \sqrt[q]{\sum_{i \in [n]} \|f_i\|_q^q}$, same for the infinite norm: $\|\mathbf{f}\|_\infty := \max_{i \in [n]} \|f_i\|_\infty$. For $\mathcal{M} \in \mathbb{C}^{n \times m}$, $\|\mathcal{M}\| := s_{\mathsf{max}}(\mathbf{A}) = \sqrt{|\lambda_{\mathsf{max}}(\mathcal{M}^\top \mathcal{M})|}$.

Gadgets. We remind the construction of the"gadget" vector and its standard definition for an integer b: $\mathbf{g}_{b,k}^\top = \left[1, b, \cdots, b^{k-1}\right] \in \mathcal{R}^k$. We extend the definition over n elements names as the gadget matrix $\mathbf{G}_{b,k,n} := \mathbf{I}_n \otimes \mathbf{g}_{b,k}^\top \in \mathcal{R}^{n \times nk}$. We denote a probabilistic "inverse" $\mathbf{G}_{b,\sigma}^{-1}$ allowing to generate a non-unique preimage $\hat{\mathbf{t}}$ of $\mathbf{t}$ such that $\mathbf{G}\hat{\mathbf{t}} = \mathbf{t}$ i.e $\hat{\mathbf{t}} \leftarrow \mathcal{D}_{\Lambda_{\hat{\mathbf{t}}}^\top(\mathbf{G}),\sigma}$ as explained in [39, Theorem 4.1].

2.2 Lattices

Lattices. A lattice, denoted by Λ, is a discrete subgroup of $\mathbb{R}^n$ with additive operations. It can be defined by a basis $\mathcal{B} = (\vec{b}_i)_{i \in [m]} \in \mathbb{R}^{n \times m}$ such that it is denoted by $\Lambda = \left\{ \sum_{i \in [m]} x_i \cdot \vec{b}_i \mid \vec{x} \in \mathbb{Z}^m \right\}$, where m is its dimension. We defined its dual lattice $\Lambda^* := \{x \mid \langle x, z \rangle \in \mathbb{Z} \ \forall z \in \Lambda\}$. We have $\Lambda = (\Lambda^*)^*$.

Probability Distributions. In our case, $\mathcal{D}_{\Lambda,\vec{c},\sigma}$ describe a discrete Gaussian distribution over Λ, centered in $\vec{c} \in \mathbb{R}^m$ with standard deviation σ such that for any $\vec{x} \in \mathbb{R}^n$, $\rho_{\vec{c},\sigma}(\vec{x}) := \exp\left(-\frac{\|\vec{x}-\vec{c}\|^2}{2\sigma^2}\right)$, and $\mathcal{D}_{\Lambda,\vec{c},\sigma}(\vec{x}) := \frac{\rho_{\vec{c},\sigma}(\vec{x})}{\rho_{\vec{c},\sigma}(\Lambda)}$ where $\rho_{\vec{c},\sigma}(\Lambda) = \sum_{x \in \Lambda} \rho_{\vec{c},\sigma}(\vec{x})$. We also define the sampling of a vector of an Elliptic Gaussian using an associated covariance matrix defined by a positive-definite matrix, denoted by $\boldsymbol{\Sigma} := \sqrt{\boldsymbol{\Sigma}}^\top \sqrt{\boldsymbol{\Sigma}}$. The probability is now defined by: $\rho_{\sqrt{\boldsymbol{\Sigma}}}(\vec{x}) = \exp\left(-\pi(\vec{x}-\vec{c})^\top \boldsymbol{\Sigma}^{-1}(\vec{x}-\vec{c})\right)$. If Λ is a structured lattice, we sample its vector of coefficients and embbed it as a polynomial of $\mathcal{R}$. We write directly polynomials.

Smoothing Parameter. We remind the definition of the smoothing parameter from [40, Definition 3.1] as $\eta_\varepsilon(\Lambda)$, the smallest standard deviation σ such that $\rho_{\frac{1}{\sigma}}(\Lambda^* \backslash 0) \leq \varepsilon$. The definition has been extended for covariance matrices in [43, Definition 2.3] such that if $\eta_\varepsilon(\sqrt{\boldsymbol{\Sigma}}^{-1}\Lambda) \leq 1$ therefore $\sqrt{\boldsymbol{\Sigma}} \geq \eta_\varepsilon(\Lambda)$.

Lemma 1 ([31, **Lemma 5**]). *If* $\|\boldsymbol{\Sigma}^{-1}\| \leq \frac{1}{\eta_\varepsilon(\Lambda)^2}$ *then* $\sqrt{\boldsymbol{\Sigma}} \geq \eta_\varepsilon(\Lambda)$.

Lemma 2 ([40, **Lemma 4.4**]). *Let* Λ *a n-dimensional lattice, an arbitrary vector* $\vec{c} \in \mathbb{R}^n$, $0 < \varepsilon < 1$ *and* $\sigma \geq \eta_\varepsilon(\Lambda)$. *Let* $\vec{z} \leftarrow \mathcal{D}_{\Lambda,\vec{c},\sigma}$. *Then:* $\mathbb{P}\left[\|\vec{z} - \vec{c}\| > \sigma\sqrt{n}\right] \leq \frac{1+\varepsilon}{1-\varepsilon} \cdot 2^{-n}$. *If* $0 < \varepsilon < \frac{1}{3}$ *then:* $\mathbb{P}\left[\|\vec{z} - \vec{c}\| > \sigma\sqrt{n}\right] \leq 2^{-n+1}$.

In our construction, we will use [29, Lemma 11] in the case of structured lattice with polynomials instead of vector of coefficients.

Lemma 3 (Structured version of [29, Lemma 11]). *Let* $\mathcal{R}_p^n$ *a nd-dimensional lattice over the ring* $\mathcal{R}$ *and a modulo p. Let* $\ell > 0$ *an integer.*

Define $\mathbf{\Sigma} \in \mathbb{R}^{nd \times nd}$ a covariance matrix, associated to a secret sampling. Define ℓ covariance matrices $\mathbf{\Sigma}_i \in \mathbb{R}^{nd \times nd}$, generating $c_i \in \mathcal{R}_p$ and $C_i := \mathsf{Rot}_{nd}(c_i)$.

If we suppose $\sqrt{\left(\mathbf{\Sigma}^{-1} + \sum_{i \in [\ell]} C_i^T \mathbf{\Sigma}_i^{-1} C_i\right)^{-1}} \geq \eta_\varepsilon(\mathcal{R}^n)$, then for any center $\vec{u} \in \mathbb{R}^{nd}$, then the following two distributions over $(\mathcal{R}^n)^\ell$ are at statistical 2ε:

$$\left\{ (\mathbf{z}_1, \cdots, \mathbf{z}_\ell) \,\middle|\, \mathbf{s} \leftarrow \mathcal{D}_{\mathbf{u}+\mathcal{R}^n, \sqrt{\mathbf{\Sigma}}}, \mathbf{y}_i \leftarrow \mathcal{D}_{-C_i\mathbf{u}+\mathcal{R}^n, \sqrt{\mathbf{\Sigma}_i}}, \mathbf{z}_i = c_i\mathbf{s} + \mathbf{y}_i \right\},$$

$$\left\{ (\mathbf{z}_1, \cdots, \mathbf{z}_\ell) \,\middle|\, \mathbf{s}' \leftarrow \mathcal{D}_{\mathcal{R}^n, \sqrt{\mathbf{\Sigma}}}, \mathbf{y}_i \leftarrow \mathcal{D}_{\mathcal{R}^n, \sqrt{\mathbf{\Sigma}_i}}, \mathbf{z}_i = c_i\mathbf{s}' + \mathbf{y}_i \right\}.$$

We use the simplified discrete convolution lemma, initially defined in [43].

Lemma 4 (Discrete Convolution Lemma from [29, Lemma 7]). *Let $\mathbf{\Sigma}_1, \mathbf{\Sigma}_2$ be positive-definite matrices. Let two n-dimensional lattices Λ_1 and Λ_2 such that $\Lambda_2 \subseteq \Lambda_1$. If $\sqrt{\mathbf{\Sigma}_1} \geq \eta_\varepsilon(\Lambda_1)$ and $\mathbf{\Sigma}_3^{-1} := \mathbf{\Sigma}_1^{-1} + \mathbf{\Sigma}_2^{-1}$ satisfies $\sqrt{\mathbf{\Sigma}_3} \geq \eta_\varepsilon(\Lambda_2)$ for $0 < \varepsilon < \frac{1}{2}$. Then, for arbitrary $\vec{c}_1, \vec{c}_2 \in \mathbb{R}^n$, the distribution of the convolution defined by $\mathcal{C} := \{\mathbf{x}_1 + \mathbf{x}_2 \mid \mathbf{x}_1 \leftarrow \mathcal{D}_{\Lambda_1, \vec{c}_1, \sqrt{\mathbf{\Sigma}_1}}, \mathbf{x}_2 \leftarrow \mathcal{D}_{\Lambda_2, \vec{c}_2, \sqrt{\mathbf{\Sigma}_2}}\}$ is within statistical distance 8ε of $\mathcal{D}_{\Lambda_1, \vec{c}_1+\vec{c}_2, \sqrt{\mathbf{\Sigma}_1+\mathbf{\Sigma}_2}}$.*

Assumptions. We remind the definitions of M(I)SIS and (Hint-)MLWE.

Definition 5 (M(I)SIS[32]). *Given $n > 0$, $m(n), B(n) > 0$ and a modulo $q(n)$ such that $B < q$. Consider a public and uniformly sampled matrix $\mathbf{A} \in \mathcal{R}_q^{n \times m}$, a target $\mathbf{t} \in \mathcal{R}_q^n$, the Module (Inhomogenous) Short Integer Solution $\mathsf{MISIS}_{\mathcal{R},q,n,m,\beta}$ problem asks for a solution $\mathbf{x} \in \mathcal{R}_q^m$ with $0 < \|\mathbf{x}\| \leq \beta$ that solves $\mathbf{A}\mathbf{x} = \mathbf{t}$ over $\mathcal{R}_q$. The $\mathsf{MSIS}_{\mathcal{R},q,n,m,\beta}$ assumption is the special case where $\mathbf{t} = \mathbf{0}$.*

Definition 6 (Hint- MLWE [31, Defi. 7], [32, Defi. 4.1]). *Given $n, \ell, m > 0$ and distributions χ, $\xi_1, \ldots, \xi_\ell$ over $\mathcal{R}^{m+n}$ and Γ is one over $\mathcal{R}$. Consider a random matrix $\mathbf{A} \xleftarrow{\$} \mathcal{R}_q^{m \times n}$, a secret $\mathbf{r} \leftarrow \chi$ parsed as $[\mathbf{s}^\top, \mathbf{e}^\top]$ for the MLWE sample. Consider ℓ hints as $\mathbf{z}_i = c_i\mathbf{r} + \mathbf{y}_i$, with $c_i \leftarrow \Gamma$ and $\mathbf{y}_i \leftarrow \xi_i$. The Hint- MLWE $_{\mathcal{R},q,n,m,\ell,\chi, \xi_1, \ldots, \xi_\ell, \Gamma}$ problem asks to distinguish between an honest $(\mathbf{A}, \mathbf{A}\mathbf{s} + \mathbf{e}, (c_i)_{i \in [\ell]}, (\mathbf{z})_{i \in [\ell]})$ and $(\mathbf{A}, \mathbf{b}, (c_i)_{i \in [\ell]}, (\mathbf{z})_{i \in [\ell]})$ for $\mathbf{b} \xleftarrow{\$} \mathcal{R}_q^m$.*

As shown above, the assumption is a relaxed alternative to the classical MLWE with ℓ linear hints over the secret element $\mathbf{r}$. The assumption was shown in [31] to be as hard as a MLWE instance in the restricted case where the element $\mathbf{r}$ and the masks $\mathbf{y}_i$ follow Gaussian distributions. We recall the hardness of Hint-MLWE from [24] with Elliptic Gaussians distributions.

Theorem 7 (Hardness of Hint-MLWE [24, Theorem 2]). *Let n, m, q and ℓ be positive integers and Γ a distribution over $\mathcal{R}$. We define upper bounds β_i such that $\beta_i^2 \geq \|c_i\|_1^2 > 0$, where each c_i is sampled from Γ. For the covariance matrices $\Sigma_1, \Sigma_{2,1}, \ldots, \Sigma_{2,\ell}$, we define σ as $\frac{1}{\sigma^2} = 2\left(s_1(\Sigma_1^{-1}) + \sum_{i\in[\ell]} \beta_i^2 s_1(\Sigma_{2,i}^{-1})\right)$.*

If $\sigma \geq \sqrt{2} \cdot \eta_\varepsilon(\mathcal{R}^{n+m})$, for some $0 < \varepsilon \leq \frac{1}{2}$, then there is an polynomial reduction from $\mathsf{MLWE}_{\mathcal{R},q,n,m,\mathcal{D}_\sigma}$ to $\mathsf{Hint\text{-}MLWE}_{\mathcal{R},q,n,m,\ell,\mathcal{D}_{\sqrt{\Sigma_1}},\mathcal{D}_{\sqrt{\Sigma_{2,1}}},\ldots,\mathcal{D}_{\sqrt{\Sigma_{2,\ell}}},\Gamma}$ that reduces the advantage by at most 2ε.

to an interactive protocol $(\mathcal{P}, \mathcal{V})$ with the public statement x. We say that the protocol is simulatable if the tr generated by $\mathcal{S}$ is indistinguishable from the transcript of a fully honest execution of the interactive protocol. ted to commitment scheme.

2.3   Ajtai and ABDLOP commitment schemes

Ajtai commitment scheme. We remind the original Ajtai commitment scheme inspired from [3] and describe a Lyubashevsky-like proof of opening [34].

Parameters : let $n \in \mathbb{N}$, $(\alpha_1, m_1) \in \mathbb{N}^2$ the message parameters and the others $m_2 \in \mathbb{N}, p \in \mathbb{N}, \sigma_2 \in \mathbb{R}^+$ are parametrised by n. $\mathcal{R}_p$ is a ring of degree d (a power of 2). By defining $\mathcal{M}_{\alpha,m} := \{\|\mathbf{s}\|_2 \leq \alpha \mid \mathbf{s} \in \mathcal{R}_p^m\}$, the message space is restricted to $\mathcal{M}_{\alpha_1,m_1}$ (with chosen α_1) and randomness to $\mathcal{M}_{\alpha_2,m_2}$ for $\alpha_2 := \sigma_2\sqrt{m_2 d}$.

- **Setup**$(1^\lambda) \to ck$: Given a security parameter λ, it generates a commitment key $ck = (\mathbf{A}_1, \mathbf{A}_2)$ where $\mathbf{A}_1 \xleftarrow{\$} \mathcal{R}_p^{n \times m_1}$ and $\mathbf{A}_2 \xleftarrow{\$} \mathcal{R}_{p,\mathsf{HNF}}^{n \times m_2}$.
- **Commit**$(ck, \mathbf{s}_1, \sigma) \to (\mathbf{t_A}, \delta)$: Given a message $\mathbf{s}_1 \in \mathcal{M}_{\alpha_1,m_1}$ and ck parsed as $(\mathbf{A}_1, \mathbf{A}_2)$. First, it samples $\mathbf{s}_2 \xleftarrow{\$} \mathcal{D}_{\mathcal{R}_p,\sigma_2}^{m_2}$ and verifies that $\mathbf{s}_2 \in \mathcal{M}_{\alpha_2,m_2}$. Then, it computes $\mathbf{t_A} = \mathbf{A}_1\mathbf{s}_1 + \mathbf{A}_2\mathbf{s}_2$ for the opening $\delta = (\mathbf{s}_1, \mathbf{s}_2)$.

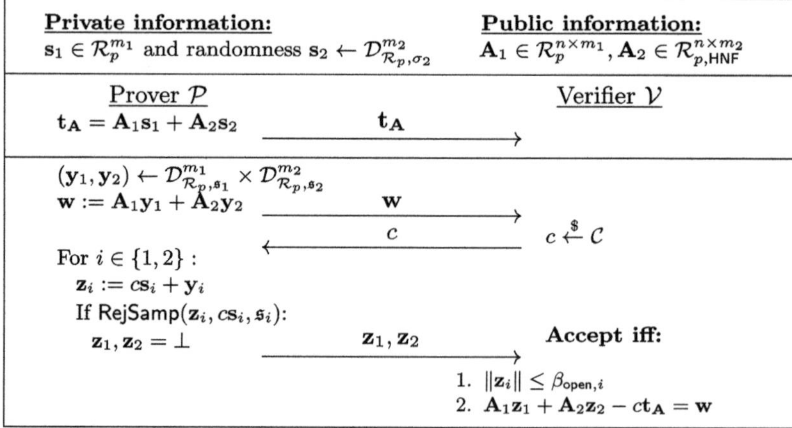

Fig. 1. Proof of Weak Opening of the Ajtai Commitment with a chosen α_i.

An honest commitment can be viewed as an element of $\mathsf{R}_{\mathsf{ajtai}}$:

$$\mathsf{R}_{\mathsf{ajtai}} := \left\{ \begin{matrix} (\mathbf{A}_1, \mathbf{A}_2, \mathbf{t_A}) \in \mathcal{R}_p^{n \times m_1} \times \mathcal{R}_{p,\mathsf{HNF}}^{n \times m_2} \times \mathcal{R}_p^n \\ (\mathbf{s}_1, \mathbf{s}_2) \in \mathcal{M}_{\alpha_1, m_1} \times \mathcal{M}_{\alpha_2, m_2} \end{matrix} \middle| \mathbf{A}_1 \mathbf{s}_1 + \mathbf{A}_2 \mathbf{s}_2 = \mathbf{t_A} \right\} .$$

The binding property is based on $\mathsf{MSIS}_{\mathcal{R}, p, n, m_1 + m_2, 2\sqrt{\alpha_1^2 + \alpha_2^2}}$. For the hiding property, an hard instance of $\mathsf{MLWE}_{\mathcal{R}, p, n, m_2, \mathcal{D}_{\sigma_2}}$ ensures indistinguishability of the hiding component $\mathbf{A}_2 \mathbf{s}_2 = \begin{bmatrix} \mathbf{I}_n & \mathbf{A}_2' \end{bmatrix} \mathbf{s}_2$. Both rely on computational difficulty.

Challenge Space. In the opening proof, we specify the same challenge space $\mathcal{C}$ than in [36] using their Lemma 15 that ensures both the stability and the smallness of element of $\mathcal{R}_p$ under a specific automorphism $\sigma_{-1} : X \mapsto X^{-1}$:

$$\mathcal{C} := \left\{ c \in \mathcal{R}_p : \|c\|_1 \leq \kappa, \sigma_{-1}(c) = c, \sqrt[2k]{\|\sigma(c^k) \cdot c^k\|_1} \leq \eta \right\} .$$

For the opening, we must have a space $\overline{\mathcal{C}} := \{c - c' : c, c' \in \mathcal{C} \text{ where } c \neq c'\} \subseteq \mathcal{R}_q^\times$.

Lemma 8 ([36, Lemma 2.6]). *Let $p = 5 \mod 8$ being a prime. For any $c \in \mathcal{R}_p$ such that $c = \sigma(c)$. Then c is invertible if and only if $c \neq 0$.*

Original Proof of Opening of Lyubashevsky [34]. The process opens to an element of $\overline{\mathsf{R}}_{\mathsf{ajtai}}$. We need to specify for a good choice of parameters that with overwhelming probability, we have $(\overline{\mathbf{s}}_1, \overline{\mathbf{s}}_2, \overline{c}) \in \overline{\mathsf{R}}_{\mathsf{ajtai}} \Rightarrow (\overline{\mathbf{s}}_1, \overline{\mathbf{s}}_2) \in \mathsf{R}_{\mathsf{ajtai}}$.

$$\overline{\mathsf{R}}_{\mathsf{ajtai}} := \left\{ \begin{matrix} (\mathbf{A}_1, \mathbf{A}_2, \mathbf{t_A}) \in \mathcal{R}_p^{n \times m_1} \times \mathcal{R}_{p,\mathsf{HNF}}^{n \times m_2} \times \mathcal{R}_p^n \\ (\overline{\mathbf{s}}_1, \overline{\mathbf{s}}_2, \overline{c}) \in \mathcal{M}_{\beta_1, m_1} \times \mathcal{M}_{\beta_2, m_2} \times \overline{\mathcal{C}} \end{matrix} \middle| \mathbf{A}_1 \overline{\mathbf{s}}_1 + \mathbf{A}_2 \overline{\mathbf{s}}_2 = \mathbf{t_A} \right\} .$$

Soundness and Completeness are shown for a good choice of parameters and standard deviations $\mathfrak{s}_i$ of the masks $\mathbf{y}_i$. The zero-knowledge property is ensured assuming rejection sampling methods on the responses $\mathbf{z}_i$. We remind the proof of opening of the Ajtai commitment scheme in Fig. 1 with $\mathfrak{s}_i := \gamma_i \eta \alpha_i$ with γ_i the rejection rate of the i-th rejection sampling. Consider a good ring $\mathcal{R}$ such that $m_i d - 1 \geq \lambda$ (the security parameter). By taking $\sigma_2 \leq \eta_\varepsilon(\mathcal{R}_p^{m_2})$, we restrict randomness to $\alpha_2 := \sigma_2 \sqrt{m_2 d}$. and set valid opening bounds $\beta_i := \mathfrak{s}_i \sqrt{2 m_i d}$.

ABDLOP Commitment Scheme [36]. The Ajtai commitment suffers from two distinct limitations. First, the message space is restricted to bounded polynomials (by α_1), second it does not permit transmitting and committing to additional elements dynamically (during the execution of the protocol). As the global aim of the framework is to prove various relations depending on each other, we need to extend the actual Ajtai commitment scheme to solve that point. In [36], the authors join to it the BDLOP construction from [9], giving the new ABDLOP commitment scheme, exploiting the same randomness $\mathbf{s}_2$ for both components. The lower part of the commitment (the BDLOP instance) is only used for dynamic messages $\mathbf{m} \in \mathcal{R}_p^\ell$ depending on the execution of the protocol, as $\mathbf{B}$ can be extended and transmitted during the execution.

$$\mathsf{R}_{\mathsf{abdlop}} := \left\{ \begin{array}{l} \left(\mathbf{A}_1, \mathbf{A}_2, \mathbf{B}, (\mathbf{t_A}, \mathbf{t_B})\right) \in \mathcal{R}_p^{n \times m_1} \times \mathcal{R}_{p,\mathsf{HNF}}^{n \times m_2} \times \mathcal{R}_{p,\mathsf{HNF}}^{\ell \times m_2} \times \mathcal{R}_p^{n+\ell} \\ (\mathbf{s}_1, \mathbf{s}_2, \mathbf{m}) \in \mathcal{M}_{\alpha_1, m_1} \times \mathcal{M}_{\alpha_2, m_2} \times \mathcal{R}_p^\ell \end{array} \left| \begin{array}{c} \mathbf{A}_1\mathbf{s}_1 + \mathbf{A}_2\mathbf{s}_2 = \mathbf{t_A} \\ \mathbf{B}\mathbf{s}_2 + \mathbf{m} = \mathbf{t_B} \end{array} \right. \right\}.$$

As explained in the original paper, for well chosen parameters, we have with overwhelming probability that $(\bar{\mathbf{s}}_1, \bar{\mathbf{s}}_2, \bar{c}) \in \overline{\mathsf{R}}_{\mathsf{abdlop}} \Rightarrow (\bar{\mathbf{s}}_1, \bar{\mathbf{s}}_2, \mathbf{t_B} - \mathbf{B}\bar{\mathbf{s}}_2) \in \mathsf{R}_{\mathsf{abdlop}}$. The opening relies essentially on the Ajtai opening $(\bar{\mathbf{s}}_1, \bar{\mathbf{s}}_2, \bar{c}) \in \overline{\mathsf{R}}_{\mathsf{ajtai}}$.

$$\overline{\mathsf{R}}_{\mathsf{abdlop}} := \left\{ \begin{array}{l} \left(\mathbf{A}_1, \mathbf{A}_2, \mathbf{B}, (\mathbf{t_A}, \mathbf{t_B})\right) \in \mathcal{R}_p^{n \times m_1} \times \mathcal{R}_{p,\mathsf{HNF}}^{n \times m_2} \times \mathcal{R}_{p,\mathsf{HNF}}^{\ell \times m_2} \times \mathcal{R}_p^{n+\ell} \\ (\bar{\mathbf{s}}_1, \bar{\mathbf{s}}_2, \bar{c}) \in \mathcal{M}_{\beta_1, m_1} \times \mathcal{M}_{\beta_2, m_2} \times \overline{\mathcal{C}} \end{array} \left| \begin{array}{c} \mathbf{A}_1\bar{\mathbf{s}}_1 + \mathbf{A}_2\bar{\mathbf{s}}_2 = \mathbf{t_A} \\ \overline{\mathbf{m}} = \mathbf{t_B} - \mathbf{B}\bar{\mathbf{s}}_2 \end{array} \right. \right\}.$$

3   Another Usage of the Gaussianisation Process

To build a rejection sampling free commit-and-prove protocol, we start by modifying the ABDLOP commitment scheme used in [36] using the Modified-Ajtai commitment joined with the Gaussianisation process from [29]. We manage to base the zero-knowledge security over a hard instance of Hint-MLWE instead of various instances of rejection sampling.

The [29] Gaussianisation Method. The [29, Section 3] method is used to construct an encoding following a Gaussian distribution centered in the secret encoding. First, it computes an encoding $\hat{\mathbf{s}}$ of an element $\mathbf{s} \in \mathcal{R}_p^m$ for $p = b^k + 1$, exploiting $\hat{\mathbf{s}} = \mathbf{G}_b^{-1}\mathbf{s} \in \mathcal{R}_p^{km}$, a decomposition in base b. In a second step, the encoding is randomized by sampling a non-influent error. At the end, it generates an additional error $\mathbf{e} \leftarrow \mathcal{D}_{\mathcal{R}^{km}, \sigma}$ to construct the final randomized encoding $\tilde{\mathbf{t}} = \hat{\mathbf{t}} + \mathfrak{p}_m \mathbf{e}$, with $\mathfrak{p}_m := X^{md} - b$.

3.1   RF-ABDLOP Commitment Scheme

Here, we adapt the idea of the Modified-Ajtai commitment scheme from [29] that commit to an encoded element $\tilde{\mathbf{s}}_1$ rather than $\mathbf{s}_1$ directly to ABDLOP, giving the Rejection-Free ABDLOP commitment scheme defined with:

$$\mathsf{R}_{\mathsf{RF}} := \left\{ \begin{array}{l} \left(\mathbf{A}_1, \mathbf{A}_2, \mathbf{B}, (\mathbf{t_A}, \mathbf{t_B})\right) \in \mathcal{R}_p^{n \times km_1} \times \mathcal{R}_{p,\mathsf{HNF}}^{n \times m_2} \times \mathcal{R}_{p,\mathsf{HNF}}^{\ell \times m_2} \times \mathcal{R}_p^{n+\ell} \\ (\mathbf{s}_1, \mathbf{s}_2, \mathbf{m}) \in \mathcal{R}_p^{m_1} \times \mathcal{M}_{\alpha_2, m_2} \times \mathcal{R}_p^\ell \end{array} \left| \begin{array}{c} \mathbf{A}_1\tilde{\mathbf{s}}_1 + \mathbf{A}_2\mathbf{s}_2 = \mathbf{t_A} \\ \mathbf{B}\mathbf{s}_2 + \mathbf{m} = \mathbf{t_B} \\ \tilde{\mathbf{s}}_1 = \mathsf{RandomizedEncode}(\mathbf{s}_1, \sigma_1) \end{array} \right. \right\}.$$

$$(1)$$

Theorem 9 (Binding). *The Rejection-Free ABDLOP commitment scheme is binding if it satisfies for $\varepsilon = \mathsf{negl}(\lambda)$, the following conditions: $\sigma_1 \geq \eta_\varepsilon(\mathcal{R}^{km_1})$, $\sigma_2 \geq \eta_\varepsilon(\mathcal{R}^{m_2})$, $km_1 d - 1 \geq \lambda$, $m_2 d - 1 \geq \lambda$ and the hardness of the assumption $\mathsf{MSIS}_{\mathcal{R},p,n,km_1+m_2,2\sqrt{\alpha_1^2+\alpha_2^2}}$ for $\alpha_1 = (b+1)\sigma_1\sqrt{km_1 d}$ and $\alpha_2 = \sigma_2\sqrt{m_2 d}$.*

Proof. Suppose having two accepted tuples $(\mathbf{s}_1, \mathbf{m}, \mathbf{s}_2)$ and $(\mathbf{s}_1', \mathbf{m}', \mathbf{s}_2')$ for a same commitment $\mathbf{t_A}, \mathbf{t_B}$. Then, we can recover $\mathbf{A}_1 \mathbf{s}_1^* + \mathbf{A}_2 \mathbf{s}_2^* := \mathbf{A}_1(\tilde{\mathbf{s}}_1 - \tilde{\mathbf{s}}_1') + \mathbf{A}_2(\mathbf{s}_2 - \mathbf{s}_2') = \mathbf{0}$. If the tuples are not equals, then we found a valid MSIS solution using Lemma 2 to bound $\|(\mathbf{s}_1^*, \mathbf{s}_2^*)\| \leq 2\|(\tilde{\mathbf{s}}_1, \mathbf{s}_2)\| \leq 2\sqrt{\|\tilde{\mathbf{s}}_1\|^2 + \|\mathbf{s}_2\|^2} \leq 2\sqrt{\alpha_1^2 + \alpha_2^2}$ as we can also bound $\|\tilde{\mathbf{s}}_1\| = \|\hat{\mathbf{s}}_1 + \mathfrak{p}_m \mathbf{e}\| \leq \|\mathfrak{p}_m\| \cdot \|\mathbf{e}\| \leq (b+1)\sigma_1\sqrt{km_1 d}$. □

Theorem 10 (Hiding). *$\mathsf{MLWE}_{\mathcal{R},p,n+\ell,m_2,\mathcal{D}_{\sigma_2}}$ implies the hiding of RF-ABDLOP.*

Proof. Hardness of MLWE implies $\begin{bmatrix} \mathbf{A}_2^\top & \mathbf{B}^\top \end{bmatrix}^\top \mathbf{s}_2 \approx_c \mathcal{U}(\mathcal{R}_p^{n+\ell})$, hence hiding. □

Security parameters will change accordingly, as the length of the $\tilde{\mathbf{s}}_1$ is k times longer $\mathbf{s}_1$. We choose parameters in order to have with overwhelming probability that $(\bar{\mathbf{s}}_1, \bar{\mathbf{s}}_2, \bar{c}) \in \overline{\mathsf{R}}_{\mathsf{RF}} \Rightarrow (\mathsf{Decode}(\bar{\mathbf{s}}_1), \mathbf{t_B} - \mathbf{B}\bar{\mathbf{s}}_2, \bar{\mathbf{s}}_2) \in \mathsf{R}_{\mathsf{RF}}$.

$$\overline{\mathsf{R}}_{\mathsf{RF}} := \left\{ \begin{array}{c} \left(\mathbf{A}_1, \mathbf{A}_2, \mathbf{B}, (\mathbf{t_A}, \mathbf{t_B}) \right) \in \mathcal{R}_p^{n \times km_1} \times \mathcal{R}_{p,\mathsf{HNF}}^{n \times m_2} \times \mathcal{R}_{p,\mathsf{HNF}}^{\ell \times m_2} \times \mathcal{R}_p^{n+\ell} \\ (\bar{\mathbf{s}}_1, \bar{\mathbf{s}}_2, \bar{c}) \in \mathcal{R}_p^{km_1} \times \mathcal{M}_{\beta_2, m_2} \times \overline{\mathcal{C}} \end{array} \middle| \begin{array}{c} \mathbf{A}_1 \bar{\mathbf{s}}_1 + \mathbf{A}_2 \bar{\mathbf{s}}_2 = \mathbf{t_A} \\ \overline{\mathbf{m}} = \mathbf{t_B} - \mathbf{B}\bar{\mathbf{s}}_2 \end{array} \right\}.$$

The proof of opening is ensured over well-chosen standard deviations of the mask $\mathbf{y}_i$. These masks follow discrete Gaussian distribution with standard deviations $\sqrt{2}\mathfrak{s}_i$. In our work, the slight modification only concerns the challenge described in Sect. 2.3, which only affects the bounds related to the lattice assumptions, resulting in the following proofs. As the transmitted $\tilde{\mathbf{s}}_i$ follows a Gaussian distribution, the zero-knowledge property is now assumed over a hard instance of Hint-MLWE, using the transmitted $\mathbf{z}_i$ as hints.

The protocol is defined as a subpart of the future framework of proof, presented in Fig. 2 for $\mathcal{F} = \varnothing$. The proof of opening follows the same construction as the one from the Ajtai commitment scheme presented in Fig. 1.

Theorem 11 (Completeness). *The proof of opening knowledge satisfies completeness if it satisfies for $\varepsilon = \mathsf{negl}(\lambda)$ the following conditions: $\sigma_1, \sqrt{2}\mathfrak{s}_1 \geq \eta_\varepsilon(\mathcal{R}^{km_1}); \sigma_2, \mathfrak{s}_2 \geq \eta_\varepsilon(\mathcal{R}^{m_2})$ for the bounds $\beta_{\mathsf{open},1} = (b+1)(\eta\sigma_1 + \sqrt{2}\mathfrak{s}_1)\sqrt{km_1 d}$ and $\beta_{\mathsf{open},2} = (\eta\sigma_2 + \mathfrak{s}_2)\sqrt{m_2 d}$.*

Proof. For the first verification step, $\|\mathbf{z}_i\| \leq \beta_{\mathsf{open},i}$ it is checked by Lemma 2 instantiated for the 4 conditions about the standard deviations. The norm of $\tilde{\mathbf{s}}_1$ is bounded the same way as in the proof of binding of the commitment scheme, for $\tilde{\mathbf{s}}_1$ and $\tilde{\mathbf{y}}_1$. The second verification step follows from the original construction. □

Theorem 12 (Knowledge Soundness). *Consider the parameters defined in Theorem 9. An extractor either produce a valid unique opening $(\bar{\mathbf{s}}_1, \bar{\mathbf{s}}_2, \bar{c}) \in \overline{\mathsf{R}}_{\mathsf{RF}}$ with non negligible probability or a $\mathsf{MSIS}_{\mathcal{R},p,n,km_1+m_2,8\eta\sqrt{\beta_{\mathsf{open},1}^2+\beta_{\mathsf{open},2}^2}}$ solution.*

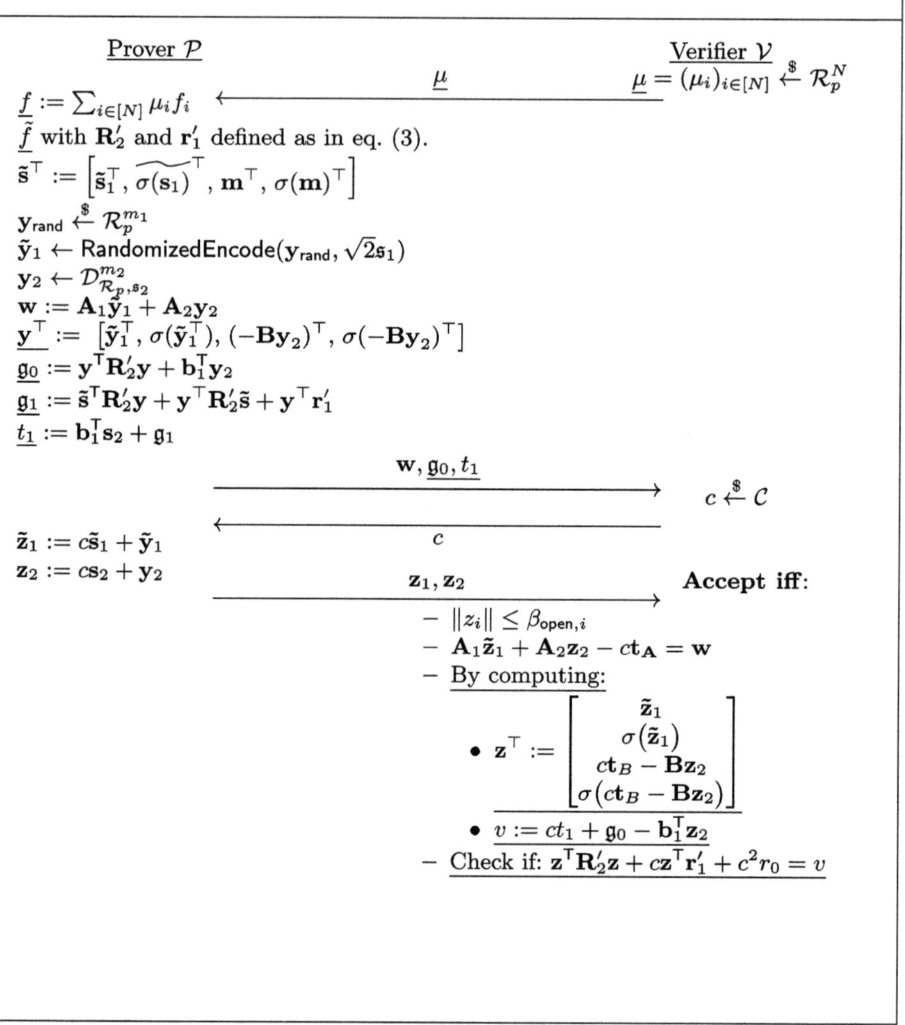

Fig. 2. Proof of knowledge of an element $(\mathbf{s}_1, \mathbf{m})$ along with the proof that for $f_i \in \mathcal{F}$ where $\mathcal{F}$ is a public set of linear and quadratic relations, we have $f_i(\psi(\mathbf{s}_1, \mathbf{m})) = 0$. The protocol can also be instantiated with $\mathcal{F} = \varnothing$ without the underlined elements of the transcript (and their construction) and the underlined verification steps.

Proof. Assuming the parameters, RF-ABDLOP is binding. Let's assume that the extractor finds two acceptable transcripts with a rewind before the challenge sampling. The two transcripts are $(\mathbf{w}, c, \tilde{\mathbf{z}}_1, \mathbf{z}_2)$ and $(\mathbf{w}, c', \tilde{\mathbf{z}}'_1, \mathbf{z}'_2)$ for a same commitment $(\mathbf{t_A}, \mathbf{t_B})$. The proof follows the one from [36, Lemma 3.1]. Define $\bar{c} := c' - c, \bar{\mathbf{z}}_i := \tilde{\mathbf{z}}'_i - \tilde{\mathbf{z}}_i$ and $\bar{\mathbf{s}}_i := \frac{\bar{\mathbf{z}}_i}{\bar{c}}$. As the transcripts were accepted, we have $\mathbf{A}_1\tilde{\mathbf{s}}_1 + \mathbf{A}_2\mathbf{s}_2 - c\mathbf{t_A} = \mathbf{A}_1\tilde{\mathbf{s}}'_1 + \mathbf{A}_2\mathbf{s}'_2 - c'\mathbf{t_A}$ that implies $\mathbf{A}_1\bar{\mathbf{s}}_1 + \mathbf{A}_2\bar{\mathbf{s}}_2 = \mathbf{t_A}$ then we will have $(\bar{\mathbf{s}}_1, \bar{\mathbf{s}}_2, \bar{c}) \in \overline{\mathsf{R}}_{\mathsf{RF}}$.

The unicity follows from [36, Lemma 3.1], take $(\bar{\mathbf{s}}_1, \bar{\mathbf{s}}_2, \bar{c}), (\bar{\mathbf{s}}'_1, \bar{\mathbf{s}}'_2, \bar{c}') \in \overline{\mathsf{R}}_{\mathsf{RF}}$ for the same public parameters $(\mathbf{A}_1, \mathbf{A}_2, \mathbf{B}, \mathbf{t_A}, \mathbf{t_B})$. Take also $\bar{\mathbf{z}}_i$ and $\bar{\mathbf{z}}'_i$ with the same construction as above. Then, we have $\mathbf{A}_1(\bar{\mathbf{s}}_1 - \bar{\mathbf{s}}'_1) + \mathbf{A}_2(\bar{\mathbf{s}}_2 - \bar{\mathbf{s}}'_2) = \mathbf{0}$, but here the MSIS is not bounded. Therefore, we proceed as in [36, Lemma 3.1], we construct another SIS solution with $\mathbf{s}^*_i := (\bar{\mathbf{s}}_i - \bar{\mathbf{s}}'_i)\bar{c} \cdot \bar{c}' = \bar{\mathbf{z}}_i\bar{c}' - \bar{\mathbf{z}}'_i\bar{c}$ where we exploit the verification step (of the norms of the responses) to prove the size of the solution: $\|(\mathbf{s}^*_1, \mathbf{s}^*_2)\| \leq 2 \cdot 2\eta \cdot 2\|(\tilde{\mathbf{z}}_1, \mathbf{z}_2)\| \leq 8\eta\sqrt{\beta^2_{\mathsf{open},1} + \beta^2_{\mathsf{open},2}}.$ $\square$

Theorem 13 (Simulatability). *The proof of opening knowledge protocol satisfies simulatability if the commitment scheme is initially hiding (assuming Theorem 10) and if for* $\varepsilon = \mathsf{negl}(\lambda)$, $\frac{1}{\varsigma^2_1} := 2\left(\frac{1}{\sigma^2_1} + \frac{\eta^2}{\mathfrak{s}^2_1}\right)$ *and* $\frac{1}{\varsigma^2_2} := 2\left(\frac{1}{\sigma^2_2} + \frac{\eta^2}{2\mathfrak{s}^2_2}\right)$ *it satisfies the following conditions:* $\sigma_1, \sqrt{2}\mathfrak{s}_1 \geq \eta_\varepsilon(\mathcal{R}^{km_1}); \mathfrak{s}_1, 2\varsigma_1 \geq \frac{\sqrt{2}}{(b-1)}$.

$$\mathcal{S}_{\underline{\mathsf{REL\text{-}AUT\text{-}}\mathcal{F}}}(\mathbf{A}_1, \mathbf{A}_2, \mathbf{B}, \underline{\mathbf{b}_1})$$

1. **Targets:** $\mathbf{t_A}, \mathbf{t_B} \xleftarrow{\$} \mathcal{R}^{n+\ell}_p$, **Challenges:** $c \xleftarrow{\$} \mathcal{C}$, $\underline{\mu = (\mu_i)_{i \in N} \xleftarrow{\$} \mathcal{R}^N_p}$;

2. $\tilde{\mathbf{s}}_1 \leftarrow \mathsf{RandomizedEncode}(\mathbf{0}, \sigma_1), \mathbf{s}_2 \leftarrow \mathcal{D}^{m_2}_{\mathcal{R}_p, \sigma_2}$;

3. $\mathbf{y}_{\mathsf{rand}} \xleftarrow{\$} \mathcal{R}^{m_1}_p, \tilde{\mathbf{y}}_1 \leftarrow \mathsf{RandomizedEncode}(\mathbf{y}_{\mathsf{rand}}, \sqrt{2}\mathfrak{s}_1), \mathbf{y}_2 \leftarrow \mathcal{D}^{m_2}_{\mathcal{R}_p, \mathfrak{s}_2}$;

4. $\tilde{\mathbf{z}}_1 = c\tilde{\mathbf{s}}_1 + \tilde{\mathbf{y}}_1, \mathbf{z}_2 = c\mathbf{s}_2 + \mathbf{y}_2$;

5. $\mathbf{w} = \mathbf{A}_1\tilde{\mathbf{z}}_1 + \mathbf{A}_2\mathbf{z}_2 - c\mathbf{t_A}$;

6. $\underline{f := \sum_{i \in [N]} \mu_i f_i}$, constructs $\tilde{f}$ with $\mathbf{R}'_2$ and $\mathbf{r}'_1$ defined as in eq. (3);

7. $\mathbf{y} := \begin{bmatrix} \tilde{\mathbf{y}}_1 \\ \sigma(\tilde{\mathbf{y}}_1) \\ -\mathbf{B}\mathbf{y}_2 \\ \sigma(-\mathbf{B}\mathbf{y}_2) \end{bmatrix}, \mathbf{z} := \begin{bmatrix} \tilde{\mathbf{z}}_1 \\ \sigma(\tilde{\mathbf{z}}_1) \\ c\mathbf{t}_B - \mathbf{B}\mathbf{z}_2 \\ \sigma(c\mathbf{t}_B - \mathbf{B}\mathbf{z}_2) \end{bmatrix}$;

8. $\underline{\mathfrak{g}_0 := \mathbf{y}^\mathsf{T}\mathbf{R}'_2\mathbf{y} + \mathbf{b}^\mathsf{T}_1\mathbf{y}_2}$;

9. $\underline{t_1 = c^{-1} \cdot \left(\mathbf{z}^\mathsf{T}\mathbf{R}'_2\mathbf{z} + c\mathbf{z}^\mathsf{T}\mathbf{r}'_1 + c^2 r_0 - \mathfrak{g}_0 + \mathbf{b}^\mathsf{T}_1\mathbf{z}_2\right)}$;

10. Output $(\mathbf{t_A}, \mathbf{t_B}, \mathbf{w}, c, \mathbf{z}_1, \mathbf{z}_2, \underline{\mu, \mathfrak{g}_0, t_1})$.

Fig. 3. Definition of $\mathcal{S}(\mathbf{A}_1, \mathbf{A}_2, \mathbf{B})$ (without underlined elements) is the simulator for RF-ABDLOP in Theorem 13, the output is defined as the set of $\mathcal{H}_6$. Definition of $\mathcal{S}_{\mathsf{REL\text{-}AUT\text{-}}\mathcal{F}}(\mathbf{A}_1, \mathbf{A}_2, \mathbf{B}, \mathbf{b}_1)$ generalizes the simulator in order to prove the simulatability in Theorem 23 for a set $\mathcal{F} = \{f_i\}_{i \in [N]}$.

$\eta_\varepsilon(\mathfrak{p}_{m_1} \mathcal{R}^{km_1})$ $; \frac{\sigma_2}{\sqrt{2}}, \mathfrak{s}_2, \varsigma_2 \geq \sqrt{2} \cdot \eta_\varepsilon(\mathcal{R}^{m_2})$ *and* $\mathsf{MLWE}_{\mathcal{R},p,n,m_2,\varsigma_2}$ *is hard (this implies* Hint-MLWE *via Theorem 7), implying indistinguishability between an output of* $\mathcal{S}$ *from Fig. 3 and an honest transcript* π *generated from* R_{RF}.

Proof. The proof essentially follows the original ones [29,31] involving Hint-MLWE The idea is to prove that the statistical distance between an output of $\mathcal{S}$ and π is negligible. Let $(\mathbf{s}_1, \mathbf{s}_2, \mathbf{m}) \in \mathsf{R}_{\mathsf{RF}}$ be the attacked message, chosen with knowledge of $\mathsf{pp} := (\mathbf{A}_1, \mathbf{A}_2, \mathbf{B})$.

4   Proving More Than Just the Knowledge

The aim of our construction is to give additional information about a committed secret message, such as proving linear or quadratic relations over the message.

The [36] framework uses the fact that the message was not modified but only hidden in a commitment scheme. As we are not sending the message unmodified but its randomized encoding version, we can no longer use the original construction. We show in this section how to modify the relations in such a way that we can prove both linear and quadratic relations over the randomized encoding that imply original relations over the initial message. Our protocol manages to extend the prove of knowledge of a committed element using the RF-ABDLOP commitment scheme, therefore, we no longer specify the rings and lengths. We refer to an instance of R_{RF} in (1) in Sect. 3.1 for the public matrices $\mathbf{A}_1, \mathbf{A}_2, \mathbf{B}$.

4.1   Transform the Relations

Relation over the Secret to One over the Encoding. A main tool of the [36] framework is the proof of both linear and quadratic relations over $\mathcal{R}_p$ of messages $\mathbf{s}_1$ and $\mathbf{m}$, which can be viewed as a matrix product over the coefficient of the matrices. We denote by $\mathcal{F}$ the full set of functions f_i to be proven w.l.o.g. equal to 0, i.e. such that $\mathbf{s}$ is in $\mathsf{Ker}\mathcal{F} := \cap_{f_i \in \mathcal{F}}\mathsf{Ker}f_i$.

A quadratic relation over the committed element can be seen as proving the equation $f(\mathbf{s}) = \mathbf{s}^\top \mathbf{R}_2 \mathbf{s} + \mathbf{r}_1^\top \mathbf{s} + r_0 = 0$ for a committed message $\mathbf{s}^\top := (\mathbf{s}_1^\top, \mathbf{m}^\top)$, a matrix $\mathbf{R}_2 \in \mathcal{R}_p^{(m_1+\ell) \times (m_1+\ell)}$ and elements $\mathbf{r}_1 \in \mathcal{R}_p^{m_1+\ell}$ and $r_0 \in \mathcal{R}_p$. The linear function restricts f to $\mathbf{R}_2$ being equal to $\mathbf{0}^{(m_1+\ell) \times (m_1+\ell)}$.

Remark 14. We can identify $\mathbf{R}_2$ as a triangular matrix (upper or lower depending on the transpose case in which we are) and, by abuse of notation, restrict the quadratic component to $\mathbf{s}^\top \mathbf{R}_2 \mathbf{s} = \mathbf{s}_1^\top \mathbf{R}_{1,1} \mathbf{s}_1 + \mathbf{s}_1^\top \mathbf{R}_{1,m} \mathbf{m} + \mathbf{m}^\top \mathbf{R}_{m,m} \mathbf{m}$ with only an upper triangular matrix $\mathbf{R}_2 := \begin{bmatrix} \mathbf{R}_{1,1} & \mathbf{R}_{1,m} \\ \mathbf{0}^{\ell \times m_1} & \mathbf{R}_{m,m} \end{bmatrix} \in \mathcal{R}_p^{(m_1+\ell) \times (m_1+\ell)}$. We extend the decomposition to the linear term $\mathbf{r}_1^\top = \begin{bmatrix} \mathbf{r}_{1,1}^\top \parallel \mathbf{r}_{1,m}^\top \end{bmatrix}$.

We define the set $\mathsf{R}_{\mathcal{F}}$ of potentially committed elements satisfying the equations $f_i(\mathbf{s}) = 0$ for each $f_i \in \mathcal{F}$ for a publicly known set of functions $\mathcal{F}$.

$$\mathsf{R}_{\mathcal{F}} := \left\{ \begin{array}{c} \left(\mathbf{A}_1, \mathbf{A}_2, \mathbf{B}, (\mathbf{t_A}, \mathbf{t_B})\right) \\ (\tilde{\mathbf{s}}_1, \mathbf{m}, \mathbf{s}_2) \end{array} \middle| \begin{array}{c} \mathbf{A}_1\tilde{\mathbf{s}}_1 + \mathbf{A}_2\mathbf{s}_2 = \mathbf{t_A} \\ \mathbf{B}\mathbf{s}_2 + \mathbf{m} = \mathbf{t_B} \\ \tilde{\mathbf{s}}_1 = \mathsf{RandomizedEncode}(\mathbf{s}_1, \sigma_1) \\ (\mathbf{s}_1, \mathbf{m}) \in \cap_{f_i \in \mathcal{F}}\mathsf{Ker}f_i \end{array} \right\}.$$

In our case, the RF-Ajtai part does not involve $\mathbf{s}_1$ but $\tilde{\mathbf{s}}_1$, therefore we can not execute the original construction in [36]. Our new procedure consists of performing the decoding protocol during the execution of the function. For a function f_i in $\mathcal{F}$, we construct a new $\tilde{f}_i$ and prove it holds for $\tilde{f}_i$ instead.

Theorem 15. *Let $\mathcal{F}$ a set of linear and quadratic functions over $(\mathbf{s}_1, \mathbf{m})$. Then, for every function $f_i \in \mathcal{F}$, the associated $\tilde{f}_i := f_i \circ \left[\mathsf{Decode}(\cdot), \mathsf{id}(\cdot)\right]$ satisfies:*

1. *$\tilde{f}_i$ acts over $\tilde{\mathbf{s}} := (\tilde{\mathbf{s}}_1, \mathbf{m})$, is computable and preserves the f_i structure,*
2. *if $(\tilde{\mathbf{s}}_1, \mathbf{m}) \in \mathsf{Ker}\tilde{f}_i$ then $(\mathbf{s}_1, \mathbf{m}) \in \mathsf{Ker}f_i$,*
3. *if $(\tilde{\mathbf{s}}_1, \mathbf{m}, \mathbf{s_2}) \in \mathsf{R}_{RF\text{-}\mathcal{F}}$ then $(\mathbf{s}_1, \mathbf{m}, \mathbf{s_2}) \in \mathsf{R}_{\mathcal{F}}$.*

$$\mathsf{R}_{RF\text{-}\mathcal{F}} := \left\{ \begin{array}{c} \left(\mathbf{A}_1, \mathbf{A}_2, \mathbf{B}, (\mathbf{t_A}, \mathbf{t_B})\right) \\ (\tilde{\mathbf{s}}_1, \mathbf{m}, \mathbf{s}_2) \end{array} \middle| \begin{array}{c} \mathbf{A}_1\tilde{\mathbf{s}}_1 + \mathbf{A}_2\mathbf{s}_2 = \mathbf{t_A} \\ \mathbf{B}\mathbf{s}_2 + \mathbf{m} = \mathbf{t_B} \\ (\tilde{\mathbf{s}}_1, \mathbf{m}) \in \cap_{f_i \in \mathcal{F}}\mathsf{Ker}\tilde{f}_i \end{array} \right\}.$$

By "preserving the structure", we mean that the function $\tilde{f}$ is still linear resp. quadratic just like f.

The following lemma proves the stability of the structure of the quadratic component performing over a (possibly randomized) encoding.

Lemma 16. *Let m, n, u, b, k positive integers. Let $p := b^k + 1$. Let $\mathbf{A} \in \mathbb{Z}^{m \times n}$, $\mathbf{x}_1 \in \mathbb{Z}_p^m$ and $\mathbf{x}_2 \in \mathbb{Z}_p^n$. Then, we have for both $\overline{\mathbf{x}}_1 \in \{\hat{\mathbf{x}}_1, \tilde{\mathbf{x}}_1\}$ and $\overline{\mathbf{x}}_2 \in \{\hat{\mathbf{x}}_2, \tilde{\mathbf{x}}_2\}$:*

$$\mathbf{x}_1^\top \mathbf{A}\mathbf{x}_2 = \overline{\mathbf{x}}_1^\top \left(\mathbf{A} \otimes \mathbf{g}_{b,k}\right)\mathbf{x}_2, \quad \mathbf{x}_1^\top \mathbf{A}\mathbf{x}_2 = \mathbf{x}_1^\top \left(\mathbf{A} \otimes \mathbf{g}_{b,k}^\top\right)\overline{\mathbf{x}}_2$$
$$\text{and } \mathbf{x}_1^\top \mathbf{A}\mathbf{x}_2 = \overline{\mathbf{x}}_1^\top \left(\mathbf{A} \otimes (\mathbf{g}_{b,k}^\top \otimes \mathbf{g}_{b,k})\right)\overline{\mathbf{x}}_2.$$

Proof. First, we have $\mathbf{G}\hat{\mathbf{t}} = \mathbf{G}\tilde{\mathbf{t}}$ because $\tilde{\mathbf{t}} = \hat{\mathbf{t}} \mod \mathfrak{p}$. The second equality comes from the linear transformation and the associativity of the matrix product. For the first one, the equality comes from $\mathbf{x}_1^\top \mathbf{A}\mathbf{x}_2 = (\mathbf{A}^\top \mathbf{x}_1)^\top \cdot \mathbf{x}_2 = \left((\mathbf{A}^\top \otimes \mathbf{g}_{b,k}^\top) \cdot \overline{\mathbf{x}}_1\right)^\top \cdot \mathbf{x}_2 = \overline{\mathbf{x}}_1^\top \cdot \left(\mathbf{A} \otimes \mathbf{g}_{b,k}\right) \cdot \mathbf{x}_2$. The generalized equality is due to the two previous equalities and the associativity of the Kronecker product. $\square$

Proof (of Theorem 15). We have $\tilde{f}_i(\tilde{\mathbf{s}}_1, \mathbf{m}) = f_i(\mathbf{G}_{b,k,dm_1}\tilde{\mathbf{s}}_1, \mathbf{m}) = f_i(\mathbf{s}_1, \mathbf{m}) = 0$, implying the second point and the computability. The third point is clearly implied by the second point. For the initial point, we mean that if f is linear (resp. quadratic), then $\tilde{f}$ is still linear (resp. quadratic). As identifying f_i as a quadratic construction showed in Remark 14, we construct $\tilde{f}(\tilde{\mathbf{s}}) = \tilde{\mathbf{s}}^\top \mathbf{R}_2' \tilde{\mathbf{s}} + \mathbf{r}_1'^\top \tilde{\mathbf{s}} + r_0$ with:

$$\mathbf{r}_1'^\top = \begin{bmatrix} \mathbf{r}_{1,1}^\top \otimes \mathbf{g}_{b,k}^\top \parallel \mathbf{r}_{1,m}^\top \end{bmatrix} \text{ and } \mathbf{R}_2' := \begin{bmatrix} \mathbf{R}_{1,1} \otimes (\mathbf{g}_{b,k}^\top \otimes \mathbf{g}_{b,k}) & \mathbf{R}_{1,m} \otimes \mathbf{g}_{b,k} \\ \mathbf{0}^{\ell \times km_1} & \mathbf{R}_{m,m} \end{bmatrix},$$

that maintains a quadratic structure over $\tilde{\mathbf{s}}$. For the linear structure, we study the quadratic component; if $\mathbf{R}_2 = \mathbf{0}^{(m_1+\ell) \times (m_1+\ell)}$, then by construction, we have $\mathbf{R}_2' = \mathbf{0}^{(km_1+\ell) \times (km_1+\ell)}$ that maintains the linear structure over $\tilde{\mathbf{s}}$. $\square$

4.2   Include the Automorphism σ

Our final goal is to prove quadratic relations over the messages committed and their automorphism $\sigma : X \mapsto X^{-1}$ over $\mathcal{R}_p$. We define the full message involved as $\mathbf{s}^\top := \left(\mathbf{s}_1^\top, \sigma(\mathbf{s}_1)^\top, \mathbf{m}^\top, \sigma(\mathbf{m})^\top\right) \in \mathcal{R}_p^{2(m_1+\ell)}$. Therefore, together with the remark 14 and the explanation above, the relation can be seen as proving the equation $f(\mathbf{s}) = \mathbf{s}^\top \mathbf{R}_2 \mathbf{s} + \mathbf{s}^\top \mathbf{r}_1 + r_0 = 0$ with the (public) matrices:

$$\mathbf{R}_2 = \begin{bmatrix} \mathbf{R}_{1,1} & \mathbf{R}_{1,\sigma(1)} & \mathbf{R}_{1,m} & \mathbf{R}_{1,\sigma(m)} \\ \mathbf{0}^{m_1 \times m_1} & \mathbf{R}_{\sigma(1),\sigma(1)} & \mathbf{R}_{\sigma(1),m} & \mathbf{R}_{\sigma(1),\sigma(m)} \\ \mathbf{0}^{\ell \times m_1} & \mathbf{0}^{\ell \times m_1} & \mathbf{R}_{m,m} & \mathbf{R}_{m,\sigma(m)} \\ \mathbf{0}^{\ell \times m_1} & \mathbf{0}^{\ell \times m_1} & \mathbf{0}^{\ell \times \ell} & \mathbf{R}_{\sigma(m),\sigma(m)} \end{bmatrix} \in \mathcal{R}_p^{2(m_1+\ell) \times 2(m_1+\ell)}, \quad (2)$$

$$\mathbf{r}_1^\top = \begin{bmatrix} \mathbf{r}_{1,1}^\top & \mathbf{r}_{1,\sigma(1)}^\top & \mathbf{r}_{1,m}^\top & \mathbf{r}_{1,\sigma(m)}^\top \end{bmatrix} \in \mathcal{R}_p^{2(m_1+\ell)}. \quad (3)$$

Remark 17 ([36, Section 4]). A naive idea is to commit both the message and its automorphism. To be more efficient, the action of σ can be computed from $(\mathbf{s}_1, \mathbf{m})$ and its commitment. It is an automorphism of $\mathcal{R}_p$, so when defining σ over an element $\mathcal{R}_p^k$, we compute it coordinate-wise. We show that it is possible to compute the automorphism of an element based on its (randomized) encoding.

Lemma 18. *Let $\sigma \in \mathsf{Aut}(\mathcal{R}_p)$ and $\mathbf{x} \in \mathcal{R}_p^{m_1}$ a secret, and $\overline{\mathbf{x}} \in \{\hat{\mathbf{x}}, \tilde{\mathbf{x}}\}$ seen as an element of $\mathcal{R}_p^{km_1}$ its encoding (resp. randomized encoding) of $\mathbf{x}$. Then, we have for each $\overline{\mathbf{x}} \in \{\hat{\mathbf{x}}, \tilde{\mathbf{x}}\}: \overline{\sigma(\mathbf{x})} = \sigma(\overline{\mathbf{x}})$ decoded in a unique $\sigma(\mathbf{x})$.*

The proof depends precisely on the choice of the Encode algorithm described in Sect. 3 and is straightforward.

Theorem 19. *Let $\mathcal{G}$ a set of linear and quadratic functions over $\mathbf{s}^\top$. Define the map $\psi : (\mathbf{x}_1, \mathbf{x}_2) \longmapsto (\mathbf{x}_1, \sigma(\mathbf{x}_1), \mathbf{x}_2, \sigma(\mathbf{x}_2))$. Then, for every function $g_i \in \mathcal{G}$, the associated function $\tilde{g}_i := g_i \circ \left[\mathsf{Decode}(\cdot), \mathsf{Decode}(\cdot), \mathsf{id}(\cdot), \mathsf{id}(\cdot)\right]$ satisfies:*

1. *$\tilde{g}_i$ acts over $\tilde{\mathbf{s}} := \psi(\tilde{\mathbf{s}}_1, \mathbf{m})$, is computable and preserves the g_i structure,*
2. *if $\psi(\tilde{\mathbf{s}}_1, \mathbf{m}) \in \mathsf{Ker}\tilde{g}_i$ then $\psi(\mathbf{s}_1, \mathbf{m}) \in \mathsf{Ker}g_i$,*
3. *if $(\tilde{\mathbf{s}}_1, \mathbf{m}, \mathbf{s}_2) \in \mathsf{R}_{RF\text{-}AUT\text{-}\mathcal{G}}$ then $\left((\mathbf{s}_1^\top \| \sigma(\mathbf{s}_1)^\top), (\mathbf{m}^\top \| \sigma(\mathbf{m})^\top), \mathbf{s}_2^\top\right)^\top \in \mathsf{R}_{\mathcal{G}}$.*

$$R_{RF\text{-}AUT\text{-}\mathcal{G}} := \left\{ \begin{array}{c} \left(\mathbf{A}_1, \mathbf{A}_2, \mathbf{B}, (\mathbf{t_A}, \mathbf{t_B})\right) \\ (\tilde{\mathbf{s}}_1, \mathbf{m}, \mathbf{s}_2) \end{array} \middle| \begin{array}{c} \mathbf{A}_1\tilde{\mathbf{s}}_1 + \mathbf{A}_2\mathbf{s}_2 = \mathbf{t_A} \\ \mathbf{B}\mathbf{s}_2 + \mathbf{m} = \mathbf{t_B} \\ \psi(\tilde{\mathbf{s}}_1, \mathbf{m}) \in \cap_{g_i \in \mathcal{G}}\mathsf{Ker}\tilde{g}_i \end{array} \right\}.$$

Proof. For 2 and 3, as $\tilde{g}_i(\psi(\tilde{\mathbf{s}}_1, \mathbf{m})) = g_i\big(\mathbf{G}_{b,k,dm_1}\tilde{\mathbf{s}}_1, \mathbf{G}_{b,k,dm_1}\sigma(\tilde{\mathbf{s}}_1), \mathbf{m}, \sigma(\mathbf{m})\big)$, we have $\tilde{g}_i(\psi(\tilde{\mathbf{s}}_1, \mathbf{m})) = g_i(\mathbf{s}_1, \sigma(\mathbf{s}_1), \mathbf{m}, \sigma(\mathbf{m})) = g_i(\psi(\mathbf{s}_1, \mathbf{m})) = 0$. We keep the same analysis done in Theorem 15 as the decoding and the identity are done coordinate-wise leading to a function with input $\left((\mathbf{s}_1^\top \| \sigma(\mathbf{s}_1)^\top), (\mathbf{m}^\top \| \sigma(\mathbf{m})^\top), \mathbf{s}_2^\top\right)^\top$. We can now see $\tilde{g}_i$ as $\tilde{g}_i(\tilde{\mathbf{s}}) = \tilde{\mathbf{s}}^\top \mathbf{R}_2' \tilde{\mathbf{s}} + \mathbf{r}_1'^\top \tilde{\mathbf{s}} + r_0$ with $\mathbf{R}_2' \in \mathcal{R}_p^{2(km_1+\ell) \times 2(km_1+\ell)}$ such that:

$$\mathbf{R}_2' := \begin{bmatrix} \mathbf{R}_{1,1} \otimes (\mathbf{g}_{b,k}^\top \otimes \mathbf{g}_{b,k}) & \mathbf{R}_{1,\sigma(1)} \otimes (\mathbf{g}_{b,k}^\top \otimes \mathbf{g}_{b,k}) & \mathbf{R}_{1,\mathbf{m}} \otimes \mathbf{g}_{b,k} & \mathbf{R}_{1,\sigma(\mathbf{m})} \otimes \mathbf{g}_{b,k} \\ \mathbf{0}^{km_1 \times km_1} & \mathbf{R}_{\sigma(1),\sigma(1)} \otimes (\mathbf{g}_{b,k}^\top \otimes \mathbf{g}_{b,k}) & \mathbf{R}_{\sigma(1),\mathbf{m}} \otimes \mathbf{g}_{b,k} & \mathbf{R}_{\sigma(1),\sigma(\mathbf{m})} \otimes \mathbf{g}_{b,k} \\ \mathbf{0}^{\ell \times km_1} & \mathbf{0}^{\ell \times km_1} & \mathbf{R}_{\mathbf{m},\mathbf{m}} & \mathbf{R}_{\mathbf{m},\sigma(\mathbf{m})} \\ \mathbf{0}^{\ell \times km_1} & \mathbf{0}^{\ell \times km_1} & \mathbf{0}^{\ell \times \ell} & \mathbf{R}_{\sigma(\mathbf{m}),\sigma(\mathbf{m})} \end{bmatrix},$$

$$\text{and } \mathbf{r}_1'^\top := \begin{bmatrix} \mathbf{r}_{1,1}^\top \otimes \mathbf{g}_{b,k}^\top & \mathbf{r}_{1,\sigma(1)}^\top \otimes \mathbf{g}_{b,k}^\top & \mathbf{r}_{1,\mathbf{m}}^\top & \mathbf{r}_{1,\sigma(\mathbf{m})}^\top \end{bmatrix} \in \mathcal{R}_p^{2(km_1+\ell)}. \tag{4}$$

Regarding the computability, it exploits Lemma 18: the (randomized) encoding, the (unique) decoding, and the automorphism σ_{-1} are commutative. $\quad\square$

In order to prove the security of the opening process, we define the associated opening relation. In particular, we choose parameters such that if $(\bar{\mathbf{s}}_1, \bar{\mathbf{s}}_2, \bar{c}) \in \overline{R}_{RF\text{-}AUT\text{-}\mathcal{G}}$ then $(\bar{\mathbf{s}}_1, \mathbf{t_B} - \mathbf{B}\bar{\mathbf{s}}_2, \bar{\mathbf{s}}_2) \in R_{RF\text{-}AUT\text{-}\mathcal{G}}$ with overwhelming probability for:

$$\overline{R}_{RF\text{-}AUT\text{-}\mathcal{G}} := \left\{ \begin{array}{c} \left(\mathbf{A}_1, \mathbf{A}_2, \mathbf{B}, (\mathbf{t_A}, \mathbf{t_B})\right) \\ (\bar{\mathbf{s}}_1, \bar{\mathbf{s}}_2, \bar{c}) \end{array} \middle| \begin{array}{c} \mathbf{A}_1\bar{\mathbf{s}}_1 + \mathbf{A}_2\bar{\mathbf{s}}_2 = \mathbf{t_A} \\ \overline{\mathbf{m}} = \mathbf{t_B} - \mathbf{B}\bar{\mathbf{s}}_2 \\ \psi(\bar{\mathbf{s}}_1, \overline{\mathbf{m}}) \in \cap_{g_i \in \mathcal{G}}\mathsf{Ker}\tilde{g}_i \end{array} \right\}.$$

4.3 Proving the Relations

One Relation Holds. We show in Theorem 19 that we manage to preserve the structure of the original relations f when constructing $\tilde{f}$. Therefore, the idea behind the preservation was to keep using the same construction from [36, Section 4.1]. Now that we can compute the automorphism over the encoding by Lemma 18, we define $\tilde{\mathbf{s}} := \psi(\tilde{\mathbf{s}}_1, \mathbf{m})$ and $\mathbf{y} := \psi(\tilde{\mathbf{y}}_1, -\mathbf{B}\mathbf{y}_2)$ both in $\mathcal{R}_p^{2(km_1+\ell)}$. We also construct the response $\mathbf{z}_m$ (computable by the verifier) as in [36] by defining $\mathbf{z}_m := c\mathbf{t_B} - \mathbf{B}\mathbf{z}_2 = c \cdot \mathbf{m} - \mathbf{B}\mathbf{y}_2$. We have:

$$\mathbf{z} := \psi(\mathbf{z}_1, \mathbf{z}_m) = c \begin{bmatrix} \tilde{\mathbf{s}}_1 \\ \sigma(\tilde{\mathbf{s}}_1) \\ \mathbf{m} \\ \sigma(\mathbf{m}) \end{bmatrix} + \begin{bmatrix} \tilde{\mathbf{y}}_1 \\ \sigma(\tilde{\mathbf{y}}_1) \\ -\mathbf{B}\mathbf{y}_2 \\ \sigma(-\mathbf{B}\mathbf{y}_2) \end{bmatrix} = c \begin{bmatrix} \tilde{\mathbf{s}}_1 \\ \widetilde{\sigma(\mathbf{s}_1)} \\ \mathbf{m} \\ \sigma(\mathbf{m}) \end{bmatrix} + \begin{bmatrix} \tilde{\mathbf{y}}_1 \\ \sigma(\tilde{\mathbf{y}}_1) \\ -\mathbf{B}\mathbf{y}_2 \\ \sigma(-\mathbf{B}\mathbf{y}_2) \end{bmatrix} = c\tilde{\mathbf{s}} + \mathbf{y}.$$

The idea of [36, Section 4.1] is to prove the vanishment of a coefficient to reduce the problem to a linear relation. The quadratic relation $\tilde{f}$ is now modified in order to have $\mathbf{z}^\top \mathbf{R}_2' \mathbf{z} + c\mathbf{z}^\top \mathbf{r}_1' + c^2 r_0 - c \cdot \mathfrak{g}_1 - \mathfrak{g}_0 = c^2 \tilde{f}(\mathbf{s})$, with two garbage polynomials $\mathfrak{g}_0 := \mathbf{y}^\top \mathbf{R}_2' \mathbf{y} \in \mathcal{R}_p$ and $\mathfrak{g}_1 := \tilde{\mathbf{s}}^\top \mathbf{R}_2' \mathbf{y} + \mathbf{y}^\top \mathbf{R}_2' \tilde{\mathbf{s}} + \mathbf{y}^\top \mathbf{r}_1' \in \mathcal{R}_p$. The prover finally commits the element $\mathfrak{g}_1$ by computing $t_1 = \mathbf{b}_1^\top \mathbf{s}_2 + \mathfrak{g}_1$ (in the BDLOP part) and transmits directly $\mathfrak{g}_0$ as it is not linked with any secret but only the randomness $\mathbf{s}_2$ freshly sampled in each execution of the protocol. The verifier checks if $\mathbf{z}^\top \mathbf{R}_2' \mathbf{z} + c\mathbf{z}^\top \mathbf{r}_1' + c^2 r_0 \overset{?}{=} c t_1 + \mathfrak{g}_0 - \mathbf{b}_1^\top \mathbf{z}_2$ implying $\tilde{f}(\mathbf{s}) = 0$.

Prove Many Relations. We simply generalize to the proof of $N = |\mathcal{F}|$ relations using the exact same construction as in [36, Section 4.2].

Lemma 20 (Induced by [36, Lemma 4.3]). *Consider a set of relations* $\mathcal{F} = \{f_i\}_{i \in [N]}$. *Define* $f := \sum_{i \in [N]} \mu_i f_i$ *with* $\mu = (\mu_i)_{i \in [N]} \overset{\$}{\leftarrow} \mathcal{R}_p^N$. *Then,*

$$\mathbb{P}\left[(\bar{\mathbf{s}}_1, \bar{\mathbf{s}}_2, \bar{c}) \notin \overline{\mathsf{R}}_{(RF\text{-}AUT\text{-})\mathcal{F}} \mid (\bar{\mathbf{s}}_1, \bar{\mathbf{s}}_2, \bar{c}) \in \overline{\mathsf{R}}_{(RF\text{-}AUT\text{-})\{f\}} \right] \leq p^{-\frac{d}{2}}.$$

We give the protocol in Fig. 2 with the following theorems ensuring security.

Theorem 21 (Completeness). *The protocol instantiated in Fig. 2 satisfies completeness if it satisfies conditions/constructions from Theorems 11 and 19.*

Proof. The proof of completeness follows from the explanations in Sect. 4 as the elements transmitted allows the right construction in order to prove $f(\mathbf{s}) = 0$ for each function in $\mathcal{F}$. The last thing to check is that the commitment scheme used is relevant and ensure completeness: we simply launch parameters and conditions from Theorems 11 and 19 . $\qquad\square$

Theorem 22 (Knowledge Soundness).
There exists an extractor of the protocol in Fig. 2 that either produces an opening $(\bar{\mathbf{s}}_1, \bar{\mathbf{s}}_2, \bar{c}) \in \overline{\mathsf{R}}_{RF\text{-}AUT\text{-}\mathcal{F}}$ with non-negligible probability or solves an instance of a $MSIS_{\mathcal{R},p,n,km_1+m_2,8\eta\sqrt{\beta_{open,1}^2+\beta_{open,2}^2}}$.

Proof. Let's assume that the extractor finds two accepted transcripts for a same commitment $(\mathbf{t}_A, \mathbf{t}_B)$ with a rewind before the sampling of the challenge $(\mu, \mathbf{w}, \mathfrak{g}_0, t_1, c, \tilde{\mathbf{z}}_1, \mathbf{z}_2)$ and $(\mu', \mathbf{w}, \mathfrak{g}_0, t_1, c', \tilde{\mathbf{z}}_1', \mathbf{z}_2')$. Regarding the elements involved in the proof of opening knowledge, the proof has already been shown in Theorem 12. The final part involves proving that the process of proving relationships cannot be compromised. Consider μ not involved as we can reduce the multiple relation to a single one by launching Lemma 20 by setting $p^{\frac{d}{2}}$ being negligible. The proof follows essentially the one from [36, Lemma 4.2] as it has the same structure, taking into account 3 rewinds in order to prove $(\bar{\mathbf{s}}_1, \bar{\mathbf{s}}_2, \bar{c}) \in \mathsf{R}_{RF\text{-}AUT\text{-}\{f\}}$. $\qquad\square$

Theorem 23 (Simulatability). *For the simulatability of the Figure 2, we consider the parameters from Theorems 11,12 and 13 and the simulator in Fig. 3 ensuring indistinguishability between an honest transcript π generated from elements of $\mathsf{R}_{RF\text{-}AUT\text{-}\mathcal{F}}$ and an output of the simulator $\mathcal{S}_{REL\text{-}AUT\text{-}\mathcal{F}}$.*

Proof. We immediately use Theorem 13 which defines the simulatability of RF-ABDLOP based on Hint-MLWE for the response $\mathbf{z}_2$. The right construction of the additional $(\mu, \mathfrak{g}_0, t_1)$ follows from the correctness and the fact that $\overline{\mathcal{C}} \subseteq \mathcal{R}_q^\times$. $\square$

5 Instantiations of the Framework

We provide a library in C that uses and extends the actual LaZer library [38], with a set of new rejection-free functions. We provide a concrete analysis of the running time of our code with the existing [36] framework launched with the [38] implementation. We obtain the results shown in Table 1 made with a compilation of our code on a laptop with Intel Core Ultra 7 165H, using gcc version 11.4.0 with the options -O3 -march=native, -mtune=native and -pthread to ensure optimality of the results. We recall the list of conditions our framework needs to follow to ensure security and compare the running time and the size of the transcript of our framework with the original ones of [36].

5.1 Overview of All the Conditions

First, the framework can be instantiated for a particular prime p s.t. $p = 5$ mod 8 in order to use Lemma 8 but since $p = b^k + 1$, this can only be done for $k = 2$. Second, the false positive probability in Lemma 20 must be negligible. Third, we need to make sure that our standard deviations respect the initial conditions mentioned in the security Theorems. By taking $\varepsilon = 2^{-\lambda}$ with the security parameter λ and as $\lambda_{kmd}(\mathbb{Q}^m) = 1$ and $\lambda_{kmd}(\mathfrak{p}_m\mathbb{Q}^m) = \sqrt{b^2 + 1} = \sqrt{p}$ for any m, we bound the smoothing parameters using [40, Lemma 3.3]. Overall, the parameters of the protocol depend on the size m_1, the degree d, the MSIS dimension n, and the size m_2 which is adjusted to achieve the MLWE security. As our security is based on MSIS and MLWE (through Hint-MLWE), we exploit the LWE-Estimator [5] to compute m_2 and n, ensuring the hardness of:

- $\mathrm{MSIS}_{\mathcal{R},p,n,km_1+m_2,8\eta\sqrt{(\eta(b+1)\sigma_1+\sqrt{2}\mathfrak{s}_1)^2(km_1d)+(\eta\sigma_2+\mathfrak{s}_2)^2(m_2d)}}$ (Know. Sound.),
- $\mathrm{MSIS}_{\mathcal{R},p,n,km_1+m_2,2\sqrt{(b+1)^2\sigma_1^2m_1kd+\sigma_2^2m_2d}}$ (Binding),
- $\mathrm{MLWE}_{\mathcal{R},p,n+\ell,m_2,\sigma_2}$ (Hiding) and $\mathrm{MLWE}_{\mathcal{R},p,n+\ell,m_2,\mathfrak{s}_2}$ (Simulatability).

5.2 Sizes and Running Time of the Framework

Size of an Instance of Our Framework. The optimisations are the same as in [36]. We use the Dilithium-based compression from [36, Appendix A.2], with variables D and γ cutting low-order bits for the transmitted elements $\mathbf{t_A}$ and $\mathbf{w}$. We use Huffman coding in order to optimise the size regarding the distribution of the responses $\mathbf{z}_i$. As the set of relations is public and unaffecting the extractor, the challenge μ can be the output of the hash-function of the previous transmissions. Unlike the challenge c, that is transmitted and quantified with the bounds η and κ defining $\mathcal{C}$ in Sect. 2.3. The rest of the size depends on the ν BDLOP commits used dynamically. Finally, a transcript π has length:

$$|\pi| := nd(\lceil \log p \rceil + 2.25 - D) + \nu d \cdot \lceil \log p \rceil + d \cdot \lceil \log(2\kappa + 1) \rceil$$
$$+ km_1 d \cdot \left(2.57 + \lceil \log(\eta(b+1)\sigma_1 + \sqrt{2}\mathfrak{s}_1) \rceil\right) + m_2 d \cdot \left(2.57 + \lceil \log(\eta\sigma_2 + \mathfrak{s}_2) \rceil\right).$$

Complexity of the Proving Phase. We study the complexity of the commitment. The prover needs to construct the elements $\tilde{\mathbf{s}}_1$ by encoding it, samples an error $\mathbf{e}$, and adds it. This is done in $\mathcal{O}(n(km_1 + m_2))$ operations over $\mathcal{R}_p$ considering optimal methods of sampling from spherical discrete Gaussian from [29, Section 3.2]. The proof is constructed with $\mathcal{O}(n(km_1 + m_2) + k^2 m_1^2)$ operations over $\mathcal{R}_p$ because the computations are bounded by $\mathbf{w}$ or $\mathfrak{g}_i$. If we consider that the multiplication in $\mathcal{O}(d \log d)$ is the heavier operation in $\mathcal{R}$, then the prover complexity is $\mathcal{O}\left(d \log d(n(km_1 + m_2) + k^2 m_1^2)\right)$ only enhancing the running time of a [36] instance by a factor at most k^2. But, we are restricted to the case $k = 2$. Then by denoting $T_{\mathcal{P},\mathsf{LNP}}$ the initial time of a single instance of the proving phase (without rejection), we can bound the proving phase of our protocol: $T_{\mathcal{P},\mathsf{new}} \leq 4T_{\mathcal{P},\mathsf{LNP}}$. Then, our framework executes in constant time and has a faster proving phase if the rejection rate is higher than 3.

Complexity of the Verifying Phase. With the same analysis, the verifier complexity is $\mathcal{O}\left(d \log d(n(km_1 + m_2) + k^2 m_1^2)\right)$, majored by the checks regarding $\mathbf{w}$ or v. We bound the running time of the prover by $T_{\mathcal{V},\mathsf{new}} \leq 4T_{\mathcal{V},\mathsf{LNP}}$.

Table 2. Sets of parameters involved in the proof frameworks. The first section of parameters are manually set, others are either computed using /scripts of [color = 0 0 1]rejection-free-framework-under-Hint-MLWE.zipour code inspired by the LaZer Library or using LWE-Estimator [5].

Sets of parameters						
Variable	Description	Set 1	Set 2	Set 3	Set 4	Set 5
λ	Security parameter	128				
d	Ring degree	64				
$\log p$	Log of the ring moduli	60	120	80	80	60
$m_{1\mathsf{LNP}}$	Number of low-norm messages	10	10	100	0	100
ℓ_{LNP}	Number of unbounded messages	2	2	0	20	20
$m_{1\mathsf{new}}$	Number of messages $= m_{1\mathsf{LNP}} + \ell_{\mathsf{LNP}}$	12	12	100	20	120
N	Number of quadratic equations to prove	3				
Security parameters						
κ	$\|c\|_\infty \leq \kappa$ with $c \in \mathcal{C}$	8				
η	$\|c\mathbf{r}\| \leq \eta\|\mathbf{r}\|$ with $c \in \mathcal{C}$ and $\mathbf{r}$ a vector of $\mathcal{R}$	140				
n	Height of $\mathbf{A}_1$ and $\mathbf{A}_2$	32	38	35	34	34
m_2	Length of the randomness $\mathbf{s}_2$	71	114	86	85	73

We manage to show a comparison of the size of the transcripts and the evolution of running time with [36] in the Appendix 2 for 5 parameter sets defined in the Fig. 2.

Acknowledgement. This work is supported by the PEPR quantique France 2030 program (ANR-22-PETQ-0008 PQ-TLS) and by the ASTRID program under the national project AMIRAL with reference ANR-21-ASTR-0016. Antoine Douteau is partially funded by the region of Normandy, France.

References

1. Agrawal, S., Kirshanova, E., Stehlé, D., Yadav, A.: Practical, round-optimal lattice-based blind signatures. In: Yin, H., Stavrou, A., Cremers, C., Shi, E. (eds.) ACM CCS 2022, pp. 39–53. ACM Press (2022). https://doi.org/10.1145/3548606.3560650

2. Aguilar-Melchor, C., et al.: HQC. Technical report, National Institute of Standards and Technology (2022). https://csrc.nist.gov/Projects/post-quantum-cryptography/round-4-submissions

3. Ajtai, M.: Generating hard instances of lattice problems (extended abstract). In: 28th ACM STOC, pp. 99–108. ACM Press (1996). https://doi.org/10.1145/237814.237838

4. Ajtai, M.: The shortest vector problem in L2 is NP-hard for randomized reductions (extended abstract). In: 30th ACM STOC, pp. 10–19. ACM Press (1998). https://doi.org/10.1145/276698.276705

5. Albrecht, M.R., Player, R., Scott, S.: On the concrete hardness of learning with errors. In: Journal of Mathematical Cryptology, vol. 9 - 3, pp. 169–203 (2015). https://doi.org/10.1515/jmc-2015-0016

6. Aragon, N., et al.: BIKE. Technical report, National Institute of Standards and Technology (2022). https://csrc.nist.gov/Projects/post-quantum-cryptography/round-4-submissions

7. Barbosa, M., et al.: Fixing and mechanizing the security proof of Fiat-Shamir with aborts and Dilithium. In: Handschuh, H., Lysyanskaya, A. (eds.) CRYPTO 2023, Part V. LNCS, vol. 14085, pp. 358–389. Springer, Cham (2023). https://doi.org/10.1007/978-3-031-38554-4_12

8. Barthe, G., et al.: Masking the GLP lattice-based signature scheme at any order. In: Nielsen, J.B., Rijmen, V. (eds.) EUROCRYPT 2018, Part II. LNCS, vol. 10821, pp. 354–384. Springer, Cham (2018). https://doi.org/10.1007/978-3-319-78375-8_12

9. Baum, C., Damgård, I., Lyubashevsky, V., Oechsner, S., Peikert, C.: More efficient commitments from structured lattice assumptions. In: Catalano, D., De Prisco, R. (eds.) SCN 18. LNCS, vol. 11035, pp. 368–385. Springer, Cham (2018). https://doi.org/10.1007/978-3-319-98113-0_20

10. Behnia, R., Chen, Y., Masny, D.: On removing rejection conditions in practical lattice-based signatures. In: Cheon, J.H., Tillich, J.P. (eds.) Post-Quantum Cryptography - 12th International Workshop, PQCrypto 2021, pp. 380–398. Springer, Cham (2021). https://doi.org/10.1007/978-3-030-81293-5_20

11. Beullens, W., Dobson, S., Katsumata, S., Lai, Y.F., Pintore, F.: Group signatures and more from isogenies and lattices: generic, simple, and efficient. DCC **91**(6), 2141–2200 (2023). https://doi.org/10.1007/s10623-023-01192-x

12. Beullens, W., Katsumata, S., Pintore, F.: Calamari and Falafl: logarithmic (linkable) ring signatures from isogenies and lattices. In: Moriai, S., Wang, H. (eds.) ASIACRYPT 2020, Part II. LNCS, vol. 12492, pp. 464–492. Springer, Cham (2020). https://doi.org/10.1007/978-3-030-64834-3_16

13. Bootle, J., Lyubashevsky, V., Nguyen, N.K., Sorniotti, A.: A framework for practical anonymous credentials from lattices. In: Handschuh, H., Lysyanskaya, A. (eds.) CRYPTO 2023, Part II. LNCS, vol. 14082, pp. 384–417. Springer, Cham (2023). https://doi.org/10.1007/978-3-031-38545-2_13

14. Bos, J.W., et al.: CRYSTALS - kyber: a CCA-secure module-lattice-based KEM. In: 2018 IEEE European Symposium on Security and Privacy, pp. 353–367. IEEE Computer Society Press (2018). https://doi.org/10.1109/EuroSP.2018.00032

15. Canetti, R., Lindell, Y., Ostrovsky, R., Sahai, A.: Universally composable two-party and multi-party secure computation. In: 34th ACM STOC, pp. 494–503. ACM Press (2002). https://doi.org/10.1145/509907.509980

16. Coron, J.S., Gérard, F., Lepoint, T., Trannoy, M., Zeitoun, R.: Improved high-order masked generation of masking vector and rejection sampling in Dilithium (2024). https://doi.org/10.46586/tches.v2024.i4.335-354

17. De Feo, L., Galbraith, S.D.: SeaSign: compact isogeny signatures from class group actions. In: Ishai, Y., Rijmen, V. (eds.) EUROCRYPT 2019, Part III. LNCS, vol. 11478, pp. 759–789. Springer, Cham (2019). https://doi.org/10.1007/978-3-030-17659-4_26

18. Devevey, J., Fallahpour, P., Passelègue, A., Stehlé, D.: A detailed analysis of Fiat-Shamir with aborts. In: Handschuh, H., Lysyanskaya, A. (eds.) CRYPTO 2023, Part V. LNCS, vol. 14085, pp. 327–357. Springer, Cham (2023). https://doi.org/10.1007/978-3-031-38554-4_11

19. Devevey, J., Fawzi, O., Passelègue, A., Stehlé, D.: On rejection sampling in Lyubashevsky's signature scheme. In: Agrawal, S., Lin, D. (eds.) ASIACRYPT 2022, Part IV. LNCS, vol. 13794, pp. 34–64. Springer, Cham (2022). https://doi.org/10.1007/978-3-031-22972-5_2

20. Devevey, J., Passelègue, A., Stehlé, D.: G+G: A fiat-shamir lattice signature based on convolved gaussians. In: Guo, J., Steinfeld, R. (eds.) ASIACRYPT 2023, Part VII. LNCS, vol. 14444, pp. 37–64. Springer, Singapore (2023). https://doi.org/10.1007/978-981-99-8739-9_2

21. Ducas, L., Durmus, A., Lepoint, T., Lyubashevsky, V.: Lattice signatures and bimodal Gaussians. In: Canetti, R., Garay, J.A. (eds.) CRYPTO 2013, Part I. LNCS, vol. 8042, pp. 40–56. Springer, Berlin, Heidelberg (2013). https://doi.org/10.1007/978-3-642-40041-4_3

22. Ducas, L., et al.: CRYSTALS-Dilithium: a lattice-based digital signature scheme. IACR TCHES 2018(1), 238–268 (2018). https://doi.org/10.13154/tches.v2018.i1.238-268, https://tches.iacr.org/index.php/TCHES/article/view/839

23. Espitau, T., Fouque, P.A., Gérard, B., Tibouchi, M.: Side-channel attacks on BLISS lattice-based signatures: exploiting branch tracing against strongSwan and electromagnetic emanations in microcontrollers. In: Thuraisingham, B.M., Evans, D., Malkin, T., Xu, D. (eds.) ACM CCS 2017, pp. 1857–1874. ACM Press (2017). https://doi.org/10.1145/3133956.3134028

24. Espitau, T., Niot, G., Prest, T.: Flood and submerse: distributed key generation and robust threshold signature from lattices. In: Reyzin, L., Stebila, D. (eds.) CRYPTO 2024, Part VII. LNCS, vol. 14926, pp. 425–458. Springer, Cham (2024). https://doi.org/10.1007/978-3-031-68394-7_14

25. Fiat, A., Shamir, A.: How to prove yourself: Practical solutions to identification and signature problems. In: Odlyzko, A.M. (ed.) CRYPTO'86. LNCS, vol. 263, pp. 186–194. Springer, Berlin, Heidelberg (1987). https://doi.org/10.1007/3-540-47721-7_12

26. Gentry, C., Peikert, C., Vaikuntanathan, V.: Trapdoors for hard lattices and new cryptographic constructions. In: Ladner, R.E., Dwork, C. (eds.) 40th ACM STOC, pp. 197–206. ACM Press (2008).https://doi.org/10.1145/1374376.1374407

27. Goldreich, O., Micciancio, D., Safra, S., Seifert, J.P.: Approximating shortest lattice vectors is not harder than approximating closest lattice vectors. Inf. Process. Lett. **71**(2), 55–61 (1999). https://doi.org/10.1016/S0020-0190(99)00083-6, https://www.sciencedirect.com/science/article/pii/S0020019099000836

28. Guo, Q., et al.: Don't reject this: key-recovery timing attacks due to rejection-sampling in HQC and BIKE. IACR TCHES **2022**(3), 223–263 (2022). https://doi.org/10.46586/tches.v2022.i3.223-263

29. Hwang, I., Seo, J., Song, Y.: Concretely efficient lattice-based polynomial commitment from standard assumptions. In: Reyzin, L., Stebila, D. (eds.) CRYPTO 2024, Part X. LNCS, vol. 14929, pp. 414–448. Springer, Cham (2024). https://doi.org/10.1007/978-3-031-68403-6_13

30. Ishai, Y., Sahai, A., Wagner, D.: Private circuits: securing hardware against probing attacks. In: Boneh, D. (ed.) CRYPTO 2003. LNCS, vol. 2729, pp. 463–481. Springer, Berlin, Heidelberg (2003). https://doi.org/10.1007/978-3-540-45146-4_27

31. Kim, D., Lee, D., Seo, J., Song, Y.: Toward practical lattice-based proof of knowledge from hint-MLWE. In: Handschuh, H., Lysyanskaya, A. (eds.) CRYPTO 2023, Part V. LNCS, vol. 14085, pp. 549–580. Springer, Cham (2023). https://doi.org/10.1007/978-3-031-38554-4_18

32. Langlois, A., Stehlé, D.: Worst-case to average-case reductions for module lattices. DCC **75**(3), 565–599 (2015). https://doi.org/10.1007/s10623-014-9938-4

33. Lyubashevsky, V.: Fiat-Shamir with aborts: applications to lattice and factoring-based signatures. In: Matsui, M. (ed.) ASIACRYPT 2009. LNCS, vol. 5912, pp. 598–616. Springer, Berlin, Heidelberg (2009). https://doi.org/10.1007/978-3-642-10366-7_35

34. Lyubashevsky, V.: Lattice signatures without trapdoors. In: Pointcheval, D., Johansson, T. (eds.) EUROCRYPT 2012. LNCS, vol. 7237, pp. 738–755. Springer, Berlin, Heidelberg (2012). https://doi.org/10.1007/978-3-642-29011-4_43

35. Lyubashevsky, V., Nguyen, N.K.: BLOOM: Bimodal lattice one-out-of-many proofs and applications. In: Agrawal, S., Lin, D. (eds.) ASIACRYPT 2022, Part IV. LNCS, vol. 13794, pp. 95–125. Springer, Cham (2022). https://doi.org/10.1007/978-3-031-22972-5_4

36. Lyubashevsky, V., Nguyen, N.K., Plançon, M.: Lattice-based zero-knowledge proofs and applications: shorter, simpler, and more general. In: Dodis, Y., Shrimpton, T. (eds.) CRYPTO 2022, Part II. LNCS, vol. 13508, pp. 71–101. Springer, Cham (2022). https://doi.org/10.1007/978-3-031-15979-4_3

37. Lyubashevsky, V., Nguyen, N.K., Seiler, G.: Shorter lattice-based zero-knowledge proofs via one-time commitments. In: Garay, J. (ed.) PKC 2021, Part I. LNCS, vol. 12710, pp. 215–241. Springer, Cham (2021). https://doi.org/10.1007/978-3-030-75245-3_9

38. Lyubashevsky, V., Seiler, G., Steuer, P.: The LaZer library: lattice-based zero knowledge and succinct proofs for quantum-safe privacy. In: Luo, B., Liao, X., Xu, J., Kirda, E., Lie, D. (eds.) ACM CCS 2024, pp. 3125–3137. ACM Press (2024). https://doi.org/10.1145/3658644.3690330

39. Micciancio, D., Peikert, C.: Trapdoors for lattices: simpler, tighter, faster, smaller. In: Pointcheval, D., Johansson, T. (eds.) EUROCRYPT 2012. LNCS, vol. 7237, pp. 700–718. Springer, Berlin, Heidelberg (2012). https://doi.org/10.1007/978-3-642-29011-4_41

40. Micciancio, D., Regev, O.: Worst-case to average-case reductions based on gaussian measures. SIAM J. Comput. **37**(1), 267–302 (2007). https://doi.org/10.1137/S0097539705447360

41. Migliore, V., Gérard, B., Tibouchi, M., Fouque, P.A.: Masking Dilithium - efficient implementation and side-channel evaluation. In: Deng, R.H., Gauthier-Umaña, V., Ochoa, M., Yung, M. (eds.) ACNS 19International Conference on Applied Cryptography and Network Security. LNCS, vol. 11464, pp. 344–362. Springer, Cham (2019). https://doi.org/10.1007/978-3-030-21568-2_17

42. Nguyen, N.K., Seiler, G.: Greyhound: Fast polynomial commitments from lattices. In: Reyzin, L., Stebila, D. (eds.) CRYPTO 2024, Part X. LNCS, vol. 14929, pp. 243–275. Springer, Cham (2024). https://doi.org/10.1007/978-3-031-68403-6_8

43. Peikert, C.: An efficient and parallel Gaussian sampler for lattices. In: Rabin, T. (ed.) CRYPTO 2010. LNCS, vol. 6223, pp. 80–97. Springer, Berlin, Heidelberg (2010). https://doi.org/10.1007/978-3-642-14623-7_5

44. Prouff, E., Rivain, M.: Masking against side-channel attacks: a formal security proof. In: Johansson, T., Nguyen, P.Q. (eds.) EUROCRYPT 2013. LNCS, vol. 7881, pp. 142–159. Springer, Berlin, Heidelberg (2013). https://doi.org/10.1007/978-3-642-38348-9_9

45. Regev, O.: On lattices, learning with errors, random linear codes, and cryptography. In: Gabow, H.N., Fagin, R. (eds.) 37th ACM STOC, pp. 84–93. ACM Press (2005). https://doi.org/10.1145/1060590.1060603

COMPASS: A Compact PASS-Lineage Accumulator with Succinct Proofs

Tao-Hsiang Chang[1,2] ![ORCID], Jen-Chieh Hsu[1] ![ORCID], Hao-Yi Hsu[1] ![ORCID], Raylin Tso[1(✉)] ![ORCID], and Masahiro Mambo[2] ![ORCID]

[1] Department of Computer Science, National ChengChi University, Taipei, Taiwan
`raylin@cs.nccu.edu.tw`
[2] School of Electrical, Information and Communication Engineering, Kanazawa University, Kanazawa, Japan

Abstract. Accumulators enable succinct membership proofs and are critical for credential revocation and other verifiable services. In the post-quantum setting, hash-based accumulators remain transparent but incur $\log n$ overhead, while RSA alternatives achieve constant costs but are not quantum-safe. Lattice-based constructions from the PASS lineage provide constant-size proofs with transparent setup, yet existing candidates either require trapdoors or yield large witnesses.

We present COMPASS, an improved lattice-based accumulator that is trapdoor-free, quantitatively analyzed, and fully implemented. Our proof-of-concept demonstrates constant-size witnesses as small as 4.3 KiB and verification times as short as 3.6 ms at $PASS_G$-style parameter points, maintaining the 128-bit BKZ-GSA security margin. Compared with our re-implementation of Maeno et al., COMPASS achieves up to 52% smaller witnesses under comparable nominal parameter scales. The design enforces strict norm bounds, weighted rejection sampling, and domain-separated challenges, and its accumulator configuration limits long-term key exposure, mitigating known PASS-lineage weaknesses.

These results show that transparent, post-quantum accumulators with compact proofs are practically attainable. COMPASS thus provides a reproducible, bandwidth-efficient foundation for credential-revocation and IoT verification systems requiring constant-time membership validation.

Keywords: post-quantum cryptography · lattice-based accumulator · verifiable credential revocation

1 Introduction

The rise of the Internet of Things (IoT) has created ecosystems in which resource-constrained devices must frequently verify or revoke membership credentials to maintain trust and security. In vehicular networks, industrial sensors, and healthcare systems, these devices operate with limited memory and computational capacity, making cryptographic tools with *constant-time verification* and *constant-size witnesses/accumulators* particularly appealing. A lightweight accumulator can efficiently compress dynamic membership lists and enable devices to update or validate membership without prohibitive cost.

Two practical scenarios illustrate this need. *Verifiable-credential revocation* systems require verifiers to check whether a credential identifier appears in a revocation list. A trapdoor-free, constant-size accumulator allows this check using short witnesses that keep bandwidth and verification time predictable on mobile or embedded platforms. Likewise, in *IoT allowlists*, gateways must confirm that a device or firmware hash belongs to an authorized set. Constant-time verification avoids the logarithmic proofs of Merkle trees and eliminates reliance on trusted setup. These examples highlight the importance of transparent, post-quantum accumulators that deliver stable proof sizes and verification cost.

The concept of a cryptographic accumulator was first introduced by Benaloh and de Mare in 1993 [3], and soon extended into dynamic and universal forms by Baric and Pfitzmann [2], Camenisch and Lysyanskaya [7], and later by Li et al. [17]. Over the years, a variety of accumulator families emerged in the *classical setting*, which can broadly be categorized as follows:

1. RSA-based accumulators relying on hidden-order groups [2,7].
2. Bilinear-pairing-based accumulators under bilinear map assumptions [6,30].
3. Other number-theoretic constructions in known-order groups [1,27].
4. Symmetric or hash-based membership structures that inspired accumulator designs, such as Merkle trees [25] and Bloom filters [4].

These categories illustrate that multiple design directions were pursued, each tailored to different application scenarios. Notably, in elliptic-curve settings, constant-size accumulator values and efficient updates were achievable [6,17], making them particularly suitable for lightweight device verification.

However, the advent of quantum computing poses an existential threat to most of these schemes. Shor's algorithm [28] efficiently breaks the hardness assumptions underlying RSA and bilinear pairings, thereby rendering entire classes of classical accumulators insecure in a post-quantum (PQ) world. As a consequence, among the traditional designs, only the *hash-based* accumulators and more recent *lattice-based* accumulators remain viable in the PQ setting [14,18,19,23].

Within the PQ landscape, hash-based accumulators retain logarithmic-size witnesses and verification time, inherited from Merkle-tree constructions [15,25]. Lattice-based designs, on the other hand, have explored both trapdoor-based [14] and trapdoorless approaches [18,19]. A few of these succeed in achieving *constant-size witnesses* [14,23], yet almost none offer *constant-time verification*, which is essential for highly constrained IoT deployments.

The notable exception is the recent work of Maeno et al. [23], who proposed the first lattice-based accumulator with both constant-time verification and constant-time updates, relying on a structured variant of the Short Integer Solution (SIS) problem, namely the Vandermonde-SIS assumption, and building upon the HPSSW signature [9,12]. While theoretically significant, their design has several limitations:

1. No practical implementation, reference code or parameter set is provided, limiting reproducibility and real-world assessment.
2. The integration between the underlying signature and the accumulator logic is relatively loose, effectively layering one primitive atop another.
3. The resulting complexity raises concerns about efficiency in constrained environments, where code size and execution time matter as much as asymptotic bounds.

1.1 Our Contributions

This work refines the lattice-based accumulator framework of Maeno et al. [23] into a compact, quantitatively analyzed, and fully implemented design. **COMPASS introduces structural, analytical, and empirical improvements** that render the scheme both deployable and theoretically cleaner.

1. **Unified accumulator-signature integration.** Prior PASS-lineage accumulators treated signature transcripts and accumulation as loosely coupled procedures. COMPASS unifies them: the same PASS-style transcript now serves directly as the accumulation witness, eliminating redundant commitments and reducing witness size by up to half under comparable nominal parameter scales. This preserves transparent setup while tightening the linkage between membership elements and the manager key.
2. **Quantitative, trapdoor-free security.** While [23] already established a straight-line V–SIS reduction, they left the bound structure implicit. We make these bounds explicit and *instantiable*, deriving closed forms for Bound_z, Bound_Z, s_y, and the resulting $\beta_{\mathrm{sig}}, \beta_{\mathrm{acc}}$, and validate them through implementation. The resulting analysis provides the first *quantitatively verifiable* trapdoor-free accumulator proof under V–SIS, with clear mapping to BKZ-GSA parameters for 128-bit security.
3. **PASS_G -based sampling and domain separation.** COMPASS replaces the earlier $\mathrm{HPSSW}/\mathrm{PASS}_{RS}$-style sampling with PASS_G-style Gaussian mechanisms, introducing explicit domain-separated oracles $(H_{c1}, H_{c2}, H_\beta)$ and weighted rejection. These ensure transcript independence and Random Oracle Model (ROM) soundness, mitigating multi-transcript and symmetry-related weaknesses recently highlighted in the PASS lineage. We implement all four algorithms (SETUP, ACCUMULATE, WITNESS, and VERIFY) with domain-separated hashes, compact coefficient packing, and a numerically stable partial Fourier backend. Two PASS_G-style parameter points were benchmarked on Google Colab (Oct. 2025): Set 1 ($N{=}512$) achieved 4.3 KiB witnesses and 3.6 ms verification; Set 2 ($N{=}1024$) achieved 8.4 KiB witnesses and 79.7 ms verification, both maintaining $\delta_{\mathsf{BKZ}} \leq 1.007$ within the 128-bit security line. These experiments constitute the first end-to-end quantitative validation of a trapdoor-free lattice accumulator.
4. **Applicability to post-quantum revocation and credential systems.** The resulting scheme fits naturally into PQ-safe credential and IoT membership frameworks that require transparent setup, bounded proofs, and predictable verification cost on constrained devices.

Together, these advances distinguish COMPASS as the first self-contained, quantitatively analyzed lattice accumulator that is both *provably secure under V–SIS* and *demonstrated in practice with reproducible BKZ-backed parameterization.*

1.2   Organization

The remainder of this paper is organized as follows. Section 2 introduces the necessary preliminaries. Section 3 presents our proposed accumulator scheme. Section 4 provides the formal security analysis. Section 5 reports implementation results and discusses practical considerations. Finally, Sect. 6 concludes the paper.

2   Preliminaries

2.1   Accumulator Schemes

Cryptographic accumulators enable committing to a (multi)set of elements with a short value and issuing succinct witnesses for membership. We follow the common four-algorithm interface (Setup, Accumulate, Witness, Verify) used in recent lattice accumulators [23]. In particular, recent work integrates a PASS-lineage signature [12,20] with a Vandermonde evaluation map to obtain short, publicly verifiable witnesses at ring dimension N and modulus q. Our scheme inherits this structure but (i) integrates the signature subroutine tightly via calibrated rejection sampling [21,22], (ii) gives concrete Gaussian bounds for parameter selection, and (iii) targets a drop-in PQ companion to EVOKE [24].

An accumulator scheme consists of four polynomial-time algorithms:

- Setup(1^λ) $\to$ (pp, **sk**): on input security parameter λ, output global parameters pp and a manager key pair, where **pk** is included in pp.
- Accumulate(pp, **sk**, $\{f_i\}_{i \in [K]}$) $\to$ (**Acc**, **Z**, $\mathcal{L}$): given parameters, secret key, and elements f_i, output an accumulator value **Acc**, aggregate state **Z**, and auxiliary list $\mathcal{L}$.
- Witness(pp, **Z**, $\mathcal{L}$, **f**$_i$) $\to w_i$: produce a witness for element **f**$_i$.
- Verify(pp, **Acc**, f_i, w_i) $\to \{0,1\}$: check whether w_i is a valid witness that f_i was accumulated into **Acc**.

Definition 1 (Correctness). *An accumulator scheme is correct if for all honestly generated* (pp, **pk**, **sk**) $\leftarrow$ Setup(1^λ), *any set* $\{f_i\}_{i \in [K]}$, *and any* $i \in [K]$, *if* (**Acc**, **Z**, $\mathcal{L}$) $\leftarrow$ Accumulate($\cdot$) *and* $w_i \leftarrow$ Witness($\cdot, i$), *then*

$$\text{Verify}(\text{pp}, \textbf{Acc}, f_i, w_i) = 1.$$

Definition 2 (Security of accumulator schem [23]). *An accumulator scheme is secure if no probabilistic polynomial-time (PPT) adversary, given* pp *and oracle access to* Accumulate *and* Witness, *can output a pair* (f, w) *such that* f *was never accumulated, yet achieved* Verify(pp, **Acc**, f, w) $= 1$, *except with negligible probability in* λ.

2.2 Vandermonde–SIS and PASS-Lineage Signature Schemes

The PASS lineage of lattice-based signatures originates from early constructions in polynomial rings by Hoffstein, Lieman, and Silverman (PASS) [11] and its refinement PASS2 [13]. The modern PASS_{RS} scheme [12] introduced rejection sampling to implement Fiat–Shamir with aborts [21,22] over structured Ring–SIS instances derived from the Partial Fourier Recovery problem. Later, MMSAT [9] extended PASS_{RS} with an aggregate-signature framework, while PASS_G [20] replaced rejection sampling with direct Gaussian sampling, leading to smaller signatures and more straightforward parameter selection.

In recent PASS variants, the signer commits using a randomized vector $\mathbf{y}$, derives challenges by hashing evaluated commitments, and computes $\mathbf{z} = \mathbf{u} + \mathbf{y}$, where $\mathbf{u}$ depends on the secret key and message. The distribution of $\mathbf{z}$ is then shaped either by rejection sampling or Gaussian sampling to ensure statistical independence from the secret. Our accumulator adopts this core structure, specialized to the Vandermonde evaluation map F_Ω in PASS_G and coupled with a membership witness aggregator.

Cyclotomic Ring and Coefficient Embedding. Assume N is a power of two and let $\Phi_{2N}(x) = x^N + 1$ be the cyclotomic polynomial. We work in the cyclotomic ring $R_q = \mathbb{Z}_q[x]/(\Phi_{2N}(x))$. Every polynomial $f(x) = \sum_{k=0}^{N-1} f_k x^k \in R_q$ is represented by its coefficient vector $\mathbf{f} = (f_0, \ldots, f_{N-1})^\top \in \mathbb{Z}_q^N$ via the coefficient embedding $\iota : R_q \to \mathbb{Z}_q^N$. Unless stated otherwise, coefficients are taken in their centered representatives in $(-\frac{q}{2}, \frac{q}{2}]$, so that $\|\cdot\|_2, \|\cdot\|_\infty$ refer to the usual norms on $\mathbb{Z}^N$. Ring multiplication in R_q corresponds to *negacyclic convolution* of coefficient vectors modulo q.

Roots and Negacyclic Fourier (Vandermonde) Matrix. Assume $2N \mid (q-1)$ and fix a primitive $2N$-th root of unity $g \in \mathbb{Z}_q$ (with $g^N \equiv -1 \,(\mathrm{mod}\ q)$). The N roots of $\Phi_{2N}(x) = x^N + 1$ in $\mathbb{Z}_q$ are the odd powers $\{\omega_j\}_{j=0}^{N-1}$ where $\omega_j := g^{2j+1}$. Define the $N \times N$ *negacyclic Fourier matrix*

$$\mathcal{F} = \left(\omega_j^k\right)_{0 \le j,k < N} \in \mathbb{Z}_q^{N \times N}.$$

For a coefficient vector $\mathbf{f} = \iota(f)$, we have $\mathcal{F}\mathbf{f} = \left(f(\omega_0), \ldots, f(\omega_{N-1})\right)^\top$. Thus $\mathcal{F}$ is a Vandermonde matrix evaluating f at all roots of $x^N + 1$.

Partial Evaluation Map. Let $\Omega = \{\omega_{j_1}, \ldots, \omega_{j_t}\}$ be any t distinct roots from $\{\omega_0, \ldots, \omega_{N-1}\}$, and let $\mathbf{F}_\Omega$ be the $t \times N$ row-submatrix of $\mathcal{F}$ indexed by $j_1, \ldots, j_t$. We define the (partial) evaluation map both as a matrix and as a function:

$$F_\Omega : R_q \to \mathbb{Z}_q^t, \qquad F_\Omega(f) := \mathbf{F}_\Omega\, \iota(f) = \left(f(\omega_{j_1}), \ldots, f(\omega_{j_t})\right).$$

This map enjoys a homomorphic property: it preserves addition and carries ring multiplication to Hadamard product: for all $a, b \in R_q$, $F_\Omega(a+b) = F_\Omega(a) + F_\Omega(b)$ and $F_\Omega(a * b) = F_\Omega(a) \odot F_\Omega(b)$.

Definition 3 (Vandermonde–SIS$_{q,t,N,\beta}^{\mathcal{K}}$ problem [20]). *Given a Vandermonde matrix $\mathbf{F}_\Omega \in \mathbb{Z}_q^{t \times N}$ drawn according to some distribution $\mathcal{K}$, find a nonzero $\mathbf{v} \in \mathbb{Z}_q^N$ such that*

$$\mathbf{F}_\Omega \mathbf{v} \equiv \mathbf{0} \quad and \quad \|\mathbf{v}\|_2 \leq \beta.$$

Here $\mathcal{K}$ samples $\Omega \subset \{g^0, \dots, g^{N-1}\}$ uniformly at random with $|\Omega| = t$, and sets $\mathbf{F}_\Omega$ to the corresponding $t \times N$ Vandermonde submatrix of the discrete Fourier transform over $\mathbb{Z}_q$.

Digital Signatures. A digital signature scheme consists of three efficient algorithms:

- KeyGen(1^λ) $\to$ (pk, sk): on input a security parameter λ, output a key pair.
- Sign(sk, m) $\to \sigma$: on input a secret key and a message m, output a signature σ.
- Verify(pk, m, σ) $\to b \in \{0, 1\}$: on input a public key, a message, and a signature, output 1 iff σ is valid.

Definition 4 (Unforgeability [22]). *A signature scheme is existentially unforgeable under chosen-message attack (EUF-CMA) if for all PPT adversaries $\mathcal{A}$, the following probability is negligible in λ:*

$$\Pr\left[\mathsf{Verify}(\mathsf{pk}, m^\star, \sigma^\star) = 1 \ \wedge \ m^\star \notin \mathcal{Q} \right] \leq \mathrm{negl}(\lambda),$$

where (pk, sk) $\leftarrow$ KeyGen(1^λ), $\mathcal{Q}$ *is the set of messages $\mathcal{A}$ queried to the signing oracle, and $(m^\star, \sigma^\star)$ is the output of $\mathcal{A}$.*

2.3   Gaussian Toolkit and Inequalities

Following [22,26], we use the unnormalized kernel $\rho_\sigma(\mathbf{x}) := \exp(-\|\mathbf{x}\|_2^2/(2\sigma^2))$ for $\mathbf{x} \in \mathbb{R}^N$ and $\sigma > 0$, and write $\rho_\sigma(\mathbb{Z}^N) := \sum_{\mathbf{y} \in \mathbb{Z}^N} \rho_\sigma(\mathbf{y})$. The discrete Gaussian $D_{\mathbf{c},\sigma}^N$ over $\mathbb{Z}^N$ with center $\mathbf{c} \in \mathbb{Z}^N$ has

$$D_{\mathbf{c},\sigma}^N(\mathbf{x}) \ = \ \rho_\sigma(\mathbf{x} - \mathbf{c}) \big/ \rho_\sigma(\mathbb{Z}^N), \qquad \mathbf{x} \in \mathbb{Z}^N,$$

and we write D_σ^N for $\mathbf{c} = \mathbf{0}$.

Norm relations. We use the standard inequalities $\|\mathbf{x}\|_\infty \leq \|\mathbf{x}\|_2 \leq \sqrt{N}\,\|\mathbf{x}\|_\infty$ for $\mathbf{x} \in \mathbb{R}^N$ (see [31]).

Negligible functions. A function negl $: \mathbb{N} \to \mathbb{R}$ is negligible if for every polynomial $p(\cdot)$ there exists N_0 such that for all $N > N_0$, $\mathrm{negl}(N) < 1/p(N)$. We use negl($N$) to denote an unspecified negligible function in the parameter N.

Lemma 1 (Rejection sampling, adapted from [22], Thm. 3.4). *Let $V \subset \mathbb{Z}^N$ be such that every $\mathbf{v} \in V$ satisfies $\|\mathbf{v}\|_2 < T$, and let $\sigma \geq C\,T\sqrt{\log N}$ for some sufficiently large constant $C > 0$. For any distribution h on V, there exists a constant $M = O(1)$ such that the following holds. Consider algorithm $\mathcal{A}$:*

1. *sample* $\mathbf{v} \leftarrow h$;
2. *sample* $\mathbf{z} \leftarrow D^N_{\mathbf{v},\sigma}$;
3. *output* $(\mathbf{z}, \mathbf{v})$ *with probability* $\min\left\{1, \frac{D^N_\sigma(\mathbf{z})}{M\,D^N_{\mathbf{v},\sigma}(\mathbf{z})}\right\}$.

Then the output distribution of $\mathcal{A}$ *is within negligible statistical distance* $\mathrm{negl}(N)/M$ *of algorithm* $\mathcal{F}$ *that:*

1. *samples* $\mathbf{v} \leftarrow h$;
2. *samples* $\mathbf{z} \leftarrow D^N_\sigma$;
3. *outputs* $(\mathbf{z}, \mathbf{v})$ *with probability* $1/M$.

Moreover, $\mathcal{A}$ *outputs with probability at least* $(1 - \mathrm{negl}(N))/M$.

Lemma 2 (Tail bounds for D^m_σ [22, Lem. 3.3]). *Let* $\mathbf{z} \leftarrow\!\!\$\, D^N_\sigma$.

1. *(Coordinate tail) There exist constants* $\alpha > 0$, $c > 1$ *such that*

$$\Pr[\,|\mathbf{z}_i| > \alpha\sigma\sqrt{\ln N}\,] \ \le\ N^{-c} \quad \text{for any coordinate } i.$$

2. *(ℓ_2 tail)*
$$\Pr[\,\|\mathbf{z}\|_2 > 2\sigma\sqrt{N}\,] \ < \ 2^{-N}.$$

Corollary 1 (ℓ_∞ tail). *From Lemma 2(1) and a union bound over coordinates,*

$$\Pr[\,\|\mathbf{z}\|_\infty > \alpha\sigma\sqrt{\ln N}\,] \ \le\ N^{1-c}.$$

Lemma 3 (Young's convolution inequality $(\ell^1 \to \ell^2)$ [10]). *Let* $a, b \in \mathbb{R}^N$, *and let* $*$ *denote linear or circular/negacyclic convolution on* $\mathbb{R}^N$. *Then*

$$\|a * b\|_2 \ \le\ \|a\|_1\,\|b\|_2.$$

Lemma 4 (Peter–Paul inequality [32]). *For any* $x, y \in \mathbb{R}^N$ *and any* $\varepsilon > 0$,

$$2\langle x, y\rangle \ \le\ \varepsilon\|x\|_2^2 + \varepsilon^{-1}\|y\|_2^2.$$

3   Proposed Scheme

We now present COMPASS accumulator scheme, which follows the constant-size paradigm introduced by Maeno, Miyaji, and Miyaji [23]. Like their work, our design achieves constant verification time, constant accumulator length, and constant witness length. Our construction, however, is more compact and efficient: both accumulator values and witnesses are smaller, verification requires fewer ring operations, and we do have a proof-of-concept (PoC) implementation that validates correctness and measures preliminary performance.

3.1 Features

COMPASS offers the following properties:

- **Constant size:** The accumulator and witnesses are of fixed length, independent of the set size K, and verification runs in constant time.
- **Compactness:** Compared to [23], the artifact sizes and verification cost are significantly reduced.
- **Implementability:** Concrete parameter definitions are included with PoC implemented, ensuring that the scheme is not only theoretically sound but also practically realizable.

3.2 Security Foundations

The security of COMPASS rests on the lattice problems underlying the PASS family of signatures [12,20], in particular the Vandermonde-SIS assumption, a structured variant of the SIS problem in cyclotomic rings. Intuitively, soundness relies on the infeasibility of producing valid transcripts $(\mathbf{z}, \mathbf{c}_1, \mathbf{c}_2)$ without knowledge of the manager's secret key. Formal reductions and probability bounds are deferred to Sect. 4.

3.3 Scheme Description

The scheme consists of four polynomial-time algorithms: Setup, Accumulate, Witness, and Verify. All parties share common public parameters pp, which specify the ring, modulus, evaluation set, hash oracles, and serialization conventions. The manager initializes the scheme, computes accumulator values, and issues witnesses to members. Verifiers later check membership relative to the published accumulator value. Notation used throughout the scheme is summarized in Table 1, and the hash/encoding primitives are given in Table 2. A member certificate $\mathsf{cert}_i \in \{0,1\}^*$ is deterministically mapped to a short ring vector via $\mathbf{f}_i \leftarrow \mathsf{HashToLat}(\mathsf{cert}_i) \in \mathcal{B}^\infty(1) \subset R_q$, implemented as $\mathsf{HashToLat} = \Phi \circ \mathsf{Hash}_C$ with fixed domain separation and a deterministic rounding/centering map Φ to $\{-1,0,1\}^N$. We write $\hat{\mathbf{f}}_i|_\Omega = F_\Omega(\mathbf{f}_i)$.

- $\mathsf{Setup}(1^\lambda) \rightarrow (\mathsf{pp}, \mathbf{sk})$: Run by the manager. Input: security parameter λ. Output: public parameters pp and secret key $\mathbf{sk}$. The Setup algorithm initializes all system parameters and fixes the derived hash oracles with explicit domain separation. This prevents cross-protocol collisions and narrows the attack surface.
- $\mathsf{Accumulate}(\mathsf{pp}, \mathbf{sk}, K, \{\mathbf{f}_i\}_{i=1}^K) \rightarrow (\mathbf{Acc}, \mathbf{Z}, \mathcal{L})$: Run by the manager. Input: public parameters, manager's secret key, integer K, and a collection of K member vectors $\mathbf{f}_i$. Output: accumulator value $\mathbf{Acc}$, aggregate state $\mathbf{Z}$, and auxiliary list $\mathcal{L}$ for witness extraction. This is the most computation-intensive algorithm. Each member's credential (or certificate) is first mapped to a short

Table 1. Notation

Symbol	Description
λ	Security parameter.
N	Ring degree.
q	Prime with $q \equiv 1 \pmod{2N}$.
g	Generator of $\mathbb{Z}_q^*$; $\omega := g^{(q-1)/(2N)}$ (a primitive $2N$-th root).
R_q	Cyclotomic ring $\mathbb{Z}_q[x]/(x^N + 1)$.
$\mathbf{a} \in R_q$	Polynomial/vector $a(x) = \sum_{i=0}^{N-1} a_i x^i$ with $a_i \in \mathbb{Z}_q$; coefficients centered in $[-q/2, q/2)$.
$\hat{\mathbf{x}}$	Full odd-index evaluation $(x(\omega), x(\omega^3), \ldots, x(\omega^{2N-1})) \in \mathbb{Z}_q^N$.
Ω	Subset of odd powers $\{\omega^{2j+1}\}$; $t = |\Omega|$.
$\mathbf{F}_\Omega$	Vandermonde submatrix in $\mathbb{Z}_q^{t \times N}$; $F_\Omega(\mathbf{x}) = \mathbf{F}_\Omega \mathbf{x} = \hat{\mathbf{x}}|_\Omega \in \mathbb{Z}_q^t$.
$\|\cdot\|_\infty, \|\cdot\|_2$	Supremum and Euclidean norms on coefficient vectors.
$\mathcal{B}^\infty(B)$	$\{\mathbf{a} \in \mathbb{Z}^N : \|\mathbf{a}\|_\infty \leq B\}$.
$\mathbf{sk} \in \mathcal{B}^\infty(1)$	Manager's secret key.
$\mathbf{pk}$	Manager's public key $F_\Omega(\mathbf{sk}) \in \mathbb{Z}_q^t$.
$\mathbf{f}_i \in \mathcal{B}^\infty(1)$	Member i's short vector (e.g., derived from a certificate).
$*$	Ring product in R_q.
$\odot$	Component-wise product on evaluations.
$\leftarrow\$$	Sampling notation; $x \leftarrow\$ [0, 1)$ is uniform on $[0, 1)$.
$D_{c,\sigma}^N$	Discrete Gaussian over $\mathbb{Z}^N$ with center c (omit if 0) and std σ.
s_y	Gaussian clipping threshold.
M	normalizing constant for weighted rejection sampling.
κ	Number of nonzeros in challenges (from FormatC).
m	Output length of Hash_{Acc} in $\mathbb{Z}_q^m$.
τ	Bit length of Hash_C's output to FormatC.
$\mathtt{ctx}$	Domain-separation context (scheme id, params digest, epoch).
$\mathsf{ser}(\cdot)$	Deterministic serialization (fixed endianness; length-prefix where needed).
$\|$	byte-string concatenation.

Table 2. Hash encoding primitives

Name	Domain	Range / Role
Hash_C	$\{0,1\}^*$	$\{0,1\}^\tau$ (input to FormatC).
FormatC	$\{0,1\}^\tau$	$\mathcal{C}_\kappa \subset \{-1,0,1\}^N$ (exactly κ nonzeros).
Hash_β	$\{0,1\}^*$	$\{\pm 1\}$ (aggregation weight).
Hash_{Acc}	$\mathbb{Z}_q^t$	$\mathbb{Z}_q^m$ (binding digest of $\hat{\mathbf{Z}}|_\Omega$).

vector $\mathbf{f}_i$ via a hash-to-lattice function. The manager then produces PASS_G-style signature-like components on these vectors, obtaining masked responses $\mathbf{z}_i$, and aggregates them into a single accumulator value. Using the dual challenges $(\mathbf{c}_{i1}, \mathbf{c}_{i2})$ binds both the manager's key and the member's vector, ensuring that neither side can be forged or reused in isolation.

- Witness($pp, \mathbf{Z}, \mathcal{L}, i$) $\rightarrow w_i$: Run by the manager. Input: accumulator state $\mathbf{Z}$, log $\mathcal{L}$, and index i. Output: witness $w_i = (\mathbf{Z}_i, \mathbf{z}_i, \mathbf{c}_{i1}, \mathbf{c}_{i2})$ for member i. Since Accumulate already performs the rejection sampling and aggregation, Witness is lightweight: it simply extracts the stored values and computes $\mathbf{Z}_i = \mathbf{Z} - \beta_i \mathbf{z}_i$.
- Verify($pp, \mathbf{Acc}, \mathbf{f}_i, w_i$) $\rightarrow b$: Run by the verifier. Input: public parameters, accumulator value $\mathbf{Acc}$, member vector $\mathbf{f}_i$, and witness w_i. Output: decision bit $b \in \{0, 1\}$ indicating membership validity. Verification proceeds in two stages. First, the verifier checks that the witness corresponds to a valid PASS_G-style transcript, i.e., that $\mathbf{z}_i$ passes the norm/acceptance checks and recomputed challenges are consistent. Second, the verifier reconstructs $\mathbf{Z}' = \mathbf{Z}_i + \beta_i \mathbf{z}_i$ and confirms that $H_{Acc}(\hat{\mathbf{Z}}'|_\Omega) = \mathbf{Acc}$, thereby proving membership.

Algorithm 1. $\mathsf{Setup}(1^\lambda)$

Require: λ
Ensure: pp, sk
1: Choose $(N, q, g, \sigma, R_q, \Omega, t, \kappa, m, \tau)$ based on λ.
2: Fix serialization $\mathsf{ser}(\cdot)$ and the base hash primitives from Table 2.
3: Sample $\mathbf{sk} \leftarrow\!\$\, \mathcal{B}^\infty(1)$; set $\mathbf{pk} \leftarrow F_\Omega(\mathbf{sk}) \in \mathbb{Z}_q^t$.
4: Define the following *derived* oracles (with fixed domain separation internally):
5:     $H_{c1}(\hat{\mathbf{y}}|_\Omega, \hat{\mathbf{f}}|_\Omega) := \mathsf{FormatC}\Big(\mathsf{Hash}_C\big(\mathsf{ser}(\texttt{"C1"}) \,\|\, \mathsf{ser}(\hat{\mathbf{y}}|_\Omega) \,\|\, \mathsf{ser}(\hat{\mathbf{f}}|_\Omega)\big)\Big)$
6:     $H_{c2}(\hat{\mathbf{y}}|_\Omega, \mathbf{pk}) := \mathsf{FormatC}\Big(\mathsf{Hash}_C\big(\mathsf{ser}(\texttt{"C2"}) \,\|\, \mathsf{ser}(\hat{\mathbf{y}}|_\Omega) \,\|\, \mathsf{ser}(\mathbf{pk})\big)\Big)$
7:     $H_\beta(\mathbf{c}_1, \mathbf{c}_2, \hat{\mathbf{f}}|_\Omega) := \mathsf{Hash}_\beta\big(\mathsf{ser}(\texttt{"BETA"}) \,\|\, \mathsf{ser}(\mathbf{c}_1) \,\|\, \mathsf{ser}(\mathbf{c}_2) \,\|\, \mathsf{ser}(\hat{\mathbf{f}}|_\Omega)\big)$
8:     $H_{Acc}(\hat{\mathbf{Z}}|_\Omega) := \mathsf{Hash}_{Acc}(\hat{\mathbf{Z}}|_\Omega)$
9: Set $pp = \big(N, q, g, \sigma, R_q, \Omega, t, \kappa, m, \tau, F_\Omega, \mathbf{pk}, H_{c1}, H_{c2}, H_\beta, H_{Acc}, \mathsf{ser}\big)$.
10: **return** $(pp, \mathbf{sk})$

---

4   Security Analysis

4.1   Correctness of the COMPASS

Theorem 1 (Correctness). *COMPASS satisfies Definition 1. Namely, for any security parameter λ, for any honestly generated $(pp, \mathbf{pk}, \mathbf{sk}) \leftarrow \mathsf{Setup}(1^\lambda)$ (with fixed oracles H_{c1}, H_{c2}, H_{Acc} and domain separation), any collection $\{\mathbf{f}_i\}_{i \in [K]}$, and any $(\mathbf{Acc}, \mathbf{Z}, \mathcal{L}) \leftarrow \mathsf{Accumulate}(pp, \mathbf{sk}, K, \{\mathbf{f}_i\})$, if $w_i \leftarrow \mathsf{Witness}(pp, \mathbf{Z}, \mathcal{L}, i)$ for any $i \in [K]$, then*

$$\mathsf{Verify}(pp, \mathbf{Acc}, \mathbf{f}_i, w_i) = 1.$$

Proof. Let $(\beta_i, \mathbf{c}_{i1}, \mathbf{c}_{i2}, \mathbf{z}_i) \in \mathcal{L}$ be the entry for member i. By construction in Accumulate, these satisfy $\mathbf{z}_i = \mathbf{sk} * \mathbf{c}_{i1} + \mathbf{f}_i * \mathbf{c}_{i2} + \mathbf{y}_i$ for some sampled $\mathbf{y}_i$, with $\|\mathbf{z}_i\|_2 \leq \mathsf{Bound}_z$. Moreover, $\beta_i = H_\beta(\mathbf{c}_{i1}, \mathbf{c}_{i2}, \hat{\mathbf{f}}_i|_\Omega)$ and $\mathbf{Z} = \sum_{j=1}^K \beta_j \mathbf{z}_j \bmod q$,

---

Algorithm 2. $\mathsf{Accumulate}(\mathsf{pp}, \mathsf{sk}, \{\mathbf{f}_i\}_{i=1}^{K})$

Require: $\mathsf{pp}, \mathsf{sk}, K, \mathbf{f}_1, \ldots, \mathbf{f}_K \in R_q$
Ensure: $\mathbf{Acc}, \mathbf{Z}, \mathcal{L} = \{(\beta_i, \mathbf{c}_{i1}, \mathbf{c}_{i2}, \mathbf{z}_i)\}_{i=1}^{K}$
 1: Precompute $\hat{\mathbf{f}}_i|_{\Omega} \leftarrow F_{\Omega}(\mathbf{f}_i)$ for all $i \in [K]$.
 2: **repeat**
 3:      $Z \leftarrow \mathbf{0}$; $\mathcal{L} \leftarrow \varnothing$
 4:      **for** $i = 1$ **to** K **do**
 5:          **repeat**
 6:              Sample $\mathbf{y}_i \leftarrow\!\!\$\, D_{\sigma}^{N}$ (resample if $\|y\|_{\infty} > s_y$) and compute $\hat{\mathbf{y}}_i|_{\Omega} \leftarrow F_{\Omega}(\mathbf{y}_i)$
 7:              $\mathbf{c}_{i1} \leftarrow H_{c1}(\hat{\mathbf{y}}_i|_{\Omega}, \hat{\mathbf{f}}_i|_{\Omega})$; $\mathbf{c}_{i2} \leftarrow H_{c2}(\hat{\mathbf{y}}_i|_{\Omega}, \mathbf{pk})$
 8:              $\mathbf{z}_i \leftarrow \mathbf{sk} * \mathbf{c}_{i1} + \mathbf{f}_i * \mathbf{c}_{i2} + \mathbf{y}_i$
 9:              **if** $\|\mathbf{z}_i\|_2 > \mathsf{Bound}_z$ **then reject**
10:              $\mathbf{u}_i \leftarrow \mathbf{sk} * \mathbf{c}_{i1} + \mathbf{f}_i * \mathbf{c}_{i2}$
11:              $p_i \leftarrow \mathsf{Prob}_{accept\_z}$;   draw $U \leftarrow\!\!\$\, [0, 1)$
12:              **if** $U > p_i$ **then reject**
13:              **accept** $\mathbf{z}_i$
14:          **until** $\mathbf{z}_i$ accepted
15:          $\beta_i \leftarrow H_{\beta}(\mathbf{c}_{i1}, \mathbf{c}_{i2}, \hat{\mathbf{f}}_i|_{\Omega})$; $\mathbf{Z} \leftarrow (\mathbf{Z} + \beta_i \mathbf{z}_i) \bmod q$
16:          $\mathcal{L} \leftarrow \mathcal{L} \cup \{(\beta_i, \mathbf{c}_{i1}, \mathbf{c}_{i2}, \mathbf{z}_i)\}$
17:      **end for**
18: **until** $\|\mathbf{Z}\|_2 \leq \mathsf{Bound}_Z$
19: $\hat{\mathbf{Z}}|_{\Omega} \leftarrow F_{\Omega}(\mathbf{Z})$; $\mathsf{Acc} \leftarrow H_{Acc}(\hat{\mathbf{Z}}|_{\Omega})$
20: **return** $(\mathbf{Acc}, \mathbf{Z}, \mathcal{L})$

Algorithm 3. $\mathsf{Witness}(\mathsf{pp}, \mathbf{Z}, \mathcal{L}, \mathbf{f}_i)$

Require: $\mathsf{pp}, \mathbf{Z}, \mathcal{L} : [K] \rightarrow (\beta, \mathbf{c}_1, \mathbf{c}_2, \mathbf{z}), \mathbf{f}_i$ where $i \in [K]$
Ensure: $w_i = (\mathbf{Z}_i, \mathbf{z}_i, \mathbf{c}_{i1}, \mathbf{c}_{i2})$
 1: $(\beta_i, \mathbf{c}_{i1}, \mathbf{c}_{i2}, \mathbf{z}_i) \leftarrow \mathcal{L}[i]$
 2: $\mathbf{Z}_i \leftarrow (\mathbf{Z} - \beta_i \mathbf{z}_i) \bmod q$ *(in R_q)*
 3: **return** $(\mathbf{Z}_i, \mathbf{z}_i, \mathbf{c}_{i1}, \mathbf{c}_{i2})$

Algorithm 4. $\mathsf{Verify}(\mathsf{pp}, \mathbf{Acc}, \mathbf{f}_i, w_i)$

Require: $\mathsf{pp}, \mathbf{Acc},$ member's $\mathbf{f}_i$ and w_i
Ensure: $b \in \{0, 1\}$ indicating membership
 1: **if** $\|\mathbf{z}_i\|_2 > \mathsf{Bound}_z$ **then return** 0
 2: $\hat{\mathbf{z}}_i|_{\Omega} \leftarrow F_{\Omega}(\mathbf{z}_i)$; $\hat{\mathbf{c}}_{i1}|_{\Omega} \leftarrow F_{\Omega}(\mathbf{c}_{i1})$; $\hat{\mathbf{c}}_{i2}|_{\Omega} \leftarrow F_{\Omega}(\mathbf{c}_{i2})$; $\hat{\mathbf{f}}_i|_{\Omega} \leftarrow F_{\Omega}(\mathbf{f}_i)$
 3: $\hat{\mathbf{y}}_i'|_{\Omega} \leftarrow \hat{\mathbf{z}}_i|_{\Omega} - \mathbf{pk} \odot \hat{\mathbf{c}}_{i1}|_{\Omega} - \hat{\mathbf{f}}_i|_{\Omega} \odot \hat{\mathbf{c}}_{i2}|_{\Omega}$ $(\bmod\, q)$
 4: $\mathbf{c}_{i1}' \leftarrow H_{c1}(\hat{\mathbf{y}}_i'|_{\Omega}, \hat{\mathbf{f}}_i|_{\Omega})$; $\mathbf{c}_{i2}' \leftarrow H_{c2}(\hat{\mathbf{y}}_i'|_{\Omega}, \mathbf{pk})$
 5: **if** $(\mathbf{c}_{i1}' \neq \mathbf{c}_{i1})$ **or** $(\mathbf{c}_{i2}' \neq \mathbf{c}_{i2})$ **then return** 0
 6: $\beta_i \leftarrow H_{\beta}(\mathbf{c}_{i1}, \mathbf{c}_{i2}, \hat{\mathbf{f}}_i|_{\Omega})$
 7: $\mathbf{Z}' \leftarrow (\mathbf{Z}_i + \beta_i \mathbf{z}_i) \bmod q$
 8: **if** $\|\mathbf{Z}'\|_2 > \mathsf{Bound}_Z$ **then return** 0
 9: $\hat{\mathbf{Z}}'|_{\Omega} \leftarrow F_{\Omega}(\mathbf{Z}')$; $\mathbf{Acc}' \leftarrow H_{Acc}(\hat{\mathbf{Z}}'|_{\Omega})$
10: **return** 1 **iff** $\mathbf{Acc}' = \mathbf{Acc}$ **else** 0

with $\|\mathbf{Z}\|_2 \leq \mathsf{Bound}_Z$. The accumulator value is $\mathbf{Acc} = H_{Acc}(\hat{\mathbf{Z}}|_\Omega)$. The witness algorithm outputs $w_i = (\mathbf{Z}_i, \mathbf{z}_i, \mathbf{c}_{i1}, \mathbf{c}_{i2})$ with $\mathbf{Z}_i = \mathbf{Z} - \beta_i \mathbf{z}_i \bmod q$.

Norm check. Since $\mathbf{z}_i$ was accepted in Accumulate, the bound $\|\mathbf{z}_i\|_2 \leq \mathsf{Bound}_z$ holds.

Challenge consistency. Applying F_Ω to the defining equation for $\mathbf{z}_i$ and using that F_Ω is a ring homomorphism (see Sect. 2), the verifier's reconstruction $\hat{\mathbf{y}}'_i|_\Omega$ equals the true $\hat{\mathbf{y}}_i|_\Omega$ in $\mathbb{Z}_q^t$, so the recomputed challenges $\mathbf{c}'_{i1}, \mathbf{c}'_{i2}$ match $\mathbf{c}_{i1}, \mathbf{c}_{i2}$.

Aggregate reconstruction. The verifier recomputes $\beta_i = H_\beta(\cdot)$ and sets $\mathbf{Z}' = \mathbf{Z}_i + \beta_i \mathbf{z}_i \bmod q = \mathbf{Z}$, hence $\|\mathbf{Z}'\|_2 = \|\mathbf{Z}\|_2 \leq \mathsf{Bound}_Z$.

Accumulator value. Finally, $\mathbf{Acc}' = H_{Acc}(\hat{\mathbf{Z}}'|_\Omega) = H_{Acc}(\hat{\mathbf{Z}}|_\Omega) = \mathbf{Acc}$.

All checks succeed, hence $\mathsf{Verify}(\mathsf{pp}, \mathbf{Acc}, \mathbf{f}_i, w_i) = 1$. $\square$

4.2   Rejection Sampling, Acceptance Probability, and Bounds

Fix $i \in [K]$ and write

$$\mathbf{u}_i := \mathbf{sk}*\mathbf{c}_{i1} + \mathbf{f}_i*\mathbf{c}_{i2}, \qquad \mathbf{z}_i := \mathbf{u}_i + \mathbf{y}_i, \quad \mathbf{y}_i \leftarrow\!\$\, D_\sigma^N.$$

(1) Acceptance Probability. By Lemma 1, if we accept $(\mathbf{z}_i, \mathbf{u}_i)$ with probability

$$\mathsf{Prob}_{accept\_z} = \min\left\{1, \frac{D_\sigma^N(\mathbf{z}_i)}{M\, D_{\mathbf{u}_i,\sigma}^N(\mathbf{z}_i)}\right\},$$

then since $D_{\mathbf{c},\sigma}^N(\mathbf{x}) = \rho_\sigma(\mathbf{x} - \mathbf{c})/\rho_\sigma(\mathbb{Z}^N)$, the normalizing $\rho_\sigma(\mathbb{Z}^N)$ cancels and

$$\frac{D_\sigma^N(\mathbf{z}_i)}{M\, D_{\mathbf{u}_i,\sigma}^N(\mathbf{z}_i)} = \frac{1}{M} \cdot \frac{\rho_\sigma(\mathbf{z}_i)}{\rho_\sigma(\mathbf{z}_i - \mathbf{u}_i)}.$$

Using $\rho_\sigma(\mathbf{x}) = \exp\left(-\|\mathbf{x}\|_2^2/(2\sigma^2)\right)$ and $\|\mathbf{z}_i - \mathbf{u}_i\|_2^2 = \|\mathbf{z}_i\|_2^2 - 2\langle\mathbf{z}_i, \mathbf{u}_i\rangle + \|\mathbf{u}_i\|_2^2$, we obtain

$$\frac{\rho_\sigma(\mathbf{z}_i)}{\rho_\sigma(\mathbf{z}_i - \mathbf{u}_i)} = \exp\left(\frac{\|\mathbf{z}_i - \mathbf{u}_i\|_2^2 - \|\mathbf{z}_i\|_2^2}{2\sigma^2}\right) = \exp\left(\frac{-2\langle\mathbf{z}_i, \mathbf{u}_i\rangle + \|\mathbf{u}_i\|_2^2}{2\sigma^2}\right).$$

Hence our acceptance rule is exactly

$$\mathsf{Prob}_{accept\_z} = \min\left\{1, \frac{\exp\left(\frac{-2\langle\mathbf{z}_i, \mathbf{u}_i\rangle + \|\mathbf{u}_i\|_2^2}{2\sigma^2}\right)}{M}\right\}, \tag{1}$$

and, by Lemma 1, the distribution of accepted $\mathbf{z}_i$ is exactly D_σ^N (independent of $\mathbf{u}_i$). We will choose M deterministically in paragraph (3) so that the inner term in (1) is ≤ 1 on all accepted samples.

(2) Bounding a Single Transcript $\mathbf{z}_i$. Because an accepted $\mathbf{z}_i$ is distributed as D_σ^N, Lemma 2(2) implies $\Pr\left[\,\|\mathbf{z}_i\|_2 > 2\sigma\sqrt{N}\,\right] < 2^{-N}$. We therefore set

$$\mathsf{Bound}_z := 2\sigma\sqrt{N},$$

so a single transcript exceeds the bound with probability at most 2^{-N}.

(3) Choosing M. Write $\mathbf{u}_i = \mathbf{sk} * \mathbf{c}_{i1} + \mathbf{f}_i * \mathbf{c}_{i2}$. By Lemma 3 with the facts $\|\mathbf{sk}\|_\infty, \|\mathbf{f}_i\|_\infty \leq 1$ and $\|\mathbf{c}_{i1}\|_1 = \|\mathbf{c}_{i2}\|_1 = \kappa$, we have

$$\|\mathbf{u}_i\|_2 \;\leq\; \|\mathbf{sk}\|_2\|\mathbf{c}_{i1}\|_1 + \|\mathbf{f}_i\|_2\|\mathbf{c}_{i2}\|_1 \;\leq\; 2\kappa\sqrt{N}.$$

Let $U := 2\kappa\sqrt{N}$ and recall that accepted $\mathbf{z}_i$ satisfy $\|\mathbf{z}_i\|_2 \leq \mathsf{Bound}_z$. By Lemma 4, for any $\varepsilon > 0$,

$$-2\langle \mathbf{z}_i, \mathbf{u}_i\rangle + \|\mathbf{u}_i\|_2^2 \;\leq\; \varepsilon\|\mathbf{z}_i\|_2^2 + (1+\varepsilon^{-1})\|\mathbf{u}_i\|_2^2.$$

Optimizing over ε gives $\varepsilon^\star = U/\mathsf{Bound}_z \in (0,1]$, yielding the bound

$$-2\langle \mathbf{z}_i, \mathbf{u}_i\rangle + \|\mathbf{u}_i\|_2^2 \;\leq\; U^2 + 2U\,\mathsf{Bound}_z.$$

Hence the acceptance ratio in (1) is at most

$$\frac{\exp\!\big((U^2 + 2U\,\mathsf{Bound}_z)/(2\sigma^2)\big)}{M}.$$

Taking

$$M \;:=\; \exp\!\left(\frac{U^2 + 2U\,\mathsf{Bound}_z}{2\sigma^2}\right) \tag{2}$$

ensures that the inner term in (1) is ≤ 1 for all accepted $\mathbf{z}_i$.

(4) Bounding the Aggregate $\mathbf{Z}$. Let $\beta_i \in \{\pm 1\}$ and recall $\mathbf{Z} = \sum_{i=1}^K \beta_i \mathbf{z}_i$. By Lemma 1, the accepted $\mathbf{z}_i$ are i.i.d. as D_σ^N. By the discrete Gaussian closure identity $(\rho_{\sigma_1} * \rho_{\sigma_2} = \rho_{\sqrt{\sigma_1^2+\sigma_2^2}})$, the sum $\mathbf{Z} = \sum_{i=1}^K \beta_i\mathbf{z}_i$ is distributed as $D_{\sigma\sqrt{K}}^N$. Applying Lemma 2(2) with σ replaced by $\sigma\sqrt{K}$ gives

$$\Pr\!\left[\,\|\mathbf{Z}\|_2 \geq 2\sigma\sqrt{KN}\,\right] \;\leq\; 2^{-N}.$$

We therefore set

$$\mathsf{Bound}_Z := 2\sigma\sqrt{KN},$$

so that the probability of exceeding the bound is at most 2^{-N}, which is negligible in the security parameter λ whenever N grows at least linearly with λ. For intuition, a comparable bound follows from treating coordinates as subgaussian and applying Chernoff or Markov's inequality with a union bound over coordinates; details are omitted.

To avoid coefficient wrap-around in R_q with overwhelming probability, it suffices to take $q \geq 8\sigma\sqrt{KN}$ and $q \equiv 1 \pmod{2N}$; then $\|\mathbf{Z}\|_\infty < q/2$ except with probability at most 2^{-N}.

(5) Truncating $\mathbf{y}_i$. We resample $\mathbf{y}_i$ until $\|\mathbf{y}_i\|_\infty \le s_y$. By comparing the discrete Gaussian tail to the continuous Gaussian tail and applying a standard Mills-ratio bound (valid for s_y/σ larger than a small constant), one obtains

$$\Pr\big[\,\|\mathbf{y}_i\|_\infty > s_y\,\big] \;\le\; 2N\,e^{-\,s_y^2/(2\sigma^2)}.$$

Choosing $s_y = \sigma\sqrt{2\big(\lambda\ln 2 + \ln(2N)\big)}$, equivalently $s_y = \alpha\,\sigma\sqrt{\ln N}$ and $\alpha = \sqrt{\frac{2(\lambda\ln 2 + \ln(2N))}{\ln N}}$, ensures

$$\Pr\big[\,\|\mathbf{y}_i\|_\infty > s_y\,\big] \;\le\; 2^{-\lambda}.$$

Since truncation removes only a $2^{-\lambda}$ fraction of proposals, the accepted $\mathbf{z}_i$ law differs from D_σ^N by at most $2^{-\lambda}$ in statistical distance, so the bounds Bound_z and Bound_Z discussed above remain valid up to an additional $K\cdot 2^{-\lambda}$ term.

4.3   Unforgeability of Signature Component

At first glance it may seem sufficient to argue accumulator security directly. However, the COMPASS construction (as in [23]) internally employs a signature-like transcript mechanism: each $(\mathbf{z}, \mathbf{c}_1, \mathbf{c}_2)$ in witness is bound to the $(\mathbf{sk}, \mathbf{f}, \mathbf{y})$. If this mechanism were forgeable, an adversary could create a valid-looking transcript without solving the underlying lattice problem, and thereby produce false membership proofs. To prevent this shortcut, we first establish *unforgeability* of the transcript-generation component under chosen-message attack. With this guarantee in place, the subsequent accumulator security proof can safely reduce to the hardness of the underlying lattice assumptions.

Signature subroutine. COMPASS' security partly relies on the unforgeability of an embedded signature-like routine, which behaves similarly to a PASS_G signature on a "message digest" $f \in \mathcal{B}^\infty(1)$ obtained from a message by the hash-to-lattice map. We assume that global system parameters $(N, q, \Omega, H_{c1}, H_{c2})$ are fixed by a system-wide setup. The signature subroutine is defined as the triple (KeyGenLite, SignLite, VerifyLite), where the manager holds secret key $\mathbf{sk}$ and public key $\mathbf{pk} = F_\Omega(\mathbf{sk})$. In short, KeyGenLite$(1^\lambda)$ samples $(\mathbf{sk}, \mathbf{pk})$ given the fixed parameters, while SignLite$(\mathbf{sk}, f)$ and VerifyLite$(\mathbf{pk}, f, \sigma)$ are extracted from textsfCOMPASS as shown in Algorithm 5,6. We analyze SignLite in the random oracle model for H_{c1}, H_{c2} with fixed domain separation as described in Sect. 3.

Theorem 2 (EUF–CMA of signature subroutine under V–SIS).
Assume the Vandermonde-$\mathrm{SIS}^{\mathcal{K}}_{q,t,N,\beta}$ problem (Definition 3) is hard. Then for any PPT adversary $\mathcal{A}$ making adaptive signing queries, the probability that $\mathcal{A}$ outputs a fresh forgery $(f^\star, \sigma^\star)$ against (KeyGenLite, SignLite, VerifyLite) is negligible, unless it can find a nonzero $\mathbf{v}$ with $\|\mathbf{v}\|_2 \le \beta$ and $\mathbf{F}_\Omega \mathbf{v} \equiv 0 \pmod{q}$. Hence, the signature subroutine is EUF–CMA secure under the V–SIS assumption.

Algorithm 5. SignLite($\mathbf{sk}, f$) relative to fixed $(N, q, \Omega, H_{c1}, H_{c2})$

1: Sample $\mathbf{y} \leftarrow D_\sigma^N$; if $\|\mathbf{y}\|_\infty > s_y$ then resample.
2: $\hat{\mathbf{y}}|_\Omega \leftarrow F_\Omega(\mathbf{y})$.
3: Compute dual challenges: $\mathbf{c}_1 \leftarrow H_{c1}(\hat{\mathbf{y}}|_\Omega, \hat{f}|_\Omega)$, $\mathbf{c}_2 \leftarrow H_{c2}(\hat{\mathbf{y}}|_\Omega, \mathbf{pk})$.
4: Compute $\mathbf{z} = \mathbf{sk} * \mathbf{c}_1 + f * \mathbf{c}_2 + \mathbf{y}$.
5: **Accept z** with probability $\mathbf{prob}_{accept\_z}$ per (1), where $\mathbf{u} = \mathbf{sk} * \mathbf{c}_1 + f * \mathbf{c}_2$.
6: **If rejected**, restart from Step 1. **If accepted**, output $(\mathbf{z}, \mathbf{c}_1, \mathbf{c}_2)$.

Algorithm 6. VerifyLite($\mathbf{pk}, f, \sigma$) relative to fixed $(N, q, \Omega, H_{c1}, H_{c2})$

1: Parse $\sigma = (\mathbf{z}, \mathbf{c}_1, \mathbf{c}_2)$.
2: If $\|\mathbf{z}\|_2 > \mathsf{Bound}_z$ then reject.
3: Compute $\hat{\mathbf{y}}'|_\Omega \leftarrow F_\Omega(\mathbf{z}) - \mathbf{pk} \odot F_\Omega(\mathbf{c}_1) - F_\Omega(f) \odot F_\Omega(\mathbf{c}_2) \pmod q$.
4: Recompute $\mathbf{c}_1', \mathbf{c}_2'$ as in SignLite.
5: Accept iff $\mathbf{c}_1' = \mathbf{c}_1$ and $\mathbf{c}_2' = \mathbf{c}_2$, otherwise reject.

Proof. Let $\mathcal{A}$ be a PPT adversary that forges a fresh $(f^\star, \sigma^\star)$ against the signature subroutine with non-negligible probability. We construct a PPT reduction $\mathcal{R}$ that, given an instance $(q, t, N, \beta, \mathbf{F}_\Omega)$ of Vandermonde–$\mathsf{SIS}_{q,t,N,\beta}^{\mathcal{K}}$, outputs a nonzero $\mathbf{v}$ with $\|\mathbf{v}\|_2 \leq \beta$ and $\mathbf{F}_\Omega \mathbf{v} \equiv \mathbf{0} \pmod q$.

Constructing $\mathcal{R}$. $\mathcal{R}$ runs KeyGenLite(1^λ) honestly to obtain $(\mathbf{pk}, \mathbf{sk})$, publishes $\mathbf{pk}$ to $\mathcal{A}$, and answers:

- *Signing queries f:* run SignLite *honestly*, including truncation $\|\mathbf{y}\|_\infty \leq s_y$ and rejection sampling per Lemma 1; return $(\mathbf{z}, \mathbf{c}_1, \mathbf{c}_2)$.
- *Random oracles H_{c1}, H_{c2}:* respond as random functions with fixed domain separation, programmed only to ensure consistency on repeated queries.

Let Q be the (polynomial) number of signing queries.

Good events definition. We define the events:

- $\mathsf{Good}_{\mathrm{rej}}$: every returned transcript is distributed as in Lemma 1.
- $\mathsf{Good}_{\mathrm{trunc}}$: every proposal satisfies $\|\mathbf{y}\|_\infty \leq s_y$.

By subsect. 4.2(1),(5), the distance from ideal D_σ^N due to rejection sampling and truncation is negligible; hence $\Pr[\neg \mathsf{Good}_{\mathrm{rej}}] \leq Q \cdot \mathsf{negl}(\lambda)$ and $\Pr[\neg \mathsf{Good}_{\mathrm{trunc}}] \leq Q \cdot 2^{-\lambda}$. Set $\mathsf{Good} := \mathsf{Good}_{\mathrm{rej}} \wedge \mathsf{Good}_{\mathrm{trunc}}$, so $\Pr[\neg \mathsf{Good}] = \mathsf{negl}(\lambda)$.

Adversary's forgery. Suppose $\mathcal{A}$ outputs a fresh forgery $(f^\star, \sigma^\star)$ where $\sigma^\star = (\mathbf{z}^\star, \mathbf{c}_1^\star, \mathbf{c}_2^\star)$ and VerifyLite($f^\star, \sigma^\star$) $= 1$. By the verification equations (Algorithm 4),

$$\hat{\mathbf{y}}^\star\big|_\Omega \equiv \hat{\mathbf{z}}^\star\big|_\Omega - \mathbf{pk} \odot \hat{\mathbf{c}}_1^\star\big|_\Omega - \hat{f}^\star\big|_\Omega \odot \hat{\mathbf{c}}_2^\star\big|_\Omega \pmod q,$$

for some (implicitly determined) $\hat{\mathbf{y}}^\star\big|_\Omega$. By linearity/multiplicativity of F_Ω,

$$\mathbf{F}_\Omega(\mathbf{sk} * \mathbf{c}_1^\star + f^\star * \mathbf{c}_2^\star + \mathbf{y}^\star - \mathbf{z}^\star) \equiv \mathbf{0} \pmod q. \tag{3}$$

Define
$$\mathbf{v} := \mathbf{sk}*\mathbf{c}_1^\star + f^\star*\mathbf{c}_2^\star + \mathbf{y}^\star - \mathbf{z}^\star \in \mathbb{Z}_q^N.$$

By (3), $\mathbf{F}_\Omega \mathbf{v} \equiv \mathbf{0} \pmod{q}$ and $\mathbf{v}$ lies in the kernel of $\mathbf{F}_\Omega$ modulo q.

Nontriviality. We show $\Pr[\mathbf{v} = \mathbf{0}]$ is negligible. If $\mathbf{v} = \mathbf{0}$, then $\mathbf{z}^\star = \mathbf{u}^\star + \mathbf{y}^\star$ with $\mathbf{u}^\star := \mathbf{sk}*\mathbf{c}_1^\star + f^\star*\mathbf{c}_2^\star$, so $(\mathbf{z}^\star, \mathbf{c}_1^\star, \mathbf{c}_2^\star)$ is distributed as an *honest* signature on $f^\star$, contradicting freshness unless $\mathcal{A}$ reproduced an existing transcript or caused an RO collision. Under ROM consistency and Good, both events have negligible probability. Hence,

$$\Pr[\mathbf{v} \neq \mathbf{0}] \geq \Pr[\mathcal{A} \text{forges}] - \mathrm{negl}(\lambda).$$

Norm bound. We recall the parameter bounds used throughout Algorithm 5,6:

- $\|\mathbf{c}_j^\star\|_1 = \kappa$; $\|\mathbf{sk}\|_2, \|f^\star\|_2 \leq \sqrt{N}$ (by construction);
- $\|\mathbf{y}^\star\|_\infty \leq s_y$ (truncation) so $\|\mathbf{y}^\star\|_2 \leq \sqrt{N}\,s_y$ from Sect. 2.3; and
- $\|\mathbf{z}^\star\|_2 \leq \mathsf{Bound}_z$ (rejection cap) from Sect. 4.2.

For the negacyclic convolution $*$ over $\mathbb{Z}_q[x]/(x^N + 1)$, Young's inequality (Lemma 3) gives $\|a*c\|_2 \leq \|a\|_2 \|c\|_1$. Hence

$$\|\mathbf{sk}*c_1^\star\|_2 \leq \kappa\sqrt{N}, \qquad \|f^\star*c_2^\star\|_2 \leq \kappa\sqrt{N}, \qquad \|y^\star\|_2 \leq \sqrt{N}\,s_y, \qquad \|z^\star\|_2 \leq \mathsf{Bound}_z.$$

By the triangle inequality,

$$\begin{aligned}
\|\mathbf{v}\|_2 &\leq \|\mathbf{sk}*\mathbf{c}_1^\star\|_2 + \|f^\star*\mathbf{c}_2^\star\|_2 + \|\mathbf{y}^\star\|_2 + \|\mathbf{z}^\star\|_2 \\
&\leq 2\kappa\sqrt{N} + \sqrt{N}\,s_y + \mathsf{Bound}_z =: \beta,
\end{aligned}$$

which is exactly the β used in the V–SIS instance.

Success probability and tightness. $\mathcal{R}$ runs in polynomial time and is straight-line (no rewinding). The only losses are the negligible probabilities from $\neg\mathsf{Good}$ and from ROM collisions, so

$$\Pr[\mathcal{R} \text{ outputs a V–SIS solution}] \geq \Pr[\mathcal{A} \text{ forges}] - \mathrm{negl}(\lambda).$$

If V–SIS$_{q,t,N,\beta}^{\mathcal{K}}$ is hard, then $\Pr[\mathcal{A} \text{ forges}]$ must be negligible. Therefore the signature subroutine is EUF–CMA secure. $\qquad\square$

Remark 1 (QROM). Our reduction neither rewinds nor forks the adversary and uses the random oracles H_{c1}, H_{c2} solely for consistency (lazy sampling) on the classical inputs revealed during signing and verification. The proof and extractor are straight-line and algebraic (no post hoc reprogramming). Therefore the same argument extends verbatim to the quantum random-oracle model (QROM), where H_{c1}, H_{c2} are implemented as quantum-accessible lazily sampled oracles. In particular, the signing simulation and the extraction of $\mathbf{v} = \mathbf{sk}*\mathbf{c}_1^\star + f^\star*\mathbf{c}_2^\star + \mathbf{y}^\star - \mathbf{z}^\star$ proceed unchanged.

4.4 Security of the COMPASS Scheme

Theorem 3 (Security of COMPASS). *Assume the Vandermonde–$SIS_{q,t,N,\beta}^{\mathcal{K}}$ problem (Definition 3) is hard. Then any PPT adversary $\mathcal{A}$ that outputs a fresh accepting witness $w^\star$ for some $f^\star$ relative to an honestly generated $(\mathsf{pp},\mathbf{pk},\mathbf{sk})$ has at most negligible advantage. Equivalently, a successful forger yields, with comparable probability, a nonzero $\mathbf{v}$ with $\|\mathbf{v}\|_2 \leq \beta$ and $\mathbf{F}_\Omega\mathbf{v} \equiv \mathbf{0} \pmod{q}$.*

Proof Let $\mathcal{A}$ interact with $(\mathsf{pp},\mathbf{pk})$ and output a fresh $(f^\star, w^\star)$ that verifies. As in Theorem 2, by Lemma 1 and the truncation analysis in §4.2(5), the accepted-transcript distribution differs from the ideal D_σ^N experiment by at most $\varepsilon_{\mathrm{tot}}(N,\lambda) = \mathrm{negl}(N) + 2^{-\lambda}$ in total variation. We analyze the ideal experiment; the overall success probability changes by at most $\varepsilon_{\mathrm{tot}}$.

Signature-consistency: reference to Theorem 2 . Let $w^\star = (\mathbf{Z}_i, \mathbf{z}^\star, \mathbf{c}_1^\star, \mathbf{c}_2^\star)$. The check in Algorithm 4 that reconstructs $\hat{\mathbf{y}}'\big|_\Omega = F_\Omega(\mathbf{z}^\star) - \mathbf{pk}\odot F_\Omega(\mathbf{c}_1^\star) - F_\Omega(f^\star)\odot F_\Omega(\mathbf{c}_2^\star)$ and verifies $(\mathbf{c}_1^\star, \mathbf{c}_2^\star)$ is exactly the signature subroutine's verification. Hence, by Theorem 2, from any successful forgery we obtain a nonzero kernel vector

$$\mathbf{v}_{\mathrm{sig}} \;=\; \mathbf{sk}*\mathbf{c}_1^\star + f^\star*\mathbf{c}_2^\star + \mathbf{y}^\star - \mathbf{z}^\star \quad \text{with} \quad F_\Omega(\mathbf{v}_{\mathrm{sig}}) \equiv \mathbf{0} \pmod{q},$$

and

$$\|\mathbf{v}_{\mathrm{sig}}\|_2 \;\leq\; 2\kappa\sqrt{N} + \sqrt{N}\, s_y + \mathsf{Bound}_z \;=:\; \beta_{\mathrm{sig}}. \tag{4}$$

Accumulator consistency. Algorithm 2 enforces $\|\mathbf{Z}\|_2 \leq \mathsf{Bound}_Z$ for the aggregate $\mathbf{Z} := \sum_i \beta_i \mathbf{z}_i$, and Algorithm 4 reconstructs $\mathbf{Z}' := \mathbf{Z}_i + \beta^\star \mathbf{z}^\star$, checks $\|\mathbf{Z}'\|_2 \leq \mathsf{Bound}_Z$, and accepts only if

$$H_{\mathrm{Acc}}\big(F_\Omega(\mathbf{Z}')\big) \;=\; \mathsf{Acc} \;=\; H_{\mathrm{Acc}}\big(F_\Omega(\mathbf{Z})\big).$$

By ROM/CRHF soundness (negligible collision probability), this implies $F_\Omega(\mathbf{Z}' - \mathbf{Z}) \equiv \mathbf{0} \pmod{q}$ except with negligible probability. Define

$$\mathbf{v}_{\mathrm{acc}} := \mathbf{Z}' - \mathbf{Z}, \quad \text{so} \quad F_\Omega(\mathbf{v}_{\mathrm{acc}}) \equiv \mathbf{0} \pmod{q}.$$

Using the enforced bounds,

$$\|\mathbf{v}_{\mathrm{acc}}\|_2 \;\leq\; \|\mathbf{Z}'\|_2 + \|\mathbf{Z}\|_2 \;\leq\; 2\,\mathsf{Bound}_Z \;=:\; \beta_{\mathrm{acc}}. \tag{5}$$

Conclusion. From a successful forgery we obtain a nonzero kernel vector $\mathbf{v}_{\mathrm{sig}}$ with bound β_{sig}; the accumulator check independently yields a (possibly zero) kernel vector $\mathbf{v}_{\mathrm{acc}}$ bounded by β_{acc}. Instantiating the Vandermonde–SIS reduction with

$$\beta := \beta_{\mathrm{sig}}$$

gives a valid solution with probability at least $\Pr[\mathcal{A}\ \text{forges}] - \varepsilon_{\mathrm{tot}} - \mathrm{negl}(\lambda)$. By hardness of Vandermonde–$\mathrm{SIS}_{q,t,N,\beta}^{\mathcal{K}}$, the forging advantage is negligible. $\square$

Remark 2 (Tighter bound β). With $\mathsf{Bound}_z = 2\sigma\sqrt{N}$, $\mathsf{Bound}_Z = 2\sigma\sqrt{KN}$ and $s_y = \sigma\sqrt{2(\lambda\ln 2 + \ln(2N))}$, we have

$$\beta_{\mathrm{sig}} = 2\kappa\sqrt{N} + \sigma\sqrt{N}\left(\sqrt{2(\lambda\ln 2 + \ln(2N))} + 2\right), \qquad \beta_{\mathrm{acc}} = 4\sigma\sqrt{KN}.$$

For PASS_G-scale parameters (e.g., $\sigma \approx 3000$), the term $2\kappa\sqrt{N}$ is negligible relative to the $\sigma\sqrt{N}$ component, and the ratio simplifies to

$$\frac{\beta_{\mathrm{acc}}}{\beta_{\mathrm{sig}}} \approx \frac{4\sqrt{K}}{\sqrt{2(\lambda\ln 2 + \ln(2N))} + 2}.$$

Hence the crossover occurs at

$$K^\star = \left(\frac{\sqrt{2(\lambda\ln 2 + \ln(2N))} + 2}{4}\right)^2,$$

which is essentially independent of σ. For $\lambda = 128$ and $N \in \{512, 1024, 2048\}$ we obtain $K^\star \in [15.7, 15.9]$. Consequently, for practical set sizes $K \geq 10^3$,

$$\beta = \min\{\beta_{\mathrm{sig}}, \beta_{\mathrm{acc}}\} = \beta_{\mathrm{sig}},$$

since β_{acc} grows as $\sqrt{K}$ while β_{sig} is K-independent.

5   Implementation and Discussion

5.1   Implementation

Our implementation is open-sourced for reproducibility[1]. We realized the prototype in Python/NumPy on Google Colab, using dense integer arrays for coefficients and SHAKE256 for all hashes. Rather than restating the algebra (already given in Sect. 2 and Sect. 3), we summarize only the main implementation decisions.

Challenge Generation. Per-element challenges are domain-separated and computed as $c_{1,i} = \mathsf{FormatC}(\mathsf{Hash}(\mathsf{ctx}_1 \,\|\, \hat{y}_i|_\Omega \,\|\, \hat{f}_i|_\Omega)))$ and $c_{2,i} = \mathsf{FormatC}(\mathsf{Hash}(\mathsf{ctx}_2 \,\|\, \hat{y}_i|_\Omega \,\|\, \mathbf{pk}))$. Here $\mathsf{FormatC}$ deterministically maps XOF output to a κ-sparse ternary vector. This makes each challenge binding to its context and reproducible for test vectors.

Acceptance Test. For each candidate $z_i = \mathsf{sk}\star c_{1,i} + f_i \star c_{2,i} + y_i$, we apply both an ℓ_2 bound and a Lyubashevsky-style probabilistic rejection with precomputed normalizer M. This ensures transcript independence of the secret key, matching the PASS family design.

---

[1] Available at https://github.com/khsjds/COMPASS.

Encoding. We use compact PASS-style challenges (c_1, c_2) (8 bytes of sign bits plus 2κ index bytes) and fixed-width mod-q packing for coefficients. Let $b_q = \lceil \log_2 q / 8 \rceil$ be the bytes per coefficient; in our runs $b_q = 4$. With the coeff-domain witness (Z_i, z_i, c_1, c_2) the size is $2N \cdot b_q + 16 + 4\kappa$ bytes.

Pipeline. The code follows the sequence: `keygen` → `accumulate` → `witness` → `serialize`/`deserialize` → `verify`, with bit-for-bit equality before and after packing.

Complexity and Optimization. The PoC evaluates F_Ω via a precomputed Vandermonde matrix $P \in \mathbb{Z}_q^{t \times N}$ and computes $\hat{f}|_\Omega = Pf \bmod q$. Time is $O(tN)$; for large q we use chunked accumulation with periodic modular reduction to avoid 64-bit overflow. An NTT backend could reduce convolution to $O(N \log N)$ and sparse back-substitution can lower the F_Ω constant factors without changing transcript formats.

PASS$_G$ -Style Parameter Points and Benchmarks. To validate our implementation and enable comparison with prior PASS-lineage work [20], we derived two parameter sets starting from PASS$_G$ baselines. Since PASS$_G$ uses small σ values yielding impractically low acceptance rates in our rejection sampling framework, we increased σ while maintaining 128-bit quantum security. Parameters (Table 3) balance witness size, security margin, and efficiency under the constraint $q \equiv 1 \pmod{2N}$.

Table 3. COMPASS performance at PASS$_G$-style parameter points (K=100, 10 runs).

Metric	Set 1	Set 2	Metric	Set 1	Set 2
Parameters			*Theory vs. Empirical*		
Max K	10,000	10,000	M	2,877	2.41
N	512	1024	Accept. (th.)	3.48×10^{-4}	0.415
t	256	512	Accept. (emp.)	3.51×10^{-4}	0.41
κ	44	36	Rel. error	$+0.9\%$	-1.2%
σ	11,336	167,771	*Verification & Security*		
q	205,207,553	4,294,957,057	Verify (ms)	3.62 ± 1.28	79.69 ± 18.15
Performance (per member)			Witness (B)	4,288	8,352
Attempts	$2,891 \pm 223$	2.43 ± 0.15	Accum. (B)	512	512
Time/att. (ms)	2.47 ± 0.16	27.95 ± 2.38	δ (BKZ)	1.0070	1.0042
Accum. (ms)	$7,155 \pm 681$	67.9 ± 6.4	Security	128-bit	128-bit

We ran 10 independent trials accumulating 100 members each (Google Colab, Oct. 2025; Python 3.9, NumPy 1.26). Both achieve 128-bit quantum security for

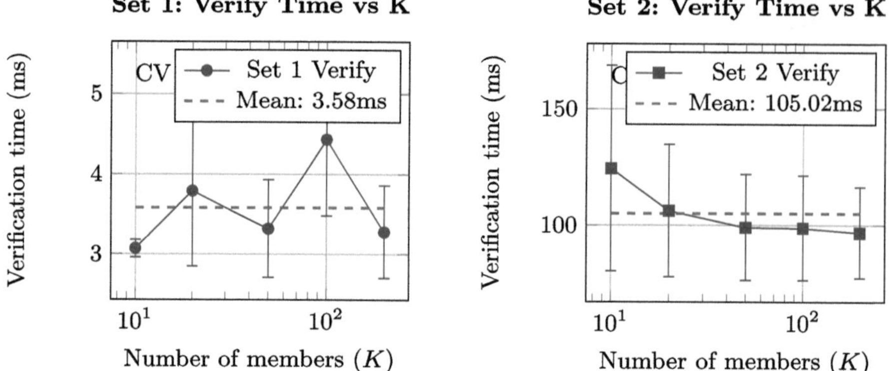

Fig. 1. Per-member verification time remains constant across accumulator sizes for both parameter sets (3 runs each), confirming sampling independence.

up to K=10,000 members under the BKZ-GSA analysis [20]: BKZ-2.0 root-Hermite factors yield δ=1.0070 (Set 1) and δ=1.0042 (Set 2), both within the 128-bit threshold.

The results reveal a fundamental tradeoff. Set 1 achieves fast verification (3.6 ms) but requires ~2,900 rejection attempts (7.2 s/member). Set 2 increases σ by 15×, yielding 41% acceptance and 105× faster accumulation (68 ms) despite 11× per-attempt overhead, but 22× slower verification (80 ms). Theoretical and empirical acceptance rates agree within 2%.

To validate that per-member costs remain independent of accumulator size, we tested both sets across $K \in \{10, 20, 50, 100, 200\}$ (3 runs each, Fig. 1). Coefficients of variation (CV=13.6% for Set 1, CV=9.7% for Set 2) confirm constant per-member metrics, consistent with theoretical sampling independence.

5.2   Discussion

While elliptic-curve based accumulators have attractive short proofs, they rely on hardness assumptions (ECDLP) that are vulnerable to quantum attacks, and thus are not PQ secure. In contrast, both our accumulator and [23] are lattice-based and remain plausible under quantum adversaries.

Efficiency and Practicality. Table 4 summarizes recent accumulator schemes (within the past three years) across lattice-, RSA-, and Merkle-based approaches. Among PQ candidates, Merkle-tree accumulators [29] incur logarithmic witnesses and verification, whereas both our construction and [23] achieve *constant*-size proofs with *constant*-time verification under lattice assumptions and transparent setup. Zhao et al. [33] also propose a lattice-based accumulator with verification independent of the set size, but requiring a trapdoor and yielding proofs of size $\tilde{O}(\lambda)$. RSA-based designs such as [16] offer constant-size/time performance but are not PQ-safe. Overall, only trapdoor-free lattice schemes combine PQ security, $O(1)$ verification, and predictable small proofs, making them particularly attractive for bandwidth- and CPU-constrained clients (e.g., IoT or VC revocation).

Table 4. Efficiency comparison of recent accumulators.

Scheme	Assumption	Proof size	Verify time	PQ-safe	Trapdoor
This work (lattice)	vSIS / PV-knapsack	$O(1)$	$O(1)$	✓	No
[33] (lattice)	Lattice	$\tilde{O}(\lambda)$	$\tilde{O}(\lambda)$	✓	Yes
[16] (RSA)	Strong-RSA	$O(1)$	$O(1)$	✗	Yes
[29] (Merkle tree)	CRHF	$O(\log n)$	$O(\log n)$	✓	No

Concrete Size Comparison. To compare structural efficiency on equal footing, we evaluate each scheme using parameter points corresponding to the *128-bit security level claimed for their underlying signature schemes*. For COMPASS, we use the two PASS_G-style configurations reported in Sect. 5.1, and for [23] we reconstruct their setting following the PASS_{RS} Set 4 parameters ($N{=}1153$, $t{=}600$, $q{=}968{,}521$) from [12], which their HPSSW variant cites as providing comparable security. All sizes were recomputed under consistent packing and serialization rules.

Since [23] did not publish an implementation nor parameter set, we re-implemented their HPSSW-style accumulator following their algorithmic description, including both $\hat{y}\,|_\Omega$ and y_i components, and measured the serialized witness size as received by a verifier. For COMPASS, the compact unified witness (Z_i, z_i, c_1, c_2) was measured using fixed-width mod-q packing as in our proof-of-concept.

Under these matched nominal conditions, COMPASS produces witnesses up to 52% smaller than those reconstructed from [23]. However, these parameters should be regarded only as nominal examples rather than safe instantiations. In the HPSSW-style configuration adopted by [23], the ratio $t/N \approx 0.52$ exceeds the $1/2$ threshold later shown by [8] to enable efficient order-2 symmetry attacks in PASS-lineage constructions. By contrast, COMPASS can freely reduce t/N to the conservative $[0.33, 0.40)$ window without harming correctness or performance: smaller t lowers both the Fourier transform cost and the total witness size, while maintaining secure and rejection-efficient sampling. Thus, Table 5 highlights the structural compactness and flexibility of COMPASS rather than any direct parity in security level.

Table 5. Witness sizes under nominal PASS-lineage parameter scales.

Scheme	N	q	Params	Witness size
Ours (Set 1)	512	205,207,553	$\kappa=44$, $\sigma=11{,}336$	4.29 KiB
Ours (Set 2)	1024	4,294,957,057	$\kappa=36$, $\sigma=167{,}771$	8.35 KiB
[23] (w/ [12] Set 4)	1153	968,521	$b(=\kappa)=36, t=600$	8.99 KiB

(i) All rows use identical serialization conventions and mod-q fixed-width packing. (ii) COMPASS witnesses correspond to the unified transcript (Z_i, z_i, c_1, c_2). (iii) The Maeno et al. entry is based on our independent re-implementation of their HPSSW-style witness (signature σ plus (Z_i, y_i)), which is inherently larger because it transmits both $\hat{y}|_\Omega$ and y_i components. (iv) These results reflect structural compactness only; performance and security-level measurements for COMPASS appear in subsection5.1.

Security Notes and Known Weaknesses of the PASS Lineage The hardness of PASS-lineage schemes such as [12,20] rests on structured lattice assumptions (vSIS / Partial Vandermonde). Recent works have identified caveats: Boudgoust et al. [5] show that certain frequency subsets Ω and parameter ratios can yield "easy instances" of Ideal-SVP, while Das-Joux [8] extend these results to practical key-recovery attacks on broader weak classes.

Mitigations in COMPASS. Our construction already incorporates two safeguards standard: (i) strict L_∞ bounds on sampled y combined with Lyubashevsky-style rejection sampling to ensure transcript security and independence from (sk, f); (ii) domain-separated challenges c_1, c_2 derived from distinct XOF contexts with explicit inclusion of Ω, the element index, and public key material.

Deployment Guidance. In addition to the above, secure use of COMPASS requires careful parameterization: choose t/N conservatively in the window $1/3 \le t/N < 1/2$ with Ω drawn uniformly at random, and conduct simple sanity checks on secret keys to avoid unusually weak samples. Unlike trapdoor-based designs, our accumulator operates with few long-lived keys, limiting exposure to the multi-transcript attacks described in [5,8]. We therefore conclude that, with conservative parameters, COMPASS remains *safe enough* for deployment in accumulator settings.

6   Conclusion

We have presented COMPASS, an improved lattice-based accumulator in the PASS lineage that retains PQ security under a transparent, trapdoor-free setup. Our construction combines compact PASS-style encodings with an accumulator-tailored design, yielding significantly smaller concrete witnesses without sacrificing verification simplicity.

Our proof-of-concept implementation, evaluated on two PASS_G-style parameter points, achieved witnesses as small as of 4.3 KiB and verification times as short as 3.6 ms while maintaining the intended 128-bit BKZ-GSA security margin. These results demonstrate the practicality of our design for credential-revocation and other low-latency verification settings. Compared with our reimplementation of Maeno et al., COMPASS achieves up to 52% smaller witnesses under comparable nominal parameter scales.

Nevertheless, recent analyses [5,8] highlight structural risks in PASS-lineage schemes when t/N exceeds $1/2$. Our formulation mitigates this by allowing conservative ratios in $[0.33, 0.40)$, which both improves soundness against symmetry attacks and reduces the computational footprint. Because accumulators require only a few long-lived keys, exposure to multi-transcript threats is inherently limited, making such conservative parameterization practical.

In summary, COMPASS advances lattice-based accumulators by demonstrating that concrete compactness, transparent setup, and tunable security can coexist in a single post-quantum framework. The presented implementation establishes a reproducible baseline for future refinements and for fully BKZ-verified parameter sets aimed at long-term deployment.

Acknowledgments. This research was supported by the National Science and Technology Council (NSTC), Taiwan (ROC), under Project Numbers NSTC 114–2634-F-004–002-MBK and NSTC 114–2221-E-004–009-MY3.

References

1. Au, M.H., Tsang, P.P., Susilo, W., Mu, Y.: Dynamic Universal Accumulators for DDH Groups and Their Application to Attribute-Based Anonymous Credential Systems. In: Fischlin, M. (ed.) CT-RSA 2009. LNCS, vol. 5473, pp. 295–308. Springer, Heidelberg (2009). https://doi.org/10.1007/978-3-642-00862-7_20
2. Barić, N., Pfitzmann, B.: Collision-free accumulators and fail-stop signature schemes without trees. In: Fumy, W. (ed.) Advances in Cryptology – EUROCRYPT '97, pp. 480–494. Springer, Berlin Heidelberg, Berlin, Heidelberg (1997)
3. Benaloh, J., de Mare, M.: One-way accumulators: a decentralized alternative to digital signatures. In: Helleseth, T. (ed.) Advances in Cryptology – EUROCRYPT '93, pp. 274–285. Springer, Berlin Heidelberg, Berlin, Heidelberg (1994)
4. Bloom, B.H.: Space/time trade-offs in hash coding with allowable errors. Commun. ACM **13**(7), 422–426 (1970). https://doi.org/10.1145/362686.362692
5. Boudgoust, K., Gachon, E., Pellet-Mary, A.: Some easy instances of ideal-SVP and implications on the partial vandermonde knapsack problem. In: Advances in Cryptology - CRYPTO 2022: 42nd Annual International Cryptology Conference, CRYPTO 2022, Santa Barbara, CA, USA, August 15–18, 2022, Proceedings, Part II, pp. 480–509. Springer-Verlag, Berlin, Heidelberg (2022). https://doi.org/10.1007/978-3-031-15979-4_17
6. Camenisch, J., Kohlweiss, M., Soriente, C.: An Accumulator Based on Bilinear Maps and Efficient Revocation for Anonymous Credentials. In: Jarecki, S., Tsudik, G. (eds.) PKC 2009. LNCS, vol. 5443, pp. 481–500. Springer, Heidelberg (2009). https://doi.org/10.1007/978-3-642-00468-1_27

7. Camenisch, J., Lysyanskaya, A.: Dynamic Accumulators and Application to Efficient Revocation of Anonymous Credentials. In: Yung, M. (ed.) CRYPTO 2002. LNCS, vol. 2442, pp. 61–76. Springer, Heidelberg (2002). https://doi.org/10.1007/3-540-45708-9_5

8. Das, D., Joux, A.: Key recovery attack on the partial vandermonde knapsack problem. In: Joye, M., Leander, G. (eds.) Advances in Cryptology - EUROCRYPT 2024, pp. 205–225. Springer Nature Switzerland, Cham (2024)

9. Doröz, Y., Hoffstein, J., Silverman, J.H., Sunar, B.: MMSAT: a scheme for multimessage multiuser signature aggregation. Cryptology ePrint Archive, Paper 2020/520 (2020). https://eprint.iacr.org/2020/520

10. Grafakos, L.: Classical fourier analysis, graduate texts in mathematics, vol. 249. Springer, New York, 3 edn. (2014). https://doi.org/10.1007/978-1-4939-1194-3

11. Hoffstein, J., Lieman, D., Silverman, J.H.: Polynomial rings and efficient public key authentication. In: Blum, M., Lee, C.H. (eds.) Proceedings of the International Workshop on Cryptographic Techniques and E-Commerce (CrypTEC '99). City University of Hong Kong Press, Hong Kong (1999)

12. Hoffstein, J., Pipher, J., Schanck, J.M., Silverman, J.H., Whyte, W.: Practical signatures from the partial fourier recovery problem. In: Boureanu, I., Owesarski, P., Vaudenay, S. (eds.) Applied Cryptography and Network Security, pp. 476–493. Springer International Publishing, Cham (2014)

13. Hoffstein, J., Silverman, J.H.: Polynomial rings and efficient public key authentication ii. In: Lam, K.Y., Shparlinski, I., Wang, H., Xing, C. (eds.) Cryptography and Computational Number Theory, pp. 269–286. Birkhäuser Basel, Basel (2001)

14. Jhanwar, M.P., Safavi-Naini, R.: Compact accumulator using lattices. In: Chakraborty, R.S., Schwabe, P., Solworth, J. (eds.) Security, Privacy, and Applied Cryptography Engineering, pp. 347–358. Springer International Publishing, Cham (2015)

15. Jiang, W., Li, J., Guo, Y., Zhang, H.: An ordered universal accumulator based on a hash chain. Applied Sciences 15(5) (2025). https://doi.org/10.3390/app15052565, https://www.mdpi.com/2076-3417/15/5/2565

16. Kemmoe, V.Y., Lysyanskaya, A.: RSA-based dynamic accumulator without hashing into primes. In: Proceedings of the 2024 on ACM SIGSAC Conference on Computer and Communications Security, pp. 4271–4285. CCS '24, Association for Computing Machinery, New York, NY, USA (2024). https://doi.org/10.1145/3658644.3690199, https://doi.org/10.1145/3658644.3690199

17. Li, J., Li, N., Xue, R.: Universal accumulators with efficient nonmembership proofs. In: Katz, J., Yung, M. (eds.) Applied Cryptography and Network Security, pp. 253–269. Springer, Berlin Heidelberg, Berlin, Heidelberg (2007)

18. Libert, B., Ling, S., Nguyen, K., Wang, H.: Lattice-Based Zero-Knowledge Arguments for Integer Relations. In: Shacham, H., Boldyreva, A. (eds.) CRYPTO 2018. LNCS, vol. 10992, pp. 700–732. Springer, Cham (2018). https://doi.org/10.1007/978-3-319-96881-0_24

19. Libert, B., Ling, S., Nguyen, K., Wang, H.: Zero-knowledge arguments for lattice-based accumulators: logarithmic-size ring signatures and group signatures without trapdoors. J. Cryptol. 36(3) (2023). https://doi.org/10.1007/s00145-023-09470-6, https://doi.org/10.1007/s00145-023-09470-6

20. Lu, X., Zhang, Z., Au, M.H.: Practical signatures from the partial fourier recovery problem revisited: A provably-secure and gaussian-distributed construction. In: Susilo, W., Yang, G. (eds.) Information Security and Privacy, pp. 813–820. Springer International Publishing, Cham (2018)

21. Lyubashevsky, V.: Fiat-Shamir with Aborts: Applications to Lattice and Factoring-Based Signatures. In: Matsui, M. (ed.) ASIACRYPT 2009. LNCS, vol. 5912, pp. 598–616. Springer, Heidelberg (2009). https://doi.org/10.1007/978-3-642-10366-7_35

22. Lyubashevsky, V.: Lattice Signatures without Trapdoors. In: Pointcheval, D., Johansson, T. (eds.) EUROCRYPT 2012. LNCS, vol. 7237, pp. 738–755. Springer, Heidelberg (2012). https://doi.org/10.1007/978-3-642-29011-4_43

23. Maeno, Y., Miyaji, H., Miyaji, A.: Lattice-based accumulator with constant time list update and constant time verification. In: El Hajji, S., Mesnager, S., Souidi, E.M. (eds.) Codes, Cryptology and Information Security, pp. 204–222. Springer Nature Switzerland, Cham (2023)

24. Mazzocca, C., Acar, A., Uluagac, S., Montanari, R.: EVOKE: efficient revocation of verifiable credentials in IoT networks. In: 33rd USENIX Security Symposium (USENIX Security 24), pp. 1279–1295. USENIX Association, Philadelphia, PA (2024). https://www.usenix.org/conference/usenixsecurity24/presentation/mazzocca

25. Merkle, R.C.: A certified digital signature. In: Brassard, G. (ed.) Advances in Cryptology – CRYPTO' 89 Proceedings, pp. 218–238. Springer, New York, New York, NY (1990)

26. Micciancio, D., Peikert, C.: Trapdoors for Lattices: Simpler, Tighter, Faster, Smaller. In: Pointcheval, D., Johansson, T. (eds.) EUROCRYPT 2012. LNCS, vol. 7237, pp. 700–718. Springer, Heidelberg (2012). https://doi.org/10.1007/978-3-642-29011-4_41

27. Papamanthou, C., Tamassia, R., Triandopoulos, N.: Authenticated hash tables based on cryptographic accumulators. Algorithmica **74**(2), 664–712 (2016)

28. Shor, P.: Algorithms for quantum computation: discrete logarithms and factoring. In: Proceedings 35th Annual Symposium on Foundations of Computer Science, pp. 124–134 (1994). https://doi.org/10.1109/SFCS.1994.365700

29. Sitouah, N., Bruschi, F., Pallotta, F.L., Mencucci, R., Sciuto, D.: An untraceable credential revocation approach based on a novel merkle tree accumulator. In: 2024 IEEE International Conference on Blockchain and Cryptocurrency (ICBC), pp. 210–214 (2024). https://doi.org/10.1109/ICBC59979.2024.10634387

30. Sudarsono, A., Nakanishi, T., Funabiki, N.: Efficient proofs of attributes in pairing-based anonymous credential system. In: Fischer-Hübner, S., Hopper, N. (eds.) Privacy Enhancing Technologies, pp. 246–263. Springer, Berlin Heidelberg, Berlin, Heidelberg (2011)

31. Vershynin, R.: High-dimensional probability: an introduction with applications in data science. Cambridge Series in Statistical and Probabilistic Mathematics, Cambridge University Press (2018)

32. Yue, D., Baldi, S., Cao, J., Li, Q., De Schutter, B.: Distributed adaptive resource allocation: an uncertain saddle-point dynamics viewpoint. IEEE/CAA J. Autom. Sinica **10**(12), 2209–2221 (2023). https://doi.org/10.1109/JAS.2023.123402

33. Zhao, Y., Yang, S., Huang, X.: Lattice-based dynamic universal accumulator: design and application. Comput. Stand. Int. **89**, 103807 (2024). https://doi.org/10.1016/j.csi.2023.103807, https://www.sciencedirect.com/science/article/pii/S0920548923000880

Isogeny-based Cryptography

Smooth Twins for Cryptographic Applications from Pell Equations

Daniel Berger[(✉)]

German Aerospace Center (DLR), Institute for the Protection of Maritime
Infrastructures, Fischkai 1, 27572 Bremerhaven, Germany
`daniel.berger@dlr.de`

Abstract. Modern isogeny-based post quantum cryptographic schemes,
like SQISign 1.0, including its variant AprèsSQI, rely on large primes p
such that $p^2 - 1$ is divisible by a large smooth factor. In order to find
those, it is easiest to find smooth twins, consecutive smooth integers,
that sum to a prime number. In this paper we develop novel methods
to make an algorithm for finding smooth twins by solving Pell equations
feasible to search for practical parameters for isogeny-based post quantum cryptography schemes. Our improvements here are twofold: First,
we improve the runtime of the algorithm from exponential to polynomial
and second, we introduce a way to enforce special requirements, that are
necessary for practical applications. Our main method, which we call
Pell-and-boost, is based on the infrastructure of orders in real quadratic
number fields and a previous technique for finding smooth twins. While
similar methods have been used in the past, they have never been powerful enough to compute parameters for practical use cases, we remedy this
and showcase how our new techniques can compute potential candidates
for primes.

Keywords: Post-quantum cryptography · isogenies · parameter
search · Pell equation · smooth integers · real quadratic number field

1   Introduction

Many isogeny-based post-quantum cryptographic schemes, such as B-SIDH [8],
the original[1] SQISign [10,11] and POKE [1] require parameters of a special form
to be efficient, which may be found from so-called *smooth twins*, two consecutive
integers that are both as smooth as possible (or at least contain a large enough
smooth factor). Explicit study of this problem for cryptographic purposes goes
back to [9]; the most effective methods for finding smooth twins are based on
sieving or computing greatest common divisors (see e.g. [24]). We instead focus
on a method, which is based on solving Pell equations.

---

[1] When we refer to SQISign, we mean SQISign 1.0. A new version, SQISign 2.0, relaxes
the requirements, only requiring $p + 1$ to be smooth, by utilizing higher dimensional
isogenies.

© The Author(s), under exclusive license to Springer Nature Switzerland AG 2026
R. Dutta et al. (Eds.): INDOCRYPT 2025, LNCS 16372, pp. 173–193, 2026.
https://doi.org/10.1007/978-3-032-13301-4_8

Utilizing Pell equations to find (all) smooth twins goes back to a method developed by Størmer [28], which was later improved by Lehmer [21]. A Pell equation is a Diophantine equation of the form $x^2 - dy^2$ with d nonsquare (sometimes squarefree). The classical method for solving this equation is via computation of the *continued fraction expansion* of $\sqrt{d}$, which is however not optimal. One of the main difficulties encountered is that the solution to a given Pell equation is often on the order of $O(\exp \sqrt{d})$ in size.

Luca and Najman utilized the same method in [22] to compute all 100-smooth twins, which took considerable computational effort. To better manage the growth of the solutions, they utilized so-called *compact representations*, which are connected to the so-called *infrastructure* in real quadratic number fields, which we will recall in Sect. 2.3 and utilize in Sect. 3.1 to give an improved algorithm to search for smooth twins of *cryptographic size.*

In [4] it was shown how the methods can be extended to increase smoothness probabilities. They used a version of Lemma 1 to translate small known pairs of consecutive smooth integers into solutions to Pell equations, computed larger solutions to the same Pell equation and translated back. This process was able to compute large consecutive smooth integers, however did not produce any whose sum is prime. They prove a lemma explaining this, which shows that their method is impractical for cryptographic applications. We comment on and extend their ideas in Sect. 3.3.

The best known smooth twins of cryptographic size were found in [3, 11, 24], some of the smoothest (not satisfying special requirements for practical applications) in [26]. While papers on computing smooth twins of cryptographic size generally mention the method of solving Pell equations, it is usually dismissed due to its exponential runtime and the inability to "enforce special requirements such as large powers of two dividing $r(r + 1)$" [24] on the numbers, which are necessary for cryptographic applications.

We show how to improve the algorithm, presenting a way to enforce such special requirements in Sect. 3.2. Furthermore we show that the algorithm does have polynomial, not exponential, runtime, when restricting the search to find primes for cryptographic applications; while the degree of that polynomial is still too large to be usable in practice, we restrict the search space further in a reasonable way to make it more manageable. To showcase the validity of our results, we use the techniques to compute a list of candidate primes.

The paper is structured as follows. In Sect. 2, we introduce the necessary background, consisting of the cryptographic requirements as well as the necessary number theory to understand our results. Section 3 contains our main contributions and introduces two improvements to a known algorithm, improving the runtime from exponential to polynomial and allowing us to enforce necessary special requirements. Additionally we introduce a probabilistic argument to further restrict the searchspace, decreasing the runtime of our improved algorithm even more. We conclude the section with a note on how to store Størmer discriminants, the main inputs that are being iterated over, in practice. Finally, in Sect. 4 we present our test results and we conclude in Sect. 5.

2 Background

In this section we introduce the cryptographic requirements, the connection between smooth twins and the Pell equation as well as real quadratic number fields and their infrastructure. Statements without proof on the latter topic are standard and may be found e.g. in [17]; we present them mostly to introduce our notation.

2.1 Cryptographic Requirements

An integer r is called B-smooth for an integer $B > 1$ if none of its prime factors are larger than B. Two consecutive integers $(r, r+1)$ are called B-*smooth twins* if $r(r+1)$ is B-smooth. Often the parameter B is implicit and thus dropped.

Isogeny-based schemes such as B-SIDH [8] (now broken [6]), SQISign [10,11] (as well as its variant AprèsSQI [7]) or POKE [1] utilize supersingular elliptic curves defined over $\mathbb{F}_{p^2}$ for some prime p. The efficiency of those schemes crucially relies on a choice of p having certain properties, most notably we need the ability to compute isogenies of degree N, requiring $N \mid (p-1)(p+1) = p^2 - 1$, so that N-torsion is defined over $\mathbb{F}_{p^2}$, allowing efficient isogeny computation. For $N = \prod_i p_i^{e_i}$, isogenies of degree N are computed as the composition of e_i isogenies of degree p_i, requiring a total of $\sum_i e_i$ prime-degree isogeny computations. An isogeny of degree p_i is computed in $O(p_i)$ time utilizing Vélu's formulas [30] or in $\widetilde{O}(\sqrt{p_i})$ time utilizing $\sqrt{\text{èlu}}$ formulas [2]. Therefore computing isogenies can only be done efficiently for smooth N.

The scheme SQISign requires 2λ bit primes p for $\lambda \in \{128, 192, 256\}$, corresponding to NIST security levels I, III and V, satisfying $2^k T \mid (p^2 - 1)$, where $T > p^{5/4}$ is an odd integer as smooth as possible and k is as large as possible. Additionally it is useful if $3^{k'} \mid T$ for k' as large as possible. More specifically we require $2^k 3^{k'} \geq 2^\lambda$. Note that there is a conflict between the size of k and k'; SQISign prioritizes the size of k. Further note that the requirement $T > p^{5/4}$ limits the size of k to $2^k < p^{3/4}$, where in practice often only $2^k \approx p^{1/4}$ is reached [11]. In the variant AprèsSQI, the requirements on T are relaxed, leading to a theoretical maximum of $2^k \approx p$; if here $k > \lambda$, no additional powers of 3 are required.

This motivates the search for primes p such that $p^2 - 1 = (p-1)(p+1)$ is as smooth as possible. This expression may be uniquely factored as $p^2 - 1 = 2d'y^2$ with $d' > 2$ smooth and squarefree and y smooth (cf. Lemma 1). So we are looking for solutions to Pell equations $x^2 - 1 = 2d'y^2$ with d' and y smooth and x prime. Note that $p-1$ and $p+1$ are even, so they are both smooth if, and only if, $((p-1)/2, (p+1)/2)$ are smooth twins and that $(p-1)/2 + (p+1)/2 = p$, so, alternatively, we are looking for smooth twins whose sum is prime.

2.2 From Smooth Twins to Pell Equations and Back

The key observation is the following

Lemma 1. *There exists a bijection between smooth twins $(r, r+1) \in \mathbb{N}^2$ and $(x, y, d') \in \mathbb{N}^3$, where y is B-smooth, $d' \neq 2$ is squarefree and B-smooth and*

$$x^2 - 2d'y^2 = 1.$$

Proof. Let $(r, r+1)$ be smooth twins. Set $x := 2r+1$, then $x^2 - 1 = 4r(r+1)$ may be written as $x^2 - 1 = 2d'y^2$ with d' being squarefree and y being B-smooth.

Conversely, let $x^2 - 2d'y^2 = 1$ with d' and y as in the lemma statement, then x is odd and setting $r := (x-1)/2$, it follows that $r(r+1) = 2d'y^2/4$ is also B-smooth.

As there are only a finite number of smooth squarefree integers, we only have to consider finitely many Pell equations. For a fixed positive integer d that is not a square, the Pell equation $x^2 - dy^2 = 1$ has a least positive nontrivial solution (x_1, y_1), which is called the *fundamental solution*. It generates all other solutions $(x_n, y_n)_{n \in \mathbb{N}}$ via $x_n + y_n\sqrt{d} = (x_1 + y_1\sqrt{d})^n$. The following result restricts which of those solutions might lead to smooth twins and, in particular, shows that there are only finitely many smooth twins for any fixed smoothness bound.

Theorem 1 (Carmichael). *If $n \neq 1, 2, 6$, then y_n has a prime divisor, that does not divide d nor y_ℓ for $\ell < n$. Furthermore any prime p must divide y_{p-1} or y_{p+1}.*

Proof. Can be found in [5].

It follows that if p is the smallest prime larger than B, we have to check at most the first $\max\{3, (p+1)/2\}$ solutions to a given Pell equation $x^2 - 2d'y^2 = 1$. Furthermore it can be shown that $(y_n)_n$ is a divisibility sequence, which further restricts the number of solutions to be considered. This gives us Algorithm 1, which serves as a starting point and will be improved in the following sections.

We call the set

$$\left\{ 2 \prod_{p \in D} p : D \subseteq S, D \neq \{2\} \right\}$$

which is iterated over by the algorithm, the set of *Størmer discriminants*. We also want to note that in practice we usually have to compute fewer solutions than Theorem 1 would suggest and obtaining the fundamental solution is the computationally most expensive part. In [4, Lemma 1] it was shown that only 2^kth solutions for some $k \in \mathbb{N}$ to Pell equations have the chance to yield smooth twins whose sum is prime, further restricting the number of solutions to be checked if we are only interested in smooth twins for cryptographic applications.

Fundamental solutions often exceed $\exp\sqrt{d}$ [14, 20] and as the continued fraction method computes them in standard decimal (respectively binary) representation, this algorithm is also exponential in runtime. Faster algorithms circumvent this by instead computing a sufficiently precise approximation of the logarithm of the fundamental solution (x_1, y_1), the so-called *regulator*[2].

$$\mathrm{R}_d = \log\left(x_1 + y_1\sqrt{d}\right).$$

---

[2] Note that this notion of regulator is non-standard and we define it here only for convenience; usually the regulator is defined in terms of the discriminant as in (2.2).

Algorithm 1: Computing smooth twins.

Data: Smoothness bound B.
Result: All smooth twins.

1 Let S be the set of all primes $p \leq B$
2 $T \leftarrow \varnothing$
3 **foreach** $D \subseteq S$ *with* $D \neq \{2\}$ **do**
4 $\quad\quad d \leftarrow 2\prod_{p \in D} p$
5 $\quad\quad$ Compute the fundamental solution (x_1, y_1) to the Pell equation
   $\quad\quad\quad x^2 - dy^2 = 1$
6 $\quad\quad I \leftarrow \{1, \ldots, (\max S + 1)/2\}$
7 $\quad\quad$ **for** $i \in I$ **do**
8 $\quad\quad\quad\quad$ Consider the ith solution (x_i, y_i)
9 $\quad\quad\quad\quad$ **if** y_i *is B-smooth* **then**
10 $\quad\quad\quad\quad\quad T \leftarrow T \cup \{((x_i - 1)/2, (x_i + 1)/2)\}$
11 $\quad\quad\quad\quad$ **else**
12 $\quad\quad\quad\quad\quad I \leftarrow I \setminus \{i, 2i, \ldots\}$
13 $\quad\quad\quad\quad$ **if** *each $p \in S \setminus D$ already occurred as divisor of some y_ℓ with $\ell < i$* **then**
14 $\quad\quad\quad\quad\quad I \leftarrow I \cap \{1, 3\}$

15 **return** T

Thus e.g. the Buchmann-McCurley algorithm (see e.g. [29]) can compute the regulator in subexponential time, however it is dependent on some generalized Riemann hypothesis. The fastest known algorithm that can unconditionally compute the regulator is presented in [12] and runs in $\mathrm{O}(\Delta^{1/6+\epsilon})$ (where Δ is the discriminant as defined in (2.1)).

Notice that the regulator corresponds, up to some computable multiplicative factor, to the number of bits in the fundamental solution. We will be using this in Sect. 3.1 to give an algorithm that checks whether the fundamental solution has b to t bits with a runtime that depends only logarithmically on d (cf. Theorem 4).

2.3 Real Quadratic Number Fields and Infrastructure

Let $d \in \mathbb{N}$ be an arbitrary nonsquare. In $\mathbb{Q}(\sqrt{d})$, we may factor the Pell equation $x^2 - dy^2 = 1$ as

$$1 = x^2 - dy^2 = \left(x + y\sqrt{d}\right)\left(x - y\sqrt{d}\right) = \mathrm{N}\left(x + y\sqrt{d}\right),$$

where $\mathrm{N}\colon \mathbb{Q}(\sqrt{d}) \to \mathbb{Z}$ is the norm function. It follows that solutions to the Pell equation correspond to nontrivial units of norm 1. The ring of integers of K is given by $\mathbb{Z}[\omega_0]$, where

$$\omega_0 = \begin{cases} \frac{1+\sqrt{d_0}}{2} & \text{if } d_0 \equiv 1 \pmod 4, \\ \sqrt{d_0} & \text{if } d_0 \equiv 2, 3 \pmod 4. \end{cases}$$

178   D. Berger

In the latter case the units correspond to integers solutions to $x^2 - d_0 y^2 = \pm 1$, but in the former case they correspond to integer solutions to $x^2 - d_0 y^2 = \pm 4$. The ring of integers $\mathbb{Z}[\omega_0]$ is also the maximal order in K, but we will be considering arbitrary orders $\mathcal{O}$, which may be uniquely written as a $\mathbb{Z}$-module $\mathcal{O} = \mathbb{Z} + \mathbb{Z} f \omega_0 = \mathbb{Z}[f\omega_0]$, where f is called the *conductor* of $\mathcal{O}$. The *discriminant* of $\mathcal{O}$ is defined as

$$\Delta := \Delta(\mathcal{O}) = \begin{cases} d & \text{if } d_0 \equiv 1 \pmod{4}, \\ 4d & \text{if } d_0 \equiv 2,3 \pmod{4}. \end{cases} \tag{2.1}$$

Note that d and Δ may at most differ by a factor of 4 and can theorefore be used interchangeably in asymptotic considerations. This uniquely determines the order and we may therefore write

$$\mathcal{O}_\Delta = \mathbb{Z} + \mathbb{Z}\frac{\Delta + \sqrt{\Delta}}{2}.$$

The group of units is given by $\mathbb{Z}[\epsilon_\Delta]$ for $\epsilon_\Delta = (x_1' + y_1'\sqrt{\Delta})/2$, which is the *fundamental unit*; (x_1', y_1') is the uniquely determined smallest solution to $x^2 - \Delta y^2 = \pm 4$ in the positive integers. The logarithm of the fundamental unit is called the regulator of the order

$$\mathrm{R}_\Delta = \log \epsilon_\Delta. \tag{2.2}$$

Similarly to a Pell equation, this may be computed via the continued fraction expansion of $\sqrt{\Delta}$. Note that $\mathrm{R}_d = k\,\mathrm{R}_\Delta$ with k chosen as in Table 1.

The fundamental solution (x_1, y_1) to a given Pell equation $x^2 - dy^2 = 1$ is directly connected to the order of conductor f and discriminant Δ, where $d = f^2 d_0$ with d_0 squarefree, and its fundamental unit (x_1', y_1') via

$$\epsilon_d = x_1 + y_1\sqrt{d} = \epsilon_\Delta^k$$

for some $k \in \mathbb{N}$, which may be explicitly computed (see Table 1).

Table 1. Solving the Pell equation from the fundamental unit.

$\begin{smallmatrix} d_0 \equiv 1 \\ (\mathrm{mod}\ 4) \end{smallmatrix}$	$\Delta \pmod{2}$	$y_1' \pmod{2}$	$\mathrm{N}(\epsilon_\Delta)$	k
✗	—	—	1	1
✗	—	—	-1	2
✓	—	0	1	1
✓	—	0	-1	2
✓	0	1	—	2
✓	1	1	1	3
✓	1	1	-1	6

The *infrastructure* of a real quadratic number field, first developed in [25], is a group-like structure (i.e. it is a "non-associative group") on equivalence classes of ideals in $\mathcal{O}$.

We call $\mathcal{O}$-ideals $\mathfrak{a}$, $\mathfrak{b}$ *equivalent*, if $\mathfrak{a} = \kappa\mathfrak{b}$ for some $\kappa \in K \setminus \{0\}$ and denote the equivalence class of $\mathfrak{a}$ by $[\mathfrak{a}]$. Setting

$$\varsigma := \begin{cases} 1 & \text{if } d_0 \not\equiv 1 \pmod 4, \\ 2 & \text{if } d_0 \equiv 1 \pmod 4, \end{cases}$$

one may write an $\mathcal{O}$-ideal $\mathfrak{a}$ uniquely as the $\mathbb{Z}$-module

$$\mathfrak{a} = S\left(\mathbb{Z}\frac{Q}{\varsigma} + \mathbb{Z}\frac{P + \sqrt{d}}{\varsigma}\right),$$

with nonnegative integers S, Q, P, subject to $\varsigma \mid Q$ and $\varsigma Q \mid (d - P^2)$. Conversely any such $\mathbb{Z}$-module defines an $\mathcal{O}$-ideal, which we also denote by $(S)(Q, P)$. We call such an ideal *primitive*, if $S = 1$ in which case we also denote it by (Q, P), and *reduced*, if it is primitive and there is no $\alpha \in \mathfrak{a}$ with $|\alpha| < \mathrm{N}(\mathfrak{a})$ and $|\overline{\alpha}| < \mathrm{N}(\mathfrak{a})$. Note that given any ideal $(S)(Q, P)$, we get a primitive representative (Q, P). One can show that if $\mathfrak{a}$ is reduced, then $\mathrm{N}(\mathfrak{a}) < \sqrt{\Delta}$ and as the norm of $(S)(Q, P)$ is given by $S^2 Q/\varsigma$, this shows that the coefficients of reduced ideals are bounded; in particular there are only finitely many. Next we want to present an algorithm for ideal reduction (i.e. the process of computing a reduced representative of a given ideal), which is a well-studied topic by itself; for an overview see [18]. We will be following [16] and use an approach closely related to continued fractions. Set $(Q_0, P_0) := (Q, P)$ and define $\rho((Q_{j-1}, P_{j-1})) = (Q_j, P_j)$ to be a step in the simple continued fraction expansion

$$\frac{P + \sqrt{d}}{Q} = q_0 + \cfrac{1}{q_1 + \cfrac{1}{\ddots + q_{j-1} + \cfrac{1}{\frac{P_j + \sqrt{d}}{Q_j}}}}.$$

By the theory of continued fractions (see for example [27]), it follows that there is some j, such that

$$\frac{P_j + \sqrt{d}}{Q_j} > 0 \text{ and } -1 < \frac{P_j - \sqrt{d}}{Q_j} < 0.$$

By [17, Theorem 5.8], it follows that (Q_j, P_j) is reduced. Using the same theorem, one can verify that if $\mathfrak{a}$ is reduced, then $\rho(\mathfrak{a})$ is also reduced. Furthermore one can show that $\rho(\mathfrak{a}) \sim \mathfrak{a}$, so repeatedly applying ρ to a given ideal $\mathfrak{a}$ results in a sequence

$$\mathfrak{a} = \rho^0(\mathfrak{a}), \rho^1(\mathfrak{a}), \rho^2(\mathfrak{a}), \ldots, \rho^j(\mathfrak{a}), \ldots$$

of pairwise equivalent ideals, that are eventually reduced. By [17, Theorem 5.18] this sequence actually contains all reduced ideals equivalent to $\mathfrak{a}$. We note that this computation mirrors the computation of the simple continued fraction expansion of a quadratic irrational, which is known to be be eventually periodic. The computation of the preperiod corresponds to the reduction of the ideal and as the simple continued fraction expansion is eventually periodic, this must also be true for the sequence of ideals. Starting with an already reduced ideal $\mathfrak{a} = \mathfrak{a}_1$, we get the set

$$\mathscr{C} := \{\mathfrak{a}_1, \mathfrak{a}_2 = \rho(\mathfrak{a}), \mathfrak{a}_3 = \rho^2(\mathfrak{a}), \ldots, \mathfrak{a}_p = \rho^{p-1}(\mathfrak{a})\},$$

with $\rho(\mathfrak{a}_p) = \mathfrak{a}$. We define $\theta_j \in K$ via $\rho^{j-1}(\mathfrak{a}) = \theta_j\mathfrak{a}$. Now $\epsilon_\Delta > 1$ by definition and thus $0 < \epsilon_\Delta^{-1} < 1$. As $\epsilon_\Delta^{-1}\mathfrak{a} = \mathfrak{a}$, [17, Theorem 5.18] now implies that there is some $m > 1$ such that $\theta_m = \epsilon_\Delta$. Certainly we must also have $m > p$ and as θ_{p+1} must be a unit and one can show that $(\theta_j)_{j\in\mathbb{N}}$ is strictly increasing, it follows that $\epsilon_\Delta = \theta_{p+1}$. As the preceding discussion is independent of the choice of $\mathfrak{a}$, we restrict ourselves to the simplest case of $\mathfrak{a} = (1)$, so that $\mathfrak{a}_m = (\theta_m)$ is principal. Note that all principal ideals $\mathfrak{a} = (\theta)$ are invertible with inverse $\overline{\mathfrak{a}} = (\overline{\theta})$. We define the *distance* δ of $\mathfrak{a}_m$ to be $\delta(\mathfrak{a}_m) := \log\theta_m$, so that $\delta(\mathfrak{a}_{p+1}) = \log\epsilon_\Delta = \mathrm{R}_\Delta$.

The set $\mathscr{C}$ now behaves very similarly to a finite cyclic group, where an application of ρ corresponds to multiplication by the generator and δ corresponds to the order of an element (do note however that distances are transcendental numbers) and the group order corresponds to the regulator. However, we also want to be able to "multiply" arbitrary elements $\mathfrak{a}_i, \mathfrak{a}_j \in \mathscr{C}$ to produce an element $\mathfrak{a}_k \in \mathscr{C}$ with $\delta(\mathfrak{a}_k) \approx \delta(\mathfrak{a}_i) + \delta(\mathfrak{a}_j)$. The product $\mathfrak{a}_i\mathfrak{a}_j$ of the ideals is in general not reduced (not even primitive), but one could just apply the reduction algorithm to obtain a reduced representative. As the coefficients of the factors as well as the reduced representative of the product are bounded by $\sqrt{\Delta}$ and the coefficients of the (unreduced) product are only bounded by Δ, this is however not optimal. Shanks already came up with a way to fix this, resulting in an algorithm (dubbed NUCOMP), where all the intermediary results are bounded by $\mathrm{O}(\sqrt{\Delta})$; a modern description of that algorithm may be found in [15]. We refer to this operation as $\mathfrak{a}_i * \mathfrak{a}_j = \mathfrak{a}_k$ and note that with our implementation we always obtain $\delta(\mathfrak{a}_i * \mathfrak{a}_j) \leq \delta(\mathfrak{a}_i) + \delta(\mathfrak{a}_j)$ and that the difference is bounded by $\mathrm{O}(\log\Delta)$; using sufficiently many applications of ρ, we can obtain an ideal with distance closest to $\delta(\mathfrak{a}_i) + \delta(\mathfrak{a}_j)$ ([17, Appendix]). The set $\mathscr{C}$, together with ρ, δ and $*$, is referred to as infrastructure. As NUCOMP is not associative, neither is $*$, so this isn't a group, however many algorithms that apply to groups can still be adapted.

Modern algorithms implementing infrastructure utilize compact representations [17, Chapter 12], which allow storing fundamental solutions ϵ_Δ requiring only $\mathrm{O}(\log\log\epsilon_\Delta \log\Delta)$ bits, as opposed to $\mathrm{O}(\log\epsilon_\Delta)$ bits. Adapting a exponentiation by squaring technique to the infrastructure results in the following

Theorem 2. *There is an algorithm that, given a distance x, computes a compact representation of θ, such that $(\theta) \in \mathscr{C}$, $\delta((\theta)) \leq x$ and $\delta(\rho((\theta))) > x$. This algorithm executes in $\mathrm{O}(\log x \log\Delta)$ elementary operations on numbers with $\mathrm{O}(\log\Delta)$ bits.*

Proof. This is essentially the algorithm CRAX from [17, Algorithm 12.4].

3   Practical Adjustments

In this section we introduce various improvements on Algorithm 1 when we are interested in finding smooth twins for cryptographic applications. The first improvement in Sect. 3.1 introduces a technique to check whether an nth solution can possibly be of the correct size to skip having to solve Pell equations that cannot yield practical solutions. In Sect. 3.2 we introduce a way to enforce useful additional requirements, which also restricts the search space. Finally in Sect. 3.3 we restrict the search space further with a probabilistic argument.

3.1   Limiting the Search to Smooth Twins of Cryptographic Size

Note that for a fixed d and thus Δ, the distance δ of an ideal (θ_j) in the infrastructure corresponds, up to a computable multiplicative constant, to the number of bits in θ_j (respectively its coordinates as a $\mathbb{Z}$-module). We may therefore use Theorem 2 to compute the first ideal (θ_m) in the infrastructure such that θ_m has b bits. By successive applications of ρ, we can then check every ideal whose generator has at most t bits. At each step we may check whether (θ_j) is the unit ideal, in which case θ_j corresponds to an nth solution to a Pell equation. To bound the number of ρ applications, we introduce the following

Theorem 3 (Khintchine-Lévy).   *For almost all real $t \in [0, 1)$ we have*

$$\lim_{n \to \infty} \sqrt[n]{\prod_{i=1}^{n} \phi_i} = \mathrm{e}^\lambda,$$

where $\phi_i = \phi_i(t)$ is the ith complete quotient in the simple continued fraction expansion of t and $\lambda = \pi^2/(12 \log 2) \approx 1.1865691104$.

Proof. Can be found in [19,23].

Applying this to the infrastructure, where the complete quotients are given by

$$\phi_i = \frac{P_i + \sqrt{d}}{Q_i},$$

it follows from this and the fact that $(Q_i)_{i \in \mathbb{N}}$ is bounded that

$$\log \theta_j = \log \left(\frac{Q_{j-1}}{Q_0} \prod_{i=1}^{j-1} \phi_i \right) \approx (j-1)\lambda \approx j-1$$

and therefore $\delta(\theta_j) \approx j$, i.e. we expect each step in the infrastructure to travery a distance of roughly 1. More precisely, if we look at two applications of ρ, we can compute a lower bound of the distance traversed. For $j \geq 2$, we get

$$\theta_j = q_{j-2}\theta_{j-1} + \theta_{j-2} \geq \theta_{j-1} + \theta_{j-2} \geq 2\theta_{j-2}$$

and thus

$$\log \theta_j \geq \log \theta_{j-2} + \log 2. \tag{3.1}$$

Combining this result with an earlier algorithm gives us

Theorem 4. *There is an algorithm that checks whether some x_n has b to t bits in $\mathrm{O}(\log b(\log d)^2) + \mathrm{O}(|t - b| \log d)$ time, where (x_n, y_n) is the nth solution to the Pell equation $x^2 - dy^2 = 1$.*

Proof. The first summand is from Theorem 2 with $x = b$. The second summand follows from (3.1).

As $\mathrm{R}_d = \log(x_1 + y_1\sqrt{d})$, we get that $x_1 \approx dy_1$ both have $\lceil \log(2)^{-1} \cdot \mathrm{R}_d \rceil + \{0, 1\}$ bits. This lets us compute bounds on R_d (respectively R_Δ) when requiring x_1 to be of cryptographic size.

Finding a lower bound for R_Δ is not too hard; [14] shows that we have

$$\mathrm{R}_\Delta \geq \log\left(\frac{1}{2}\left(\sqrt{\Delta - 4} + \sqrt{\Delta}\right)\right) =: b(\Delta), \tag{3.2}$$

where equality is achieved infinitely often. Finding a suitable upper bound is much harder, the best known result is due to Hua [13], namely

$$\mathrm{R}_\Delta < \sqrt{\frac{1}{2}\Delta}\left(\frac{1}{2}\log \Delta + 1\right) =: t(\Delta). \tag{3.3}$$

It follows that $t(\Delta)$ must be large enough and $b(\Delta)$ small enough to ensure that R_Δ can correspond to smooth twins of cryptographic size. While the lower bound this imposes on Δ is at least reasonable (albeit too small to really make a difference), the upper bound grows exponentially in Δ and is thus too large to be useful. More precisely if t_b is the upper bound on the number of bits for potential smooth twins, we obtain from (3.2) that

$$\Delta \leq \frac{(1 + e^{2\log 2 \cdot t_b})^2}{e^{2\log 2 \cdot t_b}} = 2(\cosh(2\log 2 \cdot t_b) + 1).$$

Alternatively, by applying logarithms on both sides,

$$\log_2 \Delta \leq \log_2\left(\frac{(1 + e^{2\log 2 \cdot t_b})^2}{e^{2\log 2 \cdot t_b}}\right) \approx \log_2\left(e^{2\log 2 \cdot t_b}\right) = 2t_b.$$

As from $x^2 - 1 = dy^2$ it already follows that d may have at most twice as many bits as x, this bound is trivial and not really useful.

Example 1. For NIST-I, we are looking for primes having approximately 256 bits. If we limit the search space to 250 to 260 bits. It follows that $t(\Delta) > \log 2 \cdot 250$ and $b(\Delta) \leq \log 2 \cdot 260$. This leads to $\Delta > 2490$ and $\Delta < 3.4324 \cdot 10^{156} \approx 2^{2 \cdot 260}$.

Nevertheless it allows for a more fine-grained analysis of Algorithm 1, when we are only interested in smooth twins of cryptographic size. From

$$\log \prod_{p \leq N} p = \mathrm{O}(N),$$

we deduce that the product of $O(N)$ distinct primes has at least N bits, i.e. there is some least integer $k_b \in O(b)$ such that the product of any k_b distinct primes has more than b bits. It follows that instead of having to check all $2^{\pi(B)} - 1 = O(2^{B/\log B})$ Størmer discriminants as input to Algorithm 1, it suffices to check those less than 2^b. If we assume $b < \pi(B)/2 = O(B/\log B)$ is fixed, there are

$$\sum_{\substack{D \subseteq S \\ 2\prod_{p \in D} p \leq 2^b}} 1 \leq \sum_{i=0}^{k_b} \binom{\pi(B)}{i} = O\left(\pi(B)^{k_b}\right) = O\left(\left(\frac{B}{\log B}\right)^{k_b}\right) = O\left(B^{k_b}\right)$$

$$(3.4)$$

many. Note that for cryptographic applications, b is bounded above (by say ≈ 512, corresponding to NIST security level III) and thus also k_b is bounded above (by some expression that is linear in b), leading to an algorithm that is polynomial in B, albeit with a large exponent.

Example 2. Continuing Example 1, we confirm by computation that

$$\prod_{p \leq \pi(389)} p > 2(\cosh(2\log 2 \cdot 260) + 1),$$

thus showing that the algorithm runs in $O(B^{389})$ for this case. This runtime is still too bad to be feasible in practice, motivating the need for our improvements in the next sections.

3.2   Utilizing Conductors $f > 2$

Solving of the Pell equation $x^2 - 2d'y^2 = 1$ as in Lemma 1 corresponds to computing the fundamental unit in the unique order with conductor $f \in \{1, 2\}$ in $\mathbb{Q}(\sqrt{2d'})$, depending on whether $2 \mid d'$. Instead of this, we might as well consider the Pell equation $x^2 - 2^k 3^{k'} d' y^2 = 1$ for $k \geq 1$ and $k' \geq 0$ instead, solutions to which correspond to elements of the unique order with conductor $f \in \{2^k 3^{k'}, 2^{k-1} 3^{k'}, 2^k 3^{k'-1}, 2^{k-1} 3^{k'-1}\}$ in $\mathbb{Q}(\sqrt{2^k 3^{k'} d'})$, depending on whether $k' > 0$, $2 \mid d'$ and $3 \mid d'$. This ensures that $x^2 - 1$ is divisible by $2^k 3^{k'}$. Lemma 2 shows that we can use this to encode minimal divisibility conditions in the conductor.

Lemma 2. *We have $\mathbb{Z}[f\sqrt{d_0}] \subseteq \mathbb{Z}[f'\sqrt{d_0}]$ if, and only if, $f' \mid f$.*

Proof. First assume $\mathbb{Z}[f\sqrt{d_0}] \subseteq \mathbb{Z}[f'\sqrt{d_0}]$, in particular $f\sqrt{d_0} = x' + y'f'\sqrt{d_0}$ for some $x', y' \in \mathbb{Z}$. As $\sqrt{d_0}$ is irrational, this immediately implies $f = y'f'$.

Now let's assume $y'f' = f$ for some $y' \in \mathbb{Z}$. If now $\xi = x + yf\sqrt{d_0}$ with $x, y \in \mathbb{Z}$ is arbitrary, then $\xi = x + yy'f'\sqrt{d_0} \in \mathbb{Z}[f'\sqrt{d_0}]$.

In particular Lemma 2 shows that if we write $\xi \in \mathbb{Z}[f\sqrt{d_0}] \cap \mathbb{Z}[f'\sqrt{d_0}]$ with coordinates in either $\mathbb{Z}$-module, both representations share the same x-coordinate. While this may suggest that we should instead only consider the order $\mathbb{Z}[\sqrt{d_0}]$, which is contained in all other relevant orders, recall that when restricting ourselves to solutions of cryptographic size, we can first use Theorem 4 to check whether a solution even has the chance to be of the correct size, which is computationally a lot cheaper than actually computing the size of the fundamental unit, i.e. it only depends logarithmically on Δ instead of a runtime of at least $O(\Delta^{1/6+\epsilon})$ (cf. [12]) As larger discriminants increase the chance of the fundamental unit being too large, this is worth it. If this check succeeds however, we should solve the Pell equation $x^2 - dy^2 = 1$, where we factor $2^k 3^{k'} d' = dy'^2$ with d squarefree, as algorithms for solving the Pell equation have a runtime that is at least linear in the discriminant. Furthermore, larger discriminants with a prescribed factor of $2^k 3^{k'}$ decrease the possible size of the squarefree number d' as we have an upper bound on the size of the discriminant, improving (3.4) further.

Example 3. Continuing Example 1 and Example 2, consider SQISign, which requires $2^k 3^{k'} \mid (x^2 - 1)$ for $2^k 3^{k'} \geq 2^\lambda$, this reduces the possible size of d' by $\lambda = 128$ bits, thus reducing the size of the discriminant from $\Delta < 3.4324 \cdot 10^{156} \approx 2^{2 \cdot 260}$ to $\Delta < 2.9643 \cdot 10^{79} \approx 2^{2 \cdot (260-128)} = 2^{2 \cdot 132}$ and the runtime of the algorithm from $O(B^{389})$ to $O(B^{44})$.

Following this, we will refer to nonsquare numbers of the form $2^k 3^{k'} d'$ with d' smooth and squarefree as Størmer discriminants as well.

3.3   Limiting the Number of Factors

In [4] there are some heuristic arguments that suggest searching with Størmer discriminants that are small or contain only small prime factors to be more successful. Here they only consider Størmer discriminants of the form $2d'$, so it is better to think of choosing d'. During tests with larger smoothness bounds, we instead noticed that those Størmer discriminants which have comparatively fewer prime factors are more likely to yield potential smooth twins (cf. also Table 2). This can be partially explained by the fact that numbers, even smooth ones, tend to not have too many (distinct) prime factors on average (note, though, that smooth numbers tend to have more factors than non-smooth numbers). A B-smooth number $N \in [1, X]$ is a product of primes $p_i \leq B$ with $\prod_i p_i \leq X$, so $\sum_i \log p_i \leq \log X$. As $\sum_{p \leq B} \log p = O(B)$ and by the Prime Number Theorem, the average value $\overline{\log p}$ of $\log(p)$ is

$$\overline{\log p} \sim \frac{\log B}{B} \sum_{p \leq B} \log p \sim \log(B).$$

So we expect the average number of prime factors to be

$$\frac{\log X}{\log B}. \tag{3.5}$$

This also shows that as the smoothness bound increases, the number of factors of smooth numbers tend to decrease. Note that we are only choosing the (number of) factors of d', not $2^k 3^{k'} y^2$ and that minimum size conditions on the latter, decrease the possible size of the former and thus the expected number of factors.

Factor $N \in [1, X]$ as

$$N = \prod_{p \mid N} p^{e_p} = \left(\prod_{p \mid N} p^{\lfloor e_p/2 \rfloor} \right)^2 \cdot \prod_{\substack{p \mid N \\ e_p \text{ odd}}} p,$$

with

$$d := \prod_{\substack{p \mid N \\ e_p \text{ odd}}} p$$

squarefree. We are interested in the expected size and number of factors of d. For each prime $p \mid N$ we have $p \mid d$ if, and only if e_p is odd. Heuristically, the parity of e_p is equally likely to be even or odd, as for large p we tend to have $p \in \{0, 1\}$ and for small p, where e_p is allowed to be large, it should behave like a random nonnegative integer. It follows that d is expected to have half as many factors as N and that the expected size $\mathrm{E}[d]$ of d can be computed from

$$\mathrm{E}[\log d] \approx \sum_{p \leq B} \frac{1}{2} \log \ p \sim \frac{1}{2} B,$$

which implies

$$\mathrm{E}[d] \approx e^{\frac{1}{2}B}. \tag{3.6}$$

We conclude that it makes sense to only consider those x such that the d' in $x^2 - 1 = 2^k 3^{k'} d' y^2$ has at least ℓ and at most m distinct prime factors. This way we only have to consider at most

$$\sum_{i=\ell}^{m} \binom{\pi(B)}{i} = \mathrm{O}\left(\pi(B)^m - \pi(B)^\ell \right)$$

many Størmer discriminants. Therefore the modified algorithm has a runtime of $\mathrm{O}(B^m/(\log B)^m) = \mathrm{O}(B^m)$.

Example 4. For the best currently known primes p from B-smooth twins [24], we computed the number of factors $\omega(d)$ as well as its size $\log_2(d)$ of the squarefree number d, where $dy'^2 = 2^k 3^{k'} d' = p^2 - 1$ for the corresponding Størmer discriminant $2^k 3^{k'} d'$. We also computed $\log_2(x_1)$ for the corresponding fundamental solution $x_1 + y_1 \sqrt{d}$. The results are summarized in Table 2. Interestingly, none of the primes correspond to a fundamental solution; all of them correspond to a 2nd solution. Another thing we notice is that $\log_2(d) \approx 2 \log_2(x_d) \approx \log_2(p) \approx 2\lambda$, in stark contrast to [4], which suggests choosing d as small as possible. This is only partly explained by the fact that $2^k 3^{k'}$ already contains a large square (cf. Lemma 3), a better explanation might be (3.6), although we have to recall that the considered numbers $p^2 - 1$ here only contain a large smooth part and are allowed to contain some further non-smooth factors. This presents a heuristic improvement of (3.2) and (3.3).

Table 2. Data from the Størmer discriminants for the best known primes corresponding to smooth twins.

Prime	$\omega(d)$	$\log_2(d)$	(k, k')	$\log_2(x_1)$
p_{1223}^{I}	22	246.2	$(68, 38)$	125.4
p_{1973}^{I}	21	241.5	$(76, 36)$	126.4
p_{8011}^{I}	17	223.0	$(84, 30)$	126.3
p_{5563}^{III}	26	377.3	$(56, 84)$	188.6
p_{22741}^{III}	26	378.1	$(70, 84)$	189.0
p_{47441}^{III}	25	377.0	$(98, 68)$	188.5
$p_{194581}^{\mathrm{III}}$	26	361.5	$(130, 60)$	190.2
p_{40609}^{V}	29	503.2	$(74, 114)$	251.6
p_{66343}^{V}	29	488.7	$(92, 108)$	252.4
p_{141079}^{V}	29	499.0	$(116, 90)$	249.5
p_{318233}^{V}	32	498.9	$(146, 72)$	250.1
$p_{3917}^{\mathrm{APRÈS}}$	15	247.2	$(100, 64)$	126.4
$p_{2791}^{\mathrm{APRÈS}}$	16	237.5	$(108, 14)$	124.4
$p_{3627}^{\mathrm{APRÈS}}$	15	243.5	$(120, 12)$	124.6
$p_{4441}^{\mathrm{APRÈS}}$	16	209.5	$(130, 18)$	124.0
$p_{12433}^{\mathrm{APRÈS}}$	15	253.0	$(162, 18)$	126.5
p_{1961}^{POKE}	17	232.4	$(130, 10)$	126.4
p_{1879}^{POKE}	14	203.8	$(130, 10)$	125.1
p_{1373}^{POKE}	20	236.5	$(134, 6)$	122.9
p_{1693}^{POKE}	14	194.7	$(132, 2)$	121.6
p_{1487}^{POKE}	18	226.6	$(130, 4)$	124.9

Lemma 3. *Write $2^k 3^{k'} d' = d y'^2$ with d squarefree and assume $2^k 3^{k'} \geq 2^\lambda$, then $2\log_2(y') \geq \lambda - \log_2(6)$.*

Proof. We have $2^{\lfloor k/2 \rfloor} 3^{\lfloor k'/2 \rfloor} \mid y'$ as well as $6(2^{\lfloor k/2 \rfloor} 3^{\lfloor k'/2 \rfloor})^2 \geq 2^k 3^{k'} \geq 2^\lambda$. It follows that

$$\log_2(y') \geq \left\lfloor \frac{k}{2} \right\rfloor + \log_2(3) \left\lfloor \frac{k'}{2} \right\rfloor \geq \frac{\lambda - \log_2(6)}{2}.$$

3.4   Representing Størmer discriminants

Let S be the set of all B-smooth primes as in Algorithm 1 and set $n := |S|$. Now Størmer discriminants correspond to binary numbers of length n, i.e. elements of $\mathbb{F}_2^n$ via d'. This also induces an order on the set of Størmer discriminants from the lexicographical order on $\mathbb{F}_2^*$. A given Størmer discriminant corresponds to a subset $D \subseteq S$ and can be recursively computed from any partition of D, so this naturally lends itself to a dynamic programming approach. As there are however still too many Størmer discriminants to store all of them, we present an approach that is a decent tradeoff between time and space requirements, requiring $O(n)$ space. The main benefit of this approach however, is that it becomes quite easy to skip all Størmer discriminants that are too large (cf. Sects. 3.1 to 3.3), too small or have too few or too many prime factors (cf. Sect. 3.3), which corresponds to the Hamming weight of the node.

We construct a full binary tree, where nodes are elements of $\mathbb{F}_2^*$ (also storing the corresponding Størmer discriminant) and the leaves are elements $b_1 b_2 \ldots b_n \in \mathbb{F}_2^n$, i.e. correspond to Størmer discriminants. The tree is rooted at the empty string ε. At each node s, the left child corresponds to $s\|0$ and the right child to $s\|1$ (where $\|$ is string concatenation). We may identify a non-leaf node with a leaf by appending sufficiently many 0s. During computation we store the current path from root to a leaf $(\varepsilon, b_1 = \varepsilon\|b_1, \ldots, b_1 b_2 \ldots b_n)$ in memory. The tree can be constructed in $O(n)$ time from any input string (respectively Størmer discriminant); computing and storing the current path takes $O(n)$ space. We call a node valid if every leaf in the subtree rooted at that node stores a Størmer discriminant within the specified bounds and has a Hamming weight within the specified bounds. To compute the next element from the current leaf, we go up to the parent node until we do so from a left child and consider the right child of the current parent node. If the right child encountered here is not valid, we instead discard it and continue going up; this corresponds to deleting the entire subtree rooted at that right child. From here we go left, until reaching a leaf or encountering a non-valid node, in which case we go right instead; this corresponds to deleting the entire subtree rooted at that left child. Pseudocode of the algorithm is given in Algorithm 2. Computation of the next Størmer discriminant in the sequence takes $O(n)$ moves in the tree.

Algorithm 2: Computing the next Størmer discriminant

Data: Current path $(s_0 = \varepsilon, s_1, \ldots, s_n)$ in the tree.
Result: Next Størmer discriminant.

```
 1  k ← n
 2  do
 3  │   if sₖ ends in 0 then
 4  │   │   if sₖ₋₁||1 is not valid then
 5  │   │   └   continue
 6  │   │   sₖ ← sₖ₋₁||1
 7  │   │   k ← k + 1
 8  │   │   if sₖ₋₁||0 is valid then
 9  │   │   │   sₖ ← sₖ₋₁||0
10  │   │   else
11  │   │   └   sₖ ← sₖ₋₁||1
12  │   │   if k = n then
13  │   │   └   return (s₀, …, sₙ)
14  │   k ← k − 1
15  while k > 0
```

---

4   Test Results

Note that Theorem 4, due to the cyclic nature of the infrastructure, only checks whether an nth solution to a Pell equation is of cryptographic size. For $n > 1$, as $\log_2(x_n) \approx n \cdot \log_2(x_1)$, we can derive solutions $x_1, \ldots, x_{n-1}$ from the fundamental solution x_1 having $1/j$ the required size for $j = 2, \ldots, n$. Those solutions can be boosted to the required size with a method from [3], which we dub "Pell-and-boost" in this case. Let $(r_n, r_n \pm 1)$ be smooth twins having $1/n$ of the required size; consider the polynomial $p_n(x) = 2x^n - 1$. Then $p_n(r_n)^2 - 1 = 4r_n^n(r_n^n - 1)$ has the required size and a smooth factor of size at least $n \log_2(r_n)$ bits; additionally as

$$r_n^n - 1 = \prod_{d \mid n} \Phi_d(r_n),$$

where Φ_d is the dth cyclotomic polynomial, we get that $\Phi_1(r_n) = (r_n - 1) \mid (p_n(r_n)^2 - 1)$ and for even n also $\Phi_2(r_n) = (r_n + 1) \mid (p_n(r_n)^2 - 1)$, further increasing the size of the guaranteed smooth factor by $\log_2(r_n \pm 1)$ bits. Furthermore if $2^k 3^{k'} \mid r_n$, then $2^{nk} 3^{nk'} \mid (p_n(r_n)^2 - 1)$. So if $p_n(r_n)$ is prime, it might be a suitable candidate. Note that every smooth twin generates two potential candidates. The solution x_n, which has the correct number of bits, can then be checked for primality directly (without boosting); note that this is only possible if n is a power of two [4, Lemma 1]. We present pseudocode of the final improved version of Algorithm 1 in Algorithm 3. Note that the value n_λ in the algorithm can be easily computed from a fundamental solution (respectively the

---

Algorithm 3: Computing primes for cryptographic applications.

Data: Smoothness bound B, size constraint $d' \in [b, t]$, constraint on number of factors $\omega(d') \in [b', t']$, exponent k for 2, exponent k' for 3.

Result: Potential primes P

1   Let S be the set of all primes $p \leq B$

2   $P \leftarrow \varnothing$

3   **foreach** $D \subseteq S$ with $D \neq \{2\}$ and $|D| + \{1, 2\} \in [b', t']$ *(depending on whether $k' > 0$)* **do**

4     $d \leftarrow 2^k 3^{k'} \prod_{p \in D} p$

5   **if** *there is no $n_\lambda \in \mathbb{N}$ such that $n_\lambda \, \mathrm{R}_d$ has roughly 2λ bits (cf. Theorem 4)* **then**

6   **continue**

7   Compute the fundamental solution (x_1, y_1) to the Pell equation $x^2 - dy^2 = 1$

8   Compute n_λ such that $x_1^{n_\lambda}$ has roughly 2λ bits

9     $i \leftarrow 1$

10   **while** $i < n$ **do**

11   Consider the ith solution (x_i, y_i)

12   **if** $p \leftarrow 2((x_i \pm 1)/2)^{n_\lambda - i + 1} - 1$ *is prime* **then** $P \leftarrow P \cup \{p\}$

13      $i \leftarrow i + 1$

14   **if** x_i *is prime* **then** $P \leftarrow P \cup \{x_i\}$

15 **return** P

---

regulator) and depends only on the security parameter λ. We may thus compute three values, one for each security parameter, and do the following computatioins concurrently. This way we do not have to redo the most costly computations if we want to search for parameters for a different security level. We have run this algorithm over Størmer discriminants of the form $2^{17} d'$ for $d' \in [2^{32}, 2^{128}]$ with $\omega(d') \in [4, 16]$ and d' being squarefree and 151-smooth, while searching for primes p having 240 to 260 bits, 370 to 390 bits or 500 to 520 bits for the respective security levels. This search was exhaustive for this parameter set and also included most such parameters with d' being 157-smooth. It ran on two AMD EPYC 7713 64-Core CPUs for roughly two weeks, utilizing 200 threads and found a total of 2576 potential primes, which may be found, together with additional data and all our code, in our GitHub repository https://github.com/db711/infrastructure.

When restricting to those that satisfy $2^k 3^{k'} > 2^\lambda$, only 15 for security level I, 6 for security level II and 3 for security level III remain. We provide those primes in Table 3 (see Appendix A). We see that many of those primes satisfy the extra condition $k > \lambda$, so they might be of particular interest for AprèsSQI.

5 Conclusions and Further Research

We have seen that our methods are capable of computing parameters for practical use. A more detailed analysis is quite complex, see e.g. [24, Section 2.5], which develops a cost metric for the signing procedure. In [24, Section 6] they mention SageMath and Magma scripts that were used for evaluating the primes. Testing our primes with the Magma script[3] suggests that some of them are at least competitive, particularly the following (see also Table 3):

$$p^{I}_{25561} := \texttt{0x44B73922CD9088DE97E99471DC7FFFFFFFFFFFFFFFFFFFFFFFFFFFFFFFF},$$

$$p^{III}_{18143} := \texttt{0x3FAB317A7C31C81AE51AA9CA8136D2C1C654C3A60C94C59A81FFFFFFFFFFF}$$
$$\texttt{FFFFFFFFFFFFFFFFFFFFFFFFFFFFFF},$$

$$p^{III}_{63709} := \texttt{0x1A916EBEB889097E4CFFC3E6FD4BF4F0882EE9B10BAFF3855121CCC08CD5FB}$$
$$\texttt{1F041FFFFFFFFFFFFFFFFFFFFFFFFFFFF},$$

$$p^{III}_{53693} := \texttt{0xA4F0F46410AECA97D5D6D8532CE9948B1A4F1CC82360344D3EF7872435BC1F}$$
$$\texttt{FFFFFFFFFFFFFFFFFFFFFFFFFFFFFFFF}.$$

When comparing it to other methods, one major thing we notice is that when solving a Pell equation for a given Størmer discriminant, we find not only smooth twins divisible by the factors of the Størmer discriminant, but also by additional even powers of primes. We may therefore hope to find smooth twins, which are divisible by large smooth squares (and thus possibly smoother overall), even when searching with B-smooth Størmer discriminants for very small values of B.

Heuristically, however, the largest prime factor of a number rarely appears squared, which limits our methods. It would thus be interesting further work to develop an algorithm that quickly generates random Størmer discriminants $2^k 3^{k'} d'$ with d' and $\omega(d')$ being bounded and d' being B-smooth for larger values of B (i.e. also allowing a few large prime factors to appear in the factorization of d'). This could then be used to run a randomized search.

Acknowledgements. I am grateful to the anonymous reviewers for their valuable comments that led to various improvements.

---

[3] Available at https://delta.cs.cinvestav.mx/~francisco/dftp_pushX_velu_v3.magma.

A   List of Primes

Table 3. Our computed primes, their respective bit length and associated values k, k'.

Prime p	$\lfloor \log_2 p \rfloor + 1$	(k, k')
0x863A0B7A19F0FE9011A81FFFFFFFFFFFFFFFFFFFFFFFFFFFFFFFF	212	$(130, 1)$
0x44B73922CD9088DE97E99471DC7FFFFFFFFFFFFFFFFFFFFFFFFFFFFFFFFF	231	$(128, 3)$
0x91E3B64279B3A9E1AFF2DE1772A1FFFFFFFFFFFFFFFFFFFFFFFFFFFFFFFFF	232	$(122, 5)$
0xE5CF0E089370490CEC80A5FC0BC31846E4FFFFFFFFFFFFFFFFFFFFFFFFFFF	232	$(97, 40)$
0x73CB9ECBF1368622090DFEDE300278B53FFFFFFFFFFFFFFFFFFFFFFFFFFFFF	235	$(107, 14)$
0x9661B5CA5063CCA0705A50284FFFFFFFFFFFFFFFFFFFFFFFFFFFFFFFFFFFFF	236	$(137, 9)$
0x14520E6CAE1CCE94BA53C9F12FDF7FFFFFFFFFFFFFFFFFFFFFFFFFFFFFFFFFF	237	$(128, 4)$
0x1B8973AC1BA5BF025BD31D537D7288AED9F2E79201FFFFFFFFFFFFFFFFFFFF	237	$(74, 40)$
0x1FC16FC66F3AA44BC4BD325F6A8F009EE861C681FFFFFFFFFFFFFFFFFFFFFF	237	$(82, 32)$
0x30C1AF458168586C7E08AE94A11FFFFFFFFFFFFFFFFFFFFFFFFFFFFFFFFFFFF	238	$(134, 66)$
0x378AD4390AD05647C5E46BFFFFFFFFFFFFFFFFFFFFFFFFFFFFFFFFFFFFFFFFF	238	$(155, 3)$
0xA8C226502CD3F9F819ECF4338BA81FFFFFFFFFFFFFFFFFFFFFFFFFFFFFFFFFFF	244	$(130, 40)$
0xB70493787C214ABD7496BAC281A05212DFFFFFFFFFFFFFFFFFFFFFFFFFFFFFFF	244	$(114, 14)$
0x12582F405F69EA523F20BBB580A9A68617D21DA975C1FFFFFFFFFFFFFFFFFFF	245	$(74, 56)$
0x1DD20B00718EBB52DC7FFFFFFFFFFFFFFFFFFFFFFFFFFFFFFFFFFFFFFFFFFFFF	245	$(176, 18)$
0x8B58C59A7FE20948C561B9F87371B401F3B05FFFFFFFFFFFFFFFFFFFFFFFFFFFFFFFFFFFFFFFFFFFFFFFFFFFFFFFF	356	$(210, 1)$
0x180CD7C79AAD1080FA894C7FFFFFFFFFFFFFFFFFFFFFFFFFFFFFFFFFFFFFFFFFFFFFFFFFFFFFFFFFFFFFFFFFFFFFFFF	357	$(272, 54)$
0x3FAB317A7C31C81AE51AA9CA8136D2C1C654C3A60C94C59A81FFFFFFFFFFFFFFFFFFFFFFFFFFFFFFFFFFFFFFFFFFFFF	358	$(162, 32)$
0x1A916EBEB889097E4CFFC3E6FD4BF4F0882EE9B10BAFF3855121CCC08CD5FB1F041FFFFFFFFFFFFFFFFFFFFFFFFFFFF	361	$(98, 72)$
0xA4F0F46410AECA97D5D6D8532CE9948B1A4F1CC82360344D3EF7872435BC1FFFFFFFFFFFFFFFFFFFFFFFFFFFFFFFFFFF	364	$(122, 72)$
0x1F6F90158B0D0B4AE5C055F643ADC731EE5DBB5313F2B8048BFDFEC8E13FFFFFFFFFFFFFFFFFFFFFFFFFFFFFFFFFFFFF	365	$(135, 42)$
0x32BB0C1B66AD262AE9A078BFA5FBCB926CDB88B005203A60B547FFFFFFFFFFFFFFFFFFFFFFFFFFFFFFFFFFFFFFFFFFFFFFFFFFFFFFFFFFFF	474	$(272, 18)$
0x6DB3CA5C632AC6F41B9388CE6A87FFFFFFFFFFFFFFFFFFFFFFFFFFFFFFFFFFFFFFFFFFFFFFFFFFFFFFFFFFFFFFFFFFFFFFFFFFFFFFFFFFFF	483	$(376, 68)$
0x8DAD5B8D1951F86EABF825C6E150F40D0A265A0129C78C2DC19D422EF27AC3367FFFFFFFFFFFFFFFFFFFFFFFFFFFFFFFFFFFFFFFFFFFFFFFFFFFFFFFFFF	492	$(236, 52)$

References

1. Basso, A., Maino, L.: POKé: a compact and efficient PKE from higher-dimensional isogenies. In: Fehr, S., Fouque, P.A. (eds.) Advances in Cryptology – EUROCRYPT 2025. pp. 94–123. Springer, Cham (2025). https://doi.org/10.1007/978-3-031-91124-8_4

2. Bernstein, D.J., De Feo, L., Leroux, A., Smith, B.: Faster computation of isogenies of large prime degree. Open Book Series 4(1), 39–55 (2020). https://doi.org/10.2140/obs.2020.4.39. https://msp.org/obs/2020/4-1/p04.xhtmlSciences Publishers

3. Bruno, G., Corte-Real Santos, M., Costello, C., Eriksen, J.K., Meyer, M., Naehrig, M., Sterner, B.: Cryptographic smooth neighbors. In: Guo, J., Steinfeld, R. (eds.) Advances in Cryptology – ASIACRYPT 2023, pp. 190–221. Springer, Singapore (2023). https://doi.org/10.1007/978-981-99-8739-9_7

4. Buzek, J., Hasan, J., Liu, J., Naehrig, M., Vigil, A.: Finding twin smooth integers by solving Pell equations (2022). https://doi.org/10.48550/arXiv.2211.04315. http://arxiv.org/abs/2211.04315

5. Carmichael, R.D.: On the numerical factors of the arithmetic forms $\alpha^n \pm \beta^n$. Ann. Math. **15**(1/4), 30–48. https://doi.org/10.2307/1967797. https://www.jstor.org/stable/1967797

6. Castryck, W., Decru, T.: An efficient key recovery attack on SIDH. In: Hazay, C., Stam, M. (eds.) Advances in Cryptology – EUROCRYPT 2023, pp. 423–447. Springer, Heidelberg (2023). https://doi.org/10.1007/978-3-031-30589-4_15

7. Corte-Real Santos, M., Eriksen, J.K., Meyer, M., Reijnders, K.: AprèsSQI: extra fast verification for SQIsign using extension-field signing. In: Joye, M., Leander, G. (eds.) Advances in Cryptology – EUROCRYPT 2024, pp. 63–93. Springer, Heidelberg (2024). https://doi.org/10.1007/978-3-031-58716-0_3

8. Costello, C.: B-SIDH: supersingular isogeny diffie-hellman using twisted torsion. In: Moriai, S., Wang, H. (eds.) Advances in Cryptology – ASIACRYPT 2020, pp. 440–463. Springer, Heidelberg (2020). https://doi.org/10.1007/978-3-030-64834-3_15

9. Costello, C., Meyer, M., Naehrig, M.: Sieving for twin smooth integers with solutions to the prouhet-tarry-escott problem. In: Canteaut, A., Standaert, F.X. (eds.) Advances in Cryptology – EUROCRYPT 2021, pp. 272–301. Springer, Cham (2021). https://doi.org/10.1007/978-3-030-77870-5_10

10. De Feo, L., Kohel, D., Leroux, A., Petit, C., Wesolowski, B.: SQISign: compact post-quantum signatures from quaternions and isogenies. In: Moriai, S., Wang, H. (eds.) ASIACRYPT 2020. LNCS, vol. 12491, pp. 64–93. Springer, Cham (2020). https://doi.org/10.1007/978-3-030-64837-4_3

11. De Feo, L., Leroux, A., Longa, P., Wesolowski, B.: New algorithms for the deuring correspondence. In: Hazay, C., Stam, M. (eds.) Advances in Cryptology – EUROCRYPT 2023. pp. 659–690. Springer, Heidelberg (2023). https://doi.org/10.1007/978-3-031-30589-4_23

12. de Haan, R., Jacobson, M., Williams, H.: A fast, rigorous technique for computing the regulator of a real quadratic field. Math. Comput. **76**(260), 2139–2160 (Oct 2007). https://doi.org/10.1090/S0025-5718-07-01935-7. https://www.ams.org/mcom/2007-76-260/S0025-5718-07-01935-7/

13. Hua, L.k.: On the least solution of Pell's equation (1942). https://projecteuclid.org/journals/bulletin-of-the-american-mathematical-society/volume-48/issue-10/On-the-least-solution-of-Pells-equation/bams/1183504769.pdf

14. Jacobson, M.J., , Richard F., L., , Williams, H.C.: An Investigation of Bounds for the Regulator of Quadratic Fields. Experimental Mathematics **4**(3), 211–225 (Jan 1995). https://doi.org/10.1080/10586458.1995.10504322, https://doi.org/10.1080/10586458.1995.10504322, publisher: Taylor & Francis _eprint: https://doi.org/10.1080/10586458.1995.10504322

15. Jacobson, M.J., Scheidler, R., Williams, H.C.: An improved real-quadratic-field-based key exchange procedure. J. Cryptol. **19**(2), 211–239 (2006). https://doi.org/10.1007/s00145-005-0357-6

16. Jacobson, M.J., Scheidler, R., Williams, H.C.: The efficiency and security of a real quadratic field based key exchange protocol. In: Public-Key Cryptography and Computational Number Theory, pp. 89–112. De Gruyter (2011). https://www.degruyterbrill.com/document/doi/10.1515/9783110881035.89/pdf

17. Jacobson, M.J., Williams, H.C.: Solving the pell equation. In: CMS Books in Mathematics. Springer, New York (2009). https://doi.org/10.1007/978-0-387-84923-2

18. Jacobson Jr, M.J., Sawilla, R.E., Williams, H.C.: Efficient ideal reduction in quadratic fields. Int. J. Math. Comput. Sci. **1**, 83–116 (2006). https://citeseerx.ist.psu.edu/document?repid=rep1&type=pdf&doi=621e8b4d38ed9fe240cd444938c79e161a11fa48

19. Khintchine, A.: Zur metrischen Theorie der diophantischen Approximationen. Mathematische Zeitschrift **24**(1), 706–714 (1926). https://doi.org/10.1007/BF01216806

20. Lagarias, J.C.: On the computational complexity of determining the solvability or unsolvability of the equation $X^2 - DY^2 = -1$. Trans. Am. Math. Soc. **260**(2), 485–508 (1980). https://doi.org/10.1090/S0002-9947-1980-0574794-0. https://www.ams.org/tran/1980-260-02/S0002-9947-1980-0574794-0/

21. Lehmer, D.H.: On a problem of Störmer. Illinois J. Math. **8**(1), 57–79 (1964). https://projecteuclid.org/journalArticle/Download?urlId=10.1215%2Fijm%2F1256067456

22. Luca, F., Najman, F.: On the largest prime factor of x^2-1. Math. Comput. **80**(273), 429–435 (2011). https://doi.org/10.1090/S0025-5718-2010-02381-6. https://www.ams.org/mcom/2011-80-273/S0025-5718-2010-02381-6/

23. Lévy, P.: Sur le développement en fraction continue d'un nombre choisi au hasard. Compositio mathematica **3**, 286–303

24. Santos, M.C.R., Eriksen, J.K., Meyer, M., Rodríguez-Henríquez, F.: Finding practical parameters for isogeny-based cryptography (2024). https://eprint.iacr.org/2024/1150

25. Shanks, D.: The infrastructure of a real quadratic field and its applications. In: Proceedings of the Number Theory Conference, pp. 217–224 (1972)

26. Sterner, B.: towards optimally small smoothness bounds for cryptographic-sized smooth twins and their isogeny-based applications. In: Eichlseder, M., Gambs, S. (eds.) Selected Areas in Cryptography – SAC 2024, pp. 178–202. Springer, Cham (2025). https://doi.org/10.1007/978-3-031-82852-2_8

27. Steuding, J.: Diophantine Analysis. CRC Press, Boca Raton

28. Størmer, C.: Quelques théorèmes sur l'équation de Pell et leurs applications (1897)

29. Vollmer, U.: An accelerated buchmann algorithm for regulator computation in real quadratic fields. In: Fieker, C., Kohel, D.R. (eds.) Algorithmic Number Theory, vol. 2369, pp. 148–162. Springer, Heidelberg (2002). https://doi.org/10.1007/3-540-45455-1_12

30. Vélu, J.: Isogenies entre courbes elliptiques. Comptes-Rendus de l'Academie des Sci. **273**, 238–241 (1971). https://cir.nii.ac.jp/crid/1573387449093111296

Hardened CTIDH: Dummy-Free and Deterministic CTIDH

Gustavo Banegas[1], Andreas Hellenbrand[2], and Matheus Saldanha[3(✉)]

[1] Inria and Laboratoire d'Informatique de l'Ecole polytechnique, Institut Polytechnique de Paris, Palaiseau, France
`gustavo@cryptme.in`
[2] RheinMain University of Applied Sciences, Wiesbaden, Germany
`andreas.hellenbrand@hs-rm.de`
[3] Universidade Federal de Santa Catarina, Florianópolis, Brazil
`matheus.saldanha@posgrad.ufsc.br`

Abstract. Isogeny-based cryptography has emerged as a promising post-quantum alternative, with CSIDH and its constant-time variant CTIDH offering efficient group-action protocols. dCTIDH recently introduced a efficient deterministic version. However, CTIDH and dCTIDH rely on dummy operations in differential addition chains (DACs) and Matryoshka-isogenies, which can be exploitable by fault-injection attacks. In this work, we present the first *dummy-free* implementation of dCTIDH. Our approach combines two recent ideas: `DACsHUND`, which enforces equal-length DACs within each batch without padding, and a reformulated Matryoshka structure that removes dummy multiplications and validates all intermediate points. Our analysis shows that small primes such as $3, 5$, and 7 severely restrict feasible `DACsHUND` configurations, motivating new parameter sets that exclude them. We implement dummy-free dCTIDH-2048-194 and dCTIDH-2048-205, achieving group action costs of roughly 357,000 to 362,000 $\mathbb{F}_p$-multiplications, with median evaluation times of 1.59 to 1.60 (Gcyc). These results do not surpass dCTIDH, but they outperform CTIDH by roughly 5% while eliminating dummy operations entirely. Compared to dCSIDH, our construction is more than $4\times$ faster. To the best of our knowledge, this is the first *efficient* implementation of a CSIDH-like protocol that is simultaneously deterministic, constant-time, and fully dummy-free.

Keywords: post-quantum cryptography · isogeny-based cryptography · CSIDH

Author list in alphabetical order; see https://ams.org/profession/leaders/ CultureStatement04.pdf. This work has been supported by the Federal Ministry of Research, Technology and Space (BMFTR) under the project QUDIS (ID 16KIS2089), and by the HYPERFORM consortium, funded by France through Bpifrance, and by the France 2030 program under grant agreement ANR-22-PETQ0008 PQ-TLS, Matheus Saldanha was funded by the Coordenação de Aperfeiçoamento de Pessoal de Nível Superior – Brasil (CAPES) – Finance Code 001.
Date of this document: 2025-11-04.

1   Introduction

In recent years, isogeny-based cryptography has attracted significant attention from both mathematicians and cryptographers, due to its features such as non-interactive key exchange and compact key sizes. Following the cryptanalysis of SIKE [8,13,18], research on isogeny-based key exchange has shifted toward CSIDH [9], which currently remains *unbroken*. The attacks that compromised SIKE—based on the supersingular isogeny framework—do not apply to CSIDH or its variants, thereby preserving their relevance as viable post-quantum key exchange candidates.

Despite its resilience, CSIDH is relatively slow compared to other post-quantum schemes. Furthermore, achieving secure implementations requires countermeasures against side-channel attacks, which further increase computational overhead. To address these limitations, variants such as CTIDH [1] and dCTIDH-2048 [6] have been proposed. These schemes improve performance by introducing more structured sets of isogeny paths and leveraging fixed parameter sets that simplify implementations. Both employ *isogeny batching* techniques, which simultaneously enhance security and performance. CTIDH achieves faster key exchange by introducing a new key space based on batches of isogenies, together with a constant-time algorithm for the CSIDH group action that synergizes with the new structure. Building on CTIDH, dCTIDH-2048 adopts a more deterministic approach, refining the batching technique through the introduction of *Widely Overlapping Meta-Batches* (WOMBats).

Beyond performance, dummy operations introduce an attack surface for *active* side-channels (fault injection): by targeting these redundant steps, an adversary can induce faults that desynchronize the control flow and leak secrets. This risk is not merely theoretical: Campos, Kannwischer, Meyer, Onuki, and Stöttinger [7] demonstrated fault attacks against dummy-padded isogeny computations, and more recently [12] exploited dummies in implementations of CSIDH which are constant-time. While batching in CTIDH raises the bar, their results indicate that **practical fault attacks remain feasible**, which motivates pursuing dummy-free techniques.

Contributions. In this work, we investigate in depth the use of DACsHUND and dummy-free Matryoshka isogenies (a constant-time isogeny computation that equalizes the cost of isogeny evaluations), and their combined role in enabling a fully dummy-free dCTIDH-2048 implementation. Our primary goal is to produce an optimized variant of CTIDH/dCTIDH-2048 that eliminates dummy operations while maintaining strong security properties.

1. We analyze the DACsHUND method for dummy-free differential addition chain (DAC) computations. For each prime, we enumerate all possible DAC configurations and adapt the dCTIDH-2048 greedy parameter search to enforce equal-length DACsHUNDs within each batch, thus avoiding dummy operations. We evaluate the resulting configurations under different dCTIDH-2048 settings and quantify the performance impact.

2. We implement the dummy-free Matryoshka isogeny approach and integrate its cost into the greedy search process, enabling parameter optimization that accounts for its specific constraints.
3. We propose new dCTIDH-2048 parameter sets that leverage these dummy-free techniques for improved performance. We focus on configurations that exclude the primes $3, 5, 7$, as they are less compatible with DACsHUND.

Availability of Software. Our implementation and greedy search scripts are available at

https://github.com/AndHell/hardenedCTIDH.

Related Work. Several works have sought to make CSIDH constant-time or deterministic. For instance, in [14], the authors address challenges such as point sampling and introduce the SIMBA technique. However, their approach still relies on dummy operations to compute isogenies. In parallel, other research has explored dummy-free constant-time methods, including two-point ladders and strategy-based scheduling of small-prime isogenies [15]. While these methods help mitigate timing leakage, they do not fully resolve batch-level DAC harmonization or eliminate the dummy padding inherent in Matryoshka. In CTIDH [1], the authors apply batching of isogenies using atomic blocks and Matryoshka to achieve a faster constant-time implementation of CSIDH. However, this approach is neither deterministic nor dummy-free.

Campos, Hellenbrand, Meyer, and Reijnders introduced dCTIDH-2048 [6], a deterministic variant of CTIDH. Their central innovation is the use of *WOM-Bats*, which combine overlapping batches with multiple isogenies per batch to enable efficient deterministic evaluation. Their implementation is highly optimized, both in terms of the number of finite-field operations per prime and the efficiency of those operations. Nevertheless, as the authors emphasize, dCTIDH-2048 still relies on dummy operations in both Matryoshka isogenies and DAC padding. As a result, it is not dummy-free, leaving open the challenge of combining determinism, constant-time execution, and full dummy-freeness in a single construction. To address this, dCTIDH-2048 proposed two potential directions: *DACsHUND* and *dummy-free Matryoshka isogenies*. In this work, we investigate these approaches in detail, with the goal of achieving the first fully dummy-free variant of dCTIDH-2048.

Recent work has explored radical 3-isogenies as a replacement for small-degree isogenies in CSIDH-like protocols, reporting up to a $4\times$ for key generation speedup for dCTIDH [10]. However, initiating a 3-isogeny chain still requires repeated sampling, which introduces probabilistic behavior. As a result, radical 3-isogeny chains cannot be made dummy-free and remain too costly in practice compared to 2-isogeny walks.

Addressing a related challenge in isogeny computation, Bernstein, Cottaar, and Lange [2] revisit the problem of constructing differential addition chains, introducing new algorithms that minimize both chain length and computational overhead. Their work focuses on faster methods for finding minimum-length continued-fraction differential addition chains, significantly improving over previous search strategies. In our setting, we also rely on efficient DACs and employ a greedy search procedure to identify chains that satisfy the structural constraints imposed by DACsHUND. While their algorithms target global optimality in chain length, our approach prioritizes compatibility with batching and constant-time requirements, aiming at practical dummy-free instantiations within the dCTIDH-2048 framework.

2   Background

2.1   Elliptic Curves and Isogenies

Given a finite field $\mathbb{F}_p$, an *elliptic curve* E over $\mathbb{F}_p$ is typically represented in the *Montgomery form*

$$By^2 = x^3 + Ax^2 + x,$$

where $A, B \in \mathbb{F}_p$ and $B(A^2 - 4) \neq 0$ ensures nonsingularity. This model admits particularly efficient and constant-time arithmetic based only on x-coordinates, making it well suited for isogeny-based cryptographic protocols. The group law on $E(\mathbb{F}_p)$ uses the point at infinity $\mathcal{O}$ as the identity element. For a detailed treatment of elliptic curve theory and arithmetic, see [19].

Projective Coordinates. In practice, elliptic-curve arithmetic is often performed in *projective coordinates* to avoid costly field inversions. An affine point (x, y) is represented as $(X : Y : Z)$, corresponding to $(X/Z, Y/Z)$ when $Z \neq 0$, while the point at infinity $\mathcal{O}$ is given by $(0 : 1 : 0)$. This representation replaces inversions with a few additional multiplications, making additions, doublings, and isogeny evaluations both more efficient and easier to implement in constant time.

Isogenies. An *isogeny* between elliptic curves, $\phi : E \to E'$, is a non-constant rational map that preserves the group law. While isogenies are not in general uniquely determined by their kernels, a separable isogeny is determined by its kernel—a finite subgroup of E—up to post-composition with an automorphism of the image curve. When working with Montgomery curves, these maps can be efficiently evaluated using only the x-coordinates of points, yielding major computational advantages for large-degree isogeny walks as required in CSIDH [9] and related protocols.

The use of x-only arithmetic not only simplifies the application of Vélu's formulas—the classical tool for computing an isogeny from its kernel—but also enables constant-time implementations via techniques such as the Montgomery ladder. For small odd prime degrees ℓ, the kernel is typically generated by an $\mathbb{F}_p$-rational point of order ℓ^T for some $T \geq 1$, whose image modulo ℓ defines the desired ℓ-isogeny. This allows efficient construction of the corresponding quotient curve.

The two main algorithms for evaluating such isogenies are *Vélu's formulas* [20] and $\sqrt{\text{élu}}$[1] method [3]. Both reduce the problem to computing a polynomial of the form

$$h_S(X) = \prod_{s \in S} (X - x([s]P)), \tag{1}$$

---

[1] Pronounced "square-root Vélu".

where P is a point of order ℓ, and S is an index set determined by ℓ. The main computational tasks are to determine the new Montgomery coefficient A' of the image curve E' and to evaluate the images of selected points under ϕ. These operations are conventionally denoted as `xIsog` and `xEval`, respectively: `xIsog`(P) computes the coefficients of the codomain curve corresponding to the isogeny with kernel $\langle P \rangle$, while `xEval`(P, Q) evaluates the image of a point Q under this isogeny using the precomputed kernel information.

In the classical Vélu approach, the index set is $S = \{1, 2, \ldots, (\ell - 1)/2\}$; one computes $x([s]P)$ for all $s \in S$ and forms the product $h_S(X)$. This yields essentially linear cost in ℓ: about 4ℓ $\mathbb{F}_p$-multiplications to update A' and 2ℓ per evaluated image point, i.e., overall $\widetilde{O}(\ell)$; Vélu is conceptually simple and practically optimal for small prime degrees.

For larger ℓ, the $\sqrt{}$élu algorithm applies a baby-step/giant-step decomposition on the odd-index set $S = \{1, 3, 5, \ldots, \ell - 2\}$ via $S \leftrightarrow (U \times V) \cup W$, obtaining h_S as h_W times a resultant involving h_U and a polynomial derived from h_V. This reorganizes the arithmetic to $\widetilde{O}(\sqrt{\ell})$. Vélu is a special case of $\sqrt{}$élu with $U = V = \varnothing$.

Remark 1. We note that both CTIDH and dCTIDH-2048 rely on "*Matryoshka*" isogenies, a technique to enforce uniform evaluation costs across batches of primes. Because this construction is central to CTIDH, dCTIDH-2048, and also to our work, we provide a more detailed discussion in Sect. 2.6.

2.2 CSIDH

Introduced by [9] in 2018, CSIDH is a non-interactive key exchange protocol based on the action of the ideal class group of an imaginary quadratic order on a set of supersingular elliptic curves defined over a prime field $\mathbb{F}_p$. This class group action is realized through chains of isogenies between elliptic curves, each of small odd prime degree ℓ_i dividing $p + 1$.

The protocol operates on a restricted set $\mathcal{E}$ of supersingular elliptic curves $E/\mathbb{F}_p$ whose endomorphism ring is isomorphic to $\mathbb{Z}[\sqrt{-p}]$, with all curves having exactly $p + 1$ points. For a prime p of the form

$$p + 1 = 2^f \cdot g \cdot \prod_{i=1}^{n} \ell_i,$$

where $f \geq 2$, g a small cofactor, and the ℓ_i are small, distinct odd primes, the group structure of $\mathcal{E}(\mathbb{F}_p)$ admits a torsion decomposition enabling efficient computation of ℓ_i-degree isogenies.

The underlying group action is defined as follows: a secret key is a vector $(e_1, \ldots, e_n)$ with $e_i \in [-m_i, m_i]$, representing the ideal class

$$[\mathfrak{a}] = \prod_{i=1}^{n} [\mathfrak{l}]_i^{e_i}.$$

Its action on a fixed base curve E_0 is computed as a walk in the ℓ_i-isogeny graph, with each step corresponding to an ℓ_i-isogeny in the forward or backward direction according to the sign of e_i. The resulting curve $E' = [\mathfrak{a}] * E_0$ is the public key.

CSIDH is *commutative*: for secret keys $[\mathfrak{a}]$ and $[\mathfrak{b}]$,

$$[\mathfrak{a}] * ([\mathfrak{b}] * E_0) = [\mathfrak{b}] * ([\mathfrak{a}] * E_0),$$

allowing both parties to derive the same shared secret curve without interaction.

Security relies on the *isogeny path-finding problem*: given two supersingular curves E and E' over $\mathbb{F}_p$ with the same $\mathbb{F}_p$-rational endomorphism ring $\mathcal{O}$, find an explicit $\mathbb{F}_p$-rational isogeny $\phi : E \to E'$ of smooth degree. This problem is believed to be hard for both classical and quantum algorithms when instantiated with sufficiently large p and appropriate parameters. Quantum security analysis remains active, with recent work suggesting that primes of at least 2048 bits may be required for conservative approaches. For a detailed discussion of CSIDH's security against subexponential hidden-shift attacks and the parameter choices, see the detailed analyses of isogenies and quantum algorithms in [4, 16].

2.3   Constant-Time Isogeny Diffie–Hellman (CTIDH)

The CTIDH [1] variant removes timing side channels by ensuring that all isogeny walks execute in constant time. Instead of conditionally applying an ℓ_i-isogeny based on the exponent e_i, CTIDH introduces a batching strategy with a redefined key space. A batch is defined as $\mathcal{B}_i = \{\ell_{i,1}, \ldots, \ell_{i,N_i}\}$, where all primes in $\mathcal{B}_i$ are handled collectively. An $\ell_{i,j}$-isogeny is then evaluated using the same computational routine and number of operations as the largest-degree ℓ_{i,N_i}-isogeny through *Matryoshka isogenies*, which conceal the true degree by padding smaller-degree computations with dummy steps so that every evaluation incurs the same cost as an ℓ_{i,N_i}-isogeny within each batch B_i.

To further mitigate leakage, CTIDH assigns a bound m_i to each batch, prescribing a fixed number of isogeny evaluations. When the required number of evaluations is smaller than m_i, dummy isogenies are inserted so that every batch always performs exactly m_i evaluations. This masks timing variations across batches. However, it does not protect against fault attacks, since the dummy operations themselves remain a potential target.

The dominant cost of an isogeny evaluation lies in computing its kernel polynomial, which involves scalar multiplications by different prime factors and would otherwise lead to timing variations. To mitigate this, CTIDH employs *differential addition chains* (DACs). By padding shorter chains with dummy operations, all scalar multiplications are forced to cost the same, analogous to the Matryoshka approach used for isogeny evaluations.

Finally, performance improvements arise from assigning bounds m_i to entire batches rather than to individual primes. This yields a larger combinatorial key space:

$$\#K_{N,M} = \prod_{i=1}^{B} \Phi(N_i, m_i), \quad \Phi(x, y) = \sum_{k=0}^{\min\{x,y\}} \binom{x}{k}\binom{y}{k} 2^k,$$

where $\Phi(x, y)$ counts integer vectors in $\mathbb{Z}^x$ with ℓ_1-norm at most y.

2.4 dCSIDH: Deterministic and Dummy-Free CSIDH

The dCSIDH, or `secsidh`, variant [5] was introduced as a high-security implementation of CSIDH that simultaneously achieves *determinism* and *dummy-freeness*. Unlike CTIDH, which relies on dummy operations to enforce constant-time behavior, dCSIDH eliminates both randomness and dummy padding by restricting the key space to exponents $e_i \in \{-1, 1\}$. This restriction ensures that every isogeny degree is used exactly once in a fixed direction, providing determinism in both point sampling and isogeny evaluation.

From a security perspective, determinism provides stronger protection against fault attacks. However, from a performance standpoint, this comes at a significant cost: eliminating dummy operations removes batching flexibility. Furthermore, to ensure security against quantum hidden-shift attacks, dCSIDH typically employs primes of at least 2048 bits. As a result, benchmarks show that dCSIDH runs approximately 3 to 5 times slower than the CTIDH at comparable parameter sizes.

2.5 dCTIDH-2048: Deterministic CTIDH

While CSIDH offers an elegant algebraic structure and promising post-quantum security, its reference design is vulnerable to practical implementation issues most notably timing and fault attacks. To address these challenges, dCTIDH-2048 [6] was introduced as a refinement of CTIDH, enhancing the original protocol with deterministic evaluation, stronger side-channel resistance, and improved performance.

The dCTIDH-2048 scheme is a deterministic variant of CTIDH that resolves the reliance on probabilistic point sampling and non-deterministic isogeny evaluation. Its key innovation is the introduction of *Widely Overlapping Meta-Batches* (WOMBats), which combine two complementary batching ideas: *multiple isogenies per batch* and *overlapping batches*.

In the original CTIDH, exactly one isogeny is computed per batch in order to avoid secret-dependent behavior. This constraint limits efficiency, since even if the secret key requires several isogenies from the same batch, only one can be evaluated. In contrast, if we restrict secret exponents to unitary values $e_i \in \{-1, 1\}$, then multiple isogenies of distinct degrees can be computed safely within a single batch. For a batch $\mathcal{B}_i = \{\ell_{i,1}, \ldots, \ell_{i,N_i}\}$, we can choose any number $M_i \leq N_i$ of distinct degrees, evaluating M_i isogenies via M_i calls to Matryoshka$[\ell_{i,1}, \ell_{i,N_i}]$.

This significantly reduces the number of total isogenies needed, as the key space grows combinatorially:

$$\Psi(N_i, M_i) = \binom{N_i}{M_i} \cdot 2^{M_i} \quad \text{or} \quad \Psi_{\text{dummy}}(N_i, M_i) = \sum_{j=0}^{M_i} \binom{N_i}{j} \cdot 2^j$$

if dummy isogenies are allowed (i.e. $e_i \in \{-1, 0, 1\}$).

Another approach to enlarge the key space and improve efficiency is to use batches that overlap in some of their prime factors. Suppose the first batch is $\mathcal{B}_1 = \{\ell_1, \dots, \ell_{N_1}\}$. Instead of defining $\mathcal{B}_2 = \{\ell_{N_1+1}, \dots, \ell_{N_1+N_2}\}$, we let the batches share $\omega_{1,2}$ primes:

$$\mathcal{B}_2 = \{\ell_{N_1-\omega_{1,2}+1}, \dots, \ell_{N_1+N_2-\omega_{1,2}}\}.$$

This overlapping structure amplifies the combinatorial growth of the key space without requiring a proportional increase in the number of isogeny evaluations. To ensure determinism, the bounds M_1, M_2 must satisfy $M_1 + M_2 \leq N_1 + N_2 - \omega_{1,2}$, preventing multiple isogenies from being applied to the same degree.

dCTIDH combines the two techniques above into WOMBats. A WOMBat $\mathcal{W} = \{\ell_{i,1}, \dots, \ell_{i,N}\}$ with bound M is evaluated as M overlapping batches

$$\mathcal{B}_1 = \{\ell_1, \dots, \ell_{N-M+1}\}, \quad \mathcal{B}_2 = \{\ell_2, \dots, \ell_{N-M+2}\}, \quad \dots \quad \mathcal{B}_M = \{\ell_M, \dots, \ell_N\}.$$

Each $\mathcal{B}_j$ overlaps in $N - M$ primes with its neighbors, and exactly one isogeny is computed from each, realized as a Matryoshka isogeny Matryoshka$[\ell_j, \ell_{N-M+j}]$. In this way, the WOMBat structure deterministically covers all possible distributions of M distinct isogeny degrees, while guaranteeing constant computational cost. The resulting key space of $N_\mathcal{W}$ disjoint WOMBats is

$$\prod_{i=1}^{N_\mathcal{W}} \Psi(N_i, M_i) = \prod_{i=1}^{N_\mathcal{W}} \binom{N_i}{M_i} \cdot 2^{M_i}.$$

To mitigate timing leakage, dCTIDH-2048 employs *DACs*, which pad shorter chains with dummy steps within each WOMBat to achieve constant-time scalar multiplication as previous mentioned. Consequently, even though dCTIDH-2048 eliminates *randomness* during evaluation, *the DAC and Matryoshka computations still incorporate dummy steps to maintain constant-time execution.*

2.6   Techniques in CSIDH-like Schemes

Efficient implementations of CSIDH and its variants rely on specialized techniques that simultaneously ensure constant-time execution and improve the performance of scalar multiplications and isogeny evaluations. Among the most important are *Differential Addition Chains (DACs)*, which realize scalar multiplications in constant time using only x-coordinates, and *Matryoshka isogenies*, which enable constant-time evaluation of isogenies while reducing computational cost through the exploitation of nested structures within isogeny chains.

Differential Addition Chains (DACs). Differential Addition Chains (DACs) are algorithmic frameworks for scalar multiplication on elliptic curves, particularly in the Montgomery model, where only x-coordinates are used. By avoiding full group operations and secret-dependent branching, DACs enable constant-time and side-channel-resistant implementations–an essential feature in isogeny-based cryptography where points are ephemeral and curves evolve along isogeny walks.

Definition 1 (Differential Addition Chain). *A differential addition chain for an integer n is a sequence $1 = c_0, c_1, \ldots, c_r = n$ such that for each $i \in \{1, \ldots, r\}$ there exist indices $j, k < i$ with*

$$c_i = c_j + c_k, \text{ and } c_j - c_k \in \{0, c_0, c_1, \ldots, c_{i-1}\}.$$

In other words, each new sum in the chain must correspond to a difference already present in the chain (or zero).

Example 1 A differential addition chain for 29 is 1, 2, 3, 5, 8, 13, 21, 29, since, for instance, $13 = 8 + 5$ with difference $8 - 5 = 3 \in \{1, 2, 3, 5, 8\}$.

In this work we focus on the subclass of *continued-fraction DACs* that are used in both CTIDH and dCTIDH-2048, which admit a compact bitstring encoding. For simplicity, we use the terms *DAC* and *continued-fraction DAC* interchangeably throughout.

Definition 2 (Continued-fraction DAC). *Let $(a_2, b_2, c_2), \ldots, (a_r, b_r, c_r)$ be a sequence of triples with $r \geq 3$, $(a_2, b_2, c_2) = (1, 2, 3)$, $c_r = n$, and for each $i \geq 3$:*

$$(a_i, b_i, c_i) = \begin{cases} (b_{i-1}, c_{i-1}, c_{i-1} + b_{i-1}), & or \\ (a_{i-1}, c_{i-1}, c_{i-1} + a_{i-1}), \end{cases}$$

with $c_i = a_i + b_i$. The corresponding continued-fraction DAC is then the sequence $1, 2, c_2, \ldots, c_r = n$.

Example 2. The differential addition chain shown above for 29 can also be interpreted as a continued-fraction differential addition chain. However, the sequence 1, 2, 3, 5, 8, 16, 29, while being a valid differential addition chain, is not a continued-fraction differential addition chain.

Definition 3 (Compressed Representation). *Each position in a continued-fraction DAC is uniquely defined by a single bit:*

$$\begin{cases} (b_{i-1}, c_{i-1}, c_{i-1} + b_{i-1}), & \text{if } f_i = 0, \\ (a_{i-1}, c_{i-1}, c_{i-1} + a_{i-1}), & \text{if } f_i = 1, \end{cases}$$

Hence, a continued-fraction DAC of length r with coefficients $1, 2, c_2, \ldots, c_r$ can be compactly represented by just $r - 2$ bits, namely $f_3, \ldots, f_r$.

Example 3. A continued-fraction DAC for 13 admits the compressed bitstring representation $f = 11110$.

In the Montgomery model, scalar multiplication $[k]P$ can be realized by iterating only differential operations:

$$\texttt{DIFF\_ADD}(P, Q, P - Q) \quad \text{and} \quad \texttt{xDBL}(P),$$

while tracking the differential $P-Q$. This makes the procedure fully deterministic and constant-time. Algorithm 1 illustrates how scalar multiplication by n can be carried out using a compressed DAC bitstring, relying solely on the two fundamental operations $\texttt{xDBL}$ and $\texttt{DIFF\_ADD}$.

Algorithm 1. $\texttt{DAC}$ – Scalar multiplication via a compressed DAC

Input: Point P, compressed DAC $f_3, \ldots, f_r$ for n
Output: $x([n]P)$
  1: $X_0 \leftarrow x(P)$ $\triangleright\ x(A)$ with $A = P$
  2: $X_1 \leftarrow \texttt{xDBL}(P)$ $\triangleright\ x(B) = x(2P)$
  3: $X_2 \leftarrow X_0$ $\triangleright\ x(A - B) = x(P - 2P) = x(P)$
  4: **for** $i = 3$ to r **do**
  5:      $X_3 \leftarrow \texttt{DIFF\_ADD}(X_0, X_1, X_2)$ $\triangleright\ = x(A + B)$ from $x(A), x(B), x(A-B)$
  6:      **if** $f_i = 0$ **then**
  7:          $(X_0, X_1, X_2) \leftarrow (X_1, X_3, X_0)$
  8:      **else**
  9:          $(X_0, X_1, X_2) \leftarrow (X_0, X_3, X_1)$
10:      **end if**
11: **end for**
12: **return** X_1 $\triangleright\ = x([n]P)$

Note that two DACs corresponding to different integers incur the same computational cost whenever their compressed representations have the same length, regardless of the integers themselves.

Remark 2. Compared to the classic Montgomery ladder (which is also constant-time), continued-fraction DACs compress structured additions/doublings for fixed small ℓ and integrate more naturally with batch scheduling, which is why CTIDH/dCTIDH-2048 prefer DACs for kernel generation.

In CSIDH-like protocols, DACs are used to compute kernel generators for isogenies. Secret keys are exponent vectors $(e_1, \ldots, e_n)$ indicating how many times an isogeny of a specific degree is applied. Each scalar multiplication $[\ell_i]P$ (for small primes ℓ_i) is performed using a fixed DAC, ensuring constant-time execution.

In CTIDH and dCTIDH-2048, DACs are precomputed according to the allowed exponent bounds, and scalar multiplications are often *batched* to reuse intermediate results. However, the length of these DACs–and therefore the computational cost of a multiplication by ℓ_i–depends directly on ℓ_i. To keep the isogeny degree ℓ_i secret, CTIDH enforces constant-time multiplications for all factors within a batch B. This is done by precomputing an optimal DAC for each $\ell_i \in B$ and padding it with dummy steps if necessary, so that multiplication by any cofactor from B requires the same number of operations as the largest ℓ_i in the batch.

However, since dummy padding is normally applied to maintain constant-time execution, it can leave room for active attacks, such as fault injection. The dCTIDH scheme addresses this issue with *DACsHUND*, a technique for dummy-free DAC evaluation, which we explore later in this work.

Matryoshka Isogenies. As previously mentioned, the computational cost of evaluating isogenies via Vélu's formulas or $\sqrt{\text{élu}}$ grows respectively as $\widetilde{O}(\ell)$ and $\widetilde{O}(\sqrt{\ell})$ in the isogeny degree ℓ. Since in CSIDH-like protocols one must evaluate isogenies of different prime degrees, these costs naturally vary across primes, potentially leaking information and complicating optimization. Additionally, when primes are grouped in batches, such as in CTIDH and dCTIDH-2048, isogeny evaluations must also cost the same within each batch. To address this, CTIDH introduces the notion of *Matryoshka isogenies*, a technique that enforces uniform evaluation cost across a batch of primes.

The core idea is to impose a "nested" evaluation structure on the kernel polynomial

$$h_S(X) = \prod_{s \in S} \big(X - x([s]P)\big),\tag{2}$$

where $S = \{1, 2, \ldots, (\ell - 1)/2\}$ in the Vélu case, or $S = \{1, 3, 5, \ldots, \ell - 2\}$ in the $\sqrt{\text{élu}}$ case. For Vélu's method, this amounts to cycling through the multiples $[s]P$, generating and evaluating $h_S(X)$ on the fly. Once the loop reaches $(\ell-1)/2$, one can continue appending *dummy iterations*, thereby aligning the total number of operations to that required by the largest prime $\ell_{\max}$ in the batch. In this way, any ℓ-isogeny in the batch can be evaluated at the uniform cost $\widetilde{O}(\ell_{\max})$.

The same concept extends to $\sqrt{\text{élu}}$ evaluations. In this case, the index set $S \longleftrightarrow (U \times V) \cup W$ is split into a box $U \times V$ and a leftover set W. Then, $h_S(X)$ can be computed by multiplying $h_W(X)$ with the resultant of $h_U(X)$ and a polynomial derived from V, with all sets U, V, and W having size $\widetilde{O}(\sqrt{\ell})$.

To apply a Matryoshka structure, U and V are chosen according to the smallest degree in the batch, while W is padded according to the largest. This ensures a uniform evaluation cost across primes in the batch, albeit with some efficiency loss since the parameters U, V, W are no longer optimally tuned for each ℓ.

We denote by Matryoshka$[\ell_i, \ell_j]$ a computation that performs any isogeny of degree $\ell \in [\ell_i, \ell_j]$ at the cost of ℓ_j, whether using Vélu or $\sqrt{\text{élu}}$ as appropriate. This nested framework makes it possible to batch isogeny evaluations without leaking degree information, while still achieving sublinear performance when $\sqrt{\text{élu}}$ is applicable. For further details, see [1,3].

3 DACsHUND

In dCTIDH, key generation requires computing a sequence of scalar multiplications $[\ell_i]P$ for a fixed set of primes $\ell_1, \ldots, \ell_n$. These multiplications are carried out on Montgomery curves using x-only arithmetic (xADD, xDBL) and differential addition chains (DACs). Implementations must be side-channel resistant, deterministic, and ideally batched to maximize performance.

In constant-time settings, the minimal DAC for each prime generally has a different length. To equalize the execution flow, previous approaches required padding shorter DACs with dummy operations so that all scalar multiplications within a batch completed in the same number of steps. While effective, this introduces redundancy and increases susceptibility to certain advanced fault-injection attacks. To overcome this limitation, we introduce *DACsHUND* (Differential Addition Chain Having Unnecessities Needed for Dummy-freeness), originally proposed in the future work of the dCTIDH paper, which enables dummy-free DAC execution.

Definition 4 (DACsHUND). *Let $\{\mathcal{B}_1, \ldots, \mathcal{B}_n\}$ be a family of n batches, where each batch $\mathcal{B}_i$ consists of N_i primes: $\mathcal{B}_i = \{\ell_{1,i}, \ldots, \ell_{N_i,i}\}$ with $\ell_{1,i} \leq \cdots \leq \ell_{N_i,i}$. Each prime $\ell_{j,i}$ has an associated set $\mathcal{D}_{j,i}$ of admissible DAC lengths. The configuration $\{\mathcal{B}_1, \ldots, \mathcal{B}_n\}$ is a valid DACsHUND if, for every batch $\mathcal{B}_i$, the intersection $\bigcap_{j=1}^{N_i} \mathcal{D}_{j,i}$ is non-empty.*

Intuitively, the idea is to partition the primes into batches such that all DACs in a batch share at least one common length. This eliminates the need for dummy padding while preserving constant-time execution. Algorithm 2 formalizes the batch validation procedure. This general framework not only supports dCTIDH, but can also be applied to related protocols such as CTIDH.

Example 4. Consider a batch $\mathcal{B}_1 = \{11, 13, 17, 19\}$. The corresponding DAC sets are:

$$\mathcal{D}_{1,1} = \{3, 4, 8\},$$
$$\mathcal{D}_{2,1} = \{3, 4, 5, 10\},$$
$$\mathcal{D}_{3,1} = \{4, 5, 7, 14\},$$
$$\mathcal{D}_{4,1} = \{4, 5, 6, 8, 16\}.$$

Since their intersection is $\{4\}$, this is a valid *DACsHUND* configuration. However, if prime 5 is added, its DAC set $\{1, 2\}$ leads to an empty intersection, invalidating the batch.

DACsHUND Map. The first step in building a DACsHUND configuration is to enumerate all admissible DACs for each prime in the range of interest. Instead of storing only the shortest DAC, we record every possible DAC length and its corresponding representation. This yields a map DACsHUND associating each prime p with its set of DAC lengths. For example, DACsHUND$[13] = \{3, 4, 5, 10\}$.

We adopt a straightforward *brute-force* strategy: enumerating all possible compressed DAC representations up to a prescribed length (e.g., 16), and testing

---

Algorithm 2. `IsValidDACsHUND` — Validation of DACsHUND Compatibility

Input: Batch sizes $N = (N_1, \ldots, N_B)$, number of batches B, prime list $\mathcal{P}$
Output: `True` if valid; `False` otherwise
1: Partition $\mathcal{P}$ into batches $\mathcal{P}^{(1)}, \ldots, \mathcal{P}^{(B)}$ of sizes $N_1, \ldots, N_B$
2: **for** $i = 1$ to B **do**
3:      $I \leftarrow \bigcap_{p \in \mathcal{P}^{(i)}}$ `DACsHUND`$[p]$
4:      **if** $I = \emptyset$ **then**
5:         **return** `False`
6:      **end if**
7: **end for**
8: **return** `True`

---

each candidate to verify whether it corresponds to a valid prime. Although this approach does not exploit optimized DAC search methods [2], the search space remains sufficiently small that an exhaustive traversal can be completed in a small time frame. On an AMD Ryzen 7 3800X running at 3.9 GHz, the complete brute-force exploration of all compressed DAC representations of length up to 16 (2^{16} DACs) takes approximately 0.25 s.

3.1 Searching Batch Configurations

With `DACsHUND` in place, the next step is to search for valid batch configurations. The `dCTIDH` batch search builds on the greedy strategy of `CTIDH` and is defined by three parameters: the number of batches B, the batch size vector $N = (N_1, \ldots, N_B)$ specifying the number of primes per batch, and the bound vector $M = (M_1, \ldots, M_B)$ that ensures the resulting configuration spans a sufficiently large key space.

Initialization. The standard dCTIDH greedy initialization assigns equal size to all batches ($N_i = n/B$ with $\sum N_i = n$), but this often produces invalid `DACsHUND` configurations with empty intersections. To address this, we construct an initial configuration iteratively: starting with $N = (1, \ldots, 1)$, we cycle through the batches, incrementing one N_i at a time, and accept the update only if the resulting configuration is `DACsHUND`-valid. This continues until all primes are allocated. The procedure is shown in Algorithm 3.

Greedy Search. The greedy algorithm modifies a configuration by decreasing the size N_i of one batch B_i and increasing the size of another $B_j \neq B_i$. This is repeated while exploring feasible bounds M_i for each batch. To integrate `DACsHUND`, we start with a valid initial configuration obtained via Algorithm 3 and introduce a validation step using Algorithm 2 at each modification to ensure that the new batch configuration preserves non-empty DAC intersections. If multiple DAC lengths are available, the smallest one is selected to minimize scalar multiplication cost. The cost function is thus adapted to consider the shortest valid DAC from the intersection of each batch.

Algorithm 3. `FindInitialBatchSizes` — Search for Valid Initial Configurations

Input: Number of batches B, prime list $\mathcal{P}$
Output: Batch size tuple N if valid; `None` otherwise
1: Initialize $N \leftarrow (1, \ldots, 1) \in \mathbb{Z}^B$
2: **while** $\sum_{i=1}^{B} N_i < |\mathcal{P}|$ **do**
3:     $\Delta \leftarrow$ `False`
4:     **for** $i = 1$ to B **do**
5:         Let $N' \leftarrow N$ with $N_i' \leftarrow N_i + 1$
6:         **if** IsVALIDDACsHUND$(N', B, \mathcal{P})$ **then**
7:             $N \leftarrow N'$, $\Delta \leftarrow$ `True`
8:         **end if**
9:     **end for**
10:     **if** $\Delta =$ `False` **then**
11:         **return** `None`
12:     **end if**
13: **end while**
14: **return** N

Remark 3. Small primes such as 3, 5, and 7 have very restricted DAC sizes (e.g., $\mathcal{D}_3 = \{0\}$). Their inclusion can yield inefficient configurations under `DACsHUND` constraints. For this reason, we also explore configurations excluding these primes and substituting them with larger ones to assess the performance trade-offs.

4   Dummy-Free Matryoshka

As outlined in §2.6, both CTIDH and dCTIDH-2048 employ the Matryoshka structure to conceal the true degree of an isogeny within a batch. In this setting, an isogeny of degree ℓ_k contained in a batch (ℓ_l, ℓ_r) is evaluated at the uniform cost of an ℓ_r-isogeny. As introduced in CTIDH [1], this construction proceeds as follows. One first computes the sequence of points

$$P, [2]P, \ldots, \left[\frac{\ell_r - 1}{2}\right] P,$$

and from these builds the kernel polynomial. The polynomial is factored into two parts: the *real factors*,

$$\prod_{i=0}^{(\ell_k - 1)/2} \big(x - x([i]P)\big),$$

which correspond to the actual ℓ_k-isogeny, and the *dummy factors*,

$$\prod_{i=(\ell_k - 1)/2 + 1}^{(\ell_r - 1)/2} \big(x - x([i]P)\big),$$

which pad the cost up to ℓ_r and thereby hide the true degree ℓ_k.

This dummy-based approach introduces two distinct entry points for fault-injection attacks. First, the dummy multiplications in the kernel polynomial may be distinguishable from real ones, enabling targeted faults. Second, the unused multiples

$$\left[\tfrac{\ell_k-1}{2}+1\right]P,\dots,\left[\tfrac{\ell_r-1}{2}\right]P,$$

although computed, are never required by the true kernel and thus create additional leakage channels.

To address these vulnerabilities, [6] introduced a modified Matryoshka structure. Their refinement eliminates dummy multiplications by reformulating the kernel product so that redundant terms cancel out algebraically, rather than being introduced explicitly. Furthermore, the unused multiples are validated against their expected relations, preventing an adversary from exploiting them as a source of leakage. This restructuring preserves the constant-time nature of Matryoshka while *significantly reducing its exposure to fault attacks.*

As this approach does not affect the computation of the kernel polynomial up to ℓ_l, the bounds for Vélu vs. $\sqrt{\ }$élu remain unchanged. In our implementation, up to a lower batchbound of $\ell_l <= 89$ Vélu formula are used, and $\sqrt{\ }$élu else. The remain part of the batch from ℓ_l to ℓ_r can be views as if it would be computed with Vélu.

4.1 Matryoshka 2.0

The idea described in [6, Appendix A], eliminates dummy operations entirely while retaining the same cost profile. The key observation is that for any point P, we have $x([i]P) = x([\ell - i]P)$. This symmetry allows the algorithm to verify that every multiple's x-coordinate must be computed correctly, since each will appear twice.

Algorithm 4 shows the full computation of the kernel polynomial h using the dummy-free Matryoshka approach, as described in [6]. Instead of inserting dummy factors, Matryoshka 2.0 replaces them with real multiplications of a modified form:

$$x = \tfrac{1}{2}x([i]P) - \alpha \cdot \tfrac{1}{2}x([i]P),$$

where α is chosen in constant time to be -1 if the value $x([i]P)$ has already appeared for some $j < i$, and 1 otherwise. Thus, lines 12 to 14 carry the same information as checking whether $i > \frac{\ell_k-1}{2}$ to determine if a dummy operation needs to be computed in the original version.

This achieves two crucial properties: uniformity of computation, since every iteration performs a real multiplication of the same cost, leaving no distinction between *real* and *dummy* steps; and the absence of unused data, since all multiples $x([i]P)$ are incorporated into the product, eliminating the risk of computing unnecessary points.

The original Matryoshka implementation in CTIDH (and dCTIDH-2048) uses projective space to represent x-only points as $(X : Z)$, thereby avoiding costly

Algorithm 4. Matryoshka 2.0 (based on [6])

Input: A degree ℓ_k, a batch $[\ell_l, \ldots \ell_r]$ and a point P such that $\ell_k \cdot P = \mathcal{O}$
Output: The kernel polynomial $h(x)$ for $\phi : E \to E/\langle P \rangle$

1: $b_k \leftarrow \frac{\ell_k - 1}{2}, b_l \leftarrow \frac{\ell_l - 1}{2}, b_r \leftarrow \frac{\ell_r - 1}{2}$
2: $t \leftarrow b_r - b_l$
3: Compute (x-coordinates of) $\{P, [2]P, \ldots, [b_r]P\}$.
4: $h(x) \leftarrow 1$

5: **for** $i \in [1, \ldots, b_l]$ **do** $\triangleright$ compute the *linear* part up to b_l
6:      $m \leftarrow x([i]P)$
7:      $h(x) \leftarrow h(x) \cdot (x - m)$
8: **end for**

9: **for** $i \in [b_l + 1, \ldots, b_r]$ **do**
10:      $m \leftarrow \frac{1}{2} x([i]P)$
11:      $\alpha \leftarrow 1$
12:      **for** $j \in [(b_l + 1 - t), \ldots, (i-1)\}$ **do** $\triangleright$ checks if $x[iP]$ has appeared already
13:          $\alpha \leftarrow \alpha \cdot$ cCompare$(x([i]P), x([j]P))$ $\triangleright$ returns -1 if so
14:      **end for**
15:      $h_1(x) \leftarrow h(x) \cdot (x - m)$
16:      $h_2(x) \leftarrow h(x) \cdot \alpha \cdot m$
17:      $h(x) \leftarrow h_1(x) - h_2(x)$
18: **end for**

19: **return** $h(x) \leftarrow x^{b_k - b_r} \cdot h(x)$ $\triangleright$ fix the degree

---

inversions. As in Vélu's formulas, the kernel polynomial must be evaluated at $\frac{h(1)}{h(-1)}$ to compute the codomain coefficient A'. In the projective setting, the evaluations at 1 and -1 are directly integrated into the implementation.

To adapt Algorithm 4 to projective space, we replace the affine expression $m = \frac{1}{2} x([i]P)$ with its projective equivalent. Writing $[i]P = (X_i : Z_i)$, we obtain

$$\frac{m_x}{m_z} = \frac{X_i}{2 \cdot Z_i}.$$

Accordingly, we updated in lines 15–17 with the following

$$\frac{h_x}{h_z} = \frac{h_x \cdot \big((\alpha \cdot m_x) + m_x - m_z\big)}{h_z \cdot \big((\alpha \cdot m_x) + m_x + m_z\big)}.$$

The cCompare routine must also be modified to compare projective points, increasing its cost to $2\mathbf{M}$. In practice we can implement cCompare as a projective constant-time equals and a constant-time swap between m and $-m$ to save a multiplication by -1. Finally, the degree correction step in line 19 simplifies to a constant-time sign flip of h_z.

Igonoring additions, the computation of one $\mathtt{Matryoshka}_{[\ell_l,\ell_r]}$-isogeny is thereby increased by $\sum_{i=1}^{t}(t-1+i)\cdot 2\mathbf{M}$, with $t=((\ell_r-1)/2)-((\ell_l-1)/2)$, compared to the dummy based version.

4.2   Matryoshka 1.414 ($\sqrt{\text{élu}}$)

For the Matryoshka[2] variant using $\sqrt{\text{élu}}$, *Algorithm 4 cannot be applied directly*, since not all multiples $[i]K$ required for comparison are available due to the index system that splits the computation into $U\times V\cup W$. However, we can exploit the structure of Matryoshka-$\sqrt{\text{élu}}$: the $U\times V$ component covers the kernel polynomial only up to ℓ_l, so all dummy factors necessarily appear in W. Moreover, W consists solely of even multiples of P. This enables us to validate each x-coordinate of the multiples $[2]P,[4]P,\ldots,[\frac{\ell_r-1}{2}]P$ by checking whether they match the double of their corresponding halves, that is, by verifying $\mathtt{xDBL}(x([i]P))=x([2i]P)$.

Depending on the batch size and the $\sqrt{\text{élu}}$ parameters, in some cases, not all odd halves are generated within $U\times V$. Therefore, the odd points must be computed explicitly in the range

$$max\big(bs,(\frac{(\ell_r-1)}{2}-2\cdot bs\cdot gs)/2\big),$$

where (bs,gs) denote the baby-step/giant-step parameters of $\sqrt{\text{élu}}$ for the ℓ_l-isogeny. This ensures that every even multiple in W pairs with its half, allowing for consistent validation without dummy points. Algorithm 5 summarizes the resulting *dummy-free Matryoshka algorithm* adapted to $\sqrt{\text{élu}}$.

As a result of the $\mathtt{xDBL}$ trick, the overhead of projective Matryoshka 1.414 is just $2\cdot\mathbf{M}+xDAC$ for iteration, together with the $((br-2bsgs)/2)-bs$ additionial $\mathtt{xADD}$ to compute the missing point halves.

5   Implementation

We base our implementation on the dCTIDH-2048 code from https://github.com/PaZeZeVaAt/dCTIDH, which in turn builds on the $\mathtt{secsidh}$[3] implementation [5]. This code incorporates the optimal strategies showed in [11] to accelerate kernel point computations by balancing the trade-off between pushing points through isogenies and computing kernels via DACs. In addition, it provides assembly-optimized $\mathbb{F}_p$ arithmetic for the different parameter sets.

We extend this implementation by integrating the new $\mathtt{DACsHUND}$ parameters for DAC computation and by adapting Algorithms 4 and 5 to projective space.

---

[2] We called Matryoshka 1.414 since $\sqrt{2}\approx 1.414$.

[3] Publicly available at https://github.com/kemtls-secsidh/secsidh.

Algorithm 5. Matryoshka 1.414

Input: A degree ℓ_k, a batch $[\ell_l, \ldots \ell_r]$, a point P such that $\ell_k \cdot P = \mathcal{O}$ and $\sqrt{\text{élu}}$ parameters (bs, gs) for ℓ_l

Output: The kernel polynomial $h(x)$ for $\phi : E \to E/\langle P \rangle$

1: $b_k \leftarrow \frac{\ell_k - 1}{2}, b_l \leftarrow \frac{\ell_l - 1}{2}, b_r \leftarrow \frac{\ell_r - 1}{2}$

2: $t \leftarrow b_r - b_l$

3: Compute multiples according to $\sqrt{\text{élu}}$

4: Compute odd multiples $[bs + 2]P, \ldots, [(br - 2 * bs * gs)/2]P$ if $bs < (br - 2 * bs * gs)/2$

5: $h(x) \leftarrow 1$

6: Compute $\sqrt{\text{élu}}$ using (bs, gs)

7: **for** $i \in [0, \ldots, b_r - 2 * bs * gs]$ **do**

8:      $m \leftarrow \frac{1}{2}x([2 * i + 2]P)$

9:      $\alpha \leftarrow 1$ if $i \leq b_k - 2 * bs * gs$ else -1

10:      $\alpha \leftarrow \alpha \cdot$ -cCOMPARE$(\text{xDBL}(x([i + 1]P)), x([2 * i + 2]P))$ ▷ -1 if points are equal, else 1.

11:      $h_1(x) \leftarrow h(x) \cdot (x - m)$

12:      $h_2(x) \leftarrow h(x) \cdot \alpha \cdot m$

13:      $h(x) \leftarrow h_1(x) - h_2(x)$ ▷ h is multiplied by x when $\alpha = -1$

14: **end for**

15: **return** $h(x) \leftarrow x^{b_k - b_r} \cdot h(x)$

5.1 Batch Configurations

To determine optimal parameter sets for dCTIDH-2048, we build on the configurations reported in the original dCTIDH-2048 work. In particular, we focus on the parameter sets dCTIDH-2048-194 and dCTIDH-2048-205, which serve as natural starting points and enable direct comparison with their non–dummy-free dCTIDH-2048 counterparts.

Further analysis shows that the small primes 3, 5, and 7 in the set $\{\ell_i\}$ severely restrict possible batch structures under DACsHUND constraints. To address this, we run our greedy search while excluding either 3, or 3, 5, 7 from the set $\{\ell_i\}$.

Table 1 presents the results for the dCTIDH-2048-194 and dCTIDH-2048-205 parameter sets. We evaluate configurations with between 12 and 20 batches for each parameter set. A complete run over all batch configurations requires approximately 16 h using 32 threads on a server equipped with dual AMD EPYC 7643 processors (2.3 GHz, 192 threads in total).

While the performance differences remain within $\approx 5\%$, our results indicate that the best configuration comes from skipping only the prime 3. Therefore, we implement dummy-free dCTIDH-2048 for the parameter sets dCTIDH-2048-205 and dCTIDH-2048-194 by excluding the 3-isogeny.

Table 1. Best greedy results for the dCTIDH-2048-194 and dCTIDH-2048-205 parameter sets. The column *cost* reports the estimated number of $\mathbb{F}_p$-multiplications required for a full group action.

Variant	ℓ Skipped	Batches	Isogenies	Cost
dCTIDH-2048-205	–	15	70	$327,942$
dCTIDH-2048-194	–	17	75	$334,458$
dCTIDH-2048-205	3	17	73	$327,390$
dCTIDH-2048-194	3	14	73	$332,920$
dCTIDH-2048-205	3, 5, 7	13	70	$334,846$
dCTIDH-2048-194	3, 5, 7	13	72	$341,526$

Remark 4. The greedy search only optimizes the plain cost of isogeny evaluation using optimal strategies. Therefore, it does not account for additional, albeit constant, costs in the group action, such as cofactor removal, and a final inversion to return an affine codomain, are not accounted for, explaining the differences. to the benchmarks measured in Table 2.

5.2   Performance

All benchmarks were performed on an Intel Core i7-6700 (Skylake) processor, running Debian 12 with Hyper-Threading and Turbo Boost disabled, and compiled using `gcc-12.2.0`.

Table 2 compares the results against dCSIDH as only other constant-time, dummy-free and deterministic scheme, CTIDH (from the secsidh implementation), as well as the relevant dCTIDH-2048 parameter sets.

Table 2. Results of a group action evaluation in multiplications **M**, squarings **S**, and additions **a**, and median cycle count (Gcyc) of 10,000 experiments, performed on a Skylake CPU.

variant	M	S	a	$\mathbb{F}_p$-mult.	Gcyc
CTIDH-2048	$287,207 \pm 21\%$	$83,759 \pm 9\%$	–	$370,966 \pm 17\%$	$1.652 \pm 17\%$
dCSIDH-2048 [5]	$1,315,203$	$227,501$	–	$1,542,704$	7.039
dCTIDH-2048-2048-205 [6]	$263,545$	$50,825$	$465,224$	$314,370$	1.418
dCTIDH-2048-2048-194 [6]	$266,101$	$51,258$	$469,258$	$317,359$	1.410
This work (205)	$303,058$	$54,074$	$560,276$	$357,132$	1.600
This work (194)	$307,004$	$55,215$	$553,193$	$362,219$	1.595

Table 2 compares the cost of the group action across different CSIDH implementations. As expected, dCSIDH is by far the most expensive: its fully deterministic and dummy-free design results in more than 1.5 million field multiplications and a median cost of 7.0 Gigacycles, making it impractical in comparison with other approaches.

Both parameter sets of dCTIDH-2048 (194 and 205) are more efficient, requiring about 314–317k $\mathbb{F}_p$ multiplications and completing a group action in roughly 1.410–1.418 Gigacycles. This confirms that batching and WOMBats provide a strong efficiency, albeit at the cost of dummy operations.

Our dummy-free implementation adds a small overhead compared to dCTIDH-2048: 358–362k $\mathbb{F}_p$-multiplications and 1.595–1.600 Gigacycles. This represents a slowdown of only 12–14%, while completely eliminating dummy multiplications in both DACs and Matryoshka isogenies (when we compare with dCTIDH-2048). At the same time, we still outperform the original CTIDH by about 4%, demonstrating the advantages of the WOMBat keyspace, even under the additional DACsHUND constraints.

Remark 5. Similar to dCTIDH-2048, this work focuses solely on optimizing the group action, which is just one part of a full key exchange. During key generation, One also needs to compute a torsion point of order $\prod \ell_i$, and in the key derivation step, the order of this point must be validated (which also ensures supersingularity). However, excluding the degree 3 speeds up the point search and validation by up to 20% compared to the dCTIDH-2048. Recent work by Pope, Reijnders, Robert, Sferlazza, and Smith [17] used a pairing-based approach for validation, suggesting a possible fourfold speedup. We leave the integration of pairing-based validation and point search into the dCTIDH-2048-framework as future work.

6   Conclusion

We have presented the first **dummy-free implementation of dCTIDH-2048**, combining **DACsHUND** with *dummy-free Matryoshka isogenies*. Our approach eliminates all dummy operations in both differential addition chains and isogeny evaluations, providing the first dCTIDH-2048 implementation that is deterministic, constant-time, and fully dummy-free. We showed how to adapt the greedy parameter search to incorporate these constraints, and identified viable parameter sets for dCTIDH-2048-194 and dCTIDH-2048-205, noting that very small primes such as $3, 5, 7$ are incompatible with **DACsHUND**.

In our implementation, we report results in Table 2 using the new batching strategy and the Matryoshka 1.414 variant. We show that even without dummy isogenies, our performance remains close to that of dCTIDH-2048. Moreover, we demonstrate an improvement of roughly 4% over CTIDH for both our implementations of dCTIDH-2048-2048-194 and dCTIDH-2048-2048-205.

Acknowledgements. The authors would like to thank Fabio Campos, Krijn Reijnders, and Michael Meyer for their discussions and insights on Matryoshka isogenies.

References

1. Banegas, G., et al.: CTIDH: faster constant-time CSIDH. IACR Trans. Cryptogr. Hardw. Embed. Syst. **2021**(4), 351–387 (2021)
2. Bernstein, D.J., Cottaar, J., Lange, T.: Searching for differential addition chains. Res. Number Theory **11**(2), 45 (2025)
3. Bernstein, D. J., De Feo, L., Leroux, A., Smith, B.: Faster computation of isogenies of large prime degree. Open Book Ser. **4**(1), 39–55 (2020)
4. Bernstein, D. J., Lange, T., Martindale, C., Panny, L.: Quantum circuits for the CSIDH: optimizing quantum evaluation of isogenies. In: Ishai, Y., Rijmen, V. (eds.) Advances in Cryptology - EUROCRYPT 2019 - 38th Annual International Conference on the Theory and Applications of Cryptographic Techniques, Germany, 19–23 May 2019, Proceedings, Part II, volume 11477 of Lecture Notes in Computer Science, Darmstadt, pp. 409–441. Springer (2019)
5. Optimizations and practicality of high-security CSIDH. IACR Commun. Cryptol. **1**(1), 5 (2024)
6. Campos, F., Hellenbrand, A., Meyer, M., Reijnders, K.: dCTIDH: fast & deterministic CTIDH. IACR Trans. Cryptogr. Hardw. Embed. Syst. **2025**(3), 516–541 (2025)
7. Campos, F., Kannwischer, M. J., Meyer, M., Onuki, H., Stöttinger, M.: Trouble at the CSIDH: protecting CSIDH with dummy-operations against fault injection attacks. In 17th Workshop on Fault Detection and Tolerance in Cryptography, FDTC 2020, 13 September 2020,, Milan, Italy, pp. 57–65. IEEE (2020)
8. Castryck, W., Decru, T.: An efficient key recovery attack on SIDH. In: Hazay, C., Stam, M. (eds.) Advances in Cryptology - EUROCRYPT 2023 - 42nd Annual International Conference on the Theory and Applications of Cryptographic Techniques, 23–27 April 2023, Proceedings, Part V, volume 14008 of Lecture Notes in Computer Science, Lyon, France, pp. 423–447. Springer (2023)
9. Castryck, W., Lange, T., Martindale, C., Panny, L., Renes, J.: CSIDH: an efficient post-quantum commutative group action. In: Peyrin, T., Galbraith, S.D. (eds.) Advances in Cryptology - ASIACRYPT 2018 - 24th International Conference on the Theory and Application of Cryptology and Information Security, 2–6 December 2018, Proceedings, Part III, volume 11274 of Lecture Notes in Computer Science, Brisbane, QLD, Australia, pp. 395–427. Springer (2018)
10. Chi-Domínguez, J.-J., Ochoa-Jiménez, E., Rodas, R.N.P.: Let us walk on the 3-isogeny graph: efficient, fast, and simple. IACR Trans. Cryptogr. Hardw. Embed. Syst. **2025**(4), 644–666 (2025)
11. Chi-Domínguez, J.-J., Rodríguez-Henríquez, F.: Optimal strategies for CSIDH. Adv. Math. Commun. **16**(2), 383–411 (2022)
12. Chiu, T., LeGrow, J., Xiong, W.: Practical fault injection attacks on constant time CSIDH and mitigation techniques. In: Chang, C.-H., Rührmair, U., Szefer, J., Batina, L., Regazzoni, F. (eds.) Proceedings of the 2024 Workshop on Attacks and Solutions in Hardware Security, ASHES 2024, 14– 18 October 2024, Salt Lake City, UT, USA, pp. 11–22. ACM (2024)
13. Maino, L., Martindale, C., Panny, L., Pope, G., Wesolowski, B.: A direct key recovery attack on SIDH. In: Hazay, C., Stam, M. (eds.) Advances in Cryptology - EUROCRYPT 2023 - 42nd Annual International Conference on the Theory and Applications of Cryptographic Techniques, 23–27 April 2023, Proceedings, Part V, volume 14008 of Lecture Notes in Computer Science, Lyon, France, pages 448–471. Springer, (2023)

14. Meyer, M., Campos, F., Reith, S.: On Lions and Elligators: an efficient constant-time implementation of CSIDH. In: Ding, J., Steinwandt, R. (eds.) Post-Quantum Cryptography, pp. 307–325. Springer International Publishing, Cham (2019)
15. Onuki, H., Aikawa, Y., Yamazaki, T., Takagi, T.: A constant-time algorithm of CSIDH keeping two points. IEICE Trans. Fundam. Electr., Commun. Comput. Sci. **E103-A**(10), 1174–1182 (2020)
16. Peikert, C.: He gives c-sieves on the CSIDH. In: Canteaut, A., Ishai, Y. (eds.) Advances in Cryptology - EUROCRYPT 2020 - 39th Annual International Conference on the Theory and Applications of Cryptographic Techniques, 10–14 May 2020, Proceedings, Part II, volume 12106 of Lecture Notes in Computer Science, Zagreb, Croatia, pp. 463–492. Springer (2020)
17. Pope, G., Reijnders, K., Robert, D., Sferlazza, A., Smith, B.: Simpler and faster pairings from the montgomery ladder. IACR Commun. Cryptol. **2**(2), 29 (2025)
18. Robert, D.: Breaking SIDH in polynomial time. In: Hazay, C., Stam, M. (eds.) Advances in Cryptology - EUROCRYPT 2023 - 42nd Annual International Conference on the Theory and Applications of Cryptographic Techniques, 23–27 April 2023, Proceedings, Part V, volume 14008 of Lecture Notes in Computer Science, Lyon, France, pp. 472–503. Springer (2023)
19. Silverman, J.H.: The Arithmetic of Elliptic Curves. Graduate Texts in Mathematics, 2 ed. Springer, New York (2009)
20. Vélu, J.: Isogénies entre courbes elliptiques. Comptes-Rendus de l'Académie des Sciences **273**, 238–241 (1971)

Key-Updatable Identity-Based Signature Schemes

Tobias Guggemos[1,2(✉)] and Farzin Renan[3]

[1] Ludwig-Maximilians-Universität München, Munich, Germany
`guggemos@nm.ifi.lmu.de`
[2] German Aerospace Center (DLR), Wessling, Germany
`tobias.guggemos@dlr.de`
[3] Middle East Technical University, Ankara, Turkey
`farzin.renan@gmail.com`

Abstract. Identity-based signature (IBS) schemes eliminate the need for certificate management, thereby reducing communication and computational overhead. A major challenge, however, is the efficient update or revocation of compromised keys, as existing approaches such as revocation lists or periodic key renewal incur significant network costs in dynamic settings. We address this challenge by introducing a symmetric element that enables key updates in IBS schemes through a single multicast message. Our approach achieves logarithmic network overhead in the number of keys, with constant computation and memory costs. We further propose a general framework that transforms any IBS scheme into a key-updatable IBS scheme (KUSS), and formalize the associated security requirements, including token security, forward security, and post-compromise security. The versatility of our framework is demonstrated through five instantiations based on Schnorr-type, pairing-based, and isogeny-based IBS, and we provide a detailed security analysis.

Keywords: Identity-Based Signatures · Key Revocation · Group Communication · ECC · Pairing-based · Isogeny-based Cryptography

1 Introduction

Efficient communication is critical for resource-constrained devices, including IoT sensors, smart lighting, and vehicular networks. Minimizing network overhead directly extends device lifetime and reduces maintenance costs. Traditional certificate-based authentication is ill-suited for such environments: each signature carries certificates, and verification requires scanning Certificate Revocation Lists (CRLs) maintained by Certificate Authorities (CAs). In large-scale dynamic systems, such as Car-to-Car (Car2Car) communication, this results in complex multi-CA trust infrastructures [1].

Identity-based signatures (IBS) address these challenges by binding public keys to user identities. A Trusted Third Party (TTP) generates user keys and publishes a master public key (mpk), allowing verification using only the sender's

R. Dutta et al. (Eds.): INDOCRYPT 2025, LNCS 16372, pp. 216–238, 2026.
https://doi.org/10.1007/978-3-032-13301-4_10

identity and mpk. This design makes IBS an attractive alternative to Public Key Infrastructure (PKI) in closed or dynamic groups. However, key revocation and updates remain challenging. Boneh and Franklin's generic key renewal method [2] requires secure channels and incurs network overhead proportional to private key sizes. Similarly, revocation in IBS can lead to high costs, including linear communication overhead [2], expensive pairing or lattice operations [3,4], or logarithmic growth in key and signature size [5].

Principles of Key Updates. In both PKI and IBS, once keys are distributed, they cannot simply be retracted. PKIs rely on CRLs to mark revoked certificates. In IBS, tracking invalid identities through lists would undermine the goal of avoiding third-party checks [6]. A natural alternative is re-keying: the TTP revokes the current mpk and issues fresh keys to valid users.

1.1 Related Work and Our Contribution

The generic approach for updating of IBS keys, proposed by Boneh and Franklin [2] is too costly for constrained networks, motivating several alternatives.

Hierarchical Keys. Hierarchical Identity-Based Signatures (HIBS) [5] provide flexible role-based structures, promising for sensor networks, but signature and secret key sizes grow with hierarchy depth. Optimizations achieving constant signature sizes [7] require large public keys and still do not support efficient subtree renewal. Tree-based revocation, explored initially for IBE [3] and later adapted to IBS [8], requires multiple pairing operations, making these approaches impractical for constrained settings.

Key Insulation and Updatable Encryption. Key insulation [9] and ciphertext updates [10] inspired our approach. In [9], both master and user keys are updated via tokens, with signatures embedding update information. Our method differs by introducing a symmetric update token that simultaneously updates all users' keys via a single broadcast. Security relies solely on the confidentiality of this token, supporting forward and post-compromise security with minimal overhead.

Contribution. We extend IBS with two new phases (Next, Update) and introduce a symmetric element δ. A TTP distributes a single confidential update token to all valid users, leveraging group key management for efficiency. We further generalize the transformation of any IBS scheme into a Key-Updatable Signature Scheme (KUSS) and formalize the associated security requirements, including token security (UF-KUSS-CMA), forward security, and post-compromise security.

To illustrate the generality of our approach, we apply our framework to five IBS schemes:

GG: by Galindo et al. [11], based on the Schnorr signature [12].
vBNN: by Cao et al. [13], also based on the Schnorr signature.
Hess: by Hess [14], based on bilinear pairings.

BLMQ: Barreto et al. [15], employed bilinear pairings.
CSIIBS: by Shaw et al. [17], based on CSI-FiSh signature [16].

Existing re-keying approaches exhibit significant trade-offs. Standards such as X.509 rely on revocation lists, keeping signature sizes small but imposing heavy verification and storage costs. HIBS [5] achieves logarithmic update complexity at the expense of larger signatures and verification keys. Key insulation [9,18] reduces computational costs but does not eliminate network and memory overhead. Revocable IBE schemes [3] achieve efficient logarithmic revocation but are not effectively adapted to IBS, while pairing-based schemes such as **Hess** [14] and **BLMQ** [15] incur high computational costs.

Table 1. Comparison of our approach (**gIBS**) with other re-keying approaches ($n =$ number of group members)

	$\mathcal{O}_{|\sigma|}$	$\mathcal{O}_S$	$\mathcal{O}_V$	$\mathcal{O}_{N_V}$	$\mathcal{O}_{U_{TTP}}$	$\mathcal{O}_{U_{GM}}$	$\mathcal{O}_{N_U}$	$\mathcal{O}_{|U|}$
ACE [19, 20]	1	1	$\log n$	-	1	1	n	n
X.509 [21]	1	1	n	-	1	1	n	n
OCSP [22]	1	1	1	m	1	-	-	1
vBNN-IBS [13]	1	1	n	-	1	1	n	n
Group Signature [23]	1	$\log n$	1	-	$\log n$	1	$\log n$	$\log n$
Group Signature [24]	1	1	1	-	1	1	n	n^2
IBS [25]	1	1	1	-	1	1	n	1
Inline-HIBS [5]	$\log n$	1	$\log n$	-	1	1	$\log n$	n
Pre-Computed-HIBS [5]	1	1	1	-	$\log n$	n	$\log n$	n
Key-Insulation [18]	1	1	1	-	1	1	n	n
Our Solution	**1**	**1**	**1**	-	**$\log n$**	**1**	**1**	**$\log n$**

$\mathcal{O}_{|\sigma|}$ = signature size, $\mathcal{O}_S$ = signing cost, $\mathcal{O}_V$ = verification cost, $\mathcal{O}_{N_V}$ = network overhead for verifying m signatures, $\mathcal{O}_{U_{TTP}}$ = preparation cost at TTP, $\mathcal{O}_{U_{GM}}$ = processing cost at GM, $\mathcal{O}_{N_U}$ = network overhead for a key update, $\mathcal{O}_{|U|}$ = size of the update message.

Our solution achieves constant-size signatures, lightweight signing and verification, and key update overhead that grows logarithmically with the number of users. Unlike pairing-based methods, it is broadly applicable across signature constructions, making it suitable for resource-constrained and dynamic environments. Furthermore, by leveraging hierarchical update strategies (e.g., the Logical Key Hierarchy (LKH) [26]), key updates can be efficiently transmitted in a single message. A detailed complexity comparison with state-of-the-art approaches is provided in Table 1.

Notation. The security parameter is denoted by λ. For a set $\mathbb{S}$, the notation $s \xleftarrow{\$} \mathbb{S}$ represents sampling an element s uniformly at random from $\mathbb{S}$. A probabilistic polynomial-time (PPT) algorithm A producing output x on input y is denoted as $x \xleftarrow{\$} A(y)$. For a deterministic polynomial-time (DPT) algorithm, this is represented as $x := A(y)$, or $x \leftarrow A(y)$. A function $\mathsf{negl} : \mathbb{N} \to \mathbb{R}$ is called *negligible* if, for every $k \in \mathbb{N}$, there exists an $n_0 \in \mathbb{N}$ such that for all $n \geq n_0$, the inequality $\mathsf{negl}(n) \leq \frac{1}{n^k}$ holds. The concatenation of two strings s_1 and s_2 is denoted by $s_1 \| s_2$. For a positive integer n, $\mathbb{Z}_n := \mathbb{Z}/n\mathbb{Z}$. We use the notation $\forall i \in [T]$ to denote "for all $i \in \{1, 2, \ldots, T\}$".

2 Preliminaries

2.1 Identity-Based Signature Scheme

Formally, an IBS scheme consists of three entities: the Key Generation Center (KGC), the signer, and the verifier. The formal definition of an IBS scheme, along with its corresponding security properties, is provided following the syntax in [17].

Definition 2.1 (Identity-based Signature Scheme). *An identity-based signature (IBS) scheme is a quadruple* $(\mathsf{Setup}, \mathsf{Ext}, \mathsf{Sign}, \mathsf{Ver})$ *consisting of four polynomial-time algorithms with the following syntax:*

$\mathsf{Setup}(1^\lambda) \xrightarrow{\$} (\mathsf{pp}, \mathsf{msk}, \mathsf{mpk})$: *A* PPT *algorithm executed by the KGC, takes the security parameter* λ *as input, outputs the public parameters* pp, *the master secret key* msk, *and the corresponding master public key* mpk.

$\mathsf{Ext}(\mathsf{msk}, \mathsf{id}) \xrightarrow{\$} \mathsf{usk}$: *A* PPT *algorithm executed by the KGC, takes the master secret key* msk *and a member's identity* id, *generates a user's secret key* usk *associated with* id.

$\mathsf{Sign}(\mathsf{usk}, m) \xrightarrow{\$} \sigma$: *A* PPT *algorithm that takes the user's secret key* usk *and a message* m *as input, outputs a signature* σ *for the message* m.

$\mathsf{Ver}(\mathsf{id}, m, \sigma) \to 0/1$: *A* DPT *algorithm that takes a signer's identity* id, *a message* m, *and a signature* σ, *checks the validity of the signature* σ *on the message* m.

Definition 2.2 (Correctness). *An IBS scheme satisfies correctness if for any* $\lambda \in \mathbb{N}$, *and* $\mathsf{id} \in \mathsf{ID}$, *and any message* $m \in \{0, 1\}^*$, *the following holds:*

$$Pr\left[\mathsf{Ver}(\mathsf{id}, m, \mathsf{Sign}(\mathsf{usk}, m)) = 1 \ \middle|\ \begin{array}{c} (\mathsf{pp}, \mathsf{msk}, \mathsf{mpk}) \xleftarrow{\$} \mathsf{Setup}(1^\lambda) \\ \mathsf{usk} \xleftarrow{\$} \mathsf{Ext}(\mathsf{msk}, \mathsf{id}) \end{array} \right] = 1.$$

Definition 2.3 (UF-IBS-CMASecurity). *An IBS scheme satisfies unforgeability under chosen identity and chosen message attacks* (UF-IBS-CMA) *if for all* PPT *adversary* $\mathcal{A}$, *there exists a negligible function* negl *such that*

$$Pr\left[\mathsf{Exp}_{\mathsf{IBS}, \mathcal{A}}^{\mathsf{UF\text{-}IBS\text{-}CMA}}(\lambda) = 1 \right] \leq \mathsf{negl}(\lambda),$$

where the experiment $\mathsf{Exp}_{\mathsf{IBS}, \mathcal{A}}^{\mathsf{UF\text{-}IBS\text{-}CMA}}$ *is defined as shown in Fig. 1.*

Input: The challenger $\mathcal{C}$ takes the security parameter λ as input, produces $(\mathsf{pp}, \mathsf{msk}, \mathsf{mpk}) \xleftarrow{\$} \mathsf{Setup}(1^\lambda)$. It gives the $\mathsf{pp}, \mathsf{mpk}$ to the adversary $\mathcal{A}$ while keeping msk secret to itself.

Query Phase: In this phase, $\mathcal{C}$ responds to polynomially many adaptive queries made by $\mathcal{A}$ by following the steps described below:

- **Oracle** $\mathcal{O}_{\mathcal{E}}(\mathsf{msk}, \cdot)$: Upon receiving queries on a user identity id from $\mathcal{A}$, $\mathcal{C}$ outputs the user's secret key $\mathsf{usk} \xleftarrow{\$} \mathsf{Ext}(\mathsf{msk}, \mathsf{id})$ for the given identity id.
- **Oracle** $\mathcal{O}_{\mathcal{S}}(\mathsf{usk}, \cdot)$: Upon receiving queries on a message m and a user identity id from $\mathcal{A}$, $\mathcal{C}$ returns a signature $\sigma \xleftarrow{\$} \mathsf{Sign}(\mathsf{usk}, m)$, where $\mathsf{usk} \xleftarrow{\$} \mathsf{Ext}(\mathsf{msk}, \mathsf{id})$ is the user's secret key associated to the id.

Forgery: In this phase, $\mathcal{A}$ eventually submits a message m^*, user identity id^*, and a forgery σ^*. $\mathcal{A}$ wins the game if $1 \leftarrow \mathsf{Ver}(\mathsf{id}^*, m^*, \sigma^*)$, with the restriction that id^* has not been queried to $\mathcal{O}_{\mathcal{E}}(\mathsf{msk}, \cdot)$ and (m^*, id^*) has not been queried to $\mathcal{O}_{\mathcal{S}}(\mathsf{usk}, \cdot)$.

Fig. 1. $\mathsf{Exp}_{\mathsf{IBS}, \mathcal{A}}^{\mathsf{UF\text{-}IBS\text{-}CMA}}$ Experiment.

2.2 Elliptic Curves and Isogenies

The primary comprehensive reference widely regarded for elliptic curves is Silverman [27]. Let $k := \mathbb{F}_q$ be a finite field where $q := p^n$ for some prime $p > 3$, and positive integer n. An *elliptic curve* E over a field k is a smooth, projective variety of genus 1, defined over k, with a distinguished rational point $\mathcal{O}_E$. For a positive integer ℓ, the ℓ-*tosion subgroup* of an elliptic curve E is defined as $E[\ell] := \{P \in E(\bar{k}) \mid [\ell]P = \mathcal{O}_E\}$. An elliptic curve E is called *supersingular* if it has no nontrivial p-torsion points over $\overline{\mathbb{F}}_p$, i.e., $E[p] = \{\mathcal{O}_E\}$. Specifically, if $E/\mathbb{F}_p$ is supersingular, then $\#E(\mathbb{F}_p) = p + 1$.

An *isogeny* is a surjective morphism between elliptic curves that is also a homomorphism with respect to the natural group structure on these curves. Two elliptic curves E_1 and E_2 are said to be *isogenous* over $\mathbb{F}_q$ if there exists an isogeny between them over $\mathbb{F}_q$. Any subgroup $G \subset E(\mathbb{F}_q)$ determines a unique (separable) isogeny $\phi : E \to E' := E/G$ with $\ker(\phi) = G$, up to $\mathbb{F}_q$-isomorphism.

An *endomorphism* is an isogeny from an elliptic curve E to itself. Examples include the *multiplication-by-m* map $[m] : P \mapsto [m]P$ for $m \in \mathbb{Z}$, and the *Frobenius* map $\pi : (x, y) \mapsto (x^q, y^q)$ for $E/\mathbb{F}_q$. The set of all endomorphisms on E, denoted by $\mathrm{End}(E)$, forms a ring under addition and composition, known as the *endomorphism ring* of E. For supersingular elliptic curves, $\mathrm{End}(E)$ is isomorphic to an order in a quaternion algebra, whereas the ring of $\mathbb{F}_p$-endomorphism on E, denoted by $\mathrm{End}_{\mathbb{F}_p}(E)$, is isomorphic only to an order in the imaginary quadratic field $\mathbb{Q}(\sqrt{-p})$. Thus, for a supersingular elliptic curve E over $\mathbb{F}_p$, there is a strict inclusion, such that $\mathrm{End}_{\mathbb{F}_p}(E) \subset \mathrm{End}(E)$.

The ideal class group of an order $\mathfrak{O} \subset \mathbb{Q}(\sqrt{-p})$ is defined as the quotient of the group of invertible fractional ideals $\mathcal{I}_{\mathfrak{O}}$ by the subgroup of principal fractional ideals $\mathcal{P}_{\mathfrak{O}}$, i.e., $\mathrm{Cl}(\mathfrak{O}) := \mathcal{I}_{\mathfrak{O}}/\mathcal{P}_{\mathfrak{O}}$. There is a natural action of the class group $\mathrm{Cl}(\mathfrak{O})$ on the set $\mathcal{E}ll_p(\mathfrak{O})$ of $\mathbb{F}_p$-isomorphism classes of elliptic curves defined over $\mathbb{F}_p$ with endomorphism rings isomorphic to $\mathfrak{O}$. Specifically, for a given $\mathfrak{O}$-ideal $\mathfrak{a}$, let the subgroup $S_{\mathfrak{a}}$ be defined by the intersection of the kernels of the endomorphisms in $\mathfrak{a}$, i.e., $S_\mathfrak{a} := \bigcap_{\alpha \in \mathfrak{a}} \ker(\alpha)$. As $S_\mathfrak{a}$ is a subgroup of E, we can quotient E by $S_\mathfrak{a}$ and denote the isogenous curve by $\mathfrak{a} * E := E/S_\mathfrak{a}$. The isogeny $\phi_\mathfrak{a} : E \to E/S_\mathfrak{a}$ is well-defined and unique up to $\mathbb{F}_p$-isomorphism. The class group $\mathrm{Cl}(\mathfrak{O})$ acts via the operator $*$ on the set $\mathcal{E}ll_p(\mathfrak{O})$, i.e., $* : \mathrm{Cl}(\mathfrak{O}) \times \mathcal{E}ll_p(\mathfrak{O}) \to \mathcal{E}ll_p(\mathfrak{O})$ freely and transitively.

As in [16], an embedding $\mathbb{Z}/N\mathbb{Z} \hookrightarrow \mathrm{Cl}(\mathfrak{O})$ is defined by $a \mapsto \mathfrak{g}^a$, where $\mathrm{Cl}(\mathfrak{O})$ is cyclic with generator $\mathfrak{g}$. Elements of $\mathrm{Cl}(\mathfrak{O})$ are expressed as $\mathfrak{a} = \mathfrak{g}^a$, with a sampled randomly from $\mathbb{Z}/N\mathbb{Z}$ such that $N = \#\mathrm{Cl}(\mathfrak{O})$. Using the shorthand $[a]$ for $\mathfrak{g}^a$ and $[a]E$ for $\mathfrak{g}^a * E$, it follows that $[a][b]E = [a+b]E$.

2.3   Computational Assumptions

Our analysis relies on the following foundational group properties and computational hardness assumptions. A key property underlying our reasoning is the *Latin square property*, which ensures that the operation within a group produces uniformly distributed elements:

Lemma 2.1 (Latin square property). *Let $(G, \circ)$ be a group of order $|G| = k$, and let $a \in G$. Then, for any $b, c \in G$ chosen uniformly at random,*

$$P(c = a \circ b) = \frac{1}{k}.$$

As a corollary, the operation of two randomly drawn elements from G results in a uniformly distributed random element of G. Specifically, in the group $(\mathbb{Z}_p^*, \cdot)$:

Lemma 2.2 (Randomness preservation in $(\mathbb{Z}_p^*, \cdot)$). *For any $a \in \mathbb{Z}_p^*$ and a random $b \in \mathbb{Z}_p^*$, the product $c = a \cdot b$ is uniformly distributed in $\mathbb{Z}_p^*$.*

Proof. Let p be a prime and $(\mathbb{Z}_p^*, \cdot)$ a group of order $p - 1$. For any $x \in \mathbb{Z}_p^*$,

$$P(x = x' \mid x' \xleftarrow{\$} \mathbb{Z}_p^*) = \frac{1}{p-1} = P(x = a \cdot b),$$

showing that the product of a random b with a yields a random group element $c \in \mathbb{Z}_p^*$.

Moreover, the primary computational assumption underlying our isogeny-based instantiation, introduced in Sect. 4.2, is the *Multi-Target Group Action Inverse Problem (MT-GAIP)*, which captures the difficulty of inverting group actions derived from isogenies. The CSI-FiSh signature scheme [16], and consequently **CSIIBS** [17], relies on the hardness of random instances of MT-GAIP.

*Problem 2.1 (**Multi-Target Group Action Inverse Problem (MT-GAIP)** [16]).* Suppose $E_0, E_1, \ldots, E_k$ are $k+1$ supersingular elliptic curves defined over $\mathbb{F}_p$, with $\mathrm{End}_{\mathbb{F}_p}(E_i) \cong \mathfrak{O}$ for all $0 \leq i \leq k$. Find an ideal $\mathfrak{a} \subset \mathfrak{O}$ such that $E_i = \mathfrak{a} * E_j$ for some distinct indices $i, j \in \{0, \ldots, k\}$.

3 Key-Updatable Identity-Based Signature Scheme

This section introduces the Key-Updatable Identity-Based Signature Scheme (KUSS), formalizes its framework and security properties, and demonstrates its use in designing a protocol for IBS key updates.

Symmetric Group Re-keying. In a communication group, each participant, called a Group Member (GM), holds a copy of the same symmetric group key, which is used to ensure confidentiality and authenticate group communications. Updating one member's key therefore requires updating all members' keys; this process is known as re-keying. Efficient symmetric re-keying schemes typically preserve three security properties: **i.** Forward Access Control, **ii.** Backward Access Control, and **iii.** Key Independence [28].

Although decentralized and distributed re-keying mechanisms exist [29], we focus on centralized approaches, motivated by their structural similarity to IBS key management. In such approaches, a Group Controller Key Server (GCKS) manages group membership and securely distributes key updates. Protocols such as G-IKEv2 [30], COSE [19], and key management mechanisms like the LKH [26] enable efficient key updates.

Our proposed IBS key update mechanism leverages these efficient symmetric re-keying techniques. In the remainder of this paper, we assume symmetric key updates are efficient and use LKH as an example of a centralized binary-tree approach, where update complexity scales as $\mathcal{O}(\log n)$.

Basic Idea. At the heart of IBS is the mapping of a user's identity string to the corresponding user secret key usk via a mathematical relation with the master secret key. Building on this principle, our approach–illustrated in Fig. 2–is motivated by the following observation:

Update Token δ. In many schemes, when the master secret key is renewed, the transition from msk_0 to msk_1 can be captured by a mathematical object δ, even if the keys are chosen independently. Since msk is used to derive both the master public key mpk and user secret key usk, updates to these keys can also be expressed in terms of δ (or a related element). Crucially, as long as msk_0 remains uncompromised, knowledge of δ reveals nothing about either msk_0 or msk_1.

In the context of user key updates–or more generally, key revocation–it is both secure and efficient for the TTP to distribute δ to all valid users rather than issuing entirely new keys individually. Each user can then independently update their keys to usk_1 and mpk_1. In case of key compromise, it suffices to prevent the update from reaching users outside the system. This approach also supports key revocation by distributing the update to all parties except the revoked user.

Update Accumulator Δ. When keys are updated multiple times, the cumulative difference from the original key material can still be expressed through the mathematical object δ. In all schemes considered here, the overall change is represented as the accumulation of individual updates, denoted by Δ.

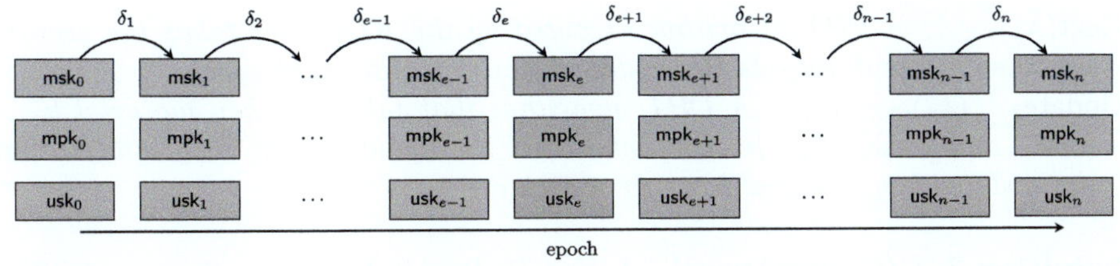

Fig. 2. Idea of updating IBS keys with an update token δ over epochs e

Example. Consider the **Hess** IBS scheme [14]. Let msk_1 be the current master secret key, and let the TTP select a fresh key msk_2 for a re-key operation. Since both msk_1 and msk_2 belong to a cyclic group $\mathbb{Z}_p^*$, there exists $\delta_1 \in \mathbb{Z}_p^*$ such that $\mathsf{msk}_2 = \delta_1 \cdot \mathsf{msk}_1$. In **Hess** IBS, the master public key is $\mathsf{mpk} = \mathsf{msk} \circ P$, and the user secret key is $\mathsf{usk} = \mathsf{msk} \circ \mathsf{H}(\mathsf{id})$, where $\mathsf{H}(\mathsf{id}) \in \langle P \rangle$ is a hash of the user's identity. Hence, transitioning from msk_1 to msk_2 updates both mpk and usk by δ_1:

	msk	mpk	usk
epoch 1	msk_1	$\mathsf{msk}_1 \circ P$	$\mathsf{msk}_1 \circ \mathsf{H}(\mathsf{id})$
epoch 2	$\delta_1 \cdot \mathsf{msk}_1$	$\delta_1 \cdot \mathsf{msk}_1 \circ P$	$\delta_1 \cdot \mathsf{msk}_1 \circ \mathsf{H}(\mathsf{id})$

After a subsequent update $\delta_2 \xleftarrow{\$} \mathbb{Z}_p^*$, the master secret becomes $\mathsf{msk}_3 = \delta_2 \cdot \mathsf{msk}_2 = \Delta_2 \cdot \mathsf{msk}_1$, where $\Delta_2 = \delta_2 \cdot \delta_1$. The **Hess** scheme allows seamless key updates because δ can be incorporated directly into user keys, without modifying signing or verification algorithms. Once the updated keys are computed, δ can be discarded. In contrast, other key-updatable schemes may require storing δ (or Δ) for signing or verification.

Formal Definition. The information each entity needs to update their key material, denoted δ, is called the *update token*. A *key update* refers to distributing this token and computing the corresponding new keys. The term *epoch* denotes the interval between successive updates. The update accumulator Δ, also called the *group shared secret*, is accessible only to non-revoked users during a given epoch and is not part of the public parameters. An IBS scheme in which the mathematical relation between the master and user keys allows key updates is called a *Key-Updatable Identity-Based Signature Scheme* (KUSS), formally defined as follows:

Definition 3.1 (Key-Updatable Identity-Based Signature). *A Key-Updatable Identity-Based Signature Scheme (**KUSS**) for message space $\mathcal{M}$ is a tuple of polynomial-time algorithms*

$$\mathsf{KUSS} := (\mathsf{Setup}, \mathsf{Ext}, \mathsf{Next}, \mathsf{Update}, \mathsf{Sign}, \mathsf{Ver}),$$

where $(\mathsf{Setup}, \mathsf{Ext}, \mathsf{Sign}, \mathsf{Ver})$ *are inherited from a standard IBS scheme (Definition 2.1), and* $(\mathsf{Next}, \mathsf{Update})$ *are additional algorithms defined as follows:*

$\mathsf{Next}(\lambda) \xrightarrow{\$} \delta$: *A **PPT** algorithm executed by the KGC that takes the security parameter λ and outputs the update token δ for the next epoch.*
$\mathsf{Update}_{\delta_{e+1}}(k_e) \to k_{e+1}$: *A **DPT** algorithm that takes the key material $k_e :=$ $(\mathsf{msk}_e, \mathsf{mpk}_e, \mathsf{usk}_e, \Delta_e)$ from epoch e and the update token δ_{e+1}, and outputs the updated key material k_{e+1} for epoch $e + 1$.*

Definition 3.2 (Correctness). *A Key-Updatable Signature Scheme (KUSS), satisfies correctness if for any $\lambda \in \mathbb{N}$, any epoch e, and $\mathsf{id} \in \mathsf{ID}$, and any message $m \in \mathcal{M}$, the following holds:*

$$\Pr\left[\mathsf{Ver}_{\mathsf{mpk}_e}(\mathsf{id}, m, \mathsf{Sign}(\mathsf{usk}_e, m)) = 1 \, \middle| \, \begin{array}{c} (\mathsf{pp}, \mathsf{msk}_0, \mathsf{mpk}_0) \xleftarrow{\$} \mathsf{Setup}(1^\lambda), \\ \mathsf{usk}_0 \xleftarrow{\$} \mathsf{Ext}(\mathsf{msk}_0, \mathsf{id}), \\ \delta_i \xleftarrow{\$} \mathsf{Next}(\lambda), \, 1 \le i \le e, \\ k_i \leftarrow \mathsf{Update}_{\delta_i}(k_{i-1}), \, \text{where} \\ k_i := (\mathsf{msk}_i, \mathsf{mpk}_i, \mathsf{usk}_i, \Delta_i), 1 \le i \le e. \end{array} \right] = 1.$$

Transformability Conditions. The KUSS is generally applicable to all IBS schemes that satisfy the following conditions:

Property 1 (Symmetric Integration). It is possible to select a shared secret Δ such that it can be integrated into the signing process with usk, i.e., in $\mathsf{Sign}(\mathsf{usk}, m)$, without increasing the number of signatures produced.

Concretely, since the key update process is identical for all users, the update token δ and the accumulator Δ can be interpreted as symmetric keys shared among the users and the TTP. Secretly updating this symmetric element implicitly updates the corresponding asymmetric keys, because signatures produced with an invalid symmetric element become invalid. In other words, any IBS scheme can be generically transformed into a KUSS by incorporating such a symmetric element into the signing process. For example, if Δ is used as a symmetric key in $\mathsf{HMAC}(\Delta, m)$ to "pre-sign" the message, then verification satisfies

$$\mathsf{Ver}(\mathsf{Sign}(\mathsf{usk}, m)) = 1 \iff \mathsf{Ver}(\mathsf{Sign}(\mathsf{usk}, \mathsf{HMAC}(\Delta, m))) = 1.$$

While Property 1 establishes that a shared secret Δ can be integrated symmetrically into the signing process, differences in security and performance may arise in a concrete implementation using a symmetric update token. For example, maintaining a secondary signing key could require additional storage and management, and the signing operation might involve extra computation. Security considerations may also differ between a secondary signing key and an updated one, though in schemes such as the **Hess** scheme [14], these differences are nominal.

If the update tokens (or the shared secret Δ) can be seamlessly incorporated into both the signing and verification keys, we classify these keys as **hybrid**. The following properties define criteria that a concrete KUSS instantiation should satisfy if such integration is possible only for signing or verification.

Property 2 (Signing Key Integration). The user secret key usk and the accumulator Δ_e can participate in a binary operation during signing, such that the updated key usk_e signs messages in the same way as the original usk in $\mathsf{Sign}(\mathsf{usk}, m)$.

Property 3 (Public Key Integration). The master public key mpk and the accumulator Δ_e can participate in a binary operation during verification, such that the updated key mpk_e verifies signatures in the same way as the original mpk in $\mathsf{Ver}_{\mathsf{mpk}}(\mathsf{id}, m, \sigma)$.

3.1 Security Model

Having introduced key updates in IBS schemes via an update token, we now formalize the security considerations. This construction can be seen as a hybrid of a standard PKI and a group key management system. Security definitions for both paradigms have been extensively studied and standardized (e.g., [31–34]). For IBS schemes, the fundamental security notion is UF-IBS-CMA (Definition 2.3). We emphasize that the introduction of the shared secret Δ and the corresponding update mechanism does not weaken the security guarantees of the constructions described in Sect. 4. In the following, we examine security properties inspired by Updatable Encryption [10], adapted to the context of Key-Updatable Identity-Based Signature Schemes (KUSS).

Token Security. Integrating a shared secret into a signature scheme does not weaken the security of the underlying construction, as it remains protected under the UF-IBS-CMA definition. Building on this foundation, we formally define the token security property for KUSS as follows.

Definition 3.3 (UF-KUSS-CMASecurity). *A KUSS scheme is secure against unforgeability under chosen-message-identity-and-epoch attacks (UF-KUSS-CMA) if for all PPT adversaries $\mathcal{A}$, there exists a negligible function* negl *such that*

$$\Pr\left[\mathsf{Exp}_{KUSS,\mathcal{A}}^{\textit{UF-KUSS-CMA}}(\lambda) = 1\right] \leq \mathsf{negl}(\lambda),$$

where the experiment $\mathsf{Exp}_{KUSS,\mathcal{A}}^{\textit{UF-KUSS-CMA}}$ *is defined as illustrated in Fig. 3.*

Forward and Post-Compromise Security. These properties capture the resilience of a KUSS against key compromises, ensuring that both past and future signatures remain secure even if a secret key is exposed in a given epoch.

(i) **Forward Security** An adversary compromising any secret key (msk, usk, or Δ) in some epoch e^* does not gain any advantage in forging signatures for previous epochs $e < e^*$. That is, all signatures and signing keys from epochs *before* the compromise retain their integrity.

(ii) **Post-Compromise Security** An adversary compromising any secret key (msk, usk, or Δ) in some epoch e^* does not gain any advantage in forging signatures for subsequent epochs $e > e^*$. That is, all signatures and keys from epochs *after* the compromise remain secure.

Input: The challenger $\mathcal{C}$, given the security parameter λ, runs $(\mathsf{pp}, \mathsf{msk}_0, \mathsf{mpk}_0) \leftarrow \mathsf{Setup}(1^\lambda)$. It provides pp and mpk_0 to the adversary $\mathcal{A}$ while keeping msk_0 secret. Finally, it initializes an empty set $\mathsf{idList} = \emptyset$.

Query Phase: In this phase, $\mathcal{C}$ responds to polynomially many adaptive queries made by $\mathcal{A}$ by following the steps described below:

- **Oracle** $\mathcal{O}_{\mathsf{Ext}}(\mathsf{msk}_e, \cdot)$: Upon receiving a query on a user identity id from $\mathcal{A}$ at epoch e, if $\mathsf{id} \notin \mathsf{idList}$, the challenger $\mathcal{C}$ generates the corresponding user secret key $\mathsf{usk}_e \xleftarrow{\$} \mathsf{Ext}(\mathsf{msk}_e, \mathsf{id})$ and updates the list as $\mathsf{idList} \leftarrow \mathsf{idList} \cup \{\mathsf{id}\}$. Otherwise, $\mathcal{C}$ aborts.
- **Oracle** $\mathcal{O}_{\mathsf{N}}(\lambda)$: On receiving queries at epoch e from $\mathcal{A}$, $\mathcal{C}$ generates a new update token for epoch $e+1$ via $\delta_{e+1} \xleftarrow{\$} \mathsf{Next}(\lambda)$.
- **Oracle** $\mathcal{O}_{\mathsf{U}}(\cdot, \cdot)$: On receiving queries from $\mathcal{A}$ to update to a new epoch, $\mathcal{C}$ uses the update token δ_{e+1} along with the current key materials $k_e := (\mathsf{msk}_e, \mathsf{mpk}_e, \mathsf{usk}_e, \Delta_e)$, (for all $\mathsf{id} \in \mathsf{idList}$), for epoch e to compute the updated key materials $k_{e+1} \leftarrow \mathsf{Update}_{\delta_{e+1}}(k_e)$, where $k_{e+1} := (\mathsf{msk}_{e+1}, \mathsf{mpk}_{e+1}, \mathsf{usk}_{e+1}, \Delta_{e+1})$. It then returns mpk_{e+1} to $\mathcal{A}$.
- **Oracle** $\mathcal{O}_{\mathsf{Sign}}(\mathsf{usk}_e, \cdot)$: On receiving queries on a message m and a user identity id at epoch e from $\mathcal{A}$, $\mathcal{C}$ responds with a signature $\sigma \xleftarrow{\$} \mathsf{Sig}(m, \mathsf{usk}_e)$, where $\mathsf{usk}_e \xleftarrow{\$} \mathsf{Ext}(\mathsf{msk}_e, \mathsf{id})$ or $\mathsf{usk}_e \leftarrow \mathsf{Update}_{\delta_e}(\cdot, \cdot, \mathsf{usk}_{e-1}, \cdot)$ is the user secret key associated to the identity id at epoch e.

Forgery: In this phase, $\mathcal{A}$ eventually submits a message m^*, a user identity id^*, and a forgery σ^*, along with e^*. $\mathcal{A}$ wins the game if $1 \leftarrow \mathsf{Ver}_{\mathsf{mpk}_{e^*}}(\mathsf{id}^*, m^*, \sigma^*)$, with the restriction that id^* and (m^*, id^*) has not been queried to $\mathcal{O}_{\mathsf{Ext}}$ and $\mathcal{O}_{\mathsf{Sign}}$, respectively, and $\mathcal{O}_{\mathsf{N}}$ has not queried at epoch e^*.

Fig. 3. $\mathsf{Exp}_{\mathsf{KUSS}, \mathcal{A}}^{\mathsf{UF\text{-}KUSS\text{-}CMA}}$ Experiment.

3.2 gIBS: A Protocol for IBS Key Distribution and Updates

RFC 4046 [28] defines a symmetric group re-keying architecture with the following relevant keys:

Key Encryption Key (KEK): A shared secret between a Group Member (GM) and the Group Controller Key Server (GCKS), e.g., established via an authenticated Diffie-Hellman exchange.

Group Key Encryption Key (GKEK): A secret shared among all group members, used solely to secure other keys. It may be part of an efficient re-keying mechanism such as LKH.

This architecture enables seamless integration of IBS and KUSS keys. User secrets (usk) and key updates (δ, Δ) can thus be integrated and securely communicated to new and existing GMs via RFC 4046 [28]:

Setup: The GCKS initializes the group by setting $(\mathsf{pp}, \mathsf{msk}_0, \mathsf{mpk}_0, \Delta_0) \xleftarrow{\$} \mathsf{Setup}(\lambda)$.

Join: A client with identity id wishes to join the communication group.

1. The GCKS generates $\delta_{e+1} \xleftarrow{\$} \mathsf{Next}(\lambda)$ to move to the next epoch and sends it to existing clients encrypted under GKEK.
2. All clients of epoch e call $(\mathsf{usk}_{e+1}, \mathsf{mpk}_{e+1}, \Delta_{e+1}) \leftarrow \mathsf{Update}_{\delta_{e+1}}(\mathsf{usk}_e, \mathsf{mpk}_e, \Delta_e)$.

3. The GCKS calls $(\mathsf{msk}_{e+1}, \mathsf{mpk}_{e+1}, \Delta_{e+1}) \leftarrow \mathsf{Update}_{\delta_{e+1}}(\mathsf{msk}_e, \mathsf{mpk}_e, \Delta_e)$.

4. The GCKS calls $\mathsf{usk}_{e+1} \xleftarrow{\$} \mathsf{Ext}(\mathsf{msk}_{e+1}, \mathsf{id})$ and sends $(\mathsf{pp}, \mathsf{usk}_{e+1}, \mathsf{mpk}_{e+1}, \Delta_{e+1})$ to the joining client secured by the corresponding KEK.

Leave: A client with id is excluded from the communication group.

1. The GCKS calls $\delta_{e+1} \xleftarrow{\$} \mathsf{Next}(\lambda)$ in order to move to the next epoch and sends a re-key message including δ_{e+1} to all clients except id secured by a new GKEK, which is efficiently done e.g. by LKH.

2. All clients of epoch e call $(\mathsf{usk}_{e+1}, \mathsf{mpk}_{e+1}, \Delta_{e+1}) \leftarrow \mathsf{Update}_{\delta_{e+1}}(\mathsf{usk}_e, \mathsf{mpk}_e, \Delta_e)$.

3. The GCKS calls $(\mathsf{msk}_{e+1}, \mathsf{mpk}_{e+1}, \Delta_{e+1}) \leftarrow \mathsf{Update}_{\delta_{e+1}}(\mathsf{msk}_e, \mathsf{mpk}_e, \Delta_e)$.

Communication: All clients of epoch e call $\sigma \xleftarrow{\$} \mathsf{Sign}(\mathsf{usk}_e, m)$ and $\mathsf{Ver}_{\mathsf{mpk}_e}(\mathsf{id}, m, \sigma)$ in order to sign and verify messages for epoch e.

4   Instantiations of KUSS

In this section, we present five instantiations of IBS schemes adapted to the KUSS framework. Specifically, Sect. 4.1 addresses elliptic curve-based IBS schemes, introducing two pairing-based and two Schnorr-type adaptations. Section 4.2, on the other hand, details an instantiation for adapting an isogeny-based IBS scheme to KUSS. Furthermore, we examine the modifications required to ensure compliance with Properties 1, 2, and 3, as established earlier.

4.1   Schnorr-Type and Pairing-Based IBS Schemes Adaptations

The integration of the shared secret into the pairing-based IBS schemes **Hess** [14] and **BLMQ** [15] is illustrated in Fig. 4, while its incorporation into the Schnorr-type IBS schemes **GG** [11] and **vBNN** [13] is presented in Fig. 5.

These schemes, all of which rely on the (elliptic curve) discrete logarithm problem, transform trivially under Property 1. In particular, the update token δ_e for an epoch e is derived by sampling a random element from $\mathbb{Z}_p^*$. The multiplicative accumulator for epoch e is then defined as $\Delta_e := \prod_{j=0}^{e} \delta_j$. Consequently, the master secret key msk_e and master public key mpk_e at an arbitrary epoch e are updated as $\mathsf{msk}_e = \Delta_e \cdot \mathsf{msk}$ and $\mathsf{mpk}_e = (\Delta_e \cdot \mathsf{msk})P$, respectively, where P denotes a generator point on the elliptic curve. By adapting the computation of usk and mpk to align with their roles in the signing and verification algorithms of the schemes, all four schemes can seamlessly incorporate a **group shared secret** without modifying the structure of the signature.

The two pairing-based KUSS schemes, **Hess+g** and **BLMQ+g**, as illustrated in Fig. 4, additionally satisfy Properties 2 and 3. Specifically, when the updated user secret key usk_e is utilized for signing, the verification algorithm requires only the corresponding updated master public key mpk_e and, in the case of **BLMQ+g**, the updated generator Q_e as well.

	Hess+g	**BLMQ+g**
Public param.	$H_1 : \{0,1\}^* \to \mathbb{G}^*, P \xleftarrow{\$} \mathbb{G}^*$ $H_2 : \{0,1\}^* \to \mathbb{Z}_p^*$ $e : \mathbb{G} \times \mathbb{G} \to \mathbb{Z}_p^*$	$H_{1,2} : \{0,1\}^* \to \mathbb{Z}_p^*$ $e : \mathbb{G}_1 \times \mathbb{G}_2 \to \mathbb{Z}_p^*$ $Q \xleftarrow{\$} \mathbb{G}_2, P := \psi(Q) \in \mathbb{G}_1$ $q := e(P,Q)$
Extract	$\mathsf{usk}_e := (\Delta_e \cdot \mathsf{msk})H_1(\mathsf{id})$	$\mathsf{usk}_e := \frac{1}{(\mathsf{msk}+H_1(\mathsf{id}))\cdot\Delta_e}P$
Sign	$x \xleftarrow{\$} \mathbb{Z}_p^*, Q \xleftarrow{\$} \mathbb{G}^*$ $r := e(Q,P)^x, h := H_2(m,r)$ $S_e := h\,\mathsf{usk}_e + xQ$ $\sigma_e := (h, S_e)$	$x \xleftarrow{\$} \mathbb{Z}_p^*, r := q^x$ $h := H_2(m,r)$ $S_e := (x+h)\mathsf{usk}_e$ $\sigma_e := (h, S_e)$
Verify	$\tilde{r} := e(S_e, P) \cdot e(H_1(\mathsf{id}), -\mathsf{mpk}_e)^h$ $h \overset{?}{=} H_2(m, \tilde{r})$	$\tilde{r} := e(S_e, H_1(\mathsf{id})\, Q_e + \mathsf{mpk}_e) \cdot q^{-h}$ $h \overset{?}{=} H_2(m, \tilde{r})$

Fig. 4. Hess and BLMQ adaptations.

In the two Schnorr-type KUSS schemes, **GG+g** and **vBNN+g**, as depicted in Fig. 5, the updated user secret key $\mathsf{usk}_e = (u_e, R)$ inherently includes a secondary Schnorr signature component, u_e. Seamlessly updating usk_e would require recomputing u_e, which is infeasible without access to the system's msk. To address this limitation, we propose incorporating Δ_e – a value known exclusively to valid group members – into the verification process through multiplication with the auxiliary value R. However, this approach does not comply with Property 3. Alternatively, the signature σ_e could be extended to include $\Delta_e R$. That said, if the verifier is a group member, they can compute $\Delta_e R$ independently, which is significantly more efficient than embedding it into the signature.

	GG+g	**vBNN+g**
Public param.	$H_{1,2} : \{0,1\}^* \to \mathbb{Z}_p^*$ $P \xleftarrow{\$} \mathbb{G}^*$	$H_{1,2} : \{0,1\}^* \to \mathbb{Z}_p^*$ $P \xleftarrow{\$} \mathbb{G}^*$
Extract	$r \xleftarrow{\$} \mathbb{Z}_p^*, R := rP$ $u_e := (r + \mathsf{msk} \cdot H_1(R, \mathsf{id})) \cdot \Delta_e$ $\mathsf{usk}_e := (u_e, R)$	$r \xleftarrow{\$} \mathbb{Z}_p^*, R := rP$ $u_e := (r + \mathsf{msk} \cdot H_1(R, \mathsf{id})) \cdot \Delta_e$ $\mathsf{usk}_e := (u_e, R)$
Sign	$x \xleftarrow{\$} \mathbb{Z}_p^*, X := xP$ $h := H_2(\mathsf{id}, m, X)$ $s_e := x + h \cdot u_e$ $\sigma_e := (s_e, R, X)$	$x \xleftarrow{\$} \mathbb{Z}_p^*, X := xP$ $h := H_2(\mathsf{id}, m, R, X)$ $s_e := x + h \cdot u_e$ $\sigma_e := (s_e, R, h)$
Verify	$c := H_1(R, \mathsf{id})$ $d := H_2(\mathsf{id}, m, X)$ $s_e P \overset{?}{=} X + d(\Delta_e R + c\,\mathsf{mpk}_e)$	$c := H_1(R, \mathsf{id})$ $\tilde{X} := s_e P - h(\Delta_e R + c\,\mathsf{mpk}_e)$ $h \overset{?}{=} H_2(\mathsf{id}, m, R, \tilde{X})$

Fig. 5. GG and vBNN adaptations.

One can show that the correctness of the above four adaptations directly follows from the correctness of their respective underlying IBS schemes.

4.2 Isogeny-Based Adaptation

To construct an isogeny-based KUSS, we build upon the IBS scheme derived from the CSI-FiSh signature developed by Shaw et al. [17].

In the isogeny-based IBS setting, the master secret key and master public key are defined as sequences of S_0 components, namely $\mathsf{msk} := \{s_i\}_{i=1}^{S_0}$ and $\mathsf{mpk} := \{E_i := [s_i]E_0\}_{i=1}^{S_0}$. Consequently, both the token and accumulator are also represented as S_0-element sequences. Unlike the multiplicative accumulator used in the elliptic curve instantiation (Sect. 4.1), the isogeny-based setting employs an additive accumulator.

For each epoch e, an update token is generated as $\boldsymbol{\delta}_e = \{\delta_{i,e}\}_{i=1}^{S_0}$ with $\delta_{i,e} \xleftarrow{\$} \mathbb{Z}_N$, and the additive accumulator is updated recursively as $\boldsymbol{\Delta}_e = \boldsymbol{\Delta}_{e-1} + \boldsymbol{\delta}_e$, where $\Delta_{i,e} = \sum_{j=0}^{e} \delta_{i,j}$ for $i \in [S_0]$. The master secret and public keys at epoch e are then defined as $\mathsf{msk}_e = \{s_i + \Delta_{i,e}\}_{i=1}^{S_0}$ and $\mathsf{mpk}_e = \{E_i^{(e)}\}_{i=1}^{S_0} := \{[\Delta_{i,e}]E_i\}_{i=1}^{S_0}$. Here, $s_i + \Delta_{i,e}$ incorporates the cumulative contribution of the accumulator into the initial secret, while $E_i^{(e)}$ denotes the corresponding updated public key component.

As described in [17], the public parameters consist of a prime p, base curve E_0, generator $\mathfrak{g}$, class number $h_{\mathcal{O}} = N$, and integers T_1, T_2, S_0, S_1 with $T_1 < S_0 = 2^{n_0} - 1$ and $T_2 < S_1 = 2^{n_1} - 1$. The hash functions are defined as $\mathsf{H}_1 : \{0,1\}^* \to [0, S_0]^{T_1 S_1}$ and $\mathsf{H}_2 : \{0,1\}^* \to [0, S_1]^{T_1 T_2}$. Based on these parameters, the extraction, signing, and verification procedures are detailed as follows:

Extraction. Given msk_e and an identity id at epoch e, the algorithm samples $r_{i,j} \xleftarrow{\$} \mathbb{Z}_N$ and computes $R_{i,j} = [r_{i,j}]E_0$ for $i \in [T_1]$, $j \in [S_1]$. The hash $\mathbf{u} = \{u_i\}_{i=1}^{T_1 S_1} := \mathsf{H}_1(\{R_{i,j}\}_{i=1,j=1}^{T_1,S_1}\|\mathsf{id})$ is evaluated. Let $s_0 := 0$. For each pair (i,j), compute

$$x_{i,j}^{(e)} := r_{i,j} - s_{u_i} - \Delta_{u_i,e} \pmod{N}.$$

Define $\mathbf{x}^{(e)} := \{x_{i,j}^{(e)}\}_{i=1,j=1}^{T_1,S_1}$, and set $\mathsf{usk}_e := (\mathbf{u}, \mathbf{x}^{(e)})$.

Signing. To sign m at epoch e, let $x_{i,0}^{(e)} := 0$ for $i \in [T_1]$. Compute $X_{i,j} := [x_{i,j}^{(e)}]E_{u_i}^{(e)}$ for $i \in [T_1]$, $j \in [S_1]$. Sample $k_{i,j} \xleftarrow{\$} \mathbb{Z}_N$ and set $K_{i,j} := [k_{i,j}]E_{u_i}^{(e)}$ for $i \in [T_1]$, $j \in [T_2]$. The hash $\mathbf{v} = \{v_{i,j}\}_{i=1,j=1}^{T_1,T_2} := \mathsf{H}_2(\{K_{i,j}\}_{i=1,j=1}^{T_1,T_2}\|m)$ is computed, and

$$z_{i,j} := k_{i,j} - x_{i,v_{i,j}}^{(e)} \pmod{N}.$$

Define $\mathbf{z} := \{z_{i,j}\}_{i=1,j=1}^{T_1,T_2}$ and $\mathbf{X} := \{X_{i,j}\}_{i=1,j=1}^{T_1,S_1}$. The signature is $\sigma = (\mathbf{v}, \mathbf{z}, \mathbf{X})$.

Verification. Given (m, σ, id), recompute $\mathbf{u} = \mathsf{H}_1(\{X_{i,j}\}_{i=1,j=1}^{T_1,S_1}\|\mathsf{id})$. Then, for all (i,j), compute

$$K'_{i,j} := \begin{cases} [z_{i,j}]E_{u_i}^{(e)}, & v_{i,j} = 0, \\ [z_{i,j}]X_{i,v_{i,j}}, & v_{i,j} \neq 0. \end{cases}$$

Finally, compute $\mathbf{v}' := \mathsf{H}_2(\{K'_{i,j}\}_{i=1,j=1}^{T_1,T_2}\|m)$ and accept iff $\mathbf{v} = \mathbf{v}'$.

Correctness. For all $(\mathsf{pp}, \mathsf{msk}_0, \mathsf{mpk}_0) \leftarrow \mathsf{Setup}(1^\lambda)$, any message $m \in \mathcal{M}$, and any signature $\sigma = (\mathbf{v}, \mathbf{z}, \mathbf{X}) \leftarrow \mathsf{Sign}(\mathsf{usk}_e, m)$ with $\mathsf{usk}_e = (\mathbf{u}, \mathbf{x}^{(e)})$ from Ext or Update, it suffices to show that $K'_{i,j} = K_{i,j}$ for all $i \in [T_1]$, $j \in [T_2]$.

First, recover $\mathbf{u}$ via $\mathsf{H}_1(\{X_{i,j}\}_{i=1,j=1}^{T_1,S_1}\|\mathsf{id})$, noting that $X_{i,j} = R_{i,j}$ for $i \in [T_1]$, $j \in [S_1]$. Indeed,

$$X_{i,j} = [x_{i,j}^{(e)}]E_{u_i}^{(e)} = [r_{i,j} - s_{u_i} - \Delta_{u_i,e}][s_{u_i} + \Delta_{u_i,e}]E_0 = [r_{i,j}]E_0 = R_{i,j}.$$

Next, for $\mathbf{v}$, consider

$$K'_{i,j} = \begin{cases} [z_{i,j}]E_{u_i}^{(e)}, & v_{i,j} = 0, \\ [z_{i,j}]X_{i,v_{i,j}}, & v_{i,j} \neq 0, \end{cases} \tag{$*$}$$

which evaluates to $[k_{i,j}]E_{u_i}^{(e)}$ in both cases. Hence, $\{K'_{i,j}\}_{i=1,j=1}^{T_1,T_2} = \{K_{i,j}\}_{i=1,j=1}^{T_1,T_2}$ and

$$\mathbf{v}' = \mathsf{H}_2(\{K'_{i,j}\}_{i=1,j=1}^{T_1,T_2}\|m) = \mathbf{v},$$

confirming the correctness of the construction.

Our proposed scheme **CSIIBS+g** satisfies Properties 1, 2, and 3. Specifically, for an arbitrary epoch e, the signature generated using the user secret key usk_e in KUSS is indistinguishable from the signature generated using the user secret key usk in IBS. This holds because

$$\mathsf{usk}_e = (\mathbf{u}, \mathbf{x}^{(e)}) = (\mathbf{u}, \{x_{i,j}^{(e)}\}_{i=1,j=1}^{T_1,S_1}) = (\mathbf{u}, \{r_{i,j} - s_{u_i} - \Delta_{u_i,e}\}_{i=1,j=1}^{T_1,S_1}),$$

whereas the user secret key in IBS is defined as $\mathsf{usk} := (\mathbf{u}, \mathbf{x}) = (\mathbf{u}, \{r_{i,j} - s_{u_i}\}_{i=1,j=1}^{T_1,S_1})$. Since $\mathbf{x}$ and $\mathbf{x}^{(e)}$ are randomized, they are indistinguishable. Consequently, the signing algorithm $\mathsf{Sig}(\mathsf{usk}_e, \cdot)$ in KUSS operates as though it were using $\mathsf{Sig}(\mathsf{usk}, \cdot)$ in IBS, producing a signature of the form

$$\sigma = (\mathbf{v}, \mathbf{z}, \mathbf{X}) = (\{v_{i,j}\}_{i=1,j=1}^{T_1,T_2}, \{z_{i,j}\}_{i=1,j=1}^{T_1,T_2}, \{X_{i,j}\}_{i=1,j=1}^{T_1,S_1}).$$

It follows that the construction satisfies Property 2. Finally, to verify the signature we just need $\mathsf{mpk}_e = \{E_i^{(e)}\}_{i=1}^{S_0}$ to compute $(*)$ thereby satisfying Property 3.

5   Security Proof

In this section, we analyze the security of the constructions introduced in Sect. 4. In particular, we provide detailed proofs for UF-KUSS-CMA security for one pre- and one post-quantum scheme (i.e. **BLMQ+g** and **CSIIBS+g**). We also outline proof strategies for establishing forward security and post-compromise security, illustrating the resilience of these constructions against the corresponding adversarial models.

5.1   Token Security

Theorem 5.1. *Suppose that the **BLMQ** scheme is unforgeable under adaptively chosen-message-and-identity attacks (UF-IBS-CMA), then the **BLMQ+g** scheme is unforgeable under adaptively chosen-message-identity-and-epoch attacks (UF-KUSS-CMA).*

Proof. Assume there exists an adversary $\mathcal{A}$ that succeeds in the experiment $\mathsf{Exp}^{\mathsf{UF\text{-}KUSS\text{-}CMA}}_{\mathbf{BLMQ+g},\mathcal{A}}(\lambda)$ with non-negligible probability ε and within time t. Let $\mathcal{B}$ be an adversary participating in the experiment $\mathsf{Exp}^{\mathsf{UF\text{-}IBS\text{-}CMA}}_{\mathbf{BLMQ},\mathcal{B}}(\lambda)$. On input λ, the challenger computes $(\mathsf{pp}, \mathsf{msk}, \mathsf{mpk})$ by invoking the Setup algorithm of **BLMQ** scheme and forwards $(\mathsf{pp}, \mathsf{mpk})$ to $\mathcal{B}$. Then, $\mathcal{B}$ forwards $(\mathsf{pp}, \mathsf{mpk}_0, \Delta_0)$ to $\mathcal{A}$, where $\Delta_0 \xleftarrow{\$} \mathbb{Z}_p^*$ initializes the epoch accumulator. On input $(\mathsf{pp}, \mathsf{mpk}_0, \Delta_0)$, $\mathcal{A}$ produces a forgery tuple $(\mathsf{id}^*, m^*, \sigma^*, e^*)$, where $\sigma^* = (h^*, S_{e^*}^*)$ satisfying the following:

$$\tilde{r} = e(S_{e^*}^*, \mathsf{H}_1(\mathsf{id}^*)Q_{e^*} + \mathsf{mpk}_{e^*}) \cdot q^{-h^*},$$
$$h^* = \mathsf{H}_2(m^*, \tilde{r}).$$

Upon receiving the forgery from $\mathcal{A}$, the adversary $\mathcal{B}$ outputs a forgery $(\mathsf{id}^*, m^*, \tilde{\sigma})$, where $\tilde{\sigma} = (h^*, \tilde{S})$ and $\tilde{S} = \Delta_{e^*} S_{e^*}^*$. The signature $\tilde{\sigma}$ is a valid forgery for the message m^* associated with id^* under the mpk of the **BLMQ** scheme as $\tilde{r}$ can still be recovered as follows:

$$\tilde{r} = e(\tilde{S}, \mathsf{H}_1(\mathsf{id}^*)Q + \mathsf{mpk}) \cdot q^{-h^*}$$
$$= q^{(x^* + h^* - h^*)} = q^{x^*}$$

Thus, $\tilde{\sigma}$ constitutes a valid forged signature produced by $\mathcal{B}$ within time $t' \leq t + t_m$ with non-negligible probability ε, where t_m is the time required to compute Δ_{e^*} operations in $\mathbb{G}$. This result contradicts the UF-IBS-CMA security of the **BLMQ** scheme. $\qquad\square$

Theorem 5.2. *If the **CSIIBS**, as introduced in [17], is unforgeable under adaptively chosen-message-and-identity attacks (UF-IBS-CMA), then the **CSIIBS+g** scheme, introduced in Sect. 4.2, is unforgeable under adaptively chosen-message-identity-and-epoch attacks (UF-KUSS-CMA), according to Definition 3.3.*

Proof. Assume there exists an efficient adversary $\mathcal{A}$ that breaks the UF-KUSS-CMA security of **CSIIBS+g** (Definition 3.3) with non-negligible probability. We show that this implies the existence of a forger $\mathcal{B}$ that breaks the UF-IBS-CMA security of **CSIIBS**. To do so, $\mathcal{B}$ leverages $\mathcal{A}$'s success in the experiment $\mathsf{Exp}_{\mathsf{CSIIBS+g},\mathcal{A}}^{\mathsf{UF\text{-}KUSS\text{-}CMA}}(\lambda)$ to participate in $\mathsf{Exp}_{\mathsf{CSIIBS},\mathcal{B}}^{\mathsf{UF\text{-}IBS\text{-}CMA}}(\lambda)$ (Definition 2.3). Specifically, $\mathcal{B}$ acts as the challenger for $\mathcal{A}$, simulating all oracles and responses using its own oracle access.

Setup. In the $\mathsf{Exp}_{\mathsf{CSIIBS},\mathcal{B}}^{\mathsf{UF\text{-}IBS\text{-}CMA}}(\lambda)$ experiment, the challenger computes $(\mathsf{pp}, \mathsf{msk}_0, \mathsf{mpk}_0)$ by invoking the Setup algorithm of **CSIIBS**, and forwards $(\mathsf{pp}, \mathsf{mpk}_0)$ to $\mathcal{B}$. $\mathcal{B}$ forwards $(\mathsf{pp}, \mathsf{mpk}_0, \Delta_0)$ to $\mathcal{A}$, where $\Delta_0 := \mathbf{0}$ initializes the epoch accumulator. It also initializes two empty lists: $\mathsf{idList}, \mathsf{mList}$, and an epoch sequence epSeq, with $\mathsf{epSeq} := (\Delta_0) = (\Delta_i)_{i=0}^{0}$.

Simulation of Queries. $\mathcal{A}$ can issue polynomially many adaptive queries to the following oracles: $\mathcal{O}_{\mathsf{Ext}}, \mathcal{O}_{\mathsf{N}}, \mathcal{O}_{\mathsf{U}}$, and $\mathcal{O}_{\mathsf{Sign}}$. $\mathcal{B}$ simulates these oracles using its own access to the $\mathsf{Exp}_{\mathsf{CSIIBS},\mathcal{B}}^{\mathsf{UF\text{-}IBS\text{-}CMA}}(\lambda)$ experiment:

- *Simulating* $\mathcal{O}_{\mathsf{Ext}}(\mathsf{msk}_e, \cdot)$: For a queried identity id, if $\mathsf{id} \in \mathsf{idList}$, return $\perp$. Otherwise, $\mathcal{B}$ queries its extracting oracle with id to obtain a secret key usk. For epoch e, it computes the epoch-specific user secret key $\mathsf{usk}_e := (\mathbf{u}, \mathbf{x}^{(e)} = \{x_{i,j} - \Delta_{u_i,e}\}_{i=1,j=1}^{T_1,S_1})$, and updates $\mathsf{idList} \leftarrow \mathsf{idList} \cup \{\mathsf{id}\}$.
- *Simulating* $\mathcal{O}_{\mathsf{N}}(\lambda, \cdot)$: Upon receiving a query to this oracle at epoch e, $\mathcal{B}$ samples a new random update token δ_{e+1} for epoch $e+1$.
- *Simulating* $\mathcal{O}_{\mathsf{U}}(\cdot, \cdot)$: Upon receiving a query to this oracle at epoch e on input δ_{e+1} and $k_e = (\mathsf{msk}_e, \mathsf{mpk}_e, \mathsf{usk}_e, \Delta_e)$, $\mathcal{B}$ computes $\Delta_{e+1} = \delta_{e+1} + \Delta_e$, $\mathsf{mpk}_{e+1} = \{[s_i + \Delta_{i,e} + \delta_{i,e+1}]E_0\}_{i=1}^{S_0}$, and $\mathsf{usk}_{e+1} = (\mathbf{u}, \mathbf{x}^{(e+1)}) = (\mathbf{u}, \{r_{i,j} - s_{u_i} - \Delta_{u_i,e} - \delta_{u_i,e+1}\}_{i=1,j=1}^{T_1,S_1})$ for all $\mathsf{id} \in \mathsf{idList}$. Finally, $\mathcal{B}$ updates the sequence $\mathsf{epSeq} \leftarrow (\Delta_i)_{i=0}^{e+1}$.
- *Simulating* $\mathcal{O}_{\mathsf{Sign}}(\mathsf{usk}_e, \cdot)$: Upon receiving a query to this oracle on input a message $m \in \mathcal{M}$ at epoch e, $\mathcal{B}$ produces a signature $\sigma = (\mathbf{v}, \mathbf{z}, \mathbf{X})$ on m under mpk using its oracle. $\mathcal{B}$, then modifies $\mathbf{z}$ by setting $z_{i,j} \leftarrow z_{i,j} - \Delta_{u_i,e} \mod N$ for which $v_{i,j} = 0$. $\mathcal{B}$ return the modified signature $\sigma' = (\mathbf{v}, \mathbf{z}', \mathbf{X})$ to $\mathcal{A}$, and updates the list $\mathsf{mList} \leftarrow \mathsf{mList} \cup \{m\}$.

Extracting the Forgery: $\mathcal{A}$ eventually submits a forgery $(\mathsf{id}^*, m^*, \sigma^*, e^*)$ to $\mathcal{B}$, where $\sigma^* = (\mathbf{v}^*, \mathbf{z}^*, \mathbf{X}^*)$. $\mathcal{B}$ modifies the signature σ^* to frame a forgery under the mpk by retrieving Δ_{e^*} from epSeq and setting $z_{i,j}^* \leftarrow z_{i,j}^* + \Delta_{u_i,e^*} \mod N$ for which $v_{i,j} = 0$. Thereby, $\mathcal{B}$ creates a forged signature $\bar{\sigma}^* = (\mathbf{v}^*, \bar{\mathbf{z}}^*, \mathbf{X}^*)$, and outputs $(\mathsf{id}^*, m^*, \bar{\sigma}^*)$ as a forgery to its challenger.

For the analysis, assume that the probability with which $\mathcal{A}$ wins in the experiment $\mathsf{Exp}_{\mathsf{CSIIBS+g},\mathcal{A}}^{\mathsf{UF\text{-}KUSS\text{-}CMA}}(\lambda)$ is non-negligible. We shall now demonstrate that $\mathcal{B}$ provides a perfect simulation of the oracle $\mathcal{O}_{\mathsf{Sign}}$ on message m. The signature $\sigma = (\mathbf{v}, \mathbf{z}, \mathbf{X})$ on m under $\mathsf{mpk} = \{E_i\}_{i=1}^{S_0}$ received by $\mathcal{B}$ from its own signing oracle consists of $\mathbf{v} = \{v_{i,j}\}_{i=1,j=1}^{T_1,T_2}$, $\mathbf{z} = \{z_{i,j}\}_{i=1,j=1}^{T_1,T_2} = \{k_{i,j} - x_{i,v_{i,j}}\}_{i=1,j=1}^{T_1,T_2}$, and $\mathbf{X} = \{X_{i,j}\}_{i=1,j=1}^{T_1,S_1} = \{[x_{i,j}]E_{u_i}\}_{i=1,j=1}^{T_1,S_1}$, where $\{v_{i,j}\}_{i=1,j=1}^{T_1,T_2} = \mathsf{H}_2(K_{i,j}\|m)$ with $K_{i,j} = [k_{i,j}]E_{u_i}$ for all $k_{i,j} \in \mathbb{Z}_N$, $i \in [T_1]$, and $j \in [T_2]$. By modifying $\mathbf{z}$ via $z'_{i,j} := z_{i,j} - \Delta_{u_i,e} \mod N$ for which $v_{i,j} = 0$, we obtain $\mathbf{z}' := \{z'_{i,j}\}_{i=1,j=1}^{T_1,T_2}$. We still retrieve $\mathbf{v} = \{v_{i,j}\}_{i=1,j=1}^{T_1,T_2}$ under mpk_e as follows:

- $[z'_{i,j}]E_{u_i}^{(e)} = [k_{i,j} - x_{i,v_{i,j}} - \Delta_{u_i,e}]E_{u_i}^{(e)} = [k_{i,j} - \Delta_{u_i,e}]E_{u_i}^{(e)} = [k_{i,j}]E_{u_i}$,
- $[z'_{i,j}]X_{i,v_{i,j}} = [k_{i,j} - x_{i,v_{i,j}}][x_{i,v_{i,j}}]E_{u_i} = [k_{i,j}]E_{u_i}$,

in case $v_{i,j} = 0$, and $v_{i,j} \neq 0$, respectively. Thus, $\sigma = (\mathbf{v}, \mathbf{z}', \mathbf{X})$ is a signature on m under the mpk_e associated to identity id.

Similarly, the forgery of $\mathcal{B}$ is computed from the forgery of $\mathcal{A}$. Let $\mathcal{A}$ submit a valid forgery $(\mathsf{id}^*, m^*, \sigma^*, e^*)$, where the signature is $\sigma^* = (\{v_{i,j}^*\}_{i=1,j=1}^{T_1,T_2}, \{z_{i,j}^*\}_{i=1,j=1}^{T_1,T_2}, \{X_{i,j}^*\}_{i=1,j=1}^{T_1,S_1})$. Then, $\mathcal{B}$ modifies the signature σ^* to frame a forgery under the mpk by setting $\bar{z}_{i,j}^* := z_{i,j}^* + \Delta_{u_i,e^*} \mod N$ for which $v_{i,j}^* = 0$, thereby under mpk the hash value $\mathbf{v}^* = \{v_{i,j}^*\}_{i=1,j=1}^{T_1,T_2}$ can be still retrieved as follows:

- $[\bar{z}_{i,j}^*]E_{u_i} = [k_{i,j}^* - x_{i,v_{i,j}^*}^* + \Delta_{u_i,e^*}]E_{u_i} = [k_{i,j}^* + \Delta_{u_i,e^*}]E_{u_i} = [k_{i,j}^*]E_{u_i}^{(e^*)}.$
- $[\bar{z}_{i,j}^*]X_{i,v_{i,j}^*}^* = [k_{i,j}^* - x_{i,v_{i,j}^*}^*][x_{i,v_{i,j}^*}^*]E_{u_i}^{(e^*)} = [k_{i,j}^*]E_{u_i}^{(e^*)}.$

in case $v_{i,j}^* = 0$, and $v_{i,j}^* \neq 0$, respectively. Thus, $\bar{\sigma}^* = (\mathbf{v}^*, \bar{\mathbf{z}}^*, \mathbf{X}^*)$ is a valid signature on m^* under the mpk for identity id^*. Moreover, $\mathcal{B}$ queries identical messages as $\mathcal{A}$ while responding to signing queries for $\mathcal{A}$, and therefore whenever $\mathcal{A}$ wins in the experiment $\mathsf{Exp}_{\mathsf{CSIIBS+g},\mathcal{A}}^{\mathsf{UF\text{-}KUSS\text{-}CMA}}(\lambda)$, then $\mathcal{B}$ wins in the experiment $\mathsf{Exp}_{\mathsf{CSIIBS},\mathcal{B}}^{\mathsf{UF\text{-}IBS\text{-}CMA}}(\lambda)$. $\qquad\square$

5.2   Forward Security

Forward security in KUSS is achieved if and only if the following conditions are met:

1. A client's entry triggers a transition to epoch $e + 1$ by invoking Next and Update.
2. Upon entry, the new client receives only Δ_{e+1} but not δ_{e+1}.

The second condition ensures forward security in cases of re-entry. Specifically, if a device lacks knowledge of previously used Δ, disclosing δ would pose no risk. However, if the device had been a group member two sessions before, knowledge of δ could enable the reconstruction of the previous Δ value. Consider the following scenario:

$$\Delta_0 \xleftarrow{\$} \mathbb{Z}_p^*, \quad \delta_{1,2} \xleftarrow{\$} \mathbb{Z}_p^*,$$
$$\Delta_1 = \Delta_0 \cdot \delta_1,$$
$$\Delta_2 = \Delta_0 \cdot \delta_1 \cdot \delta_2,$$

where Δ_0, Δ_1, and Δ_2 represent the shared secrets for epochs 0, 1, and 2, respectively.

Now, for schemes introduced in Sect. 4.1, assume a client who was a group member in epoch 0, but not in epoch 1, re-joins the group in epoch 2. This client knows Δ_0 and Δ_2, but lacks knowledge of Δ_1, δ_1, and δ_2. If this client learns δ_2, it could compute: $\frac{\Delta_2}{\Delta_0 \cdot \delta_2} = \delta_1$ and learn $\Delta_1 = \Delta_0 \cdot \delta_1$. However, with only Δ_0 and Δ_2, reconstructing Δ_1 would require: $\frac{\Delta_2}{\delta_2} = \Delta_1$ and $\Delta_0 \cdot \delta_1 = \Delta_1$. According to Lemma 2.2, the probability of obtaining such δ_1 or δ_2 is $\epsilon = \frac{1}{p-1}$. Thus, by updating Δ upon a client's entry and refraining from disclosing the current δ to new (or re-entering) members, forward security is preserved in KUSS.

In the case of **CSIIBS+g**, introduced in Sect. 4.2, the member is still unable to compute $\boldsymbol{\Delta}_{e-1}$ since the probability of finding such $\boldsymbol{\delta}_{e-1}$ or $\boldsymbol{\delta}_e$ tokens such that

$$\boldsymbol{\Delta}_{e-1} = \boldsymbol{\Delta}_{e-2} + \boldsymbol{\delta}_{e-1} \mod N$$
$$= \boldsymbol{\Delta}_e - \boldsymbol{\delta}_e \mod N$$

is computationally infeasible: $\frac{1}{h(\mathfrak{O})} = \frac{1}{N} \approx \frac{1}{p^{1/2}} \approx \frac{1}{2^\lambda}$.

5.3 Post-compromise Security

Theorem 5.3 (KUSS Post-Compromise Security). *Assuming the hardness of the One-More Discrete Logarithm Problem (1MDLP) [35], it is computationally infeasible for an adversary to compute δ_{e+1} from mpk_{e+1} in the $\mathbf{X} + \mathbf{g}$ construction, where $\mathbf{X} \in \{\mathbf{Hess}, \mathbf{BLMQ}, \mathbf{GG}, \mathbf{vBNN}\}$.*

Proof. Let $\mathcal{A}$ have access to the Oracle $\mathcal{O}_\mathsf{N}$, which transitions the global state at arbitrary epoch e to epoch $e + 1$, and to $\mathcal{O}_\mathsf{U}(\cdot, \mathsf{mpk}_e, \cdot, \cdot)$, which returns a $\mathsf{mpk}_{e+1} = \delta_{e+1}\mathsf{mpk}_e$ (where $\delta_{e+1} \xleftarrow{\$} \mathbb{Z}_p^*$). In considered schemes introduced in Sect. 4.1, $\mathsf{mpk} = \mathsf{msk}\, P$, where $\mathsf{msk} \xleftarrow{\$} \mathbb{Z}_p^*$ and P is the subgroup generator of an elliptic curve. Thus, we have:

$$\mathsf{mpk}_{e+1} = (\delta_{e+1} \cdot \Delta_e \cdot \mathsf{msk}_0)P$$
$$= (\delta_{e+1} \cdot \mathsf{msk}_e)P$$
$$= \mathsf{msk}_{e+1}\, P$$

According to Lemma 2.2, given msk_e the probability $P(k \xleftarrow{\$} \mathbb{Z}_p^* | \mathsf{msk}_{e+1} = k \cdot \mathsf{msk}_e) = \frac{1}{p-1}$. The output of $\mathcal{O}_\mathsf{U}(\cdot, \mathsf{mpk}_e, \cdot, \cdot)$ fits the definition of *random target points* in the Chall Oracle in [31], where $Y \xleftarrow{\$} \mathbb{G}$. Hence, the advantage for $\mathcal{A}$ to compute δ_{e+1} from mpk_{e+1} is bounded by the 1MDLP [35]. $\square$

In a similar argument, in the case of **CSIIBS+g**, all signatures and keys from epochs following the compromise remain secure. Let the adversary $\mathcal{A}$ have access to the oracle $\mathcal{O}_\mathsf{N}$, which advances the global state from epoch e to epoch

$e + 1$, and to $\mathcal{O}_\mathsf{U}(\cdot, \mathsf{mpk}_e, \cdot, \cdot)$, which returns $\mathsf{mpk}_{e+1} = \langle [\boldsymbol{\delta}_{e+1}], \mathsf{mpk}_e \rangle$, where $\boldsymbol{\delta}_{e+1} \xleftarrow{\$} (\mathbb{Z}/N\mathbb{Z})^{S_0}$. Then, we have

$$\begin{aligned}
\mathsf{mpk}_{e+1} &= \{E_i^{(e+1)}\}_{i=1}^{S_0} \\
&= \{[s_i + \Delta_{i,e} + \delta_{i,e+1}]E_0\}_{i=1}^{S_0} \\
&= \{[\delta_{i,e+1}]E_i^{(e)}\}_{i=1}^{S_0} \\
&= \langle [\boldsymbol{\delta}_{e+1}], \mathsf{mpk}_e \rangle.
\end{aligned}$$

However, Problem 2.1 implies that finding such a token $\boldsymbol{\delta}_{e+1} = \{\delta_{i,e+1}\}_{i=1}^{S_0}$ satisfying $\mathsf{mpk}_{e+1} = \langle [\boldsymbol{\delta}_{e+1}], \mathsf{mpk}_e \rangle$ is computationally infeasible.

5.4   Known Weaknesses in BLMQ+g

Let a client be expelled at epoch $e - 1$ and rejoin at epoch $e + 1$. In **BLMQ+g**, a signature from epoch e has the form $\sigma_e = (h, S_e)$, where $S_e = (x + h)\, \mathsf{usk}_e$, or equivalently, $S_e = (x + h) \cdot \Delta_e^{-1} \mathsf{usk}$, with h being the hash of the message and $x \xleftarrow{\$} \mathbb{Z}_p^*$. Since the client does not know Δ_e, the signature appears perfectly random. However, the client knows $S_{e-1} = (x+h) \cdot \Delta_{e-1}^{-1} \mathsf{usk}$ and $S_{e+1} = (x+h) \cdot \Delta_{e+1}^{-1} \mathsf{usk}$. If, and only if, the client knows both Δ_{e+1} and Δ_{e-1}, she can compute: $S_{e-1}^* := \Delta_{e-1} S_{e-1}$, and $S_{e+1}^* := \Delta_{e+1} S_{e+1}$. If $S_{e-1}^* = S_{e+1}^*$, she can deduce that the signature was generated with the same original usk (even without knowing usk).

In practice, this vulnerability allows a replay attack using messages signed with previously used Δ's from other users in the system. Let an attacker know the shared secret Δ_e from a recorded signature σ_e at some epoch $e < e^*$, and have access to the current Δ_{e^*}. The attacker can calculate a valid signature $\sigma_{e^*} := (h, (\Delta_e \cdot \Delta_{e^*}^{-1})S_e)$. Although the attacker cannot change the message or identity, it can evolve the signature to a new epoch. However, such a replay attack can be easily mitigated with state-of-the-art mechanisms, such as sequence numbers or timestamps, included in the content of every message m. It is worth noting that such a replay attack is possible in all the original schemes. Thus, the KUSS transformation mathematically prevents this attack for all schemes except **BLMQ**.

6   Conclusion

Reliable key updates for asymmetric keys are critical in the event of key compromise or member revocation within a communication group. Efficient key updates are particularly important for sender authentication in groups comprising resource-constrained clients. Existing mechanisms, such as X.509 certificates or traditional IBS schemes, address revocation but often incur substantial computational, networking, or management overhead. Efficient symmetric re-keying mechanisms, in contrast, offer practical solutions.

In this work, we introduced a shared secret to facilitate efficient key updates in IBS schemes. Leveraging symmetric re-keying within group infrastructures, our protocol, gIBS, extends classical IBS schemes to group settings, providing provable verification of authentication keys and ensuring both forward and backward secrecy of the shared update token. Our KUSS framework provides a formal foundation for analyzing key secrecy, independence, and existential unforgeability under adaptive attacks. The proposed transformation, applied to pairing-based and isogeny-based schemes, maintains correctness while incurring minimal computational overhead and reducing network load, especially when combined with efficient key distribution mechanisms like LKH. This work demonstrates the generality of the approach and its applicability beyond IoT, including post-quantum settings. Future work includes extending security proofs to Schnorr-like schemes, validating compatibility with other elliptic curve IBS constructions, and exploring applications to alternative cryptographic settings, such as lattices.

References

1. Whyte, W., Weimerskirch, A., Kumar, V., Hehn, T.: A security credential management system for V2V communications. In: 2013 IEEE Vehicular Networking Conference, pp. 1–8. IEEE (2013)
2. Boneh, D., Franklin, M.: Identity-based encryption from the weil pairing. In: Annual International Cryptology Conference, pp. 213–229. Springer (2001)
3. Boldyreva, A., Goyal, V., Kumar, V.: Identity-based encryption with efficient revocation. In: Proceedings of the 15th ACM Conference on Computer and Communications Security, pp. 417–426. ACM (2008)
4. Chen, J., Lim, H.W., Ling, S., Wang, H., Nguyen, K.: Revocable identity-based encryption from lattices. In: Australasian Conference on Information Security and Privacy, pp. 390–403. Springer (2012)
5. Gentry, C., Silverberg, A.: Hierarchical ID-based cryptography. In: International Conference on the Theory and Application of Cryptology and Information Security, pp. 548–566. Springer (2002)
6. Shamir, A.: Identity-based cryptosystems and signature schemes. In: Workshop on the Theory and Application of Cryptographic Techniques, pp. 47–53. Springer (1984)
7. Au, M.H., Liu, J.K., Yuen, T.H., Wong, D.S.: Practical Hierarchical Identity Based Encryption and Signature Schemes Without Random Oracles. Cryptology ePrint Archive (2006)
8. Seo, J.H., Emura, K.: Revocable identity-based encryption revisited: security model and construction. In: International Workshop on Public Key Cryptography, pp. 216–234. Springer (2013)
9. Dodis, Y., Katz, J., Xu, S., Yung, M.: Strong key-insulated signature schemes. In: International Workshop on Public Key Cryptography, pp. 130–144. Springer (2002)
10. Lehmann, A., Tackmann, B.: Updatable encryption with post-compromise security. In: Annual International Conference on the Theory and Applications of Cryptographic Techniques, pp. 685–716. Springer (2018)
11. Galindo, D., Garcia, F.D.: A Schnorr-like lightweight identity-based signature scheme. In: International Conference on Cryptology in Africa, pp. 135–148. Springer (2009)

12. Schnorr, C.-P.: Efficient signature generation by smart cards. J. Cryptol. **4**(3), 161–174 (1991)
13. Cao, X., Kou, W., Dang, L., Zhao, B.: IMBAS: identity-based multi-user broadcast authentication in wireless sensor networks. Comput. Commun. **31**(4), 659–667 (2008)
14. Hess, F.: Efficient identity based signature schemes based on pairings. In: International Workshop on Selected Areas in Cryptography, pp. 310–324. Springer (2002)
15. Barreto, P.S.L.M., Libert, B., McCullagh, N., Quisquater, J.-J.: Efficient and provably-secure identity-based signatures and signcryption from bilinear maps. In: International Conference on the Theory and Application of Cryptology and Information Security, pp. 515–532. Springer (2005)
16. Beullens, W., Kleinjung, T., Vercauteren, F.: CSI-FiSh: efficient isogeny based signatures through class group computations. In: International Conference on the Theory and Application of Cryptology and Information Security, pp. 227–247. Springer (2019)
17. Shaw, S., Dutta, R.: Identification scheme and forward-secure signature in identity-based setting from isogenies. In: International Conference on Provable Security, pp. 309–326. Springer (2021)
18. Ohtake, G., Hanaoka, G., Ogawa, K.: An efficient strong key-insulated signature scheme and its application. In: European Public Key Infrastructure Workshop, pp. 150–165. Springer (2008)
19. Palombini, F., Tiloca, M.: RFC 9594: Key Provisioning for Group Communication Using Authentication and Authorization for Constrained Environments (ACE). RFC Editor (2024)
20. Tiloca, M., Park, J., Palombini, F.: Key management for OSCORE groups in ACE. Internet-Draft draft-ietf-ace-key-groupcomm-oscore-02, Work in Progress (2019)
21. Cooper, D., Santesson, S., Farrell, S., Boeyen, S., Housley, R., Polk, W.: Internet X.509 Public Key Infrastructure Certificate and Certificate Revocation List (CRL) Profile. RFC 5280 (2008)
22. Malpani, A., Galperin, S., Adams, C.: X.509 Internet Public Key Infrastructure Online Certificate Status Protocol-OCSP. RFC 6960 (2013)
23. Libert, B., Peters, T., Yung, M.: Scalable group signatures with revocation. In: Annual International Conference on the Theory and Applications of Cryptographic Techniques, pp. 609–627. Springer (2012)
24. Zhou, S., Lin, D.: Shorter verifier-local revocation group signatures from bilinear maps. In: International Conference on Cryptology and Network Security, pp. 126–143. Springer (2006)
25. Felde, N.G., Grundner-Culemann, S., Guggemos, T.: Authentication in dynamic groups using identity-based signatures. In: 2018 14th International Conference on Wireless and Mobile Computing, Networking and Communications (WiMob), pp. 1–6. IEEE (2018)
26. Liu, Z., Lai, Y., Ren, X., Bu, S.: An efficient LKH tree balancing algorithm for group key management. In: 2012 International Conference on Control Engineering and Communication Technology, pp. 1003–1005. IEEE (2012)
27. Silverman, J.H.: The Arithmetic of Elliptic Curves, vol. 106. Springer (2009)
28. Baugher, M., Canetti, R., Dondeti, L., Lindholm, F.: Multicast Security (MSEC) Group Key Management Architecture. RFC 4046 (2005)
29. Challal, Y., Seba, H.: Group key management protocols: a novel taxonomy. Int. J. Inf. Technol. **2**(1), 105–118 (2005)
30. Smyslov, V., Weis, B.: Group Key Management using IKEv2. Internet-Draft draft-ietf-ipsecme-g-ikev2-23, Work in Progress (2025)

31. Bellare, M., Namprempre, C., Neven, G.: Security proofs for identity-based identification and signature schemes. J. Cryptol. **22**(1), 1–61 (2009)
32. Bellare, M., Yee, B.: Forward-security in private-key cryptography. In: Cryptographers' Track at the RSA Conference, pp. 1–18. Springer (2003)
33. Itkis, G.: Forward security, adaptive cryptography: time evolution. In: Handbook of Information Security, vol. 3, pp. 927–944. Wiley (2006)
34. Kim, Y., Perrig, A., Tsudik, G.: Simple and fault-tolerant key agreement for dynamic collaborative groups. In: Proceedings of the 7th ACM Conference on Computer and Communications Security, pp. 235–244. ACM (2000)
35. Koblitz, N., Menezes, A.: Another look at non-standard discrete log and Diffie-Hellman problems. J. Math. Cryptol. **2**(4), 311–326 (2008)

Beyond Sequential Walks: Parallelizing the GA-Dlog Problem

Sudeshna Karmakar, Abul Kalam, and Santanu Sarkar$^{(\boxtimes)}$

Indian Institute of Technology Madras, Chennai, India
sudeshnakarmakar.dgp@gmail.com, abulkalam.sunny@gmail.com,
sarkar.santanu.bir1@gmail.com

Abstract. The Group Action Discrete Logarithm (GA-dlog) problem is a central hardness assumption in isogeny-based cryptography, forming the basis of protocols such as CSIDH and its variants. The current state-of-the-art algorithm of May and Ostuzzi (PKC 2025) combines large-scale precomputation with collision-finding walks. However, a key limitation of their approach is that its theoretical success probability is loosely bounded. In this work, we first revisit the May–Ostuzzi framework and provide a refined analysis of its success probability, proving tighter bounds that closely match experimental behaviour. Next, we propose a new algorithmic framework that supports parallelization in both the Precomputation and Online Phases. By decomposing the group into structured components and restricting random walks to coset-like subsets, our method distributes efficiently across multiple cores. For a typical instantiation with group size N and Δ available cores, the May–Ostuzzi algorithm with a hint of size $N^{1/3}$ requires precomputation time $\frac{N^{2/3}}{\Delta}$ (using parallel computation) and online time $N^{1/3}$. Our method improves these costs when the group contains a direct-summand subgroup of order $N_1 \leq \Delta^{3/2}$, reducing the hint size to $\left\lceil \frac{N_1}{\Delta} \right\rceil \frac{N^{1/3}}{N_1^{1/3}}$, the precomputation cost to $\left\lceil \frac{N_1}{\Delta} \right\rceil \frac{N^{2/3}}{N_1^{2/3} \Delta}$, and the online cost to $\frac{N^{1/3}}{N_1^{1/3}}$. This yields an efficient reduction in both time and memory requirements under realistic parallel computation models.

Keywords: GA-dlog · parallelization · random walks · May-Ostuzzi algorithm

1 Introduction

The discrete logarithm problem (DLP) is a fundamental hardness assumption in modern cryptography, serving as the security foundation for numerous protocols in key exchange, encryption, and authentication. Its significance was first demonstrated by the introduction of the Diffie–Hellman (DH) key exchange protocol in 1976 [DH76]. The problem is mathematically formulated as follows:

Let $(G, \cdot) = \langle g \rangle$ be a finite cyclic group of order n. For any $y \in G$, we can write $y = g^v = f_g(v)$ for some $v \in \mathbb{Z}_n$. The Discrete Logarithm Problem is defined as the task of determining the unique exponent $v = f_g^{-1}(y) \pmod{n}$. This exponent $v \in \mathbb{Z}_n$ is called the discrete logarithm of y with respect to the base g, and is denoted by $\log_g(y)$.

The presumed hardness of the discrete logarithm problem serves as the foundation for a wide range of cryptographic protocols. A classical example is the Diffie–Hellman key exchange [DH76], which enables two parties to establish a shared secret key over an insecure channel without directly transmitting it. Building on this principle, ElGamal encryption [ElG85] provides a public-key encryption scheme. While the DLP also forms the basis of several digital signature schemes like Digital Signature Algorithm [NIS92], ensuring authenticity and integrity of messages. The problem's significance extends further to elliptic curve cryptography (ECC) [Kob87], which achieves equivalent levels of security with much smaller key sizes, thereby offering efficiency advantages. In addition, DLP-based hardness assumptions are integral to a variety of advanced cryptographic protocols, such as password-authenticated key exchange (PAKE) [BM92], [BPR00] illustrating the central role of the DLP in modern cryptography.

The discrete logarithm problem, while foundational to numerous cryptographic protocols, still lacks a formal unconditional proof of computational hardness. Several algorithms are known for solving it: for general finite groups, methods such as Baby-Step Giant-Step [Sha71] and Pollard's rho [Pol75] run in $\mathcal{O}(\sqrt{n})$ time, while the Pohlig–Hellman algorithm [PH06] is effective when the group order has small prime factors. Van Oorschot and Wiener established a parallel collision search paradigm for the discrete logarithm problem in cyclic groups [vOW94, OW99]. For finite fields, the Number Field Sieve (NFS) and the Function Field Sieve (FFS) achieve subexponential complexity and remain the most powerful known approaches for large-scale instances. Although these methods are far more efficient than naive trial multiplication, no polynomial-time algorithm is known on classical computers, leaving the DLP widely regarded as computationally intractable.

Although the discrete logarithm problem remains hard for classical computers, Shor's quantum algorithm [Sho94, Sho97], which uses the quantum period-finding technique, can solve it in polynomial time, compromising the security of classical schemes like the Diffie–Hellman (DH) key exchange. This shift has led to the Group Action Discrete Logarithm (GA-dlog) problem, a natural generalization of the DLP that forms the foundation of isogeny-based cryptography [BY91, Cou06, Sto10, JDF11, DFKS18] and the key exchange protocol CSIDH [CLM+18]. The problem is mathematically formulated as follows:

Let $\mathcal{X}$ be a finite set without any intrinsic group structure, and fix a distinguished element $x \in \mathcal{X}$, referred to as the *origin*. Consider a finite abelian group $(\mathcal{G}, \cdot)$ acting on $\mathcal{X}$ via a group action $\star$, such that

$$\mathcal{X} = \{g \star x \mid g \in \mathcal{G}\},$$

with the additional assumption that $|\mathcal{X}| = |\mathcal{G}| = N$. Thus, every element of $\mathcal{X}$ can be uniquely represented as the action of some group element on the origin.

Let $\mathcal{G}$ be generated by a finite set $\mathbf{g} = \{g_1, \ldots, g_n\}$. For an integer vector $\mathbf{v} = (v_1, \ldots, v_n) \in \mathbb{Z}^n$, we define

$$\mathbf{g}^{\mathbf{v}} = g_1^{v_1} \cdot g_2^{v_2} \cdots g_n^{v_n}.$$

This yields the exponentiation map

$$\varphi_{\mathbf{g}} : \mathbb{Z}^n \to \mathcal{G}, \qquad \mathbf{v} \mapsto \mathbf{g}^{\mathbf{v}},$$

whose kernel $\Lambda = \ker(\varphi_{\mathbf{g}})$ is an n-dimensional integer lattice. Consequently, we obtain a bijection

$$f_{\mathbf{g},x} : \mathbb{Z}^n/\Lambda \longrightarrow \mathcal{X}, \qquad \mathbf{v} \mapsto \mathbf{g}^{\mathbf{v}} \star x.$$

Given $y \in \mathcal{X}$ of the form $y = \mathbf{g}^{\mathbf{v}} \star x$, the GA-dlog problem is to invert $f_{\mathbf{g},x}$, i.e., to compute the unique

$$\mathbf{v} = f_{\mathbf{g},x}^{-1}(y) \pmod{\Lambda},$$

which in turn corresponds to the group element $\mathbf{g}^{\mathbf{v}} \in \mathcal{G}$.

In this framework, the lattice $\Lambda = \ker(\varphi_{\mathbf{g}})$ is the central structural parameter that governs modular reduction and efficient computation. In the discrete logarithm setting, the group order is generally assumed to be known, and for elliptic curves this can be efficiently determined using Schoof's polynomial-time algorithm [Sch95]. Similarly, in the group action discrete logarithm setting, algorithms assume knowledge of a basis of the lattice Λ. This knowledge allows reduction of exponents modulo Λ. Without this reduction, exponents can grow exponentially, making evaluations computationally infeasible. In the CSIDH setting, a basis of Λ can be computed in quantum polynomial time [Hal05] or, classically, in subexponential time [HM89]. A concrete computations reported for CSIDH-512 in [BKV19].

Building on the GA-dlog problem, several post-quantum cryptographic protocols have been developed. CSIDH [CLM+18] enables efficient non-interactive key exchange using commutative isogenies on supersingular elliptic curves, while CTIDH [BBC+21] extends this approach with twisted Edwards curves for improved performance. Signature schemes such as CSI-FiSh [BKV19] and SeaSign [FG18] use class group actions combined with Fiat–Shamir techniques to provide secure digital signatures, and the CRS Scheme employs a common reference string to support secure computations and zero-knowledge proofs. Together, these protocols illustrate the practical applicability of GA-dlog for quantum-resistant cryptography.

Solving the GA-dlog problem extends classical discrete logarithm techniques using collision-finding and precomputation. The Galbraith–Hess–Smart (GHS) [GHS02] algorithm is a generic collision-based method with complexity $\mathcal{O}(N^{1/2})$ and polynomial memory, while multiple-instance algorithms improve efficiency when solving several GA-dlog instances simultaneously. Although Shor's algorithm does not directly apply due to the lack of group structure on $\mathcal{X}$, subexponential quantum methods such as Kuperberg's algorithm [Kup04] provide the best-known quantum attacks, and reductions to the Group Action Computational Diffie–Hellman (GA-CDH) problem offer alternative strategies.

In PKC 2025, May et al. further adapt classical precomputation techniques by precomputing s random walks of length t, storing only endpoints as *"hint"*, with parameters chosen as $s = t = N^{1/3}$. Their method achieves time complexity $\mathcal{O}(N^{1/3})$ in *Online Phase* and in *Precomputation Phase* time complexity is $\mathcal{O}(N^{2/3})$ and space $\mathcal{O}(N^{1/3})$, covering a substantial portion of the domain. This approach is adapted from the framework of Corrigan-Gibbs and Kogan [CGK18] and the Legendre PRF [MZ21] setting, closely following the precomputation strategies established therein.

Attacks of this type, relying on large-scale precomputation, are realistically feasible only for exceptionally powerful adversaries, such as nation-state agencies, that possess the capability to perform years-long precomputations. Such a precomputation may run over several years, using billions of devices to generate a reusable *"hint"*. Once such a precomputation is completed, the resulting hint can significantly accelerate subsequent GA-dlog computations in the *Online Phase*.

In such a setting, an adversary would almost certainly seek to maximize the efficiency of both the *Precomputation* and *Online Phases* by exploiting parallelism and multi-core architectures. Parallel computation is not merely a convenience here but a necessity, as it allows the adversary to distribute the workload across billions of cores and thereby reduce the time-to-attack to a realistic window. One of the main contributions of this work is to use parallel computations. We propose an algorithm for solving the GA-dlog problem with precomputation that enables parallel computation in the *Online Phase*.

Our Contributions. This work makes two main contributions:

(i) May et al. showed that if we consider s walks of length t, the success probability of the *Online Phase* is at least $\frac{st^2}{8N}$. However, this bound is rather loose and significantly underestimates practical performance. For instance, with parameters $s = t = \lfloor \frac{1}{2} N^{1/3} \rfloor$, their analysis gives a lower bound of $\frac{1}{64}$, implying up to 64 repetitions on average to solve a GA-dlog instance. However, our experiments with group size 2^{18} showed that only about 6 repetitions were required in practice. We therefore provide a refined analysis that yields a tighter bound on the success probability. For example, under the same parameters, the bound improves from 64 to 12.5. Similarly, for $s = t = \lfloor \frac{1}{3} N^{1/3} \rfloor$ and $s = t = \lfloor \frac{1}{4} N^{1/3} \rfloor$, our analysis improves the bounds from 216 and 512 repetitions to 33.62 and 73.37, respectively.

(ii) We also design a new parallel-GA algorithm that uses multiple cores in both the *Precomputation* and *Online Phases*. For a typical instance with a group of size N and Δ cores, if the group admits a direct-summand subgroup of order $N_1 \leq \Delta^{3/2}$, the algorithm achieves the following improvements:
- the hint size reduces from $N^{1/3}$ to $\lceil \frac{N_1}{\Delta} \rceil \frac{N^{1/3}}{N_1^{1/3}}$,
- the precomputation cost reduces from $\frac{N^{2/3}}{\Delta}$ to $\lceil \frac{N_1}{\Delta} \rceil \frac{N^{2/3}}{N_1^{2/3}\Delta}$, and

- the online cost reduces from $N^{1/3}$ to $\frac{N^{1/3}}{N_1^{1/3}}$.

Organization of the Paper. The paper is organized as follows. In Sect. 2, we revisit the May–Ostuzzi algorithm and provide a refined analysis of the success probability of the *Online Phase*. In Sect. 3, we present our parallel algorithm for solving the GA-dlog problem. Finally, Sect. 4 concludes the paper with a summary of our results.

Notations. Throughout this paper, we follow the notation of May et al. [MO25], as our focus is primarily on analyzing the May–Ostuzzi algorithm.

2 Revisiting May-Ostuzzi

In this section, we revisit the attacks on single instances introduced by May and Ostuzzi [MO25], which are grounded in the use of pseudo-random walks. Their approach generally divides the computation of the target exponent vector into two separate phases:

Precomputation Phase: In this initial stage, a list, referred to as the *hint*, is generated that stores the endpoints of a family of pseudo-random walks, along with their associated GA-dlogs. Notably, these walks are constructed independently of the specific target instance $y \in \mathcal{X}$. Because of this independence from the target instance, this phase is often referred to as the precomputation or offline phase.

Online Phase: In this phase we generate a random walk using the target instance $y \in \mathcal{X}$ and at each steps check whether it collides with an element in the hint list obtained in the *Precomputation Phase* to recover the solution.

In the following, we discuss these phases in details.

Precomputation Phase. In the *Precomputation Phase*, we generate a family of s random walks

$$W^{(i)} = (x_j^{(i)})_{0 \leq j \leq t}, \quad i = 1, 2, \ldots, s,$$

each of length t, derived solely from the given elements $\mathbf{g}$ and x. The endpoints of these walks are stored in a list $\mathcal{L}$, which will later act as "*hint*" for detecting collisions with the walk constructed in the *Online Phase*, thereby leading to recovery of the solution.

Each walk $W^{(i)}$ starts at $x_0^{(i)} = \mathbf{g}^{\mathbf{w}_0^{(i)}} \star x$, where $\mathbf{w}_0^{(i)} \in \mathbb{Z}^n/\Lambda$ chosen uniformly at random, and proceeds according to a pseudorandom function (PRF)

$$h : \mathcal{X} \longrightarrow \mathbb{Z}^n/\Lambda$$

via the recursion

$$x_{j+1}^{(i)} := \mathbf{g}^{h(x_j^{(i)})} \star x_j^{(i)}, \quad j \geq 0$$

with the corresponding GA-dlog values

$$\mathbf{w}_{j+1}^{(i)} := h(x_j^{(i)}) + \mathbf{w}_j^{(i)} = \mathbf{w}_0^{(i)} + \sum_{k=0}^{j} h(x_k^{(i)}), \quad j \geq 0.$$

For each of these s random walks $W^{(i)}$ of length t, only the endpoint $x_t^{(i)}$, together with its associated GA-dlog $\mathbf{w}_t^{(i)}$, is stored in the list $\mathcal{L}$.

Online Phase. In the *Online Phase*, we address a given GA-dlog instance of the form

$$y = \mathbf{g}^\mathbf{v} \star x.$$

We initiate a random walk $W = (x_j)_{0 \leq j \leq 2t}$ of length $2t$, constructed analogously to those in the *Precomputation Phase*. Specifically, we start from $\mathbf{w}_0 = 0^n \in \mathbb{Z}^n/\Lambda$, which yields the starting point of the walk

$$x_0 = \mathbf{g}^{\mathbf{w}_0} \star y = y.$$

As the walk proceeds, we check whether any element x_j of the online walk collides with an endpoint $x_t^{(\ell)}$ contained in the hint list $\mathcal{L}$ of some precomputed walk $W^{(\ell)}$. If no collision occurs within $2t$ steps, a fresh random starting point $\mathbf{w}_0 \in \mathbb{Z}^n/\Lambda$ is chosen and the process is repeated. Once a collision is found between x_j in the online walk and $x_t^{(\ell)} \in \mathcal{L}$ in the precomputation list, the GA-dlog of y can be recovered as $\mathbf{v} \equiv \mathbf{w}_t^{(\ell)} - \mathbf{w}_j \pmod{\Lambda}$ as established in Proposition 1. A high-level description of May-Ostuzzi algorithm is demonstrated in Fig. 1.

Proposition 1. *[MO25]*

(i) *Let $x_j = x_k^{(\ell)}$ be the first collision between the online walk W and the pre-computed walk $W^{(\ell)}$. Then, from that point onwards, the two walks coincide.*

(ii) *Let $x_j = x_t^{(\ell)}$ denote the colliding endpoints of the walks W and $W^{(\ell)}$, respectively. From this collision, we recover the desired GA-dlog of y as*

$$\mathbf{v} = \mathbf{w}_t^{(\ell)} - \mathbf{w}_j \pmod{\Lambda},$$

where $\mathbf{w}_t^{(\ell)}$ and $\mathbf{w}_j$ are the GA-dlogs corresponding to $x_t^{(\ell)}$ and x_j, respectively.

The following theorem states the success probability of the algorithm.

Theorem 1. *[MO25] Let $(\mathcal{G}, \mathcal{X}, \star)$ be a regular group action with $N = |\mathcal{G}| = |\mathcal{X}|$. For parameters $s, t \in \mathbb{N}$ satisfying $4st^2 \leq N$, the Precomputation Phase performs $\mathcal{O}(st)$ steps to generate a hint $\mathcal{L}$ of size $\mathcal{O}(s)$. Using $\mathcal{L}$, the Online Phase solves a GA-dlog instance in $\mathcal{O}(t)$ steps with success probability at least $\frac{st^2}{8N}$.*

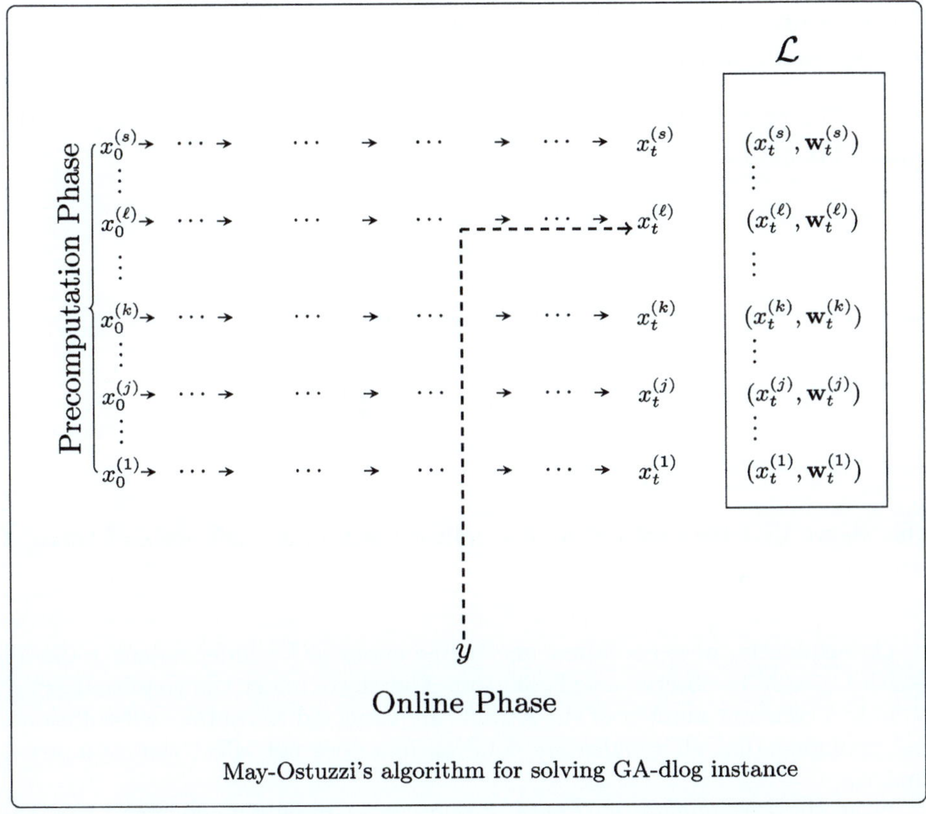

Fig. 1. May-Ostuzzi's algorithm for solving GA-dlog instance.

For parameter selection, May et al. [MO25] recommend choosing

$$s = t = \left\lfloor \tfrac{1}{2} N^{1/3} \right\rfloor$$

in their PRECOMPUTE-GA algorithm. With this choice, the success probability of the *Online Phase* is at least $\frac{1}{64}$. Consequently, in the average case one needs to repeat the *Online Phase* at most 64 times to solve a GA-dlog instance.

However, in practice the performance is significantly better. For example, for a group of order $N = 2^{18}$, we set $s = t = \lfloor \tfrac{1}{2} N^{1/3} \rfloor = 32$. In our experiments, solving 1000 GA-dlog instances using the May–Ostuzzi algorithm required, on average, only 5.77 repetitions of the *Online Phase*. This demonstrates that the theoretical lower bound on the success probability given in the theorem is quite loose compared to the experimental behaviour. Motivated by this discrepancy, we therefore proceed to derive a refined bound through a tighter analysis.

Let us first analyze the probability that a random walk has a collision within itself in the following proposition.

Proposition 2. *Consider a random walk W of length t. Then the probability that W has a collision within itself is at most $\frac{t^2}{N}$.*

Proof. The probability that W contains no collision is $\prod_{i=0}^{t}\left(1 - \frac{i}{N}\right)$, so the probability of at least one collision is

$$p = 1 - \prod_{i=0}^{t}\left(1 - \frac{i}{N}\right).$$

Using $1 - x \le e^{-x} \le 1 - x/2$ for $x \le 1$, we obtain

$$-p = \prod_{i=0}^{t}\left(1 - \frac{i}{N}\right) - 1 \ \ge \ \prod_{i=0}^{t} e^{-\frac{2i}{N}} - 1$$

$$\approx e^{-\frac{t^2}{N}} - 1 \ \ge \ 1 - \frac{t^2}{N} - 1 \ = \ -\frac{t^2}{N}.$$

This shows that the probability of a collision within a single walk of length t satisfies

$$p \le \frac{t^2}{N}.$$

Consequently, in expectation one needs about N/t^2 independent walks of length $t \le \sqrt{N}$ to observe a self-collision. Therefore, under the condition $st^2 \le N$, only a constant number of the s walks are expected to contain self-collisions, and assuming that all s walks are collision-free does not affect the asymptotic analysis.

From the *Precomputation Phase*, suppose we sample s independent random walks of length t, subject to $st^2 \le N$. Let X denote the random variable counting the total number of distinct elements across all s walks.

Consider two walks W_1 and W_2. Define X' as the number of steps taken in W_2 before encountering the first collision with W_1, i.e., the first index at which an element of W_2 also belongs to W_1. Clearly, X' is a non-negative, integer-valued random variable. By the tail-sum formula,

$$\mathbb{E}[X'] \ = \ \sum_{m=0}^{t-1} \mathbb{P}[X' > m] \ = \ \sum_{m=0}^{t-1}\left(1 - \frac{t}{N}\right)^{m} \ = \ \frac{N}{t}\left[1 - (1 - \tfrac{t}{N})^t\right].$$

As shown in Fig. 2, the expected number of common elements between two walks W_1 and W_2 is at most $t - \mathbb{E}[X']$. Hence, for s independent walks the expected number of distinct elements satisfies

$$\mathbb{E}[X] \ \ge \ st - \binom{s}{2}\left(t - \mathbb{E}[X']\right) \ \approx \ st - \frac{s^2 t}{2} + \frac{s^2 N}{2t}\left[1 - (1 - \tfrac{t}{N})^t\right].$$

Applying $(1 - x)^n \le 1 - nx + \frac{n(n-1)}{2}x^2$ for $x \in (0, 1)$ gives

$$\mathbb{E}[X] \ \ge \ st - \frac{s^2 t}{2} + \frac{s^2 N}{2t}\left[\frac{t^2}{N} - \frac{t(t-1)}{2}\frac{t^2}{N^2}\right] \ \approx \ st - \frac{s^2 t^3}{4N}.$$

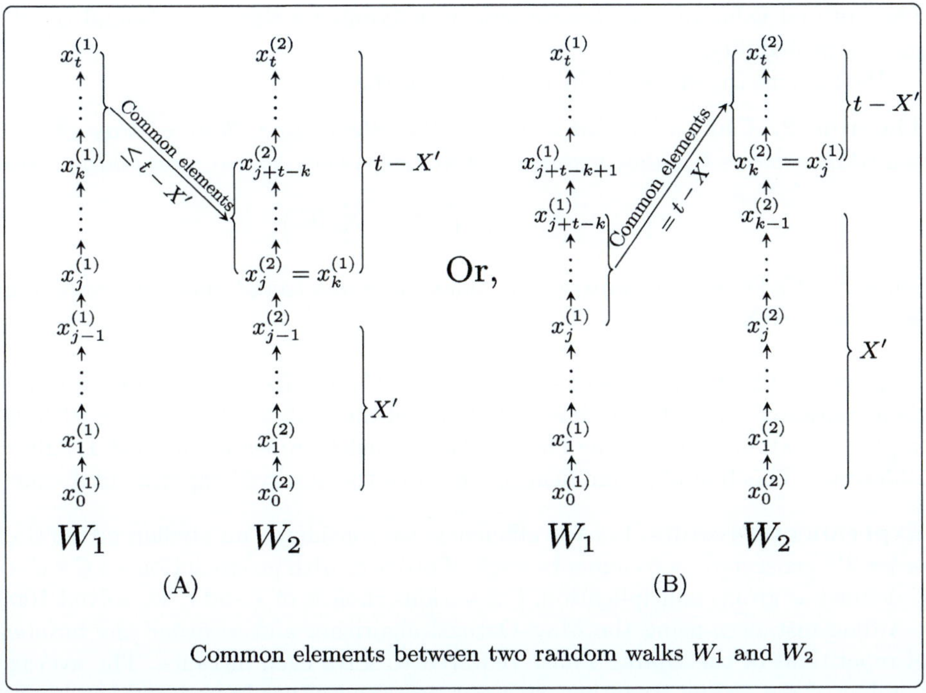

Fig. 2. Common elements between two random walks W_1 and W_2

Now fix $\gamma \in (0,1)$ depending on $\frac{st^2}{N}$ and assume $st^2 \leq 4(1-\gamma)N$. By Markov's inequality,

$$\mathbb{P}[X < \gamma st] = \mathbb{P}[st - X \geq (1-\gamma)st] \leq \frac{st - \mathbb{E}[X]}{(1-\gamma)st} \leq \frac{st^2}{4(1-\gamma)N}.$$

Therefore, with s random walks we cover at least γst distinct elements of $\mathcal{X}$ in the *Precomputation Phase* with probability

$$\mathbb{P}[X \geq \gamma st] \geq 1 - \frac{st^2}{4(1-\gamma)N}.$$

Let C_t denote the event that the walk in the *Online Phase* collides with at least one of the s precomputed random walks. Then

$$\mathbb{P}[C_t \mid X \geq \gamma st] \geq 1 - \left(1 - \frac{\gamma st}{N}\right)^t = \frac{\gamma st^2}{N} - \frac{t(t-1)}{2}\frac{\gamma^2 s^2 t^2}{N^2} \approx \frac{\gamma st^2}{N} - \frac{\gamma^2 s^2 t^4}{2N^2}.$$

Therefore, the *Online Phase* solves the GA-dlog problem with success probability at least

$$\mathbb{P}[C_t \cap X \geq \gamma st] = \mathbb{P}[X \geq \gamma st] \cdot \mathbb{P}[C_t \mid X \geq \gamma st]$$

$$\geq \left(1 - \frac{st^2}{4(1-\gamma)N}\right) \cdot \left(\frac{\gamma st^2}{N} - \frac{\gamma^2 s^2 t^4}{2N^2}\right).$$

The optimal value of γ is determined by maximizing this lower bound on the success probability.

We summarize the result in the following theorem.

Theorem 2. *Using a hint list $\mathcal{L}$ of size $\mathcal{O}(s)$, the Online Phase of May–Ostuzzi algorithm solves a GA-dlog instance in $\mathcal{O}(t)$ steps with success probability at least*

$$\left(1 - \frac{st^2}{4(1-\gamma)N}\right) \cdot \left(\frac{\gamma st^2}{N} - \frac{\gamma^2 s^2 t^4}{2N^2}\right),$$

where $\gamma \in (0,1)$ is the parameter that maximizes this bound under the constraint $st^2 \leq 4(1-\gamma)N$.

Now consider the parameters $s = t = \lfloor \frac{1}{2}N^{1/3} \rfloor$ and fix $\gamma = 0.82$. Our analysis then guarantees that the success probability of the *Online Phase* is at least 0.08, implying that, on average, the *Online Phase* must be repeated at most $1/0.08 \approx$ 12.5 times. This bound is significantly closer to the observed experimental value.

Experimental results. For the efficiency, we considered an abelian group $\mathcal{G}$ of order 2^{18} generated by 6 elements, each of order 8, with group action $\star : \mathcal{G} \times \mathcal{G} \to \mathcal{G}$ defined as group multiplication. For various choices of s and t, we solved 1000 GA-dlog instances using the May–Ostuzzi algorithm and recorded the number of repetitions of the *Online Phase* required to solve each instance. The average number of repetitions, together with our proposed upper bound and the bound of May et al., are reported in Table 1. The results indicate that our bound is consistently tighter than the one given in Theorem 1 by May et al.

Table 1. Comparison of average repetitions of the *Online Phase* with theoretical upper bounds.

(s,t)	Avg. repetitions (experiment)	Our upper bound	Bound by May et al. [MO25]
$(32, 32)$	5.77	12.5	64
$(21, 21)$	18.38	33.62	216
$(16, 16)$	40.04	73.37	512
$(12, 12)$	91.52	137.58	1000

In the following section, we present an algorithm for solving the GA-dlog problem that incorporates precomputation and supports parallel computation.

3   Parallelization of May-Ostuzzi Algorithm

We begin by recalling the group and the associated group action underlying the GA-dlog problem. Let

$$\mathcal{G} = \langle \mathbf{g} \rangle = \langle g_1, g_2, \ldots, g_n \rangle$$

be a finite abelian group of order $|\mathcal{G}| = N$. Consider the regular group action of $\mathcal{G}$ on the set

$$\mathcal{X} = \{g \star x \mid g \in \mathcal{G}\},$$

where $x \in \mathcal{X}$ is fixed as the origin. Moreover, it follows that $|\mathcal{X}| = |\mathcal{G}| = N$.

For our analysis, we assume that

$$\mathcal{G} = \mathcal{G}_1 \oplus \mathcal{G}_2,$$

where $|\mathcal{G}_1| = N_1$ and $|\mathcal{G}_2| = N_2$. The parameters N_1 and N_2 are treated as optimization variables, chosen in accordance with the number of available processing cores. Such subgroups $\mathcal{G}_1$ and $\mathcal{G}_2$ can be constructed by selecting generating sets

$$\mathbf{g}_1 = \{\mathbf{g}^{\mathbf{u}_{11}}, \mathbf{g}^{\mathbf{u}_{12}}, \cdots, \mathbf{g}^{\mathbf{u}_{1n_1}}\}, \text{ and } \mathbf{g}_2 = \{\mathbf{g}^{\mathbf{u}_{21}}, \mathbf{g}^{\mathbf{u}_{22}}, \cdots, \mathbf{g}^{\mathbf{u}_{2n_2}}\},$$

where, n_1 and n_2 are sizes of generating lists $\mathbf{g}_1$ and $\mathbf{g}_2$, respectively, such that

$$\langle \mathbf{g}_1 \rangle = \mathcal{G}_1, \quad |\mathcal{G}_1| = N_1, \quad \text{and} \quad \langle \mathbf{g}_2 \rangle = \mathcal{G}_2, \quad |\mathcal{G}_2| = N_2.$$

Similarly as previous, define the exponentiation maps

$$\varphi_{\mathbf{g}_1} : \mathbb{Z}^{n_1} \to \mathcal{G}_1, \quad \mathbf{v}_1 \mapsto \mathbf{g}_1^{\mathbf{v}_1}, \qquad \varphi_{\mathbf{g}_2} : \mathbb{Z}^{n_2} \to \mathcal{G}_2, \quad \mathbf{v}_2 \mapsto \mathbf{g}_2^{\mathbf{v}_2},$$

and let $\Lambda_1 = \ker(\varphi_{\mathbf{g}_1})$, $\Lambda_2 = \ker(\varphi_{\mathbf{g}_2})$. Then, by the direct sum property, every element of $\mathcal{G}$ has a unique representation $\mathbf{g}_1^{\mathbf{v}_1} \cdot \mathbf{g}_2^{\mathbf{v}_2}$ with $\mathbf{v}_i \in \mathbb{Z}^{n_i}/\Lambda_i$ $(i = 1, 2)$.

Consequently, the GA-dlog problem for a given instance $y \in \mathcal{X}$ transfers to find a pair $(\mathbf{v}_1, \mathbf{v}_2) \in \mathbb{Z}^{n_1}/\Lambda_1 \times \mathbb{Z}^{n_2}/\Lambda_2$ such that

$$y = \left(\mathbf{g}_1^{\mathbf{v}_1} \cdot \mathbf{g}_2^{\mathbf{v}_2}\right) \star x.$$

Further suppose that $\mathbf{v}_1$ admits a decomposition of the form

$$\mathbf{v}_1 = \bar{\alpha} - \bar{\beta}, \qquad \bar{\alpha}, \bar{\beta} \in \mathbb{Z}^{n_1}/\Lambda_1.$$

Then the problem can be reformulated as

$$\mathbf{g}_1^{\bar{\beta}} \star y = (\mathbf{g}_1^{\bar{\alpha}} \cdot \mathbf{g}_2^{\mathbf{v}_2}) \star x.$$

For a fixed $\mathbf{v}_1 \in \mathbb{Z}^{n_1}/\Lambda_1$, there exist exactly N_1 valid pairs $(\bar{\alpha}, \bar{\beta}) \in (\mathbb{Z}^{n_1}/\Lambda_1)^2$ satisfying $\mathbf{v}_1 = \bar{\alpha} - \bar{\beta}$. It suffices to identify one such pair in order to construct a solution to the GA-dlog problem.

To analyze this further, we study the structure of subsets of $\mathcal{X}$ defined by $\mathbf{g}_1$. For $\alpha \in \mathbb{Z}^{n_1}/\Lambda_1$, consider the coset-like set

$$\mathbf{g}_1^\alpha \mathcal{X} = \{(\mathbf{g}_1^\alpha \cdot g) \star x \mid g \in \mathcal{G}_2\}.$$

These subsets naturally partition $\mathcal{X}$ according to the $\mathcal{G}_1$-component. Moreover, each subset has the same cardinality, given by $|\mathbf{g}_1^\alpha \mathcal{X}| = |\mathcal{G}_2| = N_2$.

We now define a random walk $W = (x_j)_{0 \leq j \leq t}$ on $\mathcal{X}$, starting from a random point x_0 and driven by a pseudorandom function $h : \mathcal{X} \to \mathbb{Z}^{n_2}/\Lambda_2$, via the recursive rule

$$x_{j+1} = \mathbf{g}_2^{h(x_j)} \star x_j, \qquad j \geq 0.$$

The proposition below establishes a key invariance property of this walk:

Proposition 3. *Let $W = (x_j)_{j \geq 0}$ be the random walk defined above, with initial point $x_0 \in \mathbf{g}_1^\alpha \mathcal{X}$ for some $\alpha \in \mathbb{Z}^{n_1}/\Lambda_1$. Then the entire walk remains within $\mathbf{g}_1^\alpha \mathcal{X}$.*

Proof. We proceed by induction. Assume $x_j \in \mathbf{g}_1^\alpha \mathcal{X}$. Then there exists $g \in \mathcal{G}_2$ such that

$$x_j = (\mathbf{g}_1^\alpha \cdot g) \star x.$$

By the walk definition,

$$\begin{aligned}
x_{j+1} &= \mathbf{g}_2^{h(x_j)} \star x_j \\
&= \mathbf{g}_2^{h(x_j)} \star \left((\mathbf{g}_1^\alpha \cdot g) \star x \right) \\
&= (\mathbf{g}_2^{h(x_j)} \cdot \mathbf{g}_1^\alpha \cdot g) \star x \\
&= (\mathbf{g}_1^\alpha \cdot (\mathbf{g}_2^{h(x_j)} \cdot g)) \star x.
\end{aligned}$$

Since $\mathbf{g}_2^{h(x_j)} \cdot g \in \mathcal{G}_2$, it follows that $x_{j+1} \in \mathbf{g}_1^\alpha \mathcal{X}$. This completes the inductive step, and hence the walk remains within $\mathbf{g}_1^\alpha \mathcal{X}$ for all j.

Remark 1. Consider the projection morphisms

$$\pi_1 : \mathcal{G} \to \mathcal{G}_1, \text{ and } \pi_2 : \mathcal{G} \to \mathcal{G}_2.$$

A natural choice for generating sets $\mathbf{g}_1$ and $\mathbf{g}_2$ of the subgroups $\mathcal{G}_1$ and $\mathcal{G}_2$, respectively, is obtained by projecting the generators of $\mathcal{G}$:

$$\mathbf{g}_1 = \{\pi_1(g_1), \pi_1(g_2), \cdots, \pi_1(g_n)\}, \text{ and } \mathbf{g}_2 = \{\pi_2(g_1), \pi_2(g_2), \cdots, \pi_2(g_n)\}.$$

In this setting, we have $n_1 = n_2 = n$, $\Lambda = \Lambda_1 \cap \Lambda_2$ and $\mathbb{Z}^{n_1}/\Lambda_1 \oplus \mathbb{Z}^{n_2}/\Lambda_2 = \mathbb{Z}^n/\Lambda$, which makes explicit how to obtain the actual solution in $\mathbb{Z}^n/\Lambda$ from a solution in $\mathbb{Z}^{n_1}/\Lambda_1 \oplus \mathbb{Z}^{n_2}/\Lambda_2$.

We are now set to discuss our algorithm to solve GA-dlog problem. Similar to May–Ostuzzi algorithm, our algorithm also proceeds in two stages: *Precomputation Phase* and *Online Phase*. Both the phases are discussed as follows:

Precomputation Phase. In this phase, we randomly select a subset $\mathcal{A} \subseteq \mathbb{Z}^{n_1}/\Lambda_1$ of size δ_α, where δ_α is an optimization parameter. For each $\alpha \in \mathcal{A}$, we construct s instance-independent random walks of length t within the set $\mathbf{g}_1^\alpha \mathcal{X}$, as defined earlier.

Fix $\alpha \in \mathcal{A}$ and consider one such random walk $W_\alpha^{(i)}$. Following the May–Ostuzzi precomputation step, we begin by sampling $\mathbf{w}_0^{(i)} \in \mathbb{Z}^{n_2}/\Lambda_2$ uniformly at random, which determines the starting point

$$x_0^{(i)} = (\mathbf{g}_1^\alpha \cdot \mathbf{g}_2^{\mathbf{w}_0^{(i)}}) \star x \in \mathbf{g}_1^\alpha \mathcal{X}.$$

The walk then evolves in $\mathbf{g}_1^\alpha \mathcal{X}$ according to the recursive rule

$$x_{j+1}^{(i)} = \mathbf{g}_2^{h(x_j^{(i)})} \star x_j^{(i)}, \qquad j \geq 0.$$

Furthermore, we define the cumulative exponent vector

$$\mathbf{w}_{j+1}^{(i)} := h(x_j^{(i)}) + \mathbf{w}_j^{(i)} = \mathbf{w}_0^{(i)} + \sum_{k=0}^{j} h(x_k^{(i)}), \quad j \geq 0.$$

Observe that, each point of the walk admits the representation

$$x_{j+1}^{(i)} = (\mathbf{g}_1^{\boldsymbol{\alpha}} \cdot \mathbf{g}_2^{\mathbf{w}_{j+1}^{(i)}}) \star x,$$

showing that $x_{j+1}^{(i)}$ has GA-dlog $(\boldsymbol{\alpha}, \mathbf{w}_{j+1}^{(i)}) \in \mathbb{Z}^{n_1}/\Lambda_1 \times \mathbb{Z}^{n_2}/\Lambda_2$.

Accordingly, for each $\boldsymbol{\alpha} \in \mathcal{A}$, we generate s random walks of length t inside $\mathbf{g}_1^{\boldsymbol{\alpha}} \mathcal{X}$, and for each walk we store the tuple $(x_t^{(i)}, \boldsymbol{\alpha}, \mathbf{w}_t^{(i)})$ in the hint list $\mathcal{L}$. Since the walks are independent across different $\boldsymbol{\alpha}$ and among themselves, the construction of $\mathcal{L}$ admits straightforward parallelization on a multicore architecture. This allows the *Precomputation Phase* to be carried out more efficiently, with the total running time reduced in proportion to the number of available cores.

The pseudocode for the precomputation is presented in PRECOMPUTE-PARALLEL-GA (Algorithm 2), with its subroutine detailed in WALK (Algorithm 1).

Algorithm 1: WALK

Input: $(\mathbf{g}_1, \mathbf{g}_2, x) \in \mathcal{G}_1^{n_1} \times \mathcal{G}_2^{n_2} \times \mathcal{X}, \boldsymbol{\alpha} \in \mathcal{A}$, walk length t, PRF $h : \mathcal{X} \to \mathbb{Z}^{n_2}/\Lambda_2$

1   Choose a random $\mathbf{w}_0 \in \mathbb{Z}^{n_2}/\Lambda_2$;
2   $x_0 \leftarrow (\mathbf{g}_1^{\boldsymbol{\alpha}} \cdot \mathbf{g}_2^{\mathbf{w}_0}) \star x$;
3   **for** $j = 1, 2, \cdots, t$ **do**
4      $x_j \leftarrow \mathbf{g}_2^{h(x_{j-1})} \star x_{j-1}$;
5      $\mathbf{w}_j \leftarrow h(x_{j-1}) + \mathbf{w}_{j-1}$;
6   **return** $(x_t, \boldsymbol{\alpha}, \mathbf{w}_t)$

Online Phase. In this phase, we randomly select a subset $\mathcal{B} \subseteq \mathbb{Z}^{n_1}/\Lambda_1$ of size δ_β, subject to the condition

$$\delta_\alpha \cdot \delta_\beta \geq N_1.$$

This guarantees, with constant probability, the existence of a valid pair $(\bar{\boldsymbol{\alpha}}, \bar{\boldsymbol{\beta}}) \in \mathcal{A} \times \mathcal{B}$ such that $\mathbf{v}_1 = \bar{\boldsymbol{\alpha}} - \bar{\boldsymbol{\beta}}$.

For each $\boldsymbol{\beta} \in \mathcal{B}$, we generate a random walk W_β based on the given instance y. Specifically, we first choose $\mathbf{w}_0 \in \mathbb{Z}^{n_2}/\Lambda_2$ uniformly at random, which defines the starting point

$$x_0 = (\mathbf{g}_1^{\boldsymbol{\beta}} \cdot \mathbf{g}_2^{\mathbf{w}_0}) \star y.$$

The walk then evolves via the recursive rule

$$x_{j+1} = \mathbf{g}_2^{h(x_j)} \star x_j, \qquad j \geq 0,$$

Algorithm 2: PRECOMPUTE-PARALLEL-GA

Input: $\mathbf{g} = \{g_1, g_2, \cdots, g_n\} \in \mathcal{G}^n$, $x \in \mathcal{X}$, $N := |\mathcal{G}|$, $\Delta :=$ Total number of available cores

1. Choose subgroups $\mathcal{G}_1$ and $\mathcal{G}_2$ of $\mathcal{G}$ such that $\mathcal{G} = \mathcal{G}_1 \oplus \mathcal{G}_2$;
2. Set $N_1 \leftarrow |\mathcal{G}_1|$ and $N_2 \leftarrow |\mathcal{G}_2|$;
3. Define the projection morphisms: $\pi_1 : \mathcal{G} \to \mathcal{G}_1$ and $\pi_2 : \mathcal{G} \to \mathcal{G}_2$;
   // As in Remark 1
4. Construct $\mathbf{g}_1 = \{\pi_1(g_1), \pi_1(g_2), \cdots, \pi_1(g_n)\}$ and $\mathbf{g}_2 = \{\pi_2(g_1), \pi_2(g_2), \cdots, \pi_2(g_n)\}$;
5. Set $n_1 \leftarrow |\mathbf{g}_1|$ and $n_2 \leftarrow |\mathbf{g}_2|$;
   // Note that $n_1 = n_2 = n$
6. Generate PRF $h : \mathcal{X} \to \mathbb{Z}^{n_2}/\Lambda_2$;
   // Λ_2 corresponds with $\mathcal{G}_2$
7. Set $\delta_\alpha = \lceil \frac{N_1}{\Delta} \rceil$ and choose $s, t \in \mathbb{N}$ such that $4st^2 \leq N_2$; // E.g. $s = \lfloor \frac{1}{2} N_2^{\frac{1}{3}} \rfloor = t$
8. Choose a random subset $\mathcal{A} \subseteq \mathbb{Z}^{n_1}/\Lambda_1$ of size δ_α;
   // Λ_1 corresponds with $\mathcal{G}_1$
9. $\mathcal{L} \leftarrow \varnothing$;
10. **for** $\alpha \in \mathcal{A}$ **do**
11.   **for** $i = 1, 2, \cdots, s$ **do**
            // Use parallel computation
12. $z \leftarrow \text{WALK}(\alpha)$;
13. $\mathcal{L} \leftarrow \mathcal{L} \cup \{z\}$;
14. Sort $\mathcal{L}$ with respect to the first entry;
15. **return** $\mathcal{L}$

---

with the corresponding cumulative exponent vector

$$\mathbf{w}_{j+1} = h(x_j) + \mathbf{w}_j.$$

Suppose there exists a valid pair $(\bar{\alpha}, \bar{\beta}) \in \mathcal{A} \times \mathcal{B}$. Then, by construction, $\bar{\beta} + \mathbf{v}_1 \equiv \bar{\alpha} \pmod{\Lambda_1}$. In this case, the starting point of the walk $W_{\bar{\beta}}$ is given by

$$
\begin{aligned}
x_0 &= (\mathbf{g}_1^{\bar{\beta}} \cdot \mathbf{g}_2^{\mathbf{w}_0}) \star y \\
&= (\mathbf{g}_1^{\bar{\beta}} \cdot \mathbf{g}_2^{\mathbf{w}_0}) \star \left((\mathbf{g}_1^{\mathbf{v}_1} \cdot \mathbf{g}_2^{\mathbf{v}_2}) \star x \right) \\
&= (\mathbf{g}_1^{\bar{\beta}+\mathbf{v}_1} \cdot \mathbf{g}_2^{\mathbf{w}_0+\mathbf{v}_2}) \star x \\
&= (\mathbf{g}_1^{\bar{\alpha}} \cdot \mathbf{g}_2^{\mathbf{w}_0+\mathbf{v}_2}) \star x \ \in \mathbf{g}_1^{\bar{\alpha}} \mathcal{X}.
\end{aligned}
$$

By Proposition 3, the walk $W_{\bar{\beta}}$ is therefore confined to the set $\mathbf{g}_1^{\bar{\alpha}} \mathcal{X}$. Consequently, it must eventually collide with one of the precomputed walks $W_{\bar{\alpha}}^{(\ell)}$ associated with $\bar{\alpha} \in \mathcal{A}$ and for some j we obtain

$$x_j = x_t^{(\ell)},$$

where $x_t^{(\ell)}$ is the endpoint of $W_{\bar{\alpha}}^{(\ell)}$, and the corresponding tuple $(x_t^{(\ell)}, \bar{\alpha}, \mathbf{w}_t^{(\ell)})$ lies in the hint list $\mathcal{L}$. This implies

$$(\mathbf{g}_1^{\bar{\beta}} \cdot \mathbf{g}_2^{\mathbf{w}_j}) \star \left((\mathbf{g}_1^{\mathbf{v}_1} \cdot \mathbf{g}_2^{\mathbf{v}_2}) \star x \right) = (\mathbf{g}_1^{\bar{\alpha}} \cdot \mathbf{g}_2^{\mathbf{w}_t^{(\ell)}}) \star x.$$

From this identity, we recover a solution to the GA-dlog problem:

$$\mathbf{v}_1 \equiv \bar{\alpha} - \bar{\beta} \pmod{\Lambda_1}, \qquad \mathbf{v}_2 \equiv \mathbf{w}_t^{(\ell)} - \mathbf{w}_j \pmod{\Lambda_2}.$$

Notice that, the walks W_β corresponding to different $\beta \in \mathcal{B}$ are mutually independent and can therefore be executed in parallel. Suppose that a total of Δ processing cores are available. For efficiency, we set $\delta_\beta = \min\{\Delta, N_1\}$ and $\delta_\alpha = \lceil N_1/\Delta \rceil$. In this way, the walks corresponding to each $\beta \in \mathcal{B}$ can be computed in parallel across the Δ cores, thereby improving the efficiency of the *Online Phase*. A high-level description of our parallel-GA algorithm is depicted in Fig. 3.

The pseudocode for the *Online Phase* is provided in Algorithm 3, denoted as ONLINE-PARALLEL-GA-DLOG.

Algorithm 3: ONLINE-PARALLEL-GA-DLOG

Input: $(\mathbf{g}_1, \mathbf{g}_2, x) \in \mathcal{G}_1^{n_1} \times \mathcal{G}_2^{n_2} \times \mathcal{X}$, GA-dlog instance $y \in \mathcal{X}$, precomputed hint $\mathcal{L}$, walk length t, PRF $h : \mathcal{X} \to \mathbb{Z}^{n_2}/\Lambda_2$, $\Delta :=$ Total number of available cores

1   Set $\delta_\beta = \min\{\Delta, N_1\}$;
2   Choose a random subset $\mathcal{B} \subseteq \mathbb{Z}^{n_1}/\Lambda_1$ of size δ_β;
3   **for** *all $\beta \in \mathcal{B}$* **do**
     // Each β takes one core (parallel computation)
4   Choose a random $\mathbf{w}_0 \in \mathbb{Z}^{n_2}/\Lambda_2$;
5      $x_0 \leftarrow (\mathbf{g}_1^{\beta} \cdot \mathbf{g}_2^{\mathbf{w}_0}) \star y$;
6   **for** $j = 1, 2, \cdots, 2t$ **do**
7         $x_j \leftarrow \mathbf{g}_2^{h(x_{j-1})} \star x_{j-1}$;
8         $\mathbf{w}_j \leftarrow h(x_{j-1}) + \mathbf{w}_{j-1}$;
9   **if** $(x_j, \alpha, \mathbf{w}_t^{(\ell)}) \in \mathcal{L}$ *for some $(\alpha, \ell) \in \mathcal{A} \times \{1, 2, \cdots, s\}$* **then**
10           $\mathbf{v}_1 \leftarrow \alpha - \beta \pmod{\Lambda_1}$, $\mathbf{v}_2 \leftarrow \mathbf{w}_t^{(\ell)} - \mathbf{w}_j \pmod{\Lambda_2}$;
11   **return** $(\mathbf{v}_1, \mathbf{v}_2)$

12 **return** FAIL

Remark 2. As in the May–Ostuzzi algorithm, we set the length of all random walks in the *Online Phase* to $2t$. If ONLINE-PARALLEL-GA-DLOG does not yield a collision between any of the online walks W_β, where $\beta \in \mathcal{B}$, and the first components of the elements in the hint list $\mathcal{L}$, the *Online Phase* is restarted. In particular, a new subset $\mathcal{B} \subseteq \mathbb{Z}^{n_1}/\Lambda_1$ of size $\delta_\beta (= \min\{\Delta, N_1\})$ is randomly

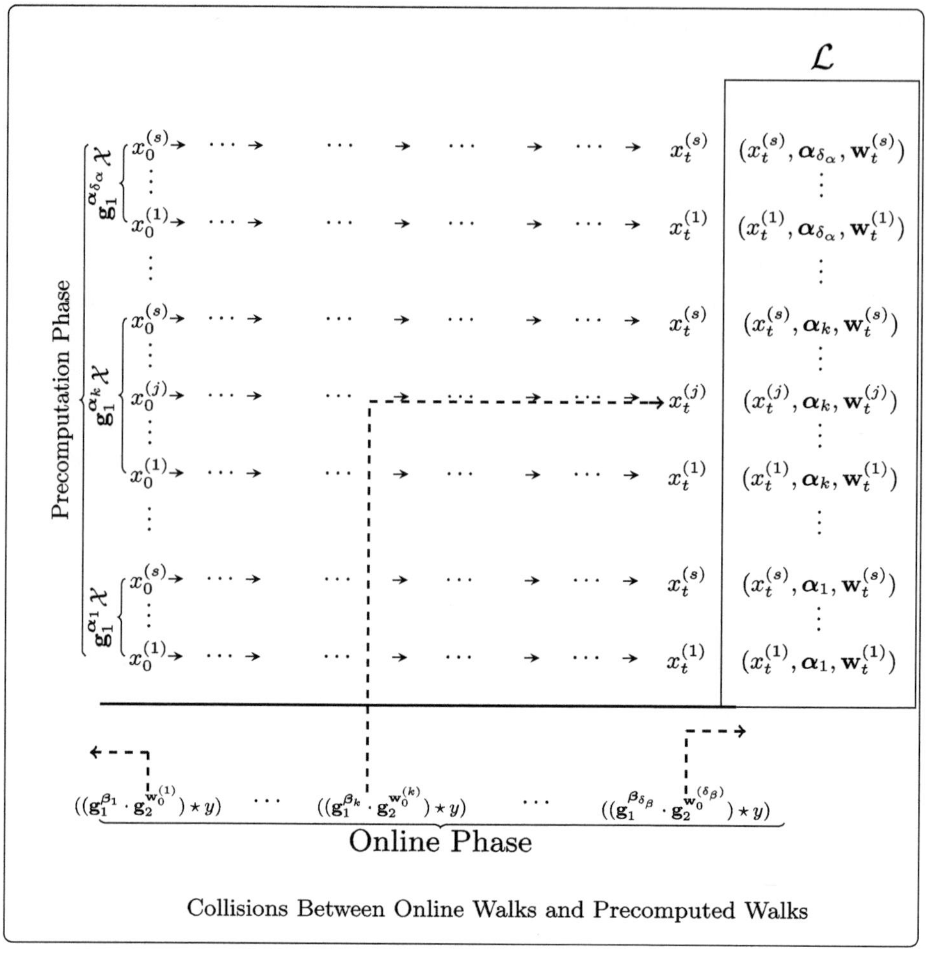

Fig. 3. Collisions Between Online Walks and Precomputed Walks

selected, and for each $\beta \in \mathcal{B}$ a fresh online walk W_β of length $2t$ is generated. This rerandomization increases the success probability of the *Online Phase*, as discussed in the following.

In PRECOMPUTE-PARALLEL-GA, we randomly select a subset $\mathcal{A} \subseteq \mathbb{Z}^{n_1}/\Lambda_1$ of size $\delta_\alpha (= \lceil \frac{N_1}{\Delta} \rceil)$. For each $\alpha \in \mathcal{A}$, we generate s instance-independent random walks of length t in order to construct the hint list $\mathcal{L}$. Therefore, the total time complexity for generating the hint list using parallel computation is $\mathcal{O}(\frac{\delta_\alpha st}{\Delta}) = \mathcal{O}\left(\lceil \frac{N_1}{\Delta} \rceil \frac{st}{\Delta}\right)$, while the space complexity is $\mathcal{O}(\delta_\alpha s) = \mathcal{O}\left(\lceil \frac{N_1}{\Delta} \rceil s\right)$.

In ONLINE-PARALLEL-GA-DLOG, we randomly select a subset $\mathcal{B} \subseteq \mathbb{Z}^{n_1}/\Lambda_1$ of size $\delta_\beta \leq \Delta$. On each core, we generate a random online walk W_β corresponding to an element $\beta \in \mathcal{B}$. The algorithm ONLINE-PARALLEL-GA-DLOG successfully solves the GA-dlog instance if there exists a valid pair $(\bar{\alpha}, \bar{\beta}) \in \mathcal{A} \times \mathcal{B}$ and the online walk $W_{\bar{\beta}}$ collides with one of the precomputed walks associated with $\bar{\alpha}$ within the first t steps.

The probability that such a valid pair exists is

$$p_1 = 1 - \left(1 - \frac{\delta_\alpha}{N_1}\right)^{\delta_\beta} \approx 1 - \left(1 - \frac{1}{\delta_\beta}\right)^{\delta_\beta} \geq 1 - \frac{1}{e}$$

If a valid pair $(\bar{\alpha}, \bar{\beta})$ exists in $\mathcal{A} \times \mathcal{B}$, then all valid walks lie in the subset $g_1^{\bar{\alpha}} \mathcal{X}$, which has size N_2. Therefore, from Sect. 2, the probability that the online walk $W_{\bar{\beta}}$ collides with one of the precomputed walks associated with $\bar{\alpha}$ within the first t steps, given that $(\bar{\alpha}, \bar{\beta})$ is a valid pair, is

$$p_2 \geq \left(1 - \frac{st^2}{4(1-\gamma)N_2}\right) \cdot \left(\frac{\gamma st^2}{N_2} - \frac{\gamma^2 s^2 t^4}{2N_2^2}\right),$$

where $\gamma \in (0, 1)$ satisfies $st^2 \leq 4(1-\gamma)N_2$ and is chosen to maximize the lower bound of p_2.

Hence, the overall success probability of ONLINE-PARALLEL-GA-DLOG within $\mathcal{O}(t)$ steps is at least

$$p_1 \cdot p_2 \geq \left(1 - \frac{1}{e}\right) \cdot \left(1 - \frac{st^2}{4(1-\gamma)N_2}\right) \cdot \left(\frac{\gamma st^2}{N_2} - \frac{\gamma^2 s^2 t^4}{2N_2^2}\right).$$

We formalize the result in the theorem below.

Theorem 3. *Let $s, t \in \mathbb{N}$ satisfy $st^2 \leq N_2$, and suppose Δ cores are available. Then* PRECOMPUTE-PARALLEL-GA *produces a hint list $\mathcal{L}$ of size $\mathcal{O}\left(\lceil \frac{N_1}{\Delta} \rceil s\right)$ within time complexity $\mathcal{O}\left(\lceil \frac{N_1}{\Delta} \rceil \frac{st}{\Delta}\right)$. Using $\mathcal{L}$, the* ONLINE-PARALLEL-GA-DLOG *algorithm solves a GA-dlog instance in $\mathcal{O}(t)$ steps with success probability at least*

$$\left(1 - \tfrac{1}{e}\right) \cdot \left(1 - \tfrac{st^2}{4(1-\gamma)N_2}\right) \cdot \left(\tfrac{\gamma st^2}{N_2} - \tfrac{\gamma^2 s^2 t^4}{2N_2^2}\right),$$

where $\gamma \in (0, 1)$ satisfies $st^2 \leq 4(1-\gamma)N_2$ and is chosen to maximize the bound.

A Comparison with May–Ostuzzi Algorithm. To compare our work with the algorithm of May–Ostuzzi, recall that their method requires a precomputation cost of $\mathcal{O}(N^{2/3})$, and both memory and online costs of $\mathcal{O}(N^{1/3})$.

Suppose Δ cores are available and consider $s = t = N_2^{1/3} = \frac{N^{1/3}}{N_1^{1/3}}$. The space complexity of our method is then

$$\mathcal{O}\left(\lceil \tfrac{N_1}{\Delta} \rceil s\right) = \mathcal{O}\left(\lceil \tfrac{N_1}{\Delta} \rceil \tfrac{N^{1/3}}{N_1^{1/3}}\right).$$

Thus, whenever there exists a direct-summand subgroup of order N_1 such that $1 < N_1 < \Delta^{3/2}$, our method requires less memory than May–Ostuzzi.

The precomputation cost of our algorithm is given by

$$\mathcal{O}\left(\lceil \tfrac{N_1}{\Delta} \rceil \tfrac{st}{\Delta}\right) = \mathcal{O}\left(\lceil \tfrac{N_1}{\Delta} \rceil \tfrac{N^{2/3}}{N_1^{2/3}\Delta}\right).$$

Even if the May–Ostuzzi algorithm's *Precomputation Phase* is parallelized over Δ cores, its complexity reduces to only $\mathcal{O}\left(\tfrac{N^{2/3}}{\Delta}\right)$, which remains more expensive than our precomputation cost as long as $1 < N_1 < \Delta^3$.

For the *Online Phase*, the complexity of our ONLINE-PARALLEL-GA-DLOG algorithm under parallel execution is $\mathcal{O}(t) = \mathcal{O}(N_2^{1/3}) = \mathcal{O}\left(\tfrac{N^{1/3}}{N_1^{1/3}}\right)$, which performs better than the May-Ostuzzi *Online Phase*.

Therefore, if the group $\mathcal{G}$ has a direct-summand subgroup of order at most $\Delta^{3/2}$, our approach provides improved time and memory performance compared to the May–Ostuzzi algorithm under multicore setting.

A detailed comparison of the two algorithms in the parallelized setting is provided in Table 2.

Table 2. Complexity comparison for the May–Ostuzzi algorithm and our parallel approach, assuming Δ available cores and N_1 as the order of a direct-summand subgroup of $\mathcal{G}$.

Cost	May–Ostuzzi [MO25]	Our Algorithm
Memory	$N^{1/3}$	$\lceil \tfrac{N_1}{\Delta} \rceil \tfrac{N^{1/3}}{N_1^{1/3}}$
Precomputation	$\tfrac{N^{2/3}}{\Delta}$	$\lceil \tfrac{N_1}{\Delta} \rceil \tfrac{N^{2/3}}{N_1^{2/3}\Delta}$
Online Phase	$N^{1/3}$	$\tfrac{N^{1/3}}{N_1^{1/3}}$

Remark 3. Following the approach of May and Ostuzzi, our complexity analysis accounts only for the number of steps in the random walks, without explicitly incorporating the cost of exponent reductions via the lattice or the associated multiplications.

Application to CSIDH-512. For the CSIDH-512 setting, we consider elliptic curves defined over the field $\mathbb{F}_p$, where p is a 512-bit prime and the system involves $n = 74$ small primes. The group order is given by [BKV19]

$$N = 3 \times 37 \times 1407181 \times 51593604295295867744293584889$$

$$\times 31599414504681995853008278745587832204909 \approx 2^{257}.$$

Assume $\Delta = 2^{20} (\approx \text{one million})$ cores are available. We may then select a direct-summand subgroup $\mathcal{G}_1$ of order

$$N_1 = 3 \times 37 \times 1407181 \approx 2^{27} < \Delta^{3/2},$$

and the complementary subgroup $\mathcal{G}_2$ of order, $N/N_1 \approx 2^{230}$ such that $\mathcal{G} = \mathcal{G}_1 \oplus \mathcal{G}_2$.

With parameters $s = t = N^{1/3}$, the May–Ostuzzi algorithm produces a hint list of size $N^{1/3} \approx 2^{86}$ at a precomputation cost of $\frac{N^{2/3}}{\Delta} \approx 2^{151}$ (using parallelization). This hint allows solving a GA-dlog instance in about $N^{1/3} \approx 2^{86}$ operations during the *Online Phase*.

In comparison, by setting $s = t = N_2^{1/3}$, our proposed algorithm produces a smaller hint of size $\left\lceil \frac{N_1}{\Delta} \right\rceil \frac{N^{1/3}}{N_1^{1/3}} \approx 2^{84}$, while reducing the precomputation cost to $\left\lceil \frac{N_1}{\Delta} \right\rceil \frac{N^{2/3}}{N_1^{2/3}\Delta} \approx 2^{140}$ under parallelization. Moreover, the *Online Phase* then requires only $\frac{N^{1/3}}{N_1^{1/3}} \approx 2^{77}$ operations, offering an efficiency improvement over the original May–Ostuzzi approach.

Experimental Verification. We implemented our parallel GA algorithms, PRECOMPUTE-PARALLEL-GA and ONLINE-PARALLEL-GA-DLOG, on abelian groups $\mathcal{G} = \langle g_1, g_2, \ldots, g_6 \rangle$ of various orders, specifically $8^6, 9^6, 10^6, 11^6$, and 12^6. Each group was generated using six generators of equal order, with group multiplication serving as the group action.

Given 24 available cores, we set $|\mathcal{G}_1| := N_1 \approx 24^{3/2} \approx 118$, following the approach discussed previously. Accordingly, we partitioned the group as $\mathcal{G}_1 = \langle g_1, g_2 \rangle$ and $\mathcal{G}_2 = \langle g_3, g_4, g_5, g_6 \rangle$, where $|\mathcal{G}_2| := N_2 = |\mathcal{G}|/N_1$. The parameters were set as $s = t = \left\lfloor \frac{1}{2} N_2^{1/3} \right\rfloor$.

For each group, we generated 100 GA-dlog instances, all of which were successfully solved by our parallel-GA algorithm. While the theoretical upper bound on the number of repetitions in the *Online Phase* is 19.69, the experimental average number of repetitions observed across these instances is reported in Table 3.

Table 3. Average number of repetitions of the *Online Phase* for 100 GA-dlog instances per group

Group Order	Average *Online Phase* Repetitions
8^6	5.89
9^6	6.17
10^6	6.94
11^6	7.17
12^6	7.21

All experiments were implemented using SageMath 10.4 and executed on a 24-core Intel i9 processor clocked at 4.1 GHz.

4   Conclusion

In this work, we revisited the May–Ostuzzi algorithm for solving the GA-dlog problem and showed that its theoretical lower bound on the success probability is overly conservative. Our refined analysis provides tighter bounds that are consistent with experimental results, giving a more accurate understanding of the algorithm's practical performance.

We further introduced a new parallel-GA framework that exploits the multicore architecture. By enabling parallelization across Δ cores, and provided that the group $\mathcal{G}$ admits a direct-summand subgroup of order $N_1 \leq \Delta^{3/2}$, our algorithm achieves reduced time and memory costs under realistic parallel models. In particular, our method reduces the hint size from $N^{1/3}$ to $\left\lceil \frac{N_1}{\Delta} \right\rceil \frac{N^{1/3}}{N_1^{1/3}}$, the precomputation cost from $\frac{N^{2/3}}{\Delta}$ to $\left\lceil \frac{N_1}{\Delta} \right\rceil \frac{N^{2/3}}{N_1^{2/3}\Delta}$, and the online cost from $N^{1/3}$ to $\frac{N^{1/3}}{N_1^{1/3}}$. In the CSIDH-512 parameter setting, with 2^{20} cores, our approach reduces the hint size from 2^{86} to 2^{84}, the precomputation cost from 2^{151} to 2^{140} operations and the online cost from 2^{86} to 2^{77}.

These results highlight that the hardness of the GA-dlog problem under parallel computation deserves further investigation, especially in the context of isogeny-based cryptosystems such as CSIDH. Future work may explore extending our framework, and adapting the techniques to other group-action-based assumptions.

Acknowledgments. Abul Kalam is supported by the Prime Minister's Research Fellowship (PMRF)–Project No.: SB22231582MAPMRF008668. We would like to thank Madhurima Mukhopadhyay for introducing the May-Ostuzzi algorithm, Debajyoti De for insightful discussions on CSIDH, and Luca De Feo along with the anonymous reviewers of INDOCRYPT 2025 for their valuable feedback and constructive suggestions, which have greatly enhanced the clarity and quality of this paper.

References

[BBC+21] Banegas, G., Bernstein, D.J., Campos, F., Chou, T., Lange, T., Meyer, M., Smith, B., Sotáková, J.: CTIDH: faster constant-time CSIDH. Cryptology ePrint Archive, Paper 2021/633 (2021). https://eprint.iacr.org/2021/633

[BKV19] Beullens, W., Kleinjung, T., Vercauteren, F.: CSI-FiSh: efficient isogeny based signatures through class group computations. In: Galbraith, S.D., Moriai, S. (eds.) ASIACRYPT 2019. LNCS, vol. 11921, pp. 227–247. Springer, Cham (2019). https://doi.org/10.1007/978-3-030-34578-5_9

[BM92] Bellovin, S.M., Merritt, M.: Encrypted key exchange: password-based protocols secure against dictionary attacks. In: Proceedings 1992 IEEE Computer Society Symposium on Research in Security and Privacy, pp. 72–84 (1992). https://doi.org/10.1109/RISP.1992.213269

[BPR00] Bellare, M., Pointcheval, D., Rogaway, P.: Authenticated key exchange secure against dictionary attacks. In: Preneel, B. (ed.) EUROCRYPT 2000. LNCS, vol. 1807, pp. 139–155. Springer, Heidelberg (2000). https://doi.org/10.1007/3-540-45539-6_11

[BY91] Brassard, G., Yung, M.: One-way group actions. In: Menezes, A.J., Vanstone, S.A. (eds.) CRYPTO 1990. LNCS, vol. 537, pp. 94–107. Springer, Heidelberg (1991). https://doi.org/10.1007/3-540-38424-3_7

[CGK18] Corrigan-Gibbs, H., Kogan, D.: The discrete-logarithm problem with preprocessing. In: Nielsen, J.B., Rijmen, V. (eds.) EUROCRYPT 2018. LNCS, vol. 10821, pp. 415–447. Springer, Cham (2018). https://doi.org/10.1007/978-3-319-78375-8_14

[CLM+18] Castryck, W., Lange, T., Martindale, C., Panny, L., Renes, J.: CSIDH: an efficient post-quantum commutative group action. In: Peyrin, T., Galbraith, S. (eds.) ASIACRYPT 2018. LNCS, vol. 11274, pp. 395–427. Springer, Cham (2018). https://doi.org/10.1007/978-3-030-03332-3_15

[Cou06] Couveignes, J.M.: Hard homogeneous spaces. Cryptology ePrint Archive, Paper 2006/291 (2006). https://eprint.iacr.org/2006/291

[DFKS18] De Feo, L., Kieffer, J., Smith, B.: Towards practical key exchange from ordinary isogeny graphs. In: Peyrin, T., Galbraith, S. (eds.) ASIACRYPT 2018. LNCS, vol. 11274, pp. 365–394. Springer, Cham (2018). https://doi.org/10.1007/978-3-030-03332-3_14

[DH76] Diffie, W., Hellman, M.: New directions in cryptography. IEEE Trans. Inf. Theory 22(6), 644–654 (1976). https://doi.org/10.1109/TIT.1976.1055638

[ElG85] ElGamal, T.: A public key cryptosystem and a signature scheme based on discrete logarithms. In: Blakley, G.R., Chaum, D. (eds.) CRYPTO 1984. LNCS, vol. 196, pp. 10–18. Springer, Heidelberg (1985). https://doi.org/10.1007/3-540-39568-7_2

[FG18] De Feo, L., Galbraith, S.D.: SeaSign: compact isogeny signatures from class group actions. Cryptology ePrint Archive, Paper 2018/824 (2018). https://eprint.iacr.org/2018/824. https://doi.org/10.1007/978-3-030-17659-4_26

[GHS02] Galbraith, S.D., Hess, F., Smart, N.P.: Extending the GHS weil descent attack. In: Knudsen, L.R. (ed.) EUROCRYPT 2002. LNCS, vol. 2332, pp. 29–44. Springer, Heidelberg (2002). https://doi.org/10.1007/3-540-46035-7_3

[Hal05] Hallgren, S.: Fast quantum algorithms for computing the unit group and class group of a number field. In: Proceedings of the Thirty-Seventh Annual ACM Symposium on Theory of Computing, STOC '05, pp. 468–474. Association for Computing Machinery, New York (2005). https://doi.org/10.1145/1060590.1060660

[HM89] Hafner, J.L., McCurley, K.S.: A rigorous subexponential algorithm for computation of class groups. J. Am. Math. Soc. 2, 837–850 (1989). https://api.semanticscholar.org/CorpusID:51781105

[JDF11] Jao, D., De Feo, L.: Towards quantum-resistant cryptosystems from supersingular elliptic curve isogenies. In: Yang, B.-Y. (ed.) PQCrypto 2011. LNCS, vol. 7071, pp. 19–34. Springer, Heidelberg (2011). https://doi.org/10.1007/978-3-642-25405-5_2

[Kob87] Koblitz, N.: Elliptic curve cryptosystems. Math. Comput. 48(177), 203–209 (1987)

[Kup04] Kuperberg, G.: A subexponential-time quantum algorithm for the dihedral hidden subgroup problem (2004). https://arxiv.org/abs/quant-ph/0302112

[MO25] May, A., Ostuzzi, M.: Multiple group action dlogs with(out) precomputation. In: Jager, T., Pan, J. (eds.) Public-Key Cryptography – PKC 2025, pp. 364–387. Springer, Cham (2025). https://doi.org/10.1007/978-3-031-91826-1_12

[MZ21] May, A., Zweydinger, F.: Legendre PRF (multiple) key attacks and the power of preprocessing. Cryptology ePrint Archive, Paper 2021/645 (2021). https://eprint.iacr.org/2021/645

[NIS92] CORPORATE NIST. The digital signature standard. Commun. ACM 35(7), 36–40 (1992). https://doi.org/10.1145/129902.129904

[OW99] van Oorschot, P.C., Wiener, M.J.: Parallel collision search with cryptanalytic applications. J. Cryptol. 12(1), 1–28 (1999). https://doi.org/10.1007/PL00003816

[PH06] Pohlig, S., Hellman, M.: An improved algorithm for computing logarithms over gf(p) and its cryptographic significance (corresp.). IEEE Trans. Inf. Theor. 24(1), 106–110 (2006). https://doi.org/10.1109/TIT.1978.1055817

[Pol75] Pollard, J.M.: A monte carlo method for factorization. BIT 15(3), 331–334 (1975). https://doi.org/10.1007/BF01933667

[Sch95] Schoof, R.: Counting points on elliptic curves over finite fields. J. de théorie des nombres de Bordeaux 7(1), 219–254 (1995). https://www.numdam.org/item/JTNB_1995__7_1_219_0/

[Sha71] Shanks, D.: Class number, a theory of factorization, and genera (1971). https://api.semanticscholar.org/CorpusID:116204965

[Sho94] Shor, P.W.: Algorithms for quantum computation: discrete logarithms and factoring. In: Proceedings 35th Annual Symposium on Foundations of Computer Science, pp. 124–134 (1994). https://doi.org/10.1109/SFCS.1994.365700

[Sho97] Shor, P.W.: Polynomial-time algorithms for prime factorization and discrete logarithms on a quantum computer. SIAM J. Comput. 26(5), 1484–1509 (1997). https://doi.org/10.1137/S0097539795293172

[Sto10] Stolbunov, A.: Constructing public-key cryptographic schemes based on class group action on a set of isogenous elliptic curves (2010). https://www.aimsciences.org/article/id/e8001706-6615-4b24-b499-8ea9d348dabb. https://doi.org/10.3934/amc.2010.4.215

[vOW94] van Oorschot, P.C., Wiener, M.J.: Parallel collision search with application to hash functions and discrete logarithms. In Proceedings of the 2nd ACM Conference on Computer and Communications Security, CCS '94, pp. 210–218. Association for Computing Machinery, New York (1994). https://doi.org/10.1145/191177.191231

Multivariate and Lattice-based Cryptography

Multivariate Encryptions with LL' Perturbations

- Is it Possible to Repair HFE in Encryption? When 0 Makes a Difference-

Jacques Patarin[1,2] and Pierre Varjabedian[1(✉)]

[1] THALES, Meudon, France
{jacques.patarin,pierre.varjabedian}@thalesgroup.com
[2] Université Paris-Saclay, UVSQ, CNRS, Laboratoire de mathématiques de Versailles, 78000 Versailles, France

Abstract. We will present here new multivariate encryption algorithms. This is interesting since few multivariate encryption schemes currently exist, while there exist many more multivariate signature schemes. Our algorithms will combine several ideas, in particular the idea of the LL' perturbation originally introduced, but only for signatures, by Gouget and Patarin in Vietcrypt 2006. In this paper, the LL' perturbation will be used for encryption and will differ greatly from the paper by Gouget and Patarin. As we will see, our algorithms resist all known attacks (in particular Gröbner attacks and MinRank attacks) and have reasonable computation times.

Keywords: Post-Quantum public key encryption · Post-Quantum key exchange · Multivariate cryptography · HFE

1   Introduction

Multivariate Cryptography is one of the main families of algorithms used to design post-quantum public key cryptography. We can trace its origin with the article [MI88] that introduced the so-called C^* scheme. The scheme was broken in [Pat95] but it led to multiple algorithms. With multivariate cryptography it is supposedly possible to perform encryption, signatures and authentication. Yet, in recent years almost all known multivariate encryption algorithms have been broken due to the discovery and improvement of attacks such as Min-Rank attacks. Due to this, at present most of the algorithms that are based on the MQ problem can only be used for signatures or authentication. For example the UOV [KPG99] family is currently largely represented in the recent NIST standardization process.

Another major multivariate algorithm is Hidden Field Equations (HFE) [Pat96]. Until recently HFE was one of the main candidate for multivariate cryptography, especially since it was one of the few able to be used for encryption (contrary to UOV mentioned previously) as well as signatures. Since its origin it was suggested to add perturbations in order to reinforce its security.

R. Dutta et al. (Eds.): INDOCRYPT 2025, LNCS 16372, pp. 263–284, 2026.
https://doi.org/10.1007/978-3-032-13301-4_12

HFE has a lot of variants. In the recent article [CMRPV24] the most important perturbations are presented with their current cryptanalysis properties. In [CMRPV24], the authors showed that it is hard to use HFE for encryption, indeed the perturbations that they studied have a high cost in order to avoid all attacks. However, the authors of [CMRPV24] have proposed a variant that seems to work for signatures. Other attempts around HFE or more generally around multivariate cryptography in encryption have been made cf. for example [SDP16, IPS+18, BM09, CBD+09]. They generally have failed to resist all of the attacks [CCFST24], [BM09] or have been found to be far less effective than expected. This is why it is considered to be much easier to make multivariate signature schemes than encryption schemes.

The main novelty of this article is that we will be using the LL' perturbation introduced in [GP06] in a specific and original way for encryption. Our LL' can also be viewed as a particular case of the other perturbation $\hat{+}$ [FmRPP22], but as we will see, our LL' is much more efficient than $\hat{+}$.

In this paper, we will focus on studying the behaviour of LL' with HFE. However, this perturbation could be used with other multivariate scheme.

1.1 Related Work

An attack against a previous version of our scheme has been published in [STV25]. The attack consisted in identifying the ciphertexts that were successfully decrypted in order to obtain information on the kernel of the linear form involved in our variant. Afterward the authors were able to get rid of them and perform an MinRank attack. However in our new protocol it is not possible to recover the messages that were successfully decrypted, as explained in Sect. 3.6, which mitigates the attack.

2 Vanilla HFE

Here by "vanilla" HFE or "Nude" HFE we mean the basic HFE scheme, *i.e.* without any additional perturbation.

The main idea for algorithms based on the difficulty to solve a quadratic system (Multivariate Quadratic problem) is to find structured quadratic polynomials that make the problem easy to solve. Then, we hide their structure via secret linear maps. For example let F be a quadratic polynomial such that the equation in x, $F(x) = y$ is easy to solve. Such polynomial F will be called the central map of the system. Let T, S be two bijective linear maps. The public key will be $T \circ F \circ S$, that way the public key will hopefully mimic the behaviour of a random quadratic map. Here the private key will be: T, S, F, hence with the private key and the hypothesis that F is easily "invertible" we can make an encryption/signature algorithm.

HFE uses such structure. Let q be a power of a prime number and d, n two integers, HFE utilizes the fact that the field $\mathbb{F}_{q^n}$ can be seen as the vector space

$\mathbb{F}_q^n$. Then an univariate polynomial $F(x)$ of $\mathbb{F}_{q^n}$ can also be seen as a multivariate polynomial $F(x_1, \ldots, x_n)$. For vanilla HFE $F(x)$ will be a polynomial like this:

$$F(x) = \sum_{i,j \leq d} a_{ij} x^{q^i + q^j}$$

We see that the monomials are quadratic in $\mathbb{F}_q$ (x^{q^i} is linear in $\mathbb{F}_q$ as power of the Froebenius morphism, hence $x^{q^i + q^j} = x^{q^i} x^{q^j}$ is a quadratic term) and that the degree is limited by $D = 2 * q^d$. This is to ensure that Cantor-Zassenhaus [CZ81] algorithm can find a solution in a reasonable time. Then let T, S be two bijective linear maps $\mathbb{F}_q^n \rightarrow \mathbb{F}_q^n$. Let $\phi : \mathbb{F}_{q^n} \rightarrow \mathbb{F}_q^n$ be an isomorphism from the field extension of $\mathbb{F}_q$ to the vector space. Then the public key will be:

$$T \circ \phi \circ F \circ \phi^{-1} \circ S.$$

However most of the time in literature the isomorphism ϕ is not written but only implied. So the public key could also be written as:

$$T \circ F \circ S.$$

The private key will be:

$$T, S, F.$$

Due to the fact that the central map is quadratic there exist a matrix representation. This representation is used in the implementation of HFE and in the cryptanalysis. Hence, when there's no ambiguity we will note the corresponding matrix representation in bold with the same letter. For example, the matrix representation of F will be $\mathbf{F}$ (size $n \times n$) and the matrix representation of S and T will be $\mathbf{S}$, $\mathbf{T}$. In our case

$$\mathbf{F} = \begin{pmatrix} \mathbf{A} & 0 \\ 0 & 0 \end{pmatrix}$$

where $\mathbf{A}$ is a matrix of size $d = \log_q(D)$. Hence, the matrix representation of F is sparse.

From an attacker's point of view the goal would either be to invert the system directly (via Gröbner basis for example) or recover the linear map T or S (via a MinRank attack for example). Both will be using the fact that the degree of the central map F is limited.

When the level of security is higher than 80bits, vanilla HFE requires large D, therefore unrealistic computation time. This is why we need perturbations from the main algorithm in order to achieve higher level of security.

Here, we will only describe how HFE works in encryption mode but it could also be used in signature.

- Let Alice be the one who sends a message and Bob the one who receives it.
- Let m be the message, or a session key that Alice wants to send encrypted to Bob.
- Let $P = T \circ F \circ S$ be the public key of Bob.

- Alice sends $m' = P(s||m)$ to Bob where s is a salt.
- Bob solves in x the equation $P(x) = m'$.
- For this Bob uses his private key to obtain the equation $f(S(x)) = T^{-1}(m')$.
- F being a univariate polynomial of $\mathbb{F}_q^n$ he uses for example Cantor-Zassenhauss or Berlekamp algorithm to solve the equation to find $S(x)$ and then x and then m since $x = s||m$.

Remark 1. As P is a system of multivariate polynomial in $X_1, \ldots, X_n$ $P(X_1, \ldots, X_n) = P_1(X_1, \ldots, X_n) \ldots P_n(X_1, \ldots, X_n)$. Let $x \in \mathbb{F}_q^n$ The notation $P(x)$ means $P(x) = P_1(x_1, \ldots, x_n) \ldots P_n(x_1, \ldots, x_n)$. We will use this notation in the rest of the article.

2.1 HFE $\hat{+}$

We will briefly present here the variant $\hat{+}$ of HFE. This variant is key to understand ours, as both are very similar in their effects. However we will later show that our variant becomes more effective because we change the method of decryption.

As with vanilla HFE we define

$$F(x) = \sum_{i,j \leq d} a_{ij} x^{q^i + q^j}$$

Let t be the $\hat{+}$ parameter and let $p_1, \ldots, p_t$ be random quadratic forms i.e. $p_i : \mathbb{F}_q^n \to \mathbb{F}_q : \sum_{j,k \leq n} a_{i,j,k} x_j x_k$. We can use the trace to give a univariate form hence we define $\tilde{p} : \mathbb{F}_{q^n} \to \mathbb{F}_q : \mathrm{Tr}(\sum_{j,k \leq n} \alpha_{i,j,k} x^{q^k + q^j})$. Hence, we define the central map of HFE$\hat{+}$ as $H = F + \sum_i r_i \tilde{p}_i$ where $r_i \in \mathbb{F}_{q^n}$.

The public key is defined in a similar fashion as

$$P = T \circ H \circ S.$$

with S and T defined in Sect. 2

Decryption Protocol. Recall the previous protocol

- Let Alice be the one who sends a message and Bob the one who receives it.
- Let m be the message, or a session key that Alice wants to send encrypted to Bob.
- Let $P = T \circ H \circ S$ be the public key of Bob.
- Alice sends $m' = P(s||m)$ to Bob where s is a salt.
- Bob uses his private key to obtain the equation $H(S(x)) = T^{-1}(m')$
- Here Bob cannot directly invert with Cantor-Zassenhauss as the degree of the $\tilde{p}_i$ is too high.
- $\tilde{p}_i$ can take at most q values hence Bob makes an exhaustive search on the possible values of the $\tilde{p}_i$ and fixes $\tilde{p}_i = \beta_i$.

- Bob solves $F(S(x)) = T^{-1}(m') - \sum_i r_i \beta_i$. If he made the right hypothesis this equation can be solved by Cantor-Zassenhauss as the degree is now bounded by d otherwise he continue the exhaustive search.
- Then Bob keeps the solution x where $\tilde{p}_i(x) = \beta_i$.

In other words, in order to decrypt one has to make an exhaustive search on the values of the $\tilde{p}_i$. Hence the cost of decryption of HFE$\hat{+}$ is q^t times the cost of decryption of HFE.

3   LL' Perturbation

3.1   Our LL' Perturbation in Encryption

Let $P_1, \ldots, P_n$ be the public key equations corresponding to a multivariate central map (e.g. HFE). $P_1, \ldots, P_n$ are quadratic polynomials in $(X_1, \ldots X_n)$. Let r be an integer such that $1 \leq r \leq n$. Let L_i, $1 \leq i \leq r$ and L'_i, $1 \leq i \leq r$ be $2r$ secret linear forms in $x_1, \ldots x_n$. Our "outside" LL' consists in adding secret linear combination $\sum_i \alpha_{i,j} L_i L'_i$ to the equations P_j, $1 \leq j \leq n$.

Therefore $P'_j = P_j + \sum_i \alpha_{i,j} L_i L'_i$.

Then for our LL' variant the public key will $P'_1, \ldots, P'_n$ while $P_1, \ldots, P_n$ will remain secret.

3.2   Analog Description

Let $P_1, \ldots, P_n$ be the public key that correspond to a multivariate central map (in our case HFE). Let S and T be (as above) the two secret linear transformations. Let $\tilde{F}_1, \ldots, \tilde{F}_n$ be the secret quadratic polynomials used for decryption. For example for HFE, $\tilde{F}_1, \ldots, \tilde{F}_n = \phi \circ F \circ \phi^{-1}$. $\tilde{F}_1, \ldots, \tilde{F}_n$ are the secret polynomials in $Z_1, \ldots, Z_n$ and $S(X_1, \ldots, X_n) = (z_1, \ldots, z_n)$. Our "inside" LL' consists in adding secret linear combinations to the A_j, $1 \leq j \leq n$ so $\tilde{H}_j = \tilde{F}_j + \sum_i \alpha_{i,j} L_i L'_i$.

In fact our "inside" and "outside" transformation are similar (since we can move from one to the other description by just modifying S and T), but for the analysis we will sometimes alternatively use "inside" or "outside".

3.3   Decryption Protocol

The main idea of the LL' perturbation is that over $\mathbb{F}_2$ we have a probability $1/4$ that $L_i L'_i \neq 0$. In order to decrypt, we will evaluate the expected number of products $L_i L'_i \neq 0$ (it will be $r/4$ on average, but it can be much less as we will see below). Then, after exhaustive search on the indices i such that $L_i L'_i \neq 0$ i.e. $L_i L'_i = 1$ we will solve the system like a normal HFE system.

Recall the previous notations, we consider the Decryption protocol for our HFE LL'

- Let Alice be the one who sends a message and Bob the one who receives it
- Let m be the message, or a session key that Alice wants to send encrypted to Bob.

- Let $P = T \circ H \circ S$ be the public key of Bob
- Alice sends multiple $m'_j = P(s||m_j)$ $(1 \leq j \leq k)$ to Bob where s is a salt (we will later discuss the values required for the number k).
- Bob select an indice j and tries to solve in x the equation $P(x) = m'_j$.
- For this Bob uses his private key to obtain the equation $H(S(x)) = T^{-1}(m'_j)$.
- Here Bob cannot directly invert with Cantor-Zassenhauss as the degree of the $\tilde{p}_i$ is too high.
- Bob makes an exhaustive search fixes b_k indices $i_1, \ldots, i_{b_k}$ and fixes the value of the corresponding $L_{i_j} L'_{i_j}$ as $\beta_{i_j} \in \mathbb{F}_q$. All other $L_i L'_i$ are supposed to be 0.
- For each guesses, Bob tries to solve $F(S(x)) = T^{-1}(m'_j) - \sum_{i_j}^{i_k} \alpha_{i_j} \beta_{i_j}$ if the hypothesis is right then this equation can be solved efficiently using algorithms such as Cantor-Zassenhauss as the degree is now bounded by d.
- Then Bob keeps the solution x such that $P(x) = m'_j$ (in other words those that correspond to the hypothesis made by bob).
- if it doesn't work then Bob asks Alice for a new set of message.

In our scheme we bound the exhaustive search hence we are able to be much more efficient than $\hat{+}$. Overall $\hat{+}$ makes an exhaustive search on all of the q^t possible values while we require an exhaustive search on q^{b_k} values with $b_k < r$. We require to send far more messages however. The discussion on the number of messages required to be sent and the appropriate bound will be answered in Sect. 5.

3.4   Variants

- In this paper, $x_1, \ldots x_n$ will be over the small field GF(2). Yet, this perturbation LL' can work on other very small Field (GF(3) for example) however it will be less efficient as the probability to cancel a linear form will decrease.
- Usually, we will have public equations of degree 2. In appendix we will also present a variant with equations of degree 3.
- It is also possible to use very simple linear combination for $L_i L'_i$. Instead of a general linear combination of the form: $\sum_i \alpha_{i,j} L_i L'_i$ we can imagine to just add $L_i L'_i$ to P_i. At present, this variant does not seems to reduce the security significantly, but it does not accelerate the decryption either (since the most difficult part of the decryption in the inversion of f and the fact that we have to repeat this operation multiple times).
- In this paper, we will use our LL' perturbation with HFE central map. Other central maps may also be used, for example C^*, cf. [MI88]. However, so far all attempts to repair C^* have failed. Therefore we are more confident with HFE and a specific analysis will probably be required on C^*.

Remark 2. A central map (as HFE for example) is really needed, *i.e.* LL' cannot be used alone. This is because a random linear combination of LL' equation will have a non-negligible probability to have a small rank and from this property it is possible to find **S** or **T**.

These variants are only general development ideas and have not been subject to a full extent analysis. Hence they will not be part of the final specification.

3.5   How to Make Our LL' Less Expensive in Computations

As said above, about $r/4$ products $L_i L_i'$ are non zero. However sometimes much less products are non zero. In order to be in such a good situation for decryption, one possibility is to send many messages or session keys and hope that at least one of them has only a small number of $L_i L_i' \neq 0$. We will see below (Sects. 5 and 6) how many messages to send on average in order to have a good probability for this to occur.

3.6   Transmission of a Session Key

Usually a public key encryption algorithm is used not to encrypt a plaintext message, but to encrypt a session key that will be used later. Then we can imagine that (with the same public key) we will encrypt potential session keys $k_1, \ldots k_l$ such that from any of these value k_i it is possible to find all the other values k_j (simple transformation can be used to do so). The receiver will decrypt one of the keys k_α. This is not yet the chosen session key; it is only meant to serve as an intermediate value. Otherwise, it could leak informations on the number of $L_i L_i' = 0$ obtained with this key. A great number of informations like this could lead to the discovery of the kernel of the $L_i L_i'$, which in turn could allow an opponent to use a projection to make the perturbation vanish and perform a regular attack. More details on how to exploit this leakage can be found in [STV25].

To prevent such information leakage, the recipient uses the key k_α to recompute all the keys $k_1, \ldots, k_l$. He then choses at random one of the key k_i and he will send back a hash of k_i (or alternatively, he will use a key derived by k_i and send an encrypted text with this key). He will also check that the message that were sent were well computed, to do so he select some of the messages randomly and check their validity. This mechanism ensures that the sender does not know which ciphertext was decrypted, and prevents any correlation between ciphertexts and successful decryption.

The protocol is constructed in the following way:

- Alice wants to send the a session key K to bob.
- Alice generates K_o at random, then she generates k keys of the form $K_i = \sigma^i(K_o)$ where σ is a public permutation.
- Alice sends Bob, messages of the form $M_i = (P(K_i), i)$.
- Bob decrypts one of the message K_j.
- Bob recomputes the messages and selects at random some of the messages sent and checks that Alice respected the protocol.
- If the protocol is respected then Bob select one of the messages K_i and sends i to Alice.
- The session key is now K_i.

When Bob checks if Alice respected the protocol he only needs to encrypt back the messages and check if their encryption is compatible with the messages sent by Alice. When the number of messages is low (less than 100) he can chose to check them all. However for higher numbers of messages he can chose to encrypt k of them.

Let's evaluate this number. Let p be the proportion of faulty messages, k the number of messages that Bob will randomly check to have been well computed and m the number of messages that Alice sent. Then the probability for an attacker to obtain informations is $prob = p\Pi_{i=0}^{k-1}\frac{m-mp-i}{m-i}$ In order to obtain significant information an attacker needs to obtain $d+1$ vectors where all of the LL' vanishes. But here the complexity will depend on the number of LL' that were allowed to be non zero in the decryption. If we allow no LL' to be non zero then we can try to guess a vector space where all LL' vanishes we only need to fraud $d+1$ times in a row and then perform a min rank attack c.f hybrid attack. The complexity would be $\frac{1}{prob^{d+1}}$.

But in our case for security above 128 bit we allow LL' to be non zero at most 1 time. Then in order to perform this attack and try to recover a vector space that will make some of the LL' vanish, he needs to make hypothesis on the right non zero LL'. Then the complexity would be $\frac{r}{prob^{d+1}}$

Remark 3. If no sent message has a small number of $L_iL_i' \neq 0$ then the receiver will have to ask for a new set of messages. In order to not leak any information when this occur, it is possible to randomly ask for new messages even if we were able to decrypt the system.

3.7   Transmission of a Specific Message m

Although, public key encryption algorithms are usually used to exchange session keys, our algorithm can also be used to transmit a specific message m. This variant is generally less convenient but still feasible. It is mandatory that an opponent does not have much more equations on the cleartext than the number of variables of the cleartext (otherwise the system becomes much easier to solve with a Gröbner basis attack). However, to ensure that at least one ciphertext yields a small number of $L_iL_i' \neq 0$, multiple ciphertexts must be generated for the same message.

To prevent information leakage between these ciphertexts while using the same public key, we transform the message M using several encodings $M_1, \ldots, M_k$ with the help of a public function (such as an AES); the role of this public transformation is to break linearity and prevent correlation between ciphertexts. For example, Let $K_1, \ldots, K_k$ be keys of an AES-128, then $M_i = AES(K_i, M)$. Then, the M_i will be encrypted with the public key of our HFE LL' and the keys K_i will be made public. The properties of the AES makes the M_i indistinguishable from a random values. Therefore, no additional information is leaked by encrypting multiple variants of the same message. The protocol is constructed in the following way:

- Alice wants to send the message M to bob.
- She sends Bob k messages of the form $M_i = P(AES(K_i, M))$ where K_i are public and P is the public key of HFE LL'.
- Bob successfully decrypts one of the messages of indices j. Hence he recovers M.
- Now that he obtains the message M Bob can check that Alice respected the protocol and that the encryption of the other values where done correctly by Alice.

With this protocol, Alice cannot know which message Bob managed to decrypt. Thus cannot use that knowledge to attempt a key-recovery attack.

3.8   Differences Between Our LL' and the LL' of [GP06]

In [GP06] a perturbation LL' was also described. However, the description differed greatly from ours and it was used in signature rather than encryption. In [GP06] n products $L_i L_i'$ are used. They were simply added to the public key without any linear combination: $P_i' = P_i + L_i L_i'$. In this paper the main difference is that we only have r products $L_i L_i'$ and that

$$P_j' = P_j + \sum_i \alpha_{i,j} L_i L_i'$$

or

$$A_j' = A_j + \sum_i \alpha_{i,j} L_i L_i'.$$

The second important difference is that in [GP06] not all of the public equations have to be satisfied, unlike in this paper where all the equations are valid equations.

3.9   LL' as a Particular Case of $\hat{+}$

The perturbation $\hat{+}$ was introduced in [FmRPP22]. With $\hat{+}$ we have r random quadratic equations that are linearly combined on the public (or secret) quadratic equation. In LL' we combine r product $L_i L_i'$. Therefore LL' can be seen as a special case of $\hat{+}$.

However, our perturbation will be much less costly in terms of computation time than $\hat{+}$. This is because $L_i L_i'$ is 0 with a probability 3/4 unlike 1/2 for $\hat{+}$.

Recall all the previous notations, Instead of adding a linear combination of random quadratic form $\hat{p}_i(x)$ we will add linear combination of $L_i L_i'$ where: $L_i(x) = \sum_i a_i x^{q^i}$.

It shows clearly the link between the two perturbations. Again we see that $\hat{+}$ is more general than LL'. However we will later show that it does not appear to induce any weaknesses to LL' compared to $\hat{+}$. Indeed our tests confirmed that the cryptanalysis of both perturbations is very similar.

However, in Sect. 4.5 we present a specific attack for LL' that is more effective than used on $\hat{+}$. Nevertheless, overall this new attack is not very efficient.

4 Cryptanalysis of LL'

4.1 General Considerations

In this Section we will study how LL' resists against the classical known attacks in multivariate cryptography (such as MinRank and Gröbner attacks). We may be afraid by the fact that in our schemes, sometimes almost all, or even all, the terms $L_i L_i'$ are 0 on a given ciphertext. We will show that an attacker cannot easily exploit this property.

In MinRank attacks we study the public equations (including the hidden terms $L_i L_i'$) independently from its values, so the fact that some $L_i L_i' = 0$ does not intervene. In other words, MinRank targets the structure of the central map and especially rank defects but it does not target the values of the polynomials.

On Gröbner attacks we try to solve the public equations with all the terms $L_i L_i'$, and the fact that we want to decrypt even when all these terms are 0 does not really help since these terms are hidden. This was confirmed on many simulations that we did (cf. Sect. 4.6).

Remark 4. It looks really difficult for an attacker to be able to recover the equations without the $L_i L_i'$ from cleartext/ciphertext pairs. This is because he does not know which message will be decrypted or which session key will be used (*i.e.* which messages or session keys with almost all $L_i L_i' = 0$).

We will now give more details about how LL' can resist all known attacks. Since LL' can be seen as a variant of $\hat{+}$ we will conduct a similar analysis as done in [FmRPP22] and compare $\hat{+}$ and LL'. First of all we will distinguish two types of attacks, Direct attacks (Gröbner Basis) and key recovery attacks (MinRank attacks).

The study of $\hat{+}$ as well as our simulations showed that LL' and $\hat{+}$ have a very similar behaviour against both types of attacks. So we will consider that both perturbations have the same impact on the degree of regularity of the polynomial. This is largely due to the fact that the evaluation of the degree of regularity is linked to the rank of the central map, and our tests showed that the ranks are equal. So it should follow the following formula:

$$d_{reg} = (d + r)/2 + 2$$

where r is the number of product of linear forms added to the public equations. As LL' is a variant of $\hat{+}$ we know that the previous formula is an upper bound of the degree of regularity.

For the MinRank attacks analysis we should distinguish two forms of MinRank that we will note MinRank **T** [BFP11], MinRank **S** [TPD21]. The goal of the MinRank will be to recover one of the linear map of the private key, once one of them is found the rest of the key is "easy" to find. Hence, MinRank **T** will denote the fact that we will try to recover the matrix **T** first and MinRank **S** the fact that we will try to recover the matrix **S**. Let us define first the MinRank problem.

Definition 1. *Let n, m, r, k be $\in \mathbb{N}$ and let $\mathbf{G}_1, \mathbf{G}_2, \ldots \mathbf{G}_k$ $n \times m$ be matrices over the field $\mathbb{F}$. The MinRank problem consists to find $u_1, u_2, \ldots u_k$ over $\mathbb{F}$ such that $\mathrm{rank}(\sum_{i=1}^{k} u_i \mathbf{G}_i) \leq r$.*

We need to rewrite the problem of finding one of the two matrices as a MinRank instance. The goal will be to use the public key matrices: $\mathbf{P}'_1 \ldots \mathbf{P}'_n$ as the matrices $\mathbf{G}_1, \mathbf{G}_2, \ldots \mathbf{G}_n$ (N.B for MinRank $\mathbf{S}$ we will not directly use the public key matrices) and to use some of the entries of the matrices $\mathbf{T}$ or $\mathbf{S}$ as the solution vector.

We should also remind here the matrix writing of the HFE LL' map. Recall here that $H = F + Q$ where F is the central map of a HFE and Q the LL' modifier.

Lemma 1. *Let $\mathbf{S}, \mathbf{T} \in M_{n \times n}(\mathbb{F}_q)$ be the linear map of the private key. Then the public key P' can be written*

$$P' = (\mathbf{P}'_1, \ldots \mathbf{P}'_n) = (\mathbf{SM}_n \mathbf{H}^{*0} \mathbf{M}_n^t \mathbf{S}^t, \ldots, \mathbf{SM}_n \mathbf{H}^{*n} \mathbf{M}_n^t \mathbf{S}^t) \mathbf{M}_n^{-1} \mathbf{T}$$

where $\mathbf{H}^{*i}$ is the matrix representation of the q^ith power of the secret polynomial H and $\mathbf{M}_n$ denotes the matrix representation of the map ϕ.

4.2   MinRank T

Let's start with MinRank $\mathbf{T}$, this method can be found in [BFP11].

Let q, n, D be standard HFE parameters, $(\mathbf{P}'_1, \ldots \mathbf{P}'_n)$ the public key and $\mathbf{T}, \mathbf{S}, \mathbf{H}$ the secret key as defined earlier. Then we have

$$(\mathbf{P}'_1, \ldots, \mathbf{P}'_n) = (\mathbf{SM}_n \mathbf{H}^{*0} \mathbf{M}_n^t \mathbf{S}^t, \ldots, \mathbf{SM}_n \mathbf{H}^{*n} \mathbf{M}_n^t \mathbf{S}^t) \mathbf{M}_n^{-1} \mathbf{T}.$$

So we can write

$$(\mathbf{P}'_1, \ldots \mathbf{P}'_n) \mathbf{T}^{-1} \mathbf{M}_n = (\mathbf{SM}_n \mathbf{H}^{*0} \mathbf{M}_n^t \mathbf{S}^t, \ldots, \mathbf{SM}_n \mathbf{H}^{*n-1} \mathbf{M}_n^t \mathbf{S}^t).$$

We will write $\mathbf{U} = \mathbf{T}^{-1} \mathbf{M}_n$ and $\mathbf{W} = \mathbf{SM}_n$. Then we have

$$(\mathbf{P}'_1, \ldots, \mathbf{P}'_n) \mathbf{U} = (\mathbf{WH}^{*0} \mathbf{W}^t, \ldots, \mathbf{WH}^{*n} \mathbf{W}^t).$$

Let $(u_{0,0}, u_{1,0}, \ldots, u_{n-1,0})$ be the first column of $\mathbf{U}$ then we have

$$\sum_{i=0}^{n-1} u_{0,i} \mathbf{P}'_i = \mathbf{WHW}^t.$$

Due to the addition of random polynomials the rank of $\mathbf{H}$ is likely very high. However if we write these polynomials in the field extension then like for $\hat{+}$. There exist $\beta_1, \ldots, \beta_r$ coefficient in $\mathbb{F}_{q^n}$ and $L_1, L'_1, \ldots, L_n, L'_n$ linear forms, such that the central map $h(x)$ verifies the equation:

$$h(x) = f(x) + \sum_{i=0}^{r} \beta_i L_i L'_i$$

where $f(x)$ is vanilla HFE central map. If we want to fully keep a univariate representation we can write, $L_i(x) = Tr(\sum_i a_i x^{q^i})$ (where Tr is the trace function of a field extension). Hence, there exist a linear map $\Pi : \mathbb{F}_{q^n} \to \mathbb{F}_{q^n}$ such that $\forall i, \Pi(\beta_i) = 0$. This map has a degree at least r [FmRPP22]. So we can further write:

$$(\mathbf{P}'_1, \ldots, \mathbf{P}'_n)\mathbf{U}\mathbf{\Pi} = (\mathbf{W}\mathbf{H}'^{*0}\mathbf{W}^t, \ldots, \mathbf{W}\mathbf{H}'^{*n}\mathbf{W}^t).$$

where $\mathbf{H}' = \mathbf{H}\mathbf{\Pi}$ where then the rank of the right member is at most $d + r$ Let $\mathbf{U}'$ be such that $\mathbf{U}' = \mathbf{U}\mathbf{\Pi}$. Hence

$$\text{rank}(\sum_{i=0}^{n-1} u'_{0,i}\mathbf{P}'_i) = \log_q(D) + r$$

which is small so finding the first column of $\mathbf{U}$ reduces to solving a Min-Rank instance with $k = n$ and the rank is $\log_q(D) + r$ on the matrices $\mathbf{P}'_1, \ldots, \mathbf{P}'_n$. As we can see the addition of r to the rank is due to the degree of the linear map Π. But the degree of this map is due to the randomness of the β_i (in the univariate representation of the perturbation), the fact that we took special quadratic polynomial in the form $L_i L'_i$ instead of random quadratic p_i (like in HFE $\hat{+}$) does not seems to change the complexity of the attack.

4.3 MinRank S

MinRank $\mathbf{S}$ was first proposed by Tao *et al.* [TPD21].

Retaining the notations $\mathbf{U}$ and $\mathbf{W}$ from the previous attack, we have $(\mathbf{P}'_1, \ldots \mathbf{P}'_{n-1}) = (\mathbf{W}\mathbf{H}^{*0}\mathbf{W}^t, \ldots, \mathbf{W}\mathbf{H}^{*n-1}\mathbf{W}^t)\mathbf{U}^{-1}$. Then we obtain

$$(\mathbf{W}^{-1}\mathbf{P}'_1\mathbf{W}^{-1,t}, \ldots, \mathbf{W}^{-1}\mathbf{P}'_{n-1}\mathbf{W}^{-1,t}) = (\mathbf{H}^{*0}, \ldots, \mathbf{H}^{*n-1})\mathbf{U}^{-1}.$$

This leads us to study the matrix whose line i is the first line of the matrix H^{*i} in other word it's the first line of the representative matrix of the central map after applying the Froebenius morphism i times. Let's call this matrix $\mathbf{G}$

We can decompose $\mathbf{H}$ in the following way: $\mathbf{H} = \mathbf{F} + \mathbf{Q}$. With $\mathbf{F}$ being the central map of a vanilla HFE and $\mathbf{Q}$ the perturbation $LL'/ \hat{+}$. The Froebenius operation being linear we can study the effect of the operation on $\mathbf{F}$ and $\mathbf{Q}$ separately. The form of $\mathbf{F}$ is sparse then the matrix we will obtain will be of the form: $\begin{pmatrix} \mathbf{A}_1 \\ 0 \\ \mathbf{A}_2 \end{pmatrix}$

On the other the matrix by $\mathbf{Q}$ will be of no particular form: $\begin{pmatrix} \mathbf{q}_0 \\ \vdots \\ \mathbf{q}_n \end{pmatrix}$. However,

we can write that $\mathbf{q}_i$ is the first row of the matrix $\sum_{j=0}^{t} \beta_j^{q^i} \mathbf{Q}_j$. Indeed $\mathbf{Q}_j$ is the representative matrix of the polynomial $\hat{p}_j(x)$ or $Tr(L_j(x)L'_j(x))$ (respectively in the case of $\hat{+}$ and LL') whose image is in $\mathbb{F}_q$. Hence the polynomial is unchanged by the Froebenius morphism.

$$\mathbf{Z} = (\mathbf{U}^{-1})^t \times \mathbf{G}$$

,

$$\mathbf{Z} = (\mathbf{U}^{-1})^t \times \left(\begin{pmatrix} \mathbf{A}_1 \\ 0 \\ \mathbf{A}_2 \end{pmatrix} + \begin{pmatrix} \mathbf{q}_0 \\ \vdots \\ \mathbf{q}_n \end{pmatrix} \right)$$

.

It means that the rank of the right matrix is at most r and thus the rank of the matrix $\begin{pmatrix} \mathbf{A}_1 \\ 0 \\ \mathbf{A}_2 \end{pmatrix} + \begin{pmatrix} \mathbf{q}_0 \\ \vdots \\ \mathbf{q}_n \end{pmatrix}$ is at most $d + r$.

Furthermore, using this and the matrix equation of HFE.

Theorem 1. *Let* $\mathbf{P}'_1, \ldots, \mathbf{P}'_n$ *be matrices of the public key and* $\mathbf{W}$ *the matrix previously defined. If one notes* $(w_0^{-1}, w_1^{-1}, \ldots w_{n-1}^{-1})$ *the first row of the matrix* $\mathbf{W}^{-1}$, *and* $b_i = (w_0^{-1}, w_1^{-1}, \ldots w_{n-1}^{-1})\mathbf{P}_i$, *then the matrix* $\mathbf{V}$ *whose rows are the* b_i *has a rank at most* $d + r$.

Proof. From the previous result we know that the rank of $\mathbf{V}\mathbf{W}^{-1^t}$ is bounded by $d + r$, hence the rank of $\mathbf{V}$ is bounded by $d + r$.

For both MinRank once we recover the entries of $\mathbf{T}$ or $\mathbf{S}$ we can recover of the totality of the key like in [BFP11, TPD21].

As mentioned we have essentially reused the cryptanalysis of HFE $\hat{+}$ [FmRPP22]. We have successfully simulated this result using a modified version of the code that was published alongside the article [BBC+22]. We did not effectively perform the attack, in other word we did not effectively solve the MinRank. But we computed the rank that appeared in the equation previously mentioned, which confirms the analysis we have performed. So our results showed the augmentation of rank as we have announced.

We performed our tests on a toy example taking: $n = 20, d = 7, r = 1, 2, 3$ and using $\hat{+}$ with similar parameters $n = 20, d = 7, t = 1, 2, 3$.

In the end, we computed the rank of the matrix $\mathbf{V}\mathbf{W}^{-1^t}$ and the matrix $\mathbf{U}' = \mathbf{U}\Pi$. and found that their rank coincided with the theory we have presented.

4.4   Resolution of MinRank

The MinRank problem is solved using the support minor method as described in [BBB+22, BBC+22]. The complexity of the resolution does not depend on the type of MinRank ($\mathbf{T}$ or $\mathbf{S}$) but only on the target rank, the Field characteristic and the degree of the extension.

$$\mathcal{O}\left((d + r)(n - 1)^4 \binom{2(d + r) + 1}{d + r}^2 \right).$$

This gives us a good idea of the number of LL' that needs to be added. We then have a least:

$r = 12$ for 80 bits of security
$r = 17$ for 100 bits of security
$r = 24$ for 128 bits of security
$r = 40$ for 192 bits of security
$r = 56$ for 256 bits of security

(These numbers are obtained with the pessimistic assumption that $d = 0$ which is never the case, therefore in the parameters section these numbers will be slightly reduced).

4.5 Hybrid Method

We can note that the fact that $L_i L_i'$ is more often 0 than a regular quadratic form Q_i it means that there are more linear map Π that cancels the perturbation. Hence, one could try to randomly compose the public key with a linear map Π_r (r stands for random) in hope that the perturbation LL' would vanish (or at least part of it).

Our attack works the following way: we select random linear map Π. Then we compose the public key P with Π in order to obtain $\tilde{P} = P \circ \Pi$. Then we try to obtain the private key of the system by solving the MinRank problem induced by $\tilde{P}$. We continue this process until we have obtained the private key.

If the projection Π cancels some of the $L_i L_i'$ it will induce a decrease of the rank. For example if one $L_i L_i'$ is cancelled it means that the rank of the first coordinate of $P(\Pi)T^{-1}$ is $d + r - 1$. And overall if we make k $L_i L_i'$ vanish the rank would become $d + r - k$.

Let's evaluate the number of tries required in order to remove at least one of the $L_i L_i'$. In other words, the probability to find Π such that their exist $1 \leq j \leq t$ such that $L_j(\Pi) = 0$ or $L_j'(\Pi) = 0$. It means that all the vectors of the basis of the image of Π are included in the kernel of L_j or L_j'. Let r be the rank of Π then the probability for Π to cancel $L_i L_i'$ is $p = 2/q^r - 1/q^{2r}$

So the probability for Π to cancel at least k $L_i L_i'$ is $prob_k = \sum_{j=k}^{t} \binom{t}{k} p^k (1 - p)^{t-k}$. In this case the target rank for the MinRank attack would become $d+r-k$. The rank of Π should be at least $d + r - k + 1$ so that we can obtain a rank defect.

So overall the complexity of the attack will become:

$$\mathcal{O}\left(\frac{1}{prob_k}(d + r - k)(n - 1)^4 \binom{2(d + r - k) + 1}{d + r - k}^2 \right).$$

We will show in Sect. 6 that this attack is less effective than the regular MinRank attacks indeed $prob_k$ decrease faster than the complexity of MinRank with a lower rank. This hybrid attack is however the only specific attack, that we know, where LL' is weaker than $\hat{+}$. As the number of cancelling of the polynomial (product of linear forms for LL', random quadratic form for $\hat{+}$) we add is the only major structural difference with $\hat{+}$.

4.6   Direct Attacks (Gröbner)

In this section we describe the behaviour of direct attacks using Gröbner basis against HFE LL'. To study the complexity of the attack, the good notion is the degree of regularity. So we will focus on the behaviour of the degree of regularity of HFE LL' central map. We will use the definition of Ding and Schmidt [DS13].

Definition 2. *We define $B = \mathbb{F}[X_1, \ldots, X_n]/\langle X_1{}^q, \ldots, X_n{}^q \rangle$ and B_d its degree d subspace. Let P be a set of homogeneous polynomial $P = \{P_1, \ldots, P_m\} \subset B_2^m$.*
Let ψ_d be the map $\psi_d : B_d^m \to B_{d+2}$ defined as

$$\psi(b_1, \ldots, b_m) = \sum_{i=1}^{m} b_i P_i$$

Then

$$R_d(P_1, \ldots, P_m) := \ker(\psi_d).$$

Further let $T_d(P_1, \ldots, P_m)$ be the subspace of trivial relations generated by the elements

$$\{b(P_i e_j - P_j e_i)|1 \leq i < j \leq m, b \in B_{d-2}\},$$

and

$$\{b(P_i^{q-1})e_i|1 \leq i \leq m, b \in B_{d-2(q-1)}\}.$$

Here e_i means the i-th unit vector consisting of all zeros except 1 at the i-th position $e_i = (0, \ldots, 0, 1, 0, \ldots, 0)$. The degree of regularity of a homogeneous quadratic set is then

$$Dreg(P_1, \ldots, P_m) := min\{d | R_{d-2}(P_1, \ldots, P_m)/T_{d-2}(P_1, \ldots, P_m) \neq \{0\}\}.$$

The work of Joux and Faugère [FJ03] showed the link between the rank and the degree of regularity. Analysis of $\hat{+}$ showed that the degree of regularity was increased by the action of $\hat{+}$ (cf. [FmRPP22]). Our tests confirmed this result and showed that LL' had a very similar impact. The computation of the Table 1 have been made on a toy example using $q = 3, D = 7, n = 10$. It shows that both of perturbations have a similar impact against Gröbner.

Table 1. Time to solve the public key of HFE $\hat{+}$ compared with HFE LL' using a Gröbner basis

Number of perturbation	HFE $\hat{+}$	HFE LL'
0	0.86	0.86
1	1.75	1.65
2	2.17	2.13
3	2.11	2.14
4	2.09	2.09

Note that the results are the average for 10 Gröbner inversion so the fact that the time may slightly decrease is simply due to the usage of small parameters. We have a bound for the degree of regularity of HFE. Ding *et al.* [DK12] showed that the degree of regularity D_{reg} is such that. $D_{reg} \leq \frac{(q-1)d}{2} + 2$. It leads to an evaluation of the complexity of the resolution of the Gröbner basis of the system.

In order to find a Gröbner basis for a system we are using algorithms from Jean-Charles Faugère (F4, F5 [Fau99] [Fau02]).

As we saw LL' have a good impact against Gröbner attacks. However, we could wonder if there exists vulnerabilities that depends in the choice of the ciphertext. For example, let f be the public key of a HFE LL' and let $x = \{x_1, \ldots x_n\}$ and $Y = f(x)$ be such that $\forall i \leq r \ L_i(X)L_i'(X) = 0$. Then we could wonder if the resolution of the equation $Y = f(X)$ with Gröbner basis is easier. We performed tests on a toy example (n = 15, r = 3, d = 7) and found that it was not different from the general case. Meaning that the fact that $\forall i \leq rL_i(X)L_i'(X) = 0$ does not induce any vulnerabilities.

We conducted our tests in the following way.

We generated a LL' private and public key. So we have $P_1', \ldots, P_n' = P_1 + Q_1, \ldots, P_n + Q_n$ where

$$Q_i = \sum_j^r a_{i,j} L_j L_j'.$$

Because as testers we know all the $L_i L_i'$ we can select a vector $x =' x_1, \ldots, x_n)$ such that $\forall i, L_i(x)L_i'(x) = 0$ we then note $Y = Y_1, \ldots, Y_n = P_1'(x), \ldots, P_n'(x) = P_1(x), \ldots, P_n(x)$. Such Y is then the image of a vanilla *HFE* (the attacker cannot directly access to this induced HFE scheme) so we test if the degree of regularity of the system. $P_1', \ldots, P_n'$ is the same from the degree of regularity of the system $P_1' - Y_1, \ldots, P_n' - Y_n$ and if it's the same from $P_1' - Y_1', \ldots, P_n' - Y_n'$ where Y_i' are randomly chosen.

Table 2. Time to solve a HFE LL' when the ciphertext is randomly chosen, compared with a ciphertext such that $Y = f(X)$ be such that $\forall i \leq r \ L_i(X)L_i'(X) = 0$

With random ciphertext	with chosen ciphertext
17.3	17.03

As we can see in Table 2 on average we take about 17sec to solve both. Each time we compared the same public key, first with a random ciphertext and then with a chosen one. We also found out that from a given public key no major differences were found, for example even if we have a time to solve that greatly differs from the average we still find close resolution time for a random ciphertext and a chosen one (for example 20.3 sec, 20.6 sec). From a theoretical perspective, this is not a great surprise as the algorithm that find a Gröbner basis do not involve the x_i's in the first place. The algorithm of Buchberger involves the reduction of S-polynomials, even if it depends on the choice of Y it does not depend on the choice of X. Then the degree of regularity of such system should not change hence the complexity of the resolution of the system via Gröbner Basis will not change.

4.7 Other Attacks

In order to break HFE other attacks exists but none of them can achieve the same efficiency as MinRank or direct attacks. We can however mention differential attacks [FGS05, DGS07]. These attacks were at the time efficient due to the choice of parameters that was at the time too optimistic. However for public keys of degree 2, due to the Gröbner basis and MinRank attacks, parameters are nowadays chosen to a point that differential attacks are irrelevant for our specific case. For a HFE polynomial of degree 3 (or higher), differential attacks are still important.

Remark 5. Overall we can observe that there is no major security differences between HFE $\hat{+}$ and HFE LL', indeed adding t quadratic form in $\hat{+}$ and t product of linear form in LL' will result in the same security. However, $\hat{+}$ is slower than LL' therefore for the same computation cost we can add more product of linear forms than random quadratic forms resulting in a security improvement.

4.8 Why the L_i and $L_i L_i'$ Must Be Kept Secret

If the L_i and L_i' linear form are made public then some efficient attacks are possible. We present here two attacks.

- With Gröbner attacks: Sometimes all the products $L_i L_i'$ are 0 on a message sent. Then when the $2r$ values L_i are public, an attacker can make an exhaustive search on r forms L_i or L_i' among these $2r$ forms, say $L_1, \ldots, L_r$, and then solve with a Gröbner basis algorithm the system of public equation plus $L_1 = 0, \ldots, L_r = 0$. Here, the Gröbner algorithm will be efficient due to these extra equations and to the fact that all the perturbations $L_i L_i'$ vanished. *i.e.* from a ciphertext Y the attacker will find the corresponding cleartext X. Moreover, this attack still works even if wrong linear forms $L_i"$ are given in addition to the right L_i and L_i'. This attack can also easily be extended when a small number of $L_i L_i'$ are non zero.
- With MinRank: Once we know some L_i we can determine their kernel K_i and their image I_i. So we can find a projection Π_K such that $L_i \circ \Pi_K = 0$ or a projection Π_I such that $\Pi_I \circ L_i = 0$. Should we compose the projection Π_K it means we can perform an attack on the matrix $\mathbf{S}$ and find a rank of d instead of $d + r$. Similarly, we can compose the projection Π_I it means we can perform an attack on the matrix $\mathbf{T}$ and find a rank of d instead of $d + r$.

5   Number of Encryption in Function of the Number of Copies Sent

In this section we will answer the question of the average number of messages (or number of session keys) required to be able to decrypt at least one of messages sent depending of the number of $L_i L_i' = 1$ allowed.

In the following we will keep the parameters we obtained in Sect. 4.4 to avoid MinRank attacks *i.e.*:

$r = 12$ for 80 bits of security
$r = 17$ for 100 bits of security
$r = 24$ for 128 bits of security
$r = 40$ for 192 bits of security
$r = 56$ for 256 bits of security

The probability to have exactly α equalities $L_i L_i' = 1$ is given by the binomial distribution:

$$\left(\frac{3}{4}\right)^{r-\alpha} \left(\frac{1}{4}\right)^{\alpha} \binom{r}{\alpha}.$$

6   Examples of Parameters and Performances with Public Equations of Degree 2

In this section we will present some explicit examples for our scheme HFE LL' with parameters and an evaluation of the performance.

We will present parameters for five level of expected security: 80, 100, 128, 192, 256 bits of security. We realised our tests of performances using the code for GeMSS (more precisely the reference implementation as it allows for flexible parameters choices). Although the code was designed for signatures, it gives us a reasonable indication for the time to decrypt (as the process of decryption is similar to the one to sign). The code does not directly model the perturbation LL' (and we cannot modify it for our own purposes) so we multiply the number of cycles we obtained by the average number of decryption required for each level of security. This gives us an evaluation of the performance that gives us a good idea of the order of magnitude of the number of cycles required to decrypt our system.

We chose the parameters in order to be protected from the attacks we previously presented.

For clarity we have chosen here only one parameter for the maximal number of $L_i L_i' = 1$ but as shown in the tables of Sect. 5 other trade-off are possible. Here, we have chosen a HFE central map with public key of degree 2 and a secret polynomial of degree 17. With such central map the rank, without perturbation, is 5 (and not 0), it explains why in Tables 3 and 4 we used a parameter r slightly smaller than in Sect. 5.

So we have to obtain parameters such that:

$$2^\lambda \leq \left((d+r)(n-1)^4 \binom{2(d+r)+1}{d+r}^2 \right).$$

where λ is the security parameter and such that:

$$2^\lambda \leq \binom{n+d_{reg}}{n}^\omega,$$

with

$$d_{reg} \approx \frac{d+r}{2} + 2.$$

Table 3. Computation time of a HFE LL' scheme, q is the characteristic, n the size of the field extension, D the degree of the central map, r the number of $L_i L_i'$, and e the maximal number of $L_i L_i' = 1$. The total time is an estimation.

Security bits	(q, n, D, r, e)	Decrypt one message (Cycles)	Avg. nbr. of decryptions	Total cycles	Total time (s)
80	$(2, 100, 17, 10, 0)$	2M	17.7	35.5M	0.011
100	$(2, 110, 17, 14, 0)$	2M	56	112M	0.035
128	$(2, 138, 17, 22, 1)$	2M	1700	3400M	1.06
192	$(2, 200, 17, 38, 1)$	3M	168883	506649	158.32
256	$(2, 266, 17, 52, 2)$	4M	28530507	114122030	35663.13

Table 4. Sizes of the ciphertext send for a HFE LL' schemes, q is the characteristic, n the size of the field extension, D the degree of the central map, r the number of $L_i L_i'$, and e the maximal number of $L_i L_i' = 1$.

Security bits	(q, n, D, r, e)	$|pk|$ (KB)	$|ciphertext|$ of one ciphertext(b)	number of messages	Total size
80	$(2, 100, 17, 10, 0)$	45	100	17.7	1770.1
100	$(2, 110, 17, 14, 0)$	61	110	56.12	6173.2
128	$(2, 138, 17, 22, 1)$	163	138	73	10198.2
192	$(2, 200, 17, 38, 1)$	497	200	4330	866066.0
256	$(2, 266, 17, 52, 2)$	1171	266	20689	5503274

In Table 3 we can see that up until 128 bits we have reasonable decryption time. Benchmarking was done on an Intel Core i7-10850H 3.2GHz CPU with 32GB of RAM. We can note that the computations are easily parallelisable. We could also use a different central map than HFE. For example we could use a C^* [MI88] central map, that is known to be faster than HFE. However, as we already mentioned, all attempts to repair C^* have failed. Therefore we are much more confident with HFE LL' than with C^* LL' where a specific analysis would be required.

As far as we know the cryptanalysis of HFE LL' does not differ much from the results obtained for HFE $\hat{+}$ as we showed earlier. Hence, we are confident with the current parameters according to the current state of attacks as we can see in Table 5 (the complexity of the attack is computed using the parameters as stated in Table 3).Unless a new idea is discovered and a vulnerability exploits the particular form of the polynomial we add, we should not expect any major changes for our parameters.

Table 5. Complexity of MinRank attack and direct attacks on our parameters (the complexity of Gröbner attack is optimistic)

level of security	MinRank	Direct attacks	Hybrid
80	84	80.9	83
100	101	103	105
128	134	129	137
192	200	199	209
256	258	252	272

7   Conclusion

Very few efficient and secure multivariate encryption algorithms exist at present. This is mainly due to the fast progress of MinRank attacks (to recover the secret key) or Gröbner attacks (to inverse the system). New type of attack (MinRank on S, [TPD21]) and new method of resolution (support minor [BBB+22]) have threatened the security of HFE and also threatened the security of most multivariate encryption algorithm (many of them are based on variants of HFE). For example, Vanilla HFE is not efficient when we want more than 80 bits of security.

In this paper, we have presented a new perturbation, called LL', to enforce the security of a multivariate encryption scheme, for example HFE. LL' perturbation was previously used only in signature schemes, and in a very different way [GP06]. With our perturbation LL' we have been able to design efficient schemes up to 128 bits of security. (For 128 bits of expected security we need about 1 s to decrypt with public equations of degree 2, and the encryption is very fast, as usually in multivariate cryptography). It may also be possible to have more security bits (by using public equations of degree 3 or 4), but then the public key become very large. It could also be possible to use C^*LL' instead of HFELL' since the Matsumot-Imai C^* is very fast this variant looks promising. However, so far all the attempt to repair C^* have failed. Therefore, we have to be cautious about C^*LL' and further analysis is required.

Due to the recent introduction of LL' algorithm, we do not recommend using these schemes right now since many multivariate scheme have been broken in the past. However it is interesting to notice that new ideas can still be found to reinforce the security of multivariate scheme. It is also interesting to notice that, again, it seems possible to use multivariate schemes for encryption.

References

[BBB+22] Bardet, M., Briaud, P., Bros, M., Gaborit, P., Tillich, J.-P.:. Revisiting algebraic attacks on MinRank and on the rank decoding problem. Cryptology ePrint Archive, Report 2022/1031 (2022). https://eprint.iacr.org/2022/1031

[BBC+22] Baena, J., Briaud, P., Cabarcas, D., Perlner, R.A., Smith-Tone, D., Verbel, J.A.: Improving support-minors rank attacks: applications to GeMSS and rainbow. In: Dodis, Y., Shrimpton, T. (eds.) CRYPTO 2022, Part III. LNCS, vol. 13509, pp. 376–405. Springer, Heidelberg (2022). https://doi.org/10.1007/978-3-031-15982-4_13

[BFP11] Bettale, L., Faugère, J.-C., Perret, L.: Cryptanalysis of multivariate and odd-characteristic HFE variants. In: Catalano, D., Fazio, N., Gennaro, R., Nicolosi, A. (eds.) PKC 2011. LNCS, vol. 6571, pp. 441–458. Springer, Heidelberg (2011). https://doi.org/10.1007/978-3-642-19379-8_27

[BM09] Billet, O., Macario-Rat, G.: Cryptanalysis of the square cryptosystems. In: Matsui, M. (ed.) ASIACRYPT 2009. LNCS, vol. 5912, pp. 451–468. Springer, Heidelberg (2009). https://doi.org/10.1007/978-3-642-10366-7_27

[CBD+09] Clough, C., Baena, J., Ding, J., Yang, B.-Y., Chen, M.: Square, a new multivariate encryption scheme. In: Fischlin, M. (ed.) CT-RSA 2009. LNCS, vol. 5473, pp. 252–264. Springer, Heidelberg (2009). https://doi.org/10.1007/978-3-642-00862-7_17

[CCFST24] Cartor, M., Cartor, R., Furue, H., Smith-Tone, D.: Improved cryptanalysis of HFERP. In: Tang, Q., Teague, V. (eds.) PKC 2024. LNCS, vol. 14601, pp. 413–440. Springer, Cham (2024). https://doi.org/10.1007/978-3-031-57718-5_14

[CMRPV24] Cogliati, B., Macariot-Rat, G., Patarin, J., Varjabedian, P.: State of the art of HFE variants. In: Saarinen, M.J., Smith-Tone, D. (eds.) PQCrypto 2024. LNCS, vol. 14772, pp. 144–167. Springer, Cham (2024). https://doi.org/10.1007/978-3-031-62746-0_7

[CZ81] Cantor, D.G., Zassenhaus, H.: A new algorithm for factoring polynomials over finite fields. Math. Comput. **36**(154), 587–592 (1981)

[DGS07] Dubois, V., Granboulan, L., Stern, J.: Cryptanalysis of HFE with internal perturbation. In: Okamoto, T., Wang, X. (eds.) PKC 2007. LNCS, vol. 4450, pp. 249–265. Springer, Heidelberg (2007). https://doi.org/10.1007/978-3-540-71677-8_17

[DK12] Ding, J., Kleinjung, T.: Degree of regularity for HFE minus (HFE-) (2012)

[DS13] Ding, J., Schmidt, D.: Solving degree and degree of regularity for polynomial systems over a finite fields. In: Fischlin, M., Katzenbeisser, S. (eds.) Number Theory and Cryptography. LNCS, vol. 8260, pp. 34–49. Springer, Heidelberg (2013). https://doi.org/10.1007/978-3-642-42001-6_4

[Fau99] Faugère, J.-C.: A new efficient algorithm for computing gröbner bases (f4). J. Pure Appl. Algebra **139**(1), 61–88 (1999)

[Fau02] Faugère, J.C.: A new efficient algorithm for computing gröbner bases without reduction to zero (F5). In: Proceedings of the 2002 International Symposium on Symbolic and Algebraic Computation, ISSAC 2002, pp. 75–83. Association for Computing Machinery, New York (2002)

[FGS05] Fouque, P.-A., Granboulan, L., Stern, J.: Differential cryptanalysis for multivariate schemes. In: Cramer, R. (ed.) EUROCRYPT 2005. LNCS, vol. 3494, pp. 341–353. Springer, Heidelberg (2005). https://doi.org/10.1007/11426639_20

[FJ03] Faugère, J.-C., Joux, A.: Algebraic cryptanalysis of hidden field equation (HFE) cryptosystems using Gröbner bases. In: Boneh, D. (ed.) CRYPTO 2003. LNCS, vol. 2729, pp. 44–60. Springer, Heidelberg (2003). https://doi.org/10.1007/978-3-540-45146-4_3

[FmRPP22] Faugère, J.-C., Rat, G.M., Patarin, J., Perret, L.: A new perturbation for multivariate public key schemes such as HFE and UOV. Cryptology ePrint Archive, Report 2022/203 (2022). https://eprint.iacr.org/2022/203

[GP06] Gouget, A., Patarin, J.: Probabilistic multivariate cryptography. In: Nguyen, P.Q. (ed.) VIETCRYPT 2006. LNCS, vol. 4341, pp. 1–18. Springer, Heidelberg (2006). https://doi.org/10.1007/11958239_1

[IPS+18] Ikematsu, Y., Perlner, R., Smith-Tone, D., Takagi, T., Vates, J.: HFERP - a new multivariate encryption scheme. In: Lange, T., Steinwandt, R. (eds.) PQCrypto 2018. LNCS, vol. 10786, pp. 396–416. Springer, Cham (2018). https://doi.org/10.1007/978-3-319-79063-3_19

[KPG99] Kipnis, A., Patarin, J., Goubin, L.: Unbalanced oil and vinegar signature schemes. In: Stern, J. (ed.) EUROCRYPT 1999. LNCS, vol. 1592, pp. 206–222. Springer, Heidelberg (1999). https://doi.org/10.1007/3-540-48910-X_15

[MI88] Matsumoto, T., Imai, H.: Public quadratic polynomial-tuples for efficient signature-verification and message-encryption. In: Barstow, D., et al. (eds.) EUROCRYPT 1988. LNCS, vol. 330, pp. 419–453. Springer, Heidelberg (1988). https://doi.org/10.1007/3-540-45961-8_39

[Pat95] Patarin, J.: Cryptanalysis of the Matsumoto and Imai public key scheme of Eurocrypt'88. In: Coppersmith, D. (ed.) CRYPTO 1995. LNCS, vol. 963, pp. 248–261. Springer, Heidelberg (1995). https://doi.org/10.1007/3-540-44750-4_20

[Pat96] Patarin, J.: Hidden fields equations (HFE) and isomorphisms of polynomials (IP): two new families of asymmetric algorithms. In: Maurer, U. (ed.) EUROCRYPT 1996. LNCS, vol. 1070, pp. 33–48. Springer, Heidelberg (1996). https://doi.org/10.1007/3-540-68339-9_4

[SDP16] Szepieniec, A., Ding, J., Preneel, B.: Extension field cancellation: a new central trapdoor for multivariate quadratic systems. In: Takagi, T. (ed.) PQCrypto 2016. LNCS, vol. 9606, pp. 182–196. Springer, Cham (2016). https://doi.org/10.1007/978-3-319-29360-8_12

[STV25] Smith-Tone, D., Valenzuela, C.: Cryptanalysis of the best HFE-LL' constructions. Cryptology ePrint Archive, Paper 2025/1362 (2025)

[TPD21] Tao, C., Petzoldt, A., Ding, J.: Efficient key recovery for All HFE signature variants. In: Malkin, T., Peikert, C. (eds.) CRYPTO 2021. LNCS, vol. 12825, pp. 70–93. Springer, Cham (2021). https://doi.org/10.1007/978-3-030-84242-0_4

Efficient Identity-Based Inner Product Functional Encryption from RLWE

Anushree Belel[1(✉)] and Junji Shikata[2]

[1] Department of Computer Engineering and Mathematics, Rovira i Virgili University, Tarragona, Spain
anushree.belel@urv.cat

[2] Graduate School of Environment and Information Sciences, Yokohama National University, Yokohama, Japan
shikata-junji-rb@ynu.ac.jp

Abstract. In this work, we propose an efficient lattice-based construction of *identity-based inner product functional encryption* (IBIPFE). Our scheme builds upon the recent IBIPFE construction of Edamura et al. (PKC 2025), incorporating the compact *approximate preimage sampling* technique of Yu et al. (CRYPTO 2023) to achieve short keys and ciphertexts. The construction relies on the RLWE assumption, guarantees a tight security reduction, and achieves *adaptive indistinguishability* (Adp-IND) in the *quantum random oracle model* (QROM). To avoid the noise re-randomization, we adopt the multi-hint extended RLWE assumption (Mera et al., PKC 2022), enabling efficient instantiation over ideal lattices. Moreover, by applying the generic transformation of Edamura et al., we upgrade our Adp-IND secure IBIPFE to also ensure *adaptive simulation-based* (Adp-SIM) security. Due to its compactness and strong security guarantees, our IBIPFE is particularly well-suited for fine-grained access control in cloud-based environments, where efficient enforcement of access policies over encrypted data is essential. Compared to earlier constructions, our design reduces both storage requirements and communication costs, making it a promising candidate for privacy-preserving cloud-based data services.

Keywords: Identity-based Inner Product Functional Encryption · Lattice · Adaptive security · Simulation-based security

1 Introduction

Functional Encryption (FE) [9] generalizes the traditional public key encryption by enabling more fine-grained access control on encrypted data, where users learn only a specific function of the message instead of the original one, thereby avoiding the conventional all-or-nothing approach. This characteristic renders FE

A. Belel—This work was done while Anushree Belel was a postdoc at Yokohama National University, Japan.

well-suited for modern cloud based environments. For instance, a cloud service provider can run analytics or filtering operations on encrypted data - such as extracting statistical summaries or applying machine learning models - without accessing the underlying sensitive data. This facilitates secure outsourcing of computational tasks such as querying encrypted databases, secure search, and privacy preserving data analysis.

A bit more formally, in a functional encryption (FE) scheme, given a ciphertext $\mathsf{Encrypt}(X)$ and a secret key SK_F associated with a function F, the holder of the key learns only the value of $F(X)$ and gains no additional information about X. The notion of security for FE are broadly categorized into indistinguishability-based (IND) and simulation-based (SIM) models. In simple terms, IND-based security guarantees that if an adversary receives multiple secret keys SK_{F_i} satisfying $F_i(X_0^\star) = F_i(X_1^\star)$, it will not be able to differentiate between the ciphertexts $\mathsf{CT}_{X_0^\star}$ and $\mathsf{CT}_{X_1^\star}$. On the other hand, SIM-based security assures that even with access to $\mathsf{CT}_{X^\star}$ and various function keys SK_{F_i}, an adversary learns nothing beyond the output values $F_i(X^\star)$. Therefore, SIM-based security is considered stronger than the IND-based one, and the adaptive version of SIM security is regarded as the best possible guarantee.

While FE schemes for general circuits [14,15] have been developed, they rely on heavy duty cryptographic tools such as indistinguishability obfuscation and multilinear maps. Due to this bottleneck, recent research has focused on crafting FE schemes for restricted functionalities - like inner product, keyword search, and boolean formula. Among these, one of the most foundational research directions is inner-product functional encryption (IPFE), first introduced by Abdalla et al. [1]. In such a scheme, given a ciphertext $\mathsf{CT}_{\mathbf{x}}$ corresponding to an l-dimensional vector $\mathbf{x}$ and a secret key $\mathsf{SK}_{\mathbf{y}}$ associated with another l-dimensional vector $\mathbf{y}$, the decryption reveals the inner product $\langle \mathbf{x}, \mathbf{y} \rangle$. To incorporate access control into IPFE, Abdalla et al. [2] introduced the concept of *identity-based inner product functional encryption* (IBIPFE), enabling only authorized users with specific identities to learn the inner product. In this framework, the encryptor embeds the plaintext vector $\mathbf{x}$ and an identity ID into the ciphertext $\mathsf{CT}_{\mathsf{ID},\mathbf{x}}$. Similarly, each secret key $\mathsf{SK}_{\mathsf{ID},\mathbf{y}}$ is tied to a vector $\mathbf{y}$ and an identity ID. Decryption reveals the inner product $\langle \mathbf{x}, \mathbf{y} \rangle$ only when the identity associated with the secret key matches the one specified in the ciphertext.

IPFE. In 2015, Abdalla et al. [1] formally proposed the notion of IPFE and presented *selective* (Sel-IND) secure instantiations based on the hardness of the *decisional Diffie-Hellman* (DDH) and the *learning with errors* (LWE) problems. Later, in 2016, Agrawal et al. [5] developed IPFE schemes with *adaptive* (Adp-IND) security based on the hardness of the DDH, LWE, and the *decision composite residuosity* (DCR) problems. More precisely, in lattice framework, they introduced two LWE-based protocols enabling inner product computation either over integers or modulo a prime p. The adaptive security of these schemes is based on a variant of LWE, known as *multi-hint extended* LWE (mhe-LWE), which ensures indistinguishability of samples even with partial noise hints, along with the Leftover Hash Lemma (LHL) to simulate uniform ciphertext components. Subsequently, in 2017, Agrawal et al. [27] enhanced the performance of the LWE-based schemes from [5] by relaxing the security requirements. Later, in 2019, Wang et al. [29] optimized the schemes from [5], preserving the same level of Adp-IND security. In 2020, Agrawal et al. [4] further transformed the LWE-based modulo p scheme [5,29], presenting a generic conversion from a $2l$-dimensional IND-secure IPFE to an l-dimensional SIM-secure variant with stateful key generation. Hence, the SIM-secure schemes incur higher overhead compared to their IND-secure counterparts. Concurrently, Lin et al. [21] proposed a more efficient Adp-SIM secure LWE-based IPFE by converting an $(l+1)$-dimensional IND-secure scheme into an l-dimensional SIM-secure one. However, their security model is considered weaker than that of [4], as the simulator is allowed access to more information. In 2022, Mera et al. [25] *first* proposed two IPFE schemes based on Ring-LWE (RLWE), offering security guarantees under both *selective* (Sel-IND) and *adaptive* (Adp-IND) notions. Their selectively secure variant is based on the hardness of mhe-RLWE problem, while the adaptively secure one relies on mhe-RLWE and employs the Ring-LHL in its security proof. As RLWE offers greater efficiency and smaller key sizes than LWE due to its structured algebraic form, the scheme from [25] stands out as the most efficient known construction.

IBE and IBIPFE. Gentry et al. [16] were the *first*, in 2008, to design an *identity-based encryption* (IBE) scheme based on the hardness of the LWE problem in the *random oracle model* (ROM), known as GPV-IBE. Since its introduction, researchers have explored lattice-based IBE from multiple dimensions - such as adapting it to the *quantum* ROM (QROM) [19,32], eliminating the dependency on ROM [3,18,30] and incorporating enhanced security guarantees [3,13,17,26]. In practice, most of these IBE schemes still exhibit inefficiencies, especially when implemented with ideal lattices and within the ROM. To address practical concerns, in 2014, Ducas et al. [11] proposed the *first* efficient lattice-based IBE design (DLP-IBE) built on NTRU lattices. Later, in 2021, Bert et al. [6,7] presented highly optimized implementations using both ideal and module lattice frameworks. Although the DLP-IBE scheme is the most efficient known lattice-based IBE, it suffers from issues like loose security reduction. In 2024, Tomita et al. [28] presented a new IBE scheme that matches DLP-IBE in efficiency and offers tight security based on RLWE assumption in the QROM. Particularly, they instantiated GPV-IBE using Yu et al.'s [31] approximate preimage sampling and leveraged the tight security proof by Katsumata et al. [19]. Instead of using noise re-randomization in the security proof, they relied on the *multi-hint extended* RLWE (mhe-RLWE) assumption introduced by Mera et al. [25].

To prevent the inherent leakage in standard IPFE schemes, Abdalla et al. [2] in 2020 enhanced IPFE by introducing fine-grained access control mechanisms, including *identity-based* IPFE (IBIPFE) and *attribute-based* IPFE (ABIPFE). They provided two instantiations of IBIPFE based on the hardness of the LWE problem. Both designs integrate existing LWE-based IBE with LWE-based IPFE. The first combines the IPFE scheme from Wang et al. [29] with the GPV-IBE, achieving Adp-IND security in the ROM. The second uses the same IPFE alongwith the IBE from Agrawal et al. [3], and achieves Sel-IND security in the standard model. Later, in 2021, Lai et al. [20] presented an LWE-based Adp-IND secure IBIPFE in the standard model. They introduced a novel lattice-based two-stage sampling method generalizing the existing GPV-style sampling. Their security proof leverages the idea of adding smudging noise to secret keys instead of ciphertexts to address the challenges in lifting Adp-IND security to the standard model. However, these existing Adp-IND secure designs suffer from loose security reductions. In 2025, Edamura et al. [12] *first* presented tightly secure LWE-based IBIPFE scheme achieving Adp-IND security in QROM. They refined Abdalla et al.'s ROM-based IBIPFE [2] to enable inner product computation modulo a prime p. The core idea behind this modification is borrowed from Agrawal et al. [5], who similarly adapted their LWE-based IPFE scheme for integers to work modulo a prime p. In security analysis, they used the proof technique of Katsumata et al. [19] to establish the tight adaptive security of their design. They also provided a generic transformation from IND-based $(l + 1)$-dimensional IBIPFE to SIM-based l-dimensional IBIPFE. Leveraging this tranformation, they achieved the *first* LWE-based IBIPFE with Adp-SIM security and a tight reduction in the QROM. Based on the above state-of-the-art, a natural question arises:

Is it possible to design a tightly secure IBIPFE that improves upon the efficiency of existing approaches without compromising on security?

Our Contribution. We respond positively to the question posed above by presenting an IBIPFE scheme that is more efficient than existing works in terms of asymptotic performance. In particular, we propose a design for IBIPFE in the QROM featuring compact master keys, secret keys, and ciphertexts, while maintaining tight reduction and strong security guarantees based on the hardness of the RLWE problem. Our construction draws inspiration from the IBE of Tomita et al. [28] and the IBIPFE of Edamura et al. [12]. In our security analysis, we rely on the mhe-RLWE assumption introduced by Mera et al. [25], rather than employing the noise re-randomization approach proposed by Katsumata et al. [19]. Informally, our work leads to the following findings.

1. Assuming the RLWE and mhe-RLWE assumptions, Our scheme IBIPFE = (Setup, KeyGen, Encrypt, Decrypt) is tightly secure in the sense of Adp-IND notion in the ROM / QROM.
2. Using a standard transformation technique from [12], our Adp-IND secure scheme can be extended to achieve Adp-SIM security under the same assumptions, along with the existence of a pseudo-random function (PRF) family.

Table 1 exhibits that our scheme achieves significantly better storage efficiency with $|\mathsf{MPK}|, |\mathsf{MSK}| = \mathcal{O}(n \log Q)$ — linear in size, whereas previous designs incur quadratic overhead, requiring $\mathcal{O}(n^2 \log^2 Q)$ or worse. Also, $|\mathsf{SK}|$ in our design is $\mathcal{O}(n \log Q + l \log Q)$, which improves upon the $\mathcal{O}(n \log^2 Q + l \log Q)$ size in prior works. The $|\mathsf{CT}|$ is also much smaller at $\mathcal{O}(ln \log Q)$, in contrast to $\mathcal{O}(n \log^2 Q + l \log Q)$ in earlier works. As summarized in Table 2, our scheme

Table 1. Comparison in terms of storage efficiency, communication performance, and security aspects

Sch.	$|\mathsf{MPK}|$	$|\mathsf{MSK}|$	$|\mathsf{SK}|$	$|\mathsf{CT}|$
[2]	$\mathcal{O}(n^2 \log^2 Q)$	$\mathcal{O}(n^2 \log^2 Q)$	$\mathcal{O}(n \log^2 Q + l \log Q)$	$\mathcal{O}(n \log^2 Q + l \log Q)$
	$\mathcal{O}(n^2 \log^2 Q)$	$\mathcal{O}(n^2 \log^2 Q)$	$\mathcal{O}(n \log^2 Q + l \log Q)$	$\mathcal{O}(n \log^2 Q + l \log Q)$
[20]	$\mathcal{O}(n^2 l \log^2 Q)$	$\mathcal{O}(n^2 \log^2 Q)$	$\mathcal{O}(n \log^2 Q + l \log Q)$	$\mathcal{O}(n \log^2 Q + l \log Q)$
[12]	$\mathcal{O}(n^2 \log^2 Q)$	$\mathcal{O}(n^2 \log^2 Q)$	$\mathcal{O}(n \log^2 Q + l \log Q)$	$\mathcal{O}(n \log^2 Q + l \log Q)$
	$\mathcal{O}(n^2 \log^2 Q)$	$\mathcal{O}(n^2 \log^2 Q)$	$\mathcal{O}(n \log^2 Q + l \log Q)$	$\mathcal{O}(n \log^2 Q + l \log Q)$
Our Work	$\mathcal{O}(n \log Q)$	$\mathcal{O}(n \log Q)$	$\mathcal{O}(n \log Q + l \log Q)$	$\mathcal{O}(l n \log Q)$
	$\mathcal{O}(n \log Q)$	$\mathcal{O}(n \log Q)$	$\mathcal{O}(n \log Q + l \log Q)$	$\mathcal{O}(l n \log Q)$

$|\mathsf{MPK}|, |\mathsf{MSK}|, |\mathsf{SK}|, |\mathsf{CT}|$: indicate the respective size of the master public / secret key, secret key, and ciphertext, n, Q : parameters corresponding to underlying hardness assumption, l: vector length

Table 2. Comparison in terms of security aspects

Sch.	Security	Reduction loss	Assumption	Model
[2]	Sel-IND	$\mathcal{O}(1)$	LWE	Standard
	Adp-IND	$\mathcal{O}(Q_{\mathsf{Hash}}^2)$	LWE	ROM
[20]	Adp-IND	$\mathcal{O}(Q_{\mathsf{Hash}})$	LWE	Standard
[12]	Adp-IND	$\mathcal{O}(1)$	LWE	QROM
	Adp-SIM	$\mathcal{O}(1)$	LWE	QROM
Our Work	Adp-IND	$\mathcal{O}(1)$	RLWE	QROM
	Adp-SIM	$\mathcal{O}(1)$	RLWE	QROM

Q_{Hash}: correspond to the number of times the random oracle are queried

satisfies tight Adp-SIM security in the QROM under the RLWE assumption, as it can be derived from the Adp-IND security through an existing transformation.

Technical Overview. We begin by outlining the core idea behind our construction. The scheme we propose is a modification of Edamura et al.'s IBIPFE [12], utilizing the compact preimage sampling developed by Yu et al. [31]. To motivate our approach, we begin with a brief introduction to the IBIPFE from [12].

<u>Edamura et al.'s IBIPFE</u>: The IBIPFE [12] construction defines the master public key as a fat matrix $\mathbf{A} \in \mathbb{Z}_Q^{n \times m}$ as well as a cryptographic hash function Hash : $\mathcal{IDS} \to \mathbb{Z}_Q^{n \times l}$ (where $\mathcal{IDS}$: identity space, l: vector length), and the corresponding master secret key is a trapdoor $\mathbf{T_A}$ associated with $\mathbf{A}$. Using this trapdoor, it is possible to efficiently generate a short preimage $\mathbf{z}_{\mathsf{ID}} \in \mathbb{Z}^{m \times l}$ satisfying $\mathbf{A}\mathbf{z}_{\mathsf{ID}} = \mathbf{u}_{\mathsf{ID}}$ for any target vector $\mathbf{u}_{\mathsf{ID}} = \mathsf{Hash}(\mathsf{ID}) \in \mathbb{Z}_Q^{n \times l}$. For the first key generation associated with identity $\mathsf{ID} \in \{0,1\}^\star$, the corresponding

secret key is given by $\mathsf{SK}_{\mathsf{ID},\mathbf{y}} = (\mathbf{y}, \mathbf{z}_{\mathsf{ID}} \cdot \mathbf{y})$ where $\mathbf{y} \in \mathbb{Z}_p^l$. To encrypt a vector $\mathbf{x} \in \mathbb{Z}_p^l$ for an identity ID, the ciphertext is formed as $\mathbf{c}_1 = \mathbf{A}^\top \mathbf{r} + \mathbf{e}_1 \in \mathbb{Z}_Q^m$ and $\mathbf{c}_2 = \mathbf{u}_{\mathsf{ID}}^\top \mathbf{r} + \mathbf{e}_2 + p^{k-1} \cdot \mathbf{x} \in \mathbb{Z}_Q^l$, where $\mathbf{r}$ is uniformly sampled, $Q = p^k$ and $\mathbf{e}_1$, $\mathbf{e}_2$ are small error vectors.

Edamura et al. [12] established that their IBIPFE scheme achieves tight security in the QROM. We present a high-level overview of the security argument in the ROM. To handle a random oracle query for an identity ID, the reduction procedure selects a uniformly random short matrix $\mathbf{z}_{\mathsf{ID}} \in \mathbb{Z}^{m \times l}$ and returns the oracle output as $\mathbf{u}_{\mathsf{ID}} = \mathbf{A}\mathbf{z}_{\mathsf{ID}}$. Provided that $\mathbf{z}_{\mathsf{ID}}$ has sufficient entropy, the resulting matrix $\mathbf{u}_{\mathsf{ID}}$ is statistically close to uniform over $\mathbb{Z}_Q^{n \times l}$, allowing the reduction to consistently answer both random oracle queries with $\mathbf{u}_{\mathsf{ID}}$ and secret key queries with $(\mathbf{y}, \mathbf{z}_{\mathsf{ID}} \cdot \mathbf{y})$ for any $\mathbf{y} \in \mathbb{Z}_p^l$. It is important to observe that the reduction has access to the secret key $\mathbf{z}_{\mathsf{ID}^\star}$ corresponding to the challenge identity $\mathsf{ID}^\star$. Leveraging this, it can generate a simulated challenge ciphertext by computing $\mathbf{c}_1^\star = \mathbf{A}^\top \mathbf{r} + \mathbf{e}_1 \in \mathbb{Z}_Q^m$, $\mathsf{ct}^\star = \mathbf{z}_{\mathsf{ID}^\star}^\top \mathbf{c}_1^\star + p^{k-1} \cdot \mathbf{x}^\star \in \mathbb{Z}_Q^l$. A crucial observation is that this simulation no longer requires the standard LWE instance $(\mathbf{u}_{\mathsf{ID}^\star}, \mathbf{u}_{\mathsf{ID}^\star}^\top \mathbf{r} + \mathbf{e}_1)$. Instead, the proof uses a noise re-randomization technique proposed by Katsumata et al. [19] to accurately simulate the distribution of the ciphertext, particularly the noise component $\mathbf{e}_1$.

Lightweight Scheme Based on Approximate Preimage Sampling of Yu et al.: The IBIPFE scheme by Edamura et al. [12] relies on the preimage sampling technique, which inherently results in inefficiency. The issue arises since the matrix $\mathbf{A}$ needs to have a width of $m = \mathcal{O}(n \log Q)$ to support preimage sampling. To enhance practical efficiency, Chen et al. [10] proposed a relaxed variant of preimage sampling, referred to as approximate preimage sampling. In this relaxed model, rather than finding an exact solution $\mathbf{z}_{\mathsf{ID}}$ satisfying $\mathbf{A}\mathbf{z}_{\mathsf{ID}} = \mathbf{u}_{\mathsf{ID}} \in \mathbb{Z}_Q^n$, one samples an approximate solution $\mathbf{z}'_{\mathsf{ID}}$ such that $\mathbf{A}\mathbf{z}'_{\mathsf{ID}} = \mathbf{u}_{\mathsf{ID}} + \widetilde{\mathbf{z}}$ with $\widetilde{\mathbf{z}} \in \mathbb{Z}_Q^n$ being a short error vector. Yu et al. [31] recently devised a compact approximate preimage sampling method by employing an almost square matrix inplace of short and fat matrix.

During KeyGen, we repeatedly apply the approximate trapdoor technique discussed earlier to reduce the overhead associated with preimage sampling. In particular, we sample approximate solution $\mathbf{z}'_{\mathsf{ID},i}$ satisfying $\mathbf{A}\mathbf{z}'_{\mathsf{ID},i} = \mathbf{u}_{\mathsf{ID},i} + \widetilde{\mathbf{z}}_i$ where $\mathbf{u}_{\mathsf{ID},i} = \mathsf{Hash}(\mathsf{ID}\|i)$ for $i \in [l]$. To encrypt a vector $\mathbf{x}^\star = (x_1^\star, x_2^\star, \ldots, x_l^\star) \in \mathbb{Z}_p^l$ under identity $\mathsf{ID}^\star$, a short random vector $\mathbf{r} \in \mathbb{Z}_Q^n$ is used inplace of a uniform one to keep the decryption error term $\widetilde{\mathbf{z}}_i^\top \mathbf{r}$ small.

$$\mathsf{ct}_i^\star = \mathbf{z}_{\mathsf{ID}^\star,i}^\top \mathbf{c}_1^\star + p^{k-1} x_i^\star = \mathbf{z}_{\mathsf{ID}^\star,i}^\top (\mathbf{A}^\top \mathbf{r} + \mathbf{e}_1) + p^{k-1} x_i^\star$$
$$= (\mathbf{A}\mathbf{z}_{\mathsf{ID}^\star,i})^\top \mathbf{r} + \mathbf{z}_{\mathsf{ID}^\star,i}^\top \mathbf{e}_1 + p^{k-1} x_i^\star = \mathbf{u}_{\mathsf{ID}^\star,i}^\top \mathbf{r} + \widetilde{\mathbf{z}}_i^\top \mathbf{r} + \mathbf{z}_{\mathsf{ID}^\star,i}^\top \mathbf{e}_1 + p^{k-1} x_i^\star$$

Edamura et al. [12] established the tight security of their scheme by employing the noise re-randomization technique introduced by Katsumata et al. [19]. In contrast, when we attempt to simulate the ciphertext component $\mathsf{ct}_i^\star$, we encounter an additional challenge: the term $\mathbf{z}_{\mathsf{ID}^\star,i}$, which arises in our setting, is an approximate preimage — not exact. As a result, evaluating $\mathbf{A}\mathbf{z}_{\mathsf{ID}^\star,i}$ introduces an extra error term $\widetilde{\mathbf{z}}_i^\top \mathbf{r}$. This deviation cannot be adequately addressed using the standard noise re-randomization method. Consequently, an alternative simulation strategy is required to complete the proof.

Our Approach: Leveraging the *Multi-Hint Extended RLWE* Assumption:
To resolve the above challenge, we replace the noise re-randomization technique with the multi-hint extended RLWE assumption (mhe-RLWE) by Mera et al. [25], which ensures RLWE remains hard even when partial hints about the secret $\mathbf{r}$ and the error $\mathbf{e}_1$ is revealed. This enables the exact simulation of the ciphertext component $\mathsf{ct}_i^\star$ using an approximate preimage vector $\mathbf{z}_{\mathsf{ID}^\star,i}$ alongwith suitable hints, eliminating the need for re-randomization. Furthermore, our proof technique aligns well with the structure of ideal lattices and naturally extends to the QROM setting as in [19].

High-Level Overview: When we instantiate our scheme using the approximate preimage sampling of Yu et al. [31], the ciphertext takes the following form:

$$\mathbf{c}_1^\star = \mathbf{a}'r + \mathbf{e}' \mod Q, \quad \mathsf{ct}_i^\star = u_{\mathsf{ID}^\star,i}r + f_i + p^{k-1}x_i^\star, i \in [l]$$

where $\mathbf{a}' \in \mathcal{R}_Q^2, r \xleftarrow{\$} \mathcal{D}_{\mathbb{Z},\delta Q}, \mathbf{e}' \xleftarrow{\$} \mathcal{D}_{\mathbb{Z}^2,\delta Q}, u_{\mathsf{ID}^\star,i} = \mathsf{Hash}(\mathsf{ID}^\star||i) \in \mathcal{R}_Q, f_i \xleftarrow{\$} \mathcal{D}_{\mathbb{Z},\zeta Q}$. We carefully sample the ciphertext $\mathsf{ct}_i^\star$ as described above, so that it can be written as

$$\mathsf{ct}_i^\star = (\mathbf{c}_1^\star - \mathbf{e}')^\top \mathbf{z}_{\mathsf{ID}^\star,i} + f_i + r(\widetilde{z}_i^\star + z_{0,i}^\star) + p^{k-1}x_i^\star$$
$$= (c_1^\star)^\top \mathbf{z}_{\mathsf{ID}^\star,i} + \left((\widetilde{z}_i^\star + z_{0,i}^\star)r + g_i\right) - \left(\mathbf{e}'^\top \mathbf{z}_{\mathsf{ID}^\star,i} + h_i\right) + p^{k-1}x_i^\star$$

by substituting f_i with $(g_i - h_i)$.

We use the mhe-RLWE assumption as follows : Given $(\mathbf{a}', \mathbf{c}_1^\star, \mathsf{hint})$ (note that hints are leaked through $\{\mathsf{ct}_i^\star\}_{i\in[l]}$), where $\mathbf{c}_1^\star = \mathbf{a}'r + \mathbf{e}' \mod Q$ or $\mathbf{c}_1^\star \xleftarrow{\$} \mathcal{R}_Q^2$, $\mathsf{hint} = \{\widetilde{z}_i^\star + z_{0,i}^\star, r(\widetilde{z}_i^\star + z_{0,i}^\star) + g_i, \mathbf{z}_{\mathsf{ID}^\star,i}, \mathbf{e}'^\top \mathbf{z}_{\mathsf{ID}^\star,i} + h_i\}_{i\in[l]}$, the adversary cannot distinguish between these two distributions. We perform a non-trivial and technically involved analysis to demonstrate that, despite the presence of structured leakage in the challenge tuple, the distribution of the ciphertext remains pseudorandom under the mhe-RLWE assumption.

2   Preliminaries

Notation. Let κ stand for the security parameter. By $y \xleftarrow{\$} \mathcal{Y}$ we indicate that y is sampled uniformly from probability distribution $\mathcal{Y}$. Let $[a]$ represent the set $\{1, 2, \ldots, a\}$ for any $a \in \mathbb{N}$ and $\mathcal{U}(\mathcal{S})$ denote the uniform distribution over $\mathcal{S}$ for a finite set $\mathcal{S}$. A map $f : \mathbb{N} \to \mathbb{R}$ is said to be a *negligible* function of n if it is $\mathcal{O}(n^{-d})$ for each $d > 0$ and it is represented as $\mathsf{negl}(n)$. For two random variables Y and Z over $\mathcal{S}$, the statistical distance $\Delta(Y, Z)$ between Y and Z is interpreted as $\Delta(Y, Z) = \sum_{s \in \mathcal{S}} \mathsf{Pr}[Y = s] - \mathsf{Pr}[Z = s]$. Two distributions Y and Z are statistically close when $\Delta(Y, Z) = \mathsf{negl}(\kappa)$. Let $\perp$ stand for null value and $\langle \cdot, \cdot \rangle$ represent inner product of two vectors. A vector is represented using bold lowercase letter and written in column format. For a vector $\mathbf{y} \in \mathbb{R}^n$, we denote its standard Euclidean norm by $\|\mathbf{y}\|$ and its supremum norm by $\|\mathbf{y}\|_\infty$. A matrix is denoted by bold uppercase letter. Given a matrix $\mathbf{Y} \in \mathbb{R}^{n \times m}$, let $\mathbf{Y}^\top$ represent its transpose and $\widetilde{\mathbf{Y}}$ denote the result of applying Gram-Schmidt orthogonalization to $\mathbf{Y}$. The largest singular value of $\mathbf{Y}$ is defined as $s_1(\mathbf{Y}) = \max_{\|\mathbf{z}\| \neq 0} \|\mathbf{Y}\mathbf{z}\| / \|\mathbf{z}\|$. The notation $\boldsymbol{\Sigma} \succ 0$ signifies that the symmetric matrix $\boldsymbol{\Sigma} \in \mathbb{R}^{m \times m}$ is positive definite, meaning that $\mathbf{y}^\top \boldsymbol{\Sigma} \mathbf{y} > 0$ holds for all non-zero vectors $\mathbf{y} \in \mathbb{R}^m \setminus \{\mathbf{0}\}$. We say $\boldsymbol{\Sigma}_1 \succ \boldsymbol{\Sigma}_2$ if the difference $\boldsymbol{\Sigma}_1 - \boldsymbol{\Sigma}_2 \succ 0$. Given a scalar s, the notation $\boldsymbol{\Sigma} \succ s$ means that the matrix $\boldsymbol{\Sigma} - s \cdot \mathbf{I}_m \succ 0$ where $\mathbf{I}_m$ denotes the identity matrix of size $m \times m$. The symbol $\sqrt{\boldsymbol{\Sigma}}$ is used to represent any matrix square root of $\boldsymbol{\Sigma}$, provided the context makes it clear.

2.1   Lattices

Let $\mathbf{C} = [\mathbf{c}_1|\mathbf{c}_2|\cdots|\mathbf{c}_n] \in \mathbb{R}^{m \times n}$ be a matrix whose columns $\mathbf{c}_i$ are linearly independent. The lattice spanned by $\mathbf{C}$ is given by $\Lambda(\mathbf{C}) = \{\mathbf{C}\mathbf{y} \mid \mathbf{y} \in \mathbb{Z}^n\}$, lattice of dimension n with basis $\mathbf{C}$. The dual lattice of Λ is defined as $\Lambda^\star = \{\mathbf{v} \in \mathsf{span}(\Lambda) \mid \langle \mathbf{u}, \mathbf{v} \rangle \in \mathbb{Z}, \forall \mathbf{u} \in \Lambda\}$.

In the context of lattice-based cryptography, q-ary lattices (q: positive integer) play a central role. Given a matrix $\mathbf{D} \in \mathbb{Z}_q^{n \times m}$, the q-ary lattice of $\mathbf{D}$ is defined as $\Lambda_q^\perp(\mathbf{D}) = \{\mathbf{u} \in \mathbb{Z}^m : \mathbf{D}\mathbf{u} = \mathbf{0} \mod q\}$. This lattice is of dimension m and satisfies $(q \cdot \mathbb{Z})^m \subseteq \Lambda_q^\perp \subseteq \mathbb{Z}^m$. For any $\mathbf{x} \in \mathbb{Z}_q^n$, the associated lattice coset is given by $\Lambda_q^{\mathbf{x}}(\mathbf{D}) = \{\mathbf{w} \in \mathbb{Z}^m : \mathbf{D}\mathbf{w} = \mathbf{x} \mod q\}$.

2.2   Gaussians

The Gaussian function ρ maps $\mathbb{R}^m$ to the interval $(0, 1]$ and is given by $\rho(\mathbf{y}) = \exp(-\pi \cdot \langle \mathbf{y}, \mathbf{y} \rangle)$. If we apply a linear transformation via an invertible matrix $\mathbf{C}$, the resulting function is $\rho_{\mathbf{C}}(\mathbf{y}) = \rho(\mathbf{C}^{-1}\mathbf{y}) = \exp(-\pi \mathbf{y}^\top \boldsymbol{\Sigma}^{-1} \mathbf{y})$, where $\boldsymbol{\Sigma} = \mathbf{C}\mathbf{C}^\top$. As the function $\rho_{\mathbf{C}}$ is fully characterized by the matrix $\boldsymbol{\Sigma}$, we also denote it by $\rho_{\sqrt{\boldsymbol{\Sigma}}}$. Given a lattice Λ and a shift vector $\mathbf{a} \in \mathsf{span}(\Lambda)$, the discrete Gaussian distribution $\mathcal{D}_{\Lambda+\mathbf{a}, \sqrt{\boldsymbol{\Sigma}}}$ is supported on $\Lambda + \mathbf{a}$ and is defined by: for $\mathbf{y} \in \Lambda + \mathbf{a}$,

$$\mathcal{D}_{\Lambda+\mathbf{a}, \sqrt{\boldsymbol{\Sigma}}}(\mathbf{y}) = \frac{\rho_{\sqrt{\boldsymbol{\Sigma}}}(\mathbf{y})}{\rho_{\sqrt{\boldsymbol{\Sigma}}}(\Lambda + \mathbf{a})}$$

In the special case where $\Lambda = \mathbb{Z}^n$ and $\mathbf{a} = \mathbf{0}$, we denote the distribution by $\mathcal{D}_{\sqrt{\boldsymbol{\Sigma}}}$. If $\boldsymbol{\Sigma} = \sigma^2 I_n$ for positive $\sigma \in \mathbb{R}$, we denote it as $\mathcal{D}_\sigma$.

Given a lattice Λ and $\epsilon \in (0, 1)$, the smoothing parameter is given by $\eta_\epsilon(\Lambda) = \min\{ s > 0 \mid \rho(s \cdot \Lambda^\star) \leq 1 + \epsilon \}$. The inequality $\sqrt{\boldsymbol{\Sigma}} \geq \eta_\epsilon(\Lambda)$ holds whenever $\rho_{\sqrt{\boldsymbol{\Sigma}^{-1}}}(\Lambda^\star) \leq (1 + \epsilon)$.

The following are useful properties of the discrete Gaussian distribution.

Lemma 1 ([22]). *For any* $\kappa > 0$, $\Pr_{\mathbf{y} \xleftarrow{\$} \mathcal{D}_\sigma}[\|\mathbf{y}\|_\infty > \sqrt{\kappa}\sigma] \leq 2\exp(-\kappa/2)$

Lemma 2 ([24]). *Let q, k, m be natural numbers with $q \geq 2$ and $k < m \leq$ poly(λ). Define the matrix $\mathbf{A} = (\mathbf{I}_k \| \bar{\mathbf{A}})$ where $\bar{\mathbf{A}} \xleftarrow{\$} \mathcal{R}_q^{k \times (m-k)}$. The distribution of $\mathbf{A}\mathbf{z}$ mod q (with $\mathbf{z} \xleftarrow{\$} \mathcal{D}_\sigma^m$ and $\sigma > 2nq^{k/m+2/(nm)}$) is $2^{-\Omega(n)}$-close to uniform over $\mathcal{R}_q^k$ with probability $(1 - 2^{-\Omega(n)})$ over the choice of $\bar{\mathbf{A}}$.*

Lemma 3 ([25]). *Consider the sublattice $L(\mathbf{C}) \subseteq \mathbb{Z}^n$ of dimension k where the columns of $(n \times k)$-matrix $\mathbf{C}$ form a basis. Given a positive definite matrix $\mathbf{\Sigma} \in \mathbb{R}^{n \times n}$, and setting $\mathbf{\Sigma}' = \sigma'^2 \mathbf{C}\mathbf{C}^\top$, sampling $\mathbf{e}$ from the discrete Gaussian distribution $\mathcal{D}_{\sqrt{(\mathbf{\Sigma}+\mathbf{\Sigma}')}}$ is indistinguishable from first sampling $\mathbf{e}_1 \leftarrow \mathcal{D}_{\sqrt{\mathbf{\Sigma}}}$, independently sampling $\mathbf{e}_2 \leftarrow \mathcal{D}_{\sqrt{\mathbf{\Sigma}'}}$, and setting $\mathbf{e} = \mathbf{e}_1 + \mathbf{e}_2$, provided that all eigen values of the matrix $\mathbf{\Gamma}_{\mathbf{\Sigma},\mathbf{\Sigma}'} = \sqrt{\sigma'^2 \mathbf{I}_k - \sigma'^4 \mathbf{C}^\top (\mathbf{\Sigma} + \mathbf{\Sigma}')^{-1} \mathbf{C}}$ are greater than the smoothing parameter $\eta_\epsilon(\mathbb{Z}^k)$.*

2.3   Ideal Lattices

We define $\mathcal{R}$ as the polynomial ring $\mathcal{R} = \mathbb{Z}[X]/\phi$ where ϕ is irreducible polynomial. To simplify both the theory and practical implementation, we set $\phi = x^n + 1$, with n being a power of 2. We use the conventional notation $\mathcal{R}_q$ to indicate the quotient ring $\mathbb{Z}_q[x]/\phi$. We choose q so that the polynomial ϕ, which has degree n, decomposes into n distinct linear factors over $\mathbb{Z}_q$, i.e., $\phi = \prod_{j \in [n]} \phi_j$, with each ϕ_j being linear. As a result, by applying the Chinese Remainder Theorem (CRT), the ring $\mathcal{R}_q$ decomposes into n ideals and can be expressed as $\mathcal{R}_q \cong \prod_{j \in [n]} \mathcal{R}_q/\phi_j$. As each quotient $\mathcal{R}_q/\phi_j$ is isomorphic to $\mathbb{Z}_q$, this yields an isomorphism between $\mathcal{R}_q$ and $\mathbb{Z}_q^n$.

For a polynomial $p = \sum_{i=0}^{n-1} p_i X^i \in \mathcal{R}$, we define $\mathsf{coeff}(p) = (p_0, p_1, \ldots, p_{n-1})^\top \in \mathbb{Z}^n$ as its coefficient vector. We also associate to p the anti-circulant matrix

$$\mathsf{Rot}(p) = \begin{pmatrix} p_0 & -p_{n-1} & \cdots & -p_1 \\ p_1 & p_0 & \cdots & -p_2 \\ \vdots & \vdots & \ddots & \vdots \\ p_{n-1} & p_{n-2} & \cdots & p_0 \end{pmatrix}$$

which consists of the coefficient vectors of $p, pX, \ldots, pX^{n-1}$ arranged as columns, i.e., $\mathsf{Rot}(p) = [\mathsf{coeff}(p), \mathsf{coeff}(pX), \ldots, \mathsf{coeff}(pX^{n-1})] \in \mathbb{Z}^{n \times n}$. For a polynomial $p \in \mathcal{R}$, let $\bar{p} = p(X^{-1})$. Then, $\bar{p} = p_0 - \sum_{i=1}^{n-1} p_{n-i} X^i$. More generally, for $k \in \mathbb{Z}_n^\star$, we define $\sigma_k(p) = p(X^k)$. For any two polynomials $p, \hat{p} \in \mathcal{R}$, we have

- $\mathsf{Rot}(p) + \mathsf{Rot}(\hat{p}) = \mathsf{Rot}(p + \hat{p})$, $\mathsf{Rot}(p) \cdot \mathsf{Rot}(\hat{p}) = \mathsf{Rot}(p\hat{p})$, $\mathsf{Rot}(\bar{p}) = \mathsf{Rot}(p)^\top$

2.4 Identity-Based Inner Product Functional Encryption

An identity-based inner product functional encryption (IBIPFE) scheme [2] with identity space $\mathcal{IDS}$ for evaluating inner products modulo a prime p is defined by a tuple of Algorithms (Setup, Encrypt, KeyGen, Decrypt), as described below:

- Setup$(1^\kappa, 1^l) \to$ (MPK, MSK): On input the security parameter κ and vector length l, a trusted authority executes this algorithm to generate master public/secret key pair (MPK, MSK). It makes MPK public while keeps MSK secret to itself.
- Encrypt(MPK, ID, $\mathbf{x}$) $\to$ CT$_{\text{ID},\mathbf{x}}$: Given the master public key MPK, an identity ID $\in \mathcal{IDS}$, and a vector $\mathbf{x} \in \mathbb{Z}_p^l$, an encryptor computes the corresponding ciphertext CT$_{\text{ID},\mathbf{x}}$.
- KeyGen(MSK, ID, $\mathbf{y}$) $\to$ SK$_{\text{ID},\mathbf{y}}$: On input the master secret key MSK, an identity ID $\in \mathcal{IDS}$, and a vector $\mathbf{y} \in \mathbb{Z}_p^l$, the trusted authority executes this algorithm to generate the corresponding secret key SK$_{\text{ID},\mathbf{y}}$.
- Decrypt(MPK, CT$_{\text{ID},\mathbf{x}}$, SK$_{\text{ID}',\mathbf{y}}$) $\to \langle \mathbf{x}, \mathbf{y} \rangle / \perp$: Taking input the master public key MPK, ciphertext CT$_{\text{ID},\mathbf{x}}$, and secret key SK$_{\text{ID}',\mathbf{y}}$, a decryptor outputs the inner product $\langle \mathbf{x}, \mathbf{y} \rangle \in \mathbb{Z}_p$ or $\perp$.

Correctness. For all (MPK, MSK) $\leftarrow$ Setup$(1^\kappa, 1^l)$, CT$_{\text{ID},\mathbf{x}} \leftarrow$ Encrypt (MPK, ID, $\mathbf{x}$), SK$_{\text{ID}',\mathbf{y}} \leftarrow$ KeyGen(MSK, ID$'$, $\mathbf{y}$), output $\perp$ if ID$' \neq$ ID. Else, Decrypt (MPK, CT$_{\text{ID},\mathbf{x}}$, SK$_{\text{ID}',\mathbf{y}}$) $= \langle \mathbf{x}, \mathbf{y} \rangle$ mod p holds with high probability.

$$
\begin{array}{c}
\text{(MPK, MSK)} \leftarrow \text{Setup } (1^\kappa, l) \\
((\text{ID}_0^\star, \mathbf{x}_0^\star), (\text{ID}_1^\star, \mathbf{x}_1^\star)) \leftarrow \mathcal{A}^{\text{KeyGen}(\text{MSK},\cdot,\cdot)}(\text{MPK}) \\
\text{CT}^\star \leftarrow \text{Encrypt}(\text{MPK}, \text{ID}_\alpha^\star, \mathbf{x}_\alpha^\star); \alpha \xleftarrow{\$} \{0, 1\} \\
\widehat{\alpha} \leftarrow \mathcal{A}^{\text{KeyGen}(\text{MSK},\cdot,\cdot)}(\text{MPK}, \text{CT}^\star)
\end{array}
$$

Fig. 1. The Adp-IND security game

Security [2]. We describe the *adaptive-indistinguishability* (Adp-IND) security notion of IBIPFE that is modelled as a game played between a challenger $\mathcal{C}$ and an adversary $\mathcal{A}$ as summarized in *Fig.* 1.

- **Setup.** At the outset of the game, the challenger $\mathcal{C}$ produces (MPK, MSK) $\leftarrow$ Setup$(1^\kappa, l)$ and issues MPK to the adversary $\mathcal{A}$.
- **Query Phase 1.** During the experiment, $\mathcal{A}$ asks secret key queries corresponding to vector $\mathbf{y}$ and identity ID polynomially many times and $\mathcal{C}$ responds with the corresponding secret key SK$_{\text{ID},\mathbf{y}} \leftarrow$ KeyGen(MSK, ID, $\mathbf{y}$).
- **Challenge.** In this phase, $\mathcal{A}$ submits a pair (ID$_0^\star$, $\mathbf{x}_0^\star$), (ID$_1^\star$, $\mathbf{x}_1^\star$) to $\mathcal{C}$. Then, $\mathcal{C}$ randomly picks a bit $\alpha \xleftarrow{\$} \{0, 1\}$, computes CT$^\star =$ CT$_{\text{ID}_\alpha^\star, \mathbf{x}_\alpha^\star} \leftarrow$ Encrypt (MPK, ID$_\alpha^\star$, $\mathbf{x}_\alpha^\star$) and provides the challenge ciphertext CT* to $\mathcal{A}$.

- **Query Phase 2.** Similar to Query Phase 1.
  It is required that any secret key query $(\mathsf{ID}, \mathbf{y})$ should satisfy $\mathsf{Func}\,(\mathsf{ID}_0^\star, \mathbf{x}_0^\star, \mathsf{ID}, \mathbf{y}) = \mathsf{Func}\,(\mathsf{ID}_1^\star, \mathbf{x}_1^\star, \mathsf{ID}, \mathbf{y})$ throughout the game where

$$
\mathsf{Func}(\mathsf{ID}^\star, \mathbf{x}^\star, \mathsf{ID}, \mathbf{y}) = \begin{cases} \langle \mathbf{x}^\star, \mathbf{y} \rangle & \text{if } \mathsf{ID}^\star = \mathsf{ID} \\ \bot & \text{Otherwise} \end{cases}
$$

- **Guess.** Finally, $\mathcal{C}$ outputs a guess $\widehat{\alpha}$ for α and wins the game if $\widehat{\alpha} = \alpha$.

The advantage of $\mathcal{A}$ in this game is defined as $\mathsf{Adv}_{\mathsf{IBIPFE},\mathcal{A}}^{\mathsf{Adp-IND}}(\kappa) = |\Pr[\widehat{\alpha} = 1 | \alpha = 0] - \Pr[\widehat{\alpha} = 1 | \alpha = 1]|$. We say that an IBIPFE scheme satisfies *adaptive indistinguishability* (Adp-IND) security notion if $\mathsf{Adv}_{\mathsf{IBIPFE},\mathcal{A}}^{\mathsf{Adp-IND}}(\kappa) \leq \mathsf{negl}(\kappa)$.

Remark 1. If $\mathsf{ID}_0^\star \neq \mathsf{ID}_1^\star$, the adversary is not allowed to obtain secret keys for $(\mathsf{ID}_0^\star, \mathbf{y})$ or $(\mathsf{ID}_1^\star, \mathbf{y})$ for any $\mathbf{y}$. This restriction is necessary because access to these specific secret keys would allow the adversary to easily distinguish between encryption of $\mathbf{x}_0$ under $\mathsf{ID}_0^\star$ and encryption of $\mathbf{x}_1$ under $\mathsf{ID}_1^\star$. If $\mathsf{ID}_0^\star = \mathsf{ID}_1^\star = \mathsf{ID}^\star$, let $\mathbf{y}_{\mathsf{ID}^\star,1}, \ldots, \mathbf{y}_{\mathsf{ID}^\star,Q_{\mathsf{ID}^\star}} \in \mathbb{Z}_p^l$ represent all vectors for which $\mathcal{A}$ obtains secret keys for $(\mathsf{ID}^\star, \mathbf{y}_{\mathsf{ID}^\star,i})$ during the security game. To ensure $\mathsf{Func}(\mathsf{ID}_0^\star, \mathbf{x}_0^\star, \mathsf{ID}, \mathbf{y}) = \mathsf{Func}(\mathsf{ID}_1^\star, \mathbf{x}_1^\star, \mathsf{ID}, \mathbf{y})$, the set $\{\mathbf{y}_{\mathsf{ID}^\star,1}, \ldots, \mathbf{y}_{\mathsf{ID}^\star,Q_{\mathsf{ID}^\star}}\}$ contains at most $(l-1)$ linearly independent vectors. Unlike in identity-based encryption (IBE), if $\mathsf{ID}_0^\star = \mathsf{ID}_1^\star = \mathsf{ID}$ holds, the challenge ciphertext does not hide information of $\mathsf{ID}^\star$ since $\mathcal{A}$ is allowed to obtain secret keys associated with $\mathsf{ID}^\star$.

We now describe the *adaptive-simulation* (Adp-SIM) security notion [4] of IBIPFE by two security games outlined in *Fig.* 2, where the left side represents the real security game $\mathsf{SIM}_{\mathsf{Real}}$ and the right side corresponds to the ideal security game $\mathsf{SIM}_{\mathsf{Ideal}}$.

- In $\mathsf{SIM}_{\mathsf{Real}}$, the adversary $\mathcal{A}$ interacts with a challenger $\mathcal{C}$, who executes the real algorithms (Setup, Encrypt, KeyGen) of an IBIPFE protocol. The adversary $\mathcal{A}$ can access an oracle KeyGen on input $(\mathsf{ID}, \mathbf{y})$ and receive as output the corresponding secret key $\mathsf{SK}_{\mathsf{ID},\mathbf{y}} \leftarrow \mathsf{KeyGen}(\mathsf{MSK}, \mathsf{ID}, \mathbf{y})$.
- In $\mathsf{SIM}_{\mathsf{Ideal}}$, the adversary $\mathcal{A}$ interacts with a simulator $\mathcal{S}$, who executes the PPT algorithms $(\mathsf{Setup}^\star, \mathsf{KeyGen}_0^\star, \mathsf{Encrypt}, \mathsf{KeyGen}_1^\star)$. The adversary has access to two oracles: $\mathsf{KeyGen}_0^\star, \mathsf{KeyGen}_1^\star$.
  - $\mathsf{KeyGen}_0^\star$: Given input $(\mathsf{ID}, \mathbf{y})$, this oracle outputs a secret key $\mathsf{SK}_{\mathsf{ID},\mathbf{y}}^\star \leftarrow \mathsf{KeyGen}_0^\star (\mathsf{MPK}^\star, \mathsf{MSK}^\star, \mathsf{ID}, \mathbf{y}, \mathsf{st})$.
  - $\mathsf{KeyGen}_1^\star$: Given input $(\mathsf{ID}, \mathbf{y}, \mathsf{Func}(\mathsf{ID}^\star, \mathbf{x}^\star, \mathsf{ID}, \mathbf{y}))$, this oracle outputs a secret key $\mathsf{SK}_{\mathsf{ID},\mathbf{y}}^\star \leftarrow \mathsf{KeyGen}_1^\star(\mathsf{MPK}^\star, \mathsf{MSK}^\star, \mathsf{ID}, \mathbf{y}, \mathsf{Func}(\mathsf{ID}^\star, \mathbf{x}^\star, \mathsf{ID}, \mathbf{y}), \mathsf{st})$. Moreover, let $\mathcal{V} = (\mathbf{y}_{\mathsf{ID}^\star,i}, v_i = \langle \mathbf{x}^\star, \mathbf{y}_{\mathsf{ID}^\star,i} \rangle)_{i \in [Q_{\mathsf{ID}^\star}]}$, where $\{\mathbf{y}_{\mathsf{ID}^\star,i}\}_{i \in [Q_{\mathsf{ID}^\star}]}$ represent all vectors on which $\mathcal{A}$ has queried the oracle $\mathsf{KeyGen}_0^\star$ associated with $\mathsf{ID}^\star$.

  Note that, we may refer to $\mathcal{A}$'s oracle access to $\mathsf{KeyGen}_0^\star$ as pre-challenge key queries and those to $\mathsf{KeyGen}_1^\star$ as post-challenge key queries.

We say that an IBIPFE scheme satisfies *adaptive-simulation* (Adp-SIM) security notion if $\mathsf{Adv}_{\mathsf{IBIPFE},\mathcal{A}}^{\mathsf{Adp-SIM}}(\kappa) = |\Pr[\widehat{\alpha} = 1 | \mathsf{SIM}_{\mathsf{Real}}] - \Pr[\widehat{\alpha} = 1 | \mathsf{SIM}_{\mathsf{Ideal}}]| \leq \mathsf{negl}(\kappa)$.

$$
\boxed{
\begin{array}{ll}
\textbf{SIM}_{\textbf{Real}} & \textbf{SIM}_{\textbf{Ideal}} \\[4pt]
(\mathsf{MPK},\mathsf{MSK}) \leftarrow \mathsf{Setup}(1^\kappa,1^l) & (\mathsf{MPK}^\star,\mathsf{MSK}^\star) \leftarrow \mathsf{Setup}^\star(1^\kappa,1^l) \\[4pt]
(\mathsf{ID}^\star,\mathbf{x}^\star) \leftarrow \mathcal{A}^{\mathsf{KeyGen}(\mathsf{MSK},\cdot,\cdot)}(\mathsf{MPK}) & (\mathsf{ID}^\star,\mathcal{V}) \leftarrow \mathcal{A}^{\mathsf{KeyGen}_0^\star(\mathsf{MPK}^\star,\mathsf{MSK}^\star,\cdot,\cdot,\mathsf{st})}(\mathsf{MPK}^\star) \\[4pt]
\mathsf{CT}^\star \leftarrow \mathsf{Encrypt}(\mathsf{MPK},\mathsf{ID}^\star,\mathbf{x}^\star) & \mathsf{CT}^\star \leftarrow \mathsf{Encrypt}^\star(\mathsf{MPK}^\star,\mathsf{ID}^\star,\mathcal{V}) \\[4pt]
\widehat{\alpha} \leftarrow \mathcal{A}^{\mathsf{KeyGen}(\mathsf{MSK},\cdot,\cdot)}(\mathsf{MPK},\mathsf{CT}^\star) & \widehat{\alpha} \leftarrow \mathcal{A}^{\mathsf{KeyGen}_1^\star(\mathsf{MPK}^\star,\mathsf{MSK}^\star,\cdot,\cdot,\cdot,\mathsf{st})}(\mathsf{MPK}^\star,\mathsf{CT}^\star)
\end{array}
}
$$

Fig. 2. The Adp-SIM security game

2.5 Approximate Trapdoor Mechanisms in Ideal Lattice Structures

Definition 1 (Approximate Trapdoor). *Let m_1 and Q be positive integers. A string $\mathsf{td}_{\mathbf{a}}$ is said to be (α,β)-approximate trapdoor for a vector $\mathbf{a} \in \mathcal{R}_Q^{m_1}$ if there exists a PPT algorithm $\mathsf{AppSampPre}$ that takes as input $\mathbf{a}, \mathsf{td}_{\mathbf{a}}$, a polynomial $u \in \mathcal{R}_Q$, auxiliary data aux, and outputs $\mathbf{z} \in \mathcal{R}^m$ that satisfy*

- $\|\mathbf{z}\| \le \beta$
- *there exists some $\widetilde{z} \in \mathcal{R}$ with $\|\widetilde{z}\| \le \alpha$ such that $\mathbf{a}^{\mathsf{T}}\mathbf{z} = u + \widetilde{z} \mod Q$*

Yu et al. introduced the compact approximate trapdoors in the context of ideal lattices. The result is summarized in the following lemma.

Lemma 4 ([31]). *Let $m_1, p, q, Q = pq$ be positive integers with $m_1 > n$. Suppose that $\mathbf{a} \in \mathcal{R}_Q^{m_1}$ and $\mathsf{td}_{\mathbf{a}} \in \mathcal{R}^{m_1}$ satisfy $\mathbf{a}^{\mathsf{T}}\mathsf{td}_{\mathbf{a}} = p \mod Q$. Let $\mathsf{aux} = \sigma$ be a positive real number such that $\sigma^2 \ge \eta_\epsilon(\mathbb{Z}^n)^2(q^2+1)(s_1(\mathsf{Rot}(\mathsf{td}_{\mathbf{a}}))^2+1)$. Then, there exists a PPT algorithm satisfying*

1. *$\mathsf{td}_{\mathbf{a}}$ is $(\sqrt{n(p^2-1)/12}, \sqrt{nm_1}\sigma)$-approximate trapdoor.*
2. *The two distributions below are statistically indistinguishable.*

$$
\left\{
\begin{array}{l}
(\mathbf{a},\mathbf{z},u,\widetilde{z}) \\
u \xleftarrow{\$} \mathcal{R}_Q, \quad \mathbf{z} \leftarrow \mathsf{AppSampPre}(\mathbf{a},\mathsf{td}_{\mathbf{a}},u,\sigma) \\
\widetilde{z} := u - \mathbf{a}^{\mathsf{T}}\mathbf{z} \mod Q
\end{array}
\right\}
\quad
\left\{
\begin{array}{l}
(\mathbf{a},\mathbf{z},u,\widetilde{z}) \\
\mathbf{z} \xleftarrow{\$} \mathcal{D}_\sigma^{m_1}, \quad \widetilde{z} \xleftarrow{\$} \mathcal{R}_p, \\
u := \mathbf{a}^{\mathsf{T}}\mathbf{z} + \widetilde{z} \mod Q
\end{array}
\right\}
$$

2.6 Hardness Assumption

Definition 2 (Ring Learning with Errors (RLWE) [23]). *Let $\mathcal{R} = \mathbb{Z}[x]/(x^n+1)$ with n a power of 2. Assume that $m_1, Q > 0$ are positive integers and χ is a distribution defined over $\mathcal{R}$. Let $\mathcal{R}_Q = \mathcal{R}/(Q \cdot \mathcal{R})$. Given $r \xleftarrow{\$} \chi$,*

$\mathbf{a}, \mathbf{b} \xleftarrow{\$} \mathcal{R}_Q^{m_1}$, $\mathbf{e} \xleftarrow{\$} \chi^{m_1}$, *the* RLWE-*problem decides if* $\mathbf{b} = \mathbf{a}r + \mathbf{e}$ *or an arbitrary element of* $\mathcal{R}_Q^{m_1}$. *The advantage of distinguisher* $\mathcal{A}$ *for* RLWE$_{Q,m_1,\chi}$ *is defined as* $\mathsf{Adv}_{\mathcal{A}}^{\mathsf{RLWE}_{Q,m_1,\chi}}(\kappa) = |\Pr[\mathcal{A}(\mathbf{a}, \mathbf{a}r + \mathbf{e} \mod Q) \to 1] - \Pr[\mathcal{A}(\mathbf{a}, \mathbf{b}) \to 1]|.$

Definition 3 (Multi-Hint Extended RLWE (mhe-RLWE) [25]). *Let* $\mathcal{R} = \mathbb{Z}[x]/(x^n + 1)$ *with* n *a power of* 2. *Assume that* n, m_1, l *are positive integers and* χ *is a distribution defined over* $\mathcal{R}$. *Let* $\mathcal{R}_Q = \mathcal{R}/(Q \cdot \mathcal{R})$. *For each* $i \in [l]$, *let* $\nu_i \in \mathcal{R}$ *and* $\boldsymbol{\xi}_i \in \mathcal{R}^{m_1}$ *be random members chosen in advance by the adversary such that* $\|\nu_i\|_\infty$ *and* $\|\boldsymbol{\xi}_i\|_\infty$ *are both bounded by a constant* $C > 0$. *Given* $\mathbf{a}, \mathbf{b} \xleftarrow{\$} \mathcal{R}_Q^{m_1}$, $r, g_1, g_2, \ldots, g_l \xleftarrow{\$} \chi$, $\mathbf{e} \xleftarrow{\$} \chi^{m_1}$, $h_1, h_2, \ldots, h_l \xleftarrow{\$} \chi$, $\mathsf{hint} = \{\nu_i, r\nu_i + g_i, \boldsymbol{\xi}_i, \mathbf{e}^\top \boldsymbol{\xi}_i + h_i\}_{i \in [l]}$, *the* mhe-RLWE *problem decides if* $\mathbf{b} = \mathbf{a}r + \mathbf{e}$ *or an arbitrary element of* $\mathcal{R}_Q^{m_1}$. *The advantage of distinguisher* $\mathcal{A}$ *for* $\mathsf{mhe} - \mathsf{RLWE}_{Q,m_1,\chi,l,C}$ *is defined as* $\mathsf{Adv}_{\mathcal{A}}^{\mathsf{mhe}-\mathsf{RLWE}_{Q,m_1,\chi,l,C}}(\kappa) = |\Pr[\mathcal{A}(\mathbf{a}, \mathbf{a}r + \mathbf{e} \mod Q, \mathsf{hint}) \to 1] - \Pr[\mathcal{A}(\mathbf{a}, \mathbf{b}, \mathsf{hint}) \to 1]|.$

Lemma 5 ([25]). *Given any Quantum Polynomial Time* (QPT) *adversary* $\mathcal{A}$, *natural numbers* n, Q, m, l, *and positive reals* $\sigma, \delta, C, \epsilon$ *satisfying*

$$\sigma \sqrt{1 - \frac{(\sqrt{2 + l}\sigma n C)^2}{\delta^2}} > \eta_\epsilon(\mathbb{Z}^{n+nl}),$$

there exists a QPT *algorithm* $\mathcal{A}_1$ *such that*

$$\mathsf{Adv}_{\mathcal{A}}^{\mathsf{mhe}-\mathsf{RLWE}_{Q,m,\mathcal{D}_\delta,l,C}}(\kappa) \leq \mathsf{Adv}_{\mathcal{A}_1}^{\mathsf{RLWE}_{Q,m,\mathcal{D}_\sigma}}(\kappa) + \mathsf{negl}(\kappa).$$

3   Our IBIPFE Scheme

Parameters. Our IBIPFE scheme uses the following parameters.

- Let p be prime, n, $Q = p^m = p \cdot p^{m-1} = p \cdot q$ be positive integers.
- Let δ be a positive real number used as discrete Gaussian parameter.
- Let α, β be positive real numbers.
- The identity space is given by $\{0,1\}^{l_{\mathsf{ID}}}$, where $l_{\mathsf{ID}} = l_{\mathsf{ID}}(\lambda)$ represents the identity length.
- A hash function $\mathsf{Hash} : \{0,1\}^{l_{\mathsf{ID}} + \lfloor \log l \rfloor} \to \mathcal{R}_Q$ is used and modelled as a (quantum) random oracle in the security analysis.
- A function $\mathsf{Expand} : \{0,1\}^{256} \to \mathcal{R}_Q$ is used to map a seed to an element in $\mathcal{R}_Q$.

- Let w_1 and w_{-1} be positive integers satisfying $w_1 + w_{-1} \leq n$. We define

$$\mathcal{T}(n, \omega_1, \omega_{-1}) := \left\{ z \in \mathcal{R} \mid z \text{ has } \begin{array}{l} \omega_1 \text{ coefficients equal to } 1; \\ \omega_{-1} \text{ coefficients equal to } -1; \\ n - \omega_1 - \omega_{-1} \text{ coefficients equal to } 0. \end{array} \right\}$$

- Let α be the parameter that determines the quality of the trapdoor such that $\sqrt{s_1(\mathsf{Rot}(f\bar{f} + g\bar{g}))} \leq \alpha \|(f, g)\| = \alpha\sqrt{2(w_1 + w_{-1})}$, where $f, g \in \mathcal{T}(n, w_1, w_{-1})$ is the trapdoor.
- Let σ be the width for approximate preimages, given by

$$\sigma \geq r\alpha\sqrt{2(w_1 + w_{-1})(1 + q^2)} = r\alpha\sqrt{2(w_1 + w_{-1})(1 + p^{2(m-1)})}$$

where $r = \eta_\epsilon(\mathbb{Z}^n)$.
- Define β as the acceptance bound of $\|(z_0 + \tilde{z}, \gamma z_1, \gamma z_2)\|$ where (z_0, z_1, z_2) is the approximate preimage, $\tilde{z}$ is the approximate error, $\gamma = \frac{\sqrt{\sigma^2 + (p^2 - 1)/12}}{\sigma}$ such that $\|z_0 + \tilde{z}\| \approx \gamma\|z_1\| \approx \gamma\|z_2\|$.

Algorithm 1 : Setup $(1^\kappa, 1^l) \rightarrow (\mathsf{MPK}, \mathsf{MSK})$

1: $\mathsf{Hash} : \{0,1\}^{l_{\mathsf{ID}} + \lfloor \log l \rfloor} \rightarrow \mathcal{R}_Q$, $\mathsf{seed}_a \xleftarrow{\$} \{0,1\}^{256}$, $\bar{a} = \mathsf{Expand}(\mathsf{seed}_a)$
2: $f_1, \ldots, f_5, g_1, \ldots, g_5 \xleftarrow{\$} \mathcal{U}(\mathcal{T}(n, \omega_1, \omega_{-1}))$
3: **for** $i = 1$ to 5 **do**
4:      **for** $j = 1$ to 5 **do**
5:          Find $k \in \mathbb{Z}_n^*$ minimizing $s_1(\mathsf{Rot}(f_i\bar{f_i} + \sigma_k(g_j)\sigma_k(\bar{g_j})))$
6:          $(f, g) = (f_i, \sigma_k(g_j))$
7:          **if** $\sqrt{s_1(\mathsf{Rot}(\bar{f}f + \bar{g}g))} \leq \alpha\sqrt{2(\omega_1 + \omega_{-1})}$ **then**
8:             $b = p - (\bar{a}f + g) \mod Q$
9:             **Return** $(\mathsf{MSK}, \mathsf{MPK}) = ((f, g), (\mathsf{seed}_a, b, \mathsf{Hash}))$
10:          **end if**
11:      **end for**
12: **end for**
13: Restart

Algorithm 2 : KeyGen $(\mathsf{MSK}, \mathsf{ID}, \mathbf{y}, \mathsf{st}) \rightarrow \mathsf{SK}_{\mathsf{ID},\mathbf{y}}$

1: $\bar{a} = \mathsf{Expand}(\mathsf{seed}_a), u_{\mathsf{ID},i} = \mathsf{Hash}(\mathsf{ID}\|i), i \in [l]$

2: $\mathbf{a} = (1, \bar{a}, b)^\top, \mathsf{td}_{\mathbf{a}} = (-g, -f, 1)^\top$

3: **for** $i = 1$ to l **do**

4:      $\mathbf{z}_i = (z_{0,i}, z_{1,i}, z_{2,i})^\top \xleftarrow{\$} \mathsf{AppSampPre}(\mathbf{a}, \mathsf{td}_{\mathbf{a}}, u_{\mathsf{ID},i}, \sigma)$

5:      $\widetilde{z}_i = u_{\mathsf{ID},i} - \mathbf{a}^\top \mathbf{z}_i \mod Q$

6:      **if** $\|(z_{0,i} + \widetilde{z}_i, \gamma z_{1,i}, \gamma z_{2,i})\| > \beta$ **then**

7:         Restart

8:      **end if**

9: **end for**

10: $\mathbf{z}_{\mathsf{ID},i} = (z_{1,i}, z_{2,i})^\top \in \mathcal{R}^2, \mathbf{Z}_{\mathsf{ID}} = \begin{pmatrix} z_{1,1} & z_{1,2} & \cdots & z_{1,\ell} \\ z_{2,1} & z_{2,2} & \cdots & z_{2,\ell} \end{pmatrix}$

11: $\mathsf{SK}_{\mathsf{ID},\mathbf{y}} = (\overline{\mathbf{y}} = \mathbf{y}, \mathbf{k}_{\mathsf{ID},\mathbf{y}} = \mathbf{Z}_{\mathsf{ID}} \cdot \overline{\mathbf{y}})$

12: If this is the first key generation for $\mathsf{ID} \in \mathcal{IDS}$, Set $\overline{\mathbf{y}}_{\mathsf{ID},1} = \mathbf{y}, \mathbf{k}_{\mathsf{ID},1} = \mathbf{k}_{\mathsf{ID},\mathbf{y}}$ and store $(\mathsf{ID}, \mathbf{Z}_{\mathsf{ID}}, (\overline{\mathbf{y}}_{\mathsf{ID},1}, \mathbf{k}_{\mathsf{ID},1})) \in \mathsf{st}$

13: Otherwise, retrieve $\left(\mathsf{ID}, \mathbf{Z}_{\mathsf{ID}}, (\overline{\mathbf{y}}_{\mathsf{ID},i}, \mathbf{k}_{\mathsf{ID},i})_{i \in [Q_{\mathsf{ID}}]}\right) \in \mathsf{st}$, where $\overline{\mathbf{y}}_{\mathsf{ID},1}, \ldots, \overline{\mathbf{y}}_{\mathsf{ID},Q_{\mathsf{ID}}}$ represent all linearly independent vectors corresponding to the secret keys that have been generated for ID so far

14: **if** $\exists d_1, d_2, \ldots, d_{Q_{\mathsf{ID}}} \in \mathbb{Z}_p$ such that $\mathbf{y} = \sum\limits_{j \in [Q_{\mathsf{ID}}]} d_j \cdot \overline{\mathbf{y}}_{\mathsf{ID},j} \mod p$ **then**

15:      Output $\mathsf{SK}_{\mathsf{ID},\mathbf{y}} = (\overline{\mathbf{y}} = \sum\limits_{j \in [Q_{\mathsf{ID}}]} d_j \cdot \overline{\mathbf{y}}_{\mathsf{ID},j}, \mathbf{k}_{\mathsf{ID},\mathbf{y}} = \sum\limits_{j \in [Q_{\mathsf{ID}}]} d_j \cdot \mathbf{k}_{\mathsf{ID},j})$

16: **else if** $\nexists d_1, d_2, \ldots, d_{Q_{\mathsf{ID}}} \in \mathbb{Z}_p$ such that $\mathbf{y} = \sum\limits_{j \in [Q_{\mathsf{ID}}]} d_j \cdot \overline{\mathbf{y}}_{\mathsf{ID},j} \mod p$ **then**

17:      Output $\mathsf{SK}_{\mathsf{ID},\mathbf{y}} = (\overline{\mathbf{y}} = \mathbf{y}, \mathbf{k}_{\mathsf{ID},\mathbf{y}} = \mathbf{Z}_{\mathsf{ID}} \cdot \overline{\mathbf{y}})$

18:      Set $\overline{\mathbf{y}}_{\mathsf{ID},Q_{\mathsf{ID}}+1} = \mathbf{y}, \mathbf{k}_{\mathsf{ID},Q_{\mathsf{ID}}+1} = \mathbf{k}_{\mathsf{ID},\mathbf{y}}$ and update $\left(\mathsf{ID}, (\overline{\mathbf{y}}_{\mathsf{ID},j}, \mathbf{k}_{\mathsf{ID},j})_{j \in [Q_{\mathsf{ID}}]}\right) \in \mathsf{st}$ by $\left(\mathsf{ID}, (\overline{\mathbf{y}}_{\mathsf{ID},j}, \mathbf{k}_{\mathsf{ID},j})_{j \in [Q_{\mathsf{ID}}+1]}\right) \in \mathsf{st}$

19: **end if**

Algorithm 3 : Encrypt $(\mathsf{MPK}, \mathsf{ID}, \mathbf{x}) \rightarrow \mathsf{CT}_{\mathsf{ID},\mathbf{x}}$

1: $\bar{a} = \mathsf{Expand}(\mathsf{seed}_a), u_{\mathsf{ID},i} = \mathsf{Hash}(\mathsf{ID}\|i) : i \in [l], \mathbf{a}' = (\bar{a}, b)^\top \in \mathcal{R}_Q^2$

2: $r \xleftarrow{\$} \mathcal{D}_{\mathbb{Z},\delta Q}, \mathbf{e}' \xleftarrow{\$} \mathcal{D}_{\mathbb{Z}^2,\delta Q}, \{f_i\}_{i \in [l]} \xleftarrow{\$} \mathcal{D}_{\mathbb{Z},\zeta Q}$

3: $\widehat{\mathbf{c}} = \mathbf{a}'r + \mathbf{e}' \mod Q = (\widehat{c}_1, \widehat{c}_2)^\top \in \mathcal{R}_Q^2$

4: $\mathsf{ct}_i = u_{\mathsf{ID},i}r + f_i + p^{m-1}x_i\mathbf{1}_{\mathcal{R}} : i \in [l]$ where $\mathbf{x} = (x_1, x_2, \ldots, x_l), \mathbf{1}_{\mathcal{R}} :$ identity of $\mathcal{R}_Q$ / polynomial of degree $(n-1)$ with all coefficients equal $1 \in \mathbb{Z}_Q$.

5: **Return** $\mathsf{CT}_{\mathsf{ID},\mathbf{x}} = (\widehat{\mathbf{c}}, \{\mathsf{ct}_i\}_{i \in [l]})$

Algorithm 4 : Decrypt $(\mathsf{MPK}, \mathsf{CT}_{\mathsf{ID},\mathbf{x}}, \mathsf{SK}_{\mathsf{ID},\mathbf{y}}) \to \langle \mathbf{x}, \mathbf{y} \rangle$

1: Parse $\mathsf{CT}_{\mathsf{ID},\mathbf{x}} = (\widehat{\mathbf{c}}, \{\mathsf{ct}_i\}_{i \in [l]})$, $\mathsf{SK}_{\mathsf{ID},\mathbf{y}} = (\overline{\mathbf{y}}, \mathbf{k}_{\mathsf{ID},\mathbf{y}})$ where $\overline{\mathbf{y}} = (\overline{y}_1, \overline{y}_2, \ldots, \overline{y}_l)$

2: Compute $\mu = \sum\limits_{i \in [l]} \overline{y}_i \mathsf{ct}_i - \widehat{\mathbf{c}}^{\top} \mathbf{k}_{\mathsf{ID},\mathbf{y}} \mod \mathcal{R}_Q$

3: Output $\arg\min_{z \in \mathbb{Z}_p} \| \mu - p^{m-1} z \mathbf{1}_{\mathcal{R}} \|$

Note. To ensure the generation of a high-quality trapdoor, the Setup algorithm samples multiple candidate polynomials and selects the particular pair satisfying required norm bound. This approach is inspired by the technique of Yu et al. [31] where sampling five f_i and five g_j produces 25 candidate pairs, probabilistically amplifying the chance that at least one pair meets the condition. If the probability that a single pair meets the required criterion is p_1, then the probability that none of the 25 pairs are suitable is $(1 - p_1)^{25}$, so the probability that at least one is suitable becomes $1 - (1 - p_1)^{25}$. For a reasonable p_1, this approaches 1, ensuring that the algorithm finds a suitable pair with overwhelming probability in a small number of iterations. For correctness and security, only the selected pair participates in subsequent protocols and proofs.

If the number is < 5, the number of candidate pairs becomes small (for example, $3^2 = 9$ or $4^2 = 16$), which reduces the probability of sampling at least one high-quality pair significantly. This leads to more frequent restarts of the setup algorithm due to failure in finding a suitable trapdoor. If the number is > 5, although the probability of finding a suitable pair improves by sampling more candidates, the computational cost increases quadratically. This leads to slower setup times. Thus, the number 5 is a practical middle ground found experimentally and theoretically (by Yu et al. [31]) to give a very high success probability $1 - (1 - p_1)^{25}$ without incurring excessive computational overhead.

Correctness. Note that if $\overline{\mathbf{y}} = \mathbf{y} = (y_1, y_2, \ldots, y_l)^{\top}$ and $\mathbf{k}_{\mathsf{ID},\mathbf{y}} = \mathbf{Z}_{\mathsf{ID}} \cdot \mathbf{y}$

$$\sum_{i \in [l]} \overline{y}_i \mathsf{ct}_i = \sum_{i \in [l]} y_i u_{\mathsf{ID},i} r + \sum_{i \in [l]} y_i f_i + p^{m-1} \langle \mathbf{x}, \mathbf{y} \rangle \mathbf{1}_{\mathcal{R}}$$

$$\widehat{\mathbf{c}}^{\top} \mathbf{k}_{\mathsf{ID},\mathbf{y}} = (\widehat{c}_1, \widehat{c}_2) \begin{pmatrix} \sum\limits_{i \in [l]} z_{1,i} y_i \\ \sum\limits_{i \in [l]} z_{2,i} y_i \end{pmatrix} = \sum_{i \in [l]} y_i \widehat{\mathbf{c}}^{\top} \mathbf{z}_{\mathsf{ID},i}$$

$$= \sum_{i \in [l]} y_i u_{\mathsf{ID},i} r - \sum_{i \in [l]} y_i r (\widetilde{z}_i + z_{0,i}) + \sum_{i \in [l]} y_i \mathbf{e}'^{\top} \mathbf{z}_{\mathsf{ID},i}$$

$$\mu = \sum_{i\in[l]} y_i \mathsf{ct}_i - \widehat{\mathbf{c}}^\top \mathbf{k}_{\mathsf{ID},\mathbf{y}}$$

$$= p^{m-1}\langle \mathbf{x},\mathbf{y}\rangle \mathbf{1}_{\mathcal{R}} + \sum_{i\in[l]} y_i f_i + \sum_{i\in[l]} y_i r(\widetilde{z}_i + z_{0,i}) - \sum_{i\in[l]} y_i {\mathbf{e}'}^\top \mathbf{z}_{\mathsf{ID},i}$$

$$= p^{m-1}\langle \mathbf{x},\mathbf{y}\rangle \mathbf{1}_{\mathcal{R}} + \mathsf{noise}$$

We have $\|r\|_\infty \le \sqrt{\kappa}\delta$, $\|\mathbf{e}'\|_\infty \le \sqrt{\kappa}\delta$, $\|(\widetilde{z}_i + z_{0,i}, z_{1,i}, z_{2,i})\| \le \beta$ for each $i \in [l]$. Therefore,

$$\Big\|\sum_{i\in[l]} \overline{y}_i\{r(\widetilde{z}_i + z_{0,i}) - {\mathbf{e}'}^\top \mathbf{z}_{\mathsf{ID},i}\}\Big\|_\infty = \Big\|\sum_{i\in[l]} \overline{y}_i (r, -{\mathbf{e}'}^\top)(\widetilde{z}_i + z_{0,i}, z_{1,i}, z_{2,i})\Big\|_\infty$$

$$\le lp\sqrt{\kappa}\delta\beta$$

$$\|\mathsf{noise}\|_\infty = \Big\|\sum_{i\in[l]} \overline{y}_i f_i + \sum_{i\in[l]} \overline{y}_i\{r(\widetilde{z}_i + z_{0,i}) - {\mathbf{e}'}^\top \mathbf{z}_{\mathsf{ID},i}\}\Big\|_\infty$$

$$\le \Big\|\sum_{i\in[l]} \overline{y}_i f_i\Big\|_\infty + \Big\|\sum_{i\in[l]} \overline{y}_i\{r(\widetilde{z}_i + z_{0,i}) - {\mathbf{e}'}^\top \mathbf{z}_{\mathsf{ID},i}\}\Big\|_\infty \le lp\sqrt{\kappa}\zeta + lp\sqrt{\kappa}\delta\beta$$

For correctness, we require that $\|\mathsf{noise}\|_\infty < p^{m-1}/2$, i.e., $lp\sqrt{\kappa}(\delta\beta+\zeta) < p^{m-1}/2$

If $\overline{\mathbf{y}} = \sum_{j\in[Q_{\mathsf{ID}}]} d_j \cdot \overline{\mathbf{y}}_{\mathsf{ID},j}$ and

$$\mathbf{k}_{\mathsf{ID},\mathbf{y}} = \sum_{j\in[Q_{\mathsf{ID}}]} d_j \cdot \mathbf{k}_{\mathsf{ID},j} = \sum_{j\in[Q_{\mathsf{ID}}]} d_j \cdot (\mathbf{Z}_{\mathsf{ID}} \cdot \overline{\mathbf{y}}_{\mathsf{ID},j}) = \mathbf{Z}_{\mathsf{ID}} \cdot \Big(\sum_{j\in[Q_{\mathsf{ID}}]} d_j \cdot \overline{\mathbf{y}}_{\mathsf{ID},j}\Big) = \mathbf{Z}_{\mathsf{ID}} \cdot \overline{\mathbf{y}}$$

$$\sum_{i\in[l]} \overline{y}_i \mathsf{ct}_i = \sum_{i\in[l]} \overline{y}_i u_{\mathsf{ID},i} r + \sum_{i\in[l]} \overline{y}_i f_i + p^{m-1}\langle \mathbf{x},\overline{\mathbf{y}}\rangle \mathbf{1}_{\mathcal{R}}$$

$$\widehat{\mathbf{c}}^\top \mathbf{k}_{\mathsf{ID},\mathbf{y}} = \sum_{i\in[l]} \overline{y}_i u_{\mathsf{ID},i} r - \sum_{i\in[l]} \overline{y}_i r(\widetilde{z}_i + z_{0,i}) + \sum_{i\in[l]} \overline{y}_i {\mathbf{e}'}^\top \mathbf{z}_{\mathsf{ID},i}$$

$$\mu = \sum_{i\in[l]} \overline{y}_i \mathsf{ct}_i - \widehat{\mathbf{c}}^\top \mathbf{k}_{\mathsf{ID},\mathbf{y}}$$

$$= p^{m-1}\langle \mathbf{x},\overline{\mathbf{y}}\rangle \mathbf{1}_{\mathcal{R}} + \sum_{i\in[l]} \overline{y}_i f_i + \sum_{i\in[l]} \overline{y}_i r(\widetilde{z}_i + z_{0,i}) - \sum_{i\in[l]} \overline{y}_i {\mathbf{e}'}^\top \mathbf{z}_{\mathsf{ID},i}$$

$$= p^{m-1}\langle \mathbf{x},\overline{\mathbf{y}}\rangle \mathbf{1}_{\mathcal{R}} + \mathsf{noise}$$

Similar to the previous case, Correctness in this case also requires that $\|\mathsf{noise}\|_\infty < p^{m-1}/2$, i.e., $lp\sqrt{\kappa}(\delta\beta + \zeta) < p^{m-1}/2$.

Parameter Conditions. Our scheme parameters must satisfy the following (Table 3):

Table 3. Parameter conditions for our scheme

$p^{m-1}/2 > lp\sqrt{\kappa}(\delta\beta + \zeta)$ for correctness
$\sigma \geq 2nQ^{2/3+2/3n}$ for Lemma 2
$\sigma \geq r\alpha\sqrt{2(w_1 + w_{-1})(1 + p^{2(m-1)})}$ for Lemma 4
$\beta \geq \sqrt{n(p^2 - 1)/12} + \sqrt{3}n\sigma$ for correctness of KeyGen
$\sigma > (1 - (\sqrt{2 + l}\sigma nC)^2/\delta^2)^{-1/2}\eta_\epsilon(\mathbb{Z}^{n+nl})$ for Lemma 5

4   Security Proof in ROM

Theorem 1. *Our identity-based inner product functional encryption* (IBIPFE) *presented in Sect. 3 satisfies* Adp-IND *security in the random oracle model (*ROM*) assuming the hardness of the* $\mathsf{RLWE}_{Q,1,\chi}$ *and the* $\mathsf{mhe} - \mathsf{RLWE}_{Q,2,\mathcal{D}_\delta,l,C}$ *problem.*

Proof. We prove the security by considering a sequence of games. For each game, we denote by Event_i the event that $\mathcal{A}$ outputs a correct guess in Game_i.

Game$_0$: This is the formal security model corresponding to Adp-IND security. At the outset of the game, the challenger $\mathcal{C}$ produces $(\mathsf{MPK}, \mathsf{MSK}) \leftarrow \mathsf{Setup}(1^\kappa)$ and provides MPK to the adversary $\mathcal{A}$. Throughout the game, $\mathcal{A}$ is allowed to ask many random oracle and secret key queries, as well as a single challenge query. The challenger responds to each query as follows:

- Upon receiving a random oracle query to Hash on $\mathsf{ID}||i$ for $i \in [l]$, generates polynomials $u_{\mathsf{ID},i} \xleftarrow{\$} \mathcal{R}_Q$, records the tuple $(\{\mathsf{ID}||i\}_{i\in[l]}, \{u_{\mathsf{ID},i}\}_{i\in[l]}, \perp)$, and replies to $\mathcal{A}$ with $\{u_{\mathsf{ID},i}\}_{i\in[l]}$.
- Upon receiving first secret key query for ID, computes $\mathsf{SK}_{\mathsf{ID},\mathbf{y}} = (\overline{\mathbf{y}} = \mathbf{y}, \mathbf{k}_{\mathsf{ID},\mathbf{y}} = \mathbf{Z}_{\mathsf{ID}} \cdot \overline{\mathbf{y}})$ where $\mathbf{Z}_{\mathsf{ID}} = \begin{pmatrix} z_{1,1} & z_{1,2} & \cdots & z_{1,l} \\ z_{2,1} & z_{2,2} & \cdots & z_{2,l} \end{pmatrix}$, $\mathbf{z}_i = (z_{0,i}, z_{1,i}, z_{2,i})^\top \xleftarrow{\$}$ $\mathsf{AppSampPre}(\mathbf{a}, \mathsf{td}_\mathbf{a}, u_{\mathsf{ID},i}, \sigma)$ for $i \in [l]$. If this is not first query for ID, retrieves $(\mathsf{ID}, \mathbf{Z}_{\mathsf{ID}}, (\overline{\mathbf{y}}_{\mathsf{ID},i}, \mathbf{k}_{\mathsf{ID},i})_{i\in[Q_{\mathsf{ID}}]}) \in \mathsf{st}$, where $\{\overline{\mathbf{y}}_{\mathsf{ID},i}\}_{i\in[Q_{\mathsf{ID}}]}$ are linearly independent vectors corresponding to secret keys that have been generated for ID. If $\mathbf{y} = \sum_{j\in[Q_{\mathsf{ID}}]} d_j \cdot \overline{\mathbf{y}}_{\mathsf{ID},j} \bmod p$, returns $\mathsf{SK}_{\mathsf{ID},\mathbf{y}} = (\overline{\mathbf{y}} = \sum_{j\in[Q_{\mathsf{ID}}]} d_j \cdot \overline{\mathbf{y}}_{\mathsf{ID},j}, \mathbf{k}_{\mathsf{ID},\mathbf{y}} = \sum_{j\in[Q_{\mathsf{ID}}]} d_j \cdot \mathbf{k}_{\mathsf{ID},j})$, else returns $(\overline{\mathbf{y}} = \mathbf{y}, \mathbf{k}_{\mathsf{ID},\mathbf{y}} = \mathbf{Z}_{\mathsf{ID}} \cdot \overline{\mathbf{y}})$
- Upon receiving challenge query on $((\mathsf{ID}_0^\star, \mathbf{x}_0^\star), (\mathsf{ID}_1^\star, \mathbf{x}_1^\star))$, the challenger chooses $\alpha \xleftarrow{\$} \{0,1\}$ and returns $\mathsf{CT}^\star = (\widehat{\mathbf{c}}^\star, \{\mathsf{ct}_i^\star\}_{i\in[l]}) \leftarrow \mathsf{Encrypt}(\mathsf{MPK}, \mathsf{ID}_\alpha^\star, \mathbf{x}_\alpha^\star)$.

When the game concludes, $\mathcal{A}$ reveals a guess $\widehat{\alpha}$ for α.

Thus, we have $\left|\Pr[\mathsf{Event}_i] - \frac{1}{2}\right| = \left|\Pr[\widehat{\alpha} = \alpha] - \frac{1}{2}\right| = \mathsf{Adv}_{\mathsf{IBIPFE},\mathcal{A}}^{\mathsf{Adp-IND}}(\kappa)$.

Game$_1$: In this game, the process of responding to random oracle queries to Hash is changed. At the start of the game, for each $j \in [Q_{\mathsf{Hash}}]$, the challenger $\mathcal{C}$ samples $\mathbf{z}_{j,i} \xleftarrow{\$} \mathcal{D}_\sigma^3$, $\widetilde{z}_{j,i} \xleftarrow{\$} \mathcal{R}_p$, computes $u_{j,i} = \mathbf{a}^\top \mathbf{z}_{j,i} + \widetilde{z}_{j,i} \bmod Q$ for $i \in [l]$ and stores the tuples $\{(u_{j,i}, \mathbf{z}_{j,i}, \widetilde{z}_{j,i})_{i\in[l]}\}_{j\in[Q_{\mathsf{Hash}}]}$ in local storage. When random oracle query is made to Hash on ID_j, $\mathcal{C}$ finds the corresponding tuple $(u_{j,i}, \mathbf{z}_{j,i}, \widetilde{z}_{j,i})_{i\in[l]}$ in local storage, returns $\{u_{j,i}\}_{i\in[l]}$, and records $(\{\mathsf{ID}_j\|i\}_{i\in[l]},$ $\{u_{j,i}\}_{i\in[l]}, (\mathbf{z}_{j,i}, \widetilde{z}_{j,i})_{i\in[l]})$ for later use.
We can invoke Lemma 1 to guarantee that all $u_{j,i}$ are statistically close to uniform as in Game$_0$. Consequently, the statistical distance between the view of $\mathcal{A}$ in Game$_0$ and Game$_1$ is $Q_{\mathsf{Hash}} \cdot \mathsf{negl}(\kappa) = \mathsf{negl}(\kappa)$. Thus, we have $|\Pr[\mathsf{Event}_0] - \Pr[\mathsf{Event}_1]| = \mathsf{negl}(\kappa)$.

Game$_2$: This game is identical to Game$_1$ except for how secret key queries are handled. Specifically, the challenger $\mathcal{C}$ no longer uses $\mathsf{td}_\mathbf{a}$ to answer $\mathcal{A}$'s first secret key query associated with ID. The challenger $\mathcal{C}$ retrieves the tuple $(\{\mathsf{ID}\|i\}_{i\in[l]}, \{u_{\mathsf{ID},i}\}_{i\in[l]}, (\mathbf{z}_i, \widetilde{z}_i)_{i\in[l]})$ from local storage where $\mathbf{z}_i = (z_{0,i}, z_{1,i}, z_{2,i})^\top$ for $i \in [l]$, sets $\mathbf{Z}_{\mathsf{ID}} = \begin{pmatrix} z_{1,1} & z_{1,2} & \cdots & z_{1,l} \\ z_{2,1} & z_{2,2} & \cdots & z_{2,l} \end{pmatrix}$, $\mathsf{SK}_{\mathsf{ID},\mathbf{y}} = (\mathbf{y}, \ \mathsf{k}_{\mathsf{ID},\mathbf{y}} = \mathbf{Z}_{\mathsf{ID}} \cdot \mathbf{y})$.

We can invoke Lemma 1 to guarantee that distribution of $\{\mathbf{z}_i\}_{i\in[l]}$ generated by AppSampPre is statistically close to $\mathcal{D}_\sigma^3$ conditioned on $\{u_{\mathsf{ID},i}\}_{i\in[l]}$. Given that $\mathcal{A}$ can access up to Q_{ID} secret key queries, we have
$$|\Pr[\mathsf{Event}_1] - \Pr[\mathsf{Event}_2]| = Q_{\mathsf{ID}} \cdot \mathsf{negl}(\kappa) = \mathsf{negl}(\kappa).$$

Game$_3$: This game is identical to Game$_2$ except for how $\mathcal{C}$ computes the MPK. The challenger $\mathcal{C}$ chooses $b \xleftarrow{\$} \mathcal{R}_Q$ randomly rather than setting $b = p - (\bar{a}f + g) \bmod Q$, and omits generating trapdoor $\mathsf{td}_\mathbf{a}$. Since the trapdoor was not needed to answer $\mathcal{A}$'s secret key queries in Game$_2$, $\mathcal{C}$ can still handle those queries just as before.

We now establish that Game$_2$ and Game$_3$ are computationally indistinguishable assuming the hardness of the $\mathsf{RLWE}_{Q,1,\chi}$ problem. To prove this, we define an RLWE adversary $\mathcal{A}_1$ that leverages the distinguisher $\mathcal{A}$ as follows:

The adversary $\mathcal{A}_1$ receives an RLWE instance $(\bar{a}, b')$ where $b' = \bar{a}f + g$ for $f, g \xleftarrow{\$} \mathcal{U}(\mathcal{T}(n, w_1, w_{-1}))$ or $b' = b_1$ for $b_1 \xleftarrow{\$} \mathcal{R}_Q$. Utilizing this, $\mathcal{A}_1$ sets the MPK $= (\bar{a}, b = p - b')$ and provides MPK to $\mathcal{A}$. In this game, $\mathcal{A}_1$ simulates the challenger in Game$_2$ by answering random oracle queries, secret key queries, and challenge query. After that, $\mathcal{A}$ returns a bit $\widehat{\alpha}$. At last, $\mathcal{A}_1$ outputs 1 if $\widehat{\alpha} = \alpha$ and 0 otherwise.

If $(\bar{a}, b')$ is a valid RLWE instance (specifically, $b' = \bar{a}f + g$), the view of $\mathcal{A}$ is identical to Game$_2$. Otherwise, if $b' = b_1 \xleftarrow{\$} \mathcal{R}_Q$, the view of $\mathcal{A}$ matches Game$_3$. Thus, under the assumption that the $\mathsf{RLWE}_{Q,1,\chi}$ problem is hard, we conclude that $||\Pr[\mathsf{Event}_2] - \Pr[\mathsf{Event}_3]|| = \mathsf{negl}(\kappa)$.

Game$_4$: This game identical to **Game$_3$** except for generation of the challenge ciphertext. To generate $\mathsf{CT}^\star = (\widehat{\mathbf{c}}^\star, \{\mathsf{ct}_i^\star\}_{i\in[l]})$ for identity $\mathsf{ID}_\alpha^\star$ and message $\mathbf{x}_\alpha^\star$, $\mathcal{C}$

extracts unique tuple ($\{\mathsf{ID}^\star_\alpha\|i\}_{i\in[l]}$, $\{u_{\mathsf{ID}^\star_\alpha,i}\}_{i\in[l]}$, ($\mathbf{z}^\star_i = (z^\star_{0,i}, z^\star_{1,i}, z^\star_{2,i})^\top, \widetilde{z}^\star_i)_{i\in[l]}$)
from local storage, assigns $\mathbf{z}_{\mathsf{ID}^\star_\alpha,i} = (z^\star_{1,i}, z^\star_{2,i})^\top$, $\mathbf{Z}_{\mathsf{ID}^\star_\alpha} = \begin{pmatrix} z^\star_{1,1} & z^\star_{1,2} & \cdots & z^\star_{1,l} \\ z^\star_{2,1} & z^\star_{2,2} & \cdots & z^\star_{2,l} \end{pmatrix}$, and
computes

$$\widehat{\mathbf{c}}^\star = \mathbf{a}'r + \mathbf{e}' \mod Q$$
$$\mathsf{ct}^\star_i = (\widehat{\mathbf{c}}^\star - \mathbf{e}')^\top \mathbf{z}_{\mathsf{ID}^\star_\alpha,i} + f_i + r(\widetilde{z}^\star_i + z^\star_{0,i}) + p^{k-1}x^\star_{\alpha,i}$$

where $r \xleftarrow{\$} \mathcal{D}_{\mathbb{Z},\delta Q}$, $\mathbf{e}' \xleftarrow{\$} \mathcal{D}_{\mathbb{Z}^2,\delta Q}$, and $f_i \xleftarrow{\$} \mathcal{D}_{\mathbb{Z},\zeta Q}$ This distribution is correct
because

$$\begin{aligned}
\mathsf{ct}^\star_i &= (\widehat{\mathbf{c}}^\star - \mathbf{e}')^\top \mathbf{z}_{\mathsf{ID}^\star_\alpha,i} + f_i + r(\widetilde{z}^\star_i + z^\star_{0,i}) + p^{k-1}x^\star_{\alpha,i} \\
&= (\mathbf{a}'r + \mathbf{e}' - \mathbf{e}')^\top \mathbf{z}_{\mathsf{ID}^\star_\alpha,i} + f_i + r(\widetilde{z}^\star_i + z^\star_{0,i}) + p^{k-1}x^\star_{\alpha,i} \\
&= r\mathbf{a}'^\top \mathbf{z}_{\mathsf{ID}^\star_\alpha,i} + f_i + r(\widetilde{z}^\star_i + z^\star_{0,i}) + p^{k-1}x^\star_{\alpha,i} \\
&= r(u_{\mathsf{ID}^\star_\alpha,i} - \widetilde{z}^\star_i - z^\star_{0,i}) + f_i + r(\widetilde{z}^\star_i + z^\star_{0,i}) + p^{k-1}x^\star_{\alpha,i} \\
&= ru_{\mathsf{ID}^\star_\alpha,i} + f_i + p^{k-1}x^\star_{\alpha,i}
\end{aligned}$$

Thus, we obtain $\Pr[\mathsf{Event}_3] = \Pr[\mathsf{Event}_4]$.

Game$_5$: This game introduces an additional change to the process of generating
the challenge ciphertext. To generate $\mathsf{CT}^\star = (\widehat{\mathbf{c}}^\star, \{\mathsf{ct}^\star_i\}_{i\in[l]})$, $\mathcal{C}$ chooses $\widehat{\mathbf{c}}^\star \xleftarrow{\$} \mathcal{R}^2_Q$
and computes $\{\mathsf{ct}^\star_i\}_{i\in[l]}$ following the method in Game$_4$.

We now establish that Game$_4$ and Game$_5$ are computationally indistinguishable assuming the hardness of the mhe-RLWE problem. To prove this, we define
an mhe-RLWE adversary $\mathcal{A}_2$ that leverages the distinguisher $\mathcal{A}$ as follows:
The adversary $\mathcal{A}_2$ receives an mhe-RLWE instance $(\mathbf{a}', \widehat{\mathbf{c}}^\star, \mathsf{hint})$ where $\widehat{\mathbf{c}}^\star = \mathbf{a}'r + \mathbf{e}'$
mod Q or $\widehat{\mathbf{c}}^\star \xleftarrow{\$} \mathcal{R}^2_Q$, $\mathsf{hint} = \{\widetilde{z}^\star_i + z^\star_{0,i}, r(\widetilde{z}^\star_i + z^\star_{0,i}) + g_i, \mathbf{z}_{\mathsf{ID}^\star_\alpha,i}, \mathbf{e}'^\top \mathbf{z}_{\mathsf{ID}^\star_\alpha,i} + h_i\}_{i\in[l]}$.
Note that the hints are leaked through

$$\begin{aligned}
\mathsf{ct}^\star_i &= (\widehat{\mathbf{c}}^\star - \mathbf{e}')^\top \mathbf{z}_{\mathsf{ID}^\star_\alpha,i} + f_i + r(\widetilde{z}^\star_i + z^\star_{0,i}) + p^{k-1}x^\star_{\alpha,i} \\
&= (\widehat{\mathbf{c}}^\star)^\top \mathbf{z}_{\mathsf{ID}^\star_\alpha,i} + \left((\widetilde{z}^\star_i + z^\star_{0,i})r + g_i\right) - \left(\mathbf{e}'^\top \mathbf{z}_{\mathsf{ID}^\star_\alpha,i} + h_i\right) + p^{k-1}x^\star_{\alpha,i}
\end{aligned}$$

where we substitute f_i with $(g_i - h_i)$ where g_i, h_i are uniformly chosen from the
same distribution $\mathcal{D}_{\mathbb{Z},\delta Q}$. This is achievable if, within the framework of Lemma 3,
we take the positive definite matrices $\mathbf{\Sigma} = \mathbf{\Sigma}' = \delta^2 I_n$ satisfying $\Gamma_{\mathbf{\Sigma},\mathbf{\Sigma}'} \geq \eta_\epsilon(\mathbb{Z}^n)$
with $\epsilon = 2^{-k}$. That is, we need to assign $\zeta = \sqrt{2}\delta$ where δ is chosen to satisfy
the mhe-RLWE assumption and to guarantee that the inequality $\Gamma_{\mathbf{\Sigma},\mathbf{\Sigma}'} \geq \eta_\epsilon(\mathbb{Z}^n)$
holds.

The adversary $\mathcal{A}_2$ provides $\mathsf{MPK} = \mathbf{a}'$ to $\mathcal{A}$. Throughout the game, $\mathcal{A}_2$ simulates the challenger $\mathcal{C}$ in Game$_4$. Upon receiving challenge query from $\mathcal{A}$, $\mathcal{A}_2$
responds the same way as done in Game$_4$. After that, $\mathcal{A}$ returns a bit $\widehat{\alpha}$. At
last, $\mathcal{A}_2$ outputs 1 if $\widehat{\alpha} = \alpha$ and 0 otherwise. If $(\mathbf{a}', \widehat{\mathbf{c}}^\star, \mathsf{hint})$ is a valid mhe-
RLWE instance (specifically, $\widehat{\mathbf{c}}^\star = \mathbf{a}'r + \mathbf{e}' \mod Q$), the view of $\mathcal{A}$ is identical

to Game_4. Otherwise, if $\widehat{\mathbf{c}}^{\star} \xleftarrow{\$} \mathcal{R}_Q^2$, the view of $\mathcal{A}$ matches Game_5. Thus, under the assumption that $\mathsf{mhe} - \mathsf{RLWE}_{Q,2,\mathcal{D}_\delta,l,C}$ problem is hard, we conclude that $|\Pr[\mathsf{Event}_4] - \Pr[\mathsf{Event}_5]| = \mathsf{negl}(\kappa)$.

Game_6: In this game, the method for generating the challenge ciphertext is modified once more. The challenger $\mathcal{C}$ samples $\{\mathsf{ct}_i^{\star} \xleftarrow{\$} \mathcal{R}_Q\}_{i \in [l]}$. Thus, we conclude that $\Pr[\mathsf{Event}_6] = 0$.

We now establish that Game_5 and Game_6 are statistically indistinguishable. We show that Game5 and Game6 are statistically indistinguishable. Given our parameter selection, Lemma 2 is applicable, which guarantees that $(\widehat{\mathbf{c}}^{\star})^{\top} \mathbf{z}_{\mathsf{ID}_\alpha^{\star},i}$ for $i \in [l]$ is statistically close to a uniform distribution. As a result, the statistical distance between the adversary $\mathcal{A}$'s view in Game_5 and Game_6 is negligible in κ, meaning $|\Pr[\mathsf{Event}_5] - \Pr[\mathsf{Event}_6]| = \mathsf{negl}(\kappa)$. Thus, by assembling all these arguments, the theorem follows.

5 Security Proof in QROM

Quantum Computation. The basis states of a single qubit are represented as column vectors $|0\rangle = (1,0)^{\top}$ and $|1\rangle = (0,1)^{\top}$. An n-qubit quantum state is expressed as $|\psi\rangle = \sum\limits_{y \in \{0,1\}^n} \beta_y |y\rangle$, where each complex coefficient β_y satisfies the normalization condition $\sum\limits_{y \in \{0,1\}^n} |\beta_y|^2 = 1$. Each state $|y\rangle$ corresponds to a tensor product $|y\rangle = |y_1 y_2 \ldots y_n\rangle = |y_1\rangle \otimes |y_2\rangle \otimes \ldots \otimes |y_n\rangle$ for $y_i \in \{0,1\}$ forming the computational basis of $\mathbb{C}^{2^n}$. When a measurement is performed on the state $|\psi\rangle$ in the computational basis, the state collapses to a specific basis state $|y\rangle$ with probability $|\beta_y|^2$, and the resulting state is $|y\rangle$. The evolution of a quantum state from $|\psi\rangle$ to $|\psi'\rangle$ is carried out by applying a unitary transformation U such that $|\psi'\rangle = U|\psi\rangle$. In general, a quantum algorithm consists of a sequence of such unitary transformations followed by measurements. The runtime of a quantum algorithm $\mathcal{A}$ is defined as the total number of universal quantum gates and measurement operations it uses. If $\mathcal{A}$ includes access to a quantum oracle, each oracle call is assumed to take one unit of time. Any computation that can be performed efficiently on a classical computer can also be efficiently implemented on a quantum computer. Specifically, for any classically computable function f, there exists a unitary operator U_f such that $U_f |y, z\rangle = |y, f(y) \oplus z\rangle$, and the number of quantum gates needed to implement U_f is proportional to the size of a classical circuit that computes f.

QROM. Boneh et al. [8] introduced the Quantum Random Oracle Model (QROM) as an extension of the classical Random Oracle Model (ROM). Similar to the ROM, the QROM treats a hash function as an idealized oracle that behaves like a truly random function. However, unlike in the classical setting, the hash function in the QROM can be queried in quantum superposition, making it accessible to quantum algorithms In the QROM, a random function $H : Y \to Z$

is selected uniformly at random at the outset. An adversary is allowed to query this oracle with a quantum superposition of inputs of the form $\sum_{y,z} \beta_{y,z} |y\rangle |z\rangle$, and receives the superposition $\sum_{y,z} \beta_{y,z} |y\rangle |H(y) \oplus z\rangle$. Let $\mathcal{A}^{|H\rangle}$ represent a quantum algorithm that has quantum access to the oracle $|H\rangle$. Zhandry [32] showed that a quantum random oracle can be effectively emulated using a family of $2Q_{\mathsf{Hash}}$-wise independent hash functions for an adversary making up to Q_{Hash} quantum queries to the random oracle.

Lemma 6. *Any quantum algorithm $\mathcal{A}$ that makes quantum queries to random oracles can be efficiently simulated by another quantum algorithm $\mathcal{B}$, which produces the same output distribution without making any queries.*

Theorem 2. *Our identity-based inner product functional encryption* (IBIPFE) *presented in Sect. 3 satisfies* Adp-IND *security in the quantum random oracle model* (QROM) *assuming the hardness of the* $\mathsf{RLWE}_{Q,1,\chi}$ *and the* $\mathsf{mhe} - \mathsf{RLWE}_{Q,2,\mathcal{D}_\delta,l,C}$ *problem.*

Proof. We prove the security by considering a sequence of games. For each game, we denote by Event_i the event that $\mathcal{A}$ outputs a correct guess in Game_i. For $i \in [l]$, let $\mathsf{Samp}(\sigma, p; r_i)$ be a PPT algorithm that takes as input a Gaussian parameter σ, a natural number p, a random string $r_i \in \{0,1\}^{l_r}$, and outputs $(\mathbf{z}_i, \widetilde{z}_i)$ where $\mathbf{z}_i$ is sampled from a distribution that is statistically close to the discrete gaussian distribution $\mathcal{D}_\sigma^3$ and $\widetilde{z}_i$ sampled from $\mathcal{U}(\mathcal{R}_p)$.

Game_0: This is the real security game. At the outset of the game, the challenger $\mathcal{C}$ chooses a random function $\mathsf{Hash} : \{0,1\}^{l_{\mathsf{ID}} + \lfloor \log l \rfloor} \to \mathcal{R}_Q$. Then, it generates $(\mathsf{MPK}, \mathsf{MSK}) \leftarrow \mathsf{Setup}(1^\kappa, 1^l)$ and provides MPK to the adversary $\mathcal{A}$. Throughout the game, $\mathcal{A}$ is allowed to ask many random oracle and secret key queries, as well as a single challenge query. The challenger responds to each query as follows:

- When $\mathcal{A}$ makes a (quantum) random oracle query on a quantum state $\sum_{\mathsf{ID}||i,z} \beta_{\mathsf{ID}||i,z} |\mathsf{ID}||i\rangle |z\rangle$, $\mathcal{C}$ returns $\sum_{\mathsf{ID}||i,z} \beta_{\mathsf{ID}||i,z} |\mathsf{ID}||i\rangle |\mathsf{Hash}(\mathsf{ID}||i) \oplus z\rangle$.
- Upon receiving first secret key query for ID, computes $\mathsf{SK}_{\mathsf{ID},\mathbf{y}} = (\overline{\mathbf{y}} = \mathbf{y}, \mathbf{k}_{\mathsf{ID},\mathbf{y}} = \mathbf{Z}_{\mathsf{ID}} \cdot \overline{\mathbf{y}})$ where $\mathbf{Z}_{\mathsf{ID}} = \begin{pmatrix} z_{1,1} & z_{1,2} & \cdots & z_{1,l} \\ 2,1 & z_{2,2} & \cdots & z_{2,l} \end{pmatrix}$, $\mathbf{z}_i = (z_{0,i}, z_{1,i}, z_{2,i})^\top \xleftarrow{\$}$ $\mathsf{AppSampPre}(\mathbf{a}, \mathsf{td}_\mathbf{a}, u_{\mathsf{ID},i}, \sigma)$ for $i \in [l]$. If this is not first query for ID, retrieves $(\mathsf{ID}, \mathbf{Z}_{\mathsf{ID}}, (\overline{\mathbf{y}}_{\mathsf{ID},i}, \mathbf{k}_{\mathsf{ID},i})_{i \in [Q_{\mathsf{ID}}]}) \in \mathsf{st}$, where $\{\overline{\mathbf{y}}_{\mathsf{ID},i}\}_{i \in [Q_{\mathsf{ID}}]}$ are linearly independent vectors corresponding to secret keys that have been generated for ID. If $\mathbf{y} = \sum_{j \in [Q_{\mathsf{ID}}]} d_j \cdot \overline{\mathbf{y}}_{\mathsf{ID},j} \bmod p$, returns $\mathsf{SK}_{\mathsf{ID},\mathbf{y}} = (\overline{\mathbf{y}} = \sum_{j \in [Q_{\mathsf{ID}}]} d_j \cdot \overline{\mathbf{y}}_{\mathsf{ID},j}, \mathbf{k}_{\mathsf{ID},\mathbf{y}} = \sum_{j \in [Q_{\mathsf{ID}}]} d_j \cdot \mathbf{k}_{\mathsf{ID},j})$, else returns $(\overline{\mathbf{y}} = \mathbf{y}, \mathbf{k}_{\mathsf{ID},\mathbf{y}} = \mathbf{Z}_{\mathsf{ID}} \cdot \overline{\mathbf{y}})$.
- Upon receiving challenge query on $((\mathsf{ID}_0^\star, \mathbf{x}_0^\star), (\mathsf{ID}_1^\star, \mathbf{x}_1^\star))$, the challenger chooses $\alpha \xleftarrow{\$} \{0,1\}$ and returns $\mathsf{CT}^\star = (\widehat{\mathbf{c}}^\star, \{\mathsf{ct}_i^\star\}_{i \in [l]}) \leftarrow \mathsf{Encrypt}(\mathsf{MPK}, \mathsf{ID}_\alpha^\star, \mathbf{x}_\alpha^\star)$.

When the game concludes, $\mathcal{A}$ reveals a guess $\widehat{\alpha}$ for α. Thus, we have $\big|\Pr[\mathsf{Event}_0] -$
$\frac{1}{2}\big| = \big|\Pr[\widehat{\alpha} = \alpha] - \frac{1}{2}\big| = \mathsf{Adv}^{\mathsf{Adp-IND}}_{\mathsf{IBIPFE},\mathcal{A}}(\kappa).$

$\mathbf{Game}_1$: In this game, the process of responding to random oracle queries to Hash is changed. At the start of the game, $\mathcal{C}$ chooses a $2Q_{\mathsf{Hash}}$-wise independent hash function $\mathsf{h}_{2Q_{\mathsf{Hash}}} : \{0,1\}^{l_{\mathsf{ID}}+\lfloor \log l \rfloor} \to \{0,1\}^{l_r}$. Then, we define $\mathsf{Hash}(\mathsf{ID}\|i) = \mathbf{a}^\top \mathbf{z}_i + \widetilde{z}_i \mod Q$ where $(\mathbf{z}_i, \widetilde{z}_i) := \mathsf{Samp}(\sigma, p; \mathsf{h}_{2Q_{\mathsf{Hash}}}(\mathsf{ID}\|i))$ for $i \in [l]$, and use this Hash throughout the game.

For a given identity ID, the output distribution of $\mathsf{Hash}(\mathsf{ID}\|i)$ is identical, and its deviation from a uniform distribution is $\mathsf{negl}(\kappa)$, as stated in Lemma 1. Observe that in this game, we only modify the distribution of $u_{\mathsf{ID},i}$ for each identity, while the secret key generation process remains the same. Therefore, by applying Lemma 6, it follows that $|\Pr[\mathsf{Event}_0] - \Pr[\mathsf{Event}_1]| = \mathsf{negl}(\lambda)$.

$\mathbf{Game}_2$: This game is identical to Game_1 except for how secret key queries are handled. Specifically, the challenger $\mathcal{C}$ no longer uses $\mathsf{td}_\mathbf{a}$ to answer $\mathcal{A}$'s first secret key query associated with ID. Upon receiving a secret key query for identity ID, the challenger $\mathcal{C}$ returns $\mathsf{SK}_{\mathsf{ID},\mathbf{y}} = (\mathbf{y}, \mathbf{k}_{\mathsf{ID},\mathbf{y}} = \mathbf{Z}_{\mathsf{ID}} \cdot \mathbf{y})$ where $\mathbf{Z}_{\mathsf{ID}} = \begin{pmatrix} z_{1,1} \, z_{1,2} \cdots z_{1,l} \\ z_{2,1} \, z_{2,2} \cdots z_{2,l} \end{pmatrix}$, $\mathbf{z}_i = (z_{0,i}, z_{1,i}, z_{2,i})^\top$, $(\mathbf{z}_i, \widetilde{z}_i) := \mathsf{Samp}(\sigma, p; \mathsf{h}_{2Q_{\mathsf{Hash}}}(\mathsf{ID}\|i)) :$ $i \in [l]$ Proceeding with the same line of reasoning as in Game_2 of the proof of Theorem 1, we have $|\Pr[\mathsf{Event}_1] - \Pr[\mathsf{Event}_2]| = Q_{\mathsf{ID}} \cdot \mathsf{negl}(\kappa) = \mathsf{negl}(\kappa)$.

$\mathbf{Game}_3$: This game is identical to Game_2 except for how $\mathcal{C}$ computes the MPK. The challenger $\mathcal{C}$ chooses $b \xleftarrow{\$} \mathcal{R}_Q$ randomly rather than setting $b = p - (\overline{a}f + g)$ $\mod Q$, and omits generating trapdoor $\mathsf{td}_\mathbf{a}$. Since the trapdoor was not needed to answer $\mathcal{A}$'s secret key queries in Game_2, $\mathcal{C}$ can still handle those queries just as before.

Proceeding with the same line of reasoning as in Game_3 of the proof of Theorem 1, we have $|\Pr[\mathsf{Event}_2] - \Pr[\mathsf{Event}_3]| = \mathsf{negl}(\kappa)$ under the assumption that $\mathsf{RLWE}_{Q,1,\chi}$ holds.

$\mathbf{Game}_4$: This game is identical to Game_3 except for generation of the challenge ciphertext. To generate $\mathsf{CT}^\star = (\widehat{\mathbf{c}}^\star, \{\mathsf{ct}_i^\star\}_{i\in[l]})$ for identity $\mathsf{ID}_\alpha^\star$ and message $\mathbf{x}_\alpha^\star$, $\mathcal{C}$ extracts the unique tuple ($\{\mathsf{ID}_\alpha^\star\|i\}_{i\in[l]}$, $\{u_{\mathsf{ID}_\alpha^\star,i}\}_{i\in[l]}$, ($\mathbf{z}_i^\star = (z_{0,i}^\star, z_{1,i}^\star, z_{2,i}^\star)^\top$, $\widetilde{z}_i^\star)_{i\in[l]}$) from local storage, assigns $\mathbf{z}_{\mathsf{ID}_\alpha^\star,i} = (z_{1,i}^\star, z_{2,i}^\star)^\top$, $\mathbf{Z}_{\mathsf{ID}_\alpha^\star} = \begin{pmatrix} z_{1,1}^\star \, z_{1,2}^\star \cdots z_{1,l}^\star \\ z_{2,1}^\star \, z_{2,2}^\star \cdots z_{2,l}^\star \end{pmatrix}$, and computes

$$\widehat{\mathbf{c}}^\star = \mathbf{a}'r + \mathbf{e}' \mod Q$$
$$\mathsf{ct}_i^\star = (\widehat{\mathbf{c}}^\star - \mathbf{e}')^\top \mathbf{z}_{\mathsf{ID}_\alpha^\star,i} + f_i + r(\widetilde{z}_i^\star + z_{0,i}^\star) + p^{k-1} x_{\alpha,i}^\star$$

where $r \xleftarrow{\$} \mathcal{D}_{\mathbb{Z},\delta Q}$, $\mathbf{e}' \xleftarrow{\$} \mathcal{D}_{\mathbb{Z}^2,\delta Q}$, and $f_i \xleftarrow{\$} \mathcal{D}_{\mathbb{Z},\varsigma Q}$ Proceeding with the same line of reasoning as in Game_4 of the proof of Theorem 1, we have $\Pr[\mathsf{Event}_3] = \Pr[\mathsf{Event}_4]$.

Game$_5$: This game introduces an additional change to the process of generating the challenge ciphertext. To generate $\mathsf{CT}^\star = (\widehat{\mathbf{c}}^\star, \{\mathsf{ct}_i^\star\}_{i\in[l]})$, $\mathcal{C}$ chooses $\widehat{\mathbf{c}}^\star \xleftarrow{\$} \mathcal{R}_Q^2$ and computes $\{\mathsf{ct}_i^\star\}_{i\in[l]}$ following the method in Game$_4$.
Proceeding with the same line of reasoning as in Game$_5$ of the proof of Theorem 1, we have $|\Pr[\mathsf{Event}_4] - \Pr[\mathsf{Event}_5]| = \mathsf{negl}(\kappa)$ under the assumption that $\mathsf{mhe} - \mathsf{RLWE}_{Q,2,\mathcal{D}_\delta,l,C}$ holds.

Game$_6$: In this game, the method for generating the challenge ciphertext is modified once more. The challenger $\mathcal{C}$ samples $\{\mathsf{ct}_i^\star \xleftarrow{\$} \mathcal{R}_Q\}_{i\in[l]}$. Thus, we have $\Pr[\mathsf{Event}_6] = 0$. Proceeding with the same line of reasoning as in Game$_4$ of the proof of Theorem 1, we have $|\Pr[\mathsf{Event}_5] - \Pr[\mathsf{Event}_6]| = \mathsf{negl}(\kappa)$. Thus, by assembling all these arguments, the theorem follows.

6   Simulation Secure IBIPFE Scheme

Following the IND-to-SIM generic transformation of [12], We present an IBIPFE scheme (for computing inner products modulo a prime p) that achieves simulation security.

- Setup$(1^\kappa, 1^l) \to (\mathsf{MPK}, \mathsf{MSK})$: A trusted authority runs $(\mathsf{IND.MPK}, \mathsf{IND.MSK}) \leftarrow \mathsf{IND.Setup}\ (1^\kappa, 1^{l+1})$, randomly select an index $k \xleftarrow{\$} \mathcal{K}$ from a function family $\mathcal{F} = \{f_k\}_{k\in\mathcal{K}}$ where each function $f_k : \mathcal{IDS} \to \mathbb{Z}_p^l$, and return $\mathsf{MPK} = \mathsf{IND.MPK} = (\mathsf{seed}_a, b, \mathsf{Hash}), \mathsf{MSK} = (\mathsf{IND.MSK}, k) = ((f, g), k)$ where $\mathsf{Hash} : \{0,1\}^{l_{\mathsf{ID}} + \lfloor \log(l+1)\rfloor} \to \mathcal{R}_Q$.
- Encrypt$(\mathsf{MPK}, \mathsf{ID}, \mathbf{x}) \to \mathsf{CT}_{\mathsf{ID},\mathbf{x}}$: An encryptor executes $\mathsf{IND.CT}_{\mathsf{ID},[\mathbf{x}||0]} \leftarrow \mathsf{IND.Encrypt}(\mathsf{MPK}, \mathsf{ID}, [\mathbf{x}\ ||\ 0])$ and returns $\mathsf{CT}_{\mathsf{ID},\mathbf{x}} = \mathsf{IND.CT}_{\mathsf{ID},[\mathbf{x}||0]} = (\widehat{\mathbf{c}}, \{\mathsf{ct}_i\}_{i\in[l+1]})$.
- KeyGen$(\mathsf{MSK}, \mathsf{ID}, \mathbf{y}) \to \mathsf{SK}_{\mathsf{ID},\mathbf{y}}$: The trusted authority computes $f_k(\mathsf{ID}) = \mathbf{r}_{\mathsf{ID}} \in \mathbb{Z}_p^l$, sets $\widehat{\mathbf{y}} = [\mathbf{y}\ ||\ \langle \mathbf{r}_{\mathsf{ID}}, \mathbf{y}\rangle \mod p]$, performs $\mathsf{IND.SK}_{\mathsf{ID},\widehat{\mathbf{y}}} \leftarrow \mathsf{IND.KeyGen}(\mathsf{MSK}, \mathsf{ID}, \widehat{\mathbf{y}})$ and returns $\mathsf{SK}_{\mathsf{ID},\widehat{\mathbf{y}}} = (\widehat{\mathbf{y}}, \mathsf{IND.SK}_{\mathsf{ID},\widehat{\mathbf{y}}})$.
- Decrypt$(\mathsf{MPK}, \mathsf{CT}_{\mathsf{ID},\mathbf{x}}, \mathsf{SK}_{\mathsf{ID},\mathbf{y}}) \to \langle \mathbf{x}, \mathbf{y}\rangle$: A decryptor takes input $\mathsf{CT}_{\mathsf{ID},\mathbf{x}} = \mathsf{IND.CT}_{\mathsf{ID},[\mathbf{x}||0]}$, $\mathsf{SK}_{\mathsf{ID},\mathbf{y}} = (\widehat{\mathbf{y}}, \mathsf{IND.SK}_{\mathsf{ID},\widehat{\mathbf{y}}})$, and outputs $\mathsf{IND.Dec}(\mathsf{IND.mpk}, \mathsf{IND.CT}_{\mathsf{ID},[\mathbf{x}||0]}, \mathsf{IND.SK}_{\mathsf{ID},\widehat{\mathbf{y}}})$.

Correctness. Given that the inner product $\langle [\mathbf{x}\ ||\ 0], [\mathbf{y}\ ||\ \langle r_{\mathsf{ID}}, \mathbf{y}\rangle] \rangle = \langle \mathbf{x}, \mathbf{y}\rangle \mod p$, the correctness of IND-IBIPFE directly guarantees the correctness of the proposed scheme SIM-IBIPFE.

7   Security

Theorem 3. *If the underlying IBIPFE scheme is AD-IND secure and $\mathcal{F}$ is a family of pseudorandom functions, then the presented scheme in Sect. 6 achieves tight AD-SIM security. Specifically, for any probabilistic polynomial time (PPT)* *adversary $\mathcal{A}$ that violates the* AD-SIM *security of the proposed scheme, there exist* PPT *adversaries* $\mathcal{A}_1$ *and* $\mathcal{A}_2$ *satisfying* $\mathsf{Adv}_{\mathsf{IBIPFE},\mathcal{A}}^{\mathsf{Adp-SIM}}(\kappa) \leq \mathsf{Adv}_{\mathsf{PRF},\mathcal{A}_1}(\kappa) + \mathsf{Adv}_{\mathsf{IBIPFE},\mathcal{A}_2}^{\mathsf{Adp-IND}}(\kappa)$ *and* $\max\{\mathsf{Time}(\mathcal{A}_1), \mathsf{Time}(\mathcal{A}_2)\} = \mathsf{Time}(\mathcal{A}) + \mathsf{poly}(\lambda, l)$.

This proof directly follows from [12].

Acknowledgements. We would like to thank Dr. Toi Tomita and Dr. Shingo Sato for their valuable suggestions and insightful discussions. We also appreciate the helpful comments provided by the anonymous reviewers of INDOCRYPT 2025. This research was in part conducted under a contract of "Research and development on new generation cryptography for secure wireless communication services" among "Research and Development for Expansion of Radio Wave Resources (JPJ000254)" which was supported by the Ministry of Internal Affairs and Communications, Japan. This work was in part supported by JSPS KAKENHI Grant Number JP23K24846. This research was also supported by the project PROVTOPIA (PID2023-150098OB-I00), funded by MICIU/AEI/10.13039/501100011033 and FEDER (EU).

References

1. Abdalla, M., Bourse, F., De Caro, A., Pointcheval, D.: Simple functional encryption schemes for inner products. In: Katz, J. (ed.) PKC 2015. LNCS, vol. 9020, pp. 733–751. Springer, Heidelberg (2015). https://doi.org/10.1007/978-3-662-46447-2_33
2. Abdalla, M., Catalano, D., Gay, R., Ursu, B.: Inner-product functional encryption with fine-grained access control. In: Moriai, S., Wang, H. (eds.) ASIACRYPT 2020. LNCS, vol. 12493, pp. 467–497. Springer, Cham (2020). https://doi.org/10.1007/978-3-030-64840-4_16
3. Agrawal, S., Boneh, D., Boyen, X.: Efficient lattice (H)IBE in the standard model. In: Gilbert, H. (ed.) EUROCRYPT 2010. LNCS, vol. 6110, pp. 553–572. Springer, Heidelberg (2010). https://doi.org/10.1007/978-3-642-13190-5_28
4. Agrawal, S., Libert, B., Maitra, M., Titiu, R.: Adaptive simulation security for inner product functional encryption. In: Kiayias, A., Kohlweiss, M., Wallden, P., Zikas, V. (eds.) PKC 2020. LNCS, vol. 12110, pp. 34–64. Springer, Cham (2020). https://doi.org/10.1007/978-3-030-45374-9_2
5. Agrawal, S., Libert, B., Stehlé, D.: Fully secure functional encryption for inner products, from standard assumptions. In: Robshaw, M., Katz, J. (eds.) CRYPTO 2016. LNCS, vol. 9816, pp. 333–362. Springer, Heidelberg (2016). https://doi.org/10.1007/978-3-662-53015-3_12
6. Bert, P., Eberhart, G., Prabel, L., Roux-Langlois, A., Sabt, M.: Implementation of lattice trapdoors on modules and applications. In: Cheon, J.H., Tillich, J.-P. (eds.) PQCrypto 2021 2021. LNCS, vol. 12841, pp. 195–214. Springer, Cham (2021). https://doi.org/10.1007/978-3-030-81293-5_11
7. Bert, P., Fouque, P.-A., Roux-Langlois, A., Sabt, M.: Practical implementation of ring-SIS/LWE based signature and IBE. In: Lange, T., Steinwandt, R. (eds.) PQCrypto 2018. LNCS, vol. 10786, pp. 271–291. Springer, Cham (2018). https://doi.org/10.1007/978-3-319-79063-3_13
8. Boneh, D., Dagdelen, Ö., Fischlin, M., Lehmann, A., Schaffner, C., Zhandry, M.: Random oracles in a quantum world. In: Lee, D.H., Wang, X. (eds.) ASIACRYPT 2011. LNCS, vol. 7073, pp. 41–69. Springer, Heidelberg (2011). https://doi.org/10.1007/978-3-642-25385-0_3

9. Boneh, D., Sahai, A., Waters, B.: Functional encryption: definitions and challenges. In: Ishai, Y. (ed.) TCC 2011. LNCS, vol. 6597, pp. 253–273. Springer, Heidelberg (2011). https://doi.org/10.1007/978-3-642-19571-6_16

10. Chen, Y., Genise, N., Mukherjee, P.: Approximate trapdoors for lattices and smaller hash-and-sign signatures. In: Galbraith, S.D., Moriai, S. (eds.) ASIACRYPT 2019. LNCS, vol. 11923, pp. 3–32. Springer, Cham (2019). https://doi.org/10.1007/978-3-030-34618-8_1

11. Ducas, L., Lyubashevsky, V., Prest, T.: Efficient identity-based encryption over NTRU lattices. In: Sarkar, P., Iwata, T. (eds.) ASIACRYPT 2014. LNCS, vol. 8874, pp. 22–41. Springer, Heidelberg (2014). https://doi.org/10.1007/978-3-662-45608-8_2

12. Edamura, T., Takayasu, A.: Tight adaptive simulation security for identity-based inner-product fe in the (quantum) random oracle model. In: IACR International Conference on Public-Key Cryptography, pp. 3–36. Springer, Heidelberg (2025). https://doi.org/10.1007/978-3-031-91826-1_1

13. Emura, K., Katsumata, S., Watanabe, Y.: Identity-based encryption with security against the KGC: a formal model and its instantiation from lattices. In: Sako, K., Schneider, S., Ryan, P.Y.A. (eds.) ESORICS 2019. LNCS, vol. 11736, pp. 113–133. Springer, Cham (2019). https://doi.org/10.1007/978-3-030-29962-0_6

14. Garg, S., Gentry, C., Halevi, S., Raykova, M., Sahai, A., Waters, B.: Candidate indistinguishability obfuscation and functional encryption for all circuits. SIAM J. Comput. **45**(3), 882–929 (2016)

15. Garg, S., Gentry, C., Halevi, S., Zhandry, M.: Functional encryption without obfuscation. In: Kushilevitz, E., Malkin, T. (eds.) TCC 2016. LNCS, vol. 9563, pp. 480–511. Springer, Heidelberg (2016). https://doi.org/10.1007/978-3-662-49099-0_18

16. Gentry, C., Peikert, C., Vaikuntanathan, V.: Trapdoors for hard lattices and new cryptographic constructions. In: Proceedings of the Fortieth Annual ACM Symposium on Theory of Computing, pp. 197–206 (2008)

17. Katsumata, S., Matsuda, T., Takayasu, A.: Lattice-based revocable (hierarchical) ibe with decryption key exposure resistance. Theor. Comput. Sci. **809**, 103–136 (2020)

18. Katsumata, S., Yamada, S.: Partitioning via non-linear polynomial functions: more compact IBEs from ideal lattices and bilinear maps. In: Cheon, J.H., Takagi, T. (eds.) ASIACRYPT 2016. LNCS, vol. 10032, pp. 682–712. Springer, Heidelberg (2016). https://doi.org/10.1007/978-3-662-53890-6_23

19. Katsumata, S., Yamada, S., Yamakawa, T.: Tighter security proofs for gpv-ibe in the quantum random oracle model. J. Cryptol. **34**, 1–46 (2021)

20. Lai, Q., Liu, F.-H., Wang, Z.: New lattice two-stage sampling technique and its applications to functional encryption – stronger security and smaller ciphertexts. In: Canteaut, A., Standaert, F.-X. (eds.) EUROCRYPT 2021. LNCS, vol. 12696, pp. 498–527. Springer, Cham (2021). https://doi.org/10.1007/978-3-030-77870-5_18

21. Lin, H., Luo, J.: Succinct and adaptively secure ABE for ABP from k-Lin. In: Moriai, S., Wang, H. (eds.) ASIACRYPT 2020. LNCS, vol. 12493, pp. 437–466. Springer, Cham (2020). https://doi.org/10.1007/978-3-030-64840-4_15

22. Lyubashevsky, V.: Lattice signatures without trapdoors. In: Pointcheval, D., Johansson, T. (eds.) EUROCRYPT 2012. LNCS, vol. 7237, pp. 738–755. Springer, Heidelberg (2012). https://doi.org/10.1007/978-3-642-29011-4_43

23. Lyubashevsky, V., Peikert, C., Regev, O.: On ideal lattices and learning with errors over rings. In: Gilbert, H. (ed.) EUROCRYPT 2010. LNCS, vol. 6110, pp. 1–23. Springer, Heidelberg (2010). https://doi.org/10.1007/978-3-642-13190-5_1

24. Lyubashevsky, V., Peikert, C., Regev, O.: A toolkit for ring-LWE cryptography. In: Johansson, T., Nguyen, P.Q. (eds.) EUROCRYPT 2013. LNCS, vol. 7881, pp. 35–54. Springer, Heidelberg (2013). https://doi.org/10.1007/978-3-642-38348-9_3

25. Mera, J.M.B., Karmakar, A., Marc, T., Soleimanian, A.: Efficient lattice-based inner-product functional encryption. In: IACR International Conference on Public-Key Cryptography, pp. 163–193. Springer, Heidelberg (2022). https://doi.org/10.1007/978-3-030-97131-1_6

26. Nishimaki, R., Yamakawa, T.: Leakage-resilient identity-based encryption in bounded retrieval model with nearly optimal leakage-ratio. In: Lin, D., Sako, K. (eds.) PKC 2019. LNCS, vol. 11442, pp. 466–495. Springer, Cham (2019). https://doi.org/10.1007/978-3-030-17253-4_16

27. Phan, D.H., Bhattacherjee, S., Stehlé, D., Agrawal, S., Yamada, S.: Efficient public trace and revoke from standard assumptions. In: Proceedings of the 2017 ACM SIGSAC Conference on Computer and Communications Security. ACM (2017)

28. Tomita, T., Shikata, J.: Efficient identity-based encryption with tight adaptive anonymity from rlwe. In: International Conference on Post-Quantum Cryptography, pp. 300–321. Springer, Heidelberg . https://doi.org/10.1007/978-3-031-62743-9_10(2024)

29. Wang, Z., Fan, X., Liu, F.-H.: FE for inner products and its application to decentralized ABE. In: Lin, D., Sako, K. (eds.) PKC 2019. LNCS, vol. 11443, pp. 97–127. Springer, Cham (2019). https://doi.org/10.1007/978-3-030-17259-6_4

30. Yamada, S.: Asymptotically compact adaptively secure lattice IBEs and verifiable random functions via generalized partitioning techniques. In: Katz, J., Shacham, H. (eds.) CRYPTO 2017. LNCS, vol. 10403, pp. 161–193. Springer, Cham (2017). https://doi.org/10.1007/978-3-319-63697-9_6

31. Yu, Y., Jia, H., Wang, X.: Compact lattice gadget and its applications to hash-and-sign signatures. In: Annual International Cryptology Conference, pp. 390–420. Springer, Heidelberg (2023). https://doi.org/10.1007/978-3-031-38554-4_13

32. Zhandry, M.: Secure identity-based encryption in the quantum random oracle model. Int. J. Quant. Inf. **13**(04), 1550014 (2015)

Module Lattice Based Constant-Size Group Signature with Verifier Local Revocation and Backward Unlinkability

Komal Pursharthi[ID] and Dheerendra Mishra[(✉)][ID]

Maulana Azad National Institute of Technology, Bhopal, India
komalpursharthi.56@gmail.com, dheerendra.m@gmail.com

Abstract. In a group signature scheme, members are able to sign messages on behalf of the group in such a way that their individual identities remain confidential. Only a designated authority within the group can determine the identity of the signer. Verifier local revocation (VLR) is recognized as an efficient technique to expel malicious group members. The research community is increasingly focusing on lattice-based VLR group signatures to improve quantum resistance. One common issue with current systems is the expansion of their group public key size in correlation with the number of group members, leading to greater computational demands which are challenging for devices with limited processing power. In this study, we introduce a group signature protocol leveraging lattice-based cryptography with VLR that achieves a constant size for the group public key, signing key, and signature, irrespective of the group's size. The security of our model relies on the difficulty of the MSIS-MLWE problem. To protect against signature forgery, we use a zero-knowledge proof system to ensure signers authenticate their identity. We provide a detailed security analysis to illustrate these features, along with a comparative study demonstrating the computational efficiency of our protocol.

Keywords: Group Signature Protocol · Post Quantum Cryptography · Lattice based Cryptography · Module-SIS · Module-LWE · Security

1 Introduction

Digital signature schemes serve to authenticate and ensure the integrity of messages. They allow anyone to confirm the legitimacy of a given signature as being from the stated signer. The fundamental criteria for digital signatures include non-repudiation and resistance to forgery. Signers cannot dispute their own signatures and unauthorised parties cannot produce a valid signature. A group signature represents a modification of the concept of digital signature. This mechanism delivers a robust cryptographic approach to guarantee both the confidentiality and authenticity of group participants, facilitating group message signing. Through a group signature, a member can endorse a message on behalf of the collective without disclosing the identity of the particular signer. This capability is vital in privacy and security-focused applications, including online voting systems, cryptocurrencies, blockchain technology, shared banking accounts, digital rights management, and numerous other domains.

The notion of group signatures was pioneered by Chaum and van Heyst [13] & they proposed very first scheme. Inspired by this construction, many group signature protocols were proposed. However, these protocols have the following unappealing properties.

- Group public key size and/or signature size increases with the size of the group. This is very challenging for group with large number of members.
- To join new members, one must update at least the public key of the group.

Solutions to these problems were proposed by Camenisch and Stadler [11]. Many other improvements are also proposed to solve the above issues. However, the security of all proposed solutions relies on the difficulty of either integer factorization problem or discrete log problem. Unfortunately, both problems are challenged by the potential advent of high scalable quantum computers [3]. The feature of qubit to take infinite values than 0 and 1, enables quantum computers to have much speed than classical computers to solve integer factorization (IF) and discrete logarithm (DL) [40]. Consequently, all schemes based on IF and DL would be exhibit insecure once a large scale quantum computer becomes reality.

Researchers are dynamically proposing secure group signature schemes which do not rely on IF and DL problems, that can be anticipated immuned to attacks by quantum computers. Lattice-based cryptography is a prominent candidate for practical post-quantum security and significant research is going on to develop and analyze lattice-based group signature schemes. Most of protocols have signature size $\mathcal{O}$ (N), where N represents the max no. of users in group. In the notion of practicality, it is a significant drawback if the size of group is big. As an improvement, later new schemes proposed with signature size $\mathcal{O}$ (log N) that overcomes barrier of linear increase with respect to no. of users. Moreover, a thorough research is going-on to make the group signature size independent of the number of users. Group signature schemes are categorised into three parts as static group signature schemes, partially dynamic group signature schemes and fully dynamic group signature schemes. Static schemes has fixed no. of group users in the setup & group manager cannot add or delete members, whereas in dynamic schemes if any malicious activity is identified from any user then schemes support revocation of the user. Moreover, a new member can join the group by interacting with group manager. Partially dynamic schemes either support user enrollment or user revocation. For practical adoption of group signature schemes for security systems, the feature of user revocation and user enrollment is essential. Therefore, Researchers are putting efforts in constructing dynamic lattice based group signature protocols achieving constant signature size and group public key size with respect to no. of group users.

1.1 Motivation and Contribution

Contemporary designs of lattice-based group signatures encounter the issue where the group public key size scales with the number of group members, leading to greater computational demands as the group expands. This escalation can render implementation unfeasible on devices with restricted computing capabilities. To the best of our understanding, there is no existing quantum-resistant group signature protocol that comprehensively supports dynamicity while maintaining constant size for the group public key, signing key, and signature. Group signature protocols relying on the complexity of the MSIS-MLWE problem help reduce the size of the signatures and public keys. Hence, we have developed an MLWE-MSIS-based group signature that ensures full dynamicity while keeping the group key and signature sizes unaffected by the number of participants in group. The principal contributions of this article are listed below:

1. We have proposed the first lattice-based VLR group signature, which achieves group public key size, signing key size & signature size all independent of no. of members in the group.
2. The security of the proposed protocol is based on the difficulty of MSIS-MLWE problem. Cryptosystems constructed based on module lattices generally offer greater efficiency, since the corresponding lattices admit more compact representations, and typically benefit from the faster arithmetic of the underlying rings.
3. It achieves backward unlikability-almost full anonymity, traceability and non-frameability.
4. The comparative analysis of group signature schemes is presented to demonstrate the better computational efficiency achieved with respect to number of cryptographic operations used in our protocol. Moreover, comparative analysis of sizes of group public key, signing key and signature is also presented in comparison with existing lattice-based group signature protocols.
5. The Stern-like zero knowledge protocol (SLZK) is employed to demonstrate the correctness of the proposed protocol.

1.2 Organization

Section 2 presents the related work of group signature protocols. Section 3 presents the overview of the proposed protocol. Section 4 provides mathematical terminologies and definitions used in the paper and lemmas used in the correctness of the construction. Section 5 describes the proposed group signature protocol. After that, Sect. 6 presents the security analysis. Section 7 provides the comparative study with respect to existing protocols and finally, Sect. 8 concludes the paper.

2 Related Work

Group signature protocols have remained a pivotal research topic in asymmetric key cryptography since the seminal work by Chaum and Heyst [13]. Their initial scheme introduced an elegant way for legitimate signers to form a group and anonymously sign messages on its behalf, while still providing a tracing mechanism to identify misbehaving members in the event of a dispute. Over the last three decades, various protocols have been designed with different security models, efficiency goals, and operational frameworks [4–8,14,21,28]. Most of these constructions rely on the hardness of IF or DL problem, which are threatened by the anticipated rise of large-scale quantum computers. As a result, there has been a substantial shift toward devising quantum-secure group signature schemes.

From Classical to Lattice-Based Constructions

Gordon et al. [20] pioneered the first lattice-based group signature protocol under the Learning with Errors (LWE) assumption, introducing a method for sampling a random superlattice alongside a short basis. In 2013, Hwang et al. [22] proposed a scheme whose signatures appear uniform and hide signer identities, while still allowing a linker with a private key to detect linked signatures. Around the same time, Laguillaumie et al. [24] introduced construction with group public key of size $\mathcal{O}(\lambda^2 \log^2 \lambda \log N)$ bits and signature of size $\mathcal{O}(\lambda \log^3 \lambda \log N)$ bits, where λ is the security parameter and N is the maximum number of group users. It also extends anonymity into fully anonymous group signature which means it resists attacker having access with a signature opening oracle. In 2015, Nguyen et al. [35] proposed a simpler efficient group signature from Lattices. It has reduced both group public key and signature sizes to $\mathcal{O}(\log N)$. This feature is achieved by using a new interactive zero-knowledge proof corresponding to a simple identity-encoding function. The security of the construction is based on the hardness of SIS and LWE problem in the random oracle model.

Dynamic Group Signatures

In 2014, Langlois et al. [25] proposed a group signature protocol with verifier local revocation. It supports membership revocation, logarithmic-size signatures, and weaker security assumptions. It has feature of selfless anonymity which means an adversary cannot distinguish between signatures from two different members as long as the signing keys of the two members involved in the challenge are not exposed, even if other keys are exposed and the adversary has access to an open oracle and can request revocation tokens. Bootle et al. [9] formalized the notion of fully dynamic group signatures, providing a broad security model that encompasses group membership changes (additions and revocations) and can be applied to various designs. Soon after, Libert et al. [26] introduced a scheme that eases the process of enrolling new members. By leveraging zero-knowledge arguments, members can prove knowledge of a private short vector, and the group manager certifies the corresponding public syndrome. In 2017, Ling et al. [31] presented a fully dynamic lattice-based group signature scheme by extending the static construction of Libert et al. [27] to support additions and revocations of users. Their design yields significantly smaller signatures while maintaining strong security in the sense of Bootle et al. [9], relying on the difficulty of SIS and LWE. Later, Ling et al. [32] proposed the first constant-size lattice-based group signature, whose signature size, public key, and users' private keys are all $\mathcal{O}(\lambda)$, depending only on the security parameter and not on N. The security hinges on the RLWE and RSIS assumptions. However, this scheme is partially dynamic signature scheme.

Progress in Efficiency and Security Models

Katsumata et al. [23] constructed two group signatures without resorting to NIZK proofs, achieving CCA-selfless anonymity and full traceability in the standard model under LWE and SIS assumptions. Canard et al. [12] then presented a constant-size lattice-based group signature with forward security in the standard model, adapting the attribute-based signature approach of Tsabary [42] to remove the reliance on the random oracle model (ROM). In 2021, Sun et al. [41] constructed an efficient fully dynamic group signature with message-dependent opening, combining the scheme of Ling et al. [29] with a double-encryption approach. Their improved zero-knowledge proof reduces the soundness error to $\frac{1}{\max(n,p)+1}$ (where n is a positive integer such that $n = \mathcal{O}(\xi)$, ξ is security parameter, p is a prime number such that $p \geq \beta. \omega(\sqrt{nlogn})$, $\beta = poly(n)$), lowering the overall communication overhead. Subsequent work by Sahin and Akleylek [39] in 2022 introduced a partially dynamic protocol in the quantum random oracle model (QROM). The scheme combines the Dilithium protocol, a Stern-like zero-knowledge proof (SLZK), and Regev's encryption, yielding notable asymptotic size efficiency. In 2023, Chen and Chen [15] offered a forward-secure solution for anonymous authentication that only requires sampling discrete Gaussian vectors during key updates. By keeping private keys linear in the lattice dimension rather than quadratic, their protocol lends itself to more lightweight applications. In 2024, Gao et al. [18] integrated timestamp-based revocation with verifier-local revocation (VLR) and provided a lattice-based scheme offering almost full anonymity. Although its signing key and signature sizes depend solely on λ, the group public key size still grows logarithmic in N.

Our Proposal

Building on the rapid progress in lattice-based group signatures, this work proposes a new VLR-enabled group signature protocol. Our scheme achieves signature size, signing key size, and group public key size independent of count of users in the group. The construction leverages the hardness of MLWE-MSIS, aiming for an efficient and fully dynamic protocol that remains practical under quantum threats.

3 Overview of the Protocol

The maximum no. of group members in the setup be $N = 2^p$, with $p \in \mathbb{Z}^+$, and the no. of maximum time periods $T = 2^{\mathcal{D}}$, where $\mathcal{D} \in \mathbb{Z}^+$ is the depth of the complete subtree used in the construction. Let ξ be the security parameter. We represent polynomial ring $\frac{\mathbb{Z}[x]}{x^d+1}$ and quotient ring $\frac{\mathbb{Z}_q[x]}{x^d+1}$ respectively by R and R_q. The global parameters of the protocol are defined on the basis of ξ are {n, α, k, m , q, r} such that n = $\mathcal{O}(\xi)$ (module rank for MSIS and MLWE), $k = \omega(log(\xi))$(a positive integer), m = 2nk, $\alpha = \lceil \rho.log(m) \rceil$ (a natural number, where ρ is a gaussian parameter), q $\geq \alpha\sqrt{d.m}$ (a prime number), r= $\omega(log\ n)$ (number of iterations in zero knowledge protocol run by verifier). There are four entities in our group signature protocol; Group Manager, Opener, Group members and verifier. In this protocol, every user get assigned with termination time t_d. If t_d passes, the member will be naturally removed from the group. Our protocol applies complete subtree for time-bound keys algorithm in [17]. In proposed construction, each leaf node of tree depicts as a time node with a value from the set {1,2,3, $\cdots$ t}. At current time t, the members which has termination time 1, 2,3, $\cdots$, t-1 will be removed from the group. For unterminated members, they have to prove that they possess a valid signing node. In other form, we can say, only members who are not terminated have signing node.

For reducing the size of the revocation list $\mathcal{RL}$, we use the idea of updating $\mathcal{RL}$ in the protocol [17]. We create a new list of revoked users $\mathcal{RU}_t$ before creating updated $\mathcal{RL}_t$. The respective revocation tokens are calculated as per $\mathcal{RU}_t$ and add these tokens into $\mathcal{RL}_t$. Moreover, the tokens of terminated users will not be saved in the $\mathcal{RL}_t$. In place of that, only tokens of non-terminated users who are malicious will be saved. Thanks to this method, this helps in the dramatic reduction in the size of revocation list. Moreover, let Φ be the leaf assigned to the user according to termination time t_d then Path(Φ)= {$\phi_1, \phi_2, \cdots \phi_D$}, where ϕ_i are nodes of tree and ϕ_1 is root of tree. For every node of tree, we randomly sample 2T-1 matrices D_ϕ to denote every node of tree. According to the concept of complete subtree, we always have $\phi' \in$ Path(Φ) $\cap \mathcal{Y}_t$ if the member is not terminated, where $\mathcal{Y}_t$ is the list of signable nodes. For a user who wants to create the signature, he requires to show :(i) he is not terminated; (ii) not removed; (iii) his index is correctly ciphered as (c_1, c_2). Towards proving that (i) and (ii) are right, it is essential to show that $M.Z + [M|D'_\phi].Z'_\phi = U$ and $\mathcal{V}_1 = \mathcal{V}_2.grt[d][t] + Y$ mod q satisfies. The essence of these values and equality is well explained in the protocol. Finally, the proof of scheme adopts the zero knowledge method of [25,30]. The speciality of this protocol is that its group public key size is dependent only on maximum time periods taken in protocol, and independent of number of members in the group. This feature makes it efficient for the scenarios, when the group size grows. Also, it makes execution practical for devices with limited computational power.

4   Preliminaries

In this section, we provide the detailing of necessary definitions, lemmas and notations which we will use in our work. Table 1 describes the notations used in our paper.

Norm: For polynomial $b = b_0 + b_1x + b_2x^2 + \cdots + b_{d-1}x^{d-1} \in R = \frac{\mathbb{Z}[x]}{\langle x^d+1 \rangle}$, the notion $||b||_\infty$ is defined as max $|b_j|$ and $||a||_p = (|a_0|^p + |a_1|^p + \cdots + |a_{d-1}|^p)^{1/p}$.

Similarly, for $b = (b_0, b_1, \cdots, b_k) \in R^k$, we define $||b||_\infty = \max ||b_j||_\infty$ and $||b||_p = (||b_1||_\infty^p + ||b_2||_\infty^p + \cdots + ||b_k||_\infty^p)^{1/p}$. By default we consider $||b|| = ||b||_2$.

Table 1. Notations used in our protocol

Notations	Description
ξ	Security Parameter
n,m,k	Integers such that n $= \mathcal{O}(\xi)$, k$= \omega(log\xi)$, m$=$2nk
R	Quotient Ring $\frac{\mathbb{Z}[X]}{x^d+1}$
d	Dimension of R
R_q	Quotient Ring $\frac{\mathbb{Z}_q[X]}{x^d+1}$
ρ	Gaussian parameter
α	$\lceil \rho.log(m) \rceil$
N	Max number of users in group
(M, T_M)	Group manager's private public key pair
(O, T_O)	Opening authority's private public key pair
GPK	Group public key
GSK	Signing key

Definition 1. *Module SIS (MSIS) Problem: Chosen $A = [a_1, a_2, \cdots, a_m] \in R_q^n$ from the uniform distribution, the MSIS challenge targets to find the non-zero vector $x = (x_1, x_2, \cdots x_m) \in R$ satisfying $\Sigma_{i=1}^m a_i.x_i = 0 \bmod q$ and $0 \leq ||x|| \leq \alpha$, where $x = (x_1, x_2, \cdots, x_m) \in R^m$ [16].*

Definition 2. *Module LWE (MLWE) Problem: Let $\mathcal{M} = R^n$ be a module over the ring R. Let $s \in R_q^n$ be a vector and $\chi \subset R_q$ be error probability distribution. The MLWE distribution $B_{q,s,\chi}^{\mathcal{M}}$ is given as follows: $(c, d = s.c + \epsilon) \in R_q^n \times R_q$, where $c \in R_q^n$ is a chosen vector and $\epsilon \leftarrow \chi$ is an error vector. MLWE search problem is to find $s \in R_q^n$ for n independent samples drawn from $B_{q,s,\chi}^{\mathcal{M}}$ [16].*

Lemma 1. *Suppose integers n, $m = \mathcal{O}(n \, logq)$, and $q \geq 2$; a prime number. The execution of TrapGen(n, m, q) results a tuple (M, T_M), where $M \in R_q^{n \times m}$ is tending to uniform distribution. $T_M \in R^{m \times m}$ is a matrix of good basis of $\wedge_q^\perp(M)$ as column vectors, holding $M.T_M = 0 \bmod q$ & $||\overline{T_M}|| \leq \mathcal{O}(\sqrt{n \, logq})$ [2].*

Lemma 2. *(SampleD [19]). On input $M \in R_q^{n \times m}$, a trapdoor matrix T_M of $\wedge_q^\perp(M)$, $U \in R_q^{n \times k}$, and $\rho = \omega(\sqrt{nlog(q)log(n)})$, the algorithm SampleD$(T_M, M, U, \rho)$ results $\mathcal{Z} \in R^{m \times k}$, satisfying $M.\mathcal{Z} = U \bmod q$ such that $||\mathcal{Z}|| \leq \alpha$.*

Lemma 3. *(SampleLeft [1]) On input $M_1, M_2 \in R_q^{n \times m}$, a trapdoor matrix T_{M_1} of $\wedge_q^\perp(M_1)$, $\mathcal{U} \in R_q^{n \times k}$, and a real $\rho > ||\overline{T_{M_1}}||.\omega(\sqrt{log(2m)})$, the algorithm SampleLeft$(M_1, M_2, T_{M_1}, \mathcal{U}, \rho)$ outputs $Z_M \in R^{2m \times k}$ that satisfies $[M_1|M_2]Z_M = \mathcal{U}$ such that $||Z_M|| \leq \alpha$.*

4.1   Security Model

This part explains working of oracle queries and formal definition of security terminologies.

We introduce a sequence of oracles to align the various queries of opponent $\Re$ in various experiments as below.

JoinU(i, t_i): On input user index i & expiration value t_i, this experiment computes $(GSK_i, cert_i, t_i)$ by executing the Join process & includes i to legitimate user list LUL, where GSK_i is the signing key and $cert_i$ is the membership certificate.

Mreg(i, σ): This oracle allows opponent $\Re$ to change the value σ of reg[i], where reg[i] has registration details according to index i and σ is the random binary string of length n.

Rreg(i): This oracle allows $\Re$ to access reg[i] corresponding to index i.

GetCert(i): On input index i, GetCert experiment results certificate of member corresponding to index i, then includes member i to DUL. DUL is a collection of members whose certificates are disclosed.

ReqToken(i,t): On input index i & time t, this experiment results token grt[i][t].

DiscloseU(i): On input index i, DiscloseU experiment outputs private key GSK_i and includes i to PUL. PUL is a collection of members whose private keys are disclosed.

Rev($\mathcal{RU}_t$): On input $\mathcal{RU}_t$, this experiment generates $\mathcal{RL}_t$ as output of Revoke(GPK, MSK, t, $\mathcal{RU}_t$) (described in Sect. 5.1) and includes $\mathcal{RU}_t$ to $\mathcal{RU}$. $\mathcal{RU}$ is the overall list of terminated members. This experiment returns $\mathcal{RL}_t$.

Sign(i,m,t): On input index i, a message m and present time t, this experiment returns a signature $\mathcal{S}$ & includes (t, $\mathcal{S}$, m) to list of signature $\mathcal{SL}$.

Open($\mathcal{S}$): On input $\mathcal{S}$, this experiment returns the identity i and time of signing j.

$\mathcal{CH}_b(j_0, j_1, m^*)$: After getting the two challenge identities (j_0, j_1) from $\Re$, this oracle arbitrarily chooses j_b to output a signature $\mathcal{S}_b^* = (\Pi^*, OVK^*, c_1^*, c_2^*, sig^*) \leftarrow$ Sign$(GPK, GSK^*, cert^*, m^*, t^*)$ (described in Sect. 5.1). Here, $j_0, j_1 \in$ LUL; $j_0, j_1 \notin \mathcal{RU}$; $j_0, j_1 \notin$ DUL & both are not terminated. This experiment can be tested only once.

Definition 3. *Traceability: This security feature ensures that without a private key, no adversary can generate a valid signature. Furthermore, it guarantees the traceability of users who produce valid signatures. The attacker is unable to create valid signatures using the private key that has expired.*

A group signature protocol is traceable for any PPT opponent $\Re$, if the benefit

$$Adv_{GS,\Re}^{trace}(\xi) := |P[exp_{GS,\Re}^{trace}(\xi) = 1]|$$

is minimal in the experiment $exp_{GS,\Re}^{trace}(\xi)$ given in Table 2.

Table 2. Traceability Experiment

Experiment: $exp_{GS,\Re}^{trace}(\xi)$
1. (GPK, T_M, reg)$\leftarrow$ $KeyGen(\xi)$; LUL:= ϕ; PUL:= ϕ; $\mathcal{SL}$:= ϕ;
2. $\Re \rightarrow$ JoinU(.), DiscloseU(.), ReqToken(.), Rev(.), Sign(.), Rreg(.);
3. $(\mathcal{S}^*, m^*, t^*) \leftarrow \Re(GPK)$;
4. If Verify(GPK, t^*, $\mathcal{S}^*$, m^*, $\mathcal{RL}_{t^*}$) = 0, return 0;
5. if (i) $\wedge$ (ii) $\wedge$ (iii) $\vee$ (iv) satisfies, return 1:
(i) $(t^*, \mathcal{S}^*, m^*) \notin \mathcal{SL}$,
(ii) d $\leftarrow$ Open(GPK, T_O, t^*, m^*, $\mathcal{S}^*$),
(iii) d $\notin$ PUL$\setminus \mathcal{RU}_{t^*} \vee$ d $= \perp$,
(iv) d $\in$ PUL $\setminus \mathcal{RU}_{t^*} \wedge t_d < t^*$;
6. Otherwise return 0.

Definition 4. *Non-Frameability: According to this security requirement, no adversary can generate a valid signature, which indicates an innocent user.*

A group signature protocol is non-frameable for any PPT opponent $\Re$, if the benefit

$$Adv_{GS,\Re}^{Non-Fr}(\xi) := |P[exp_{GS,\Re}^{Non-Fr}(\xi) = 1]|$$

is minimal in the experiment $exp_{GS,\Re}^{Non-Fr}(\xi)$ given in Table 3.

Definition 5. *Backward Unlinkability-Almost Full Anonymity (BU-AFA) [18]: A protocol achieves BU-AFA for any PPT opponent $\Re$, if benefit of opponent*

$$\boldsymbol{Adv}_{GS,\Re}^{BU}(\xi) := |P[exp_{GS,\Re}^{BU}(\xi) = 1] - \tfrac{1}{2}|$$

is minimal in the experiment $exp_{GS,\Re}^{BU}(\xi)$ given in Table 4.

Table 3. Non-Frameability Experiment

Experiment: $exp_{GS,\Re}^{Non-Fr}(\xi)$
1. (GPK, T_M, reg)$\leftarrow KeyGen(\xi)$; LUL:= ϕ; PUL:= ϕ; $\mathcal{SL}$:= ϕ;
2. $\Re \rightarrow$ DiscloseU(.), ReqToken(.), Sign(.), Rreg(.);
3. $(\mathcal{S}^*, m^*, t^*, d^*) \leftarrow \Re(GPK, MSK)$;
4. If (i)$\wedge$ (ii) $\wedge$ (iii) satisfies, return 1:
$\qquad\qquad$ (i) Verify(GPK, $t^*, \mathcal{S}^*, m^*, \mathcal{RL}_{t^*}$)=1,
$\qquad\qquad$ (ii) $d^* \leftarrow$ Open(GPK, Osk, $t^*, m^*, \mathcal{S}^*$),
$\qquad\qquad$ (iii) $d^* \in$ LUL $\wedge d^* \notin PUL \wedge (t^*, \mathcal{S}^*, m^*) \notin \mathcal{SL}$;
6. Otherwise return 0.

Table 4. Backward Unlinkability Experiment

Experiment: $exp_{GS,\Re}^{BU}(\xi)$
1. (GPK, T_M, reg)$\leftarrow KeyGen(\xi)$; LUL:= ϕ; $\mathcal{RU}$:= ϕ; DUL:= ϕ;
2. $\Re \rightarrow$ JoinU(.), Open(.), Mreg(.), ReqToken(.), GetCert(.), $\mathcal{CH}_b(.)$;
3. b' $\leftarrow \Re(GPK, GSK, T_M, t)$;
4. Return 1 if b' $=$ b, and 0 otherwise.

5 Group Signature Protocol

Let us suppose that ξ is security parameter, T is maximum number of time periods and N is number of maximum users. The group signature protocol comprises five phases, namely, KeyGen, Join, Revoke, Sign and verify. The working of each phase is described below in detail.

5.1 Description of the Protocol

- **KeyGen**($1^\xi, T, N$): The steps of working of this phase are explained as follows:

Table 5. CS-TBK Algorithm [18]

Input: a tree τ and a time t.
Output: a set of elements as nodes and named as $\mathcal{Y}$.
1: Choose two empty lists $\mathcal{X}$ and $\mathcal{Y}$;
2: for (j=1, j< t, j++) do
3:$\qquad$ include Path(Φ) to $\mathcal{X}$;
4: end
5: for each element ϕ belongs to $\mathcal{X}$ do
6:$\qquad$ if $\phi_{left} \notin \mathcal{X}$ then include ϕ_{left} to $\mathcal{Y}$ end if
7:$\qquad$ if $\phi_{right} \notin \mathcal{X}$ then include ϕ_{right} to $\mathcal{Y}$ end if
8: end for
9: if $\mathcal{Y}$ is non-empty, then
10: return $\mathcal{Y}$

1. Select n $= O(\xi)$, $\alpha = \lceil \rho.log(m) \rceil$ (a natural number, where ρ is a gaussian parameter), k$= \omega(log(\xi))$, l$=\omega(log(\xi))$, m$=$2nk, q $\geq \alpha \sqrt{d.m}$, r $= \omega(logn)$.
2. Derive manager's public private key pair (M, T_M) by executing Trap-Gen(n,m,q) in which M $\in R_q^{n \times m}$ (public) and its trapdoor $T_M \in R_q^{m \times m}$(private).
3. Uniformly sample U $\in R_q^{n \times k}$.
4. Derive opener's private public key pair (O, T_O) by executing Trap-Gen(n,m,q) in which O $\in R_q^{n \times m}$ (public) and its trapdoor $T_O \in R_q^{m \times m}$(private).
5. Suppose $\mathcal{H}_0 : \{0,1\}^* \to R_q^{n \times l}$, $\mathcal{H}_1 : \{0,1\}^* \to \{1,2,3\}^r$, $\mathcal{H}_2 : \{0,1\}^* \to R_q^{n \times k}$ and $\mathcal{G} : \{0,1\}^* \to R_q^{m \times n}$ are constructed as hash functions.
6. Generate a tree τ, uniformly select $D_\phi \leftarrow R_q^{n \times m}$ respective to every node $\phi \in \tau$.
7. Results GPK $=$ (M,O, $\{D_\phi\}_{\phi \in \tau}$,U), the private key of manager MSK$= T_M$ and the tracing key of opener Osk$= T_O$.

- **Join**(GPK, MSK, t_d) Every user who wants to join need to execute this process. Group manager sends signing key & membership certificates by communicating with user.
  1. User initiates joining request to group manager with sending t_d, where t_d is a time selected randomly from the set $\{1, 2, \cdots, T\}$. Group manager provides an index d $=$ bin(count) $\in \{0,1\}^l$ (binary expansion of length l); count is the position where user is joining. Then group manager compute $U_d = \mathcal{H}_2(d)$.
  2. After that, Group manager alot a leaf Φ to termination time t_d corresponding to index d. Note that members with identical termination time could be mapped to the identical leaf.
  3. Find $\mathcal{Z} \leftarrow$ SampleD(M, T_M, U_d, ρ). We notice that $\mathcal{Z} \in R^{m \times k}$ such that M $\mathcal{Z} = U_d$ and $||\mathcal{Z}||_\infty \leq \alpha$.
  4. After that, for every $\phi \in$ Path(Φ), Group manager calculates $\mathcal{Z}_\phi \in R^{2m \times k}$ by using SampleLeft algorithm, $\mathcal{Z}_\phi \leftarrow$ SampleLeft(M, $D_\phi, T_M, U - U_d, \rho$) such that $[M|D_\phi]\mathcal{Z}_\phi = U - U_d$.
  5. Fix the signing key GSK as $\mathcal{Z}_2 \in R^{nk \times k}$ and calculate initial token as grt[d] $= M_2.\mathcal{Z}_2$ from $\mathcal{Z} = \begin{bmatrix} \mathcal{Z}_1 \\ \mathcal{Z}_2 \end{bmatrix}$ and $M = [M_1|M_2]$.
  6. Send the membership certificate $=$ (d, $\mathcal{Z}_1$) and signing key $\mathcal{Z}_2$ to the user secretly, and keep (t_d, grt[d], $\{\mathcal{Z}_\phi\}_{\phi \in Path(\Phi)}$) to reg[d] (registration details corresponding to index d).
  7. count: count$+1$.

- **Revoke**(GPK,MSK,t, $\mathcal{RU}_t$). The present time is t. Using the CS-TBK algorithm, group manager updates $\mathcal{Y}_t$, revoked member list $\mathcal{RU}_t$, & revocation token list $\mathcal{RL}_t$.
  1. Deriving Signable Nodes: For present time t, get $\mathcal{Y}_t := (\phi_1, \cdots, \phi_{num}) \leftarrow$ CS-TBK(τ, t) as described in Table 5.
  2. Deriving Revocation Token List: Group manager randomly selects $\mathcal{W}_t \in R_q^{n \times nk}$, and for every d $\in \mathcal{RU}_t$, calculates grt[d][t]$= \mathcal{W}_t.\mathcal{Z}_1 + grt[d]$ (Note that, $\mathcal{W}_t.\mathcal{Z}_1 + grt[d] = [\mathcal{W}_t|M_2].\mathcal{Z}$) and generate $\mathcal{RL}_t = (\mathcal{W}_t, \{grt[d][t]\}_{d \in \mathcal{RU}_t})$.

   3. Output $(\mathcal{Y}_t, \mathcal{RL}_t)$.

– **Sign**(GPK,GSK,cert, m,t): Suppose $t < t_d$, the message m $\in \{0,1\}^*$. The signer fetch $\mathcal{W}_t$ from $\mathcal{RL}_t$. Now, $\exists$ a node $\phi' \in Path(\varPhi) \cap \mathcal{Y}_t$, as per CS-TBK algorithm. The user follows the below process to create group signature on message m.

   1. Get a pair of signing and verification keys (OVK,OSK)$\leftarrow$ OGen(ξ), by executing a certain key generation algorithm according to security parameter ξ to obtain one time verification key and signing key.

   2. After that, user calculates $G =\mathcal{H}_0(OVK) \in R_q^{n \times l}$. Then encrypt user's identity d using OVK. In order to do that, user randomly selects s $\leftarrow \chi^n$, $\epsilon_1 \leftarrow \chi^m$, $\epsilon_2 \leftarrow \chi^l$, and calculates $(c_1, c_2) \in R_q^m \times R_q^l$ as

$$(c_1 = O^T.s + \epsilon_1, c_2 = G^T.s + \epsilon_2 + \lfloor \tfrac{q}{2} \rfloor.d).$$

   3. Computes grt[d][t]$= [\mathcal{W}_t|M_2].\mathcal{Z}$. Then select $\sigma \leftarrow \{0,1\}^n$ randomly. Sample each noise vector $y_1, y_2, \cdots y_k \leftarrow R_q^m$ according to χ^m $(\|y_i\|_\infty \leq \alpha)$ and take noise matrix Y$= (y_1, y_2, \cdots, y_k) \in R_q^{m \times k}$. Let $\mathcal{V}_2 = \mathcal{G}(M, \mathcal{W}_t, m, \sigma) \in R_q^{m \times n}$ and calculate $\mathcal{V}_1 = \mathcal{V}_2.grt[d][t] + Y$ mod q.

   4. Create a Zero Knowledge Argument of Knowledge Π to show that there is a valid witness = (d, s, ϵ_1, ϵ_2, $\mathcal{Z}$, $\mathcal{Z}_{\phi'}$, Y) by executing the scheme r $= \omega(log\ n)$ times, like the below equalities satisfies:

$$c_1 = O^T.s + \epsilon_1,$$

$$c_2 = G^T.s + \epsilon_2 + \lfloor \tfrac{q}{2} \rfloor.d,$$

$$M.\mathcal{Z} + [M|D_{\phi'}].\mathcal{Z}_{\phi'} = U,$$

$$\mathcal{V}_1 = \mathcal{V}_2.grt[d][t] + Y mod\ q.$$

   5. Show that the member is legitimate and not terminated with the public values (M, $D_{\phi'}$, $\mathcal{W}_t, U, O, G, c_1, c_2, t$) and the witness (d, s, ϵ_1, ϵ_2, $\mathcal{Z}$, $\mathcal{Z}_{\phi'}$, Y). Then construct $\Pi = (\{CMT^{(j)}\}_{j=1}^r, \text{CH}, \{RSP^{(j)}\}_{j=1}^r)$, where CH$=$ $\{Ch^{(j)}\}_{j=1}^r = \mathcal{H}_1(m, \mathcal{W}_t, c_1, c_2, t, \{CMT^{(j)}\}_{j=1}^r)$.

   6. Calculate a one -time signature sig $\leftarrow$ OSign(OSK,c_1, c_2, Π) and derive group signature as $\mathcal{S}= (\Pi$, m, OVK, $c_1, c_2, \sigma, \mathcal{V}_1$, sig).

– **Verify**(GPK, t, m, $\mathcal{S}$, $\mathcal{RL}_t$) The steps of this process are provided below.

   1. Segment $\mathcal{S}$ as (Π, m, OVK, $c_1, c_2, \sigma, \mathcal{V}_1$, sig).

   2. If OVer(OVK, sig, c_1, c_2, Π)$\rightarrow$0, return 0;

   3. Segment Π as ($\{CMT^k \}_{k=1}^r$, CH, $\{RSP^j \}_{j=1}^r$).

   4. If $CH \neq \mathcal{H}_1(m, \mathcal{W}_t, c_1, c_2, t, \{CMT^{(j)}\}_{j=1}^r)$, output 0;

   5. For k $\in$ [r], execute steps of verification of zero knowledge protocol described in Table 6 with public parameters to decide that $RSP^{(j)}$ comprises $CMT^{(j)}$ and $Ch^{(j)}$. If any condition fails to satisfy then output 0;

   6. Calculate $\mathcal{V}_2 = \mathcal{G}(M, \mathcal{W}_t, m, \sigma)$. If $\exists\ r_i$ in $\mathcal{RL}_t$ such that $Y' = \mathcal{V}_1 - \mathcal{V}_2.r_i$ mod q and $\|Y'\|_\infty \leq \alpha$ satisfies, return 0;

   7. Return 1.

- **Open**(GPK, Osk, t, m, $\mathcal{S}$) Look for the result of Verify(GPK, t, m, $\mathcal{S}$, $\mathcal{RL}_t$), & terminate if it is 0. Otherwise, the explanation of decrypting steps is presented below.
  1. Segment $\mathcal{S}$ as $\mathcal{S} = (\Pi, \text{m }, \text{OVK}, c_1, c_2, \sigma, \text{sig})$.
  2. Calculate $G = \mathcal{H}_0(OVK)$.
  3. Find $F_{OVK} \leftarrow \text{SampleD}(O, T_O, G)$ such that $O.F_{OVK} = G \bmod q$.
  4. Use F_{OVK} to decrypt (c_1, c_2) as

$$\bar{d} = \lfloor c_2 - F_{OVK}^T c_1 \rceil \in \{0,1\}^l.$$

  5. Output $\bar{d}$.

5.2   Correctness

Correctness shows that if the verification steps executed successfully then the verified sign is correct. Here, we are showing that the verified sign is created by the non- revoked user. Verifier calculates $Y_i' = \mathcal{V}_1 - \mathcal{V}_2.r_i$. We can write $\mathcal{V}_1$ from group signature as $\mathcal{V}_1 = \mathcal{V}_2.grt[d][t^*] + Y$ and then we can show that $grt[d][t^*] \notin \mathcal{RL}_{t^*}$. We have, $Y_i' = \mathcal{V}_1 - \mathcal{V}_2.r_i = \mathcal{V}_2.grt[d][t^*] + Y - \mathcal{V}_2.r_i = \mathcal{V}_2[grt[d][t^*] - r_i] + Y$.
Case 1: If output is valid then verification step says $||Y_i'|| \geq \alpha$ (1) and we know $||Y|| \leq \alpha$ (2) as per the algorithm.
Now if r_i is chosen one by one for all values of $\mathcal{RL}_{t^*}$. So, by following (1) and (2), we have $\forall\, r_i$, $grt[d][t^*] - r_i \neq 0$. Because if $grt[d][t^*] - r_i = 0$ then $||Y_i'|| \geq \alpha$ and $||Y|| \leq \alpha$ is absurd. So, $grt[d][t^*] - r_i \neq 0 \implies grt[d][t^*] \notin \mathcal{RL}_{t^*}$. Hence, if sign is valid then it is proved that it is created by non-revoked user.
Case 2: If output is invalid then according to verification steps, we get this when $||Y_i'|| \leq \alpha$. We have expression, $Y_i' = \mathcal{V}_2[grt[d][t^*] - r_i] + Y$. Let's assume $grt[d][t^*] \notin \mathcal{RL}_{t^*}$ and $Q_i = grt[d][t^*] - r_i \bmod q$. Then $Q_i \neq 0 \bmod q$. We have, $||\mathcal{V}_2.Q_i|| = ||Y_i' - Y||$. By triangular inequality, $||\mathcal{V}_2.Q_i|| \leq ||Y_i'|| + ||Y|| \leq ||Y_i'|| + \alpha$. Now, by lemma 4 in [29], we have $||\mathcal{V}_2.Q_i|| \geq 2\alpha \implies -||\mathcal{V}_2.Q_i|| \leq -2\alpha$ and $-||\mathcal{V}_2.Q_i|| \geq -||Y_i'|| - \alpha \implies -2\alpha \geq -||Y_i'|| - \alpha \implies -\alpha \geq -||Y_i'|| \implies ||Y_i'|| \geq \alpha$ but we have, $||Y_i'|| \leq \alpha$ as output is invalid. So, our assumption is wrong and hence, $grt[d][t^*] \in \mathcal{RL}_{t^*}$. So, we get output is invalid, when the user is revoked.

Now, we are proving the correctness for tracing. We have, $\bar{d} = c_2 - F_{OVK}^T c_1 = G^T.s + \epsilon_2 + \lfloor \frac{q}{2} \rfloor.d - F_{OVK}^T(O^T.s + \epsilon_1) = G^T.s + \epsilon_2 + \lfloor \frac{q}{2} \rfloor.d - (O.F_{OVK})^T.s - F_{OVK}^T(\epsilon_1) = G^T.s + \epsilon_2 + \lfloor \frac{q}{2} \rfloor.d - G^T.s - F_{OVK}^T(\epsilon_1) = \epsilon_2 + \lfloor \frac{q}{2} \rfloor.d - F_{OVK}^T(\epsilon_1)$.

Finally, d can be recovered from $\bar{d}$ following the regev decryption that $d_j = 0$ if $\bar{d}_j$ is closer to 0, else $d_j = 1$ because $\bar{d} = \lfloor \epsilon_2 + \lfloor \frac{q}{2} \rfloor.d - F_{OVK}^T(\epsilon_1) \rceil$. The open algorithm results the identity d with overwhelming chances.

5.3   Zero Knowledge Protocol

We begin by recalling that the core component of our group signature is an interactive zero knowledge protocol. This protocol enables the prover to demonstrate to the verifier that they are an authorized group member (i.e., they possess a valid private key) and have not been revoked (i.e., their revocation token does not appear in the verifier's revocation list $\mathcal{RL}_t$). To reduce the soundness error to a negligible level, the protocol shown in Table 6 is repeated $r = \omega(\log n)$ times. The idea of proving secrets in this interactive manner was proposed in [25]. The fundamental design of our interactive protocol is outlined in detail below.

Construction of Underlying Interactive Protocol. Prover takes input of public values as $\{ M, \{D_\phi\}_{\phi \in \tau}, \mathcal{W}_t, U, O, G, c_1, c_2, t, \mathcal{V}_1, \mathcal{V}_2\}$ and his goal is to prove knowledge of $Y \in \chi^{m \times k}$, $d \in \{0,1\}^l$, $s \leftarrow \chi^n$, $\mathrm{grt}[d][t] \in R_q^{n \times k}$, $\epsilon_1 \leftarrow \chi^m$, $\epsilon_2 \leftarrow \chi^l$, $\mathcal{Z} \in R_q^{m \times k}$, $\mathcal{Z}_\phi \in R_q^{2m \times k}$ such that following equations hold:

$$c_1 = O^T s + \epsilon_1 \tag{1}$$

$$c_2 = G^T s + \epsilon_2 + \lfloor \tfrac{q}{2} \rfloor . d \tag{2}$$

$$U = M\mathcal{Z} + [M|D_{\phi'}]\mathcal{Z}_{\phi'} \tag{3}$$

$$\mathcal{V}_1 = \mathcal{V}_2 . grt[d][t] + Y \bmod q \tag{4}$$

For this target, prover uses symmetric group S_n, which is group of permutations of n symbols. To handle secret objects, prover applies extending-then-permuting techniques [25]. The process of extending and permuting the secret values is described below.

- For proving the knowledge of s which belongs to χ^n, the prover first cocatenate n "dummy" entries to s to get $s^* \in \chi^{2n}$. The selection of s^* is such that, for every permutation $\Pi_s \in S_{2n}$,

$$s^* \in \chi^{2n} \iff \Pi_s(s^*) \in \chi^{2n}$$

- For proving the knowledge of ϵ_1 which belongs to χ^m, the prover first cocatenate m "dummy" entries to ϵ_1 to obtain $\epsilon_1^* \in \chi^{2m}$. The selection of ϵ_1^* is such that, for any $\Pi_\epsilon \in S_{2m}$,

$$\epsilon_1^* \in \chi^{2m} \iff \Pi_\epsilon(\epsilon_1^*) \in \chi^{2m}$$

- For proving the knowledge of ϵ_2 which belongs to χ^l, the prover append l "dummy" entries to ϵ_2 to get $\epsilon_2^* \in \chi^{2l}$. The selection of ϵ_2^* is such that, for any $\Pi_e \in S_{2l}$,

$$\epsilon_2^* \in \chi^{2l} \iff \Pi_e(\epsilon_2^*) \in \chi^{2l}$$

- For proving the knowledge of $\mathcal{Z}$ which belongs to $R_q^{m \times k}$, the prover append 2 zero columns to each column of $\mathcal{Z}$ to get $\mathcal{Z}^* \in R_q^{m \times 3k}$. The selection of $\mathcal{Z}^*$ is such that, for any $(\Pi_{z,1}, \Pi_{z,2}, \cdots, \pi_{z,3k}) \in (S_m)^{3k}$,

$$\mathcal{Z}^* \in R_q^{m \times 3k} \iff \Pi_{z,i}(z_i) \in R_q^m \ \forall \ 1 \leq i \leq 3k, \ \text{where} \ \mathcal{Z}^* = [z_1|z_2|\cdots|z_{3k}]$$

- For proving the knowledge of $\mathcal{Z}_\phi$ which belongs to $R_q^{2m\times k}$, the prover appends 2 zero columns to each column of $\mathcal{Z}_\phi$ to get $\mathcal{Z}_\phi^* \in R_q^{2m\times 3k}$. The selection of $\mathcal{Z}_\phi^*$ is such that, for any $(\Pi_{\phi_1}, \Pi_{\phi_2}, \cdots, \Pi_{\phi_{3k}}) \in (S_{2m})^{3k}$,

$$\mathcal{Z}_\phi^* \in R_q^{2m\times 3k} \iff \Pi_{\phi_i}(Z_i) \in R_q^{2m} \; \forall \, 1 \leq i \leq 3k, \; where \; \mathcal{Z}_\phi^* = [Z_1|Z_2|\cdots Z_{3k}]$$

- For proving the knowledge of $\mathcal{R} = \mathrm{grt}[d][t]$ which belongs to $R_q^{n\times k}$, the prover append 2k zero columns to each column of $\mathcal{R}$ to get $\mathcal{R}^* \in R_q^{n\times 3k}$. The selection of $\mathcal{R}^*$ is such that, for any $(\Pi_{r,1}, \Pi_{r,2}, \cdots, \Pi_{r,3k}) \in (S_n)^{3k}$,

$$\mathcal{R}^* \in R_q^{n\times 3k} \iff \Pi_{r,i}(\mathcal{R}_i) \in R_q^{n} \; \forall \, 1 \leq i \leq 3k, \; where \; \mathcal{R}^* = [\mathcal{R}_1|\mathcal{R}_2|\cdots|\mathcal{R}_{3k}]$$

- For proving the knowledge of Y which belongs to $\chi^{m\times k}$, the prover appends 2k zero columns to each column of Y to get $Y^* \in \chi^{m\times 3k}$. The selection of Y^* is such that, for any $(\Pi_{y,1}, \Pi_{y,2}, \cdots, \Pi_{y,3k}) \in (S_m)^{3k}$,

$$Y^* \in \chi^{m\times 3k} \iff \Pi_{y,i}(Y_i) \in \chi^{m}, \; \forall \, 1 \leq i \leq 3k, \; where \; Y^* = [Y_1|Y_2|\cdots Y_{3k}]$$

We now merge all the extended private components into a vector

$$X = (s^* || \epsilon_1^* || \epsilon_2^* || z_1 || z_2 || \cdots || z_{3k} || Z_1 || Z_2 || \cdots || Z_{3k} || \mathcal{R}_1 || \mathcal{R}_2 || \cdots || \mathcal{R}_{3k} || Y_1 || Y_2 || Y_{3k})$$

$$\in R_q^{D} \; where \; D = n + 2m + 2l + 12km + 3nk \tag{5}$$

Observe that, Eqns from (1) to (4) can be identically merged into single eqn $N.X = V \bmod q$ where N & V are constructed using the public values.

Table 6. Zero knowledge protocol between prover and verifier

---

1. **Commitment**: Prover samples $s_x \leftarrow R_q^D, \eta \in \overline{S}$ and randomness $\sigma_1, \sigma_2, \sigma_3$ for COM.

He transmits CMT $= (c_1, c_2, c_3)$ to verifier, where $c_1 = COM(\eta, N.s_x; \sigma_1)$, $c_2 = COM(\tau_\eta(s_x); \sigma_2)$, $c_3 = COM(\tau_\eta(X + s_x); \sigma_3)$

2. **Challenge**: Verifier transmits ch $\leftarrow \{1, 2, 3\}$ to the prover.

3. **Response**: According to ch, prover transmits RSP calculated as below:

-ch=1 : suppose $r_x = \tau_\eta(X)$, $r_\tau = \tau_\eta(s_x)$, $RSP = (r_x, r_\tau, \sigma_2, \sigma_3)$

-ch=2 : suppose $\eta_2 = \eta$, $X_2 = X + s_x$, $RSP = (\eta_2, X_2, \sigma_1, \sigma_3)$

-ch=3 : suppose $\eta_3 = \eta$, $X_3 = s_x$, $RSP = (\eta_3, X_3, \sigma_1, \sigma_2)$

Verification Getting responses, the verifier proceeds as below:

-ch =1: Verify that $r_x \in$ VALID and $c_2 = COM(r_\tau; \sigma_2)$, $c_3 = COM(r_x + r_\tau; \sigma_3)$

-ch =2: Verify that $c_1 = COM(\eta_2, N.X_2 - V, \sigma_1)$, $c_3 = COM(\tau_{\eta_2}(X_2), \sigma_3)$

-ch =3: Verify that $c_1 = COM(\eta_3, N.X_3, \sigma_1)$, $c_2 = COM(\tau_{\eta_3}(X_3), \sigma_2)$

In every scenario, verifier results 1 iff all the conditions hold.

Now, construct a set VALID as collection of all vectors in R_q^D possessing form (5). To show X $\in$ VALID using arbitrary permutations, let us check how to shuffle the co-ordinates of X. To this extent, we first construct the set: $\overline{S} = S_{2n} \times S_{2m} \times S_{2l} \times (S_m)^{3k} \times (S_{2m})^{3k} \times (S_n)^{3k} \times (S_m)^{3k}, where \, \forall \, \eta \in (\Pi_s, \Pi_\epsilon, \Pi_e, (\Pi_{z,1}, \Pi_{z,2}, \cdots \Pi_{z,3k}), (\Pi_{\theta,1}, \cdots, \Pi_{\theta,3k}), (\Pi_{r,1}, \Pi_{r,2}, \Pi_{r,3k}),$
$(\Pi_{y,1}, \Pi_{y,2}, \cdots, \Pi_{y,3k})) \in \overline{S}$, we let τ_η be a permutation.

When acting τ_η to t $\in R_q^D$, where blocks are as in (5), it transform these blocks as below:

$$s^* \to \Pi_s(s^*); \epsilon_1^* \to \Pi_\epsilon(\epsilon_1^*); \epsilon_2^* \to \Pi_e(\epsilon_2^*)$$

$$\forall \, 1 \leq i \leq 3k, z_i \to \Pi_{z,i}(z_i)$$

$$\forall \, 1 \leq i \leq 3k, Z_i \to \Pi_{\theta,i}(Z_i)$$

$$\forall \, 1 \leq i \leq 3k, \mathcal{R}_i \to \Pi_{r,i}(\mathcal{R}_i)$$

$$\forall \, 1 \leq i \leq 3k, Y_i \to \Pi_{y,i}(Y_i)$$

Now, it can be checked that $\forall \, \eta \in \overline{S}$, X $\in$ VALID $\iff \tau_\eta(X) \in$ VALID. The brief of the zero knowledge protocol between prover and verifier is presented in Table 6.

6   Security Analysis

The proposed group signature protocol achieves Backward Unlikability with Almost Full Anonymity (BU-AFA), Traceability and Non-Frameability attributes. In this section, the formal proof of each of the property is described in the random oracle model.

6.1   BU-AFA

Theorem 1. *In ROM, presented VLR group signature protocol achieves BU under the difficulty of $M - LWE_{n,q,\chi}$ challenge.*

Proof. We are taking a below series of indistinguishable games.
G_0: It is identical as Definition 5. The challenger $\mathcal{C}$ follows every step fairly. The benefit of opponent $\Re$ in this experiment is $\mathbf{Adv}_{GS,\Re}^{BU}(\xi)$ as per definition.

G_1: Here, we will create (OVK, OSK) with initialisation of the game. If $\Re$ enquiries Open experiment for $\mathcal{S} = (\Pi, \text{m}, OVK^*, c_1, c_2, \sigma, \text{sig})$, where $OVK^* = OVK$, $\mathcal{C}$ outputs a bit uniformly and terminates. Because OVK is created prior the query stage & does not depend on $\Re$'s query on Open oracle, $OVK^* = OVK$ is minimal. Hence, the chances that a correct signature having OVK is queried by $\Re$ is minimal. So, $\Re$ cannot differentiate.

G_2: We will acquire an arbitrary matrix O^* in place of creating matrix O applying TrapGen algo for this experiment. As per the lemma 1, O is statistically close to uniform distribution. Therefore, this change is indifferentiable for $\Re$.

G_3: Respective to the former game, we initiate modification in G. We arbitrarily select $G^* \in R_q^{n \times l}$. Then set $\mathcal{H}_0(OVK^*) = G^*$. We design the hash $\mathcal{H}_0$ constructed on the arbitrarily selected matrices O^* & G^*. If $\Re$ enquiries $\mathcal{H}_0(OVK^*)$, $\mathcal{H}_0$ outputs G^*. Since, the return of oracle is selected arbitrarily, this is indifferentiable for $\Re$. If $\Re$ queries $OVK \neq OVK^*$, we first find $F_{OVK} \leftarrow D_{\mathbb{Z},s_l}^{m \times l}$, such that $\mathcal{H}_0(OVK) = G = O^*.F_{OVK}$ mod q. Then $\mathcal{H}_0$ outputs G to $\Re$ and captures (OVK, F_{OVK}, G). When $\Re$ raises query $\mathcal{H}_0(OVK)$ again and again, oracle responds with the respective F_{OVK}. As per the hardness of M-SIS challenge, G is statistically close to uniform distribution. Hence, $\Re$ still is not able to differentiate the output.

G_4: Here, we selects an arbitrary uniform $\mathcal{Y} \leftarrow R_q^{n \times k}$ and assign $\mathcal{V}_1 = \mathcal{Y}$. As the token creating phase is not dependent on bit b and $(\mathcal{V}_1, \mathcal{V}_2)$ is close to distribution $R_q^{m \times k} \times R_q^{m \times n}$. If $\Re$ can differentiate between $\mathcal{V}_1$ and $\mathcal{Y}$, it demonstrates that $\Re$ can resolve the M-LWE challenge. Hence G_4 and G_3 are indifferentiable for $\Re$.

G_5: Here, we will create an implemented proof Π^* by $\mathcal{H}_1$ experiment of G_3 for the challenged $\mathcal{S}$. As the proof architecture is zero knowledge (ZK), it pursues to be indifferentiable from perspective of $\Re$.

G_6: Respective with former game, we take $(c_1^*, c_2^*) = (t_1^*, t_2^* + \lfloor \frac{q}{2} \rfloor.bin(i_b))$ in place of $(c_1^*, c_2^*) = (O^{*T}.s + \epsilon_1, G^{*T}.s + \epsilon_2 \lfloor \frac{q}{2} \rfloor.bin(i_b))$, where $t_1^* \leftarrow R_q^m$ and $t_2^* \leftarrow R_q^l$. With underlying hardness of MLWE problem, this is computationally indifferentiable for $\Re$. We can say, suppose $\mathcal{T} = [t_1^*|t_2^*] \in R_q^{m+l}$, E= $[O^*|G^*] \in R_q^{n \times (m+l)}$ and $\epsilon = [\epsilon_1|\epsilon_2] \in \chi^{m+l}$. Then differentiating G_6 and G_5 is similar to differentiating (E, $E^T.s + \epsilon$) and (E, $\mathcal{T}$).

G_7: Here, we randomly choose a pair (t_1', t_2') to change (c_1^*, c_2^*). Same, it is computationally indifferentiable for $\Re$. In this situation, $\mathcal{S}$ has been entirely free from test bit b. We can have the chances of $\Re$ conquer $Adv_{GS,\Re}^{BU}(\xi) = 0$. These games demonstrates that the protocol achieves backward unlinkability.

6.2 Traceability

Theorem 2. *In ROM, we demonstrate that the suggested protocol is traceable relying on the challenge of MSIS difficulty.*

Proof. In this scenario, the target of $\Re$ is to fake a right signature that is not traceable to any group member or track the member who is terminated or revoked.

We suppose that there is a new opponent $\mathcal{A}$ that can resolve MSIS challenge with significant chances. Initially, $\mathcal{A}$ creates a pair of (O, T_O) based on the public values $(M, O, \{D_\phi\}_{\phi \in \tau}, U)$. Then $\mathcal{A}$ communicates with $\Re$ by replying to queries of $\Re$ as given.

- **Private key queries:** On input an identity i, the experiment results GSK_i and includes i to PUL.
- **Signature queries:** On input message m & identity i, $\mathcal{S} \leftarrow$ Sign(GPK, GSK, cert, M,t) is created by $\mathcal{A}$ fairly. The ZK scheme supports that $\mathcal{S}$ & valid group signature is indifferentiable for $\Re$.
- **Random oracle queries:** An arbitrary number is responded for every query of both hashes, $\mathcal{H}_0$ and $\mathcal{H}_1$. Query for $\mathcal{H}_1$ is selected arbitrarily from $\{1,2,3\}^r$. Additionally, we represent the kth query to hash $\mathcal{H}_1$ by r_k.

Finally, $\Re$ results $(\mathcal{S}^*, m^*, t^*)$ and Verify(GPK, t^*, S^*, m^*, $\mathcal{RL}_{t^*}$) $=1$. Then, $\mathcal{A}$ can find the respective index j by tracing $\mathcal{S}^*$, and j holds one from the below:

1. j $\notin$ PUL $\setminus \mathcal{RU}_{t^*}$ $\bigvee j =\perp$;
2. j $\in$ PUL $\setminus \mathcal{RU}_{t^*}$ $\bigwedge t_d < t^*$.

Below explinition is for how $\mathcal{A}$ would utilize this forged sign.

We need that $\Re$ raise query to $\mathcal{H}_1$ on (m, $\mathcal{W}_t$, c_1, c_2, t, $\{CMT^{(k)}\}_{k=1}^r$) before raising to $\mathcal{H}_0$. With minimum chances $(\epsilon - 3^{-r})$, $\exists$ $k^* < Q_{\mathcal{H}}$ such that this oracle query has (m, $\mathcal{W}_t$, c_1, c_2, t, $\{CMT^{(k)}\}_{k=1}^r$). According to k^*, suppose $\Re$ initiates polynomially many times with identical input from starting raise. For every repeatation, the identical arbitrary number is returned by $\mathcal{A}$ for previous $k^* - 1$ queries and a fresh arbitrary value for k^{*th} query. As per the lemma 10 in [10], it says that $\mathcal{A}$ has minimum of half chances of getting 3-fork for $(m^*, \mathcal{W}_t, c_1, c_2, t^*, \{CMT^{(k)}\}_{k=1}^r)$ when executing $\Re$ more than $\frac{32Q_{\mathcal{H}}}{\epsilon-3^{-r}}$ times.

$$t_{k^*}^{(1)} = (CH_1^{(1)}, \cdots, CH_r^{(1)})$$

$$t_{k^*}^{(2)} = (CH_1^{(2)}, \cdots, CH_r^{(2)})$$

$$t_{k^*}^{(3)} = (CH_1^{(3)}, \cdots, CH_r^{(3)})$$

Then a computation can present the probability as P[$\exists$ i $\in$ {1,2,$\cdots$,r}:$\{CH_i^{(1)}, \cdots, CH_i^{(r)}\}$= {1,2,3}]=1- $(\frac{7}{9})^r$.

Now, we suppose that index i exists. We can extract 3 forgeries respective to fork branch and obtain $(RSP_i^{(1)}, RSP_i^{(2)}, RSP_i^{(3)})$. Then we obtain a right witness $(d^*, s^*, \epsilon_1^*, \epsilon_2^*, \mathcal{Z}^*, \mathcal{Z}_{\phi'}^*, Y^*)$ by the knowledge extractor. Note that q $=$ poly(n), b$= \mathcal{O}(\sqrt{n})$, and $c_1 = O^T.s^* + \epsilon_1^*, c_2 = G^T.s + \epsilon_2^* + \lfloor\frac{q}{2}\rfloor.d^*$, $M.\mathcal{Z}^* + [M|D_\phi'] = U, \mathcal{V}_1 = \mathcal{V}_2[\mathcal{W}_t|M_2]\mathcal{Z}^* + Y^* mod \, q$, where $||Y^*||_\infty \leq \alpha, ||s^*|| \leq$ b, $||\epsilon_1^*|| \leq$ b, $||\epsilon_2^*|| \leq$ b. After that, we test (c_1, c_2) is right encryption of d^*, & $d^* \leftarrow Open(GPK, MOSK, t^*, m^*, \mathcal{S})$. Hence, this is insignificant that faked signatures can be tracked to the honest member. Additionally, the ZK scheme ensures that indices are rightly encrypted & witness is right.

6.3   Non-Frameability

Theorem 3. *In ROM, we demonstrate that the suggested protocol is non-frameable on underlying hardness of M-SIS assumption.*

Proof. We suppose that a tuple $(m^*, \mathcal{S}^*)$ can be created by an opponent $\mathfrak{R}$. This pair can be tracked to an honest member i^* with significant chances ϵ. We suppose, there is another opponent $\mathcal{A}$ that can resolve MSIS challenge with minimal chances.

A tuple (MSK, MOSK) and GPK are created by $\mathcal{A}$ honestly. Then $\mathcal{A}$ sends GPK and MSK to $\mathfrak{R}$ & communicates with $\mathfrak{R}$ by responding all queries of $\mathfrak{R}$. Malicious GM can be performed by $\mathfrak{R}$ like it can join members to the group. Other than that, $\mathcal{A}$ creates & returns $\mathcal{S} = (\Pi, m, OVK, c_1, c_2, \sigma, sig)$ if $\mathfrak{R}$ queries the signature for m.

Then, $\mathfrak{R}$ returns $\mathcal{S}^* = (\Pi^*, m^*, OVK^*, c_1^*, c_2^*, \sigma^*, sig^*)$, This signature is traced and tracked member i^* who did not create signature. We have $grt[i^*] = M_2.\mathcal{Z}_2^* = M_2.\mathcal{Z}_2'$ mod q. We suppose that $\mathcal{Z}_2^* \neq \mathcal{Z}_2'$. Because $\mathfrak{R}$ queried reg$[i^*]$, it can get grt$[i^*]$ $=M_2.\mathcal{Z}_2^*$. However, $\mathfrak{R}$ does not know $\mathcal{Z}_2^*$. In the viewpoint of $\mathfrak{R}$, $\mathcal{Z}_2^*$ is selected from gauss distribution. Hence, it possess larger min entropy and $\mathcal{Z}_2^* \neq \mathcal{Z}_2'$ with high probability. The variation D$= \mathcal{Z}_2^* - \mathcal{Z}_2' \neq 0$ and $0 < ||D||_\infty \leq 2\alpha$. We can say, the M-SIS challenge is resolved. Apart from that, we suppose, $\mathfrak{R}$ also enquiries ReqToken to get token at t', then $\mathfrak{R}$ can get $grt[i^*][t']$. We have $D' = grt[i^*][t'] - grt[i^*] = \mathcal{W}_{t'}.\mathcal{Z}_1^* = \mathcal{W}_{t'}.\mathcal{Z}_1'$. We have, $\mathcal{Z}_1^* \neq \mathcal{Z}_1'$ with high chances.

7   Performance Analysis

This section presents the computation complexity, sizes of group public key, signing key and signature key of proposed protocol. Also, it provides the comparative study with respect to existing lattice-based group signature protocols. The parametric consideration is taken as ξ be security parameter, N be the max no. of members in the group, and T is the max size of time periods. The comparative chart of sizes for group public key, signing key and signature of Lattice-based group signature protocols alongwith backward unlinkability and dynamic nature is presented in Table 7. It shows our protocol attains strong security with smaller key and signature sizes alongwith having fully dynamic structure. Moreover, Our protocol attains group public key size, group signing key and signature independent of number of members in the group.

7.1   Analysis of Features

- **Anonymity:** The protocol of Ling et al. [33] uses Merkle trees to obtain FA & dynamic nature. The feature is commending, but the problem is that the size of GSK gets unmanagable. Additionally, the downside with respect to VLR is that all nodal entries in the path need to be upgraded once for every new user registered. As tokens of revocation cannot be fully disclosed to illegitimate members, it is challenging for protocol with VLR to attain full anonymity. By referring [36,38], we design our protocol to attain AFA, which is nearly attaining full anonymity.

- **Size of GSK:** The matrices associated with identity of members in the protocols of [25,29,37] has log N matrices combined together. This causes the size of GSK & no. of group members have a linear relationship. Our approach reduces GSK size to be independent of N.

- **Size of Revocation List:** Group signature methods with VLR leads to an inescapable issue of the revocation list expanding. Naturally, this would be one of the downsides of VLR-based methods. Suppose that information of member that was canceled years ago is still retained in a trading platform's database. In most real life applications, this data is superfluous. To mitigate this issue, we apply a comprehensive subtree technique, dividing group users into two categories: terminated and non-terminated. Terminated members will no longer be able to create a valid signature. In our protocol, the revocation list only include current revocation tokens for malevolent terminated users. In effect, it significantly reduces the length of the revocation list.

- **Size of GPK:** In [18], all the nodes of tree and matrices $\{A_i\}_{i=0}^{l}$ are saved as GPK which makes size of public key linearly dependent on max of time periods and no. of members in group. In our protocol, only nodes of complete subtree, public key of group manager and public key of opening authority are stored as group public key, and hence the size of GPK is linearly dependent on max of time periods only and not dependent on no. of users in the group, which makes it practicable for real scenarios where number of users in group is large. Therefore, the protocol achieves better efficiency with respect to existing fully dynamic group signature protocols.

- **Backward Unlinkability:** Our protocol attains BU, which facilitates with signatures created on distinct periods are not linked. Based on the M-SIS challenge, let us suppose, a revocation token of some member is disclosed at period t, even then an opponent cannot get any constructive output regarding the signing key and membership certificate. So, using that token, the revocation token for different time period or signature cannot be forged by the opponent.

7.2   Computation Complexity

Our group signature protocol supports VLR with BU. It provides AFA and traceability. The revocation mechanism of this protocol is efficient as it only keeps the tokens of members who have been terminated prior the expiration time, which reduces the list size of revoked users significantly. Moreover, for calculating signing key and certificate value, manager solves $M.\mathcal{Z} = U_d$ in order to get $\mathcal{Z}$. This makes the protocol highly efficient respective to [18] without compromising with the attained security attributes BU-AFA, traceability and non-frameability. Table 7 shows the comparison of GPK,GSK and signature sizes

Table 7. Attributes comparison with relevant group signature protocols

Protocols	GPK	GSK	Signature	BU	Dynamic
Gordon et al. [20]	$\mathcal{O}(\xi^2 N)$	$\mathcal{O}(\xi^2)$	$\mathcal{O}(\xi^2 N)$	×	×
Langlois et al. [25]	$\mathcal{O}(\xi^2 log(N))$	$\mathcal{O}(\xi log(N))$	$\mathcal{O}(\xi log(N))$	×	√
Nguyen et al. [35]	$\mathcal{O}(\xi^2 log^2(N))$	$\mathcal{O}(\xi^2)$	$\mathcal{O}(\xi + log^2(N))$	×	×
Libert et al. [26]	$\mathcal{O}(\xi^2 + \xi.log(N))$	$\mathcal{O}(\xi log(N))$	$\mathcal{O}(\xi log(N))$	×	√
Ling et al. [29]	$\mathcal{O}(\xi^2 log(N))$	$\mathcal{O}(\xi log(N))$	$\mathcal{O}(\xi log(N))$	×	√
Ling et al. [34]	$\mathcal{O}(\xi^2 log^2(N))$	$\mathcal{O}(\xi + log(N))$	$\mathcal{O}(\xi log(N))$	×	√
Ling et al. [33]	$\mathcal{O}(\xi^2(logN + logT))$	$\mathcal{O}(\xi^2(logN + logT)^2 logT)$	$\mathcal{O}(\xi(logN + logT))$	×	√
Perera et al. [37]	$\mathcal{O}(\xi^2 log(N))$	$\mathcal{O}(\xi log(N))$	$\mathcal{O}(\xi log(N))$	×	√
Perera et al. [36]	$\mathcal{O}(\xi^2 log(N))$	$\mathcal{O}(\xi log(N))$	$\mathcal{O}(\xi log(N))$	×	√
Zhang et al. [43]	$\mathcal{O}(\xi^2)$	$\mathcal{O}(\xi)$	$\mathcal{O}(\xi log(N))$	√	√
Gao et al. [18]	$\mathcal{O}(\xi^2(log(N) + T))$	$\mathcal{O}(\xi)$	$\mathcal{O}(\xi)$	√	√
Ours	$\mathcal{O}(\xi^2(1 + T))$	$\mathcal{O}(\xi)$	$\mathcal{O}(\xi)$	√	√

Table 8. Number of major time-consuming functions used in relevant schemes

Functions	Langlois et al. [25]	Ling et al. [29]	Perera et al. [37]	Gao et al. [18]	Ours
SampleLeft	1	1	1	3	3
TrapGen	1	1	1	2	2
Matrix Multiplication	$l+1$	$l+3$	10	8	8
$\mathcal{H}_1 : \{0,1\}^* \rightarrow \{1,2,3\}^r$	2	2	2	2	2
Matrix Addition	l	$l+2$	7	8	7
Gaussian Sampling	l	$l+1$	3	3	3
Scalar multiplication	0	0	l	$l+1$	1

of our protocol with respect to existing group signature schemes which highlights that our protocol stands out in achieving constant signature and GPK size.

Moreover, it achieves dependency of GPK size only on max time periods and independent of numbers of members in the group. Further, the comparison of number of use of major time consuming functions in relevant schemes are presented in Table 8. Schemes Langlois et al. [25] and Ling et al. [29] are more efficient but these schemes achieve only selfless anonymity. Moreover, GPK,GSK and signature size grows logarithmic with number of users as presented in Table 7. Perera et al. [37] scheme also suffer from same issues. Table 7 and 8 reflects that without increasing computation complexity, our scheme achieves advance features like constant signing key and signature size. Moreover, public key size is independent of number of users. Also, the construction supports full dynamicity and backward unlinkability.

8 Conclusion

We have designed an MLWE-MSIS based group signature protocol with backward unlinkability which supports verifier local revocation mechanism and could significantly reduce the privacy leakage threat provoked by linkability. Respective to available fully dynamic group signature protocols in the literature, our protocol achieves AFA, non-frameability and traceability security. Furthermore, proposed protocol excels in terms that group public key size, signing key size and signature size are independent of number of users in the group. In practical scenarios of cryptocurrencies, our approach could enhance efficiency while ensuring adequate privacy protection. However, the protocol with VLR is not able to achieve full anonymity due to the limitation that all the tokens cannot be revealed. Thus, for future, one can plan to explore ways to achieve full anonymity without increasing the computation overhead of proposed design.

References

1. Ajtai, M.: Generating hard instances of the short basis problem. In: Wiedermann, J., van Emde Boas, P., Nielsen, M. (eds.) ICALP 1999. LNCS, vol. 1644, pp. 1–9. Springer, Heidelberg (1999). https://doi.org/10.1007/3-540-48523-6_1
2. Alwen, J., Peikert, C.: Generating shorter bases for hard random lattices. Theory Comput. Syst. **48**, 535–553 (2011)
3. Arute, F., et al.: Quantum supremacy using a programmable superconducting processor. Nature **574**(7779), 505–510 (2019)
4. Ateniese, G., Camenisch, J., Joye, M., Tsudik, G.: A practical and provably secure coalition-resistant group signature scheme. In: Bellare, M. (ed.) CRYPTO 2000. LNCS, vol. 1880, pp. 255–270. Springer, Heidelberg (2000). https://doi.org/10.1007/3-540-44598-6_16
5. Bellare, M., Micciancio, D., Warinschi, B.: Foundations of group signatures: formal definitions, simplified requirements, and a construction based on general assumptions. In: Biham, E. (ed.) EUROCRYPT 2003. LNCS, vol. 2656, pp. 614–629. Springer, Heidelberg (2003). https://doi.org/10.1007/3-540-39200-9_38

6. Bellare, M., Shi, H., Zhang, C.: Foundations of group signatures: the case of dynamic groups. In: Menezes, A. (ed.) CT-RSA 2005. LNCS, vol. 3376, pp. 136–153. Springer, Heidelberg (2005). https://doi.org/10.1007/978-3-540-30574-3_11
7. Boneh, D., Boyen, X., Shacham, H.: Short group signatures. In: Franklin, M. (ed.) CRYPTO 2004. LNCS, vol. 3152, pp. 41–55. Springer, Heidelberg (2004). https://doi.org/10.1007/978-3-540-28628-8_3
8. Boneh, D., Shacham, H.: Group signatures with verifier-local revocation. In: Proceedings of the 11th ACM Conference on Computer and Communications Security, pp. 168–177 (2004)
9. Bootle, J., Cerulli, A., Chaidos, P., Ghadafi, E., Groth, J.: Foundations of fully dynamic group signatures. In: Manulis, M., Sadeghi, A.-R., Schneider, S. (eds.) ACNS 2016. LNCS, vol. 9696, pp. 117–136. Springer, Cham (2016). https://doi.org/10.1007/978-3-319-39555-5_7
10. Brickell, E., Pointcheval, D., Vaudenay, S., Yung, M.: Design validations for discrete logarithm based signature schemes. In: Imai, H., Zheng, Y. (eds.) PKC 2000. LNCS, vol. 1751, pp. 276–292. Springer, Heidelberg (2000). https://doi.org/10.1007/978-3-540-46588-1_19
11. Camenisch, J., Stadler, M.: Efficient group signature schemes for large groups. In: Kaliski, B.S. (ed.) CRYPTO 1997. LNCS, vol. 1294, pp. 410–424. Springer, Heidelberg (1997). https://doi.org/10.1007/BFb0052252
12. Canard, S., Georgescu, A., Kaim, G., Roux-Langlois, A., Traoré, J.: Constant-size lattice-based group signature with forward security in the standard model. In: Nguyen, K., Wu, W., Lam, K.Y., Wang, H. (eds.) ProvSec 2020. LNCS, vol. 12505, pp. 24–44. Springer, Cham (2020). https://doi.org/10.1007/978-3-030-62576-4_2
13. Chaum, D., van Heyst, E.: Group signatures. In: Davies, D.W. (ed.) EUROCRYPT 1991. LNCS, vol. 547, pp. 257–265. Springer, Heidelberg (1991). https://doi.org/10.1007/3-540-46416-6_22
14. Chen, L., Pedersen, T.P.: New group signature schemes. In: De Santis, A. (ed.) EUROCRYPT 1994. LNCS, vol. 950, pp. 171–181. Springer, Heidelberg (1995). https://doi.org/10.1007/BFb0053433
15. Chen, S., Chen, J.: Lattice-based group signatures with forward security for anonymous authentication. Heliyon 9(4) (2023)
16. Ducas, L., et al.: Crystals-dilithium: a lattice-based digital signature scheme. IACR Trans. Cryptogr. Hardw. Embed. Syst. 238–268 (2018)
17. Emura, K., Hayashi, T., Ishida, A.: Group signatures with time-bound keys revisited: a new model, an efficient construction, and its implementation. IEEE Trans. Depend. Secure Comput. 17(2), 292–305 (2017)
18. Gao, S., Chen, X., Li, H., Susilo, W., Huang, Q.: Post-quantum secure group signature with verifier local revocation and backward unlinkability. Comput. Stand. Interfaces 88, 103782 (2024)
19. Gentry, C., Peikert, C., Vaikuntanathan, V.: How to use a short basis: trapdoors for hard lattices and new cryptographic constructions. In: STOC, vol. 8, pp. 197–206 (2008)
20. Gordon, S.D., Katz, J., Vaikuntanathan, V.: A group signature scheme from lattice assumptions. In: Abe, M. (ed.) ASIACRYPT 2010. LNCS, vol. 6477, pp. 395–412. Springer, Heidelberg (2010). https://doi.org/10.1007/978-3-642-17373-8_23
21. Groth, J.: Fully anonymous group signatures without random oracles. In: Kurosawa, K. (ed.) ASIACRYPT 2007. LNCS, vol. 4833, pp. 164–180. Springer, Heidelberg (2007). https://doi.org/10.1007/978-3-540-76900-2_10
22. Hwang, J.Y., Lee, S., Chung, B.H., Cho, H.S., Nyang, D.: Group signatures with controllable linkability for dynamic membership. Inf. Sci. 222, 761–778 (2013)

23. Katsumata, S., Yamada, S.: Group signatures without NIZK: from lattices in the standard model. In: Ishai, Y., Rijmen, V. (eds.) EUROCRYPT 2019. LNCS, vol. 11478, pp. 312–344. Springer, Cham (2019). https://doi.org/10.1007/978-3-030-17659-4_11

24. Laguillaumie, F., Langlois, A., Libert, B., Stehlé, D.: Lattice-based group signatures with logarithmic signature size. In: Sako, K., Sarkar, P. (eds.) ASIACRYPT 2013. LNCS, vol. 8270, pp. 41–61. Springer, Heidelberg (2013). https://doi.org/10.1007/978-3-642-42045-0_3

25. Langlois, A., Ling, S., Nguyen, K., Wang, H.: Lattice-based group signature scheme with verifier-local revocation. In: Krawczyk, H. (ed.) PKC 2014. LNCS, vol. 8383, pp. 345–361. Springer, Heidelberg (2014). https://doi.org/10.1007/978-3-642-54631-0_20

26. Libert, B., Ling, S., Mouhartem, F., Nguyen, K., Wang, H.: Signature schemes with efficient protocols and dynamic group signatures from lattice assumptions. In: Cheon, J.H., Takagi, T. (eds.) ASIACRYPT 2016. LNCS, vol. 10032, pp. 373–403. Springer, Heidelberg (2016). https://doi.org/10.1007/978-3-662-53890-6_13

27. Libert, B., Ling, S., Nguyen, K., Wang, H.: Zero-knowledge arguments for lattice-based accumulators: logarithmic-size ring signatures and group signatures without trapdoors. In: Fischlin, M., Coron, J.-S. (eds.) EUROCRYPT 2016. LNCS, vol. 9666, pp. 1–31. Springer, Heidelberg (2016). https://doi.org/10.1007/978-3-662-49896-5_1

28. Libert, B., Peters, T., Yung, M.: Group signatures with almost-for-free revocation. In: Safavi-Naini, R., Canetti, R. (eds.) CRYPTO 2012. LNCS, vol. 7417, pp. 571–589. Springer, Heidelberg (2012). https://doi.org/10.1007/978-3-642-32009-5_34

29. Ling, S., Nguyen, K., Roux-Langlois, A., Wang, H.: A lattice-based group signature scheme with verifier-local revocation. Theor. Comput. Sci. **730**, 1–20 (2018)

30. Ling, S., Nguyen, K., Stehlé, D., Wang, H.: Improved zero-knowledge proofs of knowledge for the ISIS problem, and applications. In: Kurosawa, K., Hanaoka, G. (eds.) PKC 2013. LNCS, vol. 7778, pp. 107–124. Springer, Heidelberg (2013). https://doi.org/10.1007/978-3-642-36362-7_8

31. Ling, S., Nguyen, K., Wang, H., Xu, Y.: Lattice-based group signatures: achieving full dynamicity with ease. In: Gollmann, D., Miyaji, A., Kikuchi, H. (eds.) ACNS 2017. LNCS, vol. 10355, pp. 293–312. Springer, Cham (2017). https://doi.org/10.1007/978-3-319-61204-1_15

32. Ling, S., Nguyen, K., Wang, H., Xu, Y.: Constant-size group signatures from lattices. In: Abdalla, M., Dahab, R. (eds.) PKC 2018. LNCS, vol. 10770, pp. 58–88. Springer, Cham (2018). https://doi.org/10.1007/978-3-319-76581-5_3

33. Ling, S., Nguyen, K., Wang, H., Xu, Y.: Forward-secure group signatures from lattices. In: Ding, J., Steinwandt, R. (eds.) PQCrypto 2019. LNCS, vol. 11505, pp. 44–64. Springer, Cham (2019). https://doi.org/10.1007/978-3-030-25510-7_3

34. Ling, S., Nguyen, K., Wang, H., Xu, Y.: Lattice-based group signatures: Achieving full dynamicity (and deniability) with ease. Theor. Comput. Sci. **783**, 71–94 (2019)

35. Nguyen, P.Q., Zhang, J., Zhang, Z.: Simpler efficient group signatures from lattices. In: Katz, J. (ed.) PKC 2015. LNCS, vol. 9020, pp. 401–426. Springer, Heidelberg (2015). https://doi.org/10.1007/978-3-662-46447-2_18

36. Perera, M.N.S., Koshiba, T.: Fully secure lattice-based group signatures with verifier-local revocation. In: 2017 IEEE 31st International Conference on Advanced Information Networking and Applications (AINA), pp. 795–802. IEEE (2017)

37. Perera, M.N.S., Koshiba, T.: Achieving full security for lattice-based group signatures with verifier-local revocation. In: Naccache, D., et al. (eds.) ICICS 2018.

LNCS, vol. 11149, pp. 287–302. Springer, Cham (2018). https://doi.org/10.1007/978-3-030-01950-1_17

38. Perera, M.N.S., Nakamura, T., Hashimoto, M., Yokoyama, H., Sakurai, K.: Almost fully anonymous attribute-based group signatures with verifier-local revocation and member registration from lattice assumptions. Theor. Comput. Sci. **891**, 131–148 (2021)

39. Şahin, M.S., Akleylek, S.: A constant-size lattice-based partially-dynamic group signature scheme in quantum random oracle model. J. King Saud Univ.-Comput. Inf. Sci. **34**(10), 9852–9866 (2022)

40. Shor, P.W.: Algorithms for quantum computation: discrete logarithms and factoring. In: Proceedings 35th Annual Symposium on Foundations of Computer Science, pp. 124–134. IEEE (1994)

41. Sun, Y., Liu, Y.: An efficient fully dynamic group signature with message dependent opening from lattice. Cybersecurity **4**(1), 1–15 (2021). https://doi.org/10.1186/s42400-021-00076-8

42. Tsabary, R.: An equivalence between attribute-based signatures and homomorphic signatures, and new constructions for both. In: Kalai, Y., Reyzin, L. (eds.) TCC 2017. LNCS, vol. 10678, pp. 489–518. Springer, Cham (2017). https://doi.org/10.1007/978-3-319-70503-3_16

43. Zhang, Y., Liu, X., Hu, Y., Gan, Y., Jia, H.: Verifier-local revocation group signatures with backward unlinkability from lattices. Front. Inf. Technol. Electron. Eng. **23**(6), 876–892 (2022)

Quantum Cryptography

New Results in Quantum Analysis of LED: Featuring One and Two Oracle Attacks

Siyi Wang[1(✉)] , Kyungbae Jang[2] , Anubhab Baksi[3] ,
Sumanta Chakraborty[4] , Bryan Lee[1] , Anupam Chattopadhyay[1] ,
and Hwajeong Seo[2]

[1] Nanyang Technological University, Singapore, Singapore
`siyi002@e.ntu.edu.sg`
[2] Hansung University, Seoul, Republic of Korea
[3] Lund University, Lund, Sweden
[4] Techno International New Town, Kolkata, India

Abstract. Quantum computing has attracted substantial attention from researchers across various fields. In case of the symmetric key cryptography, the main problem is posed by the application of Grover's search. In this work, we focus on quantum analysis of the lightweight block cipher LED. This paper proposes an optimized quantum circuit for LED, minimizing the required number of qubits, quantum gates, and circuit depth. Furthermore, we conduct Grover's attack and Search with Two Oracles (STO) attack on the proposed LED cipher, estimating the quantum resources required for the corresponding attack oracles. The STO attack outperforms the usual Grover's search when the state size is less than the key size. Beyond analyzing the cipher itself (i.e., the ECB mode), this work also evaluates the effectiveness of quantum attacks on LED across different modes of operation.

Keywords: Grover's Search · Search with Two Oracles · LED Block Cipher · Modes of Operation · Resource Estimation · Quantum Cryptanalysis

1 Introduction

The Internet of Things (IoT) is revolutionizing the way devices communicate and operate through interconnected devices, resulting in a large number of applications across a variety of domains [14,33]. Nonetheless, IoT security remains a significant concern due to the large number of devices and their limited resources. Lightweight cryptography aims to address these limitations, ensuring secure communication while maintaining efficiency in terms of power, speed, and hardware resources. However, with the development of quantum computing, quantum attacks such as Grover's algorithm pose substantial risks to encryption systems, thereby significantly impacting the security of lightweight cryptography in the quantum computing era.

R. Dutta et al. (Eds.): INDOCRYPT 2025, LNCS 16372, pp. 339–364, 2026.
https://doi.org/10.1007/978-3-032-13301-4_15

Therefore, it is crucial to explore and analyze lightweight ciphers in the context of quantum attacks. Several relevant studies have already been conducted [28,29], focusing on lightweight block ciphers such as PRESENT, GIFT, and SPECK. Moreover, we have seen recent improvements on the attack on AES, namely [27,34,38].

Among various lightweight block ciphers, LED stands out due to its compact hardware design and minimal silicon footprint. Given these properties, this paper focuses on the quantum implementation and quantum attack of the LED cipher. Specifically, we propose an efficient quantum circuit for the LED cipher, minimizing the required qubits, quantum gates, and circuit depth. Furthermore, we perform Grover's attack and Search with Two Oracles (STO) attack on the proposed quantum LED cipher and estimate the quantum resources required for these attack oracles. This work also addresses several issues found in the quantum LED cipher proposed by Song et al. [39]. By comparing it with the corrected version of their implementation, it is evident that our design provides a more comprehensive and efficient solution. Additionally, this paper also addresses modes of operation in the context of quantum attack. In classical cryptography, modes of operation such as ECB, CBC, OFB, and CFB have been widely discussed and standardized since their introduction in FIPS 81 back in 1981. However, in quantum cryptography, while quantum attacks on block ciphers have been extensively explored, modes of operation have just received limited attention so far [3]. To fill this critical gap, this work proposes frameworks for ECB, CBC, and CFB modes based on the quantum LED cipher. Furthermore, we provide a thorough analysis of the quantum attack costs associated with these different modes of operation.

Contributions and Novelty

In brief, our contributions are detailed as follows[1]:

- **Efficient Quantum Circuit for LED Block Cipher and Resource Estimates.** We present a novel approach to implementing the LED block cipher as a quantum circuit. The proposed approach optimizes the quantum circuit of the LED cipher by exploring efficient sub-circuit structures, thereby enhancing the overall efficiency.
  We evaluate the performance of our design using the ProjectQ framework [40]. Our analysis includes a detailed comparison of the required quantum resources, such as qubits and circuit depth, with respect to other block cipher implementations in the literature. Compared to previous works, our implementation achieves a more efficient quantum circuit for the LED cipher.
- **Quantum Cost Estimation of One Oracle (Grover's) and Two Oracles (STO) Attack.** In this work, Grover's attack (that employs one quantum oracle) complexity is reported for LED. Moreover, as referenced

---

[1] The relevant code is available as a public repository (https://github.com/Siyi-06/Quantum_Analysis_of_LED).

in [18,32], we also investigate the requirement and the impact of the STO attack, which is an improvement over Grover's attack when the state size is less than the key size for a block cipher (and more complex as it employs two quantum oracles). For both types of attacks, we estimate the quantum resource requirements and provide a comprehensive evaluation of their efficiency.

- **Modes of Operation for Block Ciphers in Quantum.** This work explores the modes of operation in quantum attacks, proposing frameworks for ECB, CBC and CFB modes based on the proposed quantum LED cipher. Besides, it also provides a thorough analysis of Grover's attack's effectiveness across these modes.

2 Efficient Quantum Implementation of LED Cipher

In this section, we present an efficient quantum implementation of the LED cipher, focusing on minimizing the overall cost. This cipher consists of several sub-modules, which are arranged into multiple rounds of encryption. Accordingly, we will detail the efficient implementations for the five key sub-modules of quantum LED cipher in the subsequent subsections.

2.1 AddRoundKey

The quantum AddRoundKey operation is implemented using only CNOT gates, resulting in a circuit with a depth of one. This straightforward operation is simpler than other components like the S-box and MixColumn, as it follows a generic implementation method. Specifically, in the LED cipher, a 64-qubit round key is XORed with the intermediate state using 64 CNOT gates.

2.2 AddConstants

Similar to the AddRoundKey operation, the quantum AddConstants operation is also based on a generic method. Since the round constant in the LED cipher is a fixed value, as detailed in Table 12, this quantum implementation only involves applying X gates to the state qubits where the corresponding bits in the round constant are set to 1.

2.3 SubCells

The SubCells operation applies a non-linear substitution using Sbox. As mentioned in Sect. A.1, the LED cipher reuses the same Sbox as PRESENT [9]. The coordinate functions of this Sbox are:

$$y_0 = x_0 \oplus x_1 x_2 \oplus x_2 \oplus x_3$$

$$y_1 = x_0 x_1 x_2 \oplus x_0 x_1 x_3 \oplus x_0 x_2 x_3 \oplus x_1 x_3 \oplus x_1 \oplus x_2 x_3 \oplus x_3$$

$$y_2 = x_0 x_1 x_3 \oplus x_0 x_1 \oplus x_0 x_2 x_3 \oplus x_0 x_3 \oplus x_1 x_3 \oplus x_2 \oplus x_3 \oplus 1$$

$$y_3 = x_0 x_1 x_2 \oplus x_0 x_1 x_3 \oplus x_0 x_2 x_3 \oplus x_0 \oplus x_1 x_2 \oplus x_1 \oplus x_3 \oplus 1$$

For the quantum implementation, the design of an efficient quantum Sbox is crucial, as it directly impacts the performance of this operation. Given the inherent complexity and the numerous possible approaches for implementing a quantum LED Sbox, we have explored a range of strategies to identify the most efficient design. These strategies are outlined in detail below.

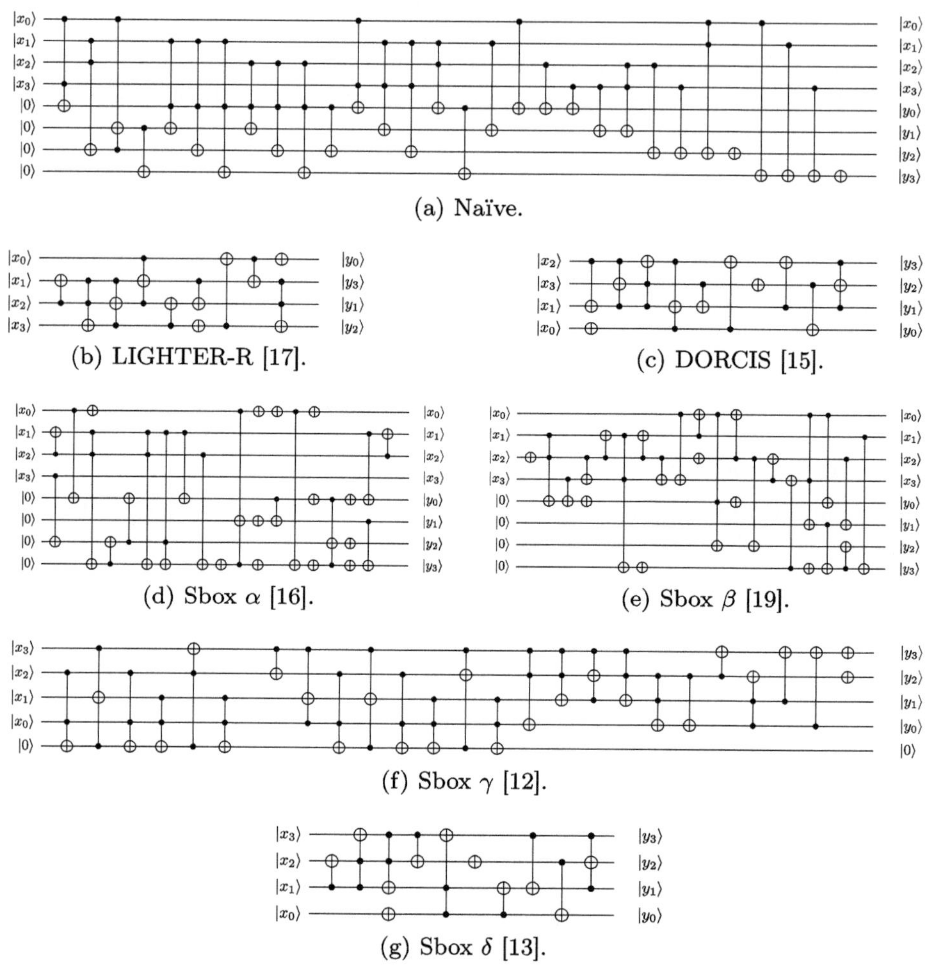

(a) Naïve.

(b) LIGHTER-R [17].

(c) DORCIS [15].

(d) Sbox α [16].

(e) Sbox β [19].

(f) Sbox γ [12].

(g) Sbox δ [13].

Fig. 1. Quantum implementations of LED Sbox.

- **Naïve.** This approach directly utilizes the Sbox coordinate functions to design the quantum circuit. While straightforward, this method does not optimize for quantum resources, resulting in higher quantum costs.
- **LIGHTER-R.** LIGHTER-R [17] is a tool designed for implementing S-boxes using reversible logic libraries, specifically targeted for 4-bit S-boxes. Using

this tool, we generated a quantum implementation of the LED Sbox, which is illustrated in Fig. 1(b).

- **DORCIS.** DORCIS [15] is a tool that optimizes quantum implementations for arbitrary 3- and 4-bit S-boxes, extending the capabilities of the LIGHTER-R. Unlike LIGHTER-R, which only handles 4-bit S-boxes and operates at a top level using Toffoli gates, DORCIS incorporates quantum decomposition with Clifford and T gates, optimizing both quantum depth and T-depth. In this paper, we employed DORCIS to generate a quantum implementation of the LED S-box, as shown in Fig. 1(c).
- **Sbox α.** In the paper [16], Courtois et al. optimized various small digital circuits, including LED S-box. It is shown in Fig. 1(d).
- **Sbox β.** In 2024, Feng et al. [19] proposed optimized implementations of lightweight cryptographic S-boxes using SAT solvers, which significantly enhance the overall cryptographic performance. Here, we implement their optimized LED S-box into quantum circuit, as shown in Fig. 1(e).
- **Sbox γ.** In the paper by Cai et al. [12], a quantum S-box for the LED cipher is proposed that utilizes 5 qubits (i.e., 1 ancilla/garbage qubit). This design is illustrated in Fig. 1(f).
- **Sbox δ.** In 2024, Chen et al. [13] proposed a SAT-based model that optimizes quantum circuits incorporating three metrics. Based on this model, we designed a more compact quantum circuit for LED Sbox, as illustrated in Fig. 1(g).

Table 1. Quantum resource requirement for LED SBox.

Method	#1qCliff	#CNOT	#Toffoli	#T	#Qubits	Circuit depth	Toffoli depth
Naïve (Fig. 1(a))	2	13	15	105	8	27	14
LIGHTER-R [17] (Fig. 1(b))	2	5	4	28	4	9	4
DORCIS [15] (Fig. 1(c))*	2	5	4	28	4	8	4
Sbox α [16] (Fig. 1(d))	12	12	5	35	8	16	5
Sbox β [19] (Fig. 1(e))	6	16	4	28	8	19	4
Sbox γ [12] (Fig. 1(f))	7	5	19	133	5	24	19
Sbox δ [13] (Fig. 1(g))	2	5	4	28	4	8	4

$\star$: Used in this work

In Table 1, the quantum costs associated with each approach are presented. Here "1qClifford" refers to the Clifford gate operating on a single qubit, such as the Hadamard, X or S gates. It can be seen that both the Sbox γ by Chen et al. and the DORCIS-generated Sbox exhibit the highest efficiency, with a circuit depth of 8 and a Toffoli depth of 4. Since the two designs are very similar, either one could be chosen. In this paper, we decided to use the DORCIS-generated implementation for our quantum S-box.

2.4 ShiftRows

The ShiftRows operation, which performs circular shifts, can be implemented using quantum swap gates, also known as "physical swap" approach. Alternatively, this operation can be achieved using the "logical swap" approach, which involves rearranging the indices of the qubits logically without employing additional quantum swap gates. As one can see, the second approach is more resource-efficient, as it avoids the requirement for extra quantum resources. Therefore, in this work, we adopt the "logical swap" approach. Details of the implementation are provided in Code 2.1.

Code 2.1. ShiftRows of quantum LED implementation.

```python
def ShiftRows(eng, state):
    new_state=[]
    # 1st row
    ## left rotate 4 places
    new_state[0:4]=state[4:8]
    new_state[4:8]=state[8:12]
    new_state[8:12]=state[12:16]
    new_state[12:16]=state[0:4]
    # 2nd row
    ## left rotate 8 places
    new_state[16:20]=state[24:28]
    new_state[20:24]=state[28:32]
    new_state[24:28]=state[16:20]
    new_state[28:32]=state[20:24]
    # 3rd row
    ## left rotate 12 places
    new_state[32:36]=state[44:48]
    new_state[36:40]=state[32:36]
    new_state[40:44]=state[36:40]
    new_state[44:48]=state[40:44]
    # 0th row
    ## left rotate 0 place
    new_state[48:64]=state[48:64]
    return new_state
```

2.5 MixColumn

Based on the specification of LED (see Sect. A.1), the MixColumnsSerial matrix (A) and the MixColumns (M) matrix can be given in binary as follows:

$$M = \begin{bmatrix}
0&0&1&0&1&0&0&0&0&1&0&0&0&1&0&0\\
1&0&0&1&0&1&0&0&0&0&1&0&0&0&1&0\\
1&1&0&0&0&0&1&0&1&0&0&1&1&0&0&1\\
0&1&0&0&0&0&0&1&1&0&0&0&1&0&0&0\\
1&0&0&1&0&1&1&0&1&0&1&0&0&1&1&0\\
1&1&0&0&1&0&1&1&1&1&0&1&1&0&1&1\\
0&1&1&0&0&1&0&1&1&1&1&0&0&1&0&1\\
0&0&1&0&1&1&0&0&0&1&0&1&1&1&0&0\\
0&1&0&1&1&1&1&1&1&1&0&1&0&0&0&1\\
1&0&1&0&0&1&1&1&1&1&1&0&1&0&0&0\\
1&1&0&1&0&0&1&1&1&1&1&1&0&1&0&0\\
1&0&1&1&1&1&1&0&1&0&1&0&0&0&1&1\\
0&1&0&0&0&1&0&0&0&1&1&1&0&1&0&1\\
0&0&1&0&0&0&1&0&0&0&1&1&1&0&1&0\\
1&0&0&1&1&0&0&1&0&0&0&1&1&1&0&1\\
1&0&0&0&1&0&0&0&1&1&1&1&1&0&1&1
\end{bmatrix}$$

$$A = \begin{bmatrix}
0&0&0&0&1&0&0&0&0&0&0&0&0&0&0&0\\
0&0&0&0&0&1&0&0&0&0&0&0&0&0&0&0\\
0&0&0&0&0&0&1&0&0&0&0&0&0&0&0&0\\
0&0&0&0&0&0&0&1&0&0&0&0&0&0&0&0\\
0&0&0&0&0&0&0&0&1&0&0&0&0&0&0&0\\
0&0&0&0&0&0&0&0&0&1&0&0&0&0&0&0\\
0&0&0&0&0&0&0&0&0&0&1&0&0&0&0&0\\
0&0&0&0&0&0&0&0&0&0&0&1&0&0&0&0\\
0&0&0&0&0&0&0&0&0&0&0&0&1&0&0&0\\
0&0&0&0&0&0&0&0&0&0&0&0&0&1&0&0\\
0&0&0&0&0&0&0&0&0&0&0&0&0&0&1&0\\
0&0&0&0&0&0&0&0&0&0&0&0&0&0&0&1\\
0&0&1&0&1&0&0&0&0&1&0&0&0&1&0&0\\
1&0&0&1&0&1&0&0&0&0&1&0&0&0&1&0\\
1&1&0&0&0&0&1&0&1&0&0&1&1&0&0&1\\
0&1&0&0&0&0&0&1&1&0&0&0&1&0&0&0
\end{bmatrix}$$

Here, we explore several methods to determine the most efficient design, which are summarized as follows:

- **Naïve (Out-of-place).** The naïve approach involves using ancilla qubits to compute the new state of a qubit while preserving its previous state in a backup qubit, which is essential for the calculation of new state of two other qubits. Following this approach, Song et al. [39] proposed the naïve implementation of the MixColumn operation in the quantum LED cipher. As shown in Table 2, this implementation requires 32 qubits and 108 CNOT gates, with a quantum depth of 10.
- **PLU Factorization.** PLU factorization decomposes a given binary matrix into a permutation matrix, a lower triangular matrix, and an upper triangular matrix. This factorization can also be applied to achieve an efficient in-place quantum implementation for MixColumn operation. By following the method detailed in [24], we use Sage^2 to obtain an in-place quantum implementation through PLU factorization.
- **Gauss-Jordan Elimination.** Gauss-Jordan elimination is used to factorize any binary matrix through elementary operations, which correspond to CNOT and SWAP gates in quantum circuits. By applying this method, we achieve an in-place quantum implementation that utilizes 16 qubits with 103 CNOT gates and 8 SWAP gates, with a quantum depth of 52.
- **XZLBZ.** The XZLBZ algorithm [42] offered an innovative approach for the in-place quantum implementation of binary matrices. At that time, it was the first tool to efficiently implement a given linear layer. The most notable result from this paper was to find an in-place implementation of the AES MixColumn with 92 CNOT gates and 30 quantum depth (although they did not optimize for quantum depth). Note that a revision of this algorithm was reported in [6].
- **YWSZZ.** This is a very recent paper [43] where the authors managed to find an in-place implementation of the AES MixColumn matrix with 91 CNOT gates and (13 classical depth, 35 quantum depth). This is where the record stands till date, to the best of our knowledge. In our case though, it turns out that XZLBZ (44) outperforms YWSZZ (47) in terms of CNOT gates.

Table 2. Quantum resource requirement for LED MixColumns.

Method	#CNOT	#SWAP	Circuit depth
Out-of-place (32 qubits)			
Naïve (used in [39])	108	0	10
In-place (16 qubits)			
PLU	103	8	62
Gauss-Jordan	103	8	52
XZLBZ [42]	45	16	19
XZLBZ [42]*	44	16	16
Modified XZLBZ [6]	50	14	17
Modified XZLBZ [6]	46	12	20
YWSZZ [43]	47	16	10

*: Used in this work

---

^{2} https://doc.sagemath.org/html/en/reference/matrices/sage/matrix/matrix2.html.

As shown in Table 2, we summarize the quantum costs for the various approaches discussed before. One may note that, the XZLBZ implementation demonstrates the highest efficiency, achieving 44 CNOT gates with 16 qubits and a circuit depth of 16. Hence, we chose this implementation as our quantum MixColumn.

The matrix A requires 14 CNOT gates in the optimal in-place implementation. This was found using the MILP tool of [8]. Thus, it would require $4 \times 14 = 52$ CNOT gates to implement M by using this matrix, and hence were not considered here. Using the same tool, we also found that the minimum number of CNOT gates that are required to implement M in-place is at least 15[3].

2.6   Architecture

Using the presented modules AddConstants, SubCells, ShiftRows, and Mix-Columns, we construct the architecture for the LED quantum circuit. Even at the architectural level, the in-place design is implemented, where the cipher-text is computed directly on the input qubits (i.e., plaintext) without allocating additional qubits.

Figure 2 shows the in-place architecture of the LED quantum circuit (here AC, SC, SR, and MC represent AddConstants, SubCells, ShiftRows, and Mix-Columns, respectively). The only differences between LED-64 and LED-128 are the number of steps (s) and the AddRoundKey operation, neither of which affects the overall architecture. Thus, the in-place architecture shown in Fig. 2 applies equally to both the LED-64 and LED-128 quantum circuits ($s = 8$ and $s = 12$, respectively).

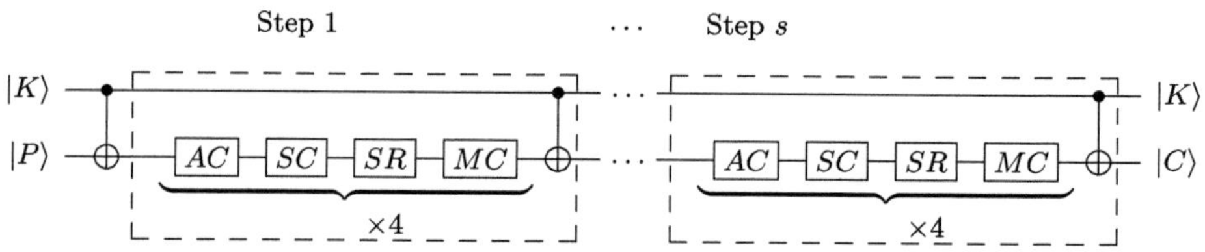

Fig. 2. In-place architecture of LED quantum circuit.

3   Consideration for Modes of Operation

In classical cryptography, modes of operation were introduced to enhance security when using block ciphers to encrypt long messages. However, in the field of quantum cryptography, this topic has not received much attention [3]. To address this gap, this work proposes several frameworks for modes of operation based on the quantum LED circuit discussed in Sect. 2. In this work, we focus on

---

[3] The program took a long time since it searched for an implementation with 15 CNOT gates, yet there was no outcome.

three widely utilized modes of operation: Electronic Codebook (ECB), Cipher Block Chaining (CBC), and Cipher Feedback (CFB).

One may note from [5, Chapter 2.2.2] that the block cipher modes can be thought of as belonging from one of the two categories, non-symmetric and symmetric. In the non-symmetric modes, both the sender and the recipient use two operations (one being the inverse of the other); whereas in symmetric modes, they use the same operation. In this paper, we take the representative modes; ECB and CBC from the non-symmetric category; and CFB from the symmetric category (analysis about the other modes, like OFB or CTR, will be similar to that of the CFB mode).

The following subsections will provide detailed implementations of the quantum frameworks for ECB, CBC, and CFB modes. For clarity and conciseness, the simplified representation of the proposed quantum LED implementation, as shown in Fig. 3, are utilized throughout the remaining part.

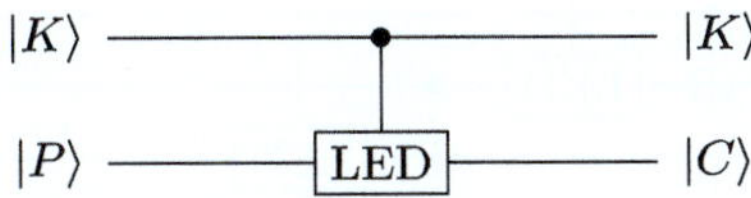

Fig. 3. Simplified representation of LED in quantum.

3.1   Electronic Codebook (ECB) Mode

This mode is straightforward, as it directly applies the quantum cipher to each block without any interdependencies between them. However, the absence of inter-block dependencies in ECB can make it very vulnerable to various attacks.

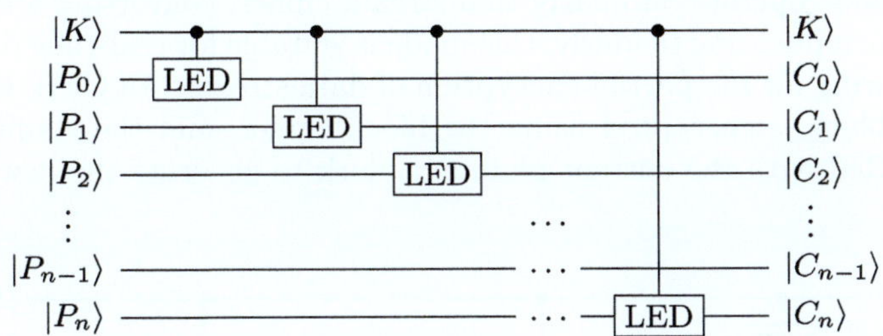

Fig. 4. Quantum framework for ECB Mode for LED.

The quantum framework for ECB mode is depicted in Fig. 4, where each plaintext block is processed independently using the quantum LED circuit. In this diagram, K denotes the key, P denotes plaintext, and C denotes ciphertext.

3.2   Cipher Block Chaining (CBC) Mode

In CBC mode, interdependencies between plaintext blocks are established by XORing each plaintext block with the ciphertext of the previous block before encryption. For the first plaintext block, the XOR operation is performed with an initialization vector (IV), which is typically a random or pseudorandom value. This chaining mechanism ensures that identical plaintext blocks produce different ciphertexts, thereby enhancing security compared to ECB mode.

The quantum framework for CBC mode is shown in Fig. 5, illustrating the process of XORing each plaintext block with the previous ciphertext block, followed by encryption using the quantum LED cipher. Here, IV denotes initialization vector.

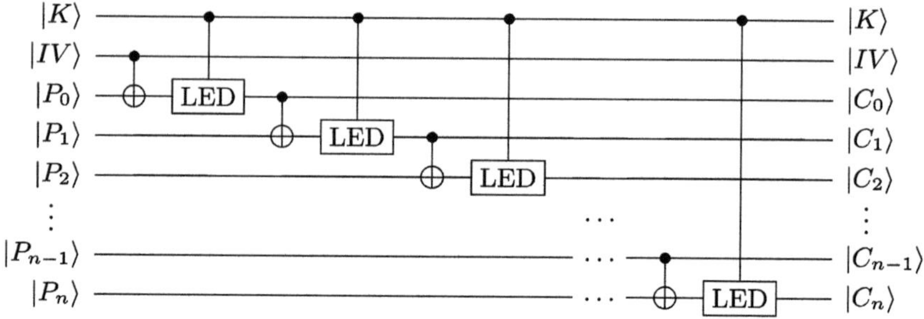

Fig. 5. Quantum framework for CBC Mode for LED.

3.3   Cipher Feedback (CFB) Mode

The CFB mode operates similarly to a stream cipher, converting a block cipher into a stream cipher. Particularly, this mode is suitable for real-time data encryption by allowing for the partial encryption of data streams. In CFB, the previous ciphertext block is encrypted using the block cipher, and the resulting output is then XORed with the current plaintext block to generate the new ciphertext block.

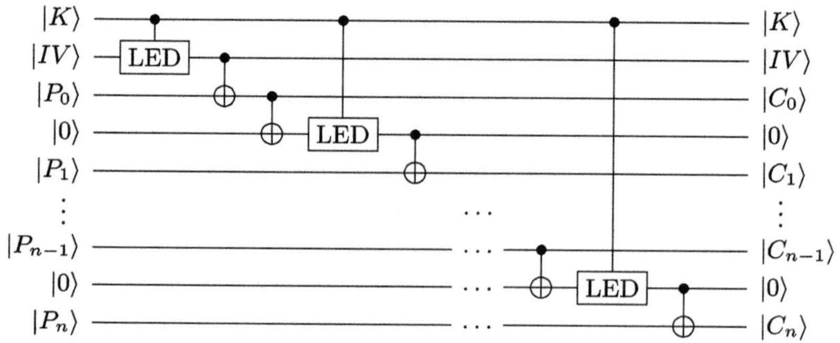

Fig. 6. Quantum framework for CFB Mode for LED.

The quantum framework for implementing CFB mode is provided in Fig. 6, where the feedback mechanism and sequential encryption process are illustrated. Although the proposed framework generates garbage qubits, all of them will be cleaned to $|0\rangle$ through the subsequent uncomputation step in Grover's attack.

In this section, we have proposed frameworks for ECB, CBC, and CFB modes based on our quantum LED implementations. Subsequently, Grover's attack estimation will be performed on the LED cipher under these proposed modes of operation frameworks to evaluate their effectiveness.

4   Results and Discussions

In this section, we conduct a thorough analysis of all the experimental results. To begin with, the issues in the previous quantum LED implementation proposed by Song et al. [39] are revised, thereby establishing a more accurate benchmark for subsequent comparisons. Following this, we compare the quantum resources required to implement the LED cipher with those needed for other block ciphers. Besides, this work provides the quantum attack resources based on the proposed quantum LED implementations, including Grover's attack oracles and STO attack oracles. At the end of this section, the effectiveness of Grover's attack oracles on various quantum modes of operation are carefully estimated, with the proposed quantum LED cipher serving as the underlying block cipher.

4.1   Revision of Previous Quantum LED Implementations

In the previous work [39], Song et al. also proposed a quantum implementation of the LED cipher. However, this work contains an issue in estimating the required number of qubits for the quantum implementation of LED cipher. In this subsection, we analyze the issue and provide a correction to ensure a more accurate comparison in subsequent content.

As shown in Table 2, the authors proposed an out-of-place naïve MixColumn. To perform one MixColumnsSerial, four MixColumn operations are required, with each MixColumn requiring 16 newly allocated output qubits, totaling 64 qubits for a single MixColumnsSerial. Since LED-64 consists of 32 rounds, each applying one MixColumnsSerial, 2048 qubits (32 rounds × 64) should be allocated for their implementation. However, their paper reports that only 142 qubits are required for the quantum LED-64 implementation without using any special techniques. This discrepancy arises from an issue in their MixColumnsSerial and naïve MixColumn implementation. To address this, we correctly allocate the output qubits for the naïve MixColumn and MixColumnsSerial in their implementation.

Table 3. Revised benchmark for the quantum implementation of LED-64

Quantum circuit	#Qubits	#Toffoli	#CNOT	#X	Circuit depth
LED-64 [39] (Buggy, taken as-is)	142	2048	19008	1438	810
LED-64 [39] (Bug-fixed by us)	2176	2048	19008	1438	1147

In Table 3, we report the corrected quantum resource requirements for their LED-64 quantum circuit. In Song et al.'s paper [39], all submodules of the quantum LED cipher are implemented without ancilla qubits, except for MixColumnsSerial. Consequently, the revised benchmark includes an additional 2048 qubits from the corrected MixColumnsSerial, bringing the total to 2176 qubits when combined with the initial 128 qubits allocated for plaintext (64 qubits) and key (64 qubits). Furthermore, the circuit depth is revised from 810 to 1147.

4.2   Experiment 1: Cost Analysis for LED in Quantum

In this subsection, we present a detailed cost analysis of the proposed quantum implementations of LED cipher. Specifically, the comparison details between our designs and various implementations of quantum lightweight block ciphers from previous researches are provided in Table 4.

Table 4. Quantum resource requirements for LED and other lightweight block ciphers.

Quantum circuit	#Qubits	#Toffoli	#CNOT	#X	Circuit depth
LED-64 (This work)	128	2048	8768	1158	753
LED-128 (This work)	192	3072	13120	1734	1127
LED-64 (Corrected) [39]	2176	2048	19008	1438	1147
PRESENT-64/80 [29]	144	2108	4683	1118	311
PRESENT-64/128 [29]	192	2232	4838	1164	311
GIFT-64/128 [29]	192	1792	1792	3261	308
GIFT-128/128 [29]	256	6144	6144	10,953	528
SIMON-64/128 [4]	192	1408	7396	1216	2643
SIMON-128/128 [4]	256	4352	17,152	4224	8427
SPECK-64/128 [26]	193	3286	9238	57	-
SPECK-128/128 [26]	257	7942	22,086	75	-
CHAM-64/128 [26]	196	2400	12,285	240	-
CHAM-128/128 [26]	268	4960	26,885	240	-

- **Our Work vs. Previous Quantum LED Design.** The proposed quantum LED-64 design in this work significantly outperforms the corrected LED-64 implementation discussed in Sect. 4.1, which represents the only previous quantum implementation of the LED cipher. Specifically, it requires fewer qubits, with a total of 128 compared to 2176 in the previous implementation, representing a reduction of approximately 94%. In terms of gate count, both designs use the same number of Toffoli gates, totaling 2048. However, our design shows clear advancements in other gate costs. Specifically, it utilizes fewer CNOT gates, with 8768 compared to 19008 in the previous implementation, resulting in a reduction of approximately 53.9%. Similarly, our design requires fewer X gates, with 1158 compared to 1438, which corresponds to a reduction of around 19.5%. Regarding circuit depth, our design has a depth of 753, whereas the previous implementation has a depth of 1147. This represents a reduction of approximately 34.4%, highlighting a significant improvement in time efficiency for our proposed design.

- **Our Work vs. Other Quantum Analysis Lightweight Block Ciphers.** Our quantum LED designs demonstrate significant advantages in efficiency when compared to other quantum lightweight block cipher implementations. For qubits, our quantum LED-64 and LED-128 designs utilize 128 and 192 qubits, respectively. Compared to all the other lightweight block ciphers mentioned in this section, our designs achieve the lowest qubit requirements. Compared to other quantum block ciphers, our proposed quantum LED designs exhibit competitive gate efficiency. For the most resource-intensive quantum gate, the Toffoli gate, both LED-64 and LED-128 maintain a relatively low count, comparable to quantum PRESENT implementations [29]. Additionally, our designs show comparable efficiency in CNOT and X gates, requiring fewer resources than ciphers such as SIMON [4] and GIFT [29]. In terms of circuit depth, our designs exhibit a relatively high cost compared to other work, ranking just below the quantum Simon designs reported in [4]. While our designs maintain competitive performance, there is still room for optimization in circuit depth.

In brief, our LED-64 and LED-128 quantum implementations demonstrate the lowest qubit requirements when compared to previous quantum block cipher implementations, while also achieving competitive gate cost and circuit depth.

4.3   Experiment 2: Grover's Attack on LED

This experiment focuses on estimating the quantum resources necessary for executing Grover's attack on the proposed LED ciphers. This process begins with a critical step—accurately estimating the quantum LED implementations. Specifically, the quantum LED-64 and LED-128 are implemented according to the methodology outlined in Sect. 2, followed by the decomposition of all involved Toffoli gates into Clifford and T gates to ensure precise resource estimation. Among various decomposition methods available [2,23,37], this paper adopts the approach introduced in Reference [2]. As shown in Fig. 7, this method utilizes 8 Clifford and 7 T gates, resulting in a T-depth of 4 and a total depth of 8 for each Toffoli gate. Based on this specific decomposition strategy, we outlined the quantum resources required for the proposed LED implementations in Table 5.

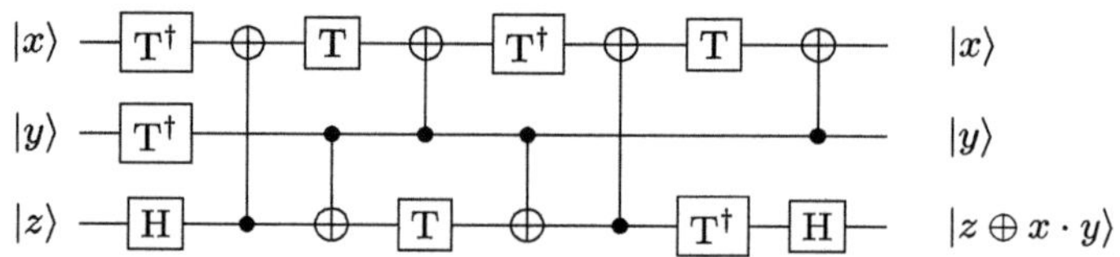

Fig. 7. Decomposition for Toffoli gate (T-depth 4, full depth 8).

In Tables 5 and 6 $TD/Td\text{-}M$ and $FD\text{-}M$ represent the trade-off performance of quantum circuits by being the product of the Toffoli/T depth and qubit count, and the full depth and qubit count, respectively. Additionally, we report the metrics $TD^2/Td^2\text{-}M$ and $FD^2\text{-}M$, which are major trade-off metrics for parallelization in Grover's search[4], emphasizing the importance of depth.

Table 5. Decomposed resource requirements for proposed quantum LED implementations.

Circuit	#CNOT	#1qCliff	#T	Toffoli depth (TD)	#Qubit (M)	Full depth (FD)	$TD\text{-}M$	$FD\text{-}M$	$TD^2\text{-}M$	$FD^2\text{-}M$
LED-64	20032	5254	14336	128	128	1649	$1.00 \cdot 2^{14}$	$1.61 \cdot 2^{17}$	$1.00 \cdot 2^{21}$	$1.3 \cdot 2^{28}$
LED-128	30016	7878	21504	192	192	2471	$1.13 \cdot 2^{15}$	$1.81 \cdot 2^{18}$	$1.69 \cdot 2^{22}$	$1.09 \cdot 2^{30}$

Based on the quantum cost of the proposed LED implementations, the corresponding Grover attack cost can be directly estimated. As detailed in Sect. A.3, recovering a k-bit key for a cipher using Grover's algorithm requires approximately $\sqrt{2^k}$ iterations of both the Grover oracle and the diffusion operator. Moreover, reference [10] provides a precise analysis, indicating that the optimal number of iterations is $\lfloor \frac{\pi}{4}\sqrt{2^k} \rfloor$. In this work, we adopt this estimate for our cost calculations.

Furthermore, we employed a widely used method [21,31] to simplify our analysis. Particularly, we disregard the diffusion operator from the cost estimation due to its minimal contribution. Consequently, the focus of Experiment 2 remains solely on the cost of the oracle, ensuring an accurate and efficient estimation process.

Next, we delve into the quantum resource analysis for Grover's attack oracle. This oracle consists of several components: the LED quantum circuit for encryption, an n-controlled NOT gate (where n represents the ciphertext size) for comparing the ciphertext with a known value, and the reverse operation of

---

[4] Grover's key search demands extreme circuit depth due to the large number of iterations. Under the constraint of circuit depth (such as the MAXDEPTH parameter introduced by NIST [35,36]), parallelization of Grover's search is required to reduce the circuit depth, but its performance is poor. As a result, the product of the squared depth and qubit count serves as the trade-off metric in Grover's parallelization. See Sect. 4.3 for more details.

the previously executed LED quantum circuit for each subsequent iteration. It is crucial to emphasize the decomposition of the n-controlled NOT gate. Based on the approach outlined in Reference [41], this gate is estimated to require $(32 \cdot n - 64)$ T gates.

Overall, the estimated costs for Grover's attack oracles on LED are summarized in Table 6. The total cost required is approximated as $\lfloor \frac{\pi}{4} \sqrt{2^k} \rfloor \times$ (Table 5 $\times$ 2) $+ \lfloor \frac{\pi}{4} \sqrt{2^k} \rfloor \times (32 \cdot n - 64)$ T gates. Since the iterations are executed sequentially, the number of qubits required remains consistent with the values presented in Table 5, with only one additional decision qubit required for ciphertext comparison.

Table 6. Quantum resource requirements for Grover's attack on LED.

Cipher	$\#T$	$\#$Gate (G)	Full depth (FD)	T-depth (Td)	$\#$Qubit (M)	G-FD	FD-M	Td-M	FD^2-M	Td^2-M
LED-64	$1.57 \cdot 2^{46}$	$1.99 \cdot 2^{47}$	$1.27 \cdot 2^{43}$	$1.57 \cdot 2^{41}$	$1.01 \cdot 2^{7}$	$1.26 \cdot 2^{91}$	$1.28 \cdot 2^{50}$	$1.58 \cdot 2^{48}$	$1.61 \cdot 2^{93}$	$1.24 \cdot 2^{90}$
LED-128	$1.13 \cdot 2^{79}$	$1.47 \cdot 2^{81}$	$1.90 \cdot 2^{75}$	$1.18 \cdot 2^{74}$	$1.00 \cdot 2^{8}$	$1.39 \cdot 2^{157}$	$1.90 \cdot 2^{83}$	$1.18 \cdot 2^{82}$	$1.80 \cdot 2^{159}$	$1.39 \cdot 2^{156}$

NIST Post-Quantum Security Levels. To assess the security of a cipher against quantum attacks, NIST [35,36] has defined security bounds for various levels:

- Level 1: The resource requirements for an attack are comparable to those for breaking AES-128: 2^{170} (stated) $\rightarrow$ $\mathbf{2^{157}}$ (state-of-the-art [27]).
- Level 3: The resource requirements for an attack are comparable to those for breaking AES-192: 2^{233} (stated) $\rightarrow$ $\mathbf{2^{221}}$ (state-of-the-art [27]).
- Level 5: The resource requirements for an attack are comparable to those for breaking AES-256: 2^{298} (stated) $\rightarrow$ $\mathbf{2^{285}}$ (state-of-the-art [27]).

Based on the cost estimates for Grover's key search against AES variants as presented by Grassl et al. [21], NIST determined the quantum attack complexities for Levels 1, 3 and 5 (corresponding to different AES variants) as 2^{170}, 2^{233} and 2^{298}, respectively (calculated as total gates multiplied by the depth of Grover's search). It is important to highlight that NIST's complexity estimates in [35] are derived from research results published in PQCrypto'16 [21]. Since that time, quantum circuits for AES have undergone continuous optimization, resulting in a significant decrease in the cost of attacks in recent years [25,27,31,44].

NIST also acknowledges that the attack complexities based on these levels are relative, given the ongoing optimizations in quantum circuits for AES (see [35, Page 17]). Therefore, if a more efficient attack is proposed, the benchmarks may need to be updated.

Recently, NIST revised the security bounds for AES [36] following the findings presented at Eurocrypt 2020 [31]. In [31], the quantum attack costs for AES-128, -192 and -256 were significantly reduced to 2^{157}, 2^{221} and 2^{285}, respectively; aligning with the updated values in [36].

It is worth noting that the newly updated costs have an issue, as the estimation in [31] was found to be incorrect (the corrected estimation is in [30]). However, the updated costs presented in [36] are also achievable in [27] through their optimized quantum circuits for AES (see Table 7). Thus, we still use the security bounds from [36] to assess the post-quantum security of LED.

In Table 7, the evaluation of post-quantum security levels for LED-64 and LED-128 is presented. As expected, LED-64 cannot achieve the required security level due to its 64-bit key length. However, LED-128 offers the same level of difficulty as breaking AES-128 using Grover's key search algorithm, thereby achieving the Level-1 post-quantum security.

Table 7. Comparison of the Grover's key search costs.

Post-quantum Security	NIST'16 [35] (based on [21])	NIST'22 [36] (based on [31])	Jang et al. [27]	Table 6 (G-FD)	
				LED-64	LED-128
Level-1 (AES-128)	2^{170}	2^{157}	2^{156}	Not achieved (2^{91})	Level 1 (2^{157})
Level-3 (AES-192)	2^{233}	2^{221}	2^{221}		
Level-5 (AES-256)	2^{298}	2^{285}	2^{286}		

4.4 Experiment 3: Estimation of Runtime and Physical Qubits for Grover's Attack on LED

With our quantum circuits for Grover's attack on variants of the LED block cipher, this section estimates the realistic quantum costs (such as physical qubits, surface code cycles, runtime, etc.) for fault-tolerant computation. Overall, our estimation follows the methodology in [1].

For fault-tolerant computing, we assume surface-code-based error correction, and the T gate incurs a high cost. In general, the T gate on a surface code is implemented by *magic state distillation* and we adopt the Reed-Muller 15-to-1 distillation method presented in [11]. To calculate the required distillation layers and code distances, we need to assume several physical parameters. We set the state injection error rate to $e_{in} = 10^{-4}$, and the physical error rate per gate, e_g, is given by $e_g = e_{in}/10$.

For Grover's attack on LED-64 and LED-128 (ECB mode, based on Table 11), the total numbers of T gates, T_g, are $1.57 \cdot 2^{46}$ and $1.13 \cdot 2^{79}$, respectively. To calculate the required distillation layers and code distances, we use Algorithm 1, based on [20] (and restated in [1]). The output error rate, e_{out}, should satisfy $e_{out} < 1/T_g$. Algorithm 1 specifies that two distillation layers are required, with $\{d_1, d_2\} = \{11, 5\}$ for LED-64 and $\{18, 7\}$ for LED-128.

Algorithm 1 Procedure for calculating the number of distillation layers and code distances.

1: **Input:** $e_{in}, e_g(= e_{in}/10), e_{out}(= 1/T_g), \varepsilon(= 1)$
2: $d \leftarrow$ empty list $[]$
3: $e \leftarrow e_{out}$
4: $i \leftarrow 0$
5: **repeat**
6: $\qquad i \leftarrow i + 1$
7: $\qquad e_i \leftarrow e$
8: $\qquad$ Find minimum d_i such that $192 d_i (100 e_g)^{\frac{d_i+1}{2}} < \frac{\varepsilon e_i}{1+\varepsilon}$
9: $\qquad e \leftarrow \sqrt[3]{e_i/(35(1+\varepsilon))}$
10: $\qquad d.\text{append}(d_i)$
11: **until** $e > e_{in}$
12: **Output:** $d = [d_1, \ldots, d_i]$

For a quantum attack on LED-64, the bottom distillation layer occupies the largest area in the surface code. With two distillation layers, generating a single magic state requires $16 \times 15 = 240$ input states. These input states are encoded using a code of distance $d_2 = 5$, corresponding to approximately $2.5 \times 1.25 \times d_2^2 \approx 78$ physical qubits per logical qubit. Consequently, the total footprint of the distillation circuit is $240 \times 78 = 18720$ physical qubits, and the round of distillation completes in $10 d_2 = 50$ surface code cycles.

The top layer of state distillation requires a code of distance $d_1 = 11$, corresponding to approximately $2.5 \times 1.25 \times d_1^2 \approx 378$ physical qubits per logical qubit. the total footprint of the distillation circuit is $16 \times 378 = 6048$ physical qubits, and the round of distillation completes in $10 d_1 = 110$ surface code cycles. However, since the physical qubits of the bottom layer can be reused in the top layer, the total number of physical qubits required for distillation is 18720.

A single magic state distillation requires $50 + 110 = 160$ surface code cycles, and we assume 200 nanoseconds per cycle in our estimation. As the top layer of distillation occupies less area than the bottom layer, pipelining allows the production of $18720/6048 \approx 3$ magic states in parallel (after the warm-up stage). To generate the magic states for $T_g = 1.57 \cdot 2^{46}$, the estimated time is $(T_g/3) \times 160 \times 200$ nanoseconds.

The ratio between the required number of T gates and the T-depth for the quantum attack (based on Table 11) is $T_g/T_d \approx 32$. Recall that with pipelining, 3 magic states can be generated in 160 code cycles. Thus, to obtain 32 magic states within the same number of cycles, we require $32/3 \approx 11$ distillation factories operating in parallel. This increases the number of physical qubits to $18720 \times 11 = 205920$, while reducing the distillation time to $(T_g/32) \times 160 \times 200$ nanoseconds ≈ 3.5 years.

We also estimate the physical overhead for Clifford gates in the surface code. For the quantum attack on LED-64, the total number of logical qubits is 129, and the total number of Clifford gates, C_g, is $1.21 \cdot 2^{47}$. The overall error rate of the circuit should satisfy $< 1/C_g$. To compute the required code distance, we determine the smallest d_c that satisfies the inequality

$$\left(\frac{e_{in}}{0.0125}\right)^{\frac{d_c+1}{2}} < 1/C_g,$$

This yields $d_c = 13$. Thus, the number of physical qubits required for Clifford gates is $129 \times 2.5 \times 1.25 \times d_c^2 \approx 68128$. However, although we do not provide a detailed proof here, the total number of cycles is determined solely by magic state production, as Clifford gates remain idle for most of the computation.

The estimated physical costs for Grover's attack on LED are summarized in Table 8. For brevity, we omit the detailed estimation for LED-128; however, it can be derived in the same manner as for LED-64. It can be observed that breaking LED-64 and LED-128 with Grover's attack will require 3.5 and 1.97×2^{34} years of computation, respectively, assuming 274048 and 907053 physical qubits are available.

Table 8. Estimated number of physical qubits and runtime for Grover's attack on LED.

		LED-64	LED-128
Distilleries	Surface code distances	$\{11, 5\}$	$\{18, 7\}$
	Factories	11	16
	Physical qubits	206250	588000
Grover	Surface code distance	13	23
	Physical qubits	68128	319053
Total	Physical qubits	274048	907053
	Surface code cycles	$1.96 \cdot 2^{48}$	$1.15 \cdot 2^{82}$
	Runtime	3.5 years	$1.97 \cdot 2^{34}$ years

4.5   Experiment 4: STO Attack on Quantum LED

In this experiment, we present a detailed analysis of the STO attack on the proposed quantum LED design, with associated quantum costs outlined in Table 9.

Table 9. Quantum resource requirements for one oracle (Grover) and two oracles (STO) attacks on LED.

Cipher	#T	#Gate (G)	Full depth (FD)	T-depth (Td)	#Qubit (M)	G-FD	FD-M	Td-M	FD^2-M	Td^2-M
LED-64 (Grover)	$1.57 \cdot 2^{46}$	$1.99 \cdot 2^{47}$	$1.27 \cdot 2^{43}$	$1.57 \cdot 2^{41}$	$1.01 \cdot 2^{7}$	$1.26 \cdot 2^{91}$	$1.28 \cdot 2^{50}$	$1.58 \cdot 2^{48}$	$1.61 \cdot 2^{93}$	$1.24 \cdot 2^{90}$
LED-64 (STO)	$1.57 \cdot 2^{46}$	$1.99 \cdot 2^{47}$	$1.27 \cdot 2^{43}$	$1.57 \cdot 2^{41}$	$1.01 \cdot 2^{7}$	$1.26 \cdot 2^{91}$	$1.28 \cdot 2^{50}$	$1.58 \cdot 2^{48}$	$1.61 \cdot 2^{93}$	$1.24 \cdot 2^{90}$
LED-128 (Grover)	$1.13 \cdot 2^{79}$	$1.47 \cdot 2^{81}$	$1.90 \cdot 2^{75}$	$1.18 \cdot 2^{74}$	$1.00 \cdot 2^{8}$	$1.39 \cdot 2^{157}$	$1.90 \cdot 2^{83}$	$1.18 \cdot 2^{82}$	$1.80 \cdot 2^{159}$	$1.39 \cdot 2^{156}$
LED-128 (STO)	$1.58 \cdot 2^{78}$	$1.03 \cdot 2^{81}$	$1.33 \cdot 2^{76}$	$1.65 \cdot 2^{74}$	$1.51 \cdot 2^{7}$	$1.36 \cdot 2^{157}$	$1.99 \cdot 2^{83}$	$1.24 \cdot 2^{82}$	$1.32 \cdot 2^{160}$	$1.02 \cdot 2^{157}$

- **STO Attack on Quantum LED-64.** The STO attack does not offer significant advantages over the Grover's oracle for attacking the LED-64 cipher. The parameter r is calculated as $r = \lceil k/n \rceil$, where k is the key size and n is the state size. A quantum attack on a block cipher requires at least r pairs of plaintext and corresponding ciphertext, where k denotes the key size and n denotes the block size. For LED-64, with both the block size and key size set to 64 bits, the required r is only 1. Hence, the Grover's attack oracle remains equally efficient when compared to the cheap oracle O_γ in STO attack. In brief, the STO attack does not provide a substantial advantage over Grover's attack oracle for the LED-64 cipher.
- **STO Attack on Quantum LED-128.** The STO attack offers significant improvement over Grover's attack for the LED-128 cipher. For LED-128, with a key size of 128 bits and a block size of 64 bits, the required number of rounds r is 2. Hence, the STO attack benefits from the cheap oracle O_γ, which is much more efficient compared to the corresponding Grover's oracle. While the STO attack requires a higher depth, it compensates by reducing the number of gates and qubits involved. This efficiency in terms of gate count and qubit cost highlights the advantages of the STO approach over Grover's attack for LED-128.

Overall, we have successfully implemented the STO attack on the proposed quantum LED ciphers, resulting in improved efficiency compared to Grover's attack. Although the STO attack does not offer significant advantages for LED-64, it is clear that for LED-128, STO attack oracle requires fewer gates and qubits, though with an increase in circuit depth. Moreover, the product of Gate and Full depth ($G\text{-}FD$) decreases, further highlighting the benefits of the STO approach in attacking LED-128.

4.6 Experiment 5: Grover's Attack on LED Across Various Modes of Operation

In the last experiment, we perform Grover's attack on the proposed quantum LED cipher across multiple modes of operation, namely ECB, CBC and CFB.

Before estimating the resources required for Grover's attack, we first analyze the quantum costs associated with the different modes of operation frameworks discussed in Sect. 3. According to these frameworks, Table 10 provides the detailed quantum resource requirements for η-block quantum LED encryption across different modes of operation, showing how the cost of each quantum framework varies with block size η.

However, regardless of the mode of operation used, Grover's attack only needs to be applied to the first block of the cipher. This is because all blocks within a specific mode generally share the same key. Consequently, as shown in Table 11, the cost estimation for the quantum attack exhibits only minor differences across different modes of operation.

In brief, the cost of Grover's attack depends solely on the first block, and the quantum resource requirements for this block vary only slightly between ECB, CBC and CFB. Therefore, there is minimal variation in the overall cost of Grover's attack across different modes of operation.

Table 10. Quantum resource requirements for LED under modes of operation with η blocks.

Quantum Circuit	#Qubits	#Toffoli	#CNOT	#X	Circuit depth
LED-64 (ECB)	$64 + 64\eta$	2048η	8768η	1158η	753η
LED-128 (ECB)	$128 + 64\eta$	3072η	13120η	1734η	1127η
LED-64 (CBC)	$128 + 64\eta$	2048η	8832η	1158η	754η
LED-128 (CBC)	$192 + 64\eta$	3072η	13184η	1734η	1128η
LED-64 (CFB)	$64 + 128\eta$	2048η	$8896\eta - 64$	1158η	$755\eta - 1$
LED-128 (CFB)	$128 + 128\eta$	3072η	$13248\eta - 64$	1734η	$1129\eta - 1$

Table 11. Quantum Resource Requirements for Grover's Attack on LED across various modes of operations.

Cipher	#T	#Gate (G)	Full depth (FD)	T-depth (Td)	#Qubit (M)	G-FD	FD-M	Td-M	FD^2-M	Td^2-M
LED-64 (ECB)	$1.57 \cdot 2^{46}$	$1.99 \cdot 2^{47}$	$1.27 \cdot 2^{43}$	$1.57 \cdot 2^{41}$	$1.01 \cdot 2^{7}$	$1.26 \cdot 2^{91}$	$1.28 \cdot 2^{50}$	$1.58 \cdot 2^{48}$	$1.61 \cdot 2^{93}$	$1.24 \cdot 2^{90}$
LED-128 (ECB)	$1.13 \cdot 2^{79}$	$1.47 \cdot 2^{81}$	$1.90 \cdot 2^{75}$	$1.18 \cdot 2^{74}$	$1.00 \cdot 2^{8}$	$1.39 \cdot 2^{157}$	$1.90 \cdot 2^{83}$	$1.18 \cdot 2^{82}$	$1.80 \cdot 2^{159}$	$1.39 \cdot 2^{156}$
LED-64 (CBC)	$1.57 \cdot 2^{46}$	$1.99 \cdot 2^{47}$	$1.27 \cdot 2^{43}$	$1.57 \cdot 2^{41}$	$1.51 \cdot 2^{7}$	$1.26 \cdot 2^{91}$	$1.91 \cdot 2^{50}$	$1.18 \cdot 2^{49}$	$1.20 \cdot 2^{94}$	$1.85 \cdot 2^{90}$
LED-128 (CBC)	$1.13 \cdot 2^{79}$	$1.47 \cdot 2^{81}$	$1.90 \cdot 2^{75}$	$1.18 \cdot 2^{74}$	$1.25 \cdot 2^{8}$	$1.39 \cdot 2^{157}$	$1.19 \cdot 2^{84}$	$1.475 \cdot 2^{82}$	$1.13 \cdot 2^{160}$	$1.74 \cdot 2^{156}$
LED-64 (CFB)	$1.57 \cdot 2^{46}$	$1.99 \cdot 2^{47}$	$1.27 \cdot 2^{43}$	$1.57 \cdot 2^{41}$	$1.51 \cdot 2^{7}$	$1.26 \cdot 2^{91}$	$1.91 \cdot 2^{50}$	$1.18 \cdot 2^{49}$	$1.20 \cdot 2^{94}$	$1.85 \cdot 2^{90}$
LED-128 (CFB)	$1.13 \cdot 2^{79}$	$1.47 \cdot 2^{81}$	$1.90 \cdot 2^{75}$	$1.18 \cdot 2^{74}$	$1.25 \cdot 2^{8}$	$1.39 \cdot 2^{157}$	$1.19 \cdot 2^{84}$	$1.475 \cdot 2^{82}$	$1.13 \cdot 2^{160}$	$1.74 \cdot 2^{156}$

5   Conclusion

In this work, we have designed efficient quantum implementation of the variants of LED block cipher (LED-64 and LED-128). By optimizing the quantum circuits through minimization of qubits, quantum gates, and circuit depth, we were able to estimate the least quantum resources required for executing the Grover's attack and the STO attack (which is more difficult than Grover's, but offers quantum advantage for LED-128). Furthermore, we have successfully implemented ECB, CBC and CFB modes for quantum LED; this kind of analysis was not done in the past, to best of our knowledge.

Acknowledgement. We gratefully acknowledge the support of 'MoE AcRF Tier 1 award RT10/23'.

A   Background

A.1   LED Block Cipher

In 2011, Guo et al. [22] introduced the LED block cipher, well-suited for efficient encryption/decryption in resource-constrained environments. LED operates on 64-bit blocks and supports key sizes of 64 and 128 bits. The cipher employs 32 rounds for the 64-bit key version and 48 rounds for the 128-bit key version. Specifically, before the first round, the LED cipher performs an initial AddRoundKey operation. Subsequently, the AddRoundKey operation is executed every four rounds. Each round consists of four steps performed sequentially: AddConstants, SubCells, ShiftRows and MixColumns. Besides, it is necessary to highlight that performing one MixColumnsSerial requires executing four MixColumn operations.

- **KeySchedule** and **AddRoundKey.** LED employs a straightforward key schedule. For LED-64, the same 64-bit user key K is used directly in each round. While for LED-128, the 128-bit user key is divided into two subkeys $(K = K0 \| K1)$, where the key in each round is alternately set to equal the left part $K0$ and the right part $K1$ of K. Each 64-bit round key is Exclusive-OR-ed with 64-bit state.
- **AddConstants.** The round constants are detailed in Table 12, which presents the constants $(rc_5, rc_4, rc_3, rc_2, rc_1, rc_0)$ encoded as byte values for each round. Particularly, rc_0 represents the least significant bit.

Table 12. Round constants used in LED.

Rounds	Constants
$1 - 24$	$01, 03, 07, 0F, 1F, 3E, 3D, 3B, 37, 2F, 1E, 3C, 39, 33, 27, 0E, 1D, 3A, 35, 2B, 16, 2C, 18, 30$
$25 - 48$	$21, 02, 05, 0B, 17, 2E, 1C, 38, 31, 23, 06, 0D, 1B, 36, 2D, 1A, 34, 29, 12, 24, 08, 11, 22, 04$

- **SubCells.** The LED cipher reuses the S-box $(C56B90AD3EF84712)$ from the PRESENT block cipher [9].
- **ShiftRows.** This operation involves cyclically left shifting the bytes in $i - th$ row of the state array by i cell positions. Specifically, the $0th$ row remains unchanged, while the $1st$ row is shifted left by 4 bits, the $2nd$ row is shifted left by 8 bits, and the $3rd$ row is shifted left by 12 bits.
- **MixColumns.** This operation processes each column of the state array as a column vector, which is then replaced by a new vector obtained by post-multiplying it by the matrix M (the MixColumns matrix). This matrix is actually obtained by another matrix, A (the MixColumnsSerial matrix), by raising it to the fourth power. Both are defined over $\mathrm{GF}(2^4)/x^4 + x + 1$ and are given by:

$$M = \begin{bmatrix} 4 & 1 & 2 & 2 \\ 8 & 6 & 5 & 6 \\ B & E & A & 9 \\ 2 & 2 & F & B \end{bmatrix} \qquad A = \begin{bmatrix} 0 & 1 & 0 & 0 \\ 0 & 0 & 1 & 0 \\ 0 & 0 & 0 & 1 \\ 4 & 1 & 2 & 2 \end{bmatrix}$$

Note that $M = A^4$ and is MDS[5]. The authors recommended to implement M by successively applying the circuit of A four times for low resource utilization.

In brief, the prudent design of LED strikes a balance between security and resource efficiency, making this cipher particularly well-suited for IoT devices and other applications with strict resource constraints.

---

[5] Also, one may note that $A = M^{64}$.

A.2   Quantum Gates

X gate, CNOT gate, Toffoli gate and SWAP gate play crucial roles in the implementation. That said, we may like to decompose the AND operations, see [15, Section II] or [7] for more details.

A.3   Grover's Attack

In quantum cryptography, Grover's attack can be utilized to perform a quantum key search, significantly reducing the time complexity of finding the correct encryption key.

Before delving into Grover's attack, it is essential to first understand Grover's algorithm. This famous quantum algorithm provides a quadratic speedup over classical algorithms for unstructured search problem. Specifically, Grover's algorithm can locate the solution to a problem within an unsorted database of N items using approximately $O(\sqrt{N})$ queries, compared to the $O(N)$ queries required by classical brute-force method.

Grover's attack is based on the principles of Grover's algorithm. Specifically, this attack begins by applying Hadamard gates to the key qubits, which creates superposition states. Next, the specific encryption quantum circuit uses these states to encrypt the plaintext. If the generated ciphertext matches the known ciphertext, this oracle then inverts the sign of the corresponding key state. Following this, the diffusion operator is applied to amplify the amplitude of the potential solution key. The combination of the oracle and diffusion operator is repeated multiple times to further enhance the amplitude of the correct key. Finally, measuring the key qubits reveals the most probable solution.

A.4   Improvement over Grover: Search with Two Oracles

In this subsection, we introduce the Search with Two Oracles (STO) attack as an enhancement of Grover's algorithm.

Initially proposed by Kimmel et al. [32], the STO attack is able to reduce the overall cost of quantum search procedures compared to directly applying the Grover's algorithm. This attack utilizes two quantum oracles: the first oracle, O_γ, is relatively inexpensive and marks both the M target items and a number of false positives; the second, O_χ, is more expensive but accurately identifies the M target items. Particularly, O_χ is identical to the one used in Grover's algorithm. For STO attack, the number of queries required remains $O(\sqrt{2^k/M})$ as that for the Grover's attack, where k denotes the key size. In brief, this attack enhances the overall efficiency by balancing the costs of O_γ and O_χ with the number of false positives. By carefully designing these oracles and managing their associated costs, the search process can be significantly optimized.

The STO attack we implement in this paper follows a similar methodology to that described in Reference [18]. In particular, our construction of the oracle O_χ employs a serial-oracle design pattern, as suggested in the same reference.

In subsequent content, we provide a thorough analysis of the quantum resources required for the proposed STO attack on LED cipher.

B   Per-Step Benchmarks

We present the per-step (4 rounds) resource benchmarks for our quantum circuits of LED-64 and LED-128 in Tables 13 and 14, respectively. Note that the initial key XOR is excluded from Tables 13 and 14.

Table 13. Quantum resources required per step (4 rounds) for LED-64.

Step	#CNOT	#NOT	#Toffoli	Toffoli depth (TD)	Circuit depth
1	1088	136	256	16	96
2	1088	148	256	16	96
3	1088	152	256	16	96
4	1088	146	256	16	96
5	1088	144	256	16	96
6	1088	148	256	16	95
7	1088	136	256	16	95
8	1088	148	256	16	96

Table 14. Quantum resources required per step (4 rounds) for LED-128.

Step	#CNOT	#NOT	#Toffoli	Toffoli depth (TD)	Circuit depth
1	1088	136	256	16	95
2	1088	148	256	16	96
3	1088	152	256	16	96
4	1088	146	256	16	96
5	1088	144	256	16	96
6	1088	148	256	16	95
7	1088	136	256	16	95
8	1088	148	256	16	96
9	1088	142	256	16	96
10	1088	146	256	16	96
11	1088	148	256	16	96
12	1088	140	256	16	95

References

1. Amy, M., Di Matteo, O., Gheorghiu, V., Mosca, M., Parent, A., Schanck, J.: Estimating the cost of generic quantum pre-image attacks on SHA-2 and SHA-3. In: Avanzi, R., Heys, H. (eds.) SAC 2016. LNCS, vol. 10532, pp. 317–337. Springer, Cham (2017). https://doi.org/10.1007/978-3-319-69453-5_18

2. Amy, M., Maslov, D., Mosca, M., Roetteler, M., Roetteler, M.: A meet-in-the-middle algorithm for fast synthesis of depth-optimal quantum circuits. IEEE Trans. Comput.-Aided Des. Integrated Circuits Syst. **32**(6), 818–830 (2013). https://doi.org/10.1109/TCAD.2013.2244643

3. Anand, M.V., Targhi, E.E., Tabia, G.N., Unruh, D.: Post-quantum security of the CBC, CFB, OFB, CTR, and XTS modes of operation. In: Takagi, T. (ed.) PQCrypto 2016. LNCS, vol. 9606, pp. 44–63. Springer, Cham (2016). https://doi.org/10.1007/978-3-319-29360-8_4

4. Anand, R., Maitra, A., Mukhopadhyay, S.: Grover on Simon. Quantum Inf. Process. **19**(9) (2020). https://doi.org/10.1007/s11128-020-02844-w

5. Baksi, A.: Classical and physical security of symmetric key cryptographic algorithms. Ph.D. thesis, School of Computer Science & Engineering, Nanyang Technological University, Singapore (2021). https://dr.ntu.edu.sg/handle/10356/152003

6. Baksi, A., et al.: Quantum implementation of linear and non-linear layers. In: IEEE International System-on-Chip Conference (SOCC) (2024)

7. Baksi, A., Jang, K.: Quantum computing fundamental and cryptographic perspective, pp. 7–20. Springer, Singapore (2024). https://doi.org/10.1007/978-981-97-0025-7_2

8. Baksi, A., Karmakar, B., Dasu, V.A.: POSTER: optimizing device implementation of linear layers with automated tools. In: Zhou, J., et al. (eds.) ACNS 2021. LNCS, vol. 12809, pp. 500–504. Springer, Cham (2021). https://doi.org/10.1007/978-3-030-81645-2_30

9. Bogdanov, A., et al.: PRESENT: an ultra-lightweight block cipher. In: Paillier, P., Verbauwhede, I. (eds.) CHES 2007. LNCS, vol. 4727, pp. 450–466. Springer, Heidelberg (2007). https://doi.org/10.1007/978-3-540-74735-2_31

10. Boyer, M., Brassard, G., Høyer, P., Tapp, A.: Tight bounds on quantum searching. Fortschritte der Physik **46**(4-5), 493–505 (1998). https://doi.org/10.1002/(sici)1521-3978(199806)46:4/5<493::aid-prop493>3.0.co;2-p

11. Bravyi, S., Kitaev, A.: Universal quantum computation with ideal clifford gates and noisy ancillas. Phys. Rev. A-Atomic Mol. Opt. Phys. **71**(2), 022316 (2005)

12. Cai, B., Gao, F., Leander, G.: Quantum attacks on two-round even-mansour. Front. Phys. **10**, 1028014 (2022)

13. Chen, J., Liu, Q., Fan, Y., Wu, L., Li, B., Wang, M.: New SAT-based model for quantum circuit decision problem: searching for low-cost quantum implementation. IACR Commun. Cryptol. **1**(1) (2024). https://doi.org/10.62056/anmmp-4c2h

14. Chen, S., Xu, H., Liu, D., Hu, B., Wang, H.: A vision of IoT: applications, challenges, and opportunities with china perspective. IEEE Internet Things J. **1**(4), 349–359 (2014)

15. Chun, M., Baksi, A., Chattopadhyay, A.: Dorcis: depth optimized quantum implementation of substitution boxes. Cryptology ePrint Archive, Paper 2023/286 (2023). https://eprint.iacr.org/2023/286

16. Courtois, N.T., Hulme, D., Mourouzis, T.: Solving circuit optimisation problems in cryptography and cryptanalysis. Cryptology ePrint Archive, Paper 2011/475 (2011). https://eprint.iacr.org/2011/475

17. Dasu, V.A., Baksi, A., Sarkar, S., Chattopadhyay, A.: Lighter-R: optimized reversible circuit implementation for sboxes. In: 2019 32nd IEEE International System-on-Chip Conference (SOCC), pp. 260–265 (2019). https://api.semanticscholar.org/CorpusID:218564036

18. Davenport, J.H., Pring, B.: Improvements to quantum search techniques for block-ciphers, with applications to AES. In: International Conference on Selected Areas in Cryptography, pp. 360–384. Springer, Cham (2020)

19. Feng, J., Wei, Y., Zhang, F., Pasalic, E., Zhou, Y.: Novel optimized implementations of lightweight cryptographic s-boxes via sat solvers. IEEE Trans. Circuits Syst. I Regul. Pap. **71**(1), 334–347 (2024). https://doi.org/10.1109/TCSI.2023.3325559

20. Fowler, A.G., Devitt, S.J., Jones, C.: Surface code implementation of block code state distillation. Sci. Rep. **3**(1), 1939 (2013)

21. Grassl, M., Langenberg, B., Roetteler, M., Steinwandt, R.: Applying Grover's algorithm to AES: quantum resource estimates. In: Takagi, T. (ed.) Post-Quantum Cryptography, pp. 29–43. Springer, Cham (2016)

22. Guo, J., Peyrin, T., Poschmann, A., Robshaw, M.: The LED block cipher. In: Preneel, B., Takagi, T. (eds.) CHES 2011. LNCS, vol. 6917, pp. 326–341. Springer, Heidelberg (2011). https://doi.org/10.1007/978-3-642-23951-9_22

23. He, Y., Luo, M.X., Zhang, E., Wang, H.K., Wang, X.F.: Decompositions of n-qubit toffoli gates with linear circuit complexity. Int. J. Theor. Phys. **56**(7), 2350–2361 (2017)

24. van Hoof, I.: Space-efficient quantum multiplication of polynomials for binary finite fields with sub-quadratic toffoli gate count. arXiv preprint arXiv:1910.02849 (2019)

25. Huang, Z., Sun, S.: Synthesizing quantum circuits of AES with lower t-depth and less qubits. Cryptology ePrint Archive, Report 2022/620 (2022). https://eprint.iacr.org/2022/620

26. Jang, K., Choi, S., Kwon, H., Kim, H., Park, J., Seo, H.: Grover on Korean block ciphers. Appl. Sci. **10**(18) (2020). https://doi.org/10.3390/app10186407. https://www.mdpi.com/2076-3417/10/18/6407

27. Jang, K., Baksi, A., Song, G., Kim, H., Seo, H., Chattopadhyay, A.: Quantum analysis of AES. Cryptology ePrint Archive, Paper 2022/683 (2022). https://eprint.iacr.org/2022/683

28. Jang, K., Choi, S., Kwon, H., Seo, H.: Grover on SPECK: quantum resource estimates. Cryptology ePrint Archive, Report 2020/640 (2020). https://eprint.iacr.org/2020/640

29. Jang, K., Song, G., Kim, H., Kwon, H., Seo, H.: Efficient implementation of present and gift on quantum computers. Appl. Sci. **11**, 4776 (2021). https://doi.org/10.3390/app11114776

30. Jaques, S., Naehrig, M., Roetteler, M., Virdia, F.: Implementing grover oracles for quantum key search on AES and LOWMC (2019). https://arxiv.org/abs/1910.01700

31. Jaques, S., Naehrig, M., Roetteler, M., Virdia, F.: Implementing Grover oracles for quantum key search on AES and LowMC. In: Canteaut, A., Ishai, Y. (eds.) EUROCRYPT 2020. LNCS, vol. 12106, pp. 280–310. Springer, Cham (2020). https://doi.org/10.1007/978-3-030-45724-2_10

32. Kimmel, S., Lin, C.Y.Y., Lin, H.H.: Oracles with costs. Communication and Cryptography, TQC (2015)

33. Lee, I., Lee, K.: The internet of things (IoT): applications, investments, and challenges for enterprises. Bus. Horiz. **58**(4), 431–440 (2015)

34. Liu, Q., Preneel, B., Zhao, Z., Wang, M.: Improved quantum circuits for AES: reducing the depth and the number of qubits. Cryptology ePrint Archive, Paper 2023/1417 (2023). https://eprint.iacr.org/2023/1417
35. NIST.: Submission requirements and evaluation criteria for the post-quantum cryptography standardization process (2016). https://csrc.nist.gov/CSRC/media/Projects/Post-Quantum-Cryptography/documents/call-for-proposals-final-dec-2016.pdf
36. NIST: Call for additional digital signature schemes for the post-quantum cryptography standardization process (2022). https://csrc.nist.gov/csrc/media/Projects/pqc-dig-sig/documents/call-for-proposals-dig-sig-sept-2022.pdf
37. Selinger, P.: Quantum circuits of t-depth one. Phys. Rev. A 87(4), 042302 (2013)
38. Shi, H., Feng, X.: Quantum circuits of AES with a low-depth linear layer and a new structure. Cryptology ePrint Archive, Paper 2024/381 (2024). https://eprint.iacr.org/2024/381
39. Song, M., Jang, K., Song, G., Kim, W., Seo, H.: Quantum circuit implementation of the led block cipher with compact qubit. J. Korea Inst. Inf. Secur. Cryptol. 33(3), 383–389 (2023)
40. Steiger, D., Häner, T., Troyer, M.: Projectq: an open source software framework for quantum computing. Quantum 2 (2016). https://doi.org/10.22331/q-2018-01-31-49
41. Wiebe, N., Roetteler, M.: Quantum arithmetic and numerical analysis using repeat-until-success circuits. arXiv preprint arXiv:1406.2040 (2014)
42. Xiang, Z., Zeng, X., Lin, D., Bao, Z., Zhang, S.: Optimizing implementations of linear layers. IACR Transactions on Symmetric Cryptology (2020)
43. Yuan, Y., Wu, W., Shi, T., Zhang, L., Zhang, Y.: A framework to improve the implementations of linear layers. IACR Trans. Symmetric Cryptol. 2024(2), 322–347 (2024). https://tosc.iacr.org/index.php/ToSC/article/view/11633
44. Zou, J., Wei, Z., Sun, S., Liu, X., Wu, W.: Quantum circuit implementations of AES with fewer qubits. In: Moriai, S., Wang, H. (eds.) ASIACRYPT 2020, pp. 697–726. Springer, Cham (2020)

One-Time Memories Secure Against Depth-Bounded Quantum Circuits

Kyosuke Sekii$^{(\boxtimes)}$ and Takashi Nishide

University of Tsukuba, Ibaraki 305-8577, Japan
`s2420537@u.tsukuba.ac.jp`, `nishide@risk.tsukuba.ac.jp`

Abstract. A one-time memory (OTM) is a useful cryptographic primitive, classically modeled after a non-interactive oblivious transfer. It is well known that secure OTMs (and more generally one-time deterministic programs) cannot exist in the standard model in either the classical or quantum setting due to Broadbent et al. (CRYPTO'13). Broadbent et al. circumvented this impossibility by assuming the existence of hardware tokens that cannot be queried in superposition. In this work, we take a different approach. Building on Liu's assumption (ITCS'23) that adversaries are limited to depth-bounded quantum circuits, we present two OTM constructions. The first is efficiently realizable and secure against adversaries restricted to constant-depth quantum circuits. The second is a feasibility result that achieves security against adversaries limited to $\mathcal{O}(\lambda^{\gamma})$-depth quantum circuits by ensuring that a successful attack would necessarily require deeper quantum computations, where λ^{γ} is a polynomial in the security parameter λ. Our results therefore extend prior work, which either relied on hardware assumptions or considered only constant-depth-bounded adversaries. As a result, by combining our proposed quantum OTMs with the framework of Broadbent et al. (CRYPTO'13), one can also realize quantum one-time programs (OTPs) for deterministic programs.

Keywords: one-time memory · quantum cryptography

1 Introduction

In 2008, Goldwasser, Kalai, and Rothblum introduced the novel concept called the one-time program (OTP), based on the notion of a one-time memory (OTM) [16]. A (stateful) OTM in [16] is a memory device that stores two messages m_0 and m_1 and operates as follows:

1. The sender stores two messages m_0 and m_1 in the OTM token, and sets a tamper-proof bit $\beta = 0$.
2. The sender sends the token to the receiver.
3. The receiver provides an input $b \in \{0, 1\}$ to the token.
4. If $\beta = 0$, the OTM sets $\beta = 1$, and outputs m_b. If $\beta \neq 0$, it performs no operation, and outputs $\perp$.

R. Dutta et al. (Eds.): INDOCRYPT 2025, LNCS 16372, pp. 365–389, 2026.
https://doi.org/10.1007/978-3-032-13301-4_16

An OTM is a memory device that permits access to exactly one of the two stored messages, m_0 or m_1, while irrevocably preventing access to the other[1]. Quantum-resistant OTMs, in particular, enable the construction of quantum-resistant OTPs [9]. OTPs allow the execution of arbitrary circuits while ensuring that, once an input is chosen and executed, the program cannot be reused with a different input. They support applications such as software copy protection and electronic money that prevents double-spending. OTMs can be implemented using hardware such as Trusted Execution Environments (TEEs) [15,33]. However, they cannot be realized without such specialized hardware in the classical settings. Since software-only implementations can be duplicated, they allow adversaries to execute the OTM on different inputs and obtain both stored values.

To address this limitation, quantum states have been considered for their intrinsic non-cloneable property [30], suggesting the possibility of software-only OTMs. However, it has been shown that even with quantum states alone, constructing OTMs, or more generally, transforming deterministic programs (including OTMs) into OTPs, remains impossible when adversaries can perform quantum computations of arbitrary depth [8,9].

These impossibility results motivate the central goal of this work: to construct OTMs, and more generally OTPs for deterministic programs, under the assumption that adversaries are limited to quantum computations of bounded depth [20]. While this is a stronger assumption beyond the standard model, it serves as a meaningful framework in that it makes relatively efficient constructions achievable and secure even on quantum computers operating in imperfect form.

1.1 Related Work

Broadbent et al. [8] constructed secure quantum OTMs using Wiesner's quantum states [29] under the assumption of stateless hardware tokens that hide internal information and disallow superposition queries. Behera et al. [5] subsequently reduced the token's query complexity and achieved noise tolerance against measurement errors. However, these tokens require more sophisticated functionality than the OTMs of [16] and still require physical delivery to the receivers, which limits their practicality.

Liu demonstrated that OTMs can be constructed from Wiesner states under the assumption of depth-bounded quantum adversaries and relatively short decoherence times of quantum states [20]. Since maintaining very deep quantum circuits is considered infeasible (NIST estimates a feasible depth of $2^{40} \sim 2^{96}$ [22]), the depth-bounded setting appears reasonable. The construction embeds the basis information θ into a quantum-resistant Time-Lock Puzzle (TLP) [7]. In [20], it is assumed that the Wiesner state decays before θ can be recovered from the TLP, so the receiver is forced to fix the choice of $b \in \{0, 1\}$ in advance and can obtain only one of m_0 or m_1, with retrieval deferred until the TLP deadline.

---

[1] This realizes the functionality of non-interactive oblivious transfer.

For the security, the deadline must exceed the decoherence time of the quantum states, which may be on the order of an hour or longer [27].

Stambler [25] proposed an OTM construction secure against adversaries restricted to depth-bounded, non-adaptive, and geometrically local quantum operations.[2] The scheme achieves information-theoretic security in the classical sense, as no restriction is placed on the classical computational power of adversaries. However, it relies on quantum random access codes whose decoding requires exponential time, limiting its practicality. In contrast, our method achieves the computationally secure OTMs while requiring only polynomial-time classical computation, making it suitable for practical deployment.

Further Related Work: OTP for Probabilistic Program. Recent works [18,19] construct OTPs from one-time probabilistic signatures [6,13] by restricting attention to probabilistic programs $f(x; r)$, where the randomness $r = \mathcal{RO}(x, \sigma)$ is derived from a valid signature σ on x and the random oracle $\mathcal{RO}$, circumventing the impossibility of transforming deterministic programs into OTPs [8,9]. In contrast, our work follows the depth-bounded adversary model [20], which enables the construction of OTPs even for deterministic programs, as shown in [9].

1.2   Related Impossibility Result: *Rewinding Attack*

As noted earlier, constructing an OTM secure against quantum polynomial time (QPT) adversaries with arbitrary-depth quantum computation is impossible [8,9]. Such an adversary can extract both m_0 and m_1 as follows: it prepares registers initialized with the sender's state (e.g., a Wiesner state), applies the OTM unitary U_{OTM} to read m_0 while leaving the input state nearly undisturbed by the Gentle Measurement Lemma, then inverts U_{OTM} to restore the original state and repeats the process with a different initialization to recover m_1. Even if the OTM read operation is specified as a classical procedure, the adversary can, in principle, implement an equivalent OTM read operation within a quantum circuit.

However, if evaluating U_{OTM} requires a quantum circuit of depth exceeding the adversary's capability, the rewinding attack becomes infeasible. The security of OTM constructions against depth-bounded adversaries relies on precisely this intuition: adversaries are limited to d-depth quantum circuits, while the depth of U_{OTM} in such OTM constructions exceeds this bound d.

---

[2] Geometrically local quantum operations restrict quantum computers to applying two-qubit gates only between adjacent qubits. Non-adjacent qubits must be brought together via SWAP gates, increasing circuit depth. Stambler conjectures that this locality constraint may be unnecessary and that security may rely solely on depth limitations.

1.3 Our Contributions

We propose two constructions for OTMs.

First Construction. Our first construction is an efficiently realizable OTM under the restriction that adversaries are limited to constant-depth quantum circuits. It combines Wiesner states [29], as in [8,20], with obfuscation techniques [12,28] based on the LWE assumption [23]. This approach removes the need for hardware tokens, relying only on classical communication and quantum-state transmission, and avoids the timing constraint of [20], thereby yielding a more practical and flexible construction.

As noted in [8], assuming perfect quantum measurements is unrealistic; addressing this for Wiesner states requires additional countermeasures. To this end, we design the obfuscated program so that its branching program efficiently tolerates small quantum measurement errors. Moreover, applying the "compute-and-compare" obfuscation of [12,28] directly would also require obfuscating pseudorandom generators (PRGs) as well, substantially increasing execution time. Our construction avoids obfuscating PRGs through a tailored program design, thereby achieving improved efficiency.

Second Construction. Building on post-quantum indistinguishability obfuscation (iO), the second construction realizes an OTM secure against $\mathcal{O}(\lambda^\gamma)$-depth quantum circuits in the quantum random oracle model, by tuning the scheme depth parameter k so that breaking it requires circuits of depth $\Omega(\lambda^{\gamma+1})$. To the best of our knowledge, this is the first such OTM secure against $\mathcal{O}(\lambda^\gamma)$-depth quantum circuits. Its security relies on the unclonability and direct product hardness of *hidden subspace states*, a primitive widely used in quantum money schemes [1,11,13]. This second construction extends security to more powerful adversaries, serving as an important feasibility result despite its reliance on heavy machinery such as iO and error correcting codes.

Comparison with Prior Works. Unlike Broadbent et al. [8], our constructions are entirely software-based and tolerate realistic quantum measurement errors without relying on hardware tokens. Compared with Liu's scheme [20], our constructions avoid the dependency on short decoherence times. Moreover, our second construction can also handle adversaries bounded by non-constant depth. Stambler's work [25] requires an exponential OTM read cost, whereas our constructions require only a polynomial cost. The work in [18,19] focus only on probabilistic programs, while our constructions support deterministic programs under the assumption of depth-bounded quantum adversaries [20], thereby circumventing the known impossibility results [8,9]. Table 1 summarizes the comparison of our constructions with prior works.

Table 1. Comparison of Our Constructions with Prior Works

Scheme	Hardware Token	Short Decoherence Time	Circuit Depth of Adversary	OTM Read Cost	Supported Programs
Broadbent et al. [8]	Required	No	Unbounded	Polynomial	Deterministic
Liu [20]	No	Required	Constant	Polynomial	Deterministic
Stambler [25]	No	No	Bounded	Exponential	Deterministic
GM24, GLR+24 [18,19]	No	No	Unbounded	Polynomial	Probabilistic
Our Const. 1	No	No	Constant	Polynomial	Deterministic
Our Const. 2	No	No	Polynomial	Polynomial	Deterministic

2   Preliminaries

2.1   Notations

We define the set of n integers $[n] := \{1, \ldots, n\}$. For a probability distribution or random variable X, sampling of x from X is denoted as $x \leftarrow X$. When x is selected uniformly at random from a set X, it is specifically denoted as $x \xleftarrow{\$} X$. The security parameter is denoted by λ. A negligible function is denoted by $\mathsf{negl}(\cdot)$, and a polynomial function is denoted by $\mathsf{poly}(\cdot)$. For a bit string x, $x[i]$ denotes its i-th bit, and for a function f, $f[i]$ denotes the function that outputs only the i-th bit of f's output, where the index of the first bit is 1. When a and b are bit strings, $\mathsf{HW}(a)$ denotes the Hamming weight of a, and $\mathsf{HD}(a, b)$ denotes the Hamming distance between a and b. We denote "probabilistic polynomial time" as PPT, and "quantum polynomial time" as QPT for short.

2.2   Quantum Computation

A pure quantum state of one qubit is represented as a unit column vector $|\psi\rangle$ in a complex Hilbert space. The orthonormal basis $\{|0\rangle, |1\rangle\}$ is referred to as the computational basis, and the orthonormal basis $\left\{ \frac{|0\rangle + |1\rangle}{\sqrt{2}}, \frac{|0\rangle - |1\rangle}{\sqrt{2}} \right\}$ is referred to as the Hadamard basis. For simplicity, we define $|+\rangle := \frac{|0\rangle + |1\rangle}{\sqrt{2}}$, $|-\rangle := \frac{|0\rangle - |1\rangle}{\sqrt{2}}$.

2.3   Wiesner States

Let $s = s_1 \cdots s_n \in \{0, 1\}^n$ and $\theta = \theta_1 \cdots \theta_n \in \{0, 1\}^n$. The n-qubit Wiesner state $|s^\theta\rangle$ is defined as $|s^\theta\rangle = |s[1]^{\theta[1]}\rangle \otimes \cdots \otimes |s[n]^{\theta[n]}\rangle$ where the i-th qubit is $|s[i]^{\theta[i]}\rangle = H^{\theta[i]}|s[i]\rangle$, H denotes the Hadamard matrix, and H^0 represents the identity matrix. For example, if $s = 0110$ and $\theta = 0011$, then $|s^\theta\rangle := |01 - +\rangle$.

It is well known that, without θ, recovering all bits of s from a Wiesner state $|s^\theta\rangle$ is difficult, even by applying any quantum operation or performing measurements in any basis. It is also known that when n is sufficiently large, it is hard to duplicate a Wiesner state without the knowledge of θ (Lemma 1).

Lemma 1 (Infeasibility of Cloning Wiesner States [10]). *For any even integer $n \geq 2$, any (unbounded) quantum adversaries $(\mathcal{A}, \mathcal{B}, \mathcal{C})$,*

$$\Pr[\mathsf{CloneGame}_{\mathcal{A},\mathcal{B},\mathcal{C}}(n) = 1] \leq 0.924^n < 2^{-0.11n},$$

where $\mathsf{CloneGame}_{\mathcal{A},\mathcal{B},\mathcal{C}}(n)$ is defined as:

1. *A challenger samples $s \leftarrow \{0,1\}^n$ and $\theta \leftarrow \{0,1\}^n$ conditioned on $\mathsf{HW}(\theta) = n/2$ uniformly at random, and sends $|s^\theta\rangle$ to $\mathcal{A}$.*
2. *$\mathcal{A}$ applies a quantum channel on $|s^\theta\rangle$ and obtains ρ_{BC} over registers B and C, which are then sent to $\mathcal{B}$ and $\mathcal{C}$ respectively.*
3. *The challenger sends θ to $\mathcal{B}$ and $\mathcal{C}$.*
4. *The game outputs 1 if and only if $\mathcal{B}$ outputs s_0 and $\mathcal{C}$ outputs s_1.*

Proof. This follows from a corollary of the stronger monogamy-of-entanglement result for coset states [14], since Wiesner states are a special case of coset states.

2.4   Depth-Bounded Quantum Circuits [20]

Depth-bounded quantum circuits [20] are defined under the assumption that adversaries can only evaluate quantum circuits of depth at most d. In our model, shallow quantum circuits are interleaved with full measurements and classical processing, following the framework of [20]. A full measurement acts on all qubits, including both input and output registers, leaving the internal state entirely classical. The model of depth-bounded quantum circuits is illustrated in Fig. 1.

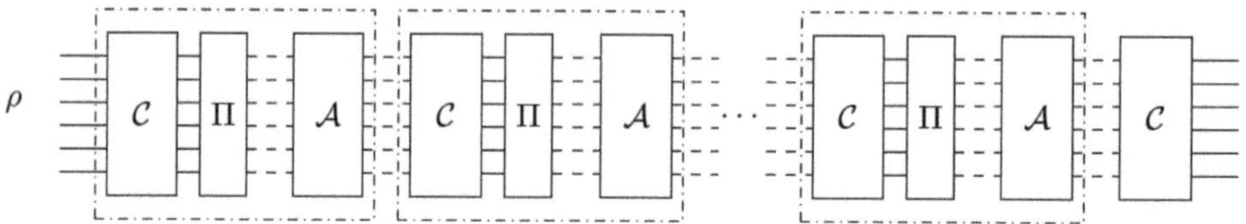

Fig. 1. d-depth bounded quantum circuits from [20]. $\mathcal{C}$ denotes quantum circuits, Π denotes measurements whereas $\mathcal{A}$ denotes classical post-processing. Dashed lines denote classical bits and solid lines denote qubits. There are polynomially many such blocks, but each $\mathcal{C}$ has depth at most d.

2.5   Quantum Random Oracle Model (QROM)

We work in the quantum random oracle model (QROM), which allows quantum circuits to access classical random oracles.

Definition 1 (Random Oracle). *A classical random oracle $\mathcal{RO}$ is defined as a unitary transformation of the form $U_f |x, y\rangle \rightarrow |x, y \oplus f(x)\rangle$ where $f : \{0,1\}^{n+1} \rightarrow \{0,1\}^m$ is a classical random function. We note that such a classical oracle can also be queried in quantum superposition.*

Unless otherwise specified, we use the term "random oracle" to mean a classical random oracle.

2.6   Measurement Error Model

In an ideal quantum computer, measuring a quantum state $|\psi\rangle$ using the projection measurement operator $\{|\psi\rangle\langle\psi|, I - |\psi\rangle\langle\psi|\}$ will always yield the outcome 0 (i.e., $|\psi\rangle$) with probability 1. However, in a quantum computer subject to a measurement error rate ϵ, the outcome of the measurement will be 0 with probability $1 - \epsilon$, and 1 with probability ϵ where ϵ is assumed to be small. In this work, we assume that when an n-qubit quantum state is measured, $n\epsilon = c$ qubits of the measurement result will be flipped, while the remaining $n - n\epsilon = n - c$ qubits are correctly measured.

2.7   One Time Memory (OTM)

An OTM scheme consists of two procedures: one for generating a token composed of quantum states and classical programs, and another for retrieving information from the token. It is parameterized by a security parameter λ and a message length ℓ_{msg}. The following definition is a variant of the one presented in [20]:

Definition 2 (Memory Token).   *A memory token scheme consists of the following quantum algorithms.*

- *Token.Gen$(1^\lambda, m_0, m_1)$ is a quantum algorithm that takes as input a security parameter λ and two messages $m_0, m_1 \in \{0,1\}^{\ell_{msg}}$, and outputs a quantum token $|\mathsf{tk}\rangle$.*
- *Token.Eval$(|\mathsf{tk}\rangle, b)$ is a quantum algorithm that takes as input a quantum token $|\mathsf{tk}\rangle$ and a bit $b \in \{0,1\}$, and outputs a message m.*

Correctness: For every security parameter λ, every message length ℓ_{msg}, every pair of messages $m_0, m_1 \in \{0,1\}^{\ell_{msg}}$ and every bit $b \in \{0,1\}$, we have

$$\Pr[\mathsf{Token.Eval}(|\mathsf{tk}\rangle, b) = m_b \mid |\mathsf{tk}\rangle \leftarrow \mathsf{Token.Gen}(1^\lambda, m_0, m_1)] = 1.$$

Next, we define "one-timeness" property of OTMs. This property states that, for depth-bounded adversary, it can either learn m_0 or m_1, but not both. Then we formalize the security via the game-based definition.

Definition 3 (One-Time Memory Secure against Depth-bounded Adversary). *Consider the following game between a challenger and a depth-bounded adversary $\mathcal{A}$.*
OTM-Game$_\mathcal{A}(1^\lambda)$:

1. *The challenger $\mathcal{C}$ samples two messages $m_0, m_1 \in \{0,1\}^{\ell_{msg}}$.*
2. *$\mathcal{C}$ samples $|\mathsf{tk}\rangle \leftarrow \mathsf{Token.Gen}(1^\lambda, m_0, m_1)$ and sends $|\mathsf{tk}\rangle$ to $\mathcal{A}$.*
3. *$\mathcal{A}$ can perform quantum computation with $|\mathsf{tk}\rangle$, but once the computation reaches depth d, $\mathcal{A}$ measures $|\mathsf{tk}\rangle$ and all quantum registers.*
4. *$\mathcal{A}$ outputs a pair m_0', m_1'.*
5. *$\mathcal{C}$ outputs 1 if $m_0' = m_0$ and $m_1' = m_1$. Otherwise, it outputs 0.*

Algorithm 1 PHard(b, sig)

    Hardcoded: s, $\theta \in \{0,1\}^n$, m_0, $m_1 \in \{0,1\}^{\ell_{msg}}$
1: cnt $= 0$
2: **for** $i \in [n]$: **if** $sig[i] \neq s[i] \wedge \theta[i] = b$ **then** cnt $\leftarrow$ cnt $+ 1$
3: **return** m_b if cnt $\leq n\epsilon$, otherwise $\perp$

We say that a quantum OTM scheme (Token.Gen, Token.Eval) *is secure, if, for any depth-bounded adversary* $\mathcal{A}$,

$$\Pr[\text{OTM-Game}_{\mathcal{A}}(1^\lambda) = 1] \leq \text{negl}(\lambda)$$

We also require that both Token.Gen and Token.Eval be computable by a d-depth quantum algorithm where $d = \mathcal{O}(1)$ or $d = \mathcal{O}(\lambda^\gamma)$ in our setting.

We recall a one-time memory in the classical stateless hardware model [5,8]. The classical hardware model assumes a device that completely hides all partial information about the hardcoded program and accepts only classical inputs.

Theorem 1 (OTM in the Classical Hardware Model [5]). *Construction 4 in the stateless hardware model, which is based on Wiesner states, is an OTM protocol secure against every QPT adversary* $\mathcal{A}$ *making a polynomial number of classical queries to the hardware token. Moreover, Construction 4 is tolerant to a measurement error rate of up to* $\epsilon = 0.14$.

Construction 4 (OTM Construction with Classical Hardware [5]). Token.Gen($1^\lambda, m_0, m_1$): The sender $\mathcal{S}$ performs the following steps:

1. Sample $\theta, s \leftarrow \{0,1\}^n$ such that $\text{HW}(\theta) = \frac{n}{2}$.
2. Prepare the Wiesner state $|s^\theta\rangle = H^\theta|s\rangle$.
3. Set $\epsilon \leftarrow \mathbb{R}$ where $0 \leq \epsilon \leq 0.14$.
4. Implement PHard (Algorithm 1) in the classical stateless hardware.
5. Send $|\text{tk}\rangle := (|s^\theta\rangle, \text{PHard})$ to the receiver.

Token.Eval($b, |\text{tk}\rangle$): Upon receiving the token $|\text{tk}\rangle$ from the sender $\mathcal{S}$, the receiver $\mathcal{R}$ selects an input bit $b \in \{0,1\}$ and executes the following:

1. Parse $|\text{tk}\rangle = (|s^\theta\rangle, \text{PHard})$.
2. Measure $(H^b)^{\otimes n}|s^\theta\rangle$ in the computational basis to obtain s'.
3. Output $m_b \leftarrow \text{PHard}(b, s')$.

2.8 Indistinguishability Obfuscation

We recall the notion of indistinguishability obfuscation (iO).

Definition 5. *An indistinguishability obfuscation scheme iO for a class of circuits* $\{\mathcal{C}_\lambda\}_{\lambda \in \mathbb{N}}$ *satisfies the following.*

Correctness. *For all* $\lambda, C \in \mathcal{C}_\lambda$, *and input* x,

$$\Pr[\tilde{C}(x) = C(x) \mid \tilde{C} \leftarrow \mathsf{iO}(1^\lambda, C)] \geq 1 - \mathsf{negl}(\lambda).$$

Security. *Let* $C_0, C_1 \in \mathcal{C}_\lambda$ *be two circuits of the same size such that* $C_0(x) = C_1(x)$ *for all inputs* x. *Then, for any QPT algorithm* $\mathcal{D}$,

$$\left| \Pr[\mathcal{D}(\tilde{C}_0, \mathsf{aux}) = 1] - \Pr[\mathcal{D}(\tilde{C}_1, \mathsf{aux}) = 1] \right| \leq \mathsf{negl}(\lambda)$$

$$\text{where } \tilde{C}_0 \leftarrow \mathsf{iO}(1^\lambda, C_0), \ \tilde{C}_1 \leftarrow \mathsf{iO}(1^\lambda, C_1).$$

2.9   Obfuscation for Compute-and-Compare Program

We briefly recall a class of programs called compute-and-compare (CC) programs and their obfuscation, referred to as *compute-and-compare obfuscation* [17,28].

Definition 6 (Compute-and-Compare (CC) Program). *Given a function* $f : \{0,1\}^{v(\lambda)} \to \{0,1\}^{w(\lambda)}$, *a target value* $y \in \{0,1\}^{w(\lambda)}$, *the following program is called a compute-and-compare program.*

$$P_{f,y}(x) = \begin{cases} 1, & \text{if } f(x) = y \\ 0, & \text{otherwise} \end{cases}$$

A distribution $\mathcal{D}_\lambda$ over f and y for CC programs is called *sub-exponentially unpredictable* if, for any QPT adversary, given the auxiliary information aux and oracle access to f, the adversary can predict the target value y only with sub-exponential probability.

We note that the CC programs to be obfuscated here are required to be *evasive* functions, as in [28], meaning informally that they output 0 with overwhelming probability when the input x is chosen at random.

Definition 7 (Compute-and-Compare Obfuscation [12,28]). *Let* CCObf *be a PPT algorithm, which takes as input a compute-and-compare program* $P_{f,y}$, *a security parameter* λ, *and outputs a classical program* $\tilde{P}_{f,y}$. *Let* $\mathcal{D}_\lambda$ *be a sub-exponentially unpredictable distribution parameterized with* λ, *and sample* $(P_{f,y}, \mathsf{aux}) \leftarrow \mathcal{D}_\lambda$. *We say that* CCObf *is an obfuscation scheme for a compute-and-compare program* $P_{f,y}$ *if it satisfies the following properties:*

Correctness. *There exists a negligible function* $\mathsf{negl}(\lambda)$ *such that, for* $P_{f,y}$,

$$\Pr[\forall x : P_{f,y}(x) = \tilde{P}_{f,y}(x) \mid \tilde{P}_{f,y} \leftarrow \mathsf{CCObf}(1^\lambda, P_{f,y})] \geq 1 - \mathsf{negl}(\lambda).$$

Security. *For every QPT adversary* $\mathcal{A}$, *there exists a PPT simulator* CCSim *that outputs a program* P' *satisfying* $P'(x) = 0$ *for all* x. $\mathcal{A}$ *can distinguish between the following two distributions with at most* $\mathsf{negl}(\lambda)$ *advantage:*

$$\left| \Pr[\mathcal{A}(\mathsf{CCObf}(1^\lambda, P_{f,y}), \mathsf{aux}) = 1] - \Pr[\mathcal{A}(\mathsf{CCSim}(1^\lambda, |f|, |y|), \mathsf{aux}) = 1] \right| \leq \mathsf{negl}(\lambda).$$

Theorem 2 ([28]). *Assuming the hardness of LWE, there exists a compute-and-compare obfuscation scheme for any class of CC programs drawn from subexponentially unpredictable distributions.*

This guarantees that $\mathsf{CCObf}(1^\lambda, P_{f,y})$ leaks no partial information about (f, y).

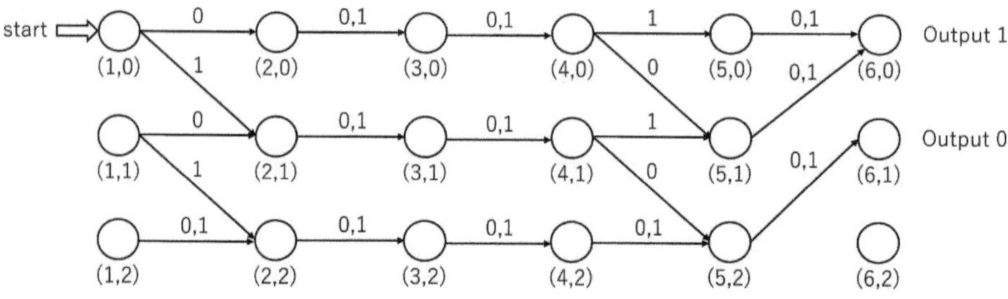

Fig. 2. An example of a branching program. If the input is 10001, the output is 0.

2.10 Branching Programs

We recall branching programs (e.g., Fig. 2) used to represent compute-and-compare programs in [12, 28], along with their obfuscation. A branching program is a directed acyclic graph in which every vertex has exactly two outgoing edges, one labeled 1 and the other labeled 0. The graph has a designated start node. To perform a computation, we begin at the start state, and, at each step, follow the edge corresponding to the bit of the input bit string labeling the current node. When a node labeled with an output value is reached, the computation outputs that value.

General branching programs representing CC programs can be obfuscated via the compute-and-compare obfuscation CCObf (Definition 7).

Theorem 3 (Obfuscation for Compute-and-Compare Branching Programs [12,28]). *Assuming the hardness of LWE, there exists compute-and-compare obfuscation CCObf for compute-and-compare branching programs.*

2.11 Hidden Subspace States

A subspace state is $A = \sum_{v \in A} |v\rangle$ where A is a subspace of the vector space $\mathbb{F}_2^n$. We will overload the notation by also writing A and $A^\perp$ to denote the corresponding membership-checking programs for A and its orthogonal complement $A^\perp$. In other words, $\mathsf{iO}(A)$ (resp. $\mathsf{iO}(A^\perp)$) checks whether its input belongs to A (resp. $A^\perp$), outputting 1 if so and 0 otherwise.

The following theorem states that any QPT adversary, given a single copy of $|A\rangle$ and membership-checking programs $(\mathsf{iO}(A), \mathsf{iO}(A^\perp))$, can clone $|A\rangle$ with only negligible probability.

Theorem 4 ($1 \to 2$ Unclonability [32]). *Consider the following game between a challenger and an adversary $\mathcal{A}$.*
$\underline{\mathsf{Exp}_{\mathcal{A}}(1^{\lambda})}$:

1. *The challenger $\mathcal{C}$ samples a random subspace $A \subseteq \mathbb{F}_2^{\lambda}$ of dimension $\lambda/2$.*
2. *$\mathcal{C}$ sends $|A\rangle$, $\mathrm{iO}(A)$, $\mathrm{iO}(A^{\perp})$ to $\mathcal{A}$.*
3. *$\mathcal{A}$ prepares and sends a (entangled) bipartite register R_1, R_2 to $\mathcal{C}$.*
4. *$\mathcal{C}$ applies the projective measurement $\{|A\rangle\langle A|, I - |A\rangle\langle A|\}$ to both R_1, and R_2, and outputs 1 if both measurements succeed. Otherwise, it outputs 0.*

Then, for any QPT adversary $\mathcal{A}$, we have $\Pr[\mathsf{Exp}_{\mathcal{A}}(1^{\lambda}) = 1] \leq \mathsf{negl}(\lambda)$.

The next theorem ensures that any QPT adversary, given $(|A\rangle, \mathrm{iO}(A), \mathrm{iO}(A^{\perp}))$, can output both vectors $v \in A$ and $w \in A^{\perp}$ only with negligible probability.

Theorem 5 (Direct Product Hardness [24]). *Consider the following game between a challenger and an adversary $\mathcal{A}$.*
$\underline{\mathsf{Exp}_{\mathcal{A}}(1^{\lambda})}$

1. *The challenger $\mathcal{C}$ samples a random subspace $A \subseteq \mathbb{F}_2^{\lambda}$ of dimension $\lambda/2$.*
2. *$\mathcal{C}$ sends $(|A\rangle, \mathrm{iO}(A), \mathrm{iO}(A^{\perp}))$ to $\mathcal{A}$.*
3. *$\mathcal{A}$ sends two vectors $v, w \in \mathbb{F}_2^{\lambda}$ to $\mathcal{C}$.*
4. *$\mathcal{C}$ outputs 1 if $(v \in A \setminus \{0^{\lambda}\} \wedge w \in A^{\perp} \setminus \{0^{\lambda}\})$, otherwise, it outputs 0.*

Then, for any QPT adversary $\mathcal{A}$, we have $\Pr[\mathsf{Exp}_{\mathcal{A}}(1^{\lambda}) = 1] \leq \mathsf{negl}(\lambda)$.

3 Efficiently Realizable OTM Secure Against Constant-Depth-Bounded Quantum Adversary

3.1 Basic Construction of OTM

We refer to an adversary who cannot execute quantum circuits of depth greater than d as a d-depth bounded adversary. Assuming d is a constant as in [20], we now construct an OTM protocol as follows:

Construction 8 (OTM with Wiesner States). $\mathsf{Token.Gen}(1^{\lambda}, m_0, m_1)$: On input $m_0, m_1 \in \{0,1\}^{\ell_{msg}}$, the sender $\mathcal{S}$ performs:

1. Sample $s, \theta \xleftarrow{\$} \{0,1\}^n$ and $f_b, y_b \xleftarrow{\$} \mathcal{D}_{\lambda}{}^3$ for $b \in \{0,1\}$ where $y_b \in \{0,1\}^{\ell_{tgt}}$.
2. Prepare the quantum state $|s^{\theta}\rangle = H^{\theta}|s\rangle$.
3. Prepare the program PMsg_b (Algorithm 2) using s, θ, f_b, y_b and m_b.
4. For each $b \in \{0,1\}$, compile $\mathsf{OPMsg}_b \leftarrow \mathsf{CCObf}(\mathsf{PMsg}_b)^4$.
5. Send $|\mathsf{tk}\rangle = (|s^{\theta}\rangle, \mathsf{OPMsg}_0, \mathsf{OPMsg}_1)$ to the receiver.

---

[3] A distribution $\mathcal{D}_{\lambda}$ over f_b and y_b is sub-exponentially unpredictable.

[4] More precisely, CCObf here should be an obfuscation scheme for a multi-bit output version, referred to as MBCC. We use the notation CCObf for simplicity, and refer to Sect. 3.3 for details.

Algorithm 2 $\mathsf{PMsg}_b(x)$

Hardcoded: $s, \theta \in \{0,1\}^n$, $f_b : \{0,1\}^n \to \{0,1\}^{\ell_{tgt}}$, $y_b \in \{0,1\}^{\ell_{tgt}}, m_b \in \{0,1\}^{\ell_{msg}}$.

Input: $x \in \{0,1\}^n$

1: Compute $f_b(x) = \begin{cases} y_b, & \text{if } x[i] = s[i] \ \forall i \in [n] \text{ with } \theta[i] = b, \\ \bot, & \text{otherwise} \end{cases}$

2: **return** m_b if $f_b(x) = y_b$, otherwise 0

✳ Here, PMsg_b denotes the code of Algorithm 2, and OPMsg_b denotes the obfuscated code of PMsg_b where PMsg_b checks the measurement outcome s' obtained in Step 2 of $\mathsf{Token.Eval}$ and outputs m_b if s' is valid.

$\mathsf{Token.Eval}(b, |\mathsf{tk}\rangle)$: On input $b \in \{0,1\}$ and a token $|\mathsf{tk}\rangle$ from the sender $\mathcal{S}$, the receiver $\mathcal{R}$ performs:

1. Parse $|\mathsf{tk}\rangle = (|s^\theta\rangle, \mathsf{OPMsg}_0, \mathsf{OPMsg}_1)$.
2. Measure $(H^b)^{\otimes n}|s^\theta\rangle$ in the computational basis to obtain s'.
3. Output $m_b = \mathsf{OPMsg}_b(s')$.

We show that Construction 8 is correct and secure against $\mathcal{O}(1)$-depth bounded adversaries.

Lemma 2 (Correctness and Security of Construction 8) *Assuming the quantum hardness of LWE, Construction 8 is an OTM secure against $\mathcal{O}(1)$-depth bounded, if all measurements are error-free ($\epsilon = 0$).*

Proof (Sketch).

Correctness. If $\mathcal{R}$ chooses $b = 0$, the measurement outcome s' matches s at all positions i where $\theta[i] = 0$ since those positions are encoded in the computational basis. When f_0 receives such s', f_0 returns y_0, leading OPMsg_0 to correctly output m_0. Similarly if $b = 1$, then $s'[i] = s[i]$ holds for all i such that $\theta[i] = 1$. In this case, $f_1(s') = y_1$, and hence $\mathsf{OPMsg}_1(s') = m_1$ as desired.

Security. We assume that $\mathcal{R}$ has noisy intermediate-scale quantum (NISQ)-level capability, modeled as an $\mathcal{O}(1)$-depth bounded adversary whose gates have fan-in at most 2 and fan-out 1. Each OPMsg_b checks $n/2$ input bits and thus the execution of this obfuscated program requires depth at least $c\log_2(n/2)$ for some constant $c > 1$, i.e., $\mathcal{O}(\log n)$, to determine its output (Fig. 3). This depth cannot be reduced since all $n/2$ bits affect the output, and the obfuscation prevents $\mathcal{R}$ from exploiting shortcuts. If $c\log_2(n/2) > d$ for a small constant d, $\mathcal{R}$ cannot execute the full quantum circuit of OPMsg_b. Consequently, $\mathcal{R}$ cannot perform a rewinding attack (see Sect. 1.2). A d-depth bounded adversary must measure all quantum states before learning the output of OPMsg_b, obtaining only classical information (Fig. 1). After measuring the Wiesner state $|s^\theta\rangle$ to obtain m_b honestly, $\mathcal{R}$ cannot find an input accepted by OPMsg_{1-b}, as the state is irreversibly collapsed and the number of possible valid inputs is $2^{n/2}$. $\qquad\square$

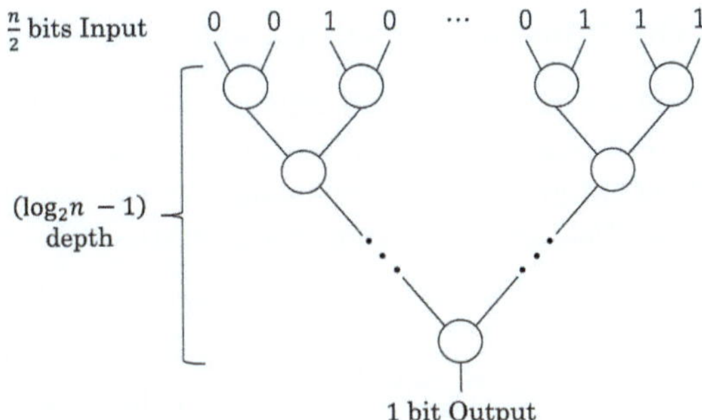

Fig. 3. Illustration of computing a program $P : \{0,1\}^{n/2} \to \{0,1\}$. Each circle represents a single (quantum) gate.

3.2 Upgrading Practicality Through Measurement Error Tolerance

So far, we have assumed error-free measurements. In practice, however, measurement errors occur with non-negligible probability at the NISQ level, and even a single error can cause the obfuscated programs or the OTM to fail. Since the security of the OTM relies on the one-time transmission of Wiesner states, the sender cannot retransmit the states in case of error.

To address this, we modify f_b in PMsg_b (Algorithm 2) to accept input strings within a small Hamming distance of the correct input, similar to **PHard** (Algorithm 1). With this modification, f_b outputs the same y_b even if a few measurement bits are flipped. We assume at most c of n bits are flipped, where $c \leq 0.14n$.

Lemma 3 (OTM with Measurement Error Tolerance). *We modify f_b in* PMsg_b *(Algorithm 2) to f'_b defined as follows:*

$$f'_b(x) = \begin{cases} y_b, & \text{if } \mathsf{HD}(x, t_b) \leq c, \\ \bot, & \text{otherwise} \end{cases} \quad \text{where } t_b[j] = \begin{cases} s[j], & \text{if } \theta[j] = b, \\ *, & \text{otherwise} \end{cases} \quad \forall j \in [n]$$

(1)

with the convention that $\mathsf{HD}(0, *) = \mathsf{HD}(1, *) = 0$.

Then Construction 8 with f_b replaced by f'_b also yields an OTM protocol that tolerates up to c measurement errors.

Proof (Sketch). Correctness of the modified PMsg_b is immediate. For security, assuming $c \leq 0.14n$, and compared with Construction 8, the fraction of inputs x for which $\mathsf{PMsg}_b(x)$ with (1) outputs m_b is roughly[5] at most $(2^{n/2} \cdot 2^{0.29n})/2^n$, which remains negligible in n, and thus $\mathsf{PMsg}_b(x)$ remains an evasive function as required. $\square$

---

[5] E.g., when $c = 0.025n$, $n = 240$, the more exact fraction is $\approx (2^{n/2} \cdot \sum_{k=0}^{c/2} \binom{n/2}{k})/2^n \approx 2^{-102}$.

3.3 Upgrading the Efficiency of Obfuscation

Now we consider how to concretely and efficiently obfuscate $\mathsf{PMsg}_b(x)$ (Algorithm 2) where $f_b(x)$ is replaced by $f_b'(x)$ (in (1)). To this end, we consider applying the obfuscation technique for multi-bit compute-and-compare (MBCC) programs in [12,28]. Here $\mathsf{PMsg}_b(x)$ corresponds to the MBCC program $P_{f_b', y_b, m_b}(x)$ defined as

$$\mathsf{PMsg}_b(x) = P_{f_b', y_b, m_b}(x) = \begin{cases} m_b, & \text{if } f_b'(x) = y_b, \\ \perp, & \text{otherwise.} \end{cases} \tag{2}$$

In [12,28], obfuscating the MBCC program $P_{f_b', y_b, m_b}(x)$ is further reduced to obfuscating ℓ_{msg} CC programs (see Definitions 6 and 7), each responsible for outputting one bit of $m_b \in \{0,1\}^{\ell_{msg}}$. More specifically, to obfuscate $P_{f_b', y_b, m_b}(x)$, it is first represented as a series of CC programs

$$P_{G_1 \circ f_b', y_{b,1}}(x), \ldots, P_{G_i \circ f_b', y_{b,i}}(x), \ldots, P_{G_n \circ f_b', y_{b,\ell_{msg}}}(x), \text{ where, for } i \in [\ell_{msg}],$$
$$\tag{3}$$

$$P_{G_i \circ f_b', y_{b,i}}(x) = \begin{cases} 1, & \text{if } (G_i \circ f_b')(x) = y_{b,i} \\ 0, & \text{otherwise} \end{cases}, \quad y_{b,i} = \begin{cases} G_i(y_b), & \text{if } m_b[i] = 1 \\ \overline{G_i(y_b)}, & \text{if } m_b[i] = 0. \end{cases}$$

Here each G_i is a pseudorandom generator (PRG)[6], with $(G_i \circ f_b')(x) = G_i(f_b'(x))$, and $\overline{G_i(y_b)}$ denoting the bitwise complement. The obfuscated $P_{f_b', y_b, m_b}(x)$ is obtained by obfuscating the ℓ_{msg} CC programs $P_{G_i \circ f_b', y_{b,i}}(x)$ in (3), so the computation of each G_i must also be included in the obfuscation. Intuitively, G_i are required to yield multiple random-looking target values $y_{b,i}$ since compute-and-compare obfuscation requires random target values although the underlying function f_b' is fixed here.

In our setting, however, each $G_i \circ f_b'$ in (3) can be replaced by an almost equivalent function $f_{b,i}$ defined in (4) of $\mathsf{PMsg}_b[i]$ (Algorithm 3), where $y_{b,1}, \ldots, y_{b,\ell_{msg}} \in \{0,1\}^{\ell_{tgt}}$ are just sampled uniformly at random without invoking G_i. This removes the need to compute G_i and improves efficiency. Thus PMsg_b in (2) consists of $(\mathsf{PMsg}_b[1], \ldots, \mathsf{PMsg}_b[\ell_{msg}])$.

Next, we discuss efficient obfuscation of each CC program $\mathsf{PMsg}_b[i]$ (Algorithm 3). Following [12,28], the program must first be converted into a small branching program (BP). If this conversion is not possible, one must resort to FHE-based obfuscation of [28], which is significantly less efficient. We note that while [28] required permutation BPs, [12] showed, via a clever trick, that non-permutation BPs can also be obfuscated efficiently.

Thus, the main task is to represent $f_{b,i}(x)$ in (4) as a compact non-permutation BP without relying on heavy machinery such as error-correcting codes for Hamming distance. Since a BP outputs a single bit while $f_{b,i}(x)$ outputs an m-bit string, we require ℓ_{tgt} BPs $(\mathsf{BPMsg}_b[i, 1], \ldots, \mathsf{BPMsg}_b[i, \ell_{tgt}])$ to

---

[6] In [28], the LWE-based PRG of [2] is used. This PRG, employing rounding in the Learning With Rounding (LWR) problem, lies in NC^1 and is strongly injective.

Algorithm 3 $\mathsf{PMsg}_b[i](x)$ (where $i \in [\ell_{msg}]$)

Hardcoded: $s, \theta \in \{0,1\}^n$, $f_{b,i} : \{0,1\}^n \to \{0,1\}^{\ell_{tgt}}$, $y_{b,i}, z_{b,i}, v_{b,i} \in \{0,1\}^{\ell_{tgt}}$, $m_b[i] \in \{0,1\}$.

Input: $x \in \{0,1\}^n$

Note: The target value $v_{b,i}$ is predetermined as $v_{b,i} = \begin{cases} y_{b,i}, & \text{if } m_b[i] = 1 \\ z_{b,i}, & \text{if } m_b[i] = 0 \end{cases}$.

1: Compute the following $f_{b,i}(x)$. Here $\overline{y}_{b,i}$ denotes the bitwise complement of $y_{b,i}$.

$$f_{b,i}(x) = \begin{cases} y_{b,i}, & \text{if } \mathsf{HD}(x, t_b) \le c, \\ \overline{y}_{b,i}, & \text{otherwise} \end{cases} \quad \text{where } t_b[j] = \begin{cases} s[j], & \text{if } \theta[j] = b, \\ *, & \text{otherwise} \end{cases} \quad \forall j \in [n] \tag{4}$$

$\triangleright$ The output bit equals $m_b[i]$ if $\mathsf{HD}(x, t_b) \le c$, and 0 otherwise
2: **return** 1 if $f_{b,i}(x) = v_{b,i}$, otherwise 0 [8]

---

represent $f_{b,i}(x)$, as in [12,28][7]. Consequently, $\mathsf{PMsg}_b(x)$ requires $\ell_{tgt} \times \ell_{msg}$ BPs in total.

We now describe how to construct a compact $\mathsf{BPMsg}_b[i,j]$ that computes the j-th output bit of $f_{b,i}(x)$. In the BP, we start at node $(1,0)$ and read the input $x \in \{0,1\}^n$ sequentially (see Fig. 2). If x is error-free, the computation proceeds along the nodes $(2,0)$, $(3,0), \ldots$, in the top row and terminates at node $(n+1, 0)$. If x contains an error, the path moves down by one row. For example, the path becomes $(1,0), (2,1), (3,1), \ldots$, if $x[1]$ is an erroneous bit. In general, if x includes k measurement errors, the computation reaches node $(n+1, k)$. At the final node, we check whether $k \le c$, which corresponds to $\mathsf{HD}(x, t_b) \le c$.

The detailed construction of $\mathsf{BPMsg}_b[i,j]$ is given below.

Construction 9 ($\mathsf{BPMsg}_b[i,j]$)**.** For $i \in [\ell_{msg}]$, $j \in [\ell_{tgt}]$, a branching program $\mathsf{BPMsg}_b[i,j]$ is constructed with the following rules.

1. For indices $h \in [n]$, $w \in \{0, \ldots, c+1\}$ and t_b defined in (4), the following rules are applied.
   - When $t_b[h] = 0$:
     - A 0-labeled edge transitions to a node at the w-th row and a 1-labeled edge transitions to a node at the $w+1$-th row. This can be illustrated pictorially as follows.

$$\begin{array}{c} \underset{(h,\,w)}{\bigcirc} \xrightarrow[\;1\;]{\;0\;} \underset{(h+1,\,w)}{\bigcirc} \\ \underset{(h+1,\,w+1)}{\bigcirc} \end{array} \qquad (5)$$

- We note, for example, that the nodes such as $(h+1, w)$ and $(h+1, w+1)$ are referred to as being in the $(h+1)$-th column of the BP[9].
  - When $t_b[h] = 1$:
    - A 1-labeled edge transitions to a node at the w-th row, and a 0-labeled edge transitions to a node at the $w+1$-th row. This is identical to (5), except that the edge labels 0 and 1 are swapped.
  - When $t_b[h] = *$:
    - A 0-labeled edge and a 1-labeled edge transition to a node at the same row. This can be illustrated pictorially as follows.

$$\underset{(h,\,w)}{\bigcirc} \xrightarrow{\;0,1\;} \underset{(h+1,\,w)}{\bigcirc}$$

2. When $h = n + 1$, the following rule is applied.
   - the first $(c+1)$ nodes transition to the node labeled "Output $y_{b,i}[j]$", and the $(c+2)$-th node transitions to the node labeled "Output $1 - y_{b,i}[j]$". This can be illustrated pictorially as follows.

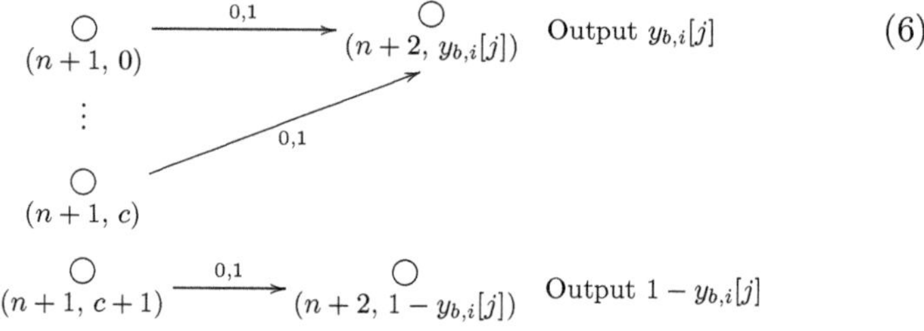

We note that the correctness of the obfuscated $\mathsf{PMsg}_b(x)$ follows from the correctness of each $\mathsf{BPMsg}_b[i, j]$.

Lemma 4 (Correctness of Construction 9). *Let* $\mathsf{BPMsg}_b[i, j]$ *be the BP generated by Construction 9. For every input* $x \in \{0, 1\}^n$, $\mathsf{BPMsg}_b[i, j](x)$ *outputs the j-th bit of* $f_{b,i}(x)$ *in (4).*

Proof. $\mathsf{BPMsg}_b[i, j]$ compares the input x with the secret string t_b in (4) bit by bit. Upon a mismatch, $\mathsf{BPMsg}_b[i, j]$ transitions from a node $(h,\, w)$ (i.e., located at column h, row w) to the node $(h+1,\, w+1)$. Since the initial state is the node $(h,\, w) = (1, 0)$, reaching a node $(n+1, k)$ where $k \leq c$, indicates that

---

the Hamming distance between the input x and t_b is at most c. Therefore, as in (6), by merging the first $(c + 1)$ nodes in the $(n + 1)$-th column into the node $(n + 2, y_{b,i}[j])$ and redirecting the transition from the node $(n + 1, c + 1)$ to the node $(n + 2, 1 - y_{b,i}[j])$, the output of $\mathsf{BPMsg}_b[i, j]$ becomes $y_{b,i}[j]$ or $\overline{y_{b,i}[j]} = 1 - y_{b,i}[j]$ accordingly. This exactly matches the j-th output bit of $f_{b,i}(x)$, establishing correctness. $\qquad\square$

We state the following lemma, which shows that there exists a compute-and-compare obfuscation for $\mathsf{BPMsg}_b[i, j]$ for $i \in [\ell_{msg}]$, $j \in [\ell_{tgt}]$.

Lemma 5 (Obfuscation for $\mathsf{BPMsg}_b[i, j]$). *Under the LWE assumption, there exists a compute-and-compare obfuscation for* $(\mathsf{BPMsg}_b[i, 1], \dots, \mathsf{BPMsg}_b[i, \ell_{tgt}])$.

Proof (Sketch). The BPs $\{\mathsf{BPMsg}_b[i, j]\}_{j \in [\ell_{tgt}]}$ generated by Construction 9 are the BPs of a CC program, and similarly to the proof sketch of Lemma 3, the fraction of inputs x for which $\{\mathsf{BPMsg}_b[i, j]\}_{j \in [\ell_{tgt}]}$ outputs $m_b[i] \in \{0, 1\}$ is negligible in n, as required. Furthermore, in our construction, $y_{b,i}$ and $z_{b,i}$ in $\mathsf{PMsg}_b[i](x)$ (Algorithm 3) can be, and indeed are chosen at random, so the target values $v_{b,i}$ (which is equal to either $y_{b,i}$ or $z_{b,i}$) are sampled from an unpredictable distribution $\mathcal{D}_\lambda$. Hence, by Theorems 2 and 3, there exists a compute-and-compare obfuscation for $\{\mathsf{BPMsg}_b[i, j]\}_{j \in [\ell_{tgt}]}$. $\qquad\square$

3.4 Main Construction

We now combine the building blocks to construct an efficiently realizable OTM protocol $(\mathsf{Token.Gen}, \mathsf{Token.Eval})$. The protocol is executed between a sender $\mathcal{S}$ and a receiver $\mathcal{R}$, and is secure against $\mathcal{O}(1)$-depth bounded quantum circuits.

Construction 10 (Efficiently Realizable OTM). $\mathsf{Token.Gen}(1^\lambda, m_0, m_1)$: On input $m_0, m_1 \in \{0, 1\}^{\ell_{msg}}$, the sender $\mathcal{S}$ performs:

1. Sample $s, \theta \xleftarrow{\$} \{0, 1\}^n$ with $\mathsf{HW}(\theta) = \frac{n}{2}$.
2. Prepare $|s^\theta\rangle = H^\theta |s\rangle$.
3. Choose appropriate $c \leftarrow \mathbb{N}$ where $c \ll n$. Sample $f_{b,i}, v_{b,i} \xleftarrow{\$} \mathcal{D}_\lambda$ where $y_{b,i}, z_{b,i} \xleftarrow{\$} \{0, 1\}^{\ell_{tgt}}$ for $b \in \{0, 1\}$ and $i \in [\ell_{msg}]$ (see (4) of $\mathsf{PMsg}_b[i](x)$ in Algorithm 3 corresponding to $\mathsf{BPMsg}_b[i, j]$).
4. Compile $\mathsf{OBPMsg}_b[i, j] \leftarrow \mathsf{CCObf}(\mathsf{BPMsg}_b[i, j])$ for $b \in \{0, 1\}$, $i \in [\ell_{msg}]$, $j \in [\ell_{tgt}]$ where $\mathsf{BPMsg}_b[i, j]$ is given in Construction 9.
5. Send $|\mathsf{tk}\rangle := (|s^\theta\rangle, \mathsf{OBPMsg}_0 = \{\mathsf{OBPMsg}_0[i, j]\}_{i \in [\ell_{msg}], j \in [\ell_{tgt}]}, \mathsf{OBPMsg}_1 = \{\mathsf{OBPMsg}_1[i, j]\}_{i \in [\ell_{msg}], j \in [\ell_{tgt}]})$ to the receiver.

$\mathsf{Token.Eval}(b, |\mathsf{tk}\rangle)$: On input $b \in \{0, 1\}$ and a token $|\mathsf{tk}\rangle$ from the sender $\mathcal{S}$, the receiver $\mathcal{R}$ performs:

1. Parse $|\mathsf{tk}\rangle$ into $(|s^\theta\rangle, \mathsf{OBPMsg}_0, \mathsf{OBPMsg}_1)$.
2. Measure $(H^b)^{\otimes n} |s^\theta\rangle$ in the computational basis to obtain s'.
3. Compute $m_b \leftarrow \mathsf{OBPMsg}_b(s')$.

Theorem 6 (Correctness and Security of Construction _10_). *Assuming the quantum hardness of LWE, Construction 10 is an OTM protocol secure against $\mathcal{O}(1)$-depth bounded adversaries, if the measurement error rate is up to $\epsilon \leq 0.14$.*

Proof (Sketch).
Correctness. Construction 10 corresponds to the error-tolerant version of Construction 8, differing only in the use of branching programs BPMsg_b. Lemma 4 shows that this difference in implementation does not affect correctness. Moreover, Lemmas 2 and 3 establish the correctness of the error-tolerant version of Construction 8. Hence, the correctness of Construction 10 follows.

Security. Lemma 5 shows that each branching program $\{\mathsf{BPMsg}_b[i,j]\}_{i \in [\ell_{msg}], j \in [\ell_{tgt}]}$ can be securely obfuscated using compute-and-compare obfuscation. It follows that BPMsg_b as a whole can also be securely obfuscated against constant-depth-bounded adversaries as in Construction 8, since it is a collection of such branching programs. Similarly, Lemmas 2 and 3 guarantee the security of the error-tolerant version of Construction 8; thus, the security of Construction 10 also follows.   □

4   Limitation of Construction 10: Bit-Flipping Attack

We have shown that Wiesner states combined with compute-and-compare obfuscation yield OTMs secure against $\mathcal{O}(1)$-depth bounded adversaries. A natural question then arises:

> *"Can one construct OTMs secure against non-constant-depth-bounded adversaries, still using Wiesner states and compute-and-compare obfuscation, by forcing the adversary to employ deeper quantum circuits via the sequential composition of obfuscated CC programs?"*

We conjecture that the answer is *no*. The relevant attack (or observation) is similar to that in [21], [3, §2.1], which present an adaptive attack on Wiesner states using either a verifying oracle or obfuscated programs that check their validity.

As an illustrative insecure construction, consider an OTM against logarithmic-depth bounded adversaries. In addition to PMsg_0 and PMsg_1, we introduce $\mathsf{PCom}_b^{(i)}$ for $i \in [k]$ and $b \in \{0,1\}$, where each $\mathsf{PCom}_b^{(i)}$ is an obfuscated CC program. $\mathsf{PCom}_b^{(1)}$ takes $x_b^{(0)} \in \{0,1\}^n$ as input and outputs $x_b^{(1)}$ if $x_b^{(0)}$ is obtained by measuring $(H^b)^{\otimes n}|s^\theta\rangle$ in the computational basis, and outputs $\perp$ otherwise. More generally, $\mathsf{PCom}_b^{(i)}$ takes $x_b^{(i-1)}$ as input and outputs $x_b^{(i)}$ only if $x_b^{(i-1)}$ is a valid output of $\mathsf{PCom}_b^{(i-1)}$. Finally, PMsg_b is modified to output m_b upon accepting a valid $x_b^{(k)}$.

At first glance, this construction appears secure against, e.g., $\mathcal{O}(\log n)$-depth bounded adversaries, since obtaining m_b seems to require evaluating a quantum circuit of depth $\Omega(k \log n)$, i.e., $m_b \leftarrow \mathsf{PMsg}_b \circ \mathsf{PCom}_b^{(k)} \circ \cdots \circ \mathsf{PCom}_b^{(1)}(x_b^{(0)})$, under the assumption that each $\mathsf{PCom}_b^{(i)}$ can be computed by logarithmic-depth circuits. However, a logarithmic-depth bounded adversary can clone the Wiesner state $|s^\theta\rangle$ as follows. Without loss of generality, we can assume that some CC program[10] in $\mathsf{PCom}_b^{(1)}$ outputs 1 on a valid encoding of $|s^\theta\rangle$. Then, to learn the i-th qubit of $|s^\theta\rangle$, the adversary applies a Pauli-X gate to the i-th qubit, feeds $|s^\theta\rangle$ into the CC program coherently, and measures the output without disturbing $|s^\theta\rangle$. If the result is 1, the i-th qubit is $|+\rangle$ or $|-\rangle$; otherwise, it is $|0\rangle$ or $|1\rangle$. Repeating this process for all n qubits of $|s^\theta\rangle$ reveals θ and $|s^\theta\rangle$, enabling cloning.

5   OTM Secure Against $\mathcal{O}(\lambda^\gamma)$-Depth Bounded Quantum Adversary with iO

We construct an OTM against $\mathcal{O}(\lambda^\gamma)$-depth bounded quantum adversaries in the QROM. This construction is partly inspired by [11, §7] and [19].

Instead of Wiesner states, we employ a *hidden subspace state* $|T_0(A_{\mathsf{Can}})\rangle = \sum_{v \in A_{\mathsf{Can}}} |T_0(v)\rangle$ where A_{Can} denotes the *canonical* $\lambda/2$ subspace of $\mathbb{F}_2^\lambda$ defined as $\mathsf{Span}(e_1, e_2, \ldots, e_{\lambda/2})$ and $e_i \in \mathbb{F}_2^\lambda$ is the standard basis vector with a 1 in the i-th coordinate and 0 in all others. The map T_0 is a full rank linear map and $T_0(A_{\mathsf{Can}})$ can be regarded as a rerandomized hidden subspace.

As also observed in [4], unlike Wiesner states, $|T_0(A_{\mathsf{Can}})\rangle$ is immune to bit-flipping attacks (Sect. 4). Intuitively, $T_0(A_{\mathsf{Can}})$ consists of $2^{n/2}$ random vectors over $\mathbb{F}_2^n$. Therefore, for any $v \in T_0(A_{\mathsf{Can}})$, the vector v' obtained by flipping a single qubit of v belongs to $T_0(A_{\mathsf{Can}})$ only with exponentially small probability.

5.1   Construction

We present an OTM construction, building on post-quantum indistinguishability obfuscation (iO) for all polynomial-size classical circuits.

$\mathsf{SampleFullRank}$ is a PPT algorithm that takes a security parameter λ and a seed as inputs, and outputs a full rank linear map $T : \mathbb{F}_2^\lambda \to \mathbb{F}_2^\lambda$. F is a post-quantum *pseudorandom function* (PRF) that takes a key and an input, and outputs the function value. Here $n = \lambda$ and m is polynomial in λ. $\mathsf{PReRand}$ is a rerandomization algorithm that takes as input an id, a vector, and a chosen bit. If $\mathsf{PReRand}$ can extract a vector in A_{Can} from the input vector, the input is considered valid, and $\mathsf{PReRand}$ outputs the next rerandomized vector along with the next rerandomized id. PMem is a memory algorithm that also takes an id, a vector, and a chosen bit as input. PMem outputs the OTM message if and only if these inputs have been rerandomized by $\mathsf{PReRand}$ exactly k times. The depth parameter k is set such that implementing $\mathsf{OPReRand}$ sequentially k times requires a quantum circuit deeper than the adversary's depth bound $\mathcal{O}(\lambda^\gamma)$.

---

[10] This corresponds to the obfuscated $\mathsf{PMsg}_b[i](x)$ (Algorithm 3) with $m_b[i] = 1$.

Construction 11 (Feasibility Construction of OTM).
Token.Gen(1^λ, m_0, m_1): The sender $\mathcal{S}$ performs the following steps:

1. Sample $K_0, K_1 \leftarrow F.\mathsf{Setup}(1^\lambda)$, $id_0 \leftarrow \{0,1\}^\lambda$.
2. Set $id_k = \underbrace{F(K_1, (F(K_1, (\dots, F(K_1, id_0)))))}_{F(K_1, \cdot) \text{ applied } k \text{ times}}$.
3. Set $T_0 = \mathsf{SampleFullRank}(1^\lambda; F(K_0, id_0))$, $T_k = \mathsf{SampleFullRank}(1^\lambda; F(K_0, id_k))$.
4. Set $|\psi\rangle = \sum_{v \in A_{\mathsf{Can}}} |T_0(v)\rangle$.
5. Compile $\mathsf{OPReRand} \leftarrow \mathsf{iO}(\mathsf{PReRand})$ where $\mathsf{PReRand}$ is in Algorithm 4.
6. Compile $\mathsf{OPMem} \leftarrow \mathsf{iO}(\mathsf{PMem})$ where PMem is in Algorithm 5.
7. Send the token $|\mathsf{tk}\rangle := (id_0, k, |\psi\rangle, \mathsf{OPReRand}, \mathsf{OPMem})$ to the receiver.

Token.Eval($|\mathsf{tk}\rangle$): The receiver $\mathcal{R}$, upon receiving the token $|\mathsf{tk}\rangle$ from $\mathcal{S}$, selects an input $b \in \{0,1\}$ and executes the following steps:

1. Parse the token $(id_0, k, |\psi\rangle, \mathsf{OPReRand}, \mathsf{OPMem})$.
2. Measure $(H^b)^{\otimes n}|\psi\rangle$ in the computational basis to obtain x_0[11]
3. For $i \in [k]$, run $(id_i, x_i) \leftarrow \mathsf{OPReRand}(id_{i-1}, x_{i-1}, b)$.
4. Output $m_b \leftarrow \mathsf{OPMem}(id_k, x_k, b) \oplus \mathcal{RO}(b \parallel x_k)$.

Algorithm 4 $\mathsf{PReRand}(id, v, b)$

    Hardcoded : K_0, K_1
1: If $v = 0^n$, output $\perp$ and halt
2: $id' = F(K_1, id)$ $\triangleright$ Generating subsequent id_{i+1} from id_i
3: $T_{\mathsf{cur}} = \mathsf{SampleFullRank}(1^\lambda; F(K_0, id))$, $T_{\mathsf{nxt}} = \mathsf{SampleFullRank}(1^\lambda; F(K_0, id'))$
4: $\omega = T_{\mathsf{cur}}^{-1}(v)$ if $b = 0$; otherwise, $\omega = T_{\mathsf{cur}}^{\top}(v)$
5: If $b = 0$ and $\omega \in A_{\mathsf{Can}}$, output $T_{\mathsf{nxt}}(\omega)$, id' and halt
6: If $b = 1$ and $\omega \in A_{\mathsf{Can}}^{\perp}$, output $(T_{\mathsf{nxt}}^{\top})^{-1}(\omega)$, id' and halt
7: Otherwise, output $\perp$

Algorithm 5 $\mathsf{PMem}(id, v, b)$

    Hardcoded : id_k, T_k, m_0, m_1
1: If $v = 0^n$ or $id \neq id_k$, output $\perp$ and halt
2: $\omega = T_k^{-1}(v)$ if $b = 0$; otherwise, $\omega = T_k^{\top}(v)$
3: If $b = 0$ and $\omega \in A_{\mathsf{Can}}$, output $m_0 \oplus \mathcal{RO}(0 \parallel v)$ and halt
4: If $b = 1$ and $\omega \in A_{\mathsf{Can}}^{\perp}$, output $m_1 \oplus \mathcal{RO}(1 \parallel v)$ and halt
5: Otherwise, output $\perp$

The correctness of Construction 11 is established as follows.

---

[11] We note that the probability that $x_0 = 0^n$ is negligible (namely $1/2^{n/2}$).

Theorem 7 (Correctness). *Construction 11 is correct.*

Proof. First, since A_{Can} is a subspace state and T_0 is a bijection, $T_0(A_{\mathsf{Can}}) = A'$ is also a subspace state where $A' = \{T_0(w) \mid w \in A_{\mathsf{Can}}\}$. In Step 2 of Token.Eval, when $b = 0$, the receiver $\mathcal{R}$ obtains a measurement outcome vector $x_0 := T_0(v_0) \in T_0(A_{\mathsf{Can}})$. Thus, in PReRand (Algorithm 4), we have $\omega = T_{\mathsf{cur}}^{-1}(x_0) = T_{\mathsf{cur}}^{-1}(T_0(v_0)) = v_0 \in A_{\mathsf{Can}}$ since $T_0 = T_{\mathsf{cur}}$. When $b = 1$, $\mathcal{R}$ obtains $x_0 \in (A')^{\perp}$[12]. Moreover, as shown in [11, §7.1], $x_0 \in (A')^{\perp}$ is equivalent to $T_0^{\top}(x_0) \in A_{\mathsf{Can}}^{\perp}$ (i.e., $x_0 \in (T_0^{\top})^{-1}(A_{\mathsf{Can}}^{\perp})$). Thus, in PReRand, we have $\omega = T_{\mathsf{cur}}^{\top}(x_0) = T_0^{\top}(x_0) \in A_{\mathsf{Can}}^{\perp}$, again since $T_0 = T_{\mathsf{cur}}$. Hence, if the input (id, v, b) to PReRand is valid, PReRand produces another valid input in the subsequent step. By iteratively executing the remaining steps of Token.Eval, $\mathcal{R}$ eventually recovers m_b correctly. $\qquad\square$

5.2 Measurement Error Tolerance

This OTM can be made immune to measurement errors. We assume up to $c \in \mathbb{N}$ errors, and apply suitable classical error-correcting codes capable of correcting up to c errors.

- ECC.Enc: Takes a bit string x and outputs a codeword C.
- ECC.Dec: Takes a codeword C and outputs the decoded bit string x if C contains at most c errors.

The sender $\mathcal{S}$ replaces $|\psi\rangle = \sum_{v \in A_{\mathsf{Can}}} |T_0(v)\rangle$ with $\sum_{v \in A_{\mathsf{Can}}} |\mathsf{ECC.Enc}(T_0(v))\rangle$ in Token.Gen, and adds ECC.Dec to $|\mathsf{tk}\rangle$. The receiver $\mathcal{R}$ then computes $x_0 \leftarrow \mathsf{ECC.Dec}(C)$ where C is the measurement outcome of $|\psi\rangle$. It is straightforward to verify that the construction remains correct. When c is sufficiently small relative to n, security follows from Theorem 8.

5.3 Security Analysis

We analyze the security of Construction 11. First, we consider the *rewinding attack* (see Sect. 1.2). When the adversary $\mathcal{A}$ is $\mathcal{O}(\lambda^{\gamma})$-depth bounded, simply setting $k = \Omega(\lambda^{\gamma+1})$ prevents $\mathcal{A}$ from outputting m_b without any intermediate measurements, i.e., it prevents $\mathcal{A}$ from relying solely on gentle measurements of $\mathcal{A}$'s registers and subsequently rewinding the computation on $|T_0(v)\rangle$.

Second, we consider the *bit-flipping attack* (see Sect. 4). Suppose that $\mathcal{A}$ attempts to guess $|T(A_{\mathsf{Can}})\rangle$ using PReRand (Algorithm 4). Here, PReRand can be used as $\mathsf{iO}(A_{\mathsf{Can}})$ since both return a fixed value when the input is a certain valid *subspace state*. However, if $\mathcal{A}$ can clone the *subspace state*, it contradicts Theorem 4. Hence, the probability that $\mathcal{A}$ succeeds in the bit-flipping attack is negligible.

We now provide a more formal security proof, reducing the security of our construction to the direct product hardness (Theorem 5). That is, we show that if $\mathcal{A}$ can obtain both m_0 and m_1 from the OTM, then we can construct a reduction that uses $\mathcal{A}$ to break this hardness assumption.

---

[12] This property originates from the subspace-state quantum money scheme [1].

Theorem 8 (Security). *Construction 11 yields an OTM protocol secure against $\mathcal{O}(\lambda^{\gamma})$-depth bounded adversaries in the QROM.*

We describe the security through a sequence of hybrids.

Hyb_0: This hybrid is identical to $\mathsf{OTM\text{-}Game}_{\mathcal{A}}(1^{\lambda})$ (Definition 3) where $(\mathsf{Token.Gen}, \mathsf{Token.Eval})$ are as defined in Construction 11.

Hyb_1: Sample a random subspace $A^* \subseteq \mathbb{F}_2^{\lambda}$ and prepare the state $|A^*\rangle$. Define the classical program as $\mathsf{P}_0 \leftarrow \mathsf{iO}(A^*)$ and $\mathsf{P}_1 \leftarrow \mathsf{iO}((A^*)^{\perp})$. Set $\mathsf{OPReRand} \leftarrow \mathsf{iO}(\mathsf{PReRand}')$, where $\mathsf{PReRand}'$ is Algorithm 6. Set $\mathsf{OPMem} \leftarrow \mathsf{iO}(\mathsf{PMem}')$, where PMem' is Algorithm 7. The modifications in this hybrids are marked in red.

Algorithm 6 $\mathsf{PReRand}'(id, v, b)$

    Hardcoded : $K_0, K_1, \mathsf{P}_0, \mathsf{P}_1$
1: If $v = 0^n$, output $\perp$ and halt
2: $id' = F(K_1, id)$
3: $T_{\mathsf{cur}} = \mathsf{SampleFullRank}(1^{\lambda}; F(K_0, id))$, $T_{\mathsf{nxt}} = \mathsf{SampleFullRank}(1^{\lambda}; F(K_0, id'))$
4: $\omega = T_{\mathsf{cur}}^{-1}(v)$ if $b = 0$; otherwise, $\omega = T_{\mathsf{cur}}^{\top}(v)$
5: If $b = 0$ and $\mathsf{P}_0(\omega) = 1$, output $T_{\mathsf{nxt}}(\omega)$, id' and halt
6: If $b = 1$ and $\mathsf{P}_1(\omega) = 1$, output $(T_{\mathsf{nxt}}^{\top})^{-1}(\omega)$, id' and halt
7: Otherwise, output $\perp$ and halt

Algorithm 7 $\mathsf{PMem}'(id, v, b)$

    Hardcoded : $id_k, T_k, m_0, m_1, \mathsf{P}_0, \mathsf{P}_1$
1: If $v = 0^n$ or $id \neq id_k$, output $\perp$ and halt
2: $\omega = T_k^{-1}(v)$ if $b = 0$; otherwise, $\omega = T_k^{\top}(v)$
3: If $b = 0$ and $\mathsf{P}_0(\omega) = 1$, output $m_0 \oplus \mathcal{RO}(0 \parallel v)$ and halt
4: If $b = 1$ and $\mathsf{P}_1(\omega) = 1$, output $m_1 \oplus \mathcal{RO}(1 \parallel v)$ and halt
5: Otherwise, output $\perp$

Finally, we modify $\mathsf{Token.Gen}$ as follows.

1. Sample $K_0, K_1 \leftarrow F.\mathsf{Setup}(1^{\lambda})$, $id_0 \leftarrow \{0,1\}^{\lambda}$.
2. Set $id_k = F(K_1, (F(K_1, (\ldots, F(K_1, id_0)))))$.
3. Set $T_0 = \mathsf{SampleFullRank}(1^{\lambda}; F(K_0, id_0))$, $T_k = \mathsf{SampleFullRank}(1^{\lambda}; F(K_0, id_k))$.
4. Sample a random subspace $A^* \subseteq \mathbb{F}_2^{\lambda}$, and set $|\psi\rangle = |T_0(A^*)\rangle$.
5. Compile $\mathsf{OPReRand}' \leftarrow \mathsf{iO}(\mathsf{PReRand}')$ where $\mathsf{PReRand}'$ is in Algorithm 4.
6. Compile $\mathsf{OPMem}' \leftarrow \mathsf{iO}(\mathsf{PMem}')$ where PMem' is in Algorithm 5.
7. Send the token $|\mathsf{tk}\rangle := (id_0, k, |\psi\rangle, \mathsf{OPReRand}', \mathsf{OPMem}')$ to the receiver.

Lemma 6. $\mathsf{Hyb}_0 \approx \mathsf{Hyb}_1$.

Proof (Sketch). In Hyb_1, the adversary $\mathcal{A}$ receives $|T_0(A^*)\rangle$ instead of $|T_0(A_{\mathsf{Can}})\rangle$. Since A^* can be expressed as $A^* = T_0'(A_{\mathsf{Can}}) = \{T_0'(w) \mid w \in A_{\mathsf{Can}}\}$ for a random full rank linear map T_0', we have $|T_0(A^*)\rangle = |T_0 T_0'(A_{\mathsf{Can}})\rangle$. By considering that $T_0 T_0'$ in Hyb_1 corresponds to T_0 in Hyb_0, we can argue that this substitution does not affect $\mathcal{A}$'s advantage as follows.

First, both $|T_0(A^*)\rangle$ and $|T_0(A_{\mathsf{Can}})\rangle$ appear as random subspace states to $\mathcal{A}$ due to the randomness of T_0.

Second, the subspace membership check $\mathsf{P}_0(\omega)$ in $\mathsf{OPReRand}'$ and OPMem' in Hyb_1 implements $(T_0')^{-1}(\omega) \overset{?}{\in} A_{\mathsf{Can}}$, which is functionally equivalent to the subspace membership check $\omega \overset{?}{\in} A_{\mathsf{Can}}$ in $\mathsf{OPReRand}$ and OPMem in Hyb_0. Specifically, in Hyb_1, given $v \in T_k T_0'(A_{\mathsf{Can}})$, checking $(T_0')^{-1}(\omega) \overset{?}{\in} A_{\mathsf{Can}}$ corresponds to $(T_k T_0')^{-1}(v) \overset{?}{\in} A_{\mathsf{Can}}$, whereas, in Hyb_0, given $v \in T_k(A_{\mathsf{Can}})$, checking $\omega \overset{?}{\in} A_{\mathsf{Can}}$ corresponds to $T_k^{-1}(v) \overset{?}{\in} A_{\mathsf{Can}}$. Hence, the two checks are functionally equivalent. A similar reasoning applies to $\mathsf{P}_1(\omega)$. Therefore, these hybrids are indistinguishable to $\mathcal{A}$ by the security of iO, and we conclude that $\mathsf{Hyb}_0 \approx \mathsf{Hyb}_1$. $\square$

Lemma 7. $\Pr[\mathsf{Hyb}_1 = 1] \leq \mathsf{negl}(\lambda)$.

Proof (Sketch). We construct a reduction $\mathcal{R}$ which, given a single copy of $|A^*\rangle$ along with $\mathsf{iO}(A^*)$, $\mathsf{iO}((A^*)^{\perp})$ in the game of Theorem 5, interacts with $\mathcal{A}$ in Hyb_1, while simulating $\mathcal{RO}$. We assume that the depth parameter k is set such that implementing $\mathsf{OPReRand}'$ requires a quantum circuit exceeding the adversary's depth bound $\mathcal{O}(\lambda^{\gamma})$, thereby preventing $\mathcal{A}$ from inputting $|T_k(A^*)\rangle$ into OPMem'. Hence, if $\mathcal{A}$ outputs both m_0 and m_1 with non-negligible probability, then by standard QROM techniques such as the compressed oracle technique [31] and the one-way to hiding lemma [26], $\mathcal{R}$ can extract $(0 \parallel v)$ and $(1 \parallel w)$ from the $\mathcal{RO}$ transcript with non-negligible probability, recovering both $T_k^{-1}(v) \in A^*$ and $T_k^{\top}(w) \in (A^*)^{\perp}$ using the hardcoded T_k. This implies that $\mathcal{R}$ breaks the direct product hardness (Theorem 5), leading to a contradiction. $\square$

Acknowledgments. This work was supported in part by JSPS KAKENHI Grant Numbers 23K18459 and 25K03116.

References

1. Aaronson, S., Christiano, P.: Quantum money from hidden subspaces. In: STOC, pp. 41–60. ACM (2012)
2. Banerjee, A., Peikert, C., Rosen, A.: Pseudorandom functions and lattices. In: Pointcheval, D., Johansson, T. (eds.) EUROCRYPT 2012. LNCS, vol. 7237, pp. 719–737. Springer, Heidelberg (2012). https://doi.org/10.1007/978-3-642-29011-4_42
3. Bartusek, J., Goyal, V., Khurana, D., Malavolta, G., Raizes, J., Roberts, B.: Software with certified deletion. Cryptology ePrint Archive, Paper 2023/265 (2023)

4. Bartusek, J., Goyal, V., Khurana, D., Malavolta, G., Raizes, J., Roberts, B.: Software with certified deletion. In: Joye, M., Leander, G. (eds.) EUROCRYPT 2024. LNCS, vol. 14654, pp. 85–111. Springer, Cham (2024). https://doi.org/10.1007/978-3-031-58737-5_4

5. Behera, A., Sattath, O., Shinar, U.: Noise-tolerant quantum tokens for MAC (2025). https://arxiv.org/abs/2105.05016

6. Ben-David, S., Sattath, O.: Quantum tokens for digital signatures. Quantum **7**, 901 (2023)

7. Bitansky, N., Goldwasser, S., Jain, A., Paneth, O., Vaikuntanathan, V., Waters, B.: Time-lock puzzles from randomized encodings. In: ITCS, pp. 345–356. ACM (2016)

8. Broadbent, A., Gharibian, S., Zhou, H.S.: Towards quantum one-time memories from stateless hardware. Quantum **5**, 429 (2021)

9. Broadbent, A., Gutoski, G., Stebila, D.: Quantum one-time programs. In: Canetti, R., Garay, J.A. (eds.) CRYPTO 2013. LNCS, vol. 8043, pp. 344–360. Springer, Heidelberg (2013). https://doi.org/10.1007/978-3-642-40084-1_20

10. Broadbent, A., Lord, S.: Uncloneable quantum encryption via oracles. In: TQC. Leibniz International Proceedings in Informatics (LIPIcs), vol. 158, pp. 4:1–4:22. Schloss Dagstuhl – Leibniz-Zentrum für Informatik (2020)

11. Cakan, A., Goyal, V., Yamakawa, T.: Anonymous public-key quantum money and quantum voting (2024). https://arxiv.org/abs/2411.04482

12. Chen, Y., Vaikuntanathan, V., Wee, H.: GGH15 beyond permutation branching programs: proofs, attacks, and candidates. In: Shacham, H., Boldyreva, A. (eds.) CRYPTO 2018. LNCS, vol. 10992, pp. 577–607. Springer, Cham (2018). https://doi.org/10.1007/978-3-319-96881-0_20

13. Coladangelo, A., Liu, J., Liu, Q., Zhandry, M.: Hidden cosets and applications to unclonable cryptography. In: Malkin, T., Peikert, C. (eds.) CRYPTO 2021. LNCS, vol. 12825, pp. 556–584. Springer, Cham (2021). https://doi.org/10.1007/978-3-030-84242-0_20

14. Culf, E., Vidick, T.: A monogamy-of-entanglement game for subspace coset states. Quantum **6**, 791 (2022)

15. Eldridge, H., Goel, A., Green, M., Jain, A., Zinkus, M.: One-time programs from commodity hardware. In: Kiltz, E., Vaikuntanathan, V. (eds.) TCC 2022. LNCS, vol. 13749, pp. 121–150. Springer, Cham (2022). https://doi.org/10.1007/978-3-031-22368-6_5

16. Goldwasser, S., Kalai, Y.T., Rothblum, G.N.: One-time programs. In: Wagner, D. (ed.) CRYPTO 2008. LNCS, vol. 5157, pp. 39–56. Springer, Heidelberg (2008). https://doi.org/10.1007/978-3-540-85174-5_3

17. Goyal, R., Koppula, V., Waters, B.: Lockable obfuscation. In: FOCS, pp. 612–621. IEEE (2017)

18. Gunn, S., Movassagh, R.: Quantum one-time protection of any randomized algorithm. In: Tauman Kalai, Y., Kamara, S.F. (eds.) CRYPTO 2025. LNCS, vol. 16001, pp. 334–349. Springer, Cham (2025). https://doi.org/10.1007/978-3-032-01878-6_11

19. Gupte, A., Liu, J., Raizes, J., Roberts, B., Vaikuntanathan, V.: Quantum one-time programs, revisited. In: STOC, pp. 213–221. ACM (2025)

20. Liu, Q.: Depth-bounded quantum cryptography with applications to one-time memory and more. In: ITCS. Leibniz International Proceedings in Informatics, vol. 251, pp. 82:1–82:18. Schloss Dagstuhl – Leibniz-Zentrum für Informatik (2023)

21. Lutomirski, A.: An online attack against Wiesner's quantum money (2010). https://arxiv.org/abs/1010.0256

22. NIST: Submission requirements and evaluation criteria for the post-quantum cryptography standardization process (2016)
23. Regev, O.: On lattices, learning with errors, random linear codes, and cryptography. In: STOC, pp. 84–93. ACM (2005)
24. Sattath, O., Wyborski, S.: Uncloneable decryptors from quantum copy-protection (2022). https://arxiv.org/abs/2203.05866
25. Stambler, L.: Quantum one-time memories from stateless hardware, random access codes, and simple nonconvex optimization (2025). https://arxiv.org/abs/2501.04168
26. Unruh, D.: Revocable quantum timed-release encryption. J. ACM (JACM) **62**(6), 1–76 (2015)
27. Wang, P., et al.: Single ion qubit with estimated coherence time exceeding one hour. Nat. Commun. **12**(1), 233 (2021)
28. Wichs, D., Zirdelis, G.: Obfuscating compute-and-compare programs under LWE. In: FOCS, pp. 600–611. IEEE (2017)
29. Wiesner, S.: Conjugate coding. ACM SIGACT News **15**(1), 78–88 (1983)
30. Wootters, W.K., Zurek, W.H.: A single quantum cannot be cloned. Nature **299**(5886), 802–803 (1982)
31. Zhandry, M.: How to record quantum queries, and applications to quantum indifferentiability. In: Boldyreva, A., Micciancio, D. (eds.) CRYPTO 2019. LNCS, vol. 11693, pp. 239–268. Springer, Cham (2019). https://doi.org/10.1007/978-3-030-26951-7_9
32. Zhandry, M.: Quantum lightning never strikes the same state twice. or: quantum money from cryptographic assumptions. J. Cryptol. **34**(1), 6 (2021)
33. Zhao, L., et al.: One-time programs made practical. In: Goldberg, I., Moore, T. (eds.) FC 2019. LNCS, vol. 11598, pp. 646–666. Springer, Cham (2019). https://doi.org/10.1007/978-3-030-32101-7_37

Practically Implementable Minimal Universal Gate Sets for Multi-qudit Systems with Cryptographic Validation

Anisha Dutta[1]([✉]) [iD], Sayantan Chakraborty[2] [iD], Chandan Goswami[3] [iD], and Avishek Adhikari[3] [iD]

[1] Data and Analytics, Tata Steel Limited, Jamshedpur 831001, India
`srianishadutta@gmail.com`
[2] Responsible AI CoE, Data and AI, Accenture ATCI, Kolkata 700156, India
[3] Department of Mathematics, Presidency University, College Street, Kolkata 700073, India

Abstract. The rapid growth of quantum technologies highlights the need for scalable models of computation that go beyond qubits and exploit the richer structure of Qudits. This paper introduces a novel and efficient approach for defining universal gate sets specifically designed for higher-dimensional Qudit systems ($N \geq 2$), addressing the limitations of traditional qubit-based approaches. We present a systematic methodology for constructing fundamental Qudit gates through the inherent structure of Qudit operators, providing a robust theoretical foundation for universal quantum computation with Qudits. Our rigorously proven universal and minimal gate set enables more efficient quantum circuit design. We demonstrate the construction of multidimensional extensions of controlled operations from these fundamental elements. To facilitate practical application, we provide a Python-based algorithm for decomposing arbitrary multi-Qudit operations, accompanied by detailed time- and space-complexity analyses. The framework is further validated through end-to-end implementations of Grover's algorithm and QKD, comparing traditional gate constructions with circuits entirely synthesized from the proposed universal gate set (Code Repository: Minimal-Universal-Multi-Qudit-Gate-Sets). These validations demonstrate not only the functional equivalence of the decomposed circuits, but also their direct relevance to the advancement of cryptographic protocols, paving the way for more efficient and secure Qudit-based quantum cryptography.

Keywords: Multi-fold Qudit · Minimal Universal Quantum gates · Reck's Decomposition · Quantum Cryptology · Grover's algorithm · QKD · Python-Cirq

1 Introduction

Quantum computing has so far been dominated by qubits, the familiar two-level systems that act as basic units of quantum information. Multi-state quantum systems (Qudits) offer inherent advantages over traditional qubits. These higher-dimensional units can carry more information per particle and open the door to faster computations, more compact circuits, and new directions for quantum communication and security. For quantum cryptography, this shift is especially important. Protocols such as quantum key distribution (QKD), quantum secret sharing, and secure multiparty computation could all benefit from the richer structure of Qudits. At the same time, they also raise new challenges, since building and controlling the right quantum gates becomes more complex as the system dimension grows. Qudits allow quantum circuits to be built more efficiently, making cryptographic algorithms faster, more compact, and potentially more secure. While universal gate sets for qubits are well established, extending this idea to Qudits opens up exciting opportunities for advanced quantum cryptography. In this work, we present a systematic and practical way to define fundamental Qudit gates. Using an algebraic framework, we prove that our proposed set of gates is both universal and minimal. We also validate the framework by showing that it correctly recovers the standard qubit gates as a special case. This foundation makes it possible to design more powerful and efficient quantum protocols, with clear benefits for quantum key distribution, secure multiparty computation, Grover's search, and quantum secret sharing. Table 1 summarizes the notations used throughout the paper.

1.1 Motivation

The pursuit of universal quantum computation, historically focused on qubit-based systems, has yielded significant insights into the structure and capabilities of quantum algorithms. However, the potential of multi-dimensional Qudit systems, with their enhanced information storage and processing capacity, remains largely untapped, especially within the cryptographic landscape. Our primary motivation stems from the recognized advantages of Qudits in various computational and cryptographic contexts. For instance, in quantum key distribution (QKD) protocols, Qudits can be realistically generated and offer the advantage of storing more information per quantum bit [24]. This capability translates to potentially higher secret key generation rates and improved resilience against eavesdropping attacks compared to traditional qubit-based QKD protocols. Furthermore, in the realm of cryptanalysis, multi-valued Grover's search algorithms [23], which can be realized with Qudits, offer a demonstrable time advantage over their qubit-based counterparts. This opens exciting possibilities for accelerating certain cryptanalytic tasks, thereby driving the need for more sophisticated quantum-resistant cryptographic solutions. By efficiently implementing Qudit-based quantum circuits, universal Qudit gates optimize quantum-cryptographic protocols and bolster their security, while simultaneously augmenting the effectiveness of quantum attacks on classical cryptosystems through Qudits' enhanced capabilities.

1.2 Prior-Art Search

While significant strides have been made in establishing universal gate sets for qubits, such as the $\{H, T, S, CNOT\}$ set [1], the extension of these principles to higher dimensional Qudit systems has been a more complex undertaking. Early contributions by Adriano Barenco et al. [2] laid the groundwork for understanding universal quantum gate sets, exploring their construction and utility. Subsequent works, such as those by Deutsch et al. [4,5] and A. Yu. Kitaev [6], deepened the theoretical understanding of

Table 1. Overview of the Notations

Notation	Description
N	Number of states or "fold" of a single Qudit system (dimension of one Qudit)
n	Number of Qudits in the system (total system dimension is N^n)
$|y\rangle$	General Qudit state vector
$|i\rangle$	Basis state vector for a Qudit
$\mathcal{H}_N$	N-dimensional complex Hilbert space
$U(N)$	Group of $N \times N$ unitary matrices
$SU(N)$	Group of $N \times N$ special unitary matrices (determinant 1)
I_N	Identity matrix of order N
E_{ij}	Matrix with 1 in the (i, j) position and zeros elsewhere
H	Hadamard gate
T	T-gate (a single-qubit phase gate with $\pi/4$ rotation)
S	S-gate (a single-qubit phase gate with $\pi/2$ rotation)
$CNOT$	Controlled-NOT gate (fundamental 2-qubit gate)
Φ	General phase shift operator (diagonal unitary matrix)
$\text{Phase}(j, \phi_j)$	Specific form of a phase gate
PHASE	Collection of general phase gates $\{\text{Phase}(j, \phi_j)\}$
$PHASE_1$	Subset of PHASE, specific phase gate $\{\text{Phase}(1, \theta)\}$
$T_{i,j}$	Positional subspace embedding operator for a 2×2 matrix
$T_{i,j}(U)$	Instance of $T_{i,j}$ embedding a 2×2 matrix U
$T_{elements}$	The set of $T_{i,j}(U)$ matrices
$R_{i,j}$	Rotational matrices in Reck's decomposition
$\mathscr{S}$	The proposed optimal universal gate set $(PHASE_1 \cup T_{elements})$
$u(m)$	Lie algebra of the unitary group $U(m)$
$su(m)$	Lie algebra of the special unitary group $SU(m)$
$\text{span}(iI_m)$	The span of imaginary identity matrices

universality for multi-qubit systems. More recent studies, including those by Laurin E. Fischer et al. [7] and Boykin et al. [9], have extended this concept to fault-tolerant systems and higher-dimensional Qudits, highlighting their potential to optimize quantum circuits and enhance computational efficiency. In parallel, efforts have been made to explore alternative perspectives on universality. For instance, works by Cafaro et al. [10] and Timothy F. Havel [11] has examined quantum gate sets through the lens of geometric algebra, while Vlasov [13] applied Clifford algebras to construct universal gates for n-qubit systems. In 2002, Shi's work [14] demonstrated the near-universality of the Toffoli and $CNOT$ gates, further emphasizing the role of minimal gate additions in achieving universality. The study of universality of single Qudit gates [19] and semi-universality of certain subset of Qudit gates [20–22] work as a stepping stone for the study of universality of Qudit gates. Building on this extensive body of work, and being motivated by the study of universal qutrit gates by Carlos Efrain Quintero Narvaez [15], unitary matrix decomposition by Chi-Kwong Li et al. [16], and qubit gates decomposition by Juha J. Vartiainen et al. [17], our work introduces an algebraic framework for proving universality that focuses on identifying the finite, elementary operations that are physically realizable, leading to a systematic decomposition and practical implementation of arbitrary Qudit operations. Our approach is inspired by the experimental realization of arbitrary unitary operators demonstrated by Michael Reck et al. [18], which provides a bridge between theoretical constructs and practical implementations.

From a cryptographic perspective, Qudits used in quantum key-distribution protocols [24,39,40] can be realistically generated and have the advantage of storing more information than qubits, and offer higher key rates over qubit based QKD. This capability translates to reduced generation time and faster computation. In view of quantum attacks on classical cryptosystems, multi-valued Grover search [23] provides a time advantage over Grover's algorithm implemented on qubits. Notably, Qudits offer stronger security guarantee in quantum secure multiparty computation [25,26]. Furthermore, Qudits enable increased information content per Qudit in quantum secret sharing [27,28], demonstrating their broad applicability and potential to advance quantum technologies. Recent progress in quantum machine learning highlights the potential of Qudits over qubits, demonstrating their capacity for more intricate feature mappings [30], enhanced model expressiveness, and improved efficiency [31], ultimately enabling more robust and powerful algorithms. Prior analytical approaches have often focused on specific dimensionalities or lacked a systematic, constructive methodology for identifying minimal universal gate sets. Furthermore, the practical implementation and algorithmic decomposition of such Qudit gates have presented distinct challenges. This research addresses these gaps by providing a unified and theoretically robust framework for Qudit gate set universality.

1.3 Contribution

Unlike earlier qubit-centric studies, our formulation directly links mathematical universality to experimentally achievable operations and explicitly quantifies decomposition complexity, bridging theoretical completeness with cryptographic implementability. The key contributions are enlisted below.

- **Proof of Minimality of Universal Gate Set:** We establish a rigorous theoretical framework for universal gate sets in higher-dimensional Qudit systems (Sect. 3.1), introducing a robust universal set $\mathscr{S} = PHASE_1 \cup T_{elements}$. Our key contribution is the identification and explicit proof of a minimal set of indispensable gates, comprising elementary phase shifts and essential $SU(2)$ subspace embeddings. This set is proven to approximate any unitary operation on an N-dimensional Qudit system with arbitrary precision, ensuring universality. The novelty of this work lies in its Lie-algebraic proof of the minimality of the gate set proven in Theorem 3. This mathematical rigor is complemented by a detailed complexity analysis of the algorithm used to construct any arbitrary Qudit gate from this universal set.

- **Practical Validation and Applications:** The quantitative functional equivalence across arbitrary Qudit dimensions through fidelity metrics has been discussed in Sect. 3.4, and Sect. 3.5 establishes superior decomposition efficiency over a widely used alternative universal gate sets [16]. To demonstrate applicability, we implemented the decomposition end-to-end and validated it on cryptographically relevant examples: multi-ary Grover search and a QKD workflow. For each protocol we constructed two parallel circuits: (i) the conventional circuit built from high-level Qudit gates, and (ii) a circuit in which every high-level gate is replaced by its decomposition into $T_{elements}$ and $PHASE_1$ elements following the algorithm in Sect. 3.2. The software implementation uses Python/Cirq primitives, so each decomposed element is directly realizable on platforms that expose only low-level rotations and phase shifts. Simulation results show functional equivalence (unstructured search amplitude amplification and matching QKD key bits) while revealing the expected trade-offs: decomposed circuits incur higher gate counts and depth but gain hardware-agnostic portability, explicit resource accounting, and finer-grained error budgeting. Quantitative validation, comparison plots and implementation notes are provided in the new Sect. 3.6. Collectively, these contributions deliver a minimal, implementable universal basis for Qudits together with the algorithms and empirical evidence necessary to adopt that basis in cryptographic protocols and comparative cryptanalysis.

1.4 Cryptographic Impact of the Minimal Qudit Gate Set

The identification of a minimal universal gate set for Qudits has direct implications for the design and efficiency of quantum cryptographic protocols. Qudit-based cryptography leverages higher-dimensional Hilbert spaces to enhance both information capacity and robustness against noise, but a long-standing obstacle has been the practical implementability of arbitrary Qudit unitaries within constrained hardware models. Our result addresses this bottleneck by providing a minimal, yet universal, set of gates that can be synthesized with provable efficiency.

- **Quantum Key Distribution Protocols:** Multi-level QKD schemes (e.g., using d-ary alphabets) require arbitrary basis changes and state encoding. The proposed gate set enables such transformations with reduced circuit depth, thereby lowering error rates and improving achievable key rates.

- **Quantum Cryptanalysis:** Qudit-enhanced Grover search has been shown to offer faster runtimes compared to qubit-based implementations. By providing an explicit constructive method to decompose arbitrary Qudit unitaries, our gate set strengthens the practicality of these algorithms, thereby accelerating quantum attacks on classical cryptosystems. This dual-use perspective is crucial: while our gate set empowers defensive primitives such as QKD, it also highlights the urgency of developing post-quantum cryptography resilient against Qudit-based adversaries.

- **Quantum Secret Sharing:** Secret sharing protocols based on d-ary linear codes rely on efficient encoding of quantum states into multi-Qudit subspaces. Recent constructions using symmetric entanglements and novel partitioning techniques [41,42] demonstrate that such encoding can be efficiently realized. Our work implies that these transformations can be compiled into circuits with polynomial overhead in the Qudit dimension, making high-arity secret sharing more practical.

- **Secure Multiparty Computation:** Qudit-based MPC protocols benefit from the ability to implement controlled rotations and higher-arity entangling operations. By proving minimality, we ensure that such gates can be realized from the smallest possible universal basis, reducing both implementation complexity and fault-tolerance overhead.

- **Algorithmic Advantages:** Beyond direct protocol design, the constructive universality result facilitates the integration of higher-arity search and error-resilient primitives in cryptographic applications, where efficiency and scalability are crucial.

The proposed gate set not only advances the theoretical understanding of universality in Qudit systems but also establishes a practically implementable foundation for next-generation quantum cryptography. Its efficiency and minimality strengthen the applicability of Qudit-based designs in QKD, secret sharing, and multiparty computation, while clarifying their implications for quantum cryptanalysis.

2   Definition and Prerequisites

To facilitate understanding of the proposed methods, this section provides the essential definitions and mathematical prerequisites relevant to the study. These concepts serve as the basis for the development and analysis presented in the following sections.

2.1   Subspace Embedding

A subspace embedding is a technique in linear algebra that maps a lower-dimensional subspace, spanned by a set of vectors or matrices, into a higher-dimensional space. This mapping preserves the essential structure and properties of the original subspace while satisfying specific conditions such as distance preservation or bounded distortion. Consider $T_{i_1,i_2,i_3}(U)$ that embeds a 3×3 matrix U into a 6×6 identity matrix:

$$U = \begin{pmatrix} a_{11} & a_{12} & a_{13} \\ a_{21} & a_{22} & a_{23} \\ a_{31} & a_{32} & a_{33} \end{pmatrix} \implies T_{0,1,2}(U) = \begin{pmatrix} a_{11} & a_{12} & a_{13} & 0 & 0 & 0 \\ a_{21} & a_{22} & a_{23} & 0 & 0 & 0 \\ a_{31} & a_{32} & a_{33} & 0 & 0 & 0 \\ 0 & 0 & 0 & 1 & 0 & 0 \\ 0 & 0 & 0 & 0 & 1 & 0 \\ 0 & 0 & 0 & 0 & 0 & 1 \end{pmatrix}.$$

2.2 Representation of Qudit

In quantum mechanics, qubits are two-dimensional quantum systems represented by the basis states $\{|0\rangle, |1\rangle\}$. While qubits form the foundation of many quantum computing architectures, the concept can be generalized to higher-dimensional systems known as Qudits [34]. Mathematically, a Qudit is a unit vector in an N- dimensional complex Hilbert space, $\mathscr{H}_N$. This space is spanned by an orthonormal basis of N states, labeled as $|0\rangle, |1\rangle, |2\rangle, \ldots, |N-1\rangle$. A general Qudit state $|\psi\rangle$, is a linear combination of these basis states $|\psi\rangle = \sum_{i=0}^{N-1} c_i |i\rangle$, where $c_i \in \mathbb{C}$ are complex coefficients satisfying the normalization condition $\sum_{i=0}^{N-1} |c_i|^2 = 1$. In its simplest representation, an N-dimensional Qudit can be visualized as a column vector $|\psi\rangle = (c_0, c_1, \ldots, c_{N-1})^T$, where each c_i corresponds to the amplitude of the i-th basis state, and the superscription T represents the transpose of a vector. Using the orthonormal basis vectors $|i\rangle$, this can be expressed as $|\psi\rangle = c_0 (1,0,\ldots,0)^T + c_1 (0,1,\ldots,0)^T + \cdots + c_{N-1} (0,0,\ldots,1)^T$. Here, each basis state $|i\rangle$ is represented as a column vector with 1 in the i-th position and 0 elsewhere: $|i\rangle = (0,0,\ldots,0,1,0,\ldots,0)^T$. This method emphasizes that Qudits provide a versatile way to represent quantum systems, offering more possibilities than qubits. Thus, the set $\{|0\rangle, |1\rangle, |2\rangle, \ldots, |N-1\rangle\}$ serves as the orthonormal basis for all N-dimensional Qudits.

2.3 Representation of Qudit Gate

Qudits, generalizations of qubits representing quantum states in a N-dimensional Hilbert space, are manipulated using *Qudit gates*, which are unitary operators acting on these N-dimensional spaces [29, 32, 37]. A single-Qudit gate is represented by an $N \times N$ unitary matrix, $U \in U(N)$, where $UU^\dagger = I_{N \times N}$ ($I_{N \times N}$ being the $N \times N$ identity matrix). For example, a single-Qudit gate acting on a three-level qutrit ($N = 3$) might be represented by the following unitary matrix:

$$U = \begin{pmatrix} e^{i\pi/4} & 0 & 0 \\ 0 & \frac{1}{\sqrt{2}} & \frac{i}{\sqrt{2}} \\ 0 & -\frac{i}{\sqrt{2}} & \frac{1}{\sqrt{2}} \end{pmatrix}.$$

This unitary matrix ensures $UU^\dagger = I_{3 \times 3}$, preserving the probabilistic interpretation of quantum mechanics. Multi-Qudit gates extend this concept to systems involving multiple Qudits, where a gate acting on an N^n-dimensional system is represented by a $N^n \times N^n$ unitary matrix. For example, a two-qutrit gate ($N = 3$, $n = 2$) is described by a 9×9 matrix. $T_{elements}$ **Subset of** $SU(N^n)$ Consider a 2×2 matrix $U = \begin{pmatrix} u_{11} & u_{12} \\ u_{21} & u_{22} \end{pmatrix} \in$

$SU(2)$. We can embed this matrix into a $N^n \times N^n$ matrix $T_{i,j}$ in the positions i and j as

$$
T_{i,j}(U) = \begin{pmatrix}
1 & 0 & \dots & 0 & \dots & 0 & \dots & 0 \\
0 & 1 & \dots & 0 & \dots & 0 & \dots & 0 \\
\vdots & \vdots & \ddots & \vdots & \ddots & \vdots & \ddots & \vdots \\
0 & 0 & \dots & u_{11} & \dots & u_{12} & \dots & 0 \\
\vdots & \vdots & \ddots & \vdots & \ddots & \vdots & \ddots & \vdots \\
0 & 0 & \dots & u_{21} & \dots & u_{22} & \dots & 0 \\
\vdots & \vdots & \ddots & \vdots & \ddots & \vdots & \ddots & \vdots \\
0 & 0 & \dots & 0 & \dots & 0 & \dots & 1
\end{pmatrix}
$$

We define $T_{elements} = \{T_{i,j}(U) | 0 \le i < j < N^n, U \in SU(2)\} \subset SU(N^n)$ as the set of $T_{i,j}$.

PHASE Qudit Gates. Diagonal unitary matrices, where all diagonal elements have a modulus of 1, can be expressed as:

$$
\Phi = \begin{pmatrix}
e^{i\phi_0} & 0 & \dots & 0 \\
0 & e^{i\phi_1} & \dots & 0 \\
\vdots & \vdots & \ddots & \vdots \\
0 & 0 & \dots & e^{i\phi_{N^n-1}}
\end{pmatrix}.
$$

These matrices can be systematically constructed using a specific form:

$$
Phase(j,\ \phi_j) = \begin{pmatrix}
1 & 0 & \dots & 0 & 0 & 0 & \dots & 0 \\
0 & 1 & \dots & 0 & 0 & 0 & \dots & 0 \\
\vdots & \vdots & \ddots & \vdots & \vdots & \vdots & \ddots & \vdots \\
0 & 0 & \dots & 1 & 0 & 0 & \dots & 0 \\
0 & 0 & \dots & 0 & e^{i\phi_j} & 0 & \dots & 0 \\
0 & 0 & \dots & 0 & 0 & 1 & \dots & 0 \\
\vdots & \vdots & \ddots & \vdots & \vdots & \vdots & \ddots & \vdots \\
0 & 0 & \dots & 0 & 0 & 0 & \dots & 1
\end{pmatrix},
$$

where j varies from 0 to $N^n - 1$, and ϕ_j varies over $[-\pi, \pi)$. For Qudits of N-fold, n-dimensional, the matrix $Phase(j,\ \phi_j)$ will have order $N^n \times N^n$. The collection of these matrices is defined as: $PHASE = \{Phase(j,\ \phi_j) | j \in \{0, 1, \dots, N^n - 1\}, \phi_i \in [-\pi, \pi)\}$. This set is capable of generating any phase-shift operator [35] for an n-dimensional N-fold Qudit gate. We now define a special subset $PHASE_1 \subset PHASE$ as follows.

$$
PHASE_1 = \left\{ \begin{pmatrix}
e^{i\theta} & 0 & \dots & 0 \\
0 & 1 & \dots & 0 \\
\vdots & \vdots & \ddots & \vdots \\
0 & 0 & \dots & 1
\end{pmatrix} \text{ where } \theta \in [-\pi, \pi) \right\} = \{Phase(1,\ \theta) |\ where\ \phi_i \in [-\pi, \pi)\}
$$

2.4 Decomposition of Unitary Matrices with Reck's Approach

Michael Reck et al. [18] demonstrated that any $N \times N$ unitary matrix A can be reduced to a diagonal matrix D with elements of modulus 1 by sequentially applying a series of $T_{i,j}$ transformations. These transformations are derived from $N \times N$ identity matrices, where the elements at positions $(i,i),(i,j),(j,i),(j,j)$ are replaced by the corresponding elements of a 2×2 unitary matrix $U_{i,j} \in U(2)$. Conceptually, the $T_{i,j}$ matrices are embeddings of $U_{i,j}$ into an N-dimensional subspace. For simplicity, we will denote $T_{i,j}(U_{i,j}) = T_{i,j}$. The resulting diagonal matrix D, being unitary and having diagonal elements of modulus 1, must take the form:

$$
D = \begin{pmatrix}
e^{i\theta_1} & 0 & \cdots & 0 \\
0 & e^{i\theta_2} & \cdots & 0 \\
\vdots & \vdots & \ddots & \vdots \\
0 & 0 & \cdots & e^{i\theta_N}
\end{pmatrix}_{N \times N},
$$

where e^{ik} represents a phase factor. Its inverse, also a diagonal unitary matrix, is given by:

$$
D^{-1} = \begin{pmatrix}
e^{-i\theta_1} & 0 & \cdots & 0 \\
0 & e^{-i\theta_2} & \cdots & 0 \\
\vdots & \vdots & \ddots & \vdots \\
0 & 0 & \cdots & e^{-i\theta_N}
\end{pmatrix}_{N \times N}
= \begin{pmatrix}
e^{i(-\theta_1)} & 0 & \cdots & 0 \\
0 & e^{i(-\theta_2)} & \cdots & 0 \\
\vdots & \vdots & \ddots & \vdots \\
0 & 0 & \cdots & e^{i(-\theta_N)}
\end{pmatrix}_{N \times N}.
$$

The process of sequentially applying $T_{i,j}$ transformations can be described mathematically as follows: $A \cdot T_{N,N-1} \cdot T_{N,N-2} \cdots T_{N,1} \cdot T_{N-1,N-2} \cdots T_{2,1} = D \implies A = D \cdot T_{2,1}^{-1} \cdot T_{3,1}^{-1} \cdot T_{3,2}^{-1} \cdots T_{N,N-1}^{-1}$. Rewriting in terms of $D^{-1} = E$, where E is a diagonal unitary matrix: $A = (T_{N,N-1} \cdot T_{N,N-2} \cdots T_{2,1} \cdot E)^{-1}$. Since E is diagonal with all elements of modulus 1, the inverse of A is also unitary. Now, consider an arbitrary unitary matrix M. Its inverse $M^{-1} = A$, can be decomposed using the above result: $M^{-1} = A = (T_{N,N-1} \cdot T_{N,N-2} \cdots T_{2,1} \cdot E)^{-1}$. Therefore, the unitary matrix M can be expressed as: $M = T_{N,N-1} \cdot T_{N,N-2} \cdots T_{2,1} \cdot E$. In general, any $N \times N$ unitary matrix M can be decomposed as: $\prod_{i,j=0, j<i}^{N-1} R_{i,j} \cdot \Phi$, where $R_{i,j} = T_{i,j}(U_{i,j})$ for some $U_{i,j} \in SU(2)$ and Φ is an $N \times N$ phase shift matrix. We can state the following theorem that describes the aforementioned decomposition.

Theorem 1. *(Reck's Decomposition [18]) Any unitary matrix of order N can be decomposed into two-dimensional subspace rotations and a phase shift Φ as follows $M = \prod_{i,j=0, j<i}^{N-1} R_{i,j} \Phi$, where $R_{i,j} = T_{i,j}(U_{i,j})$ for some $U_{i,j} \in SU(2)$.*

3 Proposed Optimal Universal Gate Set

In this section, a novel methodology for achieving universality with sets of N-fold n-dimensional qubit gates is presented. Unlike prior quantum gate synthesis methods, this framework directly supports N-valued qubit gates, ensuring universal computation with optimality of resources. The core of the protocol involves multiplication of Qudit gates, which enables constructing arbitrary unitary operations from a finite proposed gate set.

3.1   Universality for Qudit Gates

We define $\mathscr{S} = PHASE_1 \cup T_{elements}$ and present the following Theorem 2, which we will prove in the subsequent section.

Theorem 2. *The set* $\mathscr{S} = PHASE_1 \cup T_{elements}$ *serves as a minimal set of universal gates for multi-dimensional Qudits of N folds.*

To prove the Theorem 2 , we will employ Theorem 1, along with Lemma 1, Lemma 2, Corollary 2 and Theorem 3. The flow Diagram 1 depicts the Flow of the proof.

Lemma 1. *Each matrix $R_{i,j}$ in the Reck's decomposition (Theorem 1) is an element of the set $\mathscr{T}_{elements}$.*

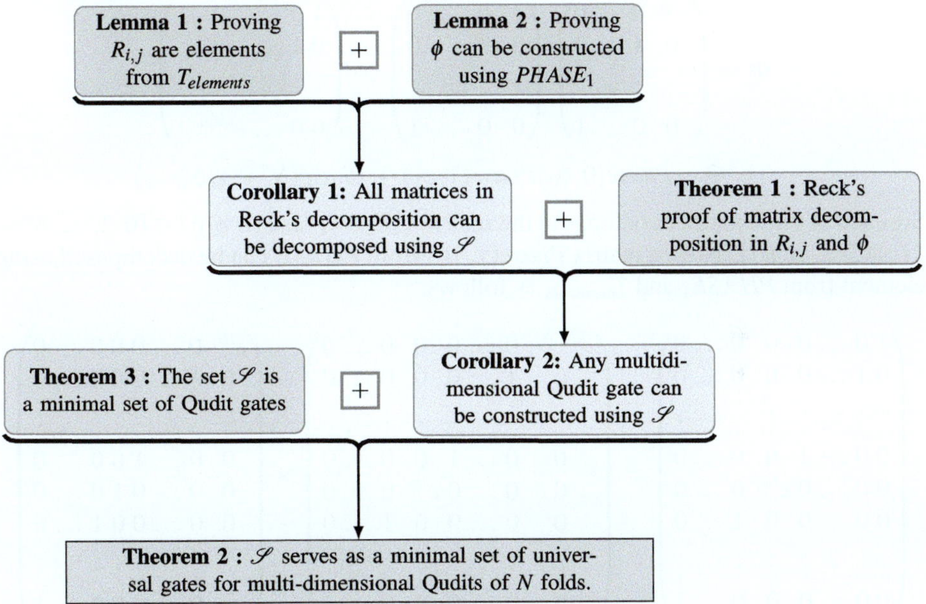

Proof. In Reck's decomposition, we express M as $M = \prod\limits_{i,j=0,j<i}^{N-1} R_{i,j}\Phi$, where $R_{i,j} = T_{i,j}(U_{i,j})$ for some $U_{i,j} \in SU(2)$.

From definition of the set $T_{elements} = \{T_{i,j}(U) | 0 \leq i < j < N^n, U \in SU(2)\}$, we see that each of the $T_{i,j}$ matrices belong to $T_{elements}$ itself. Also, from Reck's decomposition, $R_{i,j} = T_{i,j}(U)$ where $U \in SU(2)$. Thus, each of the $R_{i,j}$ can be constructed with $T_{elements}$.

Lemma 2. *Any phase shift matrix Φ in the Reck's decomposition (Theorem 1) can be constructed from elements of $PHASE_1 = \{Phase(1,\ \phi)|\ where\ \phi \in [-\pi,\pi)\}$ and $T_{elements}$.*

Proof. Let us consider the phase shift matrix Φ of size $N^n \times N^n$ defined as:

$$\Phi = \begin{pmatrix} e^{i\phi_0} & 0 & 0 & \cdots & 0 \\ 0 & e^{i\phi_1} & 0 & \cdots & 0 \\ 0 & 0 & e^{i\phi_2} & \cdots & 0 \\ \vdots & \vdots & \vdots & \ddots & \vdots \\ 0 & 0 & 0 & \cdots & e^{i\phi_{N^n-1}} \end{pmatrix}.$$

This matrix Φ is a diagonal matrix, where the diagonal entries are complex exponential, $e^{i\phi_0}, e^{i\phi_1}, e^{i\phi_2}, \ldots, e^{i\phi_{N^n-1}}$ representing phase shifts. We want to show that Φ can be represented as a product of elements from $PHASE = \{Phase(i,\ \phi_i)|\ i \in \{0,1,\ldots,N^n - 1\}$, where $\phi_i \in [-\pi,\pi)\}$. Observe that the phase matrix Φ can be expressed as of the form:

$$\Phi = \begin{pmatrix} e^{i\phi_0} & 0 & \cdots & 0 \\ 0 & 1 & \cdots & 0 \\ \vdots & \vdots & \ddots & \vdots \\ 0 & 0 & \cdots & 1 \end{pmatrix} \begin{pmatrix} 1 & 0 & \cdots & 0 \\ 0 & e^{i\phi_1} & \cdots & 0 \\ \vdots & \vdots & \ddots & \vdots \\ 0 & 0 & \cdots & 1 \end{pmatrix} \cdots \begin{pmatrix} 1 & 0 & \cdots & 0 \\ 0 & 1 & \cdots & 0 \\ \vdots & \vdots & \ddots & \vdots \\ 0 & 0 & \cdots & e^{i\phi_{N^n-1}} \end{pmatrix}$$

$$\implies \Phi = Phase(0,\phi_0)Phase(1,\phi_1)\cdots Phase(N^n - 1,\phi_{N^n-1}).$$

So, each of these matrices belongs to the set $PHASE = \{Phase(i,\ \phi_i)|\ i \in \{0,2,\ldots,N^n - 1\}$, $\phi_i \in [-\pi,\pi)\}$. Now, a matrix $Phase(j,\ \theta_j)$ from $PHASE$ can be decomposed using element from $PHASE_1$ and $T_{elements}$ as follows.

$$\begin{pmatrix} 1 & 0 & \cdots & 0 & 0 & 0 & \cdots & 0 \\ 0 & 1 & \cdots & 0 & 0 & 0 & \cdots & 0 \\ \vdots & \vdots & \ddots & \vdots & \vdots & \vdots & \ddots & \vdots \\ 0 & 0 & \cdots & 1 & 0 & 0 & \cdots & 0 \\ 0 & 0 & \cdots & 0 & e^{i\phi_j} & 0 & \cdots & 0 \\ 0 & 0 & \cdots & 0 & 0 & 1 & \cdots & 0 \\ \vdots & \vdots & \ddots & \vdots & \vdots & \vdots & \ddots & \vdots \\ 0 & 0 & \cdots & 0 & 0 & 0 & \cdots & 1 \end{pmatrix} = \begin{pmatrix} e^{i(-\phi_j)} & 0 & \cdots & 0 & 0 & 0 & \cdots & 0 \\ 0 & 1 & \cdots & 0 & 0 & 0 & \cdots & 0 \\ \vdots & \vdots & \ddots & \vdots & \vdots & \vdots & \ddots & \vdots \\ 0 & 0 & \cdots & 1 & 0 & 0 & \cdots & 0 \\ 0 & 0 & \cdots & 0 & e^{i\phi_j} & 0 & \cdots & 0 \\ 0 & 0 & \cdots & 0 & 0 & 1 & \cdots & 0 \\ \vdots & \vdots & \ddots & \vdots & \vdots & \vdots & \ddots & \vdots \\ 0 & 0 & \cdots & 0 & 0 & 0 & \cdots & 1 \end{pmatrix} \times \begin{pmatrix} e^{i\phi_j} & 0 & \cdots & 0 & 0 & 0 & \cdots & 0 \\ 0 & 1 & \cdots & 0 & 0 & 0 & \cdots & 0 \\ \vdots & \vdots & \ddots & \vdots & \vdots & \vdots & \ddots & \vdots \\ 0 & 0 & \cdots & 1 & 0 & 0 & \cdots & 0 \\ 0 & 0 & \cdots & 0 & 1 & 0 & \cdots & 0 \\ 0 & 0 & \cdots & 0 & 0 & 1 & \cdots & 0 \\ \vdots & \vdots & \ddots & \vdots & \vdots & \vdots & \ddots & \vdots \\ 0 & 0 & \cdots & 0 & 0 & 0 & \cdots & 1 \end{pmatrix}$$

Here the first matrix in the product is a member of $T_{elements}$, and the second one is from $PHASE_1 \subset PHASE$, defined above. Thus any matrix in the decomposition $\Phi = Phase(0,\phi_0)Phase(1,\phi_1)\cdots Phase(N^n - 1,\phi_{N^n-1})$ can be constructed with elements from $PHASE_1$ and $T_{elements}$, ultimately leading to any phase shift operator matrix Φ can be constructed using $PHASE_1$ and $T_{elements}$ sets only. This completes the proof.

Corollary 1. *All the matrices in Reck's decomposition (Theorem 1) can be decomposed using matrices of the set $\mathscr{S}$.*

Proof. The two types of matrices in the expansion of Reck's decomposition are $R_{i,j}$ and Φ. From Lemma 1, $R_{i,j}$ can be constructed using $T_{elements} \subset \mathscr{S}$. Also, from Lemma 2, ϕ can be constructed using $PHASE_1 \subset \mathscr{S}$. Thus, any matrix in expansion of Reck's decomposition, can be decomposed using $\mathscr{S}$.

Corollary 2. *The set $\mathscr{S} = PHASE_1 \cup T_{elements}$ forms a universal gate set for multidimensional Qudit gates of N-fold dimension.*

Proof. Any unitary matrix can be decomposed according to Reck's decomposition, with the $R_{i,j}$ and Φ matrices only (Theorem 1). Since any multidimensional Qudit gate of N- fold dimension is an N^n- order unitary matrix, it follows that such gates can be decomposed as a product of elements from $\mathscr{S} = PHASE_1 \cup T_{elements}$. This demonstrates that the set $\mathscr{S}$ is universal for constructing all multidimensional Qudit gates of N-fold dimension.

Theorem 3. *The set $\mathscr{S} = PHASE_1 \cup T_{elements}$ forms a minimal set of universal gates for multidimensional Qudit gates of N-fold and n-dimension.*

Proof. Let $\mathscr{S} = PHASE_1 \cup T_{elements}$ be a gate set intended for the generation of the unitary group $U(N^n)$, for positive integers N and n. We aim to formally demonstrate that both $PHASE_1$ and $T_{elements}$ are indispensable components of $\mathscr{S}$ for achieving universality in $U(N^n)$. For this, we rely on the distinct properties of the Lie algebras [43] generated by each subset. The Lie group generated by a set of gates G is denoted by $\text{Lie}(G)$, and its associated Lie algebra by $\text{LieAlg}(G)$.

The Lie algebra of $U(N^n)$, denoted $u(N^n)$, consists of all $N^n \times N^n$ anti-Hermitian matrices. It admits a direct sum decomposition $u(N^n) = su(N^n) \oplus \text{span}(iI_{N^n})$ where $su(N^n)$ is the Lie algebra of the special unitary group $SU(N^n)$ (comprising trace-zero anti-Hermitian matrices), and $\text{span}(iI_{N^n})$ represents the one-dimensional subalgebra associated with global phase changes in $U(N^n)$. Our argument demonstrates that $T_{elements}$ is essential for spanning $su(N^n)$ and $PHASE_1$ for spanning $\text{span}(iI_{N^n})$.

1. **Necessity of $T_{elements}$ for Generating $su(N^n)$ (Mixing Operations)**

   - Definition and Lie Algebra of $PHASE_1$: The set $PHASE_1$ is defined as composed of individual gates of the form $\text{Phase}(1,\theta) = \text{diag}(e^{i\theta},1,\dots,1) \in U(M)$ for $\theta \in [-\pi, \pi]$. The infinitesimal generator associated with $\text{Phase}(1,\theta)$ is obtained by taking the derivative with respect to θ at $\theta = 0$:

$$L(PHASE_1) = \frac{d}{d\theta}\text{diag}(e^{i\theta},1,\dots,1)\Big|(\theta = 0) = i \cdot \text{diag}(1,0,\dots,0) = iE_{11}$$

   where E_{11} is the matrix unit with a 1 in the (1,1) position and zeros elsewhere. The Lie algebra $\text{LieAlg}(PHASE_1)$ generated by $PHASE_1$ is therefore $\text{span}(iE_{11})$.
   - Properties of $\text{LieAlg}(PHASE_1)$: Any element in $\text{LieAlg}(PHASE_1)$ is a diagonal anti-Hermitian matrix. Furthermore, the set of all diagonal $N^n \times N^n$ anti-Hermitian matrices forms a Lie subalgebra of $u(N^n)$. Any sum or Lie bracket (commutator) of diagonal matrices yields a diagonal matrix. Consequently, the Lie algebra $\text{LieAlg}(PHASE_1)$ contains only diagonal matrices.

- Inability of $PHASE_1$ to Generate $su(N^n)$'s Non-Diagonal Components: The Lie algebra $su(N^n)$ for $N \geq 2$, $n \geq 1$ contains numerous non-diagonal anti-Hermitian matrices. For instance, in $su(2)$, the generators $i\sigma_x = \begin{pmatrix} 0 & i \\ i & 0 \end{pmatrix}$ and $i\sigma_y = \begin{pmatrix} 0 & 1 \\ -1 & 0 \end{pmatrix}$ are non-diagonal. Since $\mathrm{LieAlg}(PHASE_1)$ consists exclusively of diagonal matrices, it cannot generate these non-diagonal elements of $su(N^n)$. Therefore, the Lie group $\mathrm{Lie}(PHASE_1)$ cannot generate arbitrary 'mixing' operations, which are represented by non-diagonal elements in $SU(N^n)$.
- Role of $T_{elements}$ in Generating $su(N^n)$: The set $T_{elements}$ consists of $U(N^n)$ gates that are embeddings of $SU(2)$ matrices. The Lie algebra $su(2)$ contains non-diagonal trace-zero anti-Hermitian matrices $(i\sigma_x, i\sigma_y, i\sigma_z)$. A fundamental result in quantum control and universal gate sets (e.g., related to Reck's decomposition and the structure of $SU(N^n)$) establishes that $SU(N^n)$ can be generated by such embedded $SU(2)$ operations. This implies that the Lie algebra $\mathrm{LieAlg}(T_{elements})$ can generate the necessary non-diagonal elements to span $su(N^n)$.
- Conclusion on $T_{elements}$'s Necessity: Without $T_{elements}$, the gate set $\mathscr{S}$ would effectively be reduced to $PHASE_1$. As shown, $\mathrm{LieAlg}(PHASE_1)$ is a proper sub-algebra of $u(N^n)$, specifically lacking the ability to generate non-diagonal transformations in $su(N^n)$. Such transformations are critical for state mixing, entanglement, and spanning $SU(N^n)$. Thus, $T_{elements}$ is indispensable for $\mathscr{S}$ to achieve universality for $U(N^n)$.

2. Necessity of $PHASE_1$ for Generating $\mathrm{span}(iI_{N^n})$ (Global Phase Operations)

- Properties and Lie Algebra of $T_{elements}$: Each gate $T \in T_{elements}$ is an embedding of an $SU(2)$ matrix into $U(N^n)$. By definition, all $SU(2)$ matrices have a determinant of 1. Since the determinant is a multiplicative homomorphism from a Lie group to $\mathbb{C}^*$, any product of gates from $T_{elements}$ (and thus any element in $\mathrm{Lie}(T_{elements})$) must also have a determinant of 1. This implies that $\mathrm{Lie}(T_{elements}) \subseteq SU(N^n)$. The Lie algebra $\mathrm{LieAlg}(T_{elements})$ is a subalgebra of $su(N^n)$. By definition, all elements of $su(N^n)$ are trace-zero anti-Hermitian matrices. Therefore, any element $L \in \mathrm{LieAlg}(T_{elements})$ must satisfy $\mathrm{Tr}(L) = 0$.
- Inability of $T_{elements}$ to Generate $\mathrm{span}(iI_{N^n})$: The $u(N^n)$ Lie algebra includes elements of the form ixI_{N^n} for $x \in [-\pi, \pi)$. The trace of such a matrix is $\mathrm{Tr}(ixI_{N^n}) = iN^n x$. For $N, n > 0$ and $x \neq 0$, this trace is non-zero. Since all elements in $\mathrm{LieAlg}(T_{elements})$ are trace-zero, $\mathrm{LieAlg}(T_{elements})$ cannot generate any element of the form ixI_{N^n} where $x \neq 0$. Consequently, the Lie group $\mathrm{Lie}(T_{elements})$ cannot generate operations that effect a non-trivial global phase change (i.e., elements in $U(N^n) \setminus SU(N^n)$).
- Role of $PHASE_1$ in Generating $\mathrm{span}(iI_{N^n})$: The gate $\mathrm{Phase}(1, \theta)$ defined as $\mathrm{Phase}(1, \theta) = \mathrm{diag}(e^{i\theta}, 1, \ldots, 1) \in PHASE_1$ has an infinitesimal generator iE_{11}. This generator has a non-zero trace, specifically $\mathrm{Tr}(iE_{11}) = i$. This demonstrates that $\mathrm{LieAlg}(PHASE_1)$ contains elements with a non-zero trace. Importantly, the parameter θ in $PHASE_1$ is continuous ($\theta \in [-\pi, \pi)$). This continuous degree of freedom is essential for generating a continuous range of global phase shifts, which is precisely what the $U(1)$ factor in $U(N^n)$ requires (generated by iI_{N^n}).

While iE_{11} itself is not iI_{N^n}, it provides the capability to introduce non-zero trace components, which are absent in LieAlg($T_{elements}$). When combined with the $su(N^n)$ generators from $T_{elements}$, the non-zero trace component provided by $PHASE_1$ allows for the full $u(N^n)$ Lie algebra to be spanned, including the span(iI_{N^n}) part.

- Conclusion on $PHASE_1$'s Necessity: If $PHASE_1$ were removed, the gate set $\mathscr{S}$ would be reduced to $T_{elements}$. As shown, LieAlg($T_{elements}$) is a proper subalgebra of $u(M)$, specifically lacking the ability to generate operations that induce a global phase change (i.e., elements in span(iI_{N^n})). Such global phase changes are necessary to span the entirety of $U(N^n)$. Thus, $PHASE_1$ is indispensable for $\mathscr{S}$ to achieve universality for $U(N^n)$.

3. **Indispensability of both $T_{elements}$ and $Phase_1$ set of matrices**

The Lie algebra of the target group $U(N^n)$ decomposes precisely into two orthogonal components: $u(N^n) = su(N^n) \oplus \mathrm{span}(iI_{N^n})$. The gate set $T_{elements}$ generates a Lie algebra LieAlg($T_{elements}$) that is a subalgebra of $su(N^n)$. Consequently, all infinitesimal transformations generated by $T_{elements}$ are trace-zero and thus cannot contribute to the span(iI_{N^n}) component of $u(N^n)$.

The gate set $PHASE_1$ generates a Lie algebra LieAlg($PHASE_1$) that is a subalgebra of the diagonal matrices in $u(N^n)$. While LieAlg($PHASE_1$) cannot generate non-diagonal transformations (arbitrary elements of $su(M)$), it provides generators with non-zero traces (e.g., iE_{11}) and a continuous parameter space that enables the generation of the span(iI_{N^n}) component, which $T_{elements}$ cannot.

Since LieAlg($T_{elements}$) is confined to trace-zero matrices, and LieAlg($PHASE_1$) is confined to diagonal matrices (and cannot span all of $su(N^n)$), neither set alone can generate the entirety of $u(N^n)$. However, their union $\mathscr{S} = PHASE_1 \cup T_{elements}$ (or, more precisely, the Lie algebra LieAlg($\mathscr{S}$) generated by $\mathscr{S}$) is capable of spanning $u(N^n)$, as implied by the combination of established theoretical results like Reck's decomposition and the properties of phase gates in quantum optics.

Because the Lie algebras generated by $PHASE_1$ and $T_{elements}$ provide distinct and non-overlapping necessary components to span $u(N^n)$, neither $PHASE_1$ nor $T_{elements}$ can be removed from $\mathscr{S}$ without losing the ability to generate the full unitary group $U(N^n)$. This rigorously establishes the indispensability of both $PHASE_1$ and $T_{elements}$ for the universality of $\mathscr{S}$.

Proof of Theorem 2

Proof. From Corollary 2, we have that $\mathscr{S}$ is a universal gate set for any multidimensional N-fold Qudit gates. And Theorem 3 proves the minimality of the set $\mathscr{S}$ in generating any multidimensional N-fold Qudit gate. Collating Corollary 2 and Theorem 3, we have that $\mathscr{S}$ is a minimal universal gate set. This proves Theorem 2.

3.2   Algorithm for Obtaining Decomposition Matrices

We have obtained the decomposition Algorithm 3.2 for Reck's decomposition of unitary matrices. Alongside, we implemented the decomposition algorithm in a programmatic way, analyzing its space and time complexity in average case. For the sake of simplicity, we have considered the matrices' order to be N_1 instead of N^n.

This algorithm expresses an $N_1 \times N_1$ unitary matrix M as a product of $L = N_1(N_1 - 1)/2$ elementary unitary matrices R_k and a final diagonal unitary matrix Φ. The decomposition takes the form $M = R_1 \otimes R_2 \otimes \ldots \otimes R_L \otimes \Phi$. Each R_k matrix is an $N_1 \times N_1$ unitary matrix, differing from the identity matrix I_{N_1} only in a 2×2 principal submatrix. These R_k matrices are analogous to the T_{pq} matrices in the original Reck's paper, representing complex Givens rotations. The algorithm iteratively transforms the input matrix M into a diagonal matrix Φ by sequentially pre-multiplying it with the Hermitian conjugates of the desired R_k matrices.

Time Complexity Analysis:

1. **Input Validation Subtotal:** $O(N_1^3)$ (dominated by $N_1 \times N_1$ matrix multiplication of `M @ ...`).
2. **Initialization Subtotal:** $O(N_1^2)$ `U_current = M.astype(complex)` involves creating a copy of the $N_1 \times N_1$ input matrix.).
3. **Main Loops (Givens Rotations Application):** The outer loop runs $N_1 - 1$ times. The inner loop runs $N_1 - 1 - j$ times for each j. The total number of times the inner loop body executes is $O(N_1^2)$. `U_current = R_inv_matrix @ U_current`. This is a multiplication of two $N_1 \times N_1$ matrices, leading to an $O(N_1^3)$ complexity. With $O(N_1^2)$ iterations, each containing an $O(N_1^3)$ dominant matrix multiplication, the total for the main loops is $O(N_1^2) \times O(N_1^3) = O(N_1^5)$.
4. **Final Steps:** `Phi = U_current` is an $O(1)$ reference assignment. Obtaining Phi in form of $PHASE_1$ matrix, consists of N_1 many multiplications, making the whole calculation $O(N_1)$.
5. **Overall Time Complexity:** $O(N_1^3)$ (validation) + $O(N_1^2)$ (initialization) + $O(N_1^5)$ (main loops) + $O(N_1)$ (final steps). The highest order term is $O(N_1^5)$. Figure 1 shows python-based validation result of time complexity analysis.

Space Complexity Analysis:

1. **Input and Initial Copies:** `M:` and `U_current` both occupies $O(N_1^2)$ space.
2. **Temporary Matrices within Iterative Loops:** `G_block` is a temporary 2×2 matrix, consuming $O(1)$ constant space. `R_inv_matrix` is a temporary $N_1 \times N_1$. This matrix is allocated $O(N_1^2)$ space, and this memory is effectively reused across successive loop iterations, preventing cumulative memory growth.
3. **Stored `R_matrices` List:** The list dynamically stores a collection of matrices. The number of matrices accumulated within this list is approximately $\frac{N_1(N_1-1)}{2}$, which asymptotically translates to $O(N_1^2)$ matrices. Each individual matrix stored in this list is an $N_1 \times N_1$ matrix, thus demanding $O(N_1^2)$ memory. The total space consumed by the `R_matrices` list is the product of the number of stored matrices and the space required per matrix: $O(N_1^2) \times O(N_1^2) = O(N_1^4)$.
4. **Final Output `Phi`:** Represents the final output. It is characterized as a reference to `U_current`. Additionally, the creation of $T|elements$ matrix `Phi_balance` requires additional independent memory allocation proportional to N_1^2.
5. **Overall Space Complexity:** The predominant term in the algorithm's space complexity is $O(N_1^4)$, primarily due to the storage requirements of the `R_matrices` list. Figure 1 indicates space complexity analysis.

Algorithm 1 Reck's Unitary Matrix Decomposition

Require: $M \in \mathbb{U}^{N_1 \times N_1}$: An $N_1 \times N_1$ unitary matrix.
Ensure: R_{matrices}: An ordered list $[R_1, R_2, \ldots, R_L]$ of $N_1 \times N_1$ unitary matrices.
Ensure: $\Phi \in \mathbb{U}^{N_1 \times N_1}$: A diagonal $N_1 \times N_1$ unitary matrix with all diagonal elements having a modulus of 1.

1: $\varepsilon \leftarrow 10^{-9}$ ▷ *Numerical tolerance for checking approximate zero values*
2: **Initialize:**
3: $R_{\text{matrices}} \leftarrow []$ ▷ *An ordered list to store the R_k matrices*
4: $U_{\text{current}} \leftarrow M$ ▷ *The working matrix, which will be iteratively transformed into Φ*

5: **Iterative Annihilation (Column by Column, from bottom up):**
6: **for** j from 0 to $N_1 - 2$ **do**   ▷ *Column index for the pivot element $U_{current}[j,j]$*
7:      **for** i from $N_1 - 1$ **down to** $j + 1$ **do**   ▷ *Row index for $U_{current}[i,j]$ to be annihilated*
8:          $u \leftarrow U_{\text{current}}[j,j]$ ▷ *The pivot element*
9:          $v \leftarrow U_{\text{current}}[i,j]$ ▷ *The off-diagonal element to be zeroed*
10:          **if** $|v| < \varepsilon$ **then**
11:             **continue**   ▷ *Element is already numerically zero, skip transformation*
12:
13:          **Construct** 2×2 **Complex Givens Rotation Block (G_{block}):**
14:          $r \leftarrow \sqrt{|u|^2 + |v|^2}$
15:          $c \leftarrow \bar{u}/r$
16:          $s \leftarrow \bar{v}/r$
17:          $G_{\text{block}} = \begin{pmatrix} c & s \\ -\bar{s} & \bar{c} \end{pmatrix}$ ▷ *This matrix is unitary*

18:          **Construct Full** $N_1 \times N_1$ **Inverse Transformation Matrix ($R_{\text{inv\_matrix}}$):**
19:          $R_{\text{inv\_matrix}} \leftarrow I_{N_1}$ ▷ *Initialize as the $N_1 \times N_1$ identity matrix*
20:          $R_{\text{inv\_matrix}}[j,j] \leftarrow c$, $R_{\text{inv\_matrix}}[j,i] \leftarrow s$
21:          $R_{\text{inv\_matrix}}[i,j] \leftarrow -\bar{s}$, $R_{\text{inv\_matrix}}[i,i] \leftarrow \bar{c}$

22:          **Apply Transformation and Store R_k:**
23:          $U_{\text{current}} \leftarrow R_{\text{inv\_matrix}} \otimes U_{\text{current}}$ ▷ *This operation annihilates $U_{current}[i,j]$*
24:          $R_k \leftarrow R_{\text{inv\_matrix}}^{\text{H}}$ ▷ *R_k is the Hermitian conjugate of the applied transformation*
25:          Append R_k to R_{matrices}
26:
27: **Final Diagonal Matrix:**
28: $\Phi \leftarrow U_{\text{current}}$ ▷ *$U_{current}$ is now an upper triangular unitary matrix, implying it is diagonal*
29: **if** $|\Phi| \neq 1$ **then**
30:      $\Phi_{balance} \leftarrow I_{N_1}$ ▷ *$T_{elements}$ matrix to prepare $PHASE_1$ matrix from Φ*
31:      $v = 1$

32:      **for** i from 0 **to** $N_1 - 1$ **do**
33:          $v = v.\Phi[i,i]$
34:          $\Phi_{balance}[i,i] \leftarrow \Phi[i,i]$
35:
36:      $\Phi_{balance}[i,i] \leftarrow \bar{v}$
37:      Append $\Phi_{balance}$ to R_{matrices}
38:      $\Phi \leftarrow I_{N_1}$
39:      $\Phi[0,0] \leftarrow v$
40:
41: **return** $R_{\text{matrices}}, \Phi$

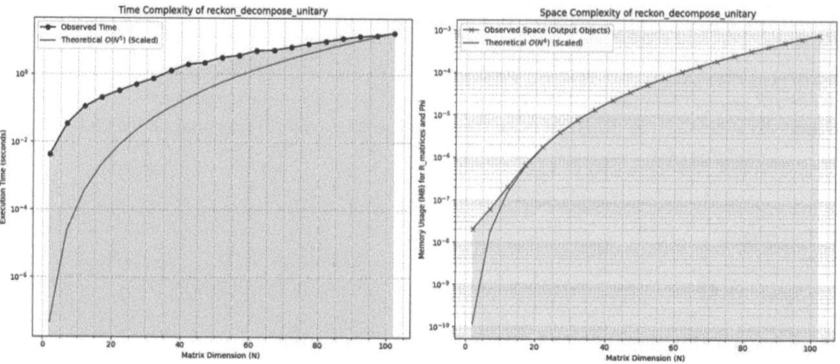

Fig. 1. Average Case Time and Space Complexity Graphs for the Algorithm compared with $O(N_1^5)$ and $O(N_1^4)$ graphs respectively converted to logarithmic scale. Graphs obtained while running for various input sizes (N_1 ranging from 2 to 100+)

3.3 Construction of Existent Universal Gate Sets Through Proposed Gate Sets

By Solovay-Kitaev theorem [1], it is established that for single qubit ($N = 2$), the set of gates $\mathscr{A} = \{H,\ T,\ S,\ CNOT\}$ constitutes a universal set of quantum gates. In our hypothesis, we propose that the set $\mathscr{S} = \{PHASE,\ SU(2^n)\}$ serves as a universal gate set for the case $N = 2$. To validate this claim, we demonstrate that the set $\mathscr{A}$ can be constructed entirely using gates from $\mathscr{S}$. The derivations are outlined as follows:

- For $N = 2$ and $n = 1$, the Hadamard gate $H = \frac{1}{\sqrt{2}} \begin{pmatrix} 1 & 1 \\ 1 & -1 \end{pmatrix}$ can be expressed as following, where the first two matrices are from $T_{elements}$, and the last is from $PHASE_1$.

$$H = \left(\begin{pmatrix} \frac{1}{\sqrt{2}} & -\frac{1}{\sqrt{2}} \\ \frac{1}{\sqrt{2}} & \frac{1}{\sqrt{2}} \end{pmatrix} \times \begin{pmatrix} -1 - i6.1e-17 & 0 \\ 0 & -1 + i6.12e-17 \end{pmatrix} \right) \times \begin{pmatrix} -1 + i6.1e-17 & 0 \\ 0 & 1 \end{pmatrix}$$

- $T = \begin{pmatrix} 1 & 0 \\ 0 & e^{i\pi/4} \end{pmatrix}$ gate can be expressed as: $T = \begin{pmatrix} e^{-i\pi/4} & 0 \\ 0 & e^{i\pi/4} \end{pmatrix} \times \begin{pmatrix} e^{i\pi/4} & 0 \\ 0 & 1 \end{pmatrix}$.

- $S = \begin{pmatrix} 1 & 0 \\ 0 & i \end{pmatrix}$ gate can be decomposed as: $S = \begin{pmatrix} e^{-i\pi/2} & 0 \\ 0 & e^{i\pi/2} \end{pmatrix} \times \begin{pmatrix} e^{i\pi/2} & 0 \\ 0 & 1 \end{pmatrix}$

- The Controlled-NOT ($CNOT$) gate for $N = 2$ can be decomposed into product of matrices from $T_{elements}$ and $PHASE_1$.

$$CNOT = \begin{pmatrix} 1 & 0 & 0 & 0 \\ 0 & 1 & 0 & 0 \\ 0 & 0 & 0 & 1 \\ 0 & 0 & 1 & 0 \end{pmatrix} = \left(\begin{pmatrix} 1 & 0 & 0 & 0 \\ 0 & 1 & 0 & 0 \\ 0 & 0 & 0 & -1 \\ 0 & 0 & 1 & 0 \end{pmatrix} \times \begin{pmatrix} -1 & 0 & 0 & 0 \\ 0 & 1 & 0 & 0 \\ 0 & 0 & 1 & 0 \\ 0 & 0 & 0 & -1 \end{pmatrix} \right) \times \begin{pmatrix} -1 & 0 & 0 & 0 \\ 0 & 1 & 0 & 0 \\ 0 & 0 & 1 & 0 \\ 0 & 0 & 0 & 1 \end{pmatrix}$$

These derivations confirm that all elements of the universal gate set $\mathscr{A}$ can be expressed using the gates of $\mathscr{S}$ in the particular case of $N = 2$. Thus, for the case of $N = 2$, $\mathscr{S}$ serves as a universal gate set for single-qubit operations.

3.4 Functional Equivalence of the Proposed Universal Gate Sets

To quantitatively validate that the decomposed gates preserve the functional behavior of their original unitary counterparts, we conducted numerical evaluations on randomly generated Qudit unitaries. For some randomly selected dimensions between 2 and 20, ten random unitaries were generated and decomposed into the sequence of rotation matrices (R_i) and phase matrices (Φ) obtained from the proposed minimal universal set $S = \mathrm{PHASE}_1 \cup T_{elements}$. The reconstructed matrices were then compared with the original unitaries using multiple fidelity and distance metrics widely employed in quantum verification. The following parameters were computed for every trial:

- **Operator fidelity:** $F = \frac{1}{m}\left|\mathrm{Tr}(U_{\mathrm{target}}^{\dagger}U_{\mathrm{approx}})\right|$, indicating global unitary overlap.
- **Trace similarity:** Normalized trace overlap between the two unitaries.
- **Frobenius error:** $\|U_{\mathrm{target}} - U_{\mathrm{approx}}\|_F$, quantifying element-wise deviation.
- **Spectral norm error:** $\|U_{\mathrm{target}} - U_{\mathrm{approx}}\|_2$, the largest singular-value deviation.
- **Eigenphase minimal arc:** Maximum phase-angle spread between eigenvalues, measured in radians.

Each metric was averaged over 10 independent trials for each Qudit dimension. All values demonstrated near-perfect agreement between the original and reconstructed matrices, with fidelity and trace-similarity consistently equal to 1.0 (up to floating-point precision), and Frobenius and spectral errors on the order of 10^{-15}. This confirms that the decomposition procedure preserves functional equivalence across dimensions. The results are summarized in Table 2.

Table 2. Functional Equivalence Validation for Random Qudit Unitaries

Qudit States	Trials Count	Operator Fidelity	Trace Similarity	Frobenius Error	Spectral Norm Error	Eigenphase Minimal Arc
2	10	$1+4\times10^{-16}$	1.000	4.32×10^{-16}	4.06×10^{-16}	0.00
8	10	$1+4\times10^{-16}$	1.000	2.94×10^{-15}	2.54×10^{-15}	0.00
9	10	$1+2\times10^{-16}$	1.000	2.82×10^{-15}	2.23×10^{-15}	8.88×10^{-17}
10	10	$1+4\times10^{-16}$	1.000	3.59×10^{-15}	3.10×10^{-15}	8.88×10^{-17}
20	10	$1+2\times10^{-16}$	1.000	7.46×10^{-15}	6.22×10^{-15}	3.55×10^{-16}

The negligible numerical errors (within machine precision) and zero-phase spread conclusively demonstrate that every decomposed gate constructed from S is functionally equivalent to its source unitary. This establishes that the proposed universal set preserves complete unitary behavior across arbitrary Qudit dimensions, thus satisfying the definition of functional equivalence for universal gate decomposition.

3.5 Efficiency over Alternative Universal Gate Sets

To assess the efficiency of the proposed Reck-based decomposition against existing frameworks, we conducted a comparative gate-count analysis with the Li–Roberts–Yin (LRY) decomposition method [16]. Both approaches decompose arbitrary unitary matrices into experimentally realizable sub-operations; however, the LRY method relies on fully controlled single-qubit operations, while Reck's formulation uses embedded two-dimensional rotations and phase shifts, leading to a more compact representation for higher-dimensional Qudits.

For each randomly generated unitary matrix of order below 30, ten independent trials were performed. The average number of resulting elementary gates was recorded for both decomposition schemes. Table 3 summarizes the outcomes. The proposed method visibly exhibits a significantly lower gate count for higher folds, with the advantage growing rapidly as dimensionality increases. This confirms that Reck's subspace-rotation approach scales more efficiently for large Qudit systems, offering a tangible reduction in synthesis complexity.

Table 3. Gate Count Comparison between Reck's & Li–Roberts–Yin's Decompositions

Qudit States	Trials Count	LRY Method (Avg. Gates)	Reck's Method (Avg. Gates)
2	10	2.0	3.0
5	10	20.0	12.0
12	10	132.0	68.0
16	10	416.0	122.0
20	10	520.0	192.0
23	10	598.0	255.0
27	10	702.0	353.0

3.6 Validation over Quantum Cryptographic Models

To establish the validity of our proposed minimal universal gate set in realistic scenarios, we performed detailed simulations of two cryptographically significant quantum models, Grover's search algorithm and the QKD protocol. These models represent complementary aspects of quantum cryptology: the former serves as a benchmark for quantum search and cryptanalysis, while the latter underpins secure communication. In each case, we constructed two parallel circuits: one employing traditional Qudit gates, and another with universal gates synthesized through Reck's decomposition.

The comparative study demonstrates that the proposed universal set faithfully reproduces the expected cryptographic behavior while providing a more generalizable and hardware-agnostic foundation. Importantly, the decomposed circuits illustrate how universal gate sets enable a consistent framework for both constructive protocols (QKD) and adversarial primitives (Grover's search).

Grover's Algorithm with Universal Decomposition: Grover's search is a classic quantum algorithm, often studied for its cryptanalytic implications because of its known quadratic speed-up in unstructured search. In the multi-Qudit setting, it requires a generalization of the oracle and diffusion operators, both of which depend heavily on the availability of Hadamard-type gates for state initialization and inversion about the mean.

In the traditional construction, the circuit begins with two 4-dimensional Qudits, each initialized using explicit Qudit Hadamard operators. The oracle (U_f) flips the phase of the marked state, and the inversion operator (U_0) reflects amplitudes about the average. These transformations are expressed as dense, high-dimensional unitary matrices, directly invoked in the simulation. In the universal decomposition construction, we replaced the Hadamard gates with their Reck-decomposed equivalents. Each Hadamard operator was expressed as a product of two sets of matrices: a diagonal phase operator Φ, and a sequence of pairwise rotation matrices R_{ij}, embedded into the multi-Qudit Hilbert space. Thus, instead of a single dense Hadamard operation, the decomposed circuit applies multiple layers of R_{ij}-based rotations and Φ-phase shifts. This decomposition ensures that the gate belongs to the minimal universal set $\mathscr{S} = PHASE_1 \cup T_{elements}$, and thereby establishes universality.

Figure 2 demonstrates outcome of simulation of Grover's Algorithms, where circuits have been created in two ways, using traditional Qudit gates and using proposed universal set of gates. While the simulation outcomes of both circuits were equivalent - amplifying the probability of the same marked state 03-the architectural distinction is fundamental. Traditional circuits rely on pre-defined, higher-level operators, whereas the decomposed version explicitly constructs the same effect using a basis of minimal gates. This reflects the central strength of the proposed framework: cryptographic models can be fully expressed using a universal and minimal gate set, eliminating the need for ad hoc gate assumptions tied to qubit-based universality. The figure illustrates that both constructions amplify the marked state with nearly identical distributions, affirming the correctness of the decomposition. Beyond correctness, this validates that cryptanalytic tasks can be generalized across dimensions while preserving efficiency. For cryptology, this means that Qudit-based Grover search can be universally synthesized and is not dependent on hardware-specific implementations of higher-level unitaries.

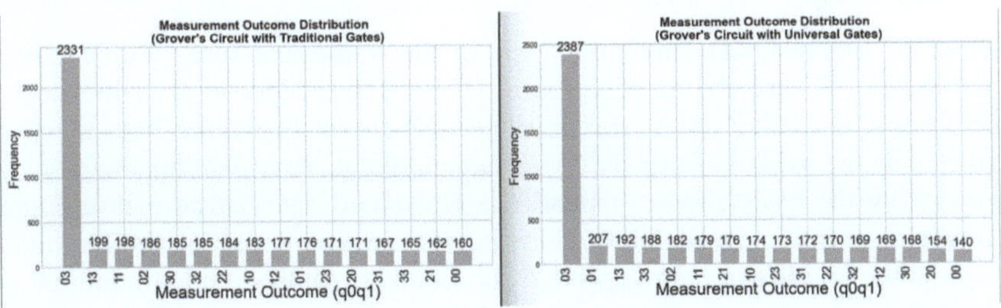

Fig. 2. Visual Comparison of Measurement Outcome Distribution for Grover's Algorithm, where Circuits are created using either Traditional or proposed Universal Gates

Quantum Key Distribution with Universal Decomposition: The BB84 QKD protocol was adapted to a Qudit setting to assess the performance of the proposed universal gate set in a defensive cryptographic application. The protocol involves three essential steps: Alice encodes random symbols in one of two bases, Bob independently measures in a chosen basis, and both parties compare basis choices to derive a shared secret key.

In traditional construction, Alice's information encoding and Bob's measurement were implemented directly with Qudit Hadamard and phase-type operators to generate rectilinear and diagonal bases. Bob's measurement outcomes were obtained by applying the corresponding inverse basis transformations before projective measurement. In the universal decomposition construction, these basis-encoding and decoding operations were built entirely from Reck-decomposed gates. Each transformation, rather than being called as a predefined unitary, was explicitly realized as a composition of Φ-phase shifts and embedded R_i rotations. Alice's encoded state and Bob's chosen basis therefore originated from circuits constructed exclusively with universal gates, without reliance on traditional Hadamard-style operators. In our implementation, the two circuits were executed in parallel for 100 simulated rounds. In 47 cases, Alice and Bob selected the same basis, leading to the establishment of 47 secret key bits. In every such case, the key bits produced by the decomposed circuits matched those from the traditional implementation, confirming the semantic equivalence of the universal gates. The following two outcomes describe the operational differences between the QKD circuits with traditional and proposed universal gates.

```
--- QKD Round 29/100 ---
Alice: Preparing bit 2 in basis 'diagonal'.
Bob: Choosing basis 'diagonal'.
0(d=3): | Qutrit_0swap2_Gate | Qu3H | Qu3H | M('diagonal_measurement')|
0(d=3): | Qu3M x3 | Qu3M x5 | Qu3M x5 | M('diagonal_measurement') |
Bob: Measured bit 2 in basis 'diagonal'.
Bob: Measured bit 2 in basis 'diagonal'.
Alice and Bob used the same basis.
  Key established for this round: 2
```

In Round 29 of the QKD simulation, within the traditional setup, Alice randomly selected the trit "2" and the diagonal basis, while Bob independently chose the same diagonal basis. On Alice's side, this resulted in a circuit beginning with the 0swap2 gate to encode the symbol, followed by a Hadamard gate to realize the diagonal basis. Bob, matching the basis choice, appended a Hadamard gate before measurement. In case of universal decomposed gates, all the gates are decomposed into unitary matrices of the proposed forms and then appended subsequently replacing the traditional gates. The 0swap2 gate has been replaced with one R_{ij}, one $\Phi_{balance}$ and one Φ matrices, while the Hadamard gate has been replaced with three R_{ij}, one $\Phi_{balance}$ and one Φ matrices. Both measurements yielded the value "2", and since the bases aligned, a key bit was successfully established, exactly as predicted by the principles of QKD.

```
--- QKD Round 67/100 ---
Alice: Preparing bit 0 in basis 'rectilinear'.
Bob: Choosing basis 'diagonal'.
0(d=3): | Qu3H | M('rectilinear_measurement') |
```

```
O(d=3):  | Qu3M | Qu3M | Qu3M | Qu3M | Qu3M | M('diagonal_measurement') |
Bob: Measured bit 1 in basis 'diagonal'.
Bob: Measured bit 2 in basis 'diagonal'.
Alice and Bob used different bases or measurement result invalid.
No key bit established for this round.
```

In Round 67 of the QKD simulation, within the traditional setup, Alice randomly selected the trit "0" and the rectilinear basis, while Bob independently chose the diagonal basis. Because of Alice's choices, this resulted in a circuit that keeps the state unaltered. Bob, using his chosen basis, appended a Hadamard gate before measurement. In case of universal decomposed gates, the only Hadamard gate in the circuit has been replaced with three R_{ij}, one $\Phi_{balance}$ and one Φ matrices. We see both measurements yielded different values of measured bit, although the choice of bases being different, this round will anyway not yield any shared key bit. The simulation logs confirm consistent agreement between the universal and traditional circuits. This shows that decomposed universal gates not only preserve functional correctness, but also maintain the cryptographic guarantees of QKD protocols.

4   Conclusion

This work establishes the universality and minimality of the proposed gate set $S = PHASE_1 \cup T_{elements}$ for higher-dimensional Qudit systems. Minimality was rigorously proven using Lie algebraic arguments, confirming the indispensability of both phase and embedded $SU(2)$ transformations. A systematic decomposition algorithm was developed for arbitrary unitary transformations, analyzed for time and space complexity, and validated through functional and cryptographic experiments. Comprehensive evaluations confirm that the proposed decomposition preserves perfect functional equivalence across dimensions and achieves superior efficiency compared to existing universal gate constructions. Implementations on Grover's search and QKD protocols further demonstrated correctness, resource transparency, and hardware independence, proving that high-level Qudit operations can be synthesized entirely from minimal universal primitives. The results reinforce three central implications for quantum cryptology: (i) universality in practice—validated through exact equivalence in decomposed circuits; (ii) hardware-agnostic portability—enabling scalable adoption across architectures; and (iii) efficiency and scalability—supporting faster QKD key generation and generalized Qudit-based search for cryptanalysis. Together, these findings present a unified, experimentally grounded framework where a single minimal universal gate set advances both defensive and adversarial dimensions of quantum cryptography. Future work will focus on physical realization and optimization of synthesis strategies to further minimize circuit depth and enhance scalability.

Acknowledgments. Prof. Avishek Adhikari receives partial support from the DST-FIST Project, funded by the Government of India, under Sanction Order SR/FST/MS-I/2019/41.

Disclosure of Interests. The authors have no competing interests to declare that are relevant to the content of this article.

References

1. Nielsen, M.A., Chuang, I.L.: Quantum Computation and Quantum Information. Cambridge university press (2010)
2. Barenco, A., et al.: Elementary gates for quantum computation. Phys. Rev. A. Am. Phys. Soc. (1995). https://doi.org/10.1103/PhysRevA.52.3457
3. Barenco, A.: A universal two-bit gate for quantum computation. In: Proceedings: Mathematical and Physical Sciences. Royal Society (1995)
4. Deutsch, D.E., Barenco, A., Ekert, A.: Universality in quantum computation. In: Proceedings of the Royal Society of London. Series A: Mathematical and Physical Sciences (1995). https://doi.org/10.1098/rspa.1995.0065
5. Deutsch, D.E., Barenco, A., Ekert, A.: Universality in quantum computation. In: Proceedings of the Royal Society of London. Series A: Mathematical and Physical Sciences. The Royal Society (1995). https://doi.org/10.1098/rspa.1995.0065
6. Yu, A.: Quantum computations: algorithms and error correction. In: Russian Mathematical Surveys (1996). https://doi.org/10.1070/RM1997v052n06ABEH002155
7. Fischer, L.E., et al.: Universal qudit gate synthesis for transmons. In: PRX Quantum. American Physical Society (2023). https://doi.org/10.1103/PRXQuantum.4.030327
8. Boykin, P.O., Mor, T., Pulver, M., Roychowdhury, V., Vatan, F.: A new universal and fault-tolerant quantum basis. Inf. Process. Lett. **75**, 101–107 (2000) https://doi.org/10.1016/S0020-0190(00)00084-3
9. Boykin, P.O., Mor, T., Pulver, M., Roychowdhury, V., Vatan, F.: On Universal and Fault-Tolerant Quantum Computing. quant-ph/9906054 (1999)
10. Cafaro, C., Mancini, S.: A geometric algebra perspective on quantum computational gates and universality in quantum computing. In: Advances in Applied Clifford Algebras. Springer Science and Business Media LLC (2011). https://doi.org/10.1007/s00006-010-0269-x
11. Havel, T.F., Doran, C.J.L.: Geometric algebra in quantum information processing. quant-ph/0004031 (2000)
12. Somaroo, S.S., Cory, D.G., Havel, T.F.: Expressing the operations of quantum computing in multiparticle geometric algebra. Phys. Lett. A. **240**, 1–7. Elsevier BV (1998). https://doi.org/10.1016/S0375-9601(98)00010-3
13. Vlasov, A.Y.: Clifford algebras and universal sets of quantum gates. Phys. Rev. A. Am. Phys. Soc. (APS) **63**, 054302 (2001). https://doi.org/10.1103/physreva.63.054302
14. Shi, Y.: Both Toffoli and Controlled-NOT need little help to do universal quantum computation. quant-ph/0205115 (2002)
15. Narvaez, C.E.Q.: Universality for Sets of Three-Valued Qubit Gates. arXiv preprint arXiv:2109.07282 (2021)
16. Li, C.-K., Roberts, R., Yin, X.: Decomposition of unitary matrices and quantum gates. arxiv:1210.7366 (2012)
17. Vartiainen, J.J., Möttöönen, M., Salomaa, M.M.: Efficient decomposition of quantum gates. Phys. Rev. Lett. **92**, 177902 (2004) https://doi.org/10.1103/physrevlett.92.177902
18. Reck, M., Zeilinger, A., Bernstein, H.J., Bertani, P.: Experimental realization of any discrete unitary operator. Phys. Rev. Lett. Am. Phys. Soc. **73**, 58 (1992). https://doi.org/10.1103/PhysRevLett.73.58
19. Sawicki, A., Karnas, K.: Universality of Single-Qudit Gates. Ann. Henri Poincaré **18**(11), 3515–3552 (2017). https://doi.org/10.1007/s00023-017-0604-z
20. Hulse, A., Liu, H., Marvian, I.: A framework for semi-universality: semi-universality of 3-Qudit SU(d)-invariant gates. arxiv:2407.21249
21. Glaudell, A.N., Ross, N.J., van de Wetering, J., Yeh, L.: Qutrit metaplectic gates are a subset of clifford+T. In: Proceedings of the Theory of Quantum Computation Conference. Schloss Dagstuhl – Leibniz-Zentrum für Informatik. https://doi.org/10.4230/LIPICS.TQC.2022.12

22. Amaro-Alcalá, D., Sanders, B.C., de Guise, H.: Randomised benchmarking for universal Qudit gates. New J. Phys. **26**, 073052 (2024). https://doi.org/10.1088/1367-2630/ad6635

23. Ivanov, S.S., Tonchev, H.S., Vitanov, N.V.: Time-efficient implementation of quantum search with Qudits. Phys. Rev. A. Am. Phys. Soc. **85**, 062321 (2012). https://doi.org/10.1103/PhysRevA.85.062321

24. Wang, S., et al.: Proof-of-principle experimental realization of a qubit-like Qudit-based quantum key distribution scheme. Quantum Sci. Technol. **3**, 025006. IOP Publishing (2018). https://doi.org/10.1088/2058-9565/aaace4

25. Lu, Y., Ding, G.: A novel quantum security multi-party extremum protocol in a d-dimensional quantum system. Phys. Scripta **99**, 095111. IOP Publishing (2024). https://doi.org/10.1088/1402-4896/ad6aee

26. Sutradhar, K., Om, H.: An efficient simulation for quantum secure multiparty computation. Sci. Rep. (2021). https://doi.org/10.1038/s41598-021-81799-z

27. Keet, A., Fortescue, B., Markham, D., Sanders, B.C.: Quantum secret sharing with Qudit graph states. Phys. Rev. A. Am. Phys. Soc. (APS) **82**, 062315 (2010). https://doi.org/10.1103/PhysRevA.82.062315

28. Tavakoli, A., Herbauts, I., Ukowski, M., Bourennane, M.: Secret sharing with a single d-level quantum system. Phys. Rev. A. Am. Phys. Soc. **92**, 030302 (2015). https://doi.org/10.1103/PhysRevA.92.030302

29. McMahon, D.: Quantum Gates and Circuits. In: Quantum Computing Explained (2007). https://doi.org/10.1002/9780470181386.ch8

30. Mandilara, A., Dellen, B., Jaekel, U., et al.: Classification of data with a Qudit, a geometric approach. Quantum Mach. Intell. **6**, 17 (2024). https://doi.org/10.1007/s42484-024-00146-3

31. Roca-Jerat, S., Román-Roche, J., Zueco, D.: Qudit machine learning. In: Machine Learning: Science and Technology. IOP Publishing (2024). https://doi.org/10.1088/2632-2153/ad360d

32. DiVincenzo, D.P.: Quantum gates and circuits. In: Proceedings of the Royal Society of London. Series A: Mathematical, Physical and Engineering Sciences. The Royal Society (1998). https://doi.org/10.1098/rspa.1998.0159

33. Feynman, R.P.: Quantum mechanical computers. In: Optics News. Optica Publishing Group (1985). https://doi.org/10.1364/ON.11.2.000011

34. Thew, R.T., Nemoto, K., White, A.G., Munro, W.J.: Qudit quantum-state tomography. Phys. Rev. A. Am. Phys. Soc. **66**, 012303 (2002). https://doi.org/10.1103/PhysRevA.66.012303

35. Howard, M., Vala, J.: Qudit versions of the qubit $\pi/8$ gate. Phys. Rev. A. Am. Phys. Soc. **86**, 022316 (2012). https://doi.org/10.1103/PhysRevA.86.022316

36. Du, F.-F., Ren, X.-M., Ma, M., Fan, G.: Qudit-based high-dimensional controlled-not gate. Opt. Lett. **49**, 1229–1232. Optica Publishing Group (2024). https://doi.org/10.1364/OL.518336

37. Wang, Y., Hu, Z., Sanders, B.C., Kais, S.: Qudits and High-Dimensional Quantum Computing. Front. Phys. (2020). https://doi.org/10.3389/fphy.2020.589504

38. Prasad, S., Scully, M.O., Martienssen, W.: A quantum description of the beam splitter. Optics Commun. **62**, 139–145 (1987). https://doi.org/10.1016/0030-4018(87)90015-0

39. Gröblacher, S., Jennewein, T., Vaziri, A., Weihs, G., Zeilinger, A.: Experimental quantum cryptography with qutrits. New J. Phys. **8**, 75 (2006). https://doi.org/10.1088/1367-2630/8/5/075

40. van de Wetering, J., Yeh, L.: Building qutrit diagonal gates from phase gadgets. In: Electronic Proceedings in Theoretical Computer Science (2023). https://doi.org/10.4204/eptcs.394.4

41. Ghosh, S., Adhikari, A.: An efficient quantum secret sharing scheme for general access structure based on a novel partitioning technique. Quantum Inf. Process. **24**, 328 (2025). https://doi.org/10.1007/s11128-025-04949-6

42. Ghosh, S., Adhikari, A.A.: $d(\geq 2)$-level (t,n) threshold quantum secret reconstruction scheme using symmetric entanglements. Quantum Inf. Process. **24**, 64 (2025). https://doi.org/10.1007/s11128-025-04679-9
43. Adhikari, M.R., Adhikari, A.: Basic Modern Algebra with Applications. Springer, New Delhi (2014). https://doi.org/10.1007/978-81-322-1599-8

Secret Sharing and Cloud Security

Traceable Bottom-Up Secret Sharing and Law and Order on Community Social Key Recovery

Rittwik Hajra, Subha Kar, Pratyay Mukherjee, and Soumit Pal[✉]

Indian Statistical Institute, Kolkata, Baranagar, India
soumitpal378@gmail.com

Abstract. A recent work by Kate et al. [EPRINT 2025] proposes a community-based social recovery scheme (SKR), where key-owners can use a subset of other community members as guardians, and in exchange, they play guardians to support other participants' key recovery. Their construction relies on a new concept called bottom-up secret sharing (BUSS). However, they do not consider a crucial feature, called traceability, which ensures that if more than a threshold number of the guardians collude, at least some colluders' identities can be traced – thereby deterring participants from colluding. In this paper, we incorporate traceability into the community social key recovery as an important feature.

We first introduce the notion of traceable BUSS, which allows tracing colluders by accessing a reconstruction box. Then, extending the work of Boneh et al. [CRYPTO 2024], we propose the first traceable BUSS construction. Finally, we show how to generically use a traceable BUSS scheme to construct a traceable SKR in the aforementioned community setting. Overall, this is the first scheme combining decentralized key management with traceability, marrying BUSS's scalability with the deterrence of traceable secret sharing.

Keywords: Bottom-Up Secret Sharing · Tracing · Social Key Recovery

1 Introduction

Secure key management in decentralized environments – such as cryptocurrency wallets and distributed communities – has become a critical challenge. Unlike passwords, cryptographic secret keys cannot be reset or recovered if lost. The stakes are enormous: it is estimated that over USD 140 billion in Bitcoin alone is permanently inaccessible due to lost keys [40]. To mitigate the risk of irrecoverable loss, users are encouraged to back up their private keys. Traditional approaches include writing down mnemonic phrases or splitting the key via threshold secret sharing (e.g., Shamir's scheme) and distributing shares to different devices or trusted individuals. However, these methods are prone to failure: physical backup can be destroyed or forgotten, and splitting keys among

R. Dutta et al. (Eds.): INDOCRYPT 2025, LNCS 16372, pp. 417–441, 2026.
https://doi.org/10.1007/978-3-032-13301-4_18

418   R. Hajra et al.

devices or custodians introduces new security vulnerabilities. In scenarios such as the death or incapacitation of a key owner, assets can still be lost if no reliable recovery mechanism is in place.

Bottom-Up Secret Sharing and Social Key Recovery. Recently Kate et al. [31] proposed a social key recovery (SKR) mechanism to address the issue. In an SKR scheme, a user (the key owner) entrusts a set of friends or community members – referred to as *guardians* – with the ability to collectively recover her secret key if needed. Typically, the secret key is split and distributed such that any $(t+1)$ out of n guardians can reconstruct the key, while any t or fewer learn nothing. This approach leverages social trust to improve reliability: even if the user loses access to her key, any qualified subset of guardians can help restore it, preventing permanent loss. Social recovery, as advocated by Buterin [16], offers significant usability benefits without relying on centralized custodians.

Central to SKR, Kate et al. [31] proposed a novel secret sharing scheme called Bottom-Up Secret Sharing (BUSS) scheme. In a BUSS scheme, there is no centralized dealer distributing shares; instead, each guardian generates its share independently, using only its own secret information and the identity of the key-owner. Intuitively, each guardian P_j computes a share σ_j as a deterministic function of (i) P_j's personal secret key and (ii) the key-owner's identity or public key. All n guardians' shares are then implicitly associated with a secret s (the key-owner's secret) by some public binding information. In the work of Kate et al. [31], the key-owner collects the guardians' contributions to interpolate a polynomial that hides s, and publishes a small public string (e.g. evaluation points of that polynomial). Later, any $t+1$ guardians can reconstruct s by recomputing their shares locally (using their own secrets) and combining them with the public string. The BUSS paradigm eliminates the need for guardians to *store any extra data beyond their own keys* - a stark contrast to naive use of Shamir's scheme [39], which would require each guardian to hold many shares for different friends and incur linear storage blowup. In a BUSS scheme, each guardian does not store any extra share for the user; instead, shares are derived on the fly from each guardian's own long-term secret.

Traceability and Non-imputability in Secret Sharing. Goyal, Song, and Srinivasan [28] recently introduced *traceable secret sharing* in the context of classical secret sharing: in a traceable scheme, any leaked share information can be traced back to the specific parties who leaked it. In a t-out-of-n threshold secret sharing scheme, suppose $f < t$ servers collude and sell their shares, possibly in an obfuscated way such that they can not be trivially traced back to the owners. In a traceable secret sharing scheme, it should be possible to trace at least one of the corrupted servers given the leaked information. Further, the tracer should be able to produce a proof that implicates the corrupted servers.

Goyal, Song and Srinivasan [28] gave the first definition and construction of a traceable secret sharing scheme. Their construction, however, is not practical as the size of the secret shares is quadratic in the size of the secret.

Boneh, Partap and Rotem [9] propose a new definition of (n, t) threshold traceable secret sharing, which allows them to give the first practical constructions from Shamir's secret sharing and Blakley's secret sharing. In their definition, the Tracer is given black-box access to a *"reconstruction box"* that has $f < t$ shares hardcoded in it. The Tracer on input the remaining $t - f$ additional shares $\mathcal{R}$ outputs the secret that can be reconstructed from the t shares it now holds. To make the definition, they have introduced two new keys, which are given to the parties along with their shares: these are *trace key* and *verification key*. The scheme is called *traceable* if, given the tracing key, the tracer can find all f parties that own one of the shares hard-coded in $\mathcal{R}$ and produce a proof that implicates these parties. The scheme is called *non-imputable* if the tracer cannot falsely accuse a party by forging a proof of their corruptness. The authors present two schemes that satisfy traceability and non-imputability, one based on Shamir's secret sharing and one based on Blakley's [6] secret sharing scheme. The schemes are practical in the sense that the share size is only twice as large as the size of the secret.

Traceable Solution of BUSS and SKR. Prior traceable secret sharing schemes were designed for traditional top-down secret sharing. The first construction by Goyal et al. [28] achieved traceability and non-imputability but at the cost of large shares (quadratic in the secret size) and complex tracing procedures based on Goldreich–Levin hardcore predicates. Subsequent works (e.g., Boneh et al. [9]) improved efficiency, but those solutions still assume a dealer distributing shares and are not directly applicable to the decentralized BUSS context. Boneh, Partap and Rotem [9] suggested the following *"An interesting open question is to devise tracing procedures for other existing secret-sharing schemes.".* No solution currently exists for introducing traceability into bottom-up, guardian-driven secret sharing – and hence into social key recovery – in a way that preserves its practical benefits, i.e., minimal guardian burden and simple, one-round operation. Thus, the following question is natural to ask:

Is it possible to build traceable bottom-up secret sharing schemes and extend the notion of traceability in social key recovery?

Our work fills this gap by presenting the first *Traceable Bottom-Up Secret Sharing* scheme and applying it to build traceable *Social Key Recovery* systems.

1.1 Our Contribution

We present the first traceable bottom-up secret sharing schemes, building on the BUSS framework defined in [31]. Our schemes are practical as they only increase the size of the secret by a factor of 2. We summarize our key contributions as follows:

1. We introduce the concept of traceability in bottom-up secret sharing and provide a concrete Traceable Bottom-Up Secret Sharing (TBUSS) scheme that satisfies the private traceability notion of Boneh, Partap, and Rotem [9].

Notably, guardians still do not store anything beyond their own keys, and share sizes remain compact.

2. Next, we propose a Non-Imputable Traceable BUSS construction, called NITBUSS.

3. Finally, we integrate our TBUSS and NITBUSS schemes with SKR to come up with Traceable and Non-Imputable SKR, respectively. We call them TSKR ("*Traceable* SKR") and NISKR ("*Non-Imputable* SKR"). The TSKR protocol preserves the "bottom-up" nature (guardians only use their own keys and public info) while adding traceability. We also outline an upgraded variant with non-imputability, obtained by using NITBUSS in place of TBUSS in the back-up and recovery process. This gives us NISKR scheme that offers the strongest level of assurance: even in a community social recovery setting, any collusion of up to $f < t + 1$ guardians will be caught with evidence, and no honest guardian can be framed. Together, these contributions lay a new foundation for traceable social key recovery, marrying the scalability of BUSS with the deterrence of traceable secret sharing.

1.2 Technical Overview

We refer the reader to go through the work of [31] and [9] for the notions of BUSS, SKR, and Traceability is Shamir's Secret Sharing, respectively. Our framework builds in layers, each adding a new security property to a standard BUSS scheme. We first show how to make BUSS traceable (TBUSS) using random evaluation points, then strengthen it with error tolerance via Reed–Solomon list decoding. Next, we hide the randomness under one-way functions to achieve non-imputability (NITBUSS). If we incorporate TBUSS in the Share and Reconstruction algorithms of SKR, we yield a *Traceable Social key Recovery* (TSKR) primitive that enforces accountability without non-imputability. Finally, we observe that simply swapping in NITBUSS in the Share/Reconstruction algorithms yields a traceable social key recovery (TSKR) primitive that achieves non-impairability (NISKR).

Traceable BUSS. We modify the construction of BUSS and propose a TBUSS construction; this modification is due to assigning random identities to the guardians. Fixed identities will lead to an inefficient tracing procedure; hence, this modification is required. We declare these random identities as their trace keys and verification keys. Suppose $f < t + 1$ guardians are corrupt and hard-coded their shares to the reconstruction box $\mathcal{R}$. If it is a good reconstruction box, then upon receiving the trace key tk along with $t + 1 - f$ extra shares, the box should output the correct secret. It is easy to see if $f = t$ and $\mathcal{R}$ are good; in that case, simple polynomial interpolation will work. So, the shares of the form $(x_i, \sigma_i)_{i=1}^{t}$ are hard-coded in the reconstruction box $\mathcal{R}$ and upon probing $\mathcal{R}$ with two additional shares (x_{t+1}, σ_{t+1}) and $(x_{t+1}, \sigma_{t+1} + 1)$ along with the public value φ, we yield a polynomial $h(X)$ from the difference of two outputs of the reconstruction box $\mathcal{R}$ on input two secrets, respectively. The public value will contribute a factor of $h_\varphi(X)$ in $h(X)$, then by finding the roots of $g(X) := h(X) \cdot (h_\varphi(X))^{-1}$ we obtain all the corrupt guardians' identities $x_1, \ldots, x_t$ as we have assumed.

For imperfect reconstruction boxes, the analysis becomes more complex. Malicious guardians may try to evade tracing by submitting incorrect or tampered shares during an unauthorized reconstruction in the hope of "confusing" the tracing mechanism. To counter this, our TBUSS construction incorporates an error-tolerance layer using Reed–Solomon list decoding. Reed–Solomon codes are a natural fit since our secret sharing is polynomial-based (a form of Reed–Solomon code). We design the tracing algorithm to treat a suspect reconstruction as a codeword and to perform list decoding to recover the candidate sets of shares that could produce that result. Even if some colluders provide erroneous values or noise, the list-decoding procedure can still pinpoint a small list of possible guardian identities, from which the true culprits are extracted. In essence, we can tolerate a bounded number of "errors" (e.g., false shares contributed to misleading tracing) and still successfully trace all the actual participants in the reconstruction. We employ the Guruswami-Sudan [29] algorithm to obtain a list of candidate polynomials G, which is polynomial in size. We perform a search on this list by factoring every polynomial to find the first polynomial whose f roots are basically corrupt guardians' identities $x_1, x_2, \ldots, x_f$.

Adding Non-imputability. A threshold secret sharing scheme is called non-imputable if the tracer cannot falsely accuse a party by forging a proof of their corruptness. To achieve non-imputability, we hide the tracing key tk using one-way functions. In the NITBUSS scheme, the random evaluation points, i.e., the trace key components, are not published in the clear. Let $\mathcal{F} : \mathbb{F} \to \mathbb{F}$ be a one-way function and instead of $(x_1, \ldots, x_N)$ we call $(u_1, \ldots, u_N)$ to be the trace key tk, where $u_i = \mathcal{F}(x_i)$ for all $i \in [N]$. The key owner, without the guardian's secret, cannot forge a consistent share for that guardian. Thus, if the tracer claims that guardian P_i's share was used, it must be that only P_i could have generated that share. Thus, no tracer or party can falsely implicate P_i since any fake evidence would fail to align with the one-way function constraint and would be rejected by the verification algorithm. We formally prove that if the one-way function $\mathcal{F}$ is hard to invert, the probability of a false accusation is negligible. Thus, non-imputability of NITBUSS essentially boils down to the hardness of inverting $\mathcal{F}$.

Traceable and Non-Imputable SKR. We integrate the TBUSS and NITBUSS primitives into a full Social Key Recovery application. The integration is seamless: we replace the backup phase of SKR with our TBUSS Share algorithm, and we replace the recovery phase of SKR with our TBUSS Reconstruction algorithm. The resulting Traceable SKR (TSKR) and Non-Imputable SKR (NISKR) protocols preserve the workflow of the original SKR – for example, a user still only needs to perform one round of communication with guardians to back up the key, and can run the recovery algorithm whenever he wishes. The TSKR protocol preserves the "bottom-up" nature (guardians only use their own keys and public info) while adding traceability. The traceability of TSKR follows from the traceability of TBUSS. There are at most N invocations of TBUSS in a TSKR, for each party P_i for $i \in [N]$, some $f < t + 1$ corrupt guardians have hard-coded their shares in a reconstruction box $\mathcal{R}_i$. Thus, we have at most N reconstruction boxes $\mathcal{R}_1, \ldots, \mathcal{R}_N$. The tracer will have access to all these reconstruction boxes and perform the tracing mechanism for every party P_i individually to come up with the corrupt guardians for party P_i for every $i \in [N]$. Note that we have just one trace key which the parties in SKR can generate themselves. So a party P_i participating in SKR only needs to store their secret sk_i, and the shares and trace keys can be computed from this secret via a hash function. By swapping in the NITBUSS variant, we obtain a Non-Imputable Social Key Recovery (NISKR) system that further guarantees no false evidence can be manufactured against honest guardians. In summary, our technical innovations enable *provably traceable and non-imputable social key recovery*, aligning key management with real-world notions of *"law and order"*.

1.3   Related Work

Social Recovery has been an important theoretical as well as practical problem. Smart Contract Based Social Recovery [16,38], Coinbase WaaS: Key-recovery using MPC [34], Account Recovery for a Privacy-Preserving Web Service [35], off-chain backup using 2FA [1], to name a few. For a comprehensive study, we refer to the SoK [17]. Bellare, Dai, and Rogaway's work on Adept Secret Sharing [4] is a key contribution that connects Social Key Recovery with traditional Secret Sharing. Alongside the usual privacy guarantee, Adept Secret Sharing introduces two extra features: authenticity and error correction. The authenticity feature lets anyone confirm that the recovered secret is indeed the right one. To make this possible, a commitment to the secret is first stored in a trustworthy public place, and later compared against the reconstructed secret.

Chor, Fiat, and Naor [19] first introduced traitor tracing for broadcast encryption, which lets a tracer identify the source of any leaked decryption keys. Many traitor tracing methods have been developed since then [7,8,11–13,18,20,23,24,27,32,33,37,41,42], but these differ from the approach taken by Boneh, Partap, and Rotem [9] and in our work. For a comprehensive review of these techniques, refer [9].

After the inception of traceable secret sharing by Goyal, Song, and Srinivasan [28], there have been many developments towards this direction. One important milestone will be the Boneh et al.'s work [9], which simplifies the syntax and definitions and presents two important traceable secret sharing schemes: Shamir and Blakely. Recently, Hoffman [30] presented a traceable version of a Chienese Remainder Based secret sharing scheme called Mignotte secret sharing scheme [36]. The work of Boneh, Partap and Rotem [9] gives rise to several important work: Traceable Verifiable Secret Sharing (VSS) [2,26], Traceable verifiable Random Function (VRF) [10], Traceable General Access Structure based secret sharing schemes [22], Traceable Threshold Encryption [5,9,15], Secret Sharing with Snitching [14,21], to name a few.

1.4   Organization of the Paper

In Sect. 2 we develop notations and the definitions and security games, used throughout the paper. In Sect. 3, we write the Traceable variant of BUSS and the full-fledged tracing algorithm along with the security theorem. Section 4 demonstrates a non-imputable variant of BUSS. In the Sect. 5 we give a traceable version of SKR. Then we compare our tracing mechanisms with the existing literature in the Sect. 6. Finally, we conclude in Sect. 7.

2   Preliminaries

2.1   Notation

We use $\mathbb{N}$ to denote the set of positive integers. For a natural number $n \in \mathbb{N}$ we denote $[n] := \{1, 2, \ldots, n\}$. Let $\mathbb{F}$ denote a finite field of a certain order unless specified by indices. A ordered tuple $(x_1, x_2, \ldots, x_n)$ is denoted by vector notation $\boldsymbol{x}_{[n]}$. Similarly for any subset $S \in [n]$, $\boldsymbol{x}_S$ and $(x_i)_{i \in S}$ are defined accordingly. For any finite set $\mathcal{D}$, we let $x \xleftarrow{\$} \mathcal{D}$ denote picking an element of $\mathcal{D}$ uniformly at random and assigning it to x. We also use the concepts of one-way function in Sect. 4, for the definition and details reader is referred to [25].

2.2   Syntax of Bottom-Up Secret Sharing (BUSS) and Social Key Recovery (SKR)

BUSS is introduced in the work of Kate et al. [31]. Consider a set of parties $P_1, P_2, \ldots, P_N$, where each party P_i has a unique identity $i \in [N]$, which enables each party to generate its share independently of both the other parties' shares and the underlying secret. In BUSS, each share is deterministically derived from the guardian's own secret key and the identity of the key-owner, ensuring that guardians are not required to store or manage any information beyond their own secret keys. The key-owner interpolates these shares into a secret-sharing polynomial and publishes a small set of public evaluation points. During recovery, any $t + 1$ guardians can locally recompute their respective shares, and together with the public values, reconstruct the original secret. Note that in the following definition, N denotes the universe size, i.e., the total number of parties in a network, but for every party $i \in [N]$, the number of guardians is fixed, i.e., $n - 1$.

Definition 1 (Bottom-Up Secret Sharing (BUSS) [31]). *For $n, t, N \in \mathbb{N}$ such that $t + 1 \leq n - 1 < N$, a bottom-up secret sharing scheme for a $(t+1)$-out of -$(n-1)$ threshold access structure over the set $[N]$ consists of the following two algorithms:*

- Share$(1^\lambda, s, \boldsymbol{\sigma}_B, B)$: *On input s, $(n-1)$ shares $\boldsymbol{\sigma}_B$, and the corresponding set of indices $B \subseteq [N] \setminus \{i\}$ such that $|B| = n-1$. Then* Share *outputs a public value φ.*
- Recon$(\varphi, \boldsymbol{\sigma}_R, R)$: *On input φ, $(t+1)$ shares $\boldsymbol{\sigma}_R$, and the corresponding set of indices $R \subseteq [N]$ such that $|R| = t+1$. Then* Recon *outputs a secret s or $\perp$.*

Correctness For any secret s, any sets R, B such that $R \subseteq B \subset [N]$, with $|R| = t + 1, |B| = n - 1$ and any σ_B:

$$\Pr[s \leftarrow \mathsf{Recon}(\varphi, \sigma_R, R) \mid \varphi \leftarrow \mathsf{Share}(1^\lambda, s, \sigma_B, B)] = 1$$

Perfect Adaptive Simulation Security   A scheme $\mathsf{BUSS} = (\mathsf{Share}, \mathsf{Recon})$ for a $(t + 1)$-out-of-$(n - 1)$ access structure is **perfect adaptive simulation secure** if for every unbounded adversary $\mathcal{A}$, there exist simulators $\mathsf{SimShare}, \mathsf{SimComb}$ such that

$$\forall \mathcal{A}, \quad \Pr[\mathsf{RealSh}_{\mathcal{A}}(\lambda) = 1] = \Pr[\mathsf{IdealSh}_{\mathcal{A}}(\lambda) = 1]$$

- $\mathsf{RealSh}_{\mathcal{A}}$
    - Receive (s, σ_C, C, B) from $\mathcal{A}$, such that $C \subseteq B$, $|C| \leq t$, and $|B| = (n - 1)$.
    - Sample $\sigma_{B \setminus C}$ uniformly at random.
    - Run $\varphi \leftarrow \mathsf{Share}(s, \sigma_B, B)$ and give φ to $\mathcal{A}$.
    - When $\mathcal{A}$ queries (Share, C') for $C' \subseteq B \setminus C$: Update $C := C \cup C'$. If $|C| > t$, abort, else, give $\sigma_{C'}$ to $\mathcal{A}$.
    - Finally, give $\sigma_{B \setminus C}$ to $\mathcal{A}$.
    - Output whatever $\mathcal{A}$ returns.
- $\mathsf{IdealSh}_{\mathcal{A}}$
    - Receive (s, σ_C, C, B) from $\mathcal{A}$, such that $C \subseteq B$, $|C| \leq t$, and $|B| = (n - 1)$.
    - Run $(\varphi, \tau) \leftarrow \mathsf{SimShare}(C)$ and give φ to $\mathcal{A}$.
    - When $\mathcal{A}$ queries (Share, C') for $C' \subseteq B \setminus C$:Update $C := C \cup C'$ and set $\sigma_{C'} := \tau_{C'}$. If $|C| > t$, abort, Else, give $\sigma_{C'}$ to $\mathcal{A}$.
    - Compute $\sigma_{B \setminus C} \leftarrow \mathsf{SimComb}(s, B, C, \sigma_C, \varphi)$ and give it to $\mathcal{A}$.
    - Output whatever $\mathcal{A}$ returns.

In the following, we revisit the SKR definition from [31]. In a decentralized community, each individual held a private key—generated securely through a process called KeyGen—that unlocked access to their digital assets. But what if someone lost their key? Rather than relying on fragile memories or risky storage, the community adopted a social recovery scheme. In this system, no one needed to remember anything beyond their own secret key. When a person backed up their key using the protocol Π_{Back}, the only result was a piece of public information—no hidden shards, no sensitive data passed around. The guardians, friends, and peers in the network didn't bear the burden of remembering secret pieces of someone else's life. It was simple and elegant. Though built for a common key generation process, the system was flexible enough to evolve, even if each person used a different method to create their key. Trust stayed local, responsibility stayed minimal, and recovery remained possible—together

Definition 2 (Social Key Recovery Scheme (SKR)[31]). *Consider a set of N parties $P_1, \ldots, P_N$, and let $\mathcal{U}$ be an access structure defined by pairs of sets (B, R), where $B \subseteq [N]$ and $R \subseteq B$. Without loss of generality, we assume that each party P_i is associated with a public identity-i. Let KeyGen denote a key generation algorithm that outputs a key pair $(\mathsf{sk}, \mathsf{pk}) \leftarrow \mathsf{KeyGen}(1^\lambda)$. A social key recovery scheme Π_{SKR} for the parties $P_1, \ldots, P_N$, the algorithm KeyGen, and the access structure $\mathcal{U}$ consists of three protocols $(\Pi_{\mathsf{Init}}, \Pi_{\mathsf{Back}}, \Pi_{\mathsf{Rec}})$ executed sequentially by subsets of the parties as follows: all parties first run Π_{Init}; subsequently, any party may initiate a single execution of Π_{Back} with a chosen subset $B \subseteq [N]$; and finally, once P_i has completed Π_{Back}, it may invoke Π_{Rec} any number of times. The syntax of the protocols is defined as follows:*

- Π_{Init}: *In this protocol, a party P_i locally generates a key pair $(\mathsf{sk}_i, \mathsf{pk}_i)$ either by computing $(\mathsf{sk}_i, \mathsf{pk}_i) \leftarrow \mathsf{KeyGen}(1^\lambda)$ or using another algorithm specified by the protocol. Each party P_i publishes pk_i. An execution is denoted by:*

$$(\mathbf{pk}_{[N]}, \mathbf{sk}_{[N]}) \leftarrow \Pi_{\mathsf{Init}}(1^\lambda, 1^N)$$

- Π_{Back}: *In this protocol, a key owner P_i who wishes to back up her secret key sk_i interacts with a set of guardians $\{P_j\}_{j \in B}$. Each guardian P_j uses secret key sk_j. The protocol concludes with a public backup string pub_i. We denote such an execution by:*

$$\mathsf{pub}_i \leftarrow \Pi_{\mathsf{Back}}(i, B, \mathbf{pk}_B, \mathbf{sk}_{B \cup \{i\}})$$

- Π_{Rec}: *If P_i wishes to recover sk_i using a recovery set R, she runs this protocol (without any secret input) with a set of guardians $\{P_j\}_{j \in R}$, each of which uses their secret key sk_j. In addition, pub_i may be used by all parties. At the end of this protocol, the key-owner may receive a private output sk_i (or $\perp$ if unsuccessful). One such execution is denoted as:*

$$\mathsf{sk}_i / \perp \leftarrow \Pi_{\mathsf{Rec}}(i, R, \mathbf{pk}_R, \mathsf{pub}_i, \mathbf{sk}_R)$$

Correctness. *For correctness we require that for any sufficiently large λ, any $i \in [N]$, any $(B, R) \in \mathcal{U}$:*

$$\Pr\left[\mathsf{sk}_i \leftarrow \Pi_{\mathsf{Rec}}(i, R, \mathbf{pk}_R, \mathsf{pub}_i, \mathbf{sk}_R) \;\middle|\; \begin{array}{l} (\mathbf{pk}_{[N]}, \mathbf{sk}_{[N]}) \leftarrow \Pi_{\mathsf{Init}}(1^\lambda, 1^N); \\ \mathsf{pub}_i \leftarrow \Pi_{\mathsf{Back}}(i, B, \mathbf{pk}_B, \mathbf{sk}_{B \cup \{i\}}) \end{array} \right] = 1$$

2.3 Syntax of Traceable Bottom-Up Secret Sharing Scheme (TBUSS)

TBUSS combines the scalability of decentralized social key recovery with the accountability of traceable secret sharing. In BUSS, each user derives backup shares for others using only their own secret key, avoiding additional storage and enabling efficient $t + 1$-out-of-$n - 1$ recovery with the help of publicly posted values. By integrating traceability, where leaked shares or reconstruction mechanisms can be traced back to the responsible parties. We mostly follow the definitions in [9]. However, we will tune their notions accordingly to cater BUSS scheme, as BUSS is not perfectly private. Although it has been shown that BUSS is Perfect Adaptive Simulation Secure [31]. But for our purpose of Tracing, we will not need those security notions.

Definition 3 (TBUSS). *For parameters $n, t, N \in \mathbb{N}$ where $t + 1 \leq n - 1 < N$, a TBUSS scheme for a $(t+1)$-out-of-$(n-1)$ threshold access structure over $[N]$ is a 4-tuple of algorithms* TBUSS $=$ (Share, Recon, Trace, Verify):

- Share$(1^\lambda, s, \boldsymbol{\sigma}_B, B) \rightarrow (\varphi, \mathsf{tk}, \mathsf{vk})$: *On input s, $(n-1)$ shares $\boldsymbol{\sigma}_B$, and the corresponding set of indices $B \subseteq [N] \backslash \{i\}$ such that $|B| = n - 1$. Then outputs a public value φ and tracing key* tk, *and verification key* vk.
- Recon$(\varphi, \boldsymbol{\sigma}_R, R) \rightarrow s$ *or* $\perp$: *On input φ, $(t+1)$ shares $\boldsymbol{\sigma}_R$, and the corresponding set of indices $R \subseteq [N]$ such that $|R| = t + 1$. Then* Recon *outputs a secret s or $\perp$.*
- Trace$^{\mathcal{R}}(\mathsf{tk}) \rightarrow (I, \pi)$: *It is a randomized algorithm that takes the tracing key* tk *and oracle access to a reconstruction box $\mathcal{R}$. It outputs a subset $I \subseteq [n-1]$ of corrupted parties and a proof π.*
- Verify$(\mathsf{vk}, I, \pi) \rightarrow \{0, 1\}$ *is a deterministic algorithm that takes the verification key* vk, *the set I, and proof π, and returns 1 if the proof is valid and 0 otherwise.*

Correctness. The correctness requirement for a TBUSS is same as BUSS, any set of $t + 1$ shares with the public value φ must be sufficient to correctly reconstruct the original secret. In other words, the reconstruction algorithm should output the correct secret whenever it is given any $t + 1$-tuple of valid shares produced by the sharing algorithm.

Definition 4 (ϵ-**Correctness**). *Let* TBUSS $=$ (Share, Recon, Trace, Verify) *be a Traceable Bottom-Up Secret Sharing scheme and let $\epsilon \in [0, 1]$ be a function of security parameter λ. We say that* TBUSS *is ϵ-correct if for every $\lambda \in \mathbb{N}$, every secret s every $0 < t + 1 \leq n$ and every pairs of sets (B, R), where $B \subseteq [N]$ and $R \subseteq B$ and $|B| = n - 1, |R| = t + 1$, it holds that*

$$\Pr\left[s' = s\right] \geq 1 - \epsilon.$$

where $(\varphi, \mathsf{tk}, \mathsf{vk}) \leftarrow$ Share$(1^\lambda, s, \boldsymbol{\sigma}_B, B)$ and $s' :=$ Recon$(\varphi, \boldsymbol{\sigma}_R, R)$,

In the context of traceable secret-sharing schemes, it is natural to strengthen the standard secrecy requirement by demanding that the tracing algorithm itself does not compromise the confidentiality of the secret. Specifically, the tracing key tk, which is designed to facilitate the identification of corrupted parties, should not reveal any additional information about the shared secret. Even if an adversary gains access to both the tracing key and fewer than $(t + 1)$ shares, they should still be unable to distinguish between different possible secrets. This intuition is formally captured in the definition of tracer secrecy provided below.

We define a Traceability game in Fig. 1. The following definition will give us the traceability advantage.

Definition 5 (Traceability). *Let* TBUSS $=$ (Share, Rec, Trace, Verify) *be a Traceable Bottom-Up Secret Sharing Scheme. Let* $\varepsilon = \varepsilon(\lambda)$ *be a function of the security parameter. We say that* TBUSS *satisfies traceability if for every PPT adversary* $\mathcal{A}$, *the following function is negligible in* λ *(See Fig. 1):*

$$\mathsf{Adv}^{\mathsf{trace}}_{\mathcal{A},\mathsf{TBUSS},\varepsilon}(\lambda) := \Pr\left[\mathsf{Game.Trace}_{\mathcal{A},\mathsf{TBUSS},\varepsilon}(\lambda) = 1\right].$$

The above definition states that, for every probabilistic polynomial time adversary, $\mathcal{A}$, the probability that it wins the game $\mathsf{Game.Trace}_{\mathcal{A},\mathsf{TBUSS},\varepsilon}(\lambda)$ defined in Fig. 1 is negligible in λ.

$\mathsf{Game.Trace}_{\mathcal{A},\mathsf{TBUSS},\varepsilon}(\lambda)$

1. $\mathcal{A}(1^{\lambda}, n-1, t+1, f)$ outputs $(\mathcal{I}, \mathsf{state})$, where $\mathcal{I} \subset [n-1]$ and $|\mathcal{I}| = f(< t+1)$ is the set of parties to corrupt.
2. $s \xleftarrow{\$} \mathcal{S}$ and $\boldsymbol{\sigma}_B \xleftarrow{\$} |\mathbb{F}|^{|B|}$.
3. $(\varphi, \mathsf{tk}, \mathsf{vk}) \leftarrow \mathsf{Share}(s, \boldsymbol{\sigma}_B, B)$
4. On input all shares of parties in $\mathcal{I}$, the adversary $\mathcal{A}(\mathsf{state}, \sigma_{i_1}, \ldots, \sigma_{i_f})$ outputs a reconstruction box $\mathcal{R}$.
5. $\mathsf{Trace}^{\mathcal{R}}(\mathsf{tk})$ outputs $(\mathcal{I}', \pi)$.
6. $\mathcal{A}$ wins if: $\mathcal{R}$ reconstructs the secret from consistent inputs with probability at least ε, i.e., $\mathcal{R}$ is good and Either $\mathcal{I} \neq \mathcal{I}'$ or $\mathsf{Verify}(\mathsf{vk}, \mathcal{I}', \pi) = 0$.

Fig. 1. The tracing game for traceable bottom-up secret sharing BUSS

In the setting of TBUSS schemes, it is essential to ensure the property of non-imputability, which guarantees that honest participants cannot be falsely accused of leaking information. This security notion asserts that even a malicious tracer (possibly attempting to manipulate the tracing mechanism) should not be able to generate a convincing proof that implicates an honest party who did not participate in any wrongdoing. In other words, the scheme must be robust against forgery of tracing evidence, thereby protecting honest users from being unjustly held accountable.

Definition 6 (Non-Imputability). *A* TBUSS $=$ (Share, Rec, Trace, Verify) *is said to be satisfied non-imputability if for every PPT adversary* $\mathcal{A}$, *the probability that it wins the game* $\mathsf{GNon\text{-}Imputability}_{\mathcal{A},\mathsf{TBUSS}}(\lambda)$ *is negligible in* λ:

$$\mathsf{Adv}^{\mathsf{ni}}_{\mathcal{A},\mathsf{NITBUSS}}(\lambda) := \Pr\left[\mathsf{GNon\text{-}Imputability}_{\mathcal{A},\mathsf{TBUSS}}(\lambda) = 1\right]$$

The above definition states that, for every PPT adversary $\mathcal{A}$ the probability that it wins the game $\mathsf{GNon\text{-}Imputability}$ is negligible in λ.

3   Traceability in BUSS

In this section, we present a traceable version of the BUSS scheme, and by abuse of notation, we call it TBUSS. We then present a tracing algorithm for TBUSS (Fig. 2).

GNon-Imputability$_{\mathcal{A},\mathsf{TBUSS}}(\lambda)$

1. $\mathcal{A}(1^\lambda, n, t) \to (i^*, s, B, \mathsf{state})$.
2. $\mathsf{Share}(s, B, \boldsymbol{\sigma}_B) \to (\varphi, \mathsf{tk}, \mathsf{vk})$.
3. On input all shares *except* for the i^*-th one and the keys tk and vk,
   $\mathcal{A}(\mathsf{state}, \boldsymbol{\sigma}_B \setminus \{\sigma_{i*}\}, \mathsf{tk}, \mathsf{vk})$ outputs $(\mathcal{I}^*, \pi)$.
4. $\mathcal{A}$ wins if $i^* \in \mathcal{I}^*$ and $\mathsf{Verify}(\mathsf{vk}, \mathcal{I}^*, \pi) = 1$.

Fig. 2. The non-imputability game for traceable bottom-up secret sharing TBUSS.

3.1   Construction of TBUSS

In BUSS, assigning fixed evaluation points (e.g., $x_i = i$) lets a malicious reconstructor $\mathcal{R}$ identify which party a share came from and reject queries involving corrupted parties. For example, if f parties are corrupted, $\mathcal{R}$ can output $\bot$ unless it receives exactly $t - f$ honest shares, making successful tracing negligible unless queries are exponential in n. To avoid this, each party is assigned a secret, random evaluation point $x_i \xleftarrow{\$} \mathbb{F}$, breaking the link between shares and party identities. Provided the field $\mathbb{F}$ is large enough, the probability of a collision $x_i = x_j$ is negligible, and any resulting correctness error can be further reduced via rejection sampling. This simple change is essential to enable efficient black-box tracing.

In this section, we provide a construction of a TBUSS, for which we borrow the underlying BUSS construction of [31]. To enable efficient tracing, we modify the original BUSS construction by replacing the fixed evaluation point with a randomly chosen one. Specifically, in the Share algorithm of the standard BUSS, each party is identified by an index i and the share is defined as $q(i) = \sigma_i$ for a suitable polynomial f. In contrast, in our traceable variant TBUSS, we assign to each party a random field element $x_i \xleftarrow{\$} \mathbb{F}$ and set $q(x_i) = \sigma_i$, we assume that $x_i \neq x_j$ for all $1 \leq i < j \leq (n-1)$. This randomization facilitates tracing while preserving the original scheme's structure. Let us construct the following TBUSS scheme for a $(t+1)$-out of-$(n-1)$ access structure over a finite field $\mathbb{F}$ in the Fig. 3. Correctness and Perfect Adaptive Simulation Security notions in our scheme TBUSS, described in Fig. 3, will follow from the work of Kate et al. [31].

Share(1^λ, s, $\boldsymbol{\sigma}_B$, B)

1. Define a polynomial q over $\mathbb{F}$ of degree $(n-1)$ such that:
   - $q(0) := s$
   - Parse $\boldsymbol{\sigma}_B \to \{(x_j, \sigma_j)\}_{j \in B}$
   - For all $j \in B$ and set $q(x_j) := \sigma_j$
2. Sample $x_{-1}, x_{-2}, \ldots, x_{-(n-t-1)} \xleftarrow{\$} \mathbb{F} \setminus \{x_j\}_{j \in B}$.
3. Compute $\varphi := (\varphi_1, \varphi_2, \ldots, \varphi_{n-t-1})$, where $\varphi_i = (x_{-i}, q(x_{-i}))$.
4. For all $j \in B$, set $\mathsf{tk}_j \leftarrow x_j$ and $\mathsf{vk}_j \leftarrow x_j$, and denote $\mathsf{tk} = (\mathsf{tk}_1, \mathsf{tk}_2, \ldots, \mathsf{tk}_{n-1})$ and $\mathsf{vk} = (\mathsf{vk}_1, \mathsf{vk}_2, \ldots, \mathsf{vk}_{n-1})$.
5. Output $(\varphi, \mathsf{tk}, \mathsf{vk})$

Recon(φ, $\boldsymbol{\sigma}_R$, R)

1. Parse φ as $(\varphi_1, \varphi_2, \ldots, \varphi_{n-t-1})$ where each $\varphi_i = (x_{-i}, q(x_{-i}))$.
2. Concatenate these $(n-t-1)$ public points with the $(t+1)$ shares $\{(x_j, \sigma_j)\}_{j \in R}$.
3. Use Lagrange interpolation over these n points to compute the unique polynomial q of degree $(n-1)$
4. Output $s := q(0)$

Fig. 3. The share and reconstruction algorithm of TBUSS.

3.2 Tracing via Polynomial Interpolation

For tracing purposes, we assume that the tracer has access to a reconstruction oracle, denoted $\mathcal{R}$, which allows query-based interaction. Suppose there are $f < t + 1$ corrupted parties whose shares have been hard-coded into $\mathcal{R}$ by the adversary. The tracer can then submit $t - f + 1$ honest shares along with the public value φ to $\mathcal{R}$, which responds with the reconstructed secret s. For simplicity, let us consider the case where t corrupted parties are controlled by the adversary, i.e., $f = t$. In this setting, the reconstruction oracle $\mathcal{R}$ receives an additional share (beyond those hardcoded) along with the public value φ as input, and returns the reconstructed secret s. This secret corresponds to the value $r(0)$ of a polynomial r, reconstructed using $(n - t - 1)$ public evaluations φ, the t hardcoded evaluations of r embedded in $\mathcal{R}$, and the one additional evaluation provided as input. Suppose the shares hardcoded in the reconstruction oracle $\mathcal{R}$ are of the form $(x_i, \sigma_i = q(x_i))_{i=1}^{t}$. Additionally, the public evaluations are given by $(x_j, \sigma_j = q(x_j))_{j=-(n-t-1)}^{-1}$ following a technique of using negative indices stated in [3], and the tracer provides one additional share (x_{t+1}, σ_{t+1}) as input. Given these inputs, the reconstruction oracle $\mathcal{R}$ is expected to output the secret s such that $s = q(0)$, where q is the unique polynomial of degree $(n-1)$ that interpolates all the provided points. In terms of Lagrange Interpolation, we will

arrive at the following expression: $s = \sum_{\substack{i=-(n-t-1) \\ i \neq 0}}^{t+1} \left(\prod_{\substack{k=-(n-t-1) \\ k \neq i, 0}}^{t+1} \frac{x_k}{x_k - x_i} \right) \sigma_i$.

Otherwise, this is not a good reconstruction box. Upon feeding the reconstruction box with $\mathcal{R}$ with (x_{t+1}, σ_{t+1}) and $(x_{t+1}, \sigma_{t+1} + 1)$ to yield s and s' respectively. Hence, we obtain, The above expression can be re-written as the following:

$$s = \sum_{\substack{i=-(n-t-1) \\ i \neq 0}}^{t} \left(\prod_{\substack{k=-(n-t-1) \\ k \neq i, 0}}^{t+1} \frac{x_k}{x_k - x_i} \right) \sigma_i + \left(\prod_{\substack{k=-(n-t-1) \\ k \neq 0}}^{t} \frac{x_k}{x_k - x_{t+1}} \right) \sigma_{t+1}$$

Now, if we feed the reconstruction box $\mathcal{R}$ with the share $(x_{t+1}, \sigma_{t+1} + 1)$, for the same x_{t+1} and σ_{t+1} as before, we obtain the following expression as before, $s' =$

$$\sum_{\substack{i=-(n-t-1) \\ i \neq 0}}^{t} \left(\prod_{\substack{k=-(n-t-1) \\ k \neq i, 0}}^{t+1} \frac{x_k}{x_k - x_i} \right) \sigma_i + \left(\prod_{\substack{k=-(n-t-1) \\ k \neq 0}}^{t} \frac{x_k}{x_k - x_{t+1}} \right) (\sigma_{t+1} + 1).$$

Now, subtracting the last two equations, we obtain the following expression,

$$(s' - s)^{-1} = \prod_{\substack{k=-(n-t-1) \\ k \neq 0}}^{t} \left(\frac{x_k - x_{t+1}}{x_k} \right)$$ Now, let us consider the polynomial

$$h(X) = \prod_{\substack{k=-(n-t-1) \\ k \neq 0}}^{t} \left(\frac{x_k - X}{x_k} \right),$$ where the variable is X.

Notice that the polynomial h has as its roots exactly the values x_i of the corrupted parties, together with the public points. Since we only need to identify the corrupted parties, we can ignore the public roots and keep the rest. Observe that the equation of s' represents an evaluation of h at the new point x_{t+1} provided to $\mathcal{R}$. If we repeat this procedure with t additional fresh points x_{t+1}, we obtain $t+1$ distinct evaluations of h. Because $\deg(h) = t$, these $t+1$ samples suffice to interpolate h uniquely. Once h is recovered, we factor it to reveal all its roots. Since the tracer's key tk already contains every party's true x_i, each root can be matched back to the corresponding corrupted party.

Although the full degree of h is actually $n - 1$, one can first simplify the expression for $h(X)$ before performing interpolation and factorization. $h(X) =$

$$\prod_{k=-(n-t-1)}^{-1} \left(\frac{x_k - X}{x_k} \right) \cdot \prod_{k=1}^{t} \left(\frac{x_k - X}{x_k} \right) = h_\varphi(X) \cdot \prod_{k=1}^{t} \left(\frac{x_k - X}{x_k} \right)$$

Note that the polynomial $h_\varphi(X)$ is public as φ is a public value and thus h_φ can be constructed publicly. Hence we arrive at the following polynomial,

$$g(X) := (s' - s)^{-1} \cdot (h_\varphi(X))^{-1} := \prod_{k=1}^{t} \left(\frac{x_k - X}{x_k} \right).$$

3.3   Tracing Imperfect Reconstruction Boxes

In the previous section, we assumed that the reconstruction oracle $\mathcal{R}$ is good, i.e., it always returns the correct secret. However, this assumption does not hold in general. In reality, the reconstruction box may not always produce the correct output. We refer to such a scenario as an imperfect reconstruction box, where $\mathcal{R}$ returns the correct reconstructed secret only with some non-negligible probability. This imperfection arises because standard polynomial interpolation can fail or yield an incorrect result when some of the provided evaluation points are erroneous. To address this issue, we observe that the task of reconstructing a polynomial of bounded degree from a set of possibly corrupted evaluation points is closely related to the list decoding problem for Reed-Solomon codes [29].

Definition 7 (The Reed-Solomon List Decoding Problem). *Let $\mathbb{F}$ be a finite field, and let $k, M, C \in \mathbb{N}$ such that $C \leq M < |\mathbb{F}|$. Given k, C and M pairs of elements $\{(x_i, \sigma_i)\}_{i=1}^{M} \in \mathbb{F}^2$, output a list G of univariate polynomials of degree at most k that agree with at least C pairs $\{(x_i, \sigma_i)\}$. In particular,*

$$G = \{g \in \mathbb{F}[X] \mid deg(g) \leq k \ \wedge \ |\{j \in [M] \mid g(x_j) = \sigma_j\}| \geq C\}$$

Theorem 1 (Guruswami-Sudan Algorithm) [29]). *Let $\mathbb{F}$ be a finite field, and let $k, M, C \in \mathbb{N}$ such that $C \leq M < |\mathbb{F}|$ and $C \geq \sqrt{kM}$. Then, there exists an algorithm $\mathcal{D}_{\mathsf{GS}}$ that solves the list decoding problem as defined in Definition 7 in time polynomial in M, k and $\log |\mathbb{F}|$.*

Since the runtime of the algorithm is polynomial, the list G it outputs will be of polynomial length. We denote $\tau(M, k, \log |\mathbb{F}|)$ to be the polynomial bounding the list length, i.e., $|G| \leq \tau$, where the input to the τ polynomial is understood from the context.

3.4 Tracing Algorithm

Our enhanced tracing procedure starts by gathering the evaluations of the polynomial $g(X)$ as outlined previously. Rather than performing exact interpolation, we apply a list-decoding algorithm to produce several candidate polynomials. We then factor each candidate in turn and verify whether all of its roots appear in the true set of x_i values held by the parties. Almost certainly, just one of these candidates will meet that requirement—its roots reveal the identities of the corrupt parties. Below, we present the formal description of this algorithm, which depends on the Guruswami–Sudan parameters M and C. We'll explain how to choose these values later in the section.

Theorem 2. *Let TBUSS be the Traceable Bottom-Up Secret Sharing scheme defined in Sect. 3.1. For every adversary $\mathcal{A}$, security parameter $\lambda \in \mathbb{N}$, parameters $M, C \in \mathbb{N}$, and $\epsilon \in \left[\sqrt{\frac{2(M+n+t^2-ft)}{p}}, 1\right]$ with $\epsilon > \sqrt{2C/M}$ and $\sqrt{fM} \leq C \leq M < p$, we have:*

$$\mathsf{Adv}^{\mathsf{trace}}_{\mathcal{A}, \mathsf{TTSS}, \varepsilon}(\lambda) \leq e^{\frac{-\epsilon^2 M}{2} \cdot (1-1/r)^2} + \frac{f \cdot (n - f - 1) \cdot \tau}{p}$$

where $|\mathbb{F}| = p$ (field size), $r = \dfrac{\epsilon^2 M}{2C}$, $n = n(\lambda)$ (size of guardian set $|B| = n-1$), $f = f(\lambda)$ (number of corruptions, $f < t+1$), $\tau = \tau(M, f, \log p)$ (list size from Guruswami-Sudan Algorithm $\mathcal{D}_{\mathsf{GS}}$).

The proof of Theorem 2 is involved and mathematically rich. While proving the Theorem 2, we had to take care of some new bad events that did not appear in the proof of Theorem 2 in the work of [9]. The detailed proof with careful analysis of these new bad events is discussed in the full version.

Learning f. If the tracing algorithm does not know the number of corrupted parties f^*, it tests values starting from $f = t - 1$ downward until successful. If it guesses $f > f^*$, any identified set of size f necessarily includes an honest party; byTheorem 2, this occurs with probability at most $\frac{f(n-f-1)\tau}{p} \leq \frac{n^2\tau}{p}$. For the correct guess $f = f^*$, the failure probability is at most $e^{-\frac{\epsilon^2 M}{8}} + \frac{n^2\tau}{p}$, which further reduces to $e^{-\lambda} + \frac{n^2\tau}{p}$ since $M \geq 8\lambda/\epsilon^2$. By a union bound, the overall probability that the tracing algorithm identifies the correct corrupted set is at least $1 - (e^{-\lambda} + 2n^3\tau/p)$ (Fig. 4).

Algorithm: $\mathsf{Trace}^{\mathcal{R}}_{\mathsf{TBUSS}}(\mathsf{tk}, f, 1^{1/\varepsilon})$

1. Parse tk as $(x_1, x_2, \ldots, x_{n-1})$
2. For $\ell = 1$ to M do:
   - For $i = f+1, \ldots, t$: sample $x_{\ell,i}, \sigma_{\ell,i} \overset{\$}{\leftarrow} \mathbb{F}$ and set $y_{\ell,i} \leftarrow (x_{\ell,i}, \sigma_{\ell,i})$.
   - Sample $x'_\ell, \sigma'_\ell, \delta_\ell \overset{\$}{\leftarrow} \mathbb{F}$. Let $y'_\ell \leftarrow (x'_\ell, \sigma'_\ell)$ and $y''_\ell \leftarrow (x'_\ell, \sigma'_\ell + \delta_\ell)$.
   - Query $\mathcal{R}$ on $(y'_\ell, y_{\ell,f+1}, \ldots, y_{\ell,t})$ and $(y''_\ell, y_{\ell,f+1}, \ldots, y_{\ell,t})$ to obtain s_ℓ and s'_ℓ respectively. This computation involves the public value φ.
   - If $s_\ell = s'_\ell$ or $\delta_\ell = 0$ or $x_{\ell,i} = x'_\ell$ for some $i \in \{f+1, \ldots, t\}$, or if $\ell > 1$, or $x'_\ell = x'_i$ for some $i \in [\ell - 1]$, and $h_\varphi(x'_\ell) = 0$ then terminate and output $\perp$. Otherwise, let $z_\ell \leftarrow \dfrac{\delta_\ell}{(s'_\ell - s_\ell) \cdot h_\varphi(x'_\ell)} \times$ $\left(\prod_{i=f+1}^{t} \dfrac{x_{\ell,i}}{x_{\ell,i} - x'_\ell}\right)$
3. Let $L = \{x'_j, z_j\}_{j \in [M]}$. Run the list decoding algorithm $\mathcal{D}_{\mathsf{GS}}$ on the following: the degree bound f, the integer C and the list L of pairs in $\mathbb{F}^2$. $\mathcal{D}_{\mathsf{GS}}$ returns a list of all candidate polynomials $G = \{g_i\}$ of degree at most f that agree with at least C evaluation points out of L.
4. For each $g_i \in G$ find all roots $w_{i,1}, \ldots, w_{i,f}$ of g_i. If for any $j \in [f]$ there is no k such that $w_{i,j} = x_k$, then remove g_i from the list G. If G is empty, then terminate and output $\perp$.
5. Let g_{j^*} be the first polynomial in G, let $\mathcal{I}$ be the set of indices $i \in [n-1]$ for which $x_k = w_{j^*,k}$ for some $k \in [f]$. Set $\pi \leftarrow \{x_i\}_{i \in \mathcal{I}}$.
6. Output $(\mathcal{I}, \pi)$

Fig. 4. The Tracing Algorithm for the Traceable Bottom-Up Secret Sharing TBUSS.

4   Non-imputability of **TBUSS**

Let us recall the Non-Imputability game defined in Sect. 2.3. To achieve non-imputability in Traceable Bottom-Up Secret Sharing (TBUSS), we enhance the scheme to prevent malicious tracers from falsely accusing honest parties. The original construction (Definition 3 and Sect. 3.1) exposes the random field elements x_j in the tracing key tk and verification key vk, allowing adversaries to trivially impute any honest party by including their x_j in a forged proof. To address this, we use a one-way function $\mathcal{F} : \mathbb{F} \to \mathbb{F}$. Intuitively, we "hide" each party's random evaluation point x_j behind a one-way function, yet still allow any extracted root to be linked back to its owner. The enhanced scheme, Non-Imputable TBUSS (NITBUSS), modifies the algorithms as follows.

4.1 Modified Algorithms for NITBUSS

In the following, we describe the Share, Trace and Verify algorithms and skip the Recon algorithm as it is easy to write following the Share algorithm.

1. Share($s, \boldsymbol{\sigma}_B, B$)
   - Define a polynomial q of degree $n - 1$ over $\mathbb{F}$ such that
     - $q(0) = s$
     - Parse $\boldsymbol{\sigma}_B \leftarrow \{(x_j, \sigma_j)\}_{j \in B}$
     - Sample $x_{-1}, x_{-2}, \ldots, x_{-(n-t-1)} \xleftarrow{\$} \mathbb{F} \setminus \{x_j\}_{j \in B}$.
     - For all $j \in B$ and set $q(x_j) := \sigma_j$
   - Compute $\varphi := (\varphi_1, \varphi_2, \ldots, \varphi_{n-t-1})$, where $\varphi_i = (x_{-i}, q(x_{-i}))$.
   - For each $j \in [n-1]$ (corresponding parties in B):
     - Compute $u_j = \mathcal{F}(x_j)$
   - Set tk $= (u_1, u_2, \ldots, u_{n-1})$ and vk $= (u_1, u_2, \ldots, u_{n-1})$.
   - Output $(\varphi, \text{tk}, \text{vk})$
2. Trace$_{\text{NITBUSS}}^{\mathcal{R}}(\text{tk}, 1^{1/\epsilon})$
   - Parse tk as $(u_1, u_2, \ldots u_n)$.
   - Run the Trace algorithm to obtain a list of candidate polynomials $G = \{g_j\}_j$ from list decoding. For each candidate polynomial g_j, do:
     - Factor g_j to extract roots $w_1, \ldots, w_f$.
     - For each root w_i:
       * Compute $u_i' = \mathcal{F}(w_i)$
       * If $u_i' \notin \{u_1, \ldots, u_f\}$, discard g_j.
   - If candidates remain, take the first valid $g^* \in G$ with roots $w_1^*, \ldots, w_f^*$.
   - For each w_i^*, find the index $k \in [n-1]$ such that $\mathcal{F}(w_i^*) = u_k$ and add k to $\mathcal{I}$.
   - Set $\pi \leftarrow (w_1^*, \ldots, w_f^*)$.
   - Output $(\mathcal{I}, \pi)$.
3. Verify($\text{vk}, \mathcal{I}, \pi$)
   - Parse vk $= (u_1, u_2, \ldots, u_{n-1})$, $\mathcal{I} = \{i_1, i_2, \ldots, i_f\} \subset [n-1]$, and $\pi = (w_1, \ldots, w_f)$
   - For each $k \in [f]$:
     - Check $\mathcal{F}(w_k) = u_{i_k}$
   - Output 1 if all checks pass; else, output 0.

The following theorem formally establishes the advantage of non-imputability of NITBUSS. We omit the proof and include it in the full version of this paper.

Theorem 3 (Non-Imputability of NITBUSS). *Let $\mathcal{F}$ be a one-way function. For every PPT adversary $\mathcal{A}$ winning the non-imputability game for NITBUSS, there exists a PPT algorithm (inverter) $\mathcal{B}$ such that for every $\lambda \in \mathbb{N}$ it holds that,*

$$\mathsf{Adv}^{\mathsf{ni}}_{\mathcal{A},\mathsf{NITBUSS}}(\lambda) = \Pr[\mathcal{F}(\mathcal{B}(\mathcal{F}(x^*))) = \mathcal{F}(x^*)]$$

where $x \xleftarrow{\$} \mathbb{F}$ and the probability is also over the random coins of $\mathcal{B}$.

5 Traceable Social Key Recovery

We propose a traceable version of the SKR scheme. The tracer has access to reconstruction boxes $\mathcal{R}$, which produce outputs upon receiving inputs in a specified format. In a $(t+1)$-out-of-$(n-1)$ scheme, if there are $f < t+1$ corruptions, the adversary may have hard-coded f inputs into the reconstruction box. In such a case, the tracer must provide the remaining $t + 1 - f$ valid inputs to complete the reconstruction and obtain the output. Suppose there are N parties, and let $\mathcal{U}$ be an access structure consisting of pairs of sets (B, R) such that $B \subseteq [N]$ and $R \subseteq B$ with $|B| = n - 1$ and $|R| = t + 1$. The TSKR scheme consists of three protocols, i.e., $\mathsf{TSKR} = (\Pi_{\mathsf{Init}}, \Pi_{\mathsf{Back}}, \Pi_{\mathsf{Rec}})$. In the Π_{Init} protocol, each party runs a key generation algorithm to produce a key pair $(\mathsf{sk}_i, \mathsf{pk}_i)$. In the Π_{Back} protocol, each party backs up their public value φ_i. Once all parties have executed both the Π_{Init} and Π_{Back} protocols, they can invoke the recovery protocol Π_{Rec} any number of times to reconstruct the secret. Note that Π_{Back} and Π_{Rec} invoke TBUSS.Share and TBUSS.Recon, respectively.

Traceable Social Key Recovery (TSKR)

$\Pi_{\mathsf{Init}}(1^\lambda, 1^N)$: Each party P_i runs the key generation algorithm $(\mathsf{sk}_i, \mathsf{pk}_i) \leftarrow \mathsf{KeyGen}(1^\lambda)$ and publishes their public key pk_i and stores their secret key sk_i.

$\Pi_{\mathsf{Back}}(i, B, \mathsf{pk}_B, \mathsf{sk}_{R \cup \{i\}})$: Party P_i runs this as a key-owner with a set of $(n-1)$ guardians $\{P_j\}_{j \in B}$ as follows:
- For each guardian P_j, P_i sends a backup request to P_j.
- Each P_j computes $x_{i,j} := H(j, \mathsf{sk}_j, \mathsf{pk}_i)$ and $\sigma_{i,j} = H(i, \mathsf{sk}_j)$ and sends $(x_j, \sigma_{i,j})$ back to the party P_i.
- On receiving $(x_{i,j}, \sigma_{i,j})$, and P_i defines $s := \mathsf{sk}_i$ computes:

$$(\varphi, \mathsf{tk}, \mathsf{vk}) \leftarrow \mathsf{TBUSS.Share}(\mathsf{sk}_i, \{(x_{i,j}, \sigma_{i,j})\}_{j \in B}, B)$$

and publishes $pub := \varphi$.

$\Pi_{\mathsf{Rec}}(i, R, \mathbf{pk}_{[N]}, \varphi, \mathsf{sk}_R)$: The party P_i wants to compute his secret with a set of $t + 1$ guardians as follows:
 - For each guardian P_j, P_i send a request to P_j for $j \in R$.
 - After receiving $(x_{i,j}, \sigma_{i,j})_{j \in R}$, P_i retrieve φ and compute

$$s' \leftarrow \mathsf{TBUSS.Recon}(\varphi, \{(x_{i,j}, \sigma_{i,j})\}_{j \in R}, R)$$

 - Then the party P_i checks the validity of the key pair (s', pk_i). If yes then privately output $\mathsf{sk}_i = s'$, else output $\perp$.

5.1   Security of TSKR

See the Traceability game for SKR in Fig. 6. This game is slightly different from the tracing game we have defined for BUSS in Fig. 1. The adversary can corrupt $f < t + 1$ guardians for every party P_k for $k \in [N]$. Let s_k be the secret key of party P_k. Now, the adversary can corrupt f guardians of party P_k to gain access to their shares σ_{k,i_j} for all $j \in [f]$ and outputs a reconstruction box $\mathcal{R}_k$ for every $k \in [N]$. The Tracer with access to $\mathcal{R}_k$ for $k \in [N]$ and with the corresponding guardians' trace key tuple $\mathsf{tk}^{(k)}$ for $k \in [N]$ outputs the list of corrupt guardians for party P_k for $k \in [N]$. Note that there is a master trace key tk for all the participating N parties, which the parties can derive by themselves. The trace key vectors $\mathsf{tk}^{(k)}$ are subsets of this master trace key tk. The parties

---

Game.Trace$_{\mathcal{A},\mathsf{TSKR},\varepsilon}(\lambda, N)$: Tracing game for TSKR.

1. $\mathcal{A}(1^\lambda, 1^N, n - 1, t + 1, f)$ outputs $(\{\mathcal{I}_k\}_{k \in [N]}, \mathsf{state})$, where $\mathcal{I}_k \subset [n - 1]$ and $|\mathcal{I}_k| = f(< t + 1)$ is the set of parties to corrupt.
2. N secrets $\{s_k\}_{k \in [N]}$ is sampled uniformly at random from $\mathcal{S}$.
3. Run: $\mathsf{Share}(s_k, \boldsymbol{\sigma}_{B_k}, B)$ output $(\varphi_k, \mathsf{tk}^{(k)}, \mathsf{vk}^{(k)})$ for $k \in [N]$.
4. On input all shares of parties in $\mathcal{I}_1, \ldots, \mathcal{I}_N$, the adversary $\mathcal{A}(\mathsf{state}, \boldsymbol{\sigma}_{\mathcal{I}_1}, \ldots, \boldsymbol{\sigma}_{\mathcal{I}_N})$ outputs reconstruction boxes $\mathcal{R}_1, \ldots, \mathcal{R}_N$ respectively.
5. $\mathsf{Trace}^{\mathcal{R}_1, \ldots, \mathcal{R}_N}(\mathsf{tk}^{(1)}, \ldots, \mathsf{tk}^{(N)})$ outputs $(\mathcal{I}'_1, \ldots, \mathcal{I}'_N, \pi_1, \ldots, \pi_N)$.
6. $\mathcal{A}$ wins if: $\mathcal{R}_1, \ldots, \mathcal{R}_N$ reconstruct the secret from consistent inputs with probability at least ε, i.e., $\mathcal{R}_1, \ldots, \mathcal{R}_N$ are good and there exists $k \in [N]$ such that either $\mathcal{I}_k \neq \mathcal{I}'_k$ or $\mathsf{Verify}(\mathsf{vk}^{(k)}, \mathcal{I}'_k, \pi_k) = 0$.

Fig. 5. The Tracing Game for the Traceable Social Key Recovery TSKR.

Definition 8 (Traceability of TSKR). *Let* $\mathsf{TSKR} = (\Pi_{\mathsf{Init}}, \Pi_{\mathsf{Back}}, \Pi_{\mathsf{Rec}})$ *be a Traceable SKR. Let* $\varepsilon = \varepsilon(\lambda, N)$ *be a function of the security parameters. We say that* TSKR *satisfies traceability if for every PPT adversary* $\mathcal{A}$, *the following function is negligible in* λ *and* N:

$$\mathsf{Adv}^{\mathsf{trace}}_{\mathcal{A},\mathsf{TSKR},\varepsilon}(\lambda, N) := \Pr\left[\mathsf{Game.Trace}_{\mathcal{A},\mathsf{TSKR},\varepsilon}(\lambda, N) = 1\right].$$

5.2   Tracing on **SKR**

First, we describe the corruption model used in our SKR scheme. Our model is similar to that of Boneh, Partap, and Rotem [9], with the key difference being that, in our setting, each party holds its own secret key sk_i, rather than a single shared secret. Any subset $\mathcal{I} \subset B$ of f colluding parties can construct a reconstruction box $\mathcal{R}$ such that, when given any additional $t - f + 1$ valid inputs, it can reconstruct the corresponding secret. We consider N such colluding groups, each constructing its own reconstruction box $\mathcal{R}_1, \ldots, \mathcal{R}_N$. For simplicity, we first assume that each colluding group consists of exactly f parties. Given any reconstruction box $\mathcal{R}^{(k)}$, if it receives $t - f + 1$ valid inputs, it can recover the associated secret. A tracer equipped with the trace key tk (as in Fig. 5) can then identify the colluding parties. So for tracing in the SKR scheme, the tracing algorithm runs over all the reconstruction boxes $\mathcal{R}_1, \ldots, \mathcal{R}_N$. After the tracing on each reconstruction box, we get the colluding sets are $\mathcal{I}_1, \ldots, \mathcal{I}_N$ then the overall colluding set is $\mathcal{I} = \cup_{k=1}^{N} \mathcal{I}_k$. We present our tracing algorithm for TSKR in Fig. 6.

Algorithm: $\mathsf{Trace}_{\mathsf{TSKR}}^{\mathcal{R}^{(1)}, \ldots, \mathcal{R}^{(N)}}(\mathsf{tk}, f)$

1. Parse tk as $(\mathsf{tk}^{(1)}, \ldots, \mathsf{tk}^{(N)})$.
2. For $k = 1, \ldots, N$:
   - With parameters $M, C \in \mathbb{N}$ and $\epsilon_1, \ldots, \epsilon_N$ run $\mathsf{Trace}_{\mathsf{TBUSS}}^{\mathcal{R}_k}(\mathsf{tk}^{(k)}) \to (\mathcal{I}_k, \pi_k)$
3. Compute $\mathcal{I}_{\mathsf{TSKR}} = (\mathcal{I}_1, \ldots, \mathcal{I}_N)$ and $\pi = (\pi_1, \ldots, \pi_k)$
4. Output $(\mathcal{I}_{\mathsf{TSKR}}, \pi)$

Fig. 6. The Tracing Algorithm for TSKR.

Theorem 4. *Let* TSKR *be the Traceable Social Key Recovery (SKR) scheme defined in Sect. 5.2. For every adversary $\mathcal{A}$, security parameters $\lambda, N \in \mathbb{N}$, parameters $M, C \in \mathbb{N}$, and $\epsilon_1, \epsilon_2, \ldots, \epsilon_N \in \left[\sqrt{\frac{2(M+n+t^2-ft)}{p}}, 1 \right]$ with $\epsilon_i > \sqrt{2C/M}$ for each $i \in [N]$ and $\sqrt{fM} \leq C \leq M < p$, we have:*

$$\mathsf{Adv}_{\mathcal{A}, \mathsf{TSKR}, \epsilon}^{\mathsf{trace}}(\lambda, N) \leq N \cdot \left(e^{\frac{-\epsilon^2 M}{2} \cdot (1-1/r)^2} + \frac{f \cdot (n - f - 1) \cdot \tau}{p} \right)$$

where $\epsilon = \min\{\epsilon_1, \ldots, \epsilon_N\}$, $|\mathbb{F}| = p$ (field size), $r = \dfrac{\epsilon^2 M}{2C}$, $n = n(\lambda)$ (size of guardian set $B = n - 1$), $f = f(\lambda)$ (number of corruptions, $f < t + 1$), $\tau = \tau(M, f, \log p)$ be the list size from the Guruswami-Sudan Algorithm $\mathcal{D}_{\mathsf{GS}}$.

Proof. It is easy to see that $\mathsf{Adv}^{\mathsf{trace}}_{\mathcal{A},\mathsf{TSKR},\epsilon}(\lambda, N) \leq N \cdot \mathsf{Adv}^{\mathsf{trace}}_{\mathcal{A},\mathsf{TBUSS},\epsilon}(\lambda)$ and by Theorem 2 we have that $\mathsf{Adv}^{\mathsf{trace}}_{\mathcal{A},\mathsf{TBUSS},\epsilon}(\lambda) \leq \left(e^{\frac{-\epsilon^2 M}{2} \cdot (1-1/r)^2} + \dfrac{f \cdot (n - f - 1) \cdot \tau}{p} \right).$
Thus, we have the theorem. $\qquad\square$

Remark 1. Note that for each $k \in [N]$ we have that $|\mathcal{I}_k| = f < t + 1$. But it is not true that $|\mathcal{I}_{\mathsf{TSKR}}| = f$ or $|\mathcal{I}_{\mathsf{TSKR}}| < t + 1$. We do not have any restriction on the size of $\mathcal{I}_{\mathsf{TSKR}}$ other than the maximum upper bound of N. It may very well happen that $|\mathcal{I}_{\mathsf{TSKR}}| = N$, i.e., all the parties $P_1, \dots, P_N$ participating in TSKR protocol are corrupt. Each party P_i can be a guardian of several other parties, thus P_i can freely choose whose share to sell, i.e., the number of corruption of P_i can be from the set $\{0, \dots, N-1\}$. So, for the corruption model of TSKR, we do not a priori declare a number of corruption, unlike TBUSS, in which the number of corruption f, is known beforehand. We also have the provision to report every set of corrupt parties $\mathcal{I}_k$ for each party P_k where $k \in [N]$, to track the corrupt guardians.

5.3   Non-imputability of SKR

We describe the NISKR construction in the full version of this paper; however, we state the formal result on non-imputability in this section.

Theorem 5 (Non-Imputability of NISKR). *Let $\mathcal{F}$ be a one-way function. For every PPT adversary $\mathcal{A}$ winning the non-imputability game for NISKR, there exists a PPT algorithm (inverter) $\mathcal{B}$ such that for every $\lambda, N \in \mathbb{N}$ it holds that,*

$$\mathsf{Adv}^{\mathsf{ni}}_{\mathcal{A},\mathsf{NISKR}}(\lambda, N) = \Pr[\mathcal{F}(\mathcal{B}(\mathcal{F}(x^*))) = \mathcal{F}(x^*)]$$

where $x \xleftarrow{\$} \mathbb{F}$ and the probability is also over the randomness $\mathcal{B}$.

The proof of the Theorem 5 is similar to the proof of the Theorem 3, hence provided in the full version of the paper.

6   Complexity and Comparison of Our Tracing Procedures

We now provide a comparative overview of existing traceable secret sharing (TSS) schemes. The scheme of Goyal, Song, and Srinivasan (GSS) [28] is based on Shamir's secret sharing and was the first to achieve public traceability. However, its practicality is limited by the fact that the share size grows by a factor of λ, making it unsuitable for large-scale deployments. Moreover, the scheme does not support weighted access structures, and its security guarantees do not extend against computationally unbounded adversaries.

Boneh, Partap, and Rotem [9] subsequently introduced traceable variants of Shamir's and Blakley's schemes. These constructions are considerably more efficient: the share size only doubles compared to the original schemes, and they work for arbitrary thresholds t. Nonetheless, their notion of traceability is private rather than public, and they do not extend to weighted access structures or offer security in the presence of unbounded adversaries.

The Mignotte-based traceable scheme (MTSS) [30] achieves the traceability notion of [9], the share size again grows only by a factor of 2, but the tracing algorithm is computationally demanding, running in time exponential in κ.

Our basic traceability notion for TBUSS satisfies the traceability guarantee introduced in the work of Boneh, Partap, and Rotem [9]. Also, this is efficient in the sense that the share size is increased by only a factor of 2. The tracing algorithm for TBUSS is also efficient as it matches the exact time complexity of the Traceable Shamir in the work of [9]. The SKR scheme has tracing complexity of $N \cdot \mathrm{poly}(\lambda, \epsilon^{-1})$ and follows the same efficiency except for a multiplicative factor of N (Table 1).

Table 1. Overview of traceable secret sharing schemes. Here, $|\mathsf{sh}|$ denotes the share size. The column 'Any $\mathcal{A}$' specifies whether security is guaranteed against computationally unbounded adversaries.

TSS scheme	$|\mathsf{sh}|$ increase	Trace Time	Public	Any $\mathcal{A}$
GSS [28]	λ	$\mathrm{poly}(\lambda, \epsilon^{-1})$	yes	no
T-Shamir [9]	2	$\mathrm{poly}(\lambda, \epsilon^{-1})$	no	no
T-Blakely [9]	2	$\mathrm{poly}(\lambda, \epsilon^{-1})$	no	no
NI-Shamir [9]	2	$\mathrm{poly}(\lambda, \epsilon^{-1})$	no	no
NI-Blakely [9]	2	$\mathrm{poly}(\lambda, \epsilon^{-1})$	no	no
MTSS [30]	2	$\mathrm{poly}(2^\kappa, \epsilon^{-1})$	no	no
SPMTSS [30]	2	$\mathrm{poly}(\lambda, \epsilon^{-1})$	semi	yes
MTSS for small t [30]	$2\lambda/t$	$\mathrm{poly}(2^\kappa, \epsilon^{-1})$	no	no
SPMTSS for small t [30]	$2\lambda/t$	$\mathrm{poly}(\lambda, \epsilon^{-1})$	semi	yes
TBUSS [This Work]	2	$\mathrm{poly}(\lambda, \epsilon^{-1})$	no	no
TSKR [This Work]	2	$N \cdot \mathrm{poly}(\lambda, \epsilon^{-1})$	no	no
NITBUSS [This Work]	2	$\mathrm{poly}(\lambda, \epsilon^{-1})$	no	no
NISKR [This Work]	2	$N \cdot \mathrm{poly}(\lambda, \epsilon^{-1})$	no	no

7   Conclusion and Future Direction

In this paper, we have proposed a tracing mechanism for the Bottom-Up Secret Sharing (BUSS) scheme. We have thoroughly analyzed the tracing advantage of BUSS. Additionally, we have shown Non-Imputability of BUSS by integrating the construction with a one-way function. Eventually, we have shown that the hardness of inversion of the one-way function yields non-imputability security of BUSS. As an application of traceable BUSS we have proposed a traceable

Social Key Recovery (TSKR) construction and provided a tracing mechanism along with non-imputability advantage. Although our Tracing Mechanism for TSKR essentially relies on the tracing mechanism of BUSS, it eventually contributes a factor of N in the tracing advantage probability. Also we have shown non-imputability of both BUSS and SKR, thereby constructing non-imputable variants of the traceable schemes BUSS and SKR, namely NITBUSS and NISKR.

In this work we have shown private traceability of BUSS and SKR, only the tracer learns the trace keys of the parties and hence can perform trace. An interesting avenue for future work is public traceability of BUSS. Also, it will be an interesting future direction to come up with a tracing mechanism for SKR independent of the tracing mechanism for BUSS.

References

1. Argent: How to recover my wallet with guardians on-chain (complete guide) (2025). https://support.argent.xyz/hc/en-us/articles/360007338877-How-to-recover-my-wallet-with-guardians-onchain-complete-guide
2. Baghery, K., Ebrahimi, E., Mirzamohammadi, O., Sedaghat, M.: Traceable verifiable secret sharing and applications. Cryptology ePrint Archive, Paper 2025/318 (2025). https://eprint.iacr.org/2025/318
3. Baird, L., et al.: Threshold signatures in the multiverse. Cryptology ePrint Archive, Paper 2023/063 (2023). https://eprint.iacr.org/2023/063
4. Bellare, M., Dai, W., Rogaway, P.: Reimagining secret sharing: creating a safer and more versatile primitive by adding authenticity, correcting errors, and reducing randomness requirements. Cryptology ePrint Archive, Paper 2020/800 (2020). https://eprint.iacr.org/2020/800
5. Bhattacharyya, R., Bormet, J., Faust, S., Mukherjee, P., Othman, H.: CCA-secure traceable threshold (ID-based) encryption and application. Cryptology ePrint Archive, Paper 2025/341 (2025). https://eprint.iacr.org/2025/341
6. Blakley, G.R.: Safeguarding cryptographic keys. In: 1979 International Workshop on Managing Requirements Knowledge (MARK), pp. 313–318 (1899). https://api.semanticscholar.org/CorpusID:38199738
7. Boneh, D., Franklin, M.: An efficient public key traitor tracing scheme. In: Wiener, M. (ed.) CRYPTO 1999, pp. 338–353. Springer, Heidelberg (1999)
8. Boneh, D., Naor, M.: Traitor tracing with constant size ciphertext. In: Proceedings of the 15th ACM Conference on Computer and Communications Security, CCS 2008, pp. 501–510. Association for Computing Machinery, New York (2008). https://doi.org/10.1145/1455770.1455834
9. Boneh, D., Partap, A., Rotem, L.: Traceable secret sharing: strong security and efficient constructions. Cryptology ePrint Archive, Paper 2024/405 (2024). https://eprint.iacr.org/2024/405
10. Boneh, D., Partap, A., Rotem, L.: Traceable verifiable random functions. Cryptology ePrint Archive, Paper 2025/312 (2025). https://eprint.iacr.org/2025/312
11. Boneh, D., Sahai, A., Waters, B.: Fully collusion resistant traitor tracing with short ciphertexts and private keys. In: Vaudenay, S. (ed.) EUROCRYPT 2006. LNCS, vol. 4004, pp. 573–592. Springer, Heidelberg (2006). https://doi.org/10.1007/11761679_34
12. Boneh, D., Shaw, J.: Collusion-secure fingerprinting for digital data. In: Coppersmith, D. (ed.) CRYPTO 1995, pp. 452–465. Springer, Heidelberg (1995)

13. Boneh, D., Zhandry, M.: Multiparty key exchange, efficient traitor tracing, and more from indistinguishability obfuscation. Algorithmica **79**(4), 1233–1285 (2017). https://doi.org/10.1007/s00453-016-0242-8

14. Bormet, J., Dziembowski, S., Faust, S., Lizurej, T., Mielniczuk, M.: Strong secret sharing with snitching. Cryptology ePrint Archive, Paper 2025/1119 (2025). https://eprint.iacr.org/2025/1119

15. Bormet, J., Hofmann, J., Othman, H.: Traceable threshold encryption without trusted dealer. Cryptology ePrint Archive, Paper 2025/342 (2025). https://eprint.iacr.org/2025/342

16. Buterin, V.: Why we need wide adoption of social recovery wallets (2021). https://vitalik.eth.limo/general/2021/01/11/recovery.html

17. Chatzigiannis, P., Chalkias, K., Kate, A., Mangipudi, E.V., Minaei, M., Mondal, M.: SoK: Web3 recovery mechanisms. Cryptology ePrint Archive, Paper 2023/1575 (2023). https://eprint.iacr.org/2023/1575

18. Chen, Y., Vaikuntanathan, V., Waters, B., Wee, H., Wichs, D.: Traitor-tracing from LWE made simple and attribute-based. In: Beimel, A., Dziembowski, S. (eds.) TCC 2018. LNCS, vol. 11240, pp. 341–369. Springer, Cham (2018). https://doi.org/10.1007/978-3-030-03810-6_13

19. Chor, B., Fiat, A., Naor, M., Pinkas, B.: Tracing traitors. IEEE Trans. Inf. Theory **46**(3), 893–910 (2000). https://doi.org/10.1109/18.841169

20. Dodis, Y., Fazio, N.: Public key trace and revoke scheme secure against adaptive chosen ciphertext attack. Cryptology ePrint Archive, Paper 2003/095 (2003). https://eprint.iacr.org/2003/095

21. Dziembowski, S., Faust, S., Lizurej, T., Mielniczuk, M.: Secret sharing with snitching. In: Proceedings of the 2024 on ACM SIGSAC Conference on Computer and Communications Security, CCS 2024, pp. 840–853. Association for Computing Machinery, New York (2024). https://doi.org/10.1145/3658644.3690296

22. Farràs, O., Guiot, M.: Traceable secret sharing schemes for general access structures. Cryptology ePrint Archive, Paper 2025/1120 (2025). https://eprint.iacr.org/2025/1120

23. Fiat, A., Tassa, T.: Dynamic traitor tracing. In: Wiener, M. (ed.) CRYPTO 1999. LNCS, vol. 1666, pp. 354–371. Springer, Heidelberg (1999). https://doi.org/10.1007/3-540-48405-1_23

24. Garg, S., Kumarasubramanian, A., Sahai, A., Waters, B.: Building efficient fully collusion-resilient traitor tracing and revocation schemes. In: Proceedings of the 17th ACM Conference on Computer and Communications Security, CCS 2010, pp. 121–130. Association for Computing Machinery, New York (2010). https://doi.org/10.1145/1866307.1866322

25. Goldreich, O.: Foundations of Cryptography. Cambridge University Press, Cambridge (2001)

26. Gong, T., Kate, A., Maji, H.K., Nguyen, H.H.: Disincentivize collusion in verifiable secret sharing. In: Fehr, S., Fouque, P.A. (eds.) EUROCRYPT 2025, pp. 34–64. Springer, Cham (2025)

27. Goyal, R., Koppula, V., Waters, B.: Collusion resistant traitor tracing from learning with errors. In: Proceedings of the 50th Annual ACM SIGACT Symposium on Theory of Computing, STOC 2018, pp. 660–670. Association for Computing Machinery, New York (2018). https://doi.org/10.1145/3188745.3188844

28. Goyal, V., Song, Y., Srinivasan, A.: Traceable secret sharing and applications. In: Malkin, T., Peikert, C. (eds.) CRYPTO 2021. LNCS, vol. 12827, pp. 718–747. Springer, Cham (2021). https://doi.org/10.1007/978-3-030-84252-9_24

29. Guruswami, V.: Improved decoding of reed-solomon and algebraic-geometric codes, vol. 45, pp. 28–37 (1998). https://doi.org/10.1109/SFCS.1998.743426
30. Hoffmann, C.: Traceable secret sharing based on the Chinese remainder theorem. Cryptology ePrint Archive, Paper 2024/811 (2024). https://eprint.iacr.org/2024/811
31. Kate, A., Mukherjee, P., Saleem, H., Sarkar, P., Roberts, B.: ANARKey: a new approach to (socially) recover keys. Cryptology ePrint Archive, Paper 2025/551 (2025). https://eprint.iacr.org/2025/551
32. Kiayias, A., Yung, M.: Self protecting pirates and black-box traitor tracing. In: Kilian, J. (ed.) CRYPTO 2001. LNCS, vol. 2139, pp. 63–79. Springer, Heidelberg (2001). https://doi.org/10.1007/3-540-44647-8_4
33. Kurosawa, K., Desmedt, Y.: Optimum traitor tracing and asymmetric schemes. In: Nyberg, K. (ed.) EUROCRYPT 1998. LNCS, vol. 1403, pp. 145–157. Springer, Heidelberg (1998). https://doi.org/10.1007/BFb0054123
34. Lindell, Y.: Cryptography and MPC in coinbase wallet as a service (WAAS) (2023). https://www.coinbase.com/en-in/blog/digital-asset-management-with-mpc-whitepaper
35. Little, R., Qin, L., Varia, M.: Secure account recovery for a privacy-preserving web service. Cryptology ePrint Archive, Paper 2024/962 (2024). https://eprint.iacr.org/2024/962
36. Mignotte, M.: How to share a secret. In: Beth, T. (ed.) Cryptography, pp. 371–375. Springer, Heidelberg (1983)
37. Naor, M., Pinkas, B.: Threshold traitor tracing. In: Krawczyk, H. (ed.) CRYPTO 1998. LNCS, vol. 1462, pp. 502–517. Springer, Heidelberg (1998). https://doi.org/10.1007/BFb0055750
38. Pedin, A.B., Siasi, N., Sameni, M.: Smart contract-based social recovery wallet management scheme for digital assets. In: Proceedings of the 2023 ACM Southeast Conference, ACMSE 2023, pp. 177–181. Association for Computing Machinery, New York (2023). https://doi.org/10.1145/3564746.3587016
39. Shamir, A.: How to share a secret. Commun. ACM **22**(11), 612–613 (1979). https://doi.org/10.1145/359168.359176
40. Times, T.N.Y.: Tens of billions worth of bitcoin have been locked by people who forgot their key (2021). https://www.nytimes.com/2021/01/13/business/tens-of-billions-worth-of-bitcoin-have-been-locked-by-people-who-forgot-their-key.html
41. Wee, H.: Functional encryption for quadratic functions from k-Lin, revisited. In: Pass, R., Pietrzak, K. (eds.) TCC 2020. LNCS, vol. 12550, pp. 210–228. Springer, Cham (2020). https://doi.org/10.1007/978-3-030-64375-1_8
42. Zhandry, M.: New techniques for traitor tracing: Size $n^{1/3}$ and more from pairings. Cryptology ePrint Archive, Paper 2020/954 (2020). https://eprint.iacr.org/2020/954

Beyond Confidentiality: Framing-Resistant Secure Vault Schemes

Meghna Sengupta(✉)

University of Edinburgh, Edinburgh, UK
meghna.816@gmail.com

Abstract. Vault schemes have recently been proposed as a cryptographic abstraction for storing sensitive and non-sensitive data in outsourced databases while preserving confidentiality guarantees. Prior work has focused primarily on privacy goals, formalized through indistinguishability notions that protect sensitive records even when an adversary has access to tokens and the database contents. In this work we argue that confidentiality alone is not sufficient: a new integrity threat arises in the form of *framing attacks,* in which a malicious server or colluding user forges records that appear to originate from an honest client.

We introduce *framing resistance* as a complementary security goal for vaults, and formalize it via a new indistinguishability game, IND-FR. We then present two constructions achieving this notion. The first binds each record to a client through digital signatures or MACs, reducing framing resistance to the unforgeability of the underlying primitive. The second eliminates the need for client-held keys by leveraging transparency logs and server-issued signed receipts, reducing framing resistance to signature unforgeability and the append-only property of the log. Both constructions are proven secure under standard assumptions.

Finally, we compare the two approaches, highlighting tradeoffs in security guarantees, trust models, and deployment complexity. Our results show that framing resistance is both achievable and practical, and that secure vaults can go *beyond confidentiality* to offer strong integrity protections suitable for real-world cloud and data-sharing environments.

Keywords: data privacy vault · LLM privacy · cloud storage · secure data storage · tokenization · vault scheme · non-framing · framing attack

1 Introduction

The need for secure mechanisms to manage sensitive and non-sensitive information in outsourced environments has become increasingly urgent, underscored by high-profile data breaches such as the Samsung internal data leak [2]. In practice, corporations have turned to privacy vault solutions such as Skyflow [3] and Hashicorp Vault [4], which provide application-oriented tools for storing and accessing data selectively. These commercial systems highlight the importance of vaults as a practical abstraction for data privacy. At the same time, vaults are playing a growing role in supporting machine learning pipelines, including large language models (LLMs), where deterministic yet semantically secure storage of training data is essential to preserve both correctness and confidentiality.

Despite their widespread use in industry, rigorous cryptographic treatment of vaults has been limited. Bhattacharyya et al. [6] introduced the first formalization of a *vault scheme*, defining algorithms for storing and accessing sensitive and non-sensitive data and providing indistinguishability-based privacy notions against honest-but-curious servers. While this work established vaults as a concrete cryptographic primitive with provable confidentiality guarantees, it left open the question of integrity: specifically, how vaults should defend against adversaries who attempt to insert or manipulate records. In the real world, the assumption that the integrity of user attribution is guaranteed by the underlying system, leaves an important gap: the threat of framing attacks.

A framing attack occurs when an adversary - such as a malicious insider, a colluding analyst and server, or a corrupt user, injects or manipulates records so that sensitive information appears to be associated with an innocent party. Unlike conventional confidentiality breaches, framing undermines the trustworthiness of the vault's contents rather than their secrecy.

Consider a concrete example: a traveler's information is stored in a government-operated vault that links non-sensitive attributes (passport number, itinerary) with sensitive attributes (biometric data, flagged security status). If a malicious analyst colludes with the cloud provider, they could insert a fabricated record that ties the traveler's passport number to a "watchlist" marker. During routine checks, the vault would then return a result indicating that the traveler is flagged, despite the fact that no such entry was ever submitted by the traveler or by legitimate authorities. This is a classic framing attack: the adversary has not learned anything secret, but has instead forged evidence that falsely implicates an honest user.

In sensitive domains such as healthcare, finance, and law enforcement, the consequences of framing can be equally severe: a patient could be falsely linked to a medical condition, a customer to fraudulent transactions, or an employee to unauthorized access. These outcomes can damage reputations, trigger unjust penalties, or erode confidence in the entire vault infrastructure.

Ensuring non-framing security, which is the guarantee that no adversary can cause the system to attribute data to an honest user unless that user explicitly stored it, is therefore essential for vault schemes to be deployable in high-stakes environments. Importantly, non-framing is complementary to existing confidentiality notions such as indistinguishability under chosen-token attacks: it protects integrity of attribution even when adversaries already observe the entire database or collude with the storage provider. Just as authenticated encryption prevents ciphertext forgeries in communication channels, non-framing vaults prevent identity forgeries in outsourced data systems.

The practical importance of non-framing secure vaults is broad. In regulated industries, compliance frameworks such as GDPR and HIPAA require not only confidentiality but also verifiable correctness of stored data. In collaborative workflows, non-framing ensures accountability, so that audit trails cannot be falsified by malicious participants. In government or enterprise settings, where analysts may access large vaults to detect patterns, non-framing guarantees that query results reflect only genuine submissions, thereby preventing the misuse of the system to fabricate incriminating evidence. Moreover, as vaults integrate with emerging paradigms such as decentralized identity systems and transparency logs, non-framing provides a natural foundation for building tamper-evident, auditable data infrastructures.

In this work, we address this integrity gap by introducing the first non-framing secure vault scheme. We formalize the notion of framing resistance, define a new security game that captures adversarial capabilities in both honest-but-curious and malicious cloud models, and present a construction that achieves strong non-framing guarantees without sacrificing usability. Our approach combines cryptographic binding of records to user identities with lightweight verification mechanisms, ensuring that vaults remain both secure and practical in real-world deployments.

1.1 Related Works

Several prior works have addressed tokenization, data privacy, and database integrity, though none directly capture the notion of framing resistance.

Tokenization systems and standards (e.g. ANSI X9.119-2 [1]) define approaches such as format-preserving or table-based tokenization, but generally focus on preserving format or replacing sensitive data without formal security models for identity attribution or injection attacks. A Cryptographic Study of Tokenization Systems [7] investigates properties such as uniqueness and mapping between tokens and sensitive data, but does not consider active adversaries who might forge or inject records under another users' identity.

Spitz: A Verifiable Database System [13] develops mechanisms to make database operations and history tamper-evident, enforcing integrity of record insertion and query execution; its design overlaps with transparency logs for ensuring records actually existed and were appended, which is a component of what is needed for framing resistance. The work The Curse of Correlations for Robust Fingerprinting of Relational Databases [9] studies attacks on fingerprinting schemes in relational data arising from correlation, which is relevant to integrity of attribution in adversarial settings.

The paper by Bhattacharyya et al. [6] introduced the vault scheme abstraction and defined privacy notions (IND-CATA, IND-SIA) under which non-sensitive and sensitive data remain hidden under token and database exposure. However, that work did not examine integrity attacks such as the insertion or forging of records - i.e., framing attacks - which are the focus of the current paper.

A distinct line of research explores the *fuzzy vault* construction [10–12], originally introduced by Juels and Sudan [10] to protect biometric or otherwise noisy data. In a fuzzy vault, the goal is to bind a secret (e.g., a cryptographic key) to a set of features such that it can later be unlocked only by a set with sufficient overlap, thereby achieving confidentiality and error-tolerant authentication. While both frameworks share the objective of securely managing sensitive information, the underlying threat models differ: fuzzy vaults aim to protect against unauthorized reconstruction of the secret under biometric noise, whereas the secure vault schemes considered here address integrity and attribution in cloud storage. The framing resistance notion introduced in this work is therefore orthogonal to the confidentiality guarantees of fuzzy vaults, focusing instead on preventing adversarial injection or manipulation of records that could falsely implicate honest users.

To summarize, while work on verifiable databases, fingerprinting, and tokenization provide useful tools and inspiration, the gap remains in formalizing and constructing vault schemes that are both private and resilient to framing - ensuring that no adversary can cause a record to be accepted under an honest user's identity unless the user indeed stored it. The present work extends this line of research by introducing the notion of *framing attacks* and formalizing the corresponding property of *framing resistance*. While prior works established how vaults can protect the secrecy of sensitive information, no previous study has considered how vaults should defend against adversaries attempting to inject or manipulate records to falsely implicate honest users.

1.2   Our Contributions

This paper addresses the integrity gap in secure vaults by introducing the notion of non-framing security and presenting two constructions that achieve it. Our contributions align with the structure of the paper:

- **Framing attacks and framing resistance (Sect. 3).** We identify *framing attacks* as a critical threat to vault schemes, distinct from but complementary to confidentiality breaches. To capture this formally, we introduce the first security notion of *framing resistance*, denoted IND-FR, and define a rigorous security game in which an adversary attempts to cause the vault to output records falsely attributed to an honest user.
- **Client-bound signatures/MACs construction (Sect. 4).** We present a non-framing vault construction in which each stored record is authenticated by the user through a signature or MAC that binds it to their identity. We prove that this scheme achieves IND-FR security in the honest-but-curious model, reducing framing resistance to the unforgeability of the underlying signature or MAC scheme.
- **Transparency log construction (Sect. 5).** To address settings where clients cannot feasibly manage cryptographic keys, we design a second variant that leverages a public append-only transparency log combined with server-issued signed receipts. We prove that this scheme achieves IND-FR security in the malicious-server model, reducing framing resistance to either breaking the server's signing scheme or violating the append-only property of the log.

– **Comparison and tradeoffs (Sect. 6).** We provide a systematic comparison of the two approaches, analyzing their security assumptions, deployment complexity, performance implications, and application contexts. This comparison highlights the complementary nature of the two variants and points toward hybrid designs that combine their strengths.

2   Notations and Preliminaries

Notations. If x is a string, $|x|$ denotes the length of the string. $x[i]$ denotes its i-th bit. If S is a set, $|S|$ denotes the size of S, and $s \leftarrow_\$ S$ denotes sampling an element uniformly at random from S and assign it to s. $\mathbb{N}$ denotes the set of positive integers $\{1, 2, \ldots\}$. For $n \in \mathbb{N}$, $[n]$ denotes the set $\{1, 2, \cdots, n\}$. Composition of two functions is denoted by $\circ$. If $\hat{F} = F \circ \phi$, then $\hat{F}(x) = F(\phi(x))$. $\{0, 1\}^n$ denotes the set of all binary strings of length n. The set of all functions with domain $\{0, 1\}^m$ and co-domain $\{0, 1\}^n$ is denoted by $\mathcal{F}_{m,n}$.

The guessing probability of a random variable X is defined as $GP(X) = \max_x \Pr[X = x]$. The min-entropy of a random variable X is defined by $H_\infty(X) \stackrel{def}{=} -\log \max_x \Pr[X = x]$. The conditional min-entropy of X conditioned on another random variable Y is defined by $H_\infty(X \mid Y) \stackrel{def}{=} -\log \sum_y \Pr[Y = y] \max_x \Pr[X = x \mid Y = y]$.

Algorithms. The algorithms considered in this paper are randomized, unless otherwise specified. If $\mathcal{A}$ is an algorithm, we let $y \leftarrow \mathcal{A}(x_1, \ldots; r)$ to denote that running $\mathcal{A}$ with input $x_1, \ldots$, random coin r and assigning the output to y. We let $y \leftarrow_\$ \mathcal{A}(x_1, \ldots)$ be the result of choosing r uniformly at random and letting $y \leftarrow \mathcal{A}(x_1, \ldots; r)$.

Security Games. The results are proven in the framework of code based games of [5]. A game G consists of a `main` oracle and zero or more stateful oracles $O_1, O_2, \cdots, O_n$. If a game G is implemented using a function f, we write $G[f]$ to denote the game. An algorithm $\mathcal{A}$ is said to participate in game G if the `main` oracle invokes algorithm $\mathcal{A}$ who can (optionally) make queries to the oracles $O_1, O_2, \cdots, O_n$. We denote by $G^{\mathcal{A}} = 1$ that an execution of G with $\mathcal{A}$ outputs 1.

Success Probability. The success probability of an algorithm $\mathcal{A}$ in game G is defined by $\mathbf{Succ}_{\mathcal{A},G} \stackrel{def}{=} \Pr[G^{\mathcal{A}} = 1]$.

In all the descriptions, uninitialized integers are assumed to be set to 0, booleans are set to *false*, the strings, sets and lists are set to be empty.

Random Oracles. An (idealized) function $\mathcal{H} : \{0, 1\}^\delta \rightarrow \{0, 1\}^\rho$ is said to be a *Random Oracle*, if for all $x \in \{0, 1\}^\delta$, the output $\mathcal{H}(x)$ is independently and uniformly distributed over $\{0, 1\}^\rho$.

EUF-CMA Security: Unforgeability of Digital Signatures. The security of a digital signature scheme $\Sigma = (\mathsf{KeyGen}, \mathsf{Sign}, \mathsf{Verify})$ is formalized via the notion of existential unforgeability under chosen-message attacks (EUF-CMA). The game proceeds as follows: the challenger first generates a keypair $(sk, pk) \leftarrow$ KeyGen(1^λ) and gives pk to the adversary $\mathcal{A}$. The adversary is allowed to make adaptive signing queries to a signing oracle Sign$(sk, \cdot)$ on messages of its choice. For each query m, the challenger returns $\sigma \leftarrow$ Sign(sk, m).

Eventually, the adversary outputs a candidate forgery (m^*, σ^*). The adversary wins the game if:

1. Verify$(pk, m^*, \sigma^*) = 1$, and
2. m^* was never submitted to the signing oracle during the interaction.

The advantage of $\mathcal{A}$ in breaking EUF-CMA security is defined as

$$\mathsf{Adv}^{\text{euf-cma}}_{\Sigma, \mathcal{A}}(\lambda) = \Pr\left[\mathsf{Verify}(pk, m^*, \sigma^*) = 1 \ \wedge \ m^* \notin Q\right],$$

where Q is the set of messages queried to the signing oracle. We say that a signature scheme Σ is EUF-CMA secure if for all PPT adversaries $\mathcal{A}$, the advantage $\mathsf{Adv}^{\text{euf-cma}}_{\Sigma, \mathcal{A}}(\lambda)$ is negligible in the security parameter λ.

2.1 Vault Schemes

We recall the notion of a *vault scheme* from [6], which models secure storage of sensitive and non-sensitive data across three parties: a user, a storage server, and an analyst.

Definition 1. *A vault scheme consists of the following probabilistic polynomial-time algorithms:*

– Param$(1^\lambda) \rightarrow PP$: generates public parameters.
– Store$(PP, ID, M, DB) \rightarrow (psswd, tok, DB')$: run by a user to store $M = (M_s, M_{ns})$, producing an access password $psswd$, a data token tok, and updated database DB'.
– Access$_{DB}(PP, tok) \rightarrow M_{ns}$: run by the analyst to access M_{ns} given tok and read access to DB.
– Retrieve$_{DB}(PP, ID, psswd) \rightarrow (M_s, M_{ns})$: run by any party with $(ID, psswd)$ to retrieve the full message M from DB.

CORRECTNESS. For all $PP \leftarrow$ Param(1^λ), any ID, and $M = (M_s, M_{ns})$, if $(psswd, tok, DB') \leftarrow$ Store(PP, ID, M, DB), then:

– Access$(PP, tok) = M_{ns}$ (access correctness),
– Retrieve$(PP, ID, psswd) = M$ (retrieve correctness).

SECURITY GOALS. The security of a vault scheme is two-fold: (1) the non-sensitive part M_{ns} should remain private without the access token; and (2) the sensitive part M_s should remain private without the master password, even when the adversary holds tok, M_{ns}, and full access to DB.

These requirements are formalized by [6] via indistinguishability notions. We recall the two central definitions below.

PRIVACY FOR NON-SENSITIVE DATA. The IND-CATA notion requires that M_{ns} remains indistinguishable without the access token. The corresponding security game is given in Fig. 1.

Game IND-CATA

1 : $b \leftarrow\$ \{0, 1\}$
2 : $PP \leftarrow\$ \mathsf{Param}(1^{\lambda})$
3 : $(state, m_0, m_1) \leftarrow \mathcal{A}^{\mathsf{Store},\mathsf{Access}}(PP, 1^{\lambda})$
4 :   **if** $m_{ns,0} = m_{ns,1}$: **return** 0
5 : $(m\text{-}tok^*, data\text{-}tok^*) \leftarrow \mathsf{Store}(PP, m_b)$
6 : $access\text{-}tok^* = \mathcal{H}(10, m\text{-}tok^*)$
7 : $b' \leftarrow \mathcal{A}^{\mathsf{Store},\mathsf{Access}}(PP, access\text{-}tok^*, state)$
8 :   **return** $(b = b')$

Fig. 1. IND-CATA security game.

PRIVACY FOR SENSITIVE DATA. The IND-SIA notion requires that even with tokens, non-sensitive data, and full database access, the adversary cannot distinguish between records differing only in the sensitive part. The corresponding security game is given in Fig. 2.

Game IND-SIA

1 : $b \leftarrow\$ \{0, 1\}$
2 : $PP \leftarrow\$ \mathsf{Param}(1^{\lambda})$
3 : $(state, m_0, m_1) \leftarrow \mathcal{A}^{\mathsf{Store},\mathsf{Access}}(PP, 1^{\lambda}, DB)$
4 : $((m\text{-}tok^*, data\text{-}tok^*), DB') \leftarrow \mathsf{Store}(PP, m_{ns,b}, DB)$
5 : $b' \leftarrow \mathcal{A}^{\mathsf{Store},\mathsf{Access}}(PP, state, DB')$
6 :   **return** $(b = b')$

Fig. 2. IND-SIA security game.

3   Framing Attacks and Framing Resistance

3.1   Framing Attacks

Existing secure vault schemes, including the construction given by [6], focus primarily on confidentiality: ensuring that sensitive attributes cannot be recovered without the appropriate access tokens, even if an adversary has complete visibility into the outsourced database. However, these definitions do not directly capture a different but equally damaging threat: the ability of an adversary to *frame* an honest user by inserting, modifying, or manipulating records so that they appear to originate from that user.

In a *framing attack*, the adversary's goal is not to learn information but to create false evidence. Concretely, the adversary may inject a fabricated record under the identifier of an honest user, or graft a record originally created for one user onto another. If the system does not provide a mechanism for verifiably binding records to the user who created them, then an honest user can later be held accountable for data that they never stored.

Example. Consider a government-operated vault that maintains records of travelers. Each record contains non-sensitive information such as a passport number or itinerary, along with sensitive attributes such as biometric markers or whether the individual is on a security watchlist. Suppose a malicious analyst colludes with the cloud provider to inject a record linking an honest traveler's passport number to a "watchlisted" flag. During subsequent queries, the vault would faithfully return this association, despite the fact that the traveler never submitted such data. The result is reputational damage, denial of services, or even unjust penalization - all without any breach of confidentiality.

This type of adversarial behavior highlights the need for vault schemes to provide not only secrecy but also *non-framing integrity*: a guarantee that no adversary can cause a record to be accepted under a user's identity unless that user explicitly stored it.

3.2   Framing Resistance

To capture protection against framing, we introduce the security notion of *framing resistance*, denoted IND-FR. Intuitively, a vault scheme is framing resistant if even a powerful adversary with complete access to the database and oracles for corrupt users cannot produce a record that is accepted as belonging to an honest user who never created it.

Intuition. Framing resistance is the vault analogue of unforgeability in digital signatures: just as unforgeability ensures that only the holder of a signing key can produce valid signatures, framing resistance ensures that only the legitimate user can produce records that the vault will accept under their identity. Whereas confidentiality notions such as IND-SIA prevent leakage of sensitive attributes, framing resistance prevents *attribution forgeries* (Figs. 3, 4 and 5).

The IND-FR Game. Let Vault = (Setup, Enroll, Store, Access, Retrieve) be a secure vault scheme. The IND-FR game between a challenger $\mathcal{C}$ and adversary $\mathcal{A}$ is defined as follows:

Game IND-FR

$1:\quad PP \leftarrow\!\$\; \mathsf{Param}(1^\lambda)$

$2:\quad (id^*, ukey^*) \leftarrow\!\$\; \mathsf{Enroll}(PP)$

$3:\quad state \leftarrow \mathcal{A}^{\mathsf{Store}(PP,\cdot),\mathsf{Access}(PP,\cdot),\mathsf{DBop}(PP,\cdot)}(PP, DB, id^*)$

$4:\quad q^* \leftarrow \mathcal{A}^{\mathsf{Store}(PP,\cdot),\mathsf{Access}(PP,\cdot),\mathsf{DBop}(PP,\cdot)}(state)$

$5:\quad R^* \leftarrow \mathsf{Access}(PP, DB, id^*, q^*)$

$6:\quad \mathbf{return}\; \big[\exists r \in R^* : (r \notin H^* \wedge r \text{ accepted as from } id^*)\big]$

Fig. 3. IND-FR security game: The adversary interacts with the system but never obtains $ukey^*$ for the honest user id^*. It wins if it can cause the vault to output a record attributed to id^* that was never honestly stored.

1. **Setup.** The challenger runs $(pp, st) \leftarrow \mathsf{Setup}(1^\lambda)$ and enrolls an honest target user: $(id^*, ukey^*) \leftarrow \mathsf{Enroll}(pp)$. The adversary is given pp and access to the initial database interface.
2. **Oracle access.** The adversary is granted adaptive access to the following oracles:
   - $\mathcal{O}_{\mathsf{store}}(id, M_{ns}, M_s)$: store an arbitrary record on behalf of any user $id \neq id^*$.
   - $\mathcal{O}_{\mathsf{acc}}(id, q)$: issue access queries on behalf of any user $id \neq id^*$.
   - $\mathcal{O}_{\mathsf{db}}(\mathsf{op}, \cdot)$: read or modify raw database state, modeling a malicious or colluding server. In the honest-but-curious model, this oracle is restricted to read-only access; in the fully malicious model, it allows arbitrary writes or deletions.

   The adversary does *not* have access to $ukey^*$, and cannot directly invoke Store on behalf of id^*.
3. **Honest history.** Optionally, the challenger may create a set of honest records for id^* by invoking $\mathsf{Store}(pp, ukey^*, id^*, M)$ on chosen messages M. Let H^* denote this honest history.
4. **Forgery attempt.** At some point, the adversary outputs a query q^* to be executed on id^*. The challenger computes $R^* \leftarrow \mathsf{Access}(pp, st, id^*, q^*)$ using the current database state (possibly modified by $\mathcal{A}$).
5. **Winning condition.** The adversary wins if there exists a record $r \in R^*$ such that:
   (a) r is accepted by the vault as belonging to id^* (i.e., passes any verification checks); and
   (b) $r \notin H^*$, i.e., it was never created by the honest user id^*.

Advantage. The adversary's framing advantage is defined as:

$$\mathsf{Adv}_{\mathcal{A}}^{\mathsf{FR}}(\lambda) = \Pr\big[\mathcal{A} \text{ wins the above game}\big].$$

A vault scheme is *IND-FR secure* in a given adversarial model (honest-but-curious or malicious server) if $\mathsf{Adv}_{\mathcal{A}}^{\mathsf{FR}}(\lambda)$ is negligible for all PPT adversaries $\mathcal{A}$.

3.3 Relationship to IND-SIA

It is natural to ask whether existing confidentiality notions, such as IND-SIA, already capture protection against framing. We emphasize that they do not.

IND-SIA in brief. The IND-SIA game assumes the adversary has complete visibility into the database and can distinguish between two possible sensitive attributes for a chosen record. The goal is to ensure that sensitive information remains hidden unless the adversary possesses the correct access token.

Why IND-SIA $\not\Rightarrow$ IND-FR. While IND-SIA captures confidentiality against database exposure, it implicitly assumes that the database contents are honestly generated and faithfully stored. It does not prevent an adversary from creating new records that appear to belong to an honest user. In other words, IND-SIA ensures secrecy of sensitive values, but does not address *integrity of attribution.*

Orthogonality. Framing resistance and sensitive-information confidentiality are complementary. A vault scheme may be IND-SIA secure but still vulnerable to framing attacks, or vice versa. To be robust in practice, a secure vault must simultaneously achieve both properties: IND-SIA to hide sensitive attributes, and IND-FR to guarantee that no user can be falsely associated with data they did not store.

4 Construction 1: Client-Bound Signatures/MACs

Overview. Our first variant for achieving non-framing security augments the baseline vault construction from [6] with *client-bound authentication* in the form of per-user signatures or MACs. The key idea is simple: every record stored in the vault must carry an unforgeable authentication tag that binds it to the identity of the user who created it. This ensures that a malicious server, colluding analyst, or corrupt user cannot inject new records under another user's identity without either compromising the user's secret key or breaking the unforgeability of the authentication scheme. In this sense, framing resistance in vaults directly parallels unforgeability in digital signatures.

4.1 Construction

Let $\Sigma = (\mathsf{KeyGen}, \mathsf{Sign}, \mathsf{Verify})$ denote a digital signature scheme secure against existential forgery under chosen-message attacks (EUF-CMA). Alternatively, Σ may be instantiated as a secure MAC scheme in settings where symmetric keys are appropriate. We modify the baseline secure vault construction as follows.

Setup. On system initialization, the authority runs $\mathsf{Param}(1^\lambda)$ to generate public parameters PP. For each enrolled user id, a signing keypair $(sk_{id}, pk_{id}) \leftarrow \mathsf{KeyGen}(1^\lambda)$ is generated. The public key pk_{id} is registered in a directory accessible to the server and other users.

$\mathsf{Store}(id, m\text{-}tok, PP, M = (M_s, M_{ns}))$

$1:$ **if** $m\text{-}tok = \perp$

$2:$ $m\text{-}tok \leftarrow\!\!\$\, \{0,1\}^\tau$

$3:$ $k_0 = \mathcal{H}(00, m\text{-}tok, M)$

$4:$ $k_1 = \mathcal{H}(01, m\text{-}tok)$

$5:$ $k_2 = \mathcal{H}(10, m\text{-}tok)$

$6:$ $k_3 = \mathcal{G}(M)$

$7:$ $C_1 = \mathsf{Enc}(k_1, M_s)$

$8:$ $\overline{M} = \mathsf{Enc}(k_2, M_{ns})$

$9:$ $(data\text{-}tok, d) = \mathsf{Tok}(k_0, \overline{M}, k_3)$

$10:$ $ts \leftarrow\!\!\$\, \mathsf{Time}(); \quad nonce \leftarrow\!\!\$\, \{0,1\}^\nu$

$11:$ $rec\_meta \leftarrow (id \,\|\, m\text{-}tok \,\|\, d \,\|\, C_1 \,\|\, ts \,\|\, nonce)$

$12:$ $\sigma \leftarrow \mathsf{Sign}(sk_{id}, rec\_meta)$

$13:$ $DB = DB \cup (id, C_1, d, \sigma, ts, nonce)$

$14:$ **return** $(m\text{-}tok, data\text{-}tok)$

$\mathsf{Access}(access\text{-}tok, data\text{-}tok)$

$1:$ $(id, C_1, d, \sigma, ts, nonce) \leftarrow \mathsf{Lookup}^{DB}(data\text{-}tok)$

$2:$ $rec\_meta \leftarrow (id \,\|\, m\text{-}tok \,\|\, d \,\|\, C_1 \,\|\, ts \,\|\, nonce)$

$3:$ **if** $\mathsf{Verify}(pk_{id}, rec_meta, \sigma) = 0 : $ **return** $\perp$

$4:$ $\overline{M} = \mathsf{DeTok}^{DB}(data\text{-}tok)$

$5:$ $M_{ns} = \mathsf{Dec}(access\text{-}tok, \overline{M})$

$6:$ **return** M_{ns}

$\mathsf{Retrieve}(m\text{-}tok, data\text{-}tok)$

$1:$ $((id, C_1, d, \sigma, ts, nonce), \overline{M}) \leftarrow \mathsf{DeTok}^{DB}(data\text{-}tok)$

$2:$ $rec\_meta \leftarrow (id \,\|\, m\text{-}tok \,\|\, d \,\|\, C_1 \,\|\, ts \,\|\, nonce)$

$3:$ **if** $\mathsf{Verify}(pk_{id}, rec_meta, \sigma) = 0 : $ **return** $\perp$

$4:$ $k_1 = \mathcal{H}(0, m\text{-}tok); \quad k_2 = \mathcal{H}(1, m\text{-}tok)$

$5:$ $M_{ns} = \mathsf{Dec}(k_2, \overline{M})$

$6:$ $M_s = \mathsf{Dec}(k_1, C_1)$

$7:$ **return** $M = (M_s, M_{ns})$

Fig. 4. Client-bound variant. Each DB entry carries an authentication tag σ over $(id, m\text{-}tok, d, C_1, ts, nonce)$, preventing framing unless the adversary forges σ or compromises sk_{id}.

Store. To store a message $M = (M_s, M_{ns})$ on behalf of id, the client proceeds as in the original construction to compute:

$$k_0 = H(00, m\text{-}tok, M),$$
$$k_1 = H(01, m\text{-}tok),$$
$$k_2 = H(10, m\text{-}tok),$$
$$k_3 = G(M),$$
$$C_1 = \mathsf{Enc}(k_1, M_s),$$
$$M' = \mathsf{Enc}(k_2, M_{ns}),$$
$$(data\text{-}tok, d) = \mathsf{Tok}(k_0, M', k_3).$$

The client then defines a record metadata string:

$$\mathsf{rec\_meta} = (id \,\|\, m\text{-}tok \,\|\, d \,\|\, C_1 \,\|\, \mathsf{ts} \,\|\, \mathsf{nonce}),$$

where ts is a timestamp and nonce is fresh randomness. The client computes a signature

$$\sigma \leftarrow \mathsf{Sign}(sk_{id}, \mathsf{rec\_meta}).$$

Finally, the vault stores the entry

$$(id, C_1, d, \sigma, \mathsf{ts}, \mathsf{nonce})$$

in the database. The client outputs $(m\text{-}tok, data\text{-}tok)$.

Access. On input $(access\text{-}tok, data\text{-}tok)$, the server retrieves the corresponding entry $(id, C_1, d, \sigma, \mathsf{ts}, \mathsf{nonce})$ and reconstructs $\mathsf{rec\_meta}$. It checks

$$\mathsf{Verify}(pk_{id}, \mathsf{rec\_meta}, \sigma) \stackrel{?}{=} 1.$$

If verification fails, the procedure returns $\perp$. Otherwise, it computes $M = \mathsf{DeTok}(d)$ and returns $M_{ns} = \mathsf{Dec}(access\text{-}tok, M)$.

Retrieve. On input $(m\text{-}tok, data\text{-}tok)$, the server retrieves the entry, verifies the signature as above, and then computes:

$$M_{ns} = \mathsf{Dec}(k_2, M'), \quad M_s = \mathsf{Dec}(k_1, C_1).$$

If verification fails, it returns $\perp$; otherwise, it outputs (M_s, M_{ns}).

4.2 Security Against Framing

We now show that this construction achieves framing resistance (IND-FR) in the honest-but-curious model. The proof reduces framing to signature (or MAC) forgery.

Theorem 1. *If the signature scheme Σ is EUF-CMA secure, then the client-bound signature vault construction is IND-FR secure in the honest-but-curious model. In particular, for any PPT adversary $\mathcal{A}$,*

$$\mathsf{Adv}_{\mathcal{A}}^{\mathsf{FR}}(\lambda) \leq \mathsf{Adv}_{\mathcal{B}}^{\mathsf{EUF\text{-}CMA}}(\lambda) + \mathsf{negl}(\lambda),$$

for some PPT forger $\mathcal{B}$.

Proof (Proof Sketch). Suppose an adversary $\mathcal{A}$ wins the IND-FR game with non-negligible probability. We construct a forger $\mathcal{B}$ against the EUF-CMA security of Σ. The challenger gives $\mathcal{B}$ a public key pk^* and signing oracle access. $\mathcal{B}$ simulates the vault for $\mathcal{A}$ as follows:

- For the honest target user id^*, $\mathcal{B}$ sets $pk_{id^*} = pk^*$ and does not know sk_{id^*}.
- Whenever the simulation requires a signature under sk_{id^*} (for honestly generated records), $\mathcal{B}$ queries its signing oracle.
- For all other users, $\mathcal{B}$ generates real keypairs and signs locally.

Eventually, $\mathcal{A}$ outputs a forgery: a record r that is accepted as belonging to id^* but was never honestly stored. This implies that r contains a valid signature σ^* on a fresh message rec_meta* under pk^* that was never queried to the signing oracle. $\mathcal{B}$ outputs (rec_meta*, σ^*) as its EUF-CMA forgery. The reduction is perfect except with negligible probability (e.g., collisions in the random oracle). Therefore, a successful framer breaks EUF-CMA, contradicting the assumed security of Σ.

Discussion. This variant provides strong framing resistance at the cost of requiring each client to maintain signing capability. In settings where per-user keys are acceptable (e.g., enterprise deployments, government systems, or federated identity frameworks), this overhead is minor compared to the security benefits. For deployments requiring keyless clients, we propose an alternative construction in the next section based on transparency logs.

5   Construction 2: Transparency Log-Based Vault Scheme

Overview. Our second variant achieves non-framing security without requiring clients to hold long-term signing keys. Instead, it leverages a *public append-only transparency log* maintained by the vault operator (or an external service) together with server-issued signed receipts. The log ensures that once a record commitment is appended, it becomes tamper-evident, and cannot later be inserted, modified, or deleted without detection. The server signs each admission into the log, providing the client and auditors with a verifiable receipt of the record's existence. This design prevents framing because any record accepted under a user's identity must be backed by a valid receipt and inclusion proof in the log, which a forger cannot generate without either breaking signature security or violating the append-only property of the log.

$\text{Store}(id, m\text{-}tok, PP, M = (M_s, M_{ns}))$

1: **if** $m\text{-}tok = \perp$
2:     $m\text{-}tok \leftarrow\!\$ \{0,1\}^\tau$
3:   $k_0 = \mathcal{H}(00, m\text{-}tok, M)$
4:   $k_1 = \mathcal{H}(01, m\text{-}tok)$
5:   $k_2 = \mathcal{H}(10, m\text{-}tok)$
6:   $k_3 = \mathcal{G}(M)$
7:   $C_1 = \text{Enc}(k_1, M_s)$
8:   $\overline{M} = \text{Enc}(k_2, M_{ns})$
9:   $(data\text{-}tok, d) = \text{Tok}(k_0, \overline{M}, k_3)$
10:   $ts \leftarrow\!\$ \text{Time}();\quad nonce \leftarrow\!\$ \{0,1\}^\nu$
11:   $com \leftarrow H(id \,\|\, m\text{-}tok \,\|\, C_1 \,\|\, d \,\|\, nonce)$
12:   $(root, idx, proof) \leftarrow \text{Log.Append}(com)$
13:   $\rho \leftarrow \text{Sign}(SK_{\text{srv}}, com \,\|\, root \,\|\, idx \,\|\, ts)$
14:   $DB = DB \cup (id, C_1, d, com, \rho, nonce, proof, ts)$
15:   **return** $(m\text{-}tok, data\text{-}tok)$

$\text{Access}(access\text{-}tok, data\text{-}tok)$

1:   $(id, C_1, d, com, \rho, nonce, proof, ts) \leftarrow \text{Lookup}^{DB}(data\text{-}tok)$
2:   **if** $\text{Verify}(PK_{\text{srv}}, com \,\|\, root \,\|\, idx \,\|\, ts, \rho) = 0 :$ **return** $\perp$
3:   **if** $\text{Log.VerifyInclusion}(com, root, proof) = 0 :$ **return** $\perp$
4:   $\overline{M} = \text{DeTok}^{DB}(data\text{-}tok)$
5:   $M_{ns} = \text{Dec}(access\text{-}tok, \overline{M})$
6:   **return** M_{ns}

$\text{Retrieve}(m\text{-}tok, data\text{-}tok)$

1:   $((id, C_1, d, com, \rho, nonce, proof, ts), \overline{M}) \leftarrow \text{DeTok}^{DB}(data\text{-}tok)$
2:   **if** $\text{Verify}(PK_{\text{srv}}, com \,\|\, root \,\|\, idx \,\|\, ts, \rho) = 0 :$ **return** $\perp$
3:   **if** $\text{Log.VerifyInclusion}(com, root, proof) = 0 :$ **return** $\perp$
4:   $k_1 = \mathcal{H}(0, m\text{-}tok);\quad k_2 = \mathcal{H}(1, m\text{-}tok)$
5:   $M_{ns} = \text{Dec}(k_2, \overline{M})$
6:   $M_s = \text{Dec}(k_1, C_1)$
7:   **return** $M = (M_s, M_{ns})$

Fig. 5. Transparency-log variant. Each record has a commitment *com* logged in an append-only log with a signed receipt ρ and inclusion proof, preventing framing unless the adversary forges ρ or breaks log append-only.

5.1 Construction

We assume the existence of:

- A collision-resistant hash function H.
- A transparency log Log with an append-only property, implemented as a Merkle tree with public roots and efficient inclusion proofs.
- A signature scheme $\Sigma = (\mathsf{KeyGen}, \mathsf{Sign}, \mathsf{Verify})$ with server signing key SK_{srv} and verification key PK_{srv}.

The vault construction is modified as follows.

Setup. The authority generates public parameters PP and initializes Log with an empty state and a signing keypair $(SK_{\mathsf{srv}}, PK_{\mathsf{srv}})$. The verification key PK_{srv} is published.

Store. To store a message $M = (M_s, M_{ns})$, the client computes the baseline encryption and tokenization steps:

$$k_0 = H(00, \textit{m-tok}, M), \quad k_1 = H(01, \textit{m-tok}),$$
$$k_2 = H(10, \textit{m-tok}), \quad k_3 = G(M),$$
$$C_1 = \mathsf{Enc}(k_1, M_s), \quad M' = \mathsf{Enc}(k_2, M_{ns}),$$
$$(\textit{data-tok}, d) = \mathsf{Tok}(k_0, M', k_3).$$

It then forms a record commitment:

$$com = H(id \,\|\, \textit{m-tok} \,\|\, C_1 \,\|\, d \,\|\, \mathsf{nonce}).$$

The server appends com to Log, obtaining $(root, idx, proof) \leftarrow \mathsf{Log.Append}$ (com). It signs a receipt:

$$\rho = \mathsf{Sign}(SK_{\mathsf{srv}}, com \,\|\, root \,\|\, idx \,\|\, \mathsf{ts}).$$

Finally, the database stores the tuple

$$(id, C_1, d, com, \rho, \mathsf{nonce}, proof, \mathsf{ts}).$$

The client receives $(\textit{m-tok}, \textit{data-tok}, \rho, proof)$.

Access.
On input $(\textit{access-tok}, \textit{data-tok})$, the vault retrieves $(id, C_1, d, com, \rho, proof, \mathsf{ts})$ and performs:

1. Verify ρ under PK_{srv}.
2. Check that com is included in $root$ using $proof$.

If either check fails, return $\bot$. Otherwise, recover $M = \mathsf{DeTok}(d)$ and return $M_{ns} = \mathsf{Dec}(\textit{access-tok}, M)$.

Retrieve. On input $(m\text{-}tok, data\text{-}tok)$, the vault retrieves the entry, verifies ρ and the inclusion proof as above, and then computes:

$$M_{ns} = \mathsf{Dec}(k_2, M'), \quad M_s = \mathsf{Dec}(k_1, C_1).$$

If verification succeeds, output (M_s, M_{ns}); else return $\bot$.

5.2   Security Against Framing

Theorem 2. *If the signature scheme Σ is EUF-CMA secure and the transparency log* Log *is append-only with collision-resistant commitments and sound inclusion proofs, then the transparency-log variant of the vault construction is IND-FR secure against malicious servers. In particular, for any PPT adversary $\mathcal{A}$,*

$$\mathsf{Adv}_{\mathcal{A}}^{\mathsf{FR}}(\lambda) \leq \mathsf{Adv}_{\mathcal{B}_1}^{\mathsf{EUF\text{-}CMA}}(\lambda) + \mathsf{Adv}_{\mathcal{B}_2}^{\mathsf{Log}}(\lambda) + \mathsf{negl}(\lambda).$$

Proof (Proof Sketch). Suppose an adversary $\mathcal{A}$ produces a forged record r that is accepted as belonging to id^* although id^* never stored it. For r to be accepted, it must contain a valid server signature ρ and a valid inclusion proof for commitment *com*. There are two cases:

- If *com* was never appended to the log, then ρ constitutes a forgery under PK_{srv}, breaking the EUF-CMA security of Σ.
- If ρ is valid but *com* was not honestly appended, then $\mathcal{A}$ has either produced a collision in H or broken the soundness/append-only guarantees of Log.

In either case, we can construct a reduction $\mathcal{B}_1$ or $\mathcal{B}_2$ that uses $\mathcal{A}$ to break either signature unforgeability or log security. Thus, the advantage of $\mathcal{A}$ is bounded above by the sum of the advantages against these primitives, plus negligible terms for simulation errors.

Discussion. This variant shifts the trust model from individual clients to the server and log infrastructure. It is particularly attractive in settings where clients are lightweight or cannot feasibly manage cryptographic keys, such as citizen-facing e-government portals or consumer health applications. While it requires additional infrastructure in the form of a transparency log and signature verification, these mechanisms are well understood, efficient, and compatible with existing cloud deployment models.

Remark (On Server Key Generation and Signature Assumptions). Our security proof assumes that the server's signing key SK_{srv} and the corresponding public key PK_{srv} are honestly generated according to the prescribed key generation algorithm of the signature scheme Σ. This is standard in most transparency-log and certificate-based systems, where the verification key is either publicly certified or audited during system initialization. If the server were allowed to choose an arbitrary or weak key, or to later substitute a new key, additional trust anchors (e.g., a public key infrastructure or log-signed key rollovers) would be required to maintain verifiability. Similarly, our reduction relies on Σ being existentially unforgeable under chosen-message attacks (EUF-CMA); stronger notions such as binding and unforgeability under key replacement (BUFF [8]) would further reinforce this assumption, but are orthogonal to our core framing-resistance definition. Incorporating BUFF-secure signatures or key-transparency mechanisms into vault deployments is a promising practical enhancement, but is beyond the scope of this work.

6   Comparison of the Two Constructions

Both constructions presented above achieve non-framing security, but they differ significantly in trust assumptions, deployment complexity, and suitability for different application domains. In this section we highlight the main tradeoffs between the *client-bound signatures/MACs* variant and the *transparency log-based* variant.

6.1   Security Assumptions

- **Client-bound signatures.** Security reduces directly to the EUF-CMA security of the underlying signature scheme (or MAC). The vault resists framing attacks as long as each user's signing key remains uncompromised.
- **Transparency logs.** Security reduces to a combination of server signature unforgeability and the append-only property of the log. The scheme remains secure even without client-held keys, but now assumes that the log infrastructure is robust and auditable.

6.2   Trust Model

- **Client-bound signatures.** Shifts trust to individual users, who must protect their signing keys. A compromised client key directly enables framing against that user.
- **Transparency logs.** Shifts trust to the vault operator (or external log service). A compromised server key or log rewrite attack could enable framing, but such events are typically easier to detect at scale through log auditing.

6.3   Deployment Complexity

- **Client-bound signatures.** Minimal infrastructure overhead: requires only key generation and distribution at enrollment. Verification of signatures is computationally lightweight.
- **Transparency logs.** Requires additional infrastructure to maintain a public append-only log, publish roots, and provide inclusion proofs. While this introduces complexity, such systems are widely deployed in practice (e.g., Certificate Transparency).

6.4   Performance Considerations

- **Client-bound signatures.** Per-record overhead is a single digital signature (or MAC), typically a few hundred bytes, and a verification operation during retrieval. This is efficient for most workloads.
- **Transparency logs.** Per-record overhead includes a log commitment, a signed receipt, and a Merkle proof. Verification requires both signature checking and Merkle proof validation, which adds latency but remains practical.

6.5   Application Scenarios

- **Client-bound signatures.** Well suited for enterprise or government systems where users can securely maintain cryptographic keys, and accountability for individual actions is paramount.
- **Transparency logs.** Better suited for citizen-facing or consumer applications, where clients are resource-constrained or cannot be expected to manage keys. Here, the log provides accountability at the system level rather than the individual level.

Summary. Both variants offer strong protection against framing, but their deployment contexts differ. The client-bound approach is simpler and offers direct per-user accountability, while the transparency log approach avoids client key management at the cost of additional infrastructure. In practice, a hybrid model combining both mechanisms may be attractive: client-bound signatures for high-assurance users, combined with a transparency log to provide global accountability and detect misbehavior by the server.

7   Conclusion

This work introduced the first formal treatment of framing resistance in secure vault schemes. We identified the threat of framing attacks, where adversaries attempt to inject or manipulate records so that they are falsely attributed to honest users, and formalized this threat via the IND-FR security game. To address it, we presented two complementary constructions. The first, based on client-bound signatures or MACs, ensures per-user accountability by binding each record cryptographically to the identity of its creator. The second, based on transparency logs, shifts the trust to an auditable infrastructure, ensuring that every accepted record is backed by a verifiable receipt and log inclusion proof.

We proved that both constructions achieve IND-FR security under standard cryptographic assumptions, and compared their tradeoffs in terms of security guarantees, deployment complexity, and application contexts. Together, these results show that secure vaults can go beyond confidentiality to provide strong integrity guarantees in adversarial cloud environments.

While the two variants address different deployment needs, our analysis suggests that hybrid approaches combining per-user authentication with global auditability may offer the best balance of accountability and resilience, and are a natural direction for further exploration.

References

1. ANSI X9.119-2-2017, Retail Financial Services - Requirements For Protection Of Sensitive Payment Card Data - Part 2: Implementing Post-Authorization Tokenization Systems. https://webstore.ansi.org/Standards/ASCX9/ansix91192017
2. ChatGPT tied to Samsung's alleged data leak. https://cybernews.com/news/chatgpt-samsung-data-leak/?fbclid=IwAR2okEfxOY1oMB0PrX0pyBMYdRDEmdzWUONdkaPWjo5auauRST-kV4SCy5U
3. Generative AI Data Privacy with Skyflow LLM Privacy Vault . https://skyflow.com/post/generative-ai-data-privacy-skyflow-llm-privacy-vault
4. Hashicorp, Vault. https://www.hashicorp.com/products/vault
5. Bellare, M., Rogaway, P.: The security of triple encryption and a framework for code-based game-playing proofs. In: Vaudenay, S., (ed.) EUROCRYPT 2006, vol. 4004 of LNCS, pp. 409–426, May/June (2006)
6. Bhattacharyya, R., Mandal, A., Sengupta, M.: Secure vault scheme in the cloud operating model. LNCS, pp. 281–303 (2024)
7. Díaz-Santiago, S., Rodríguez-Henríquez, L.M., Chakraborty, D.: A cryptographic study of tokenization systems. Int. J. Inf. Secur. **15**(4), 413–432 (2016)
8. Garg, S., Komargodski, I., Waters, B.: Buff: binding unforgeability under function families. In: EUROCRYPT 2020 (2020). https://eprint.iacr.org/2020/1525
9. Ji, T., Yilmaz, E., Ayday, E., Li, P.: The curse of correlations for robust fingerprinting of relational databases. In: Proceedings of the 24th International Symposium on Research in Attacks, Intrusions and Defenses, RAID '21, pp. 412–427, New York, NY, USA, (2021). Association for Computing Machinery
10. Juels, A., Sudan, M.: A fuzzy vault scheme. In: Proceedings of the 2002 IEEE International Symposium on Information Theory (2002)
11. Nandakumar, R., Jain, A., Pankanti, S.: Fingerprint-based fuzzy vault: implementation and performance. IEEE Trans. Inf. Forensics Secur. (2019)
12. Singh, A.P., Singh, M.: A comprehensive survey on fuzzy vault schemes for biometric security. Comput. Sci. Rev. (2020)
13. Zhang, M., Xie, Z., Yue, C., Zhong, Z.: Spitz: a verifiable database system (2020)

Author Index

A
Adhikari, Avishek   390

B
Baksi, Anubhab   45, 339
Banegas, Gustavo   194
Belel, Anushree   285
Berger, Daniel   173

C
Chakraborty, Sayantan   390
Chakraborty, Sumanta   339
Chang, Tao-Hsiang   145
Chattopadhyay, Anupam   339

D
Douteau, Antoine   119
Dutta, Anisha   390

G
Goswami, Chandan   390
Guggemos, Tobias   216

H
Hajra, Rittwik   417
Hellenbrand, Andreas   194
Hsu, Hao-Yi   145
Hsu, Jen-Chieh   145
Husnil Arif, Mohammad Ferry   3

I
Imran, Muhammad   3

J
Jang, Kyungbae   339

K
Kalam, Abul   239
Kar, Subha   417
Karmakar, Abhijit   21
Karmakar, Sudeshna   239
Kohrita, Tohru   95
Kumari, Ruby   21

L
Lee, Bryan   339

M
Mambo, Masahiro   145
Mandal, Surajit   45
Mishra, Dheerendra   312
Miyaji, Atsuko   70
Mukherjee, Pratyay   417

N
Nagai, Atsuki   70
Nikolaev, Maksim   95
Nishide, Takashi   365

O
Okada, Yurie   70

P
Pal, Debranjan   45
Pal, Soumit   417
Patarin, Jacques   263
Pursharthi, Komal   312

R
Renan, Farzin   216
Roux-Langlois, Adeline   119

S
Saldanha, Matheus   194
Sarkar, Santanu   45, 239
Saurav, Sumeet   21
Sekii, Kyosuke   365
Sengupta, Meghna   442
Seo, Hwajeong   339

Shikata, Junji   285
Silva, Javier   95

T
Tso, Raylin   145

V
Varjabedian, Pierre   263

W
Wang, Siyi   339

MIX
Papier aus verantwortungsvollen Quellen
Paper from responsible sources
FSC® C105338

If you have any concerns about our products,
you can contact us on
ProductSafety@springernature.com

In case Publisher is established outside the EU,
the EU authorized representative is:
Springer Nature Customer Service Center GmbH
Europaplatz 3, 69115 Heidelberg, Germany

Printed by Libri Plureos GmbH
in Hamburg, Germany